HERBIVORES
Their Interaction with
Secondary Plant Metabolites

Contributors

Shalom W. Applebaum
E. A. Bell
Yehudith Birk
W. M. Blaney
Lena B. Brattsten
R. F. Chapman
Frances S. Chew
Eric E. Conn
L. Fowden
James E. Gill
Jeffrey B. Harborne
Daniel H. Janzen
P. J. Lea
Irvin E. Liener
Tom J. Mabry
Doyle McKey
John C. Reese
David F. Rhoades
Trevor Robinson
James E. Rodman
Gerald A. Rosenthal
C. A. Ryan
David S. Seigler
Karel Sláma
Tony Swain
H. L. Tookey
C. H. Van Etten

HERBIVORES
Their Interaction with Secondary Plant Metabolites

Edited by

Gerald A. Rosenthal
School of Biological Sciences
University of Kentucky
Lexington, Kentucky

Daniel H. Janzen
Department of Biology
University of Pennsylvania
Philadelphia, Pennsylvania

1979

ACADEMIC PRESS, INC.
(Harcourt Brace Jovanovich, Publishers)
Orlando San Diego San Francisco New York London
Toronto Montreal Sydney Tokyo Sao Paulo

ACADEMIC PRESS, INC.
Orlando, Florida 32887

United Kingdom Edition published by
ACADEMIC PRESS, INC. (LONDON) LTD.
24/28 Oval Road, London NW1 7DX

Library of Congress Cataloging in Publication Data

Main entry under title:

Herbivores: Their interaction with secondary plant meta–
bolites.

Includes index.
1. Herbivora––Ecology. 2. Herbivora––Physiology.
3. Plant metabolites. 4. Plant defenses. I. Rosen–
thal, Gerald A. II. Janzen, Daniel H.
QH541.H37 574.5'24 79–6944
ISBN 0–12–597180–X

PRINTED IN THE UNITED STATES OF AMERICA

84 85 86 87 9 8 7 6 5 4 3 2

Contents

List of Contributors

Numbers in parentheses indicate the pages on which the author's contributions begin.

Shalom W. Applebaum (539), Department of Entomology, Faculty of Agriculture, The Hebrew University of Jerusalem, Rehovot, Israel

E. A. Bell (353), Department of Plant Sciences, King's College, University of London, London, SE24 9JF England

Yehudith Birk (539), Departments of Entomology and Agricultural Biochemistry, Faculty of Agriculture, The Hebrew University of Jerusalem, Rehovot, Israel

W. M. Blaney (161), Department of Zoology, Birkbeck College, London, England

Lena B. Brattsten (149), Departments of Biochemistry and Ecology, University of Tennessee, Knoxville, Tennessee 37916

R. F. Chapman (161), Centre for Overseas Pest Research, Ministry for Overseas Development, College House, London, England

Frances S. Chew (271), Department of Biology, Tufts University, Medford, Massachusetts 02155

Eric E. Conn (387), Department of Biochemistry and Biophysics, University of California, Davis, California 95616

L. Fowden (135), Department of Biochemistry, Rothamsted Experimental Station, Harpenden, Herts, AL5 2JQ England

James E. Gill* (501), Department of Botany, University of Texas at Austin, Austin, Texas 78712

Jeffrey B. Harborne (619), Plant Science Laboratories, The University, Reading, England

Daniel H. Janzen (331), Department of Biology, University of Pennsylvania, Philadelphia, Pennsylvania 19104

P. J. Lea (135), Department of Biochemistry, Rothamsted Experimental Station, Harpenden, Herts, AL5 2JQ England

Irvin E. Liener (567), Department of Biochemistry, College of Biological Sciences, University of Minnesota, St. Paul, Minnesota 55108

* PRESENT ADDRESS: Department of Botany, The University of Texas Health Science Center at San Antonio, San Antonio, Texas 78284

Tom J. Mabry (501), Department of Botany, University of Texas at Austin, Austin, Texas 78712

Doyle McKey* (55), Division of Biological Sciences, Univerisity of Michigan, Ann Arbor, Michigan 48109

John C. Reese† (309), Department of Entomology, University of Wisconsin, Madison, Wisconsin 53706

David F. Rhoades (3), Department of Zoology, University of Washington, Seattle, Washington 98195

Trevor Robinson (413), Department of Biochemistry, University of Massachusetts, Amherst, Massachusetts 01003

James E. Rodman (271), Department of Biology, Yale University, New Haven, Connecticut 06520

Gerald A. Rosenthal (353), School of Biological Sciences, University of Kentucky, Lexington, Kentucky 40506

C. A. Ryan (599), Department of Agricultural Chemistry, and Program in Biochemistry and Biophysics, Washington State University, Pullman, Washington 99164

David S. Seigler (449), Department of Botany, The University of Illinois, Urbana, Illinois 61801

Karel Sláma (683), Department of Insect Physiology, Institute of Entomology ČSAV, Czechoslovak Academy of Sciences, Praha 6, Czechoslovakia

Tony Swain (657), Department of Biology, Boston University, Boston, Massachusetts 02215

H. L. Tookey (471), Northern Regional Research Center, Science and Education Administration, Agricultural Research, U.S. Department of Agriculture, Peoria, Illinois 61604

C. H. Van Etten (471), Northern Regional Research Center, Science and Education Administration, Agricultural Research, U.S. Department of Agriculture, Peoria, Illinois 61604

* PRESENT ADDRESS: Wisconsin Regional Primate Research Center, Field Research Center, Edea, Cameroun, West Africa

† PRESENT ADDRESS: Department of Entomology and Applied Ecology, University of Delaware, Newark, Delaware 19711.

Preface

It is evident that the interactions between plants and the animals that consume them, as mediated by the secondary compounds (allelochemics) of plants, is a rapidly expanding field of scientific inquiry. The subject is of great importance and interest to evolutionary biologists, agriculturalists, natural product chemists, physiologists, biological chemists, ecologists, and others. Yet, the background information needed to work effectively in this area is scattered among an enormously diverse array of journals. Individuals trained in one of the classical subdisciplines of biology, or a graduate student just becoming aware of the area of animal–plant interactions, commonly ask what to read to acquire the background information to understand and enhance the current literature in this field. We have produced this work to provide such a reference.

Each author of the chemically oriented chapters was asked to provide that information which a person reading or working in animal–plant interactions would want to have as background if it were necessary to understand something of an alkaloid, a phenol, or some other secondary natural product. Each author of the ecologically oriented chapters was asked to write as though speaking to a serious scientist who is naive about animal–plant interactions but who is competent or becoming competent in some conventional subdiscipline of biology.

Each chapter is intended to stand on its own, to be an essay by a leader in the field reflecting biases and viewpoints peculiar to that person. The chapters are deliberately not edited to remove overlap of examples or redundancy of ideas expressed; in fact, there is little overlap and what is there serves as a form of emphasis. We assume that within ten years time, or perhaps less, this work will be out of date. We hope this book will be a major contributory element to this obsolescence.

Gerald A. Rosenthal
Daniel H. Janzen

HERBIVORES

Their Interaction with Secondary Plant Metabolites

Part **I**

Ecological and Evolutionary Processes

Chapter 1

Evolution of Plant Chemical Defense against Herbivores

DAVID F. RHOADES

ISBN 0-12-597180-X

I. INTRODUCTION

Stahl (1888) appears to have been the first person to suggest that some of the chemical properties of plants may have evolved for protection against attack by herbivorous animals. This idea lay dormant for many years until Fraenkel (1959, 1969) pointed out that insect host-finding and gustatory behavior was largely under the control of plant "secondary substances," compounds that, although ubiquitous among plants, had no known metabolic function. Fraenkel suggested that the adaptive significance of these substances to plants lay in their repellent properties to insects and other herbivores.

Although there is still some debate concerning the adaptive significance of plant secondary metabolites (Robinson, 1974; Seigler and Price, 1976; Seigler, 1977), evidence is accumulating to support a primarily ecological role for these metabolites both in interactions among plants and their associated biota and as protective agents against physical environmental stresses. This is not to say that secondary metabolites have only ecological significance to plants, since all chemical systems possessed by plants must necessarily be integrated into the total metabolic scheme, and multiple function is to be expected. However, research into the ecological significance of secondary metabolites has been more fruitful than that into metabolic significance, and, although some secondary compounds have been implicated as storage compounds and regulators of plant metabolism and growth (del Moral, 1972; Galbraith *et al.*, 1972; Arntzen *et al.*, 1974; McClure, 1975; Seigler, 1977), the great majority of cases in which adaptive value has been demonstrated for these substances concern ecological interactions.

II. ECOLOGICAL ADAPTIVE SIGNIFICANCE OF PLANT SECONDARY METABOLITES

A. Physical Environment

Many workers have considered that flavonoids (Caldwell, 1971; McClure, 1975) and carotenoids (Krinsky, 1971) screen plants from ultraviolet (uv) and short-wavelength visible radiation. There is evidence that the lignan- and flavonoid-based phenolic resin of creosote bush, a desert shrub, protects the plant from both uv radiation and desiccation (Rhoades, 1977a).

B. Mutualistic Interactions with Herbivores

Secondary metabolites have been shown to be adaptive to plants in mutualistic interactions with pollinators and fruit dispersers. Flavonoids, volatile terpenes, and other compounds are visual or olfactory floral attractants to pollinators (Dodson *et al.*, 1969; Brehm and Krell, 1975). "Primary" plant products such as sugars, lipids, and amino acids in nectar function as pollinator rewards (Baker and Baker, 1975; Heinrich, 1975a,b). On the other hand, many nectars also contain alkaloids and other toxic metabolites (Baker and Baker, 1975). Although at first this seems incompatible with the reward function of nectar, toxic constituents possibly serve to maximize flower constancy in pollinators. All plants pollinated by animal vectors benefit from flower-constant behavior of pollinators since nectar removed by visitors not carrying conspecific pollen is, from the plant's viewpoint, wasted. However, rare or widely dispersed plants should be more strongly selected for adaptations to maximize pollinator constancy than should common or clumped plants, since for the former a random visitor is less likely to be carrying conspecific pollen. A plausible set of floral adaptations leading to the selection of constancy or to the induction of facultative constancy in pollinators is the presentation of a large, predictable, but highly defended reward (J. C. Bergdahl and D. F. Rhoades, in preparation). Floral characteristics constituting a predictably high reward for pollinators include large nectar volume per flower or floret containing high nutrient concentrations; high-quality nectar containing diverse amino acids, sugars, and trace nutrients; long flowering season with sequential flower or floret maturation; and low variability in these properties. Defensive floral adaptations include flower asymmetry, which requires special behavior of the visitor in order to enter; relatively inaccessible placement of the reward such as at the end of a long tubular corolla, requiring special adaptations in the trophic apparatus of visitors; hairs and other devices that render entrance more difficult; and toxic constituents in the reward. Whereas these defensive adaptations should deter generalist visitors, their effect on flower-constant specialist pollinators should be less. This is due both to the evolution of counteradaptations in the specialists and, importantly, to the proximate induction of behavioral and metabolic changes, such as the use of nectar guides for gaining entrance to asymmetric flowers and induced elevated catabolism of toxins.

We can therefore expect that those animal-pollinated plants which in evolutionary time have characteristically grown widely separated from conspecific individuals should have evolved, on the one hand, particularly rich floral rewards and, on the other, rich floral defensive adaptations, whereas characteristically common or clumped plants should have evolved both sets of adaptations to a lesser degree. Positive correlation between the occurrence of nectar alkaloids and the occurrence of amino acids, lipids, and antioxidants as nectar constituents (Baker and Baker, 1975) is consistent with this scheme. Baker and Baker (1975) suggest that nectar alkaloids may be protective agents against flower-inconstant lepidopterans and favor more oligotrophic bees, which are apparently less sensitive to alkaloids than are lepidopterans.

The complex array of volatile compounds, including esters and terpenoids (Nurstein, 1970), and nonvolatile pigments, such as anthocyanins and carotenoids (van Buren, 1970; Goodwin and Goad, 1970) in fruits, probably serve as cueing agents for dispersers. Many of the biochemical changes that occur when fruits ripen, such as the accumulation of sugars (Hulme, 1970), the disappearance of alkaloids (Heftmann, 1963), and polymerization of tannins to innocuous higher molecular weight polymers (Goldstein and Swain, 1963), are well-interpreted as being designed to ensure dispersal at an appropriate stage of seed development. On the other hand, retention of alkaloids and other toxic principles in some ripe fruits, for example, those of *Blighia sapida* (Sapindaceae), *Hippomane mancinella* (Euphorbiaceae), and some *Solanum* spp. (Solanaceae) (Kingsbury, 1964), suggests an element of defense for these fruits even when seeds have reached a developmental stage suitable for dispersal. This possibly represents a mechanism for the exclusion of "low quality" dispersers (in the sense of McKey, 1975), such as small mammals and opportunistic avian frugivores, to thus maximize the probability of dispersal by "high quality" dispersers such as obligate avian frugivores (L. S. Best, unpublished). If so it is necessary that the toxicants have a greater effect on low-quality generalist frugivores than on high-quality specialist dispersers, and this is reasonable since obligate frugivores should be under greater selection to evolve appropriate detoxification mechanisms than should facultative frugivores. If it proved to be generally true that ripe fruit or nectar toxicants are more deterrent to low-quality dispersers or pollinators than to high-quality dispersers or pollinators, this would provide powerful evidence for the above schemes.

C. Antagonistic Interactions with Herbivores

Secondary compounds have been implicated as defensive agents for plants in plant–plant (allelopathic) (Rice, 1974, 1977), plant–pathogen (Bell, 1974; Deverall, 1977), and plant–herbivore interactions. There are several recent reviews of evidence for antiherbivore function of secondary substances (Schoonhoven, 1972; McKey, 1974; Freeland and Janzen, 1974; Beck and Reese, 1976; Feeny, 1976; Rhoades and Cates, 1976). Two main lines of evidence have been used to show antiherbivore activity of these substances. The first involves deterrency, and the second involves antibiosis. In the first method, the preference of herbivores for a series of plant tissues is compared with the concentration of secondary metabolites in these tissues. If the herbivore is deterred, a negative correlation should result. Deterrency by the plant substance can then be corroborated by extraction of the substance from the plant and incorporation into artificial diets. Both within-plant and between-plant comparisons are possible. Within plants, tissues often vary in their content of secondary substances (and other properties), for example, young versus mature leaves (McKey, 1974; Rhoades and Cates, 1976). Sometimes individual plants or morphs within the population varying in secondary metabolite content and preference to herbivores have been identified (Jones, 1971; Cates, 1975; Cooper-Driver and Swain, 1976). In the second method, some fitness measure such as growth rate, fecundity, or, ideally, number of viable progeny is measured for herbivores fed the various tissues, individual plants, or morphs, and the correlation between secondary metabolite content and herbivore fitness is examined. Again, the fitness experiment can be conducted using extracts and artificial diets. Ideally, both methods, using both intact plant material and artificial diets, should be used.

The following studies provide a representative, but by no means exhaustive, sample of the evidence in support of a protective function for plant secondary metabolites against herbivores. The acyanogenic morphs of *Lotus corniculatus* and *Trifolium repens* are preferred over the cyanogenic morphs by a variety of invertebrate and vertebrate herbivores (Jones, 1972, 1977a,b). Cates (1975) demonstrated a similar polymorphism in *Asarum caudatum* relative to its major herbivore, the slug *Ariolimax columbianus,* although in this case the chemical or physical basis for differential palatability is not known. Bird damage to seeds of *Sorghum halapense* is negatively correlated to the tannin content of the seeds (McMillian *et al.*, 1972). Gustafsson and Gadd (1965) found that plots of *Lupinus* individuals lacking alkaloids could be located with 100% accuracy by the presence of thrips infestation. Similarly, sheep avoid grazing varieties of lupine that contain high concentrations of alkaloids but readily graze "sweet" varieties (Arnold and Hill, 1972). Dolinger *et al.* (1973) found high intraspecific qualitative and quantitative variation of alkaloid profile in inflorescences of *Lupinus* populations exposed to high potential herbivore pressure due to long flowering season on moist sites but little intraspecific variation at dry sites, where pressure was low. Laboratory experiments established that larvae of a lycaenid butterfly, an important herbivore of *Lupinus,* had lower survival when fed flowers from the chemically variable plant populations than when fed flowers from invariant populations. Individuality of defensive chemical profile should be advantageous to plants in high-risk populations since it minimizes the effects of counteradaptation by herbivores. Edmunds and Alstad (1978) found that black pineleaf scale insects show greatly increased mortality when transferred from their host tree to a neighboring conspecific tree compared to within-tree transplants. They attribute this phenomenon to individual variation in defensive chemistry of the trees and suggest that demic differentiation of the relatively immobile scales has occurred by adaptation over many generations to the particular defensive array of their individual host trees.

The presence of a gene controlling production of bitter cucurbitacins in some individuals of *Cucumis sativus* L. reduces feeding by and increases larval mortality of the two-spotted mite by as much as 95% (DaCosta and Jones, 1971). Stephenson (1970) reports that different chlorogenic acid concentrations in potato tubers of three varieties of potatoes determine differential palatabilities to slugs. Sasamoto (in Fraenkel, 1959) showed that an increase in silica content of rice plants reduces attack by the rice stem borer *Chilo suppressalis.* Bernays and Chapman (1977), in a detailed study, concluded that food selection by *Locusta migratoria,* a largely graminivorous species, was largely controlled by deterrent plant secondary metabolites.

Decreased pest resistance of cultivated plants relative to their wild progenitors may be partly due to decreased toxicity from agricultural selection for edibility to human beings and domestic animals. The leaves of wild cabbage (*Brassica oleracea*) contain a concentration of toxic glucosinolates five times higher, on the average, than cultivated varieties (Joseffson, 1967). Other examples of cultivated varieties containing lower concentrations of toxic substances than wild forms include vetch (*Laythrus*), lima bean (*Phaseolus*), aconite (*Aconitum*), cassava (*Manihot*), betel nut (*Areca*) (Pammell, 1911), potato (*Solanum*) (Simmonds, 1976a), mango (*Mangifera*) (Singh, 1976), lettuce (*Lactuca*) (Ryder and Whitaker, 1976; Kingsbury, 1964; Clapham *et al.*, 1952), nightshade (*Solanum*) (Kingsbury, 1964), and cucurbits (Howe *et al.*, 1976). Interestingly, cultivated tobacco (*Nicotiana*) contains as much nicotine as the primitive form (Pammell, 1911).

Plant secondary substances that have been shown either to have a negative effect on herbivore fitness or to have a deterrent effect on herbivore grazing include alkaloids, pyrethrins, rotenoids, long-chain unsaturated isobutylamides, cyanogenic glycosides, phytoecdysones and juvenile hormone analogs, cardenolides and saponins, sesquiterpene lactones, nonprotein amino acids, glucosinolates and isothiocyanates, oxalates, protoanemonin, hypericin, fluoro fatty acids, selenoamino acids, 6-methoxybenzoxazoline, gossypol, condensed tannin, phenolic resin and associated phenol oxidase and proteinase inhibitors of the soybean trypsin inhibitor type (see Rhoades and Cates, 1976), phytohemagglutinin (Janzen *et al.*, 1976), and chromenes (precocenes) (Bowers *et al.*, 1976). In many of these studies the effects of the plant substances on ecologically relevant herbivores were not examined, but the weight of circumstantial evidence is impressive.

If secondary substances are the end product of energy-demanding synthesis, it is reasonable to assume that there has been positive selection for their production. Secondary chemicals are known to be produced by energy-requiring specific syntheses from smaller molecules (Soloman and Crane, 1970). Most or at least part of the pathways of synthesis have been identified for alkaloidal secondary substances of species of *Datura* (Robinson, 1968, 1974), *Senecio* (Bull *et al.*, 1968), *Papaver, Nicotiana, Lupinus* (Robinson, 1968), and *Lycopodium* (Leete, 1963) and for phenolics in a large number of plant species (Ribereau-Gayon, 1972). Plant alkaloids have been proposed to be nitrogenous waste products analogous to the nitrogenous excretory products of animals (see James, 1950), but general anabolic synthesis of alkaloids (and secondary metabolites in general) by plants, as opposed to catabolic formation of ammonia, urea, uric acid, allatoin, etc., in animals, argues against this interpretation.

Modes of action of some secondary substances are known or suspected. Feeny (1968, 1970) demonstrated that the growth of winter moth (*Operophtera brumata* L.) larvae on an artificial diet containing casein as the protein source was significantly reduced by the presence in the diet of as little as 1% fresh weight of oak leaf tannin. Pupal weight, which is positively correlated with subsequent survival and fecundity rates, was also lower in the larvae fed on the diet containing tannin. Tannin acts by forming relatively indigestible complexes with leaf proteins, thereby reducing the rate of assimilation of dietary nitrogen (Feeny, 1969). Tannins inhibit a wide variety of animal and microbial digestive enzymes (Goldstein and Swain, 1965; Van Sumere *et al.*, 1975), reduce the ability of microorganisms to break down leaf protein (Handley, 1961), inhibit fungal growth (Williams, 1963), and render herbage less palatable to sheep (Wilkins *et al.*, 1953) and less digestible to cattle (Donnelly and Anthony, 1973). Essential oils inhibit rumen microbial activity in ruminants and disrupt digestive processes (Oh *et al.*, 1967). Cyanogenic glycosides are hydrolyzed by enzymes to yield a sugar and an α-hydroxynitrile, which dissociates to release HCN, a substance potentially lethal to any organism with an electron transport system involving cytochromes (Rehr *et al.*, 1973a; Conn and Butler, 1969). Considerable information is known concerning the effect of pyrrolizidine alkaloids on the kidneys (Bull *et al.*, 1968) and on the central nervous system (Kingsbury, 1964) of vertebrates. Insect hormone analogs (Slama, 1969; Heftmann, 1970; Nakanishi *et al.*, 1972) and antihormones (Bowers *et al.*, 1976) interfere with insect metamorphosis and development. Similarly, estrogenic vertebrate hormone analogs interfere with implantation and embryo development (Allen and Kitts, 1961; Bickoff, 1962; Loper, 1968). Toxic nonprotein amino acids, instead of normal dietary amino acids are incorporated into proteins, leading to defective protein structure (Fowden *et al.*, 1967; Rehr *et al.*, 1973b). Cardenolides disturb muscle action (Roeske *et al.*, 1976). *Solanum malacoxylon* contains high concentrations of vitamin D_3. Ingestion of the plant causes livestock to become hypercalcemic and hyperphosphatemic with extensive calcification of soft tissues due to hyperactivation of Ca^{2+} and PO_4^{3-} receptor proteins in the gut wall (Wasserman *et al.*, 1976).

The importance of toxic secondary substances to plants is indirectly indicated by the intricate detoxification mechanisms evolved by animals. Most important of these is the microsomal mixed-function oxidase (MFO) system, which catalyzes such reactions as oxidation, hydroxylation, and dealkylation in both vertebrates (La Du *et al.*, 1971) and insects (Krieger *et al.*, 1971; Wilkinson, 1976). Among vertebrates MFO activity is concentrated in the liver, through which all blood returning from the gut passes before distribution to the rest of the body. In insects MFO activity is concentrated in the midgut tissue, the major region of nutrient and toxin uptake. High MFO activity is also present in the fat body of insects which appears to act as a "scrubber" for dietary toxins, commonly lipophilic, which have survived detoxification in the midgut tissue, and have entered the hemolymph. Other important detoxification enzymes include esterases and hydrases (Hollingworth, 1976) and aldo-keto reductases (Bachur, 1976). After initial metabolism, a water-solubilizing group such as sulfate, glycine, glucose (insects), or glucuronic acid (mammals) is attached, and the conjugate is excreted (Wilkinson, 1976). Cyanide is metabolized to relatively nontoxic thiocyanate ion and excreted as such in a wide variety of insects and mammals (Rothschild, 1972). In ruminants dietary chemicals are detoxified to a large extent in the rumen by microbial activity (Freeland and Janzen, 1974). The arginyl-tRNA synthetase of the bruchid beetle

Caryedes brasiliensis discriminates between L-arginine and L-canavanine (a highly toxic analog of L-arginine present in the host seeds), and canavanyl proteins are not formed (Rosenthal *et al.*, 1977). The silkworm *Bombyx mori* detoxifies the phytoecdysone ecdysterone, present in its host plants *Morus* spp., by a combination of slow absorption through the digestive tract, rapid excretion, and catabolism of absorbed ecdysterone (Hikino *et al.*, 1975). Na^+-K^+-ATPase from neuronal tissues of the monarch butterfly is approximately 300 times less sensitive to inhibition by ouabain, a cardenolide present in host milkweeds, than that from nonadapted insects (Vaughan and Jungreis, 1977).

The relative toxicity and deterrency of plant secondary metabolites to various animal taxa are in need of review and further experimentation. Selective toxicants are known, including botanical (witness pyrethrins) and synthetic compounds (Jacobson and Crosby, 1971; Hollingworth, 1976). Many of the data on selective toxicity have resulted from the intensive search for insecticides with a low toxicity to vertebrates (Wilkinson, 1976). However, in general, defensive plant secondary metabolites have a broad spectrum of activity, and when these compounds are used by plants to deter antagonists it is adaptive for them to have such broad activity since it is unlikely that plants can perceive the exact nature of all potential threats. It is no accident that the most widespread defensive plant secondary metabolites, namely, tannins and related phenolics, act mainly by inhibition of digestive processes, providing protection against all classes of enemies that acquire nutrients by the breakdown of plant tissues. This includes both herbivores and pathogens.

Of course, plant defensive substances do not equally affect all potential enemies. Specialized herbivores with narrow host ranges are often much more resistant to the effects of defensive substances in their host plants than are generalists or herbivores that do not normally eat the plant in question. Specialists display potent detoxification and tolerance mechanisms for plant defensive substances (Self *et al.*, 1964; Hikino *et al.*, 1975; Rosenthal *et al.*, 1977; Vaughan and Jungreis, 1977), use these substances as cueing agents to aid in the location of host plants (Naya and Fraenkel, 1963; Maxwell *et al.*, 1963; Chambliss and Jones, 1966; Smith, 1966; Rees, 1969; Matsumoto, 1970; Schoonhoven, 1972; Kogan, 1977; Staedler, 1977), and sequester them for their own defense, particularly in the case of insects that feed on relatively rare, herbaceous, and early successional plants (Rothschild, 1973; Ródriguez and Levin, 1976; Roeske *et al.*, 1976; Morrow *et al.*, 1976). However, these differential effects of plant secondary metabolites on animals are usually better regarded as resulting from counteradaptation in the animals due to coevolution (Section IV,G–J) rather than as reflecting fundamental differences in toxicity or deterrency of the substances to various animal taxa. This may be particularly true in comparisons of the relative toxicity or deterrency of secondary metabolites within a closely related group of animals, for instance, insects. At higher taxonomic levels of comparison, for instance, among insects, reptiles, birds, and mammals, differential toxic or deterrent effects are more likely to be partly due to fundamental differences in animal metabolism. Few data are presently available, but Swain (1977a,b) has suggested that alkaloids may be more deterrent to mammals than to reptiles, birds, or insects (but see Section V).

III. OPTIMAL DEFENSE THEORY

Spatial and temporal patterns of foraging by herbivores and other animals have received a great deal of attention and mathematical analysis from ecologists in recent years (MacArthur and Pianka, 1966; Schoener, 1971; Pulliam, 1974; Emlen and Emlen, 1975; Pyke *et al.*, 1977; Orians and Pearson, 1979). A central hypothesis of optimal foraging theory is that animals forage in the way that maximizes their inclusive fitness and that the observed foraging patterns should reflect this maximization. Operationally, this is often taken to mean that animals should forage such that assimilation of energy or nutrients per unit time or per unit expenditure is maximized. In principle, these ideas are sound, but successful analysis of many trophic interactions, particularly plant–herbivore interactions, can be obtained only when equal consideration of evolved defense by resources is included.

In contrast to optimal foraging theory, very little mathematical formulation of optimal defense theory has yet been attempted. In the case of antagonistic plant–herbivore interactions, the nonmathematical treatments of White (1969, 1974, 1976), Janzen (1973b), McKey (1974), Haukioja and Hakala (1975), Feeny (1976), Rhoades and Cates (1976), Atsatt and O'Dowd (1976), and Benz (1977) provide a useful basis for a theory of optimal defense by plants. The following hypotheses 1 and 2 and corollary hypotheses (a)–(d), which follow from 1 and 2, are either implicit or explicit in these schemes and should be explicit in any general theory of optimal defense by organisms against predation (predation in the broad sense, including true predators, herbivores, parasites, and pathogens).

1. Organisms evolve and allocate defenses in the way that maximizes individual inclusive fitness.

2. Defenses are costly, in terms of fitness, to organisms. Therefore, less well defended individuals have higher fitness than more highly defended individuals when enemies are absent. Costly defenses are more effective than less costly defenses. The cost of defense, in terms of fitness, is due to the resultant diversion of energy and nutrients from other needs.

(a) Organisms evolve defenses in direct proportion to their risk from predators and in inverse proportion to the cost of defense, other things being equal.

(b) Within an organism, defenses are allocated in direct proportion to the risk of the particular tissue and the value of that tissue in terms of fitness loss to the organism resulting from attack on that tissue and in inverse proportion to the cost of defending the particular tissue.

Optimal defense theory thus presumes optimal foraging by herbivores and other enemies since this determines the degree of risk.

(c) Defenses are costly; therefore, commitment to defense is decreased when enemies are absent and increased when organisms are subjected to attack.

(d) Defenses are costly; therefore, commitment to defense is a positive function of the total energy and nutrient budget of the organism after normalization for all contingencies to which energy and nutrients must be allocated and is negatively related to energy and nutrients allocated by the organism to other contingencies.

It follows from (d) that environmentally stressed individuals should be less well defended against predators (broad sense) than unstressed individuals.

To illustrate that a combination of optimal foraging and optimal defense theories can lead to predictions differing from those obtained from optimal foraging theory alone, I shall consider a hypothetical seed herbivore utilizing two species of seeds, one of which is large and the other small, of equal availability. Optimal foraging theory predicts that the forager should prefer to eat large seeds since this strategy maximizes mass intake rate. Thus, large seeds are at greater risk than small seeds. Therefore, according to optimal defense theory, they should evolve more effective defenses than small seeds. The net effect of differing size and evolved defenses of the seeds should be that preference by the herbivore for the two seed species should approach equality. It is assumed in this example that the defense type of each seed species is the same, varying only in degree, and that seed size is determined largely by factors other than herbivory (independent variables discussed more fully in Sections IV,G and H).

IV. OPTIMAL DEFENSE THEORY APPLIED TO PLANT–HERBIVORE INTERACTIONS

A. Hypothesis 1

Hypothesis 1 is difficult, perhaps impossible, to test directly. It is perhaps more an axiom than a hypothesis, the truth of which can be evaluated only indirectly by testing corollary hypotheses such as (a)–(d) which naturally follow from it.

B. Hypothesis 2

Evidence in support of this hypothesis is as follows. In the absence of herbivores, acyanogenic morphs of *Trifolium repens* have higher vegetative and sexual reproductive vigor than cyanogenic morphs (Foulds and Grime, 1972). Mutants of *Nicotiana* and *Datura* especially rich in alkaloids are stunted (Mothes, 1960). The growth rate of *Pinus monticola* is negatively correlated with resin total monoterpene content (Hanover, 1966). In culture, antibiotic-free strains of fungi often rapidly appear and outcompete antibiotic-producing strains (Mothes, 1976). Insect-resistant soybean cultivars produce a lower yield of seeds and accumulate nitrogen at a slower rate than insect susceptible cultivars (Tester, 1977) in the absence of herbivores. Varieties of crop plants selected for high yield are often more susceptible to insects, pathogens, and weeds (Pimentel, 1976). Further studies of growth rates, reproductive rates, competitive ability, and other fitness measures of morphs or mutants high and low in defensive substances should enhance our understanding of defense costs to plants.

For plants, the main determinant of fitness cost of defense is very probably the metabolic cost of synthesis and storage of defensive metabolites. It is clear that energy (in ATP) diverted to the synthesis of defensive substances must be subtracted from that available to the plant for all other contingencies such as production of new energy and nutrient-capturing tissues and reproduction. The same applies to plant nutrients. Inorganic nutrients tied up in the metabolic machinery for synthesizing defensive substances are unavailable for other uses. Nitrogen is a constituent of some defensive compounds, e.g., alkaloids, cyanogenics glucosinolates, and nonprotein amino acids, but these compounds usually occur at low concentrations, compared to total nitrogen, in plant tissues. Where they do occur at high concentrations, e.g., seeds, a dual defensive and nitrogen storage function is probable. Inorganic nutrients of limiting importance to plants, other than nitrogen, are rare in defensive compounds, and the majority of defensive compounds are composed solely of carbon, hydrogen, and oxygen. Therefore, the inorganic nutrient cost of plant defense probably resides largely in the metabolic machinery rather than in the defensive compounds themselves.

With our present knowledge of biosynthetic pathways it should be possible to estimate the cost of synthesis of defensive compounds directly in terms of ATP, at least in some cases. However, synthesis cost represents only part of the total cost. Sequestration of these substances by plants in special vacuoles (James, 1950; McKey, 1974; Mothes, 1976) requires additional expenditures. First there is the cost of vacuole construction. Second, it is possible that maintenance of concentration gradients of secondary substances across vacuole walls is an energy-requiring process. Thus, direct evaluation of the metabolic cost of chemical defense is not easy, particularly for substances with a dual protective and metabolic role.

That costly defenses are more effective than less costly ones has featured importantly in the coevolutionary schemes of Feeny (1976) and Rhoades and Cates (1976) (Section IV,C). This is intuitively satisfying, but to my knowledge little theoretical justification for this concept has yet been provided. It is clear that the effectiveness of defense evolved by an organism should be directly related to the degree of selective pressure exerted by enemies, and since effectiveness can be directly related to amount of defense (for instance, amount of a given defensive chemical) the concept seems, at least on a simple level, to be justified.

C. Hypotheses a and b

Feeny (1975, 1976), Rhoades and Cates (1976), and Cates and Rhoades (1977a), incorporating to a degree the ideas of McKey (1974) and Janzen (1973b), have proposed that the type of defense and degree of defensive commitment evolved by plants are directly related to the risks of discovery of plants or individual plant tissues by herbivores. Feeny (1975, 1976) has proposed the term "quantitative" defenses for those plant secondary substances such as tannins, resins, and silica that act in a dosage-dependent fashion such that degree of protection from herbivores is directly related to tissue concentration of the substance. Quantitative defensive substances such as tannins and resins are often present in high concentrations (Feeny, 1976; Rhoades and Cates, 1976), up to 60% dry weight in the case of tannins. Mechanical defenses such as tough leaves are also included in the quantitative defense category. "Qualitative" defenses, on the other hand, include plant secondary substances such as cardenolides, glucosinolates, alkaloids, cyanogenic compounds, and many others which are characteristically present in low tissue concentration (commonly <2% dry weight) and which, although less costly to plant metabolism than quantitative defenses and protect against nonadapted herbivores, provide little protection even at very high concentration from specialized herbivores that have evolved detoxification or tolerance mechanisms. Quantitative defenses are characteristic of "apparent" plants and plant tissues, which are easy for herbivores to locate, whereas qualitative defenses are characteristic of "unapparent" plants and tissues. Plant or tissue apparency, as defined by Feeny (1976), includes all characteristics that can affect the susceptibility to discovery of plants or tissues by herbivores, such as persistence, growth form (herbaceous or woody), population density, seral stage, environmental complexity, and *host-finding adaptations of all relevant herbivores.*

Rhoades and Cates (1976) point out that most qualitative defenses of "ephemeral" (≈ unapparent) plants and plant tissues are designed to interfere with internal metabolism of herbivores, whereas most "predictable and available" (≈ apparent) plants and plant tissues appear to utilize quantitative defenses such as tannins and refractory carbohydrates that are designed to reduce the digestibility of plant tissues. Approximately 80% of woody perennial dicotyledenous plant species contain tannins as opposed to 15% of annual and herbaceous perennial dicot species, showing the importance of defense by tannins in the former group (Bate-Smith and Metcalf, 1957; Rhoades and Cates, 1976). Refractory carbohydrates in which sugar residues are either methylated (Bacon and Cheshire, 1971) or acylated (Swain, 1977a) are resistant to animal or microbial carbohydrases and are common constitutents of trees and other perennials. Qualitative defenses, on the other hand, are utilized by both herbaceous and woody plants. When present in woody plants, qualitative defenses are usually concentrated in unapparent tissues, whereas the quantitative defenses of these plants are usually concentrated in apparent tissues.

McKey (1974) stressed the importance of both vulnerability of plant parts to attack by herbivores and the value of these parts to the plant in determining optimal distribution of defense. Tannins, resins, alkaloids, and other defensive substances are often concentrated at or near the plant surface, in the epidermal or subepidermal layers (McKey, 1974; Swain, 1977a), in the bark (Howes, 1953), in trichomes (Levin, 1973), or even on the epidermal surface (Rhoades, 1977a,b). Growing shoots and young leaves are more valuable to plants than mature leaves (McKey, 1974) since damage to young leaves and buds should result in a greater loss in plant fitness than an equal quantity of damage to mature leaves. Therefore, all other factors being equal, plants should invest more heavily in the defense of growing tips and young leaves than in the defense of mature leaves. For terrestrial macrophytes it seems that value to the plant should decrease in the order roots, stems, young leaves, mature leaves, reproductive structures, since each structure in this gradient is totally dependent on the function of each of the previous structures. However, other factors are often not equal. The risk of each of these structures will vary due to persistence, availability (for instance, roots are relatively unavailable to many herbivores), and proximate nutrient content (for instance, leaves contain a higher proportion of nitrogen, phosphorus, total available energy, etc., than stems due to higher metabolic activity in leaves and greater quantities of structural materials in stems). It may be metabolically more costly to defend young undifferentiated or embryonic tissues than fully differentiated or mature tissues (McKey, 1974; Orians and Janzen, 1974). During the lifetime of a plant there may be drastic changes in defense requirements of different organs. Annual plants and the above-ground parts of herbaceous perennials die after setting seed, and defensive substances are redistributed accordingly to the reproductive organs, seeds, and underground storage organs at this time (McKey, 1974). Thus, risk, value, and cost all probably contribute to the optimal within-plant distribution of defensive adaptations.

There is remarkable agreement between the theories of Feeny (1975, 1976) and Rhoades and Cates (1976) regarding most points. For instance, both theories predict that the metabolic cost to the plant and effectiveness against herbivores of qualitative defenses should be less than those of quantitative defenses. An important difference exists, however. Feeny's definition of apparency is a proximate and subjective one, subjective from the herbivore's viewpoint, and totally dependent on host-finding adaptations of the herbivore. This leads to a serious logical problem because natural selection should act on herbivores to maximize host-finding ability, and the intensity of such selection will depend on resource properties such as persistence, abundance, and seral stage. Thus, it is problematical whether any differences in subjective resource apparencies exist. Feeny's scheme is therefore based on a definition of plant apparency that renders the entire concept subject to disappearance due to the leveling effect of natural selection (see Sections IV,G and H).

To avoid this problem Rhoades and Cates (1976) defined their measure of the susceptibility to discovery of a plant or tissue by herbivores, "predictability and availability," in terms only of ultimate plant and environmental properties, namely, persistence, growth form, seral stage, abundance, and environmental complexity, and *excluding host-finding adaptations of herbivores.* Defined in this way "predictability and availability" is largely under the control of "independent variables" (Sections IV,G and H) and is minimally subject to the feedback effects of plant–herbivore coevolution. To minimize further terminological confusion I shall use Feeny's terminology with the important proviso that "apparency" will herein assume the meaning of "predictability and availability," or in other words "apparency" will be as defined by Feeny (1976) but excluding host-finding adaptations of herbivores. Feeny's concept of apparency is thus equivalent to "proximate apparency," whereas the concept of apparency as used in this chapter is equivalent to "ultimate apparency." Apparency will be examined (Sections IV,G–L) at the level of individual resources to herbivores, since plants are a composite of many resources that often vary in their apparency and utilization by herbivores. Highly unapparent resources such as young, rapidly flushed leaves often occur on highly apparent plants. Some examples of apparent and unapparent resources are listed in Fig. 2.

Further evaluation of hypotheses a and b will entail ascertaining the generality of the suggested correlations between defense type, amount (see also Sections IV,H and I) and degree of plant or tissue risk, and tissue value. Direct measurements in controlled experiments of the effects of damage to various organs on plant fitness are needed. There is little concrete evidence that it is metabolically more costly to defend some tissues (e.g., embryonic tissues) than others. This notion deserves close scrutiny by plant physiologists and developmental biologists.

D. Hypothesis c

If defenses are costly it should be advantageous for plants to be less well defended when enemies are absent than when plants are subjected to attack. We can thus expect plants to increase their commitment to defense when they are attacked by herbivores or other enemies (Haukioja and Hakala, 1975; Benz, 1977).

That plants react to invading pathogenic organisms by producing substances (phytoalexins) toxic to the invading organisms is well established (Deverall, 1977; Bell, 1974) and appears to be a general property of plants. Defensive reaction by plants to damage by herbivores is also common (Table I) and, like defensive reaction against pathogens, may be a general property of plants.

Green and Ryan (1972) showed that mechanical damage to leaves of tomato (*Lycopersicon esculentum*) and potato (*Solanum tuberosum*) leads to a systemic increase in concentration of a proteinase inhibitor in the

Table I. Effect of herbivore damage or simulated herbivore damage on plant defenses and fitness of subsequent herbivores

Plant species	[a]	[b]	Damage	Plant response[c]	Herbivore fitness	Induction time	Relaxation time	Ref.[d]
Lycopersicon esculentum	S	A	Grazing by Colorado potato beetle or mechanical damage	Systemic increase in concentration of proteinase inhibitor	?	12–100 hours	?	1
Solanum tuberosum	S	HP	Grazing by Colorado potato beetle or mechanical damage	Systemic increase in concentration of proteinase inhibitor	No effect on growth of Colorado potato beetle	12–100 hours	?	1, 2
Beta vulgaris	C	B	Infestation by beet fly	?	29–100% increase in mortality of beet fly	24 days	18 days	2, 3
Rumex obtusifolius	P	HP	Grazing by the beetle *Gastroidea viridula*	?	Slower growth rate and ovipositional avoidance by *G. viridula*	?	?	2, 4
Medicago sativa	L	HP	Attack by pea or spotted alfalfa aphids	Production of coumestrol (estrogen mimic) is stimulated	Lowered fertility in sheep	6 weeks	?	5
Senecio jacobaea	Co	HP	50% of leaves mechanically removed	Total leaf alkaloids and *N*-oxides increase 40–47% in undamaged leaves	?	2 days	?	6

(*Continued*)

Table I (*Continued*)

Plant species	[a]	[b]	Damage	Plant response[c]	Herbivore fitness	Induction time	Relaxation time	Ref.[d]
Carex aquatilis	Cy	HP	Mechanical grazing	Total phenolics and proanthocyanidins increase 30–40% and 40–50%, respectively, in undamaged foliage	?	1 year	?	7
Heteromeles arbutifolia	R	WPE	Insect damage	Cyanide content of leaves increases	?	?	?	8
Abies grandis and *Abies amabilis*	Pi	WPE	Attack by balsam wooly aphid	Production of todomatuic acid and related juvenile hormone mimics in the wood is stimulated	?	?	1 year	9
Picea sitchensis	Pi	WPE	Attack by the Sitka spruce weevil	?	Decreased larval success and ovipositional avoidance by spruce weevils	2 years	2 years	10
Pinus sylvestris	Pi	WPE	Attack by pine sawfly	Total phenolics increase in undamaged leaves	?	1 year	?	11

Pinus radiata	Pi	WPE	Attack by the European wood wasp and its symbiotic fungus	Local increase in resin and phenol content of the wood	?	?	?	12
Larix decidua	Pi	WPD	Defoliation by larch budmoth	Delayed leaf flush and smaller, tougher leaves with lower nitrogen and higher fiber and resin contents	Higher larval mortality and lower fecundity of budmoth	1 year	4–5 years	2, 13
Betula pubescens	B	WPD	Mechanical leaf damage	Total phenolics increase in neighboring undamaged leaves	Retarded pupation by the moth *Oporinia autumnata*	2 days	?	14
	B	WPD	Artificial defoliation	Delayed leaf flush and smaller leaves	Retarded pupation and lower pupal weights of *O. autumnata*	1 year	1 year	14

[a] Abbreviations: S, Solanaceae; C, Chenopodiaceae; P, Polygonaceae; L, Leguminosae; Co, Compositae; Cy, Cyperaceae; R, Rosaceae; Pi, Pinaceae; B, Betulaceae.

[b] Abbreviations: A, annual; B, biennial; HP, herbaceous perennial; WPD, woody perennial (deciduous); WPE, woody perennial (evergreen).

[c] "Undamaged leaves" or "undamaged foliage" refers to intact tissues of attacked plants.

[d] Key to references: (1) Green and Ryan (1972); (2) Benz (1977); (3) Röttger and Klingauf (1978); (4) T. Jeker, C. Mueller, and V. Volkart (unpublished); (5) Loper (1968), Loper *et al.* (1967); (6) D. F. Rhoades (unpublished); (7) S. Archer and D. F. Rhoades (unpublished); (8) Dement and Mooney (1974); (9) Puritch and Nijholt (1974); J. F. Manville (personal communication); (10) Overhulser *et al.* (1972), R. I. Gara (personal communication); (11) Thielges (1968); (12) Hillis and Inoue (1968); (13) Baltensweiler *et al.* (1977); (14) Haukioja and Niemelä (1976, 1977), P. Niemela (personal communication).

plants, and they postulated that this may reduce subsequent herbivore attack (Table I). Although Benz (1977) found that this reaction by potato has no effect on development and growth rate of the Colorado potato beetle (*Leptinotarsa decemlineata*), a specialist herbivore of potato, tomato, and related Solanaceae, effects on herbivores less highly coevolved with the plants are not excluded. Negative effects on herbivore fitness resulting from plant damage caused by previous attack or artificial grazing have been observed in dock (*Rumex obtusifolius*), alfalfa (*Medicago sativa*), beet (*Beta vulgaris*), larch (*Larix decidua*), birch (*Betula pubescens*), and spruce (*Picea sitchensis*) (Table I). In the case of alfalfa, larch, and birch these effects on herbivores have been shown to be associated with increased content of defensive substances and other changes in foliage properties (Table I). For toyon (*Heteromeles arbutifolia*), *Senecio jacobaea*, true firs (*Abies amabilis, A. grandis*), pines (*Pinus sylvestris, P. radiata*) and sedge (*Carex aquatilis*), an increase in content of defensive substances resulting from herbivore or mechanical damage has been noted, but effects on subsequent herbivores have not been examined (Table I).

Formation of plant galls is associated with substances introduced during oviposition by the adult insect or produced by the developing larva (Osborne, 1973; Carter, 1973), suggesting that the insect largely controls gall formation. However, the occurrence of large quantities of plant defensive substances in some galls (Howes, 1953) such as oak "apples," which are used as commercial sources of tannin, suggests that gall formation may in some, perhaps most, cases be due to an interaction between insect offensive stimuli and plant defensive response. Janzen (1977) takes an alternative, but not incompatible, view as to the adaptive significance of plant defensive substances in galls. He suggests that the formation of these substances is under the control of the gall-making insects and that the substances protect the gall makers from other herbivores and predators. It is possible that a continuum of adaptive significance for these gall substances exists. At one extreme they protect the plant against nonadapted gall makers, but at the other they have a negative effect on plant fitness in the face of attack by highly coevolved gall makers.

Induction times for defensive reactions by plants induced by herbivore or mechanical damage so far studied vary from as little as 12 hours to as long as 1 year or more (Table I). In the short term, plants should respond to grazing by increasing commitment to qualitative defense systems, since qualitative defensive chemicals are synthesized and sequestered at little cost to the plant, are readily transportable to the site of damage, and are effective at low concentration, at least against generalized herbivores. This may be the only type of response exhibited by unapparent plants. In the case of apparent plants, both a short-term increase in qualitative defense and a long-term increase in quantitative defense are to be expected. A short-term increase in quantitative defenses, such as tannins, is not to be expected since much larger quantities of materials must be synthesized, and much of the synthesis could take place only at the site of damage since polymeric tannins are not known to be transported. In the long term, apparent plants should increase commitment to quantitative defenses in response to herbivore damage since these defenses are effective against both generalist and specialist herbivores, and present damage by herbivores is a good predictor of future potential damage by herbivores. Thus, if short-term reaction by plants involves principally qualitative defenses, the effect on specialized herbivores is likely to be minimal compared to that on generalized herbivores. In contrast, long-term reaction involving quantitative defenses is likely to be effective against both specialist and generalist herbivores. If defenses are costly, relaxation of defense following attack is to be expected. Defensive relaxation times ranging between 18 days and 4 or 5 years have been observed (Table I).

Hypothesis (c) is relatively easy to test. Does herbivore damage to plants render them less suitable for growth and development of subsequent herbivores? If so, what physical or chemical properties of the plants are changing in response to attack? What are the induction and relaxation times of plant defensive responses? What type and extent of tissue damage are most effective in inducing defensive responses in plants? What are the relative effects of short- and long-term defensive responses on adapted and nonadapted herbivores? Are defensive metabolites synthesized *de novo* in response to attack, or are they transported to the attacked tissue type from synthesis sites or storage depots elsewhere in the plant? Is there a causal relationship between induction and relaxation of plant defense and herbivore population fluctuations?

E. Hypothesis d

A strong correlation between population fluctuation of herbivores, particularly phytophagous insects, and weather patterns is well known. This correlation largely gave rise to the concept of "density-independent" regulation of animal numbers (Andrewartha and Birch, 1954; Milne, 1957). In some cases the effect of weather has been attributed to a direct action on the insects (Klomp, 1968; Thompson *et al.*, 1976), but in others the insects have been shown to be well adapted to climatic extremes (Morris *et al.*, 1958; White, 1969, 1974, 1976), and in general the mechanism by which weather influences insect fitness remains obscure (Gunn, 1960; Dempster, 1963). White (1969, 1974, 1976) convincingly argues that the correlation is often due primarily to effects on the plants rather than on the insects. He shows that plant stress, induced particularly by drought, can lead to enhanced proximate nutritional characteristics of plants to herbivores, such as increased tissue total or soluble nitrogen or carbohydrate content, which in turn are associated with lower mortality and increased fecundity of phytophagous insects. Other stressful influences on plants that are associated with increased population numbers of insects include nutrient-poor soils, competition, pollution, disease, damage during cultural operations, old age, ice damage, lightning, and forest fires (Mattson and Addy, 1975). Forest fertilization often produces a marked increase in tree resistance to insects (Stark, 1965). Interestingly, such an increase in tree resistance has been shown in several cases to be associated with an *increase* in crude protein content of the foliage (Buttner, 1961; Smirnoff and Bernier, 1973). Thus, it is unlikely that an increase in foliage proximate nutritional quality is the cause of increased susceptibility to insects of trees grown under nutrient-deficient soil conditions.

I propose an extension of White's hypothesis that an increase in plant proximate nutritional quality is a major cause of increased susceptibility of physically stressed plants to insects. Physical stress to plants results in an imbalance between their proximate nutritional quality to herbivores (particularly available nitrogen and carbohydrate content) and the content and quality of their defensive substances. In addition to the changes in proximate chemistry associated with plant stress discussed by White (1969, 1974, 1976), a variety of conditions stressful to plants have important effects on their content of defensive substances (Table II). Concentrations of qualitative defensive substances (alkaloids, cyanogenics) in plant tissues generally increase, whereas concentrations of quantitative defensive substances (tannins, resins, and essential oils) decrease, in response to plant stress (Table II). This pattern can be explained if we assume that plants generally possess two or more defensive systems of differing cost. Under stress, plants compensate by decreasing their commitment to costly defenses and increasing their commitment to less costly but less effective defenses. These changes, together with any stress-induced increase in proximate nutritional quality, have a net positive effect on fitness of herbivores feeding on these stressed plants.

Bracken fern (*Pteridium aquilinum*), a particularly well studied plant, contains cyanogenic glycosides and tannins (Table II) in addition to several other substances of potential defensive importance, namely, ecdysones, thiaminase, silica, nontannic phenols, and lignin (Lawton, 1976). The concentrations of cyanogenics and tannin change in accordance with the proposed scheme, at least in response to shade. Investigation of stress-induced changes in defense allocation in other plants in which both costly quantitative defenses and less costly qualitative defenses are known may allow further substantiation of hypothesis (d). Such plants include *Heteromeles arbutifolia* (Dement and Mooney, 1974) and *Prunus* spp. (Kingsbury, 1964; Swain and Hillis, 1959), which contain both cyanide and tannin, and species of *Acacia* and *Prosopis,* which often contain either cyanide plus tannin or alkaloids plus tannin (Cates and Rhoades, 1977b; R. G. Cates and D. F. Rhoades, unpublished). In fact, any woody perennial plant for which a qualitative defense system is known is worthy of investigation as a potential subject for such a study. This is because it is likely that the plant will also contain tannin, since most woody perennials do. The effects of plant stress on fitness of herbivores, plant tissue content of defensive metabolites, and plant tissue proximate nutrient content are relatively easy to investigate both in the laboratory and in the field.

F. Group Approach to Optimal Defense Theory

Most of the discussion in this chapter reflects my bias toward arguments based on the effect of selection at the level of the individual. This is not necessarily the only fruitful approach to the analysis of adaptive patterns in plant defensive attributes. Atsatt and O'Dowd (1976) present an alternative viewpoint. Elaborating on the theme that members of a plant community are affected by and in many ways are functionally dependent on properties of other members of the plant community, they develop the concept of plant "guild" defense against herbivores. They also emphasize the "ways in which plant associates can function as antiherbivore resources in ecological time and discuss the possible selective value of defense guilds through evolutionary time." Their treatment is richly illustrated with examples of mechanisms whereby individual members of a guild of plant species gain protection from herbivores by the presence in the association of "insectary plants," whose nectaries aid in the maintenance of predators and parasites, and "repellent plants," that cause nondiscriminating herbivores to avoid the guild as a whole. The concept of plant "gene conservation guilds," in which gradients of constitutional resistance to herbivores minimize counteradaptation in the herbivores, is proposed (see also Pimentel and Belotti, 1976). The potential importance of these effects to agriculture is discussed. Some of the arguments contain a strong element of group selection. For instance, observing that in complex environments insects

Table II. Effect of environmental stress on concentrations of defensive substances in plants

Plant species	Plant family[a]	Growth form[b]	Defensive substance	Stress or plant condition	Changes in content of defensive substance		Ref.[c]
					Increases	Decreases	
Cynodon dactylon	G	HP	Cyanogenic compounds	Drought, frost, high temperature	*		1
Sorghum halpense	G	HP	Cyanogenic compounds	Drought	*		1
S. vulgare	G	A	Cyanogenic compounds (dhurrin)	"Unhealthy plants"	*		1
Zea mays	G	A	Cyanogenic compounds	"When normal development is impeded"	*		1
Trifolium repens	L	HP	Cyanogenic compounds	Drought	*		2
Manihot esculenta	E	HP	Cyanogenic compounds	Shade	*		3, 4
Conium maculatum	U	B	Alkaloids	Hot, dry season	*		1
Atropa belladonna	S	HP	Alkaloids	Etiolation	*		5
Ricinus spp.	E	HP	Alkaloids (ricinine)	Etiolation	*		5
Lupinus spp.	L	A, HP	Alkaloids	Etiolation	*		5
Berberis darwinii	B	WPE	Alkaloids (berberine)	Etiolation	*		6

Datura sp.	S	A	Alkaloids	Nutrient stress	*		7
Pteridium aquilinum	P	HP	Cyanogenic compounds	Shade	*		4
			Condensed tannins	Shade		*	
Prunus domestica	R	WPD	Condensed tannins	"Old trees with broken limbs and damaged trunks obviously in unhealthy condition" or shade		*	4, 8
Acer rubrum	A	WPD	Gallotannins	SO_2 pollution		*	9
Rubus spectabilis	R	WPD	Tannins	Shade		*	10
R. parviflora	R	WPD	Tannins	Shade		*	10
Pinus spp.	Pi	WPE	Essential oils	"Physiologically weakened trees"		*	11
Conifers	Pi	WPE	Resin	Lack of soil nutrients		*	12

[a] Abbreviations: G, Gramineae; L, Leguminosae; E, Euphorbiaceae; U, Umbelliferae; S, Solanaceae; B, Berberidaceae; P, Polypodiaceae; R, Rosaceae; A, Aceraceae; Pi, Pinaceae.

[b] Abbreviations: A, annual; B, biennial; HP, herbaceous perennial; WPE, woody perennial (evergreen); WPD, woody perennial (deciduous).

[c] Key to references: (1) Blohm (1962); (2) Rogers and Frykolm (1937); (3) Nestal and MacIntyre (1973); (4) Cooper-Driver *et al.* (1977); (5) James (1950); (6) Cromwell (1933); (7) Andries (1956); (8) Hillis and Swain (1959); (9) Nozzolillo (1975); (10) D. F. Rhoades (unpublished); (11) Mattson and Addy (1975); (12) Stark (1965).

sometimes oviposit on plants not suitable for normal larval growth and development, Atsatt and O'Dowd propose the concept of "attractant-decoy" plants. Attractant-decoy plants are postulated to lure away herbivores, presumably to the plant's own detriment, to the advantage of the guild as a whole, which is composed largely of distantly related plants.

Since "attractant-decoy" plants are at a selective disadvantage in the guild with respect to herbivory, it is necessary to elaborate mechanisms by which their continued presence within the guild is maintained. This is not satisfactorily accomplished by Atsatt and O'Dowd (1976). Such altruism could be maintained by reciprocal benefits to the attractant-decoy plants, perhaps partly due to repulsion of herbivores from the guild as a whole by repellent members. However, herbivory alone would not be expected to maintain altruism within a given guild since repellent plants would doubly benefit, both from their own repellency and the attractiveness of the attractant-decoy plants, whereas the attractant-decoy plants would only singly benefit. Since they are at a selective disadvantage within a guild, the numbers of altruistic members would be expected to continually decrease, leading to their extinction within the guild. The entire guild must then somehow be replaced by a similar guild containing altruistic members. Alternatively, additional benefits to the altruists, other than group protection against herbivores, must be postulated. As is often the case with arguments based on group selective influences, the arguments become complicated and *ad hoc* hypotheses must be called on. Such arguments are unlikely to find favor among proponents of evolution by maximization of individual inclusive fitness. A parsimonius interpretation based on such is that it is advantageous for individual plants to possess adaptations likely to confuse the sensory systems of insects, causing ovipositional mistakes. Ovipositional mistakes are more likely in diverse communities, so members of such communities ("guilds") do gain a measure of protection (Tahvanainen and Root, 1972; Root, 1973). Thus, in this case, arguments based on group or individual selection arrive at similar conclusions. The relative usefulness of these two approaches in unraveling the complexities of plant–herbivore interactions remains to be seen.

G. Coevolution

Smith (1975) pointed out the importance of distinguishing between selective forces that act as dependent variables and those operating as independent variables. In a coevolving pair (e.g., Fig. 1A, cycle I), characteristics of one member of the pair influence characteristics evolv-

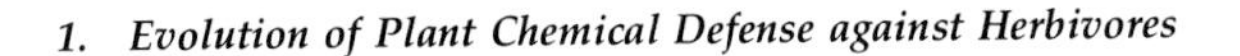

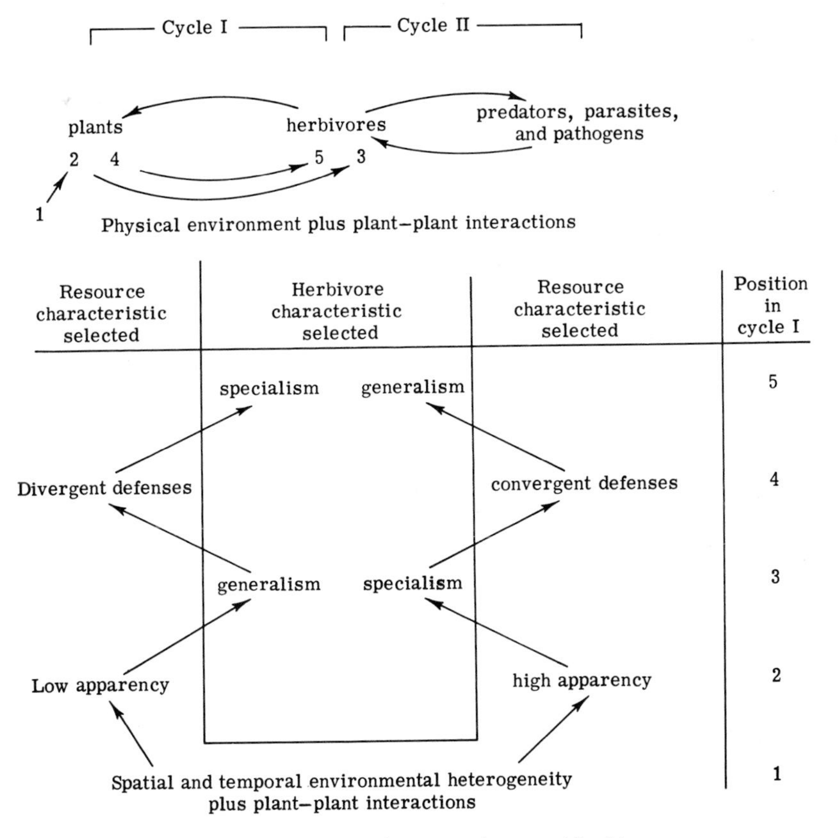

Figure 1A. Coevolution between plants and herbivores.

ing in the second member. These evolving characteristics will, in turn, feed back to affect characteristics evolving in the first member, and so on (Fig. 1A; 2–5). Characteristics 2–5 (Fig. 1A) are dependent variables, subject to feedback effects from each other and to alteration by changes in the independent variables (Fig. 1A; 1). Only physical factors (as opposed to biotic factors) can be truly independent variables. However, some biotic factors, for instance, plant–plant competition in the present case, that are minimally subject to feedback effects from the coevolving cycle are best regarded as independent variables.

To analyze a coevolving system it is important first to describe the independent variables responsible for some of the patterns in the system. These independent variables, by producing initial patterns, can thus be considered to be the driving force for the coevolving cycle, and the initial patterns provide an entry into the cycle. In this way circular

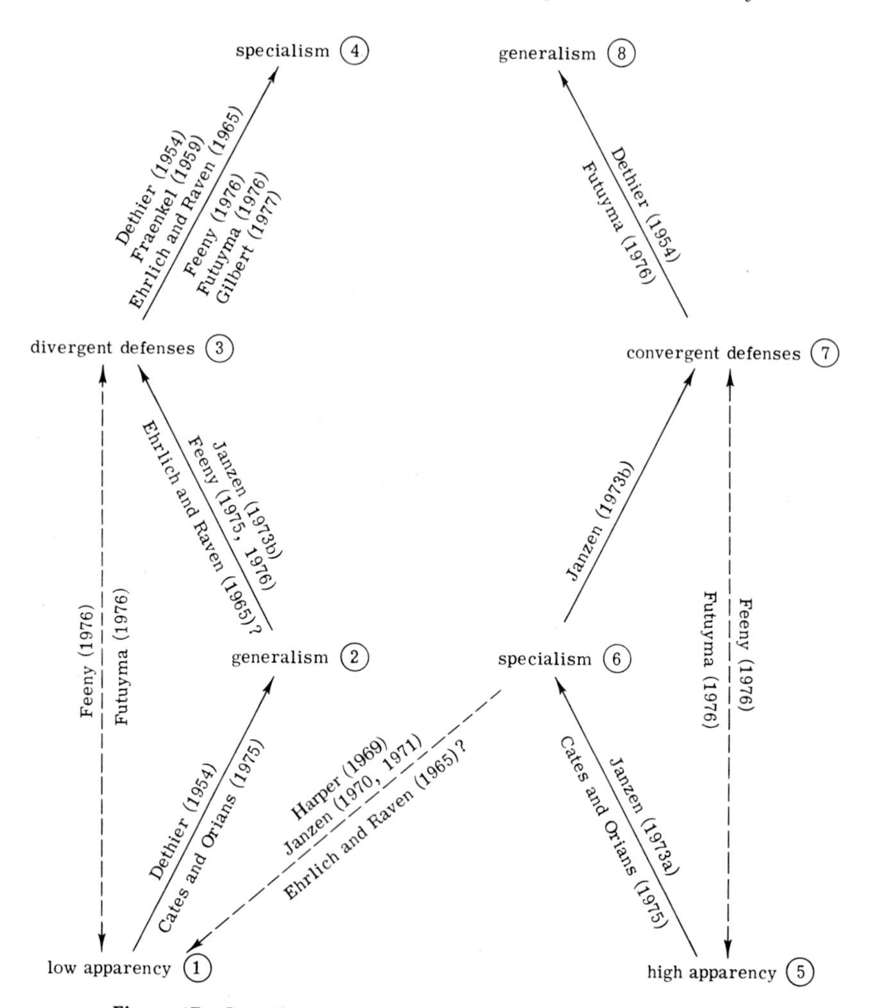

Figure 1B. Contributions to plant–herbivore coevolutionary theory.

reasoning can be avoided, and the chain of causation among all the variables can be unraveled.

H. Coevolution between Plants and Herbivores

The following, adapted from Rhoades and Cates (1976), is an attempt to unravel coevolution between plants and antagonistic herbivores (Fig. 1A, cycle I), largely ignoring coevolution between herbivores and their predators, parasites, and pathogens (cycle II) and also ignoring other cycles not depicted, e.g., between plants and mutualistic herbivores such as pollinators, the assumption being that it is useful to consider an isolated cycle. Dietary generalism and specialism are relative terms referring to opposite ends of the monophagy–polyphagy continuum.

For plant–herbivore coevolution the important independent variables are probably spatial and temporal heterogeneity in the physical environment together with competitive plant–plant interactions (Fig. 1A; 1), and the initial pattern produced by these variables is the existence of apparent and unapparent resources (see Section IV,C) (Fig. 1A; 2). For tissue apparency the most important variable is probably temporal environmental heterogeneity in water availability, temperature, and light regime since plant phenological events such as maturation of leaves and reproductive structures are so closely tied to the seasons. For plant apparency the important independent variables are temporal and spatial environmental heterogeneity, which by providing gradients in soil structure, nutrients, light regime, etc., and thus the opportunity for resource partitioning, together with competitive plant–plant interactions, have probably contributed to the evolution of plant species of various growth forms, seral stages, and abundances. Resource apparency can never be entirely free from alteration by feedback from herbivores. For instance, herbivory can be expected to alter plant abundance (Harper, 1969; Janzen, 1970). However, for present purposes it is necessary only that there be a residual contribution from the independent variables to resource apparency.

Unapparent resources should escape from their herbivores in space and time to a larger extent than should apparent resources, since it is difficult for herbivores to locate unapparent resources. Furthermore, escape in space and time should be more effective against specialist herbivores than against generalist herbivores because specialists have no alternative food source. Specialist herbivores must allocate time and energy to a search for their host, and the less apparent this resource is, the greater will be herbivore mortality during the search. For generalist herbivores, on the other hand, the apparency of individual resources is of less consequence, since generalists can opportunistically utilize whatever resource happens to be available. In other words, high apparency of resource should select for dietary specialism in herbivores, whereas low apparency should select for generalism, all other resource properties being equal (Fig. 1A; 3).

For the remainder of the discussion, plant defenses are examined mainly in terms of defensive chemistry, but the arguments are general and can be applied to mechanical and morphological defenses. If unapparent resources escape from herbivores to a greater extent than do apparent resources, other things being equal, unapparent resources should evolve defenses that are less costly to produce and store but more easily circumvented than those utilized by apparent resources, and there is evidence that this is so (Feeny, 1976; Rhoades and Cates, 1976). Escape should be more effective against specialists than against generalists. Therefore, the defenses evolved in unapparent resources should be directed particularly against generalist herbivores, and this should give rise to a divergent system of chemical defenses in such resources (Fig. 1A; 4). This is because selective forces acting on generalist herbivores should render these herbivores most capable of accommodating the "average defensive chemistry" of their host range. Thus, those resources whose defensive chemistry deviates most widely from this average will be at a selective advantage compared to those of more common chemical type. This mode of disruptive frequency-dependent selection has been termed "apostatic selection" (Clarke, 1962). The vast array of qualitative defensive chemical types utilized by herbaceous plant species and unapparent tissues of woody species and the diversity of metabolic systems within herbivores on which they act (Feeny, 1976; Rhoades and Cates, 1976) are probably the result of this divergent selective process.

Conversely, the defenses of apparent resources should be directed particularly against specialist herbivores. In the extreme case, in which each apparent resource population experiences attack by its own specialist herbivore, the defense evolved in any one resource will not be affected by the defenses evolved in other resource populations in the community. This stands in contrast to the condition for evolution of the defenses of unapparent resources. We can therefore expect parallelism or convergence of chemical defense between populations of apparent resources (Fig. 1A; 4).

The wide distribution of tannins in woody perennial dicotyledonous plant species and their high concentration in the persistent tissues of these species, in contrast to their rarity in herbaceous dicot species (Bate-Smith and Metcalf, 1957; Rhoades and Cates, 1976), is consistent with defensive convergence (or parallelism) by apparent resource populations. There are few basic chemical classes of tannins: gallotannins, elagitannins, and condensed tannins, and sometimes two or even all of these classes are present in a given woody species (Bate-Smith, 1977). Tannins act as antiherbivore substances by complexing with plant proteins and carbohydrates and also with herbivore digestive enzymes when plant tissues are injured by herbivores, to disrupt digestive processes and reduce nutrient availability (Goldstein and Swain, 1965; Feeny, 1970; Van Sumere *et al.*, 1975). All tannins complex with proteins and carbohydrates by a similar mechanism involving phenolic hydroxyl hydrogen bonding, although intervention by phenol oxidases leading to covalent binding may be involved in some cases (Loomis and Battaile, 1966; Horigome and Kandatsu, 1968; Allison, 1971; Pierpont, 1971; Van Sumere *et al.*, 1975). Rhoades (1977a,b) provided evidence that the phenolic resin of creosote bush (*Larrea* spp.), an apparent plant species, is a tanninlike protein- and starch-complexing agent that deters herbivores by disrupting digestion in a similar fashion. Concentration of the resin in the most persistent leaf age class is again suggestive of defensive convergence by apparent resources. Therefore, in contrast to unapparent resources, the functional and chemical diversity of defensive systems between apparent resource populations is lower. However, tannins are polymeric compounds, and tannin extracts typically contain a high diversity of chemical species of varying chain length, linkage pattern, and protein-complexing ability (Haslam, 1975; Ribereau-Gayon, 1972). Similarly, creosote bush resin, composed of monomeric phenols, is a complex mixture (Ródriguez *et al.*, 1972; Sakakibara *et al.*, 1976). There is thus considerable scope for individual variation of chemical profile within populations of apparent resources. Within- and between-population patterns of defensive chemistry are discussed more fully in Section IV,I.

Thus, although in the case of unapparent resources natural selection should favor the evolution of discrete and different defenses in each resource population, the defenses evolved in apparent resources should be more similar to each other. For reasons of symmetry the term "convergent defenses" will be used rather than "parallel defenses," but for present purposes these terms will be regarded as interchangeable (Fig. 1A; 4).

It is at stage 4 (Fig. 1A) that the leveling effect of natural selection becomes operative. Convergent defenses should select for dietary generalism in herbivores (Fig. 1A; 5) since if the defenses in each resource population are similar it should not be adaptive for herbivores to expend time and energy seeking a particular subset of the resources. Conversely, divergent defenses should select for dietary specialism in herbivores since if all resources are different the ability of a generalist to accommodate any one of the defenses should be less than that of a specialist herbivore, and selective advantage should accrue to specialists.

Therefore, although from considerations of resource apparency alone unapparent resources would be expected to experience attack mainly by generalists, and apparent resources mainly by specialists, the evolved defenses of the two resource groups will tend to counteract this effect via negative feedback. A further leveling may be expected from the influence of predators, pathogens, and parasites on herbivores (cycle II). Specialist herbivores, particularly those specializing on apparent resources, are in turn a more apparent resource to their enemies than are generalists. This favors selection for generalism in herbivores, to a greater extent for herbivores utilizing apparent resources than for those utilizing unapparent resources. The ratios of specialist to generalist herbivore pressure on apparent and unapparent resources should therefore tend toward equality under the combined influences of resource apparency, evolved defenses, and cycle II.

Similar arguments can be applied to fitness loss of both resource types due to herbivory. Since apparent resources are at greater risk than unapparent resources, they should evolve more effective defenses than unapparent resources. This will result in the tendency for fitness losses by both groups due to herbivory to be equalized.

Thus, there should be little difference in either herbivore-induced fitness loss or specialism of herbivory for apparent versus unapparent resources. For the above arguments to hold true, it is necessary that both the ratio of specialist to generalist herbivore pressure and fitness loss due to herbivory be slightly greater on apparent than on unapparent resources. Whether these differences should be large enough to be statistically detectable is a matter of conjecture.

It is an interesting observation that the coevolution scheme as depicted in Fig. 1A resembles a standing wave. Observable patterns (apparency, defense type) appear at the antinodes, whereas nodal patterns (dietary specialism or generalism) appear in complementary pairs, which cancel each other in the vertical direction. In other words, the leveling effect of natural selection acts for nodal patterns but not for antinodal patterns. Perhaps this is true for coevolution in general.

I. Within- and between-Population Patterns of Convergence and Divergence in Plant Defensive Chemistry

Freeland and Janzen (1974) suggested that the ability of herbivores to accommodate (detoxify or tolerate) defensive chemicals in their host plants limits the quantity of plant material of a given defensive type that can be consumed in a feeding bout. The amount of any particular resource consumed is limited to the amount containing the quantity of defensive chemicals that will not saturate the detoxification or tolerance mechanism of the herbivore for this particular defensive set. Saturation of detoxification ability or tolerance for one set of defensive chemicals does not necessarily saturate accommodation for a different set. Applied to mobile generalist herbivores this argument suggests that to maximize detoxifiable food intake a herbivore should daily seek out and take food from a variety of resources in its host range, rather than concentrate on one readily available resource (Freeland and Janzen, 1974). In other words, if a herbivore has the ability to accommodate a range of defensive chemicals, a fixed-cost (Feeny, 1976), optimal fitness will result when the full range of accommodation ability is utilized. Induction of specific detoxification (Oh *et al.*, 1967; Wilkinson and Brattsten, 1972; Freeland and Janzen, 1974; Hodgson and Tate, 1976; Allison *et al.*, 1977) or tolerance mechanisms in herbivores in response to exposure to a set of defensive chemicals is probably a common property of generalist herbivores. Induction of a large fraction of the total inducible accommodation mechanisms presumably leads to a larger net accommodation than induction of a smaller fraction.

These arguments can also be applied to the effect of defensive chemical patterns exhibited by individuals of a resource, e.g., individual plants, on fitness and preference of herbivores. Figure 2 is a representa-

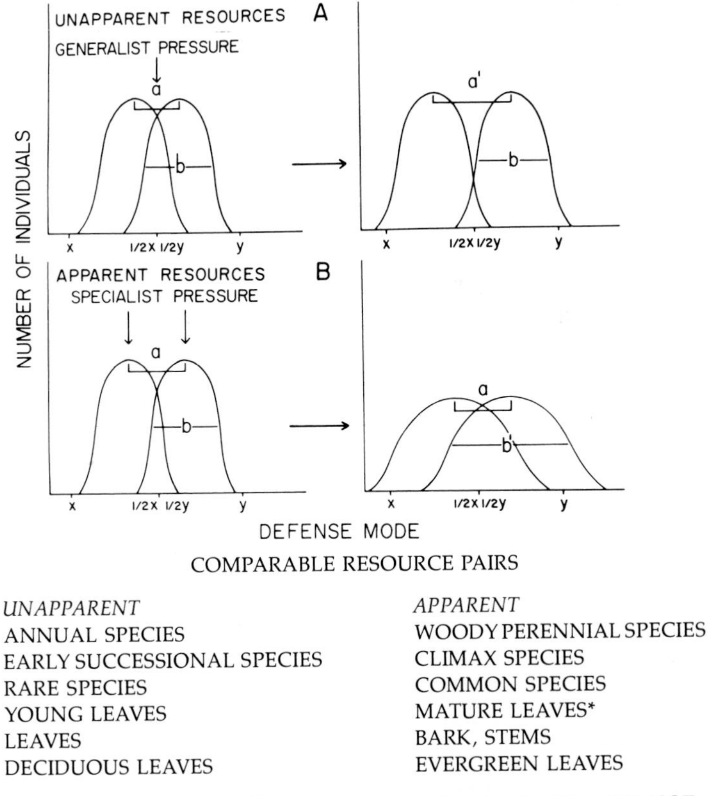

Figure 2. Within- and between-population patterns of convergence and divergence in plant defensive chemistry.

tion of two sympatric populations of unapparent resources and two sympatric populations of apparent resources under selection by herbivores (Fig. 1A; 2, 3, 4). The abscissa of Fig. 2, defense mode, represents a continuum of defense types from defense x at one extreme to defense y at the other, with admixtures of x and y in between. For instance, x and y may represent two different sets of defensive chemicals. Mean dissimilarity of defense between populations is indicated by a and a', and population variance of defense by b and b'.

It was argued above that low and high resource apparencies should select for generalism and specialism, respectively, in the herbivores (Fig. 1A; 2, 3). Furthermore, selective forces acting on herbivores should render them capable of accommodating both x and y defense sets (Fig. 2). In the case of generalist herbivores (Fig. 2A) selection should be such that accommodation ability for defensive sets x and y should tend toward equality. Specialist herbivores (Fig. 2B), on the other hand, will become more adapted to the defensive set that is dominant in their specific resource population. In either case herbivores should prefer to eat those individuals exhibiting the average defenses of the herbivores' resource ranges, (indicated by vertical arrows in Fig. 2). In this way herbivores should minimize overload of accommodation mechanisms for both x and y defensive sets.

Thus, generalist herbivore pressure on unapparent resources (Fig. 2A) should lead to a decrease in mean similarity of defense between the two populations ($a < a'$), whereas population variance of defense (b) should remain relatively unchanged. On the other hand, specialist pressure on apparent resources (Fig. 2B) should result in an increase in population variance of defense ($b < b'$), whereas mean similarity of defense between populations (a) should remain unchanged. In other words, in a community of sympatric unapparent resources, e.g., early successional annual plants, the between-species similarity of defense should be less, and the within-species similarity of defense between individuals should be greater, than the corresponding measures for sympatric apparent resources e.g., climax woody perennial plant species. These predictions can be tested.

Qualitative variation of secondary chemistry between individual plants within species and populations is a phenomenon well known to phytochemists (Hillis, 1966; Mears and Mabry, 1971; Smith, 1972; Levin, 1971; Rothschild, 1972; Manville, 1976; Mynderse and Faulkner, 1978), but its ecological significance has only recently been examined (Dolinger *et al.*, 1973; Hanover, 1975; Edmunds and Alstad, 1978) (see Section II,C). Individual chemical variation is probably a widespread and important defensive adaptation in plants. The present analysis suggests that individual variation in defensive chemistry should be greater for apparent resources than for unapparent resources.

Futuyma (1976) compared 13 primarily woody plant families to 13 primarily herbaceous families and found that the herbaceous families contained a higher total diversity of major groups of secondary compounds ($H' = 2.8$) than did the woody families ($H' = 1.7$), suggesting a lower defensive diversity in apparent than in unapparent plants. This is consistent with a higher between-resource similarity of defense in apparent than in unapparent resources. However, Futuyma's analysis considered only major chemical groups. The occurrence or nonoccurrence of alkaloids, phenolics, etc., in each family was recorded. Whether a similar pattern will result when the analysis is carried out for individual compounds in individual plants remains to be seen.

J. Contributions to the Theory of Plant–Herbivore Coevolution

I will attempt to reconstruct Fig. 1A from the diversity of hypotheses that have been advanced concerning the interrelationship of plants and herbivores (Fig. 1B) and show how these views form a coherent pattern. Doubtless, many of these ideas have been independently conceived many times, and original authorship for hypotheses is not implied for the quoted workers.

Janzen (1973a, p. 205) states that "the more predictably available a specific food item, the more a herbivore can afford to specialize in the face of interspecific competition for that general class of food item" (Fig. 1B; 5, 6). Enlarging the concept in one sense and contracting it in another, Cates and Orians (1975, p. 411) predicted that "herbivores feeding on early successional plants should be more generalized in their food and their foraging [(Fig. 1B; 1, 2)] than species utilizing later successional plants, most of which should be host-specific" (Fig. 1B; 5, 6).

Feeny (1976), observing that most herbaceous plants appeared to be defended by qualitative (divergent) defenses (Fig. 1B; 1, 3), whereas most woody plants appeared to be defended by quantitative (convergent) defenses (Fig. 1B; 5, 7), proposed that "it is the adapted specialist enemies which are likely to exert the major share of predator-related impact on herbaceous plants in ecological time" (p. 16), and, in contrast, "the effects of non-adapted enemies . . . [on herbaceous plants] are, almost by definition, likely to be trivial" (p. 23) (Fig. 1B; 1, 3, 4). Gilbert (1977, p. 405) agrees with this (Fig. 1B; 1, 3, 4). Similarly, Futuyma (1976) presents evidence that primarily woody plant families exhibit a lower density of defensive metabolites (Fig. 1B; 5, 7) than primarily herbaceous families (Fig. 1B; 1, 3) and argues that this should promote specialism in insects feeding on herbaceous plants (Fig. 1B; 1, 3, 4) and generalism in insects with woody hosts (Fig. 1B; 5, 7, 8).

It seems that Cates, Orians, and Janzen, on the one hand, and Feeny, Gilbert, and Futuyma, on the other, are predicting exactly opposite patterns of herbivore specialism on apparent and unapparent resources. Both hypotheses suffer from concentrating on only part of the truth. When the two ideas are combined, the emerging prediction is that the relative levels of pressure by specialist and generalist herbivores on apparent and unapparent resources should be similar.

Janzen (1973b) postulated that attack by a bruchid beetle species on two species of seeds could lead to character displacement of chemical defense in the seed species (Fig. 1B; 2, 3). Janzen (1973b) also hinted that within-plant defensive convergence was possible: "Insects that feed on seeds do not usually feed on leaves or roots of their host plant, and the root feeding guild is generally not found eating flowers or stems. The result is that the same chemical defense may appear in different plant parts of one species without increasing the probability of any one part being attacked" (p. 534) (Fig. 1B; 6, 7). Feeny (1975, 1976) argues that the best defense of plants against nonadapted insects may be to retain relatively unique or unusual defensive chemicals (Fig. 1B; 2, 3).

At an early stage Dethier (1954) and Fraenkel (1959) suggested that the diverse array of secondary metabolites in plants has been an important factor leading to dietary specialism and speciation in phytophagous insects (Fig. 1B; 3, 4). Dethier (1954) is widely quoted (e.g., Feeny, 1976) as originating the idea that the evolution of dietary niche width in phytophagous insects has proceeded largely from polyphagy toward monophagy. Indeed, Dethier (1954) heavily stresses factors likely to promote the evolution of dietary specialism by primitively polyphagous insects, stating that "phytophagous insects were undoubtedly originally polyphagous" and that "five lines of development from a primitive polyphagous state [to a less polyphagous state] may be postulated" (p. 41). However, Dethier (1954) also mentions, in passing, selective influences likely to promote the evolution of polyphagy. Polyphagous insects lack "complete dependence upon the fluctuating fortunes of a single plant species" (p. 37), which approximates Fig. 1B (1, 2). He also concedes that "the plant undergoes evolution and exerts a selective effect" so that phytophagous insects "enlarge their menu to include additional plant species as *convergent plant evolution* (my italics) progresses" (Dethier, 1954, p. 49) (Fig. 1B; 7, 8), although no further discussion of convergent evolution of plants is given.

It thus seems that Dethier (1954) has been, to a degree, misunderstood and happily so, since if phytophagous insects have forever evolved toward a more restricted host range (Fig. 1B; 3, 4) one wonders why there are so many polyphagous species still extant. This problem can be resolved by postulating that either (1) insufficient time has yet elapsed for all phytophagous insects to become monophagous, or (2) extinction rates are high for monophagous species and low for ancestral polyphagous species, which continually split off filial species of more restricted host range, or (3) polyphagous species are continually created *de novo*. Hypothesis (3) is absurd. Hypothesis (1) seems unlikely in view of the work of Strong (1974a,b,c) and Strong *et al.* (1977) showing rapid asymptotic accumulation of insect species, largely by indigenous insects in periods of several tens to several hundreds of years, on introduced plants and facile alteration of host plant range by laboratory selection of insects over several generations (Dethier, 1954), which demonstrates a high degree of dietary plasticity in phytophagous insects. Hypothesis (2) requires that, although each ancestral polyphagous species must continually give rise to more specialized species, a portion of the population of each ancestral species must, over evolutionary time, retain the original host range. This seems unlikely in view of the known dietary plasticity of insects and changing physical conditions and impossible when we consider that plants are evolving, speciating, and becoming extinct. The notion of general primitive polyphagy in insects leading progressively over evolutionary time to more restricted host ranges does not survive critical analysis. This idea arose because plant species vary in their defensive and other properties, which should select for dietary specialism in herbivores (Fig. 1B; 3, 4). Consideration of the influence not only of defensive diversity in plants on herbivore dietary specialism, but also of resource apparency (Fig. 1B; 1, 2; 5, 6) and defensive convergence (Fig. 1B; 7, 8) suggests that the relative levels of dietary specialist and dietary generalist pressure on plants are in dynamic equilibrium and always have been.

The arguments of Ehrlich and Raven (1965) are similarly based largely on Fig. 1B (3, 4). "The fantastic diversification of modern insects has developed in large measure as a result of a stepwise pattern of coevolutionary stages superimposed on the changing pattern of angiosperm variation" (in defensive chemistry and other properties) (Ehrlich and Raven, 1965, p. 605). Ehrlich and Raven (1965, p. 606) also suggest that herbivore pressure is, in large measure, the cause of floristic diversity. This can be interpreted to mean either that generalist herbivore pressure should lead to floristic defensive divergence and concomitant species radiation (Fig. 1B; 2, 3) or that pressure from specialized herbivores should decrease the abundance of their host plants (Fig. 1B; 6, 1), allowing a greater number of coexisting species. Ehrlich and Raven (1965) do not clearly delineate their views on this point.

Harper (1969) and Janzen (1970, 1971) also advance the view that specialist pressure is of major importance in maintaining floristic diversity by preventing the best competitors, within a given life form, from becoming common enough to eliminate the other species of that life form from the community (Fig. 1B; 6, 1). This must be partly true. However, if it is true *in toto* and if we also admit that plant apparency selects for herbivore specialism (Fig. 1B; 5, 6), observable differences in resource apparency should disappear under the leveling effect of natural selection during a single cycle (Fig. 1B; 5, 6, 1), which they manifestly do not. As pointed out previously, the evolutionary scheme assumes that real differences exist in resource apparency due to the effect of the independent variables (Fig. 1A), in spite of leveling processes such as 6, 1 in Fig. 1B. Resource apparency differences are firmly anchored by the independent variables and survive in spite of negative feedback loops at higher levels.

K. Taste, Odor, and Toxicity

Taste and odor are proximate qualities, whereas toxicity is an ultimate one. Animals avoid certain foods because they possess toxicity or some other noxious quality, not because they have a bad taste or smell. Physiologists and ecologists tend to ask different questions when confronted with cause and effect. Physiologists concentrate on the question "how?"; and how animals detect suitable food is, of course, an important question. However, ecologists generally find that "why?" questions are more useful in unraveling the evolutionary aspects of causation. Taste and odor (the "how?" aspects of food selection) are evolved responses to the fact that chemical compounds of certain molecular configuration have either noxious or beneficial qualities (the "why?" aspect) to the animal in question or are correlated with the presence in the food of substances with such qualities. For example, a noxious quality of plant tannins is that they complex with proteins. Because of this, some herbivores have evolved both oral receptor systems capable of detecting tannins in foodstuffs and the ability to react negatively to signals from these receptors. Evolution progresses such that tannins acquire a "bad taste" to these herbivores. The astringent taste of tannins to human beings is an example, and in this case the detector system is probably based on the protein-complexing property of tannins, which tan proteinaceous mucosal components and perhaps also proteins of receptor cells in the mouth. Astringency is a proximate and subjective quality. Detector systems for tannins in other animals may be similarly based on protein complexing, but to say that tannins deter herbivores because tannins are astringent is equivalent to saying that tannins deter herbivores because tannins taste astringent to human beings. This is not a very useful statement. It is better to say that tannins deter herbivores because tannins are protein-complexing agents. However, the concepts of taste and odor can be applied to animals other than human beings if defined solely in terms of observable responses of the animals or their sensory systems.

All chemical substances that taste or smell bad or good (have deterrent or attractive effects via oral or olfactory receptors) need not necessarily possess noxious or beneficial qualities. Animals should evolve detector systems capable of discriminating between only those relevant substances in their environment commonly encountered over evolutionary time. Rarely encountered substances may, by chance, mimic the chemical and physical properties of noxious or beneficial substances at the receptor level. In addition, active Batesian mimicry by plants using substances that simulate the required properties to animal receptors of noxious compounds but that in fact have no noxious qualities is possible. In such cases, the mimics would presumably use substances that are less costly to synthesize and store than noxious substances. Such Batesian convergence in taste or odor has, to my knowledge, not yet been demonstrated, but Gilbert (1975) has discussed the possible convergence by *Passiflora* spp. of leaf shape on that of common tropical plant species caused by visually hunting *Heliconius* butterflies.

Müllerian convergence of properties of plant defensive substances to animal receptors could benefit both plants and herbivores. Since arguments have already been advanced that generalist herbivore pressure should lead to divergence in noxious quality of defensive substances, we are led to the interesting possibility that the defensive substances of unapparent plants have diverged with respect to toxicity but have converged, in conjunction with animal receptors, with respect to taste or odor. Although alkaloids act on a diversity of metabolic systems in human beings (and other animals), they almost uniformly invoke a single taste—bitterness. Study of receptor responses of generalist and coevolved herbivores to a range of plant defensive substances or differing noxious quality may prove interesting.

Similar arguments can be applied to the taste and odor of beneficial substances. Nonnutritive, highly sweet proteins and terpene glycosides have been isolated from a variety of tropical fruits (Inglett and May, 1968; Eisner and Halpern, 1971; Inglett, 1975). Possibly these substances act by mimicking sugars to attract dispersers. If so it seems that these fruits must be rare and dispersed mainly by low-quality dispersers, since obligate avian frugivores are unlikely to be fooled by such a system. Gymnemic acids in the leaves of *Gymnema sylvestre* suppress sweetness of sugars to human beings and nullify the attractiveness of sugars to a variety of mammals (Eisner and Halpern, 1971). If this represents a defense, the defense is a qualitative one, and such defenses can be expected only in unapparent plant species or unapparent tissues. Defenses that camouflage the presence of free nutrients such as sugars or amino acids would seem to be necessary only in species or tissues that contain large quantities of these substances, which most plants or tissues do not, except when under physical stress.

The presence of nonnutritive sweeteners in leaves, roots, and seeds of some plant species (Inglett and May, 1968; Inglett, 1975) is much more difficult to explain as an adaptive trait. The sweetness of these substances to human beings may be merely fortuitous, as, it must be added, is possible in cases of nonnutritive sweeteners in fruits. Interestingly, sweet or bitter taste sensations in human beings can be elicited by remarkably similar substances. The flavonoid glycosides naringin and neohesperidin are bitter, but their dihydro derivatives are very sweet (Inglett, 1975). This suggests a basic similarity of structure for the receptor systems responsible for both sweet and bitter taste sensations. Conceivably, the nonnutritive sweeteners found in roots, leaves, and seeds mimic sugars to the taste receptors of human beings but mimic noxious substances to those of ecologically relevant herbivores.

In many cases animals may be attracted to or deterred by substances, not because of intrinsic properties of these substances, but because the presence of these substances is highly correlated with some other beneficial or noxious quality of the food. Specialized insects are often attracted by defensive substances in or emitted by their food plants. Except in cases in which the defensive substances are sequestered by the insect or otherwise utilized, the attractive response may not be due to any beneficial properties of the substance per se but to the fact that the plant emitting these substances, which are detoxifiable by the insects, is a known food resource.

Aposematicism by plants, using substances to warn potential herbivores of toxic quality, is possible. Volatile substances are particularly suitable for this. The patterns resulting from the interplay between Batesian and Müllerian mimicry and aposematic adaptation in taste and odor (as previously defined) of plant defensive substances, specificity of herbivores, plant apparency, and convergence and divergence in noxious quality of plants are likely to be rich.

L. Lignin as a Plant Defensive Substance

The supportive and strengthening function of lignin in plant tissues is well known (Freudenberg and Neish, 1968; Sarkanen and Ludwig, 1971). It is also widely accepted that organisms subjected to similar physical environmental stresses evolve similar adaptations to counter these stresses, other factors being equal (Orians and Solbrig, 1977). Lignin is a high molecular weight polymer of phenylpropane phenolics. Little chemical difference has been observed between lignins isolated from a variety of plant species, and only two major classes of lignins, guaiacyl lignin (present in gymnosperms, cycads, ferns, and club mosses) and guaiacyl-syringyl lignin (present in angiosperms, some conifers, grasses, and other monocotyledons) are recognized (Sarkanen and Hergert, 1971). Such convergence or parallelism in chemical structure of lignins of widely differing taxonomic origin can be viewed as resulting from selection, by the physical environment, for supportive adaptations in plants.

On the other hand, there is evidence for a defensive function of lignin. Herbivore preference for plant material and digestibility of plant material by herbivores or microorganisms are often negatively correlated with lignin or "crude fiber" content of the plant material (Burns *et al.*, 1972; Hartley and Jones, 1977). Lignin itself is relatively indigestible to herbivores and microorganisms, although it can be digested by some fungi (Abson, 1977). In addition, the presence of lignin reduces the digestibility or chemical extractability of both cellulose and protein in plant tissues (Brauns, 1952; Pirie, 1971; Kirk, 1977; Crawford, 1977). Lignin polymers contain approximately one free phenolic hydroxyl group for every three phenylpropane residues (Freudenberg, 1968), and it is possible that lignin reduces the digestibility of plant carbohydrates and proteins by hydrogen-bonded complex formation with these substances, in an analogous way to that of plant tannins. Lignin, in conjunction with tannin, has been used as a leather tanning agent (Siddiqui and Sykes, 1962), and indigestible lignin–carbohydrate complexes are known to be formed during ruminant digestion of grasses (Neilson and Richards, 1978). In plant cells lignin may be linked to cell wall carbohydrates by covalent bonds (Freudenberg, 1968). Lignin–carbohydrate and lignin–protein association in intact and disrupted plant tissues and the potential effect of these associations on tissue digestibility deserve much further study.

Thus, a dual structural and defensive role for lignin in plants is likely. The widespread occurrence of lignin in apparent plants of diverse phylogenetic origin may be due to convergence or parallelism caused by selective influences both from the physical environment and from attack by specialized herbivores (Fig. 1A; 2, 3, 4). Lignins and tannins together probably constitute the two most important classes of quantitative defensive substances in apparent plants.

M. Testing Plant Optimal Defense Theory and Plant–Herbivore Coevolutionary Theory

Evidence for and further tests of the various optimal defense hypotheses have been discussed in the relevant sections of this chapter. Many of these hypotheses are based on very limited evidence, and none can be regarded as established truth. Much research and literature analysis are necessary to determine if many of the trends or correlations described are truly general phenomena. Doubtless, some hypotheses will prove false or only partly true.

It has been argued (Section IV,H) that in undisturbed systems there should be little difference in both fitness loss to plants due to herbivory and specialism of herbivory for apparent and unapparent resources. Obtaining relevant measurements of total impact of herbivory on fitness of plants and degree of impact by specialized versus generalized herbivores on plants is not easy. In environments disturbed by man, relative plant abundance and the abundance of some herbivores, for instance, large vertebrate grazers, have been drastically and recently altered, and plants in these environments may be defensively adapted to conditions that no longer exist. It is difficult to compare herbivores that differ in their feeding method, for instance, leaf chewers, stem chewers, miners, leaf rollers, sap suckers, and gall makers. The impact of many arthropod and vertebrate herbivores that undergo cyclic changes in abundance is extremely patchy in space and time. It is becoming increasingly clear that many "species" of phytophagous insects are far from homogeneous but consist of a complex of host races, which, although morphologically similar or identical, exhibit a much narrower host range than the complex as a whole and exhibit a high degree of reproductive isolation, sometimes even when sympatric (Taksdahl, 1963; Leonard, 1974; Gilbert and Singer, 1975; Bush, 1975; Gallun *et al.*, 1975; Herrebout *et al.*, 1976; Muller, 1976; Baltensweiler *et al.*, 1977). On the other hand, the patterns of defensive chemistry predicted to occur between and within populations and species of plants and their tissues can be more readily tested (Section IV,I).

No predictions have been made here concerning the *numbers of species* of specialized and generalized herbivores utilizing apparent and unapparent resources. This is because it is the degree of impact of specialized and generalized herbivores that is important to the evolution of plant defenses rather than the number of specialized and generalized herbivore species, and there is no reason to believe that average abundance for specialized and generalized herbivore species on apparent and unapparent resources are equal. Comparing Lepidoptera species feeding on only one versus more than one plant family, Futuyma (1976) found a greater prevalence of the specialized feeders among those species that feed on herbaceous plants than among those that feed on woody hosts. However, Slansky (1976) analyzed food selection patterns of Nearctic and Palearctic butterflies using more discriminating measures of dietary niche breadth. Monophagous butterflies were defined as those feeding within one genus of plants, oligophagous as those feeding on plants in more than one genus but only in one order, and polyphagous as those feeding on plants in more than one order. Slansky found little difference in dietary specialism of butterfly species utilizing herbs, shrubs, or trees. Such studies, although valuable in their own right, may shed little light on the evolution of plant defenses unless they are accompanied by measures of relative impact of the various herbivore species on the plants or at least by estimates of relative abundance of the herbivore species.

V. CONCLUDING REMARKS

This treatment of evolution of chemical defenses of plants against herbivores interprets patterns in defensive properties of plants and herbivore characteristics in terms of their adaptive significance in ecological time, the present and recent past, and assumes that similar patterns have existed over evolutionary time. Patterns in plant defensive properties are viewed as resulting from and maintained by the interaction of selective influences exerted by the physical environment, competitive interactions, and herbivores (Figs. 1 and 2) to give a dynamic equilibrium. Plants of differing growth form, seral stage, abundance, and longevity (apparency) evolve and persist due to competitive interactions in heterogeneous environments. The resulting gradient of apparency selects for various degrees of herbivore pressure and herbivore specificity on these plants and their individual tissues. Plants evolve defenses appropriate to their risk. These evolved defenses are such as to cause differences in herbivory expected from considerations of apparency alone to disappear. Little or no reference has been made to taxonomic or phylogenetic considerations because adaptive significance knows no taxonomic or temporal boundaries. It is suggested that the broad patterns described have existed since there were plants and enemies exploiting them. On a finer scale the nature and identity of individual plants, their defenses, and herbivores have, of course, changed throughout time. Due to the total inadequacy of the fossil record, particularly with respect to plant defensive chemistry (Niklas and Giannasi, 1977; Giannasi and Niklas, 1977), we can only speculate what the nature of these changes might have been. Swain (1976, 1977b) made a heroic attempt to reconstruct the secondary chemistry of plants, properties of herbivores, and even their interactions through geological time. In view of the fact that these speculations are based entirely on properties of extant taxa, and given the known dietary plasticity of herbivores (Section IV,F) and our ignorance of the rates of evolutionary change of plant secondary chemistry, such ideas should be viewed with caution. Plant taxonomic and phylogenetic status are doubtless important in determining secondary chemistry, and this forms the basis for chemotaxonomy. However, the ideas elaborated in this chapter suggest that plant apparency to enemies is at least as important as phylogeny in determining secondary chemistry.

Swain (1976) produces evidence that man and several domestic mammalian herbivores are deterred by alkaloids in their food at a much lower concentration than are herbivorous tortoises, spiny-tailed lizards, and carnivorous green lizards and terrapins. Swain then makes the surprising extrapolation that the ability of herbivorous dinosaurs to detect alkaloids in their food was probably limited, and thus they were poisoned by alkaloid-rich late cretaceous angiosperms, leading, in large measure, to the extinction of dinosaurs. Arguments expressed in Section IV,K suggest that the high alkaloid taste tolerance of herbivorous tortoises and spiny-tailed lizards should be correlated with a high physiological tolerance for alkaloids in these reptiles. If it is acceptable to extrapolate taste tolerance of extant herbivorous reptiles to determine that of herbivorous dinosaurs, it is presumably also possible to do so for physiological tolerance. This leads to the conclusion that dinosaurs had both high taste tolerance and high physiological tolerance to alkaloids, which sheds little light on the reasons for the extinction of these animals. For the carnivorous reptiles, high taste tolerance to plant alkaloids is probably due to the fact that these animals rarely encounter such substances in their food.

This chapter examines plant–herbivore coevolution largely from the plant's viewpoint. Plants evolve defenses in response to the net selective influence exerted by their enemies. Factors that determine the life form and dietary specificity of individual herbivores have not been treated in any detail. These factors would certainly include host apparency and defensive attributes but many other factors in addition, for example, historical considerations, host morphological complexity (Schultz, 1978; Strong and Levin, 1979), and competitive interactions with other herbivores (Gilbert and Singer, 1975).

Evolution of plant defense is here examined largely in terms of defenses against herbivores. A comprehensive theory of plant defense will also include plant defenses against both pathogens and other plants. Many of the secondary metabolites active against herbivores are similarly active against pathogens and other plants. The various hypotheses of optimal defense theory (Section IV) are general and apply to all plant enemies that attempt to utilize plant tissues to the detriment of their hosts. There is already substantial evidence in support of some of these hypotheses, for example, hypothesis (c), in plant–pathogen systems. Plant pathologists should find it relatively easy to evaluate the other hypotheses. Similarly, the coevolutionary scheme (Figs. 1 and 2) could be readily expanded to include pathogens. This would merely entail substituting the term "herbivores and pathogens" for the term "herbivores" in the scheme, given that the assumptions apply equally well for plant–pathogen and plant–herbivore interactions, as many of them appear to do. If so, the deductions resulting from the scheme apply to herbivores plus pathogens rather than to herbivores alone. Inclusion of plant–plant interactions may prove more difficult. Plant–plant interactions have so far been included as independent variables (Fig. 1A). If allelopathic interactions prove to be of general importance in plant competition, the assumption that plant competition can be treated as an independent variable will be seriously weakened, particularly if it is shown that the same basic chemical types of defensive compounds are effective against plant competitors, herbivores, and pathogens. Inclusion of allelopathy in the scheme may necessitate elaboration of an additional coevolving cycle antecedent to cycle I (Fig. 1A) involving plants, their allelopathic chemicals, and other attributes interacting to produce observable differences in plant apparency.

ACKNOWLEDGMENTS

I am indebted to G. H. Orians, N. E. Beckage, R. G. Cates, A. B. Adams, and J. C. Bergdahl for providing valuable discussion and insight. In addition I thank P. R. Atsatt, G. A. Benz, J. P. Bryant, D. J. Futuyma, E. Haukioja, D. H. Janzen, A. R. P. Journet, G. A. Rosenthal, C. C. Smith, and T. C. R. White for illuminating comments on the completed manuscript, although in many cases there was insufficient time for their inclusion. Special thanks are extended to L. Erckmann for the illustrations and typing. This work was supported by National Science Foundation Grant DEB 77-03258.

REFERENCES

Abson, D., ed. (1977). "Biological Delignification." Weyerhauser Co., Seattle, Washington. (Copies available from D. Abson, Weyerhauser Co., 3400 13th Ave. S.W., Seattle, Washington 98134).

Allen, M. R., and Kitts, W. D. (1961). *Can. J. Anim. Sci.* **41,** 1–12.

Allison, M. J., Littledike, E. T., and James, L. F. (1977). *J. Anim. Sci.* **45,** 1173.

Allison, R. M. (1971). *In* "Leaf Protein: Its Agronomy, Preparation, Quality, and Use" (N. W. Pirie, ed.), pp. 78–85. IBP Handb., No. 20. Davis, Philadelphia, Pennsylvania.

Andrewartha, H. G., and Birch, L. C. (1954). "The Distribution and Abundance of Animals." Chicago Univ. Press, Chicago, Illinois.

Andries, M. (1956). Ph.D. Thesis, University of Washington, Seattle.

Arnold, G. W., and Hill, J. L. (1972). *In* "Phytochemical Ecology" (J. B. Harborne, ed.), pp. 71–101. Academic Press, New York.

Arntzen, G. O., Falkenthal, S. V., and Bobick, S. (1974). *Plant Physiol.* **53,** 304–306.

Atsatt, P. R., and O'Dowd, D. J. (1976). *Science* **193,** 24–29.

Bachur, N. R. (1976). *Science* **193,** 595–597.

Bacon, J. S. D., and Cheshire, M. V. (1971). *Biochem. J.* **124,** 555–562.

Baker, H. G., and Baker, I. (1975). *In* "Coevolution of Animals and Plants" (L. E. Gilbert and P. H. Raven, eds.), pp. 100–140. Univ. of Texas Press, Austin.

Baltensweiler, W., Benz, G., Bovey, P., and Delucchi, V. (1977). *Annu. Rev. Entomol.* **22,** 79–100.

Bate-Smith, E. C. (1977). *Phytochemistry* **16,** 1421–1426.

Bate-Smith, E. C., and Metcalf, C. R. (1957). *J. Linn. Soc. London, Bot.* **55,** 669–705.

Beck, S. D., and Reese, J. C. (1976). *Recent Adv. Phytochem.* **10,** 41–92.

Bell, E. A. (1974). *Proc. Summer Inst. Biol. Control Plant Insects Dis.* p. 403.

Benz, G. (1977). *Eucarpia/IOBC Working Group Breed. Resistance Insects Mites, Bull. SROP, 1977/8* Report, pp. 155–159.

Bernays, E. A., and Chapman, R. F. (1977). *Ecol. Entomol.* **2,** 1–18.

Bickoff, E. M. (1962). *Agric. Food Chem.* **10,** 410.

Blohm, H. (1962). "Poisonous Plants of Venezuela." Harvard Univ. Press, Cambridge, Massachusetts.

Bowers, W. S., Ohata, T., Cleere, J. S., and Marsella, P. A. (1976). *Science* **193,** 542–547.

Brauns, F. E. (1952). "The Chemistry of Lignin." Academic Press, New York.

Brehm, B. G., and Krell, D. (1975). *Science* **190,** 1221–1223.

Bull, L. B., Culvenor, C. C. J., and Dick, A. T. (1968). "The Pyrrolizidine Alkaloids." Am. Elsevier, New York.

Burns, J. C., Mochrie, R. D., and Cope, W. A. (1972). *Argron. J.* **64,** 193–195.

Bush, G. L. (1975). *In* "Evolutionary Strategies of Parasitic Insects and Mites" (P. W. Price, ed.), pp. 187–224. Plenum, New York.

Büttner, H. (1961). *Schriftenr. Landesforstverwaltung* **11,** 1–5.

Caldwell, M. M. (1971). *Photophysiology* **6,** 131.

Carter, W. (1973). "Insects in Relation to Plant Disease." Wiley, New York.

Cates, R. G. (1975). *Ecology* **56,** 391–400.

Cates, R. G., and Orians, G. H. (1975). *Ecology* **56,** 410–418.

Cates, R. G., and Rhoades, D. F. (1977a). *Biochem. Syst. Ecol.* **5,** 185–196.

Cates, R. G., and Rhoades, D. F. (1977b). *In* "Mesquite" (B. Simpson, ed.), pp. 61–83. Dowden, Hutchinson & Ross, Inc., Stroudsburg, Pennsylvania.

Chambliss, O. L., and Jones, C. M. (1966). *Science* **158,** 1392–1393.

Clapham, A. R., Tutin, T. G., and Warburg, E. F. (1952). "Flora of the British Isles." Cambridge Univ. Press, New York.

Clarke, B. (1962). *Syst. Assoc. Publ. 4, Taxonomy and Geography,* 47–70.
Conn, E. E., and Butler, G. W. (1969). *In* "Perspectives in Phytochemistry" (J. B. Harborne and T. Swain, eds.), pp. 47–64. Academic Press, New York.
Cooper-Driver, G., and Swain, T. (1976). *Nature (London)* **260,** 604.
Cooper-Driver, G., Finch, S., Swain, T., and Bernays, E. (1977). *Biochem. Syst. Ecol.* **5,** 177–185.
Crawford, R. L. (1977). *In* "Biological Delignification" (D. Abson, ed.), pp. 55–72. Weyerhauser Co., Seattle, Washington.
Cromwell, B. T. (1933). *Biochem J.* **27,** 860–872.
DaCosta, C. P., and Jones, C. M. (1971). *Science* **172,** 1145–1146.
del Moral, R. (1972). *Oecologia* **9,** 289–300.
Dement, W. A., and Mooney, H. A. (1974). *Oecologia* **13,** 62–76.
Dempster, J. P. (1963). *Biol. Rev. Cambridge Philos. Soc.* **38,** 490–529.
Dethier, V. G. (1954). *Evolution* **8,** 55–64.
Deverall, B. J. (1977). "Defense Mechanisms of Plants." Cambridge Univ. Press, London and New York.
Dodson, C. H., Dressler, R. L., Adams, R. M., and Williams, N. H. (1969). *Science* **164,** 1243–1249.
Dolinger, P. M., Ehrlich, P., Fitch, W., and Breedlove, D. (1973). *Oecologia* **13,** 191–204.
Donnelly, E. D., and Anthony, W. B. (1973). *Agron. J.* **65,** 993–994.
Edmunds, G. F. Jr., and Alstad, D. N. (1978). *Science* **199,** 941–945.
Ehrlich, P. R., and Raven, P. H. (1965). *Evolution* **18,** 586–608.
Eisner, T., and Halpern, B. P. (1971). *Science* **72,** 1362.
Emlen, J. M., and Emlen, M. G. R. (1975). *Am. Nat.* **109,** 427–436.
Feeny, P. P. (1968). *Phytochemistry* **7,** 871–880.
Feeny, P. P. (1969). *Phytochemistry* **8,** 2119–2123.
Feeny, P. P. (1970). *Ecology* **51,** 656–681.
Feeny, P. P. (1975). *In* "Coevolution of Animals and Plants" (L. E. Gilbert and P. H. Raven, eds.), pp. 1–19. Univ. of Texas Press, Austin.
Feeny, P. P. (1976). *Recent Adv. Phytochem.* **10,** 1–40.
Foulds, W., and Grime, J. (1972). *Heredity* **28,** 181–187.
Fowden, L., Lewis, D., and Tristam, H. (1967). *Adv. Enzymol.* **29,** 89–163.
Fraenkel, G. S. (1959). *Science* **129,** 1466–1470.
Fraenkel, G. S. (1969). *Entomol. Exp. Appl.* **12,** 473–486.
Freeland, W. J., and Janzen, D. H. (1974). *Am. Nat.* **108,** 269–289.
Freudenberg, K. (1968). *In* "The Constitution and Biosynthesis of Lignin" (K. Freudenberg and A. C. Neish, eds.), pp. 47–116. Springer-Verlag, Berlin and New York.
Freudenberg, K., and Neish, A. C., eds. (1968). "The Constitution and Biosynthesis of Lignin." Springer-Verlag, Berlin and New York.
Futuyma, D. J. (1976). *Am. Nat.* **110,** 285–292.
Galbraith, J. L., Horn, D. H. S., Ito, S., Kodama, M., and Sasse, J. M. (1972). *Agric. Biol. Chem.* **36,** 2393.
Gallun, R. L., Starks, K. J., and Guthrie, W. D. (1975). *Annu. Rev. Entomol.* **20,** 337–357.
Giannasi, D. E., and Niklas, K. J. (1977). *Science* **197,** 765–767.
Gilbert, L. E. (1975). *In* "Coevolution of Animals and Plants" (L. E. Gilbert and P. H. Raven, eds.), pp. 210–240. Univ. of Texas Press, Austin.
Gilbert, L. E. (1977). *Colloq. Int. C.N.R.S.* **265,** 399–413.
Gilbert, L. E., and Raven, P. H., eds. (1975). "Coevolution of Animals and Plants." Univ. of Texas Press, Austin.
Gilbert, L. E., and Singer, M. C. (1975). *Annu. Rev. Ecol. Syst.* **6,** 365–397.

Goldstein, J. L., and Swain, T. (1963). *Phytochemistry* **2,** 371–383.
Goldstein, J. L., and Swain, T. (1965). *Phytochemistry* **4,** 185–192.
Goodwin, T. W., and Goad, L. J. (1970). *In* "The Biochemistry of Fruits and Their Products" (A. C. Hulme, ed.), Vol. 1, pp. 305–368. Academic Press, New York.
Green, T. R., and Ryan, C. A. (1972). *Science* **175,** 776–777.
Gunn, D. L. (1960). *Annu. Rev. Entomol.* **5,** 279–300.
Gustafsson, A., and Gadd, I. (1965). *Hereditas* **53,** 15–39.
Handley, W. R. C. (1961). *Plant Soil* **15,** 37–73.
Hanover, J. W. (1966). *Heredity* **21,** 73–84.
Hanover, J. W. (1975). *Annu. Rev. Entomol.* **20,** 75–95.
Harborne, J. B., ed. (1972). "Phytochemical Ecology." Academic Press, New York.
Harborne, J. B., and Van Sumere, C. F., eds. (1975). "The Chemistry and Biochemistry of Plant Proteins." Academic Press, New York.
Harper, J. L. (1969). *Brookhaven Symp. Biol.* **22,** 48–61.
Hartley, R. D., and Jones, E. C. (1977). *Phytochemistry* **16,** 1531–1534.
Haslam, E. (1975). *In* "The Flavonoids" (J. B. Harborne, T. J. Mabry, and H. Mabry, eds.), Part 1, pp. 505–559. Academic Press, New York.
Haukioja, E., and Hakala, T. (1975). *Rep. Kevo Subarct. Res. Stn.* **12,** 1–9.
Haukioja, E., and Niemelä, P. (1976). *Rep. Kevo Subarct. Res. Stn.* **13,** 44–47.
Haukioja, E., and Niemalä, P. (1977). *Ann. Zool. Fenn.* **14,** 48–52.
Heftmann, E. (1963). *Annu. Rev. Plant Physiol.* **14,** 225–248.
Heftmann, E. (1970). *Recent Adv. Phytochem.* **3,** 211–217.
Heinrich, B. (1975a). *In* "Coevolution of Animals and Plants" (L. E. Gilbert and P. H. Raven, eds.), pp. 141–158. Univ. of Texas Press, Austin.
Heinrich, B. (1975b). *Annu. Rev. Ecol. Syst.* **6,** 139–170.
Herrebout, W. M., Kuijten, P. J., and Wiebes, J. T. (1976). *In* "The Host Plant in Relation to Insect Behavior and Reproduction" (T. Jermy, ed.), pp. 91–94. Plenum, New York.
Hikino, H., Ohizumi, Y., and Takemoto, T. (1975). *J. Insect Physiol.* **21,** 1953–1963.
Hillis, W. E. (1966). *Phytochemistry* **5,** 541.
Hillis, W. E., and Inoue, T. (1968). *Phytochemistry* **7,** 13–22.
Hillis, W. E., and Swain, T. (1959). *J. Sci. Food. Agric.* **10,** 135–144.
Hodgson, E., and Tate, L. G. (1976). *In* "Insecticide Biochemistry and Metabolism" (C. F. Wilkinson, ed.), pp. 115–148. Plenum, New York.
Hollingworth, R. M. (1976). *In* "Insecticide Biochemistry and Metabolism" (C. F. Wilkinson, ed.), pp. 431–506. Plenum, New York.
Horigome, T., and Kandatsu, M. (1968). *Agric. Biol. Chem.* **32,** 1093–1102.
Howe, W. L., Sanborn, J. R., and Rhodes, A. M. (1976). *Environ. Entomol.* **5,** 1043–1047.
Howes, F. N. (1953). "Vegetable Tanning Materials." Butterworth, London.
Hulme, A. C., ed. (1970). "The Biochemistry of Fruits and Their Products," Vol. 1. Academic Press, New York.
Inglett, G. E. (1975). *In* "The Chemistry and Biochemistry of Plant Proteins" (J. B. Harborne and C. F. Van Sumere, eds.), pp. 265–280. Academic Press, New York.
Inglett, G. E., and May, J. F. (1968). *Econ. Bot.* **22,** 326–332.
Jacobson, J., and Crosby, D. G., eds. (1971). "Naturally Occurring Insecticides." Dekker, New York.
James, W. D. (1950). *In* "The Alkaloids," (R. H. F. Manske and H. L. Holmes, eds.), Vol. 6, pp. 16–90, Academic Press, New York.
Janzen, D. H. (1970). *Am. Nat.* **104,** 501–528.
Janzen, D. H. (1971). *Am. Nat.* **105,** 97–112.
Janzen, D. H. (1973a). *Syst. Assoc. Spec. Vol.* **5,** 201–211.

Janzen, D. H. (1973b). *Pure Appl. Chem.* **34,** 529–538.
Janzen, D. H. (1977). *Am. Nat.* **111,** 691–713.
Janzen, D. H., Juster, H. B., and Liener, I. E. (1976). *Science* **192,** 795–796.
Jermy, T., ed. (1976). "The Host Plant in Relation to Insect Behavior and Reproduction." Plenum, New York.
Jones, D. A. (1971). *Science* **173,** 945.
Jones, D. A. (1972). *In* "Phytochemical Ecology" (J. B. Harborne, ed.), pp. 103–122. Academic Press, New York.
Jones, D. A. (1977a). *Heredity* **39,** 27–44.
Jones, D. A. (1977b). *Heredity* **39,** 45–65.
Josefsson, E. (1967). *Phytochemistry* **6,** 1617–1627.
Kingsbury, J. M. (1964). "Poisonous Plants of the United States and Canada." Prentice-Hall, Englewood Cliffs, New Jersey.
Kirk, T. K. (1977). *In* "Biological Delignification" (D. Abson, ed.), pp. 31–54. Weyerhauser Co., Seattle, Washington.
Klomp, H. (1968). *In* "Insect Abundance" (T. R. E. Southwood, ed.), p. 98. Blackwell, Oxford.
Kogan, M. (1977). *Proc. Int. Congr. Entomol., 15th, 1976,* pp. 211–227.
Krieger, R., Feeny, P. P., and Wilkinson, C. (1971). *Science* **172,** 579–580.
Krinsky, N. I. (1971). *In* "Carotenoids" (O. Isler, ed.), p. 669. Birkhaeuser, Basel.
La Du, B. N., Mandel, H. G., and Way, E. L. (1971). "Fundamentals of Drug Metabolism and Drug Disposition." Williams & Wilkins, Baltimore, Maryland.
Lawton, J. H. (1976). *Bot. J. Linn. Soc.* **73,** 187–216.
Leete, E. (1963). *In* "Biogenesis of Natural Compounds" (P. Bernfeld, ed.), pp. 739–798. Pergamon, Oxford.
Leonard, D. E. (1974). *Annu. Rev. Entomol.* **19,** 197–229.
Levin, D. A. (1971). *Am. Nat.* **105,** 157–181.
Levin, D. A. (1971). *Q. Rev. Biol.* **48,** 3–15.
Loomis, W. D., and Battaile, J. (1966). *Phytochemistry* **5,** 423–438.
Loper, G. M. (1968). *Crop Sci.* **8,** 104–108.
Loper, G. M., Hanson, C. H., and Graham, J. H. (1967). *Crop Sci.* **7,** 189–192.
MacArthur, R. H., and Pianka, E. R. (1966). *Am. Nat.* **100,** 603–609.
McClure, J. W. (1975). *In* "The Flavonoids" (J. B. Harborne, T. J. Mabry, and H. Mabry, eds.), pp. 970–1055. Academic Press, New York.
McKey, D. (1974). *Am. Nat.* **108,** 305–320.
McKey, D. (1975). *In* "Coevolution of Animals and Plants" (L. E. Gilbert and P. H. Raven, eds.), pp. 159–191. Univ. of Texas Press, Austin.
McMillian, W. W., Wiseman, B. R., Burns, R. E., Harris, H. B., and Greene, G. L. (1972). *Agron. J.* **64,** 821–822.
Manville, J. F. (1976). *Can. J. Chem.* **54,** 2365–2371.
Matsumoto, Y. (1970). *In* "Control of Insect Behavior by Natural Products" (D. L. Wood, R. M. Silverstein, and M. Nakajima, eds.), pp. 133–156. Academic Press, New York.
Mattson, W. J., and Addy, N. D. (1975). *Science* **190,** 515–522.
Maxwell, F. G., and Harris, F. A., eds. (1974). "Proceedings of the Summer Institute on Biological Control of Plant Insects and Diseases." Univ. of Mississippi Press, Jackson.
Maxwell, F. G., Jenkins, J. N., Keller, J. C., and Parrot, W. L. (1963). *J. Econ. Entomol.* **56,** 449–454.
Mears, J. A., and Mabry, T. J. (1971). *In* "Chemotaxonomy of the Leguminosae" (J. B. Harborne, D. Boulter, and B. L. Turner, eds.), pp. 73–178, Academic Press, New York.
Milne, A. (1957). *Can. Entomol.* **89,** 193–213.

Morris, R. F., Miller, C. A., Greenbank, D. D., and Mott, D. G. (1958). *Proc. Int. Congr. Entomol., 10th, 1956* Vol. 4, pp. 137–150.

Morrow, P. A., Bella, T. E., and Eisner, T. (1976). *Oecologia* **24,** 193–206.

Mothes, K. (1960). *Alkaloids (N.Y.)* **6,** 1–29.

Mothes, K. (1976). *Recent Adv. Phytochem.* **10,** 385–405.

Muller, F. P. (1976). *In* "The Host Plant in Relation to Insect Behavior and Reproduction" (T. Jermy, ed.), pp. 187–190. Plenum, New York.

Mynderse, J. S., and Faulkner, D. J. (1978). *Phytochemistry* **17,** 237–240.

Nakanishi, K., Koreeda, M., and Imai, S. (1972). *In* "Some Recent Developments in the Chemistry of Natural Products" (S. Rangaswami and N. V. Subba Rao, eds.), pp. 194–213. Prentice-Hall, New Delhi.

Naya, J. K., and Fraenkel, G. (1963). *Ann. Entom. Soc. Am.* **56,** 174–178.

Neilson, M. J., and Richards, G. N. (1978). *J. Sci. Fd. Agric.* **29,** 513–519.

Nestel, B., and MacIntyre, R., eds. (1973). "Chronic Cassava Toxicity," Monogr. IDRC-010e. Int. Dev. Res. Cent., Ottawa.

Niklas, K. J., and Giannasi, D. E. (1977). *Science* **196,** 877–878.

Nozzolillo, C. (1975). *Contrib. Pap. 15th Annu. Meet., Phytochem. Soc. North Am.*

Nurstein, H. E. (1970). *In* "The Biochemistry of Fruits and Their Products" (A. C. Hulme, ed.), Vol. 1, pp. 239–268. Academic Press, New York.

Oh, H. K., Sakai, T., Jones, M. B., and Longhurst, W. M. (1967). *Appl. Microbiol.* **15,** 777–784.

Orians, G. H., and Janzen, D. H. (1974). *Am. Nat.* **108,** 581–592.

Orians, G. H., and Pearson, N. P. (1979). *In* "Analysis of Ecological Systems" (D. Horn, ed.), Ohio State Univ. Press, Columbus (in press).

Orians, G. H., and Solbrig, O. T. (1977). *In* "Convergent Evolution in Warm Deserts" (G. H. Orians and O. T. Solbrig, eds.), pp. 225–255. Dowden, Hutchinson & Ross, Inc., Stroudsburg, Pennsylvania.

Osborne, D. J. (1973). *In* "Insect/Plant Relationships" (H. F. van Emden, ed.), pp. 33–42. Wiley, New York.

Overhulser, D., Gara, R. I., and Johnsey, R. (1972). *Ann. Entomol. Soc. Am.* **65,** 1423–1424.

Pammell, L. H. (1911). "A Manual of Poisonous Plants." Torch Press, Cedar Rapids, Iowa.

Pierpoint, W. S. (1971). *Rep., Rothampsted Exp. Stn., Harpenden, Engl.* p. 199.

Pimentel, D. (1976). *Bull. Entomol. Soc. Am.* **22,** 20–26.

Pimentel, D., and Bellotti, A. C. (1976). *Am. Nat.* **110,** 877–888.

Pirie, N. W., ed. (1971). "Leaf Protein: Its Agronomy, Preparation, Quality and Use," IBP Handb., No. 20. Davis, Philadelphia, Pennsylvania.

Pulliam, H. R. (1974). *Am. Nat.* **108,** 59–74.

Puritch, G. S., and Nijholt, W. W. (1974). *Can. J. Bot.* **52,** 585–587.

Pyke, G. H., Pulliam, H. R., and Charnov, E. L. (1977). *Qt. Rev. Biol.* **52,** 137–154.

Rees, C. J. C. (1969). *Entomol. Exp. Appl.* **12,** 565–583.

Rehr, S. S., Feeny, P. P., and Janzen, D. H. (1973a). *J. Anim. Ecol.* **1,** 63–67.

Rehr, S. S., Janzen, D. H., and Feeny, P. P. (1973b). *Science* **181,** 81–82.

Rhoades, D. F. (1977a). *Biochem. Syst. Ecol.* **5,** 281–290.

Rhoades, D. F. (1977b). *In* "Creosote Bush" (T. J. Mabry, J. H. Hunziker, and D. R. Difeo, Jr., eds.), pp. 135–175. Dowden, Hutchinson, & Ross, Inc. Stroudsburg, Pennsylvania.

Rhoades, D. F., and Cates, R. G. (1976). *Recent Adv. Phytochem.* **10,** 168–213.

Ribereau-Gayon, P. (1972). "Plant Phenolics." Oliver & Boyd, Edinburgh.

Rice, E. L. (1974). "Allelopathy." Academic Press, New York.

Rice, E. L. (1977). *Biochem. Syst. Ecol.* **5,** 201–206.

Robinson, T. (1968). "The Biochemistry of Alkaloids." Springer-Verlag, Berlin and New York.
Robinson, T. (1974). *Science* **184,** 430–435.
Rodríguez, E., and Levin, D. A. (1976). *Recent Adv. Phytochem.* **10,** 214–270.
Rodríguez, E., Valesi, A., Hunziker, J., and Mabry, T. J. (1972). *Phytochemistry* **11,** 2821–2827.
Roeske, C. N., Sieber, J. N., Brower, L. P., and Moffitt, C. M. (1976). *Recent Adv. Phytochem.* **10,** 93–167.
Rogers, C. F., and Frykolm, O. C. (1937). *J. Agric. Res.* **55,** 533–537.
Root, R. B. (1973). *Ecol. Monogr.* **43,** 95–120.
Rosenthal, G. A., Janzen, D. H., and Dahlman, D. L. (1977). *Science* **196,** 658–660.
Rothschild, M. (1972). *In* "Phytochemical Ecology" (J. B. Harborne, ed.), pp. 2–12. Academic Press, New York.
Rothschild, M. (1973). *In* "Insect/Plant Relationships" (H. F. van Emden, ed.), pp. 59–83. Wiley, New York.
Röttger, U., and Klinghauf, F. (1979). "Aenderung im Stoffwechsel von Zuckerrübenblattern durch Befall mit *Pegomyia betae* Curt. (Muscidae: Anthomyidae)." *Z. Angew. Entomol.* (in press).
Ryder, E. J., and Whitaker, T. W. (1976). *In* "Evolution of Crop Plants" (N. W. Simmonds, ed.), pp. 39–41. Longmans, Green, New York.
Sakakibara, M., Difeo, D. R., Jr., Nakatani, N., Timmermann, B., and Mabry, T. J. (1976). *Phytochemistry* **15,** 727–731.
Sarkanen, K. V., and Hergert, H. L. (1971). *In* "Lignins: Occurrence, Formation, Structure, and Reactions" (K. V. Sarkanen and C. H. Ludwig, eds.), pp. 43–94. Wiley, New York.
Sarkanen, K. V., and Ludwig, C. H., eds. (1971). "Lignins: Occurrence, Formation, Structure, and Reactions." Wiley, New York.
Schoener, T. W. (1971). *Annu. Rev. Ecol. Syst.* **2,** 369–404.
Schoonhoven, L. M. (1972). *Recent Adv. Phytochem.* **5,** 197–224.
Schultz, J. C. (1978). "Competition, Predation and the Structure of Phytophyllous Insect Communities." Unpublished thesis, Univ. of Washington, Seattle.
Seigler, D. S. (1977). *Biochem. Syst. Ecol.* **5,** 195–199.
Seigler, D. S., and Price, P. H. (1976). *Am. Nat.* **110,** 101–105.
Self, L., Guthrie, F., and Hodgson, E. (1964). *J. Insect. Physiol.* **10,** 907–909.
Siddiqui, Z. A. and Sykes, R. L. (1962). *In* "Seminar on Vegetable Tannins" (S. Rajadurai and K. U. Bhanu, eds.), pp. 137–144. Central Leather Res. Inst., Madras.
Simmonds, N. W. (1976a). *In* "Evolution of Crop Plants" (N. W. Simmonds, ed.), pp. 81–83. Longmans, Green, New York.
Simmonds, N. W., ed. (1976b). "Evolution of Crop Plants." Longmans, Green, New York.
Singh, L. B. (1976). *In* "Evolution of Crop Plants" (N. W. Simmonds, ed.), pp. 3–9. Longmans, Green, New York.
Slama, K. (1969). *Entomol. Exp. Appl.* **12,** 721–728.
Slansky, F., Jr. (1976). *J. N.Y. Entomol. Soc.* **84,** 91–105.
Smirnoff, W. A., and Bernier, B. (1973). *Can. J. For. Res.* **3,** 112–115.
Smith, B. D. (1966). *Nature (London)* **212,** 213–214.
Smith, C. C. (1975). *In* "Coevolution of Animals and Plants" (L. E. Gilbert and P. H. Raven, eds.), pp. 53–77. Univ. of Texas Press, Austin.
Smith, R. H. (1972). *U.S., For. Serv., Pac. Southwest For. Range Exp. Stn., Res. Pap.* **56,** 1–10.
Soloman, M. J., and Crane, F. A. (1970). *J. Pharm. Sci.* **59,** 1670–1672.

Staedler, E. (1977). *Proc. Int. Congr. Entomol., 15th, 1976* pp. 228–248.
Stahl, E. (1888). *Jena. Z. Med. Naturwiss.* **22,** 557–684.
Stark, R. W. (1965). *Annu. Rev. Entomol.* **10,** 303–324.
Stephenson, J. W. (1970). *Rep., Rothampsted Exp. Stn., Harpenden, Engl.* p. 249.
Strong, D. R., Jr. (1974a). *Proc. Natl. Acad. Sci., U.S.A.* **73,** 2766–2769.
Strong, D. R., Jr. (1974b). *Ann. Mo. Bot. Gard.* **61,** 692–701.
Strong, D. R., Jr. (1974c). *Science* **185,** 1064–1066.
Strong, D. R., Jr., and Levin, D. A. (1979). *Am. Nat.* (in press).
Strong, D. R., Jr., McCoy, E. D., and Rey, J. R. (1977). *Ecology* **58,** 167–175.
Swain, T. (1976). *In* "Morphology and Biology of Reptiles" (A. d'A. Bellairs and C. B. Cox, eds.), pp. 107–122. Academic Press, New York.
Swain, T. (1977a). *Annu. Rev. Plant Physiol.* **28,** 479–501.
Swain, T. (1977b). *Proc. Int. Congr. Entomol., 15th, 1976,* pp. 249–256.
Swain, T., and Hillis, W. E. (1959). *J. Sci. Food Agric.* **10,** 63–68.
Tahvanainen, J. O., and Root, R. B. (1972). *Oecologia* **10,** 321–346.
Taksdahl, G. (1963). *Ann. Entomol. Soc. Am.* **56,** 69–74.
Tester, C. F. (1977). *Phytochemistry* **16,** 1899–1901.
Thielges, B. A. (1968). *Can. J. Bot.* **46,** 724–725.
Thompson, W. A., Cameron, P. J., Wellington, W. G., and Vertinsky, I. B. (1976). *Res. Popul. Ecol.* **18,** 1–13.
van Buren, J. (1970). *In* "The Biochemistry of Fruits and Their Products" (A. C. Hulme, ed.), Vol. 1, pp. 269–304. Academic Press, New York.
van Emden, H. F., ed. (1973). "Insect/Plant Relationships." Wiley, New York.
Van Sumere, C. F., Albrecht, J., Dedonder, A., De Pooter, H., and Pé, I. (1975). *In* "The Chemistry and Biochemistry of Plant Proteins" (J. B. Harborne and C. F. Van Sumere, eds.), Chapter 8. Academic Press, New York.
Vaughan, G. L., and Jungreis, A. M. (1977). *J. Insect Physiol.* **23,** 585–589.
Wallace, J. W., and Mansell, R. L., eds. (1976). "Biochemical Interaction Between Plants and Insects," *Recent Adv. Phytochem.,* Vol. 10. Plenum, New York.
Wasserman, R. H., Henion, J. D., Haussler, M. R., and McCain, T. A. (1976). *Science* **194,** 853–854.
White, T. C. R. (1969). *Ecology* **50,** 905–909.
White, T. C. R. (1974). *Oecologia* **16,** 279–301.
White, T. C. R. (1976). *Oecologia* **22,** 119–134.
Wilkins, H. L., Bates, R. P., Hanson, P. R., Lindahl, I. L., and Davis, R. E. (1953). *Agron. J.* **45,** 335–336.
Wilkinson, C. F., ed. (1976). "Insecticide Biochemistry and Metabolism." Plenum, New York.
Wilkinson, C. F., and Brattsten, L. B. (1972). *Drug Metab. Rev.* **1,** 153–161.
Williams, A. H. (1963). *In* "Enzyme Chemistry of Phenolic Compounds" (J. B. Pridham, ed.), pp. 87–95. Macmillan, New York.

Chapter **2**

The Distribution of Secondary Compounds within Plants

DOYLE McKEY

ISBN 0-12-597180-X

I. INTRODUCTION

Plants produce a diverse set of chemicals that are toxic to varying degrees to plant pathogens and herbivores and, are therefore, of potential selective advantage in deterring these enemies. Deterrence of herbivores and pathogens appears to be the principal advantage conferred by many of these compounds (Swain, 1974, 1977). If this is the case, the plant's metabolic behavior, which affects the performance of this "function"—for example, the physiological mechanisms influencing their distribution within the plant—should also have been subject to strong selection. The objectives of this chapter are to outline the selective factors that might influence the distribution of toxic secondary compounds within plants, to examine the physiological processes by which these distributions are produced, and to review the empirical evidence that some particular patterns of distribution are somewhat general.

Plants do not produce sufficient quantities of toxic compounds to provide complete protection of all their tissues. Instead, secondary compounds occur in varying amounts in different parts of the plant. The selective factors influencing distribution of these compounds within the plant are thus largely those of allocation of a resource. The patterns of compromise between conflicting selective forces acting on the allocation of "chemical-defense investment" are likely to bear great similarity to those visualized in other classical cost–benefit problems in evolutionary ecology (Rapport and Turner, 1977).

But the allocation of secondary compounds has another dimension. This dimension is a result of the fact that chemical defenses are not all of a kind, but as a group are chemically very diverse, exhibiting qualitative differences in attributes that may be important to the performance of defensive functions. The chemical defenses of a single plant may include representatives of several very different structural classes. Each of these compounds may possess a set of properties that suit it for particular roles in defense. The apportionment of the many types of compounds at a plant's disposal, while subject to the same kinds of compromises visualized in many problems of evolutionary ecology, is a problem demanding knowledge of the biochemical and physiological properties unique to each kind of compound.

Selective forces imposed by herbivores will certainly form a major influence on the evolution of patterns of toxic-compound allocation. However, defense compounds emerge from the internal physiology of the plant, and the importance of their metabolic behavior within the plant has not been sufficiently appreciated in existing concepts about their distribution within plants. The principal shortcoming is the tendency to look at the toxic product in isolation from the metabolic events associated with its accumulation. The chemical-defense phenotype that is being acted on by natural selection is not simply the toxic product itself, isolated from its connections with plant metabolism and responsive as a unit to selection. Rather, the phenotype being acted upon by selection is a very complex set of physiological events: the biosynthetic pathways leading to the toxic product and to its catabolism, as well as components associated with the manipulation of the toxic compound, e.g., organelles involved in its compartmentation, adjustments in other systems vulnerable to toxic effects of the compound, and mechanisms for its translocation.

There is another consideration that must be dealt with: Do these toxic compounds confer selective advantages unrelated to deterrence of herbivores? The answer to this question will greatly affect our expectations about their patterns of distribution within plants. This question has been discussed elsewhere (Swain 1974, 1977) and will be only briefly considered here. Swain (1974, 1977) has emphasized that the essential internal functions ascribed to secondary products are by and large not important enough to account for the diversity and quantity of these compounds. It is interesting that a common function ascribed to secondary metabolites is that of inhibiting other functions (e.g., Gross, 1975). However, a common characteristic of regulatory systems is that they are metabolically cheap and specific, features not reported for secondary compounds. Many kinds of secondary compounds occur in plants at concentrations many times greater than the components whose activities they are supposed to regulate. Also, their regulatory activities would in some cases be decidedly nonspecific. For example, the proposal that tannins in plants may function as gibberellin antagonists (Corcoran *et al.*, 1972) must be evaluated in light of the fact that tannins are generalized protein-complexing agents. Any advantage of tannins as regulatory effectors might well be offset by their capacity to interact nonspecifically with plant enzymes. While this question certainly cannot be definitively answered here, I have proceeded on the assumption that secondary compounds cannot be assigned a general role in essential internal physiology.

The ultimate origins of secondary metabolic pathways may be traced to "primary" pathways (Swain, 1974). Is it possible that some existing pathways have conserved original primary functions? While I prefer to bypass the distinction between "original" and "acquired" functions of secondary metabolites as formulated by Del Moral (1972), it may be pointed out that the origins of defensive compounds are likely to be diverse. Synthetic sequences for the formation of some types of alkaloids have analogies in metabolic pathways common in plants (Swain, 1974), and these could have arisen via altered substrate specificity of existing enzymes. In such cases, a toxic compound may have been produced fortuitously, and suddenly, by a pathway with completely primary function. Just as suddenly, the new compound may have lost the essential function of its homologue. Its distribution within the plant would then be free to respond to selection acting on its new defensive function. In other cases, e.g., lectins, the evolution of toxic molecules may have been a long and gradual sequence of events leading from proteins important in self-recognition systems mediating intercell interactions in plant development, to proteins that recognize and interact with cells of foreign organisms (cf. Clarke *et al.*, 1975). In such cases, we might expect a confusing variety of functions to be performed by existing representatives of these compounds, and a corresponding variety of intraplant distributions.

The approach often employed by ecologists—examining compounds largely in isolation from the metabolic events associated with them—has typified the examination of these compounds by chemists. Historically, the secondary compounds have been the province of organic chemists. The concern of these investigators has been to determine the structure of new compounds and to elucidate the reaction mechanisms by which plants produce them (Floss *et al.*, 1974). The exploration of "biological" aspects of secondary compounds for a long time concerned not questions of function—investigation of which would have required the enzymologist or plant physiologist—but rather the utility of secondary compounds in assessing phyletic relationships. In addition to producing progress in the task of constructing phylogenies, studies motivated by the rationale of chemotaxonomy have resulted in an enormously useful body of information on the biosynthesis of secondary compounds. These have proceeded from analysis of reaction mechanisms to design of tracer studies using presumed precursors to verify proposed synthetic sequences, but have usually stopped short of the isolation of enzymes catalyzing the reactions observed (cf. Floss *et al.*, 1974, for alkaloids).

Conspicuously lacking in the studies of these compounds for a long time was the involvement of biochemists and plant physiologists (Floss *et al.*, 1974). Molecules that seemed to be waste products or at least appeared to serve no essential internal functions may have been regarded by biochemists as uninteresting subjects for study. For example, why should one expect interesting regulatory mechanisms in the production of waste products? Secondary compounds appeared to be metabolic sludge, and investigations of them at the physiological level had the prospect of being dead-end studies of dead-end products. Studies began to emerge, however, demonstrating that the view of secondary compounds as end-products that drop out of metabolism once produced was incorrect (Robinson, 1974; Ellis and Towers, 1970; Ellis, 1971). These findings were accorded great significance, for if secondary compounds underwent active metabolism, the view that they were waste products became difficult to justify (Swain, 1974). Furthermore, the fact that they often accumulated to high concentrations in the plant, even though there existed mechanisms for their degradation, could only mean that their accumulation was subject to physiological regulation (Ellis, 1974), a sure sign of their utility to the plant.

The realization that secondary compounds are often not end products, but subject to further metabolism within the plant, opened the possibility that many secondary compounds perform essential or useful internal-physiological functions (Siegler and Price, 1976). There seems to be a tendency to regard recent information on the metabolic activity of secondary compounds as being in conflict with the view that the prime selective advantage they confer is deterrence of pathogens and herbivores. The perceived conflict appears to be a product of the compound-in-isolation approach to thinking about defense chemicals. When the toxic product is conceived of as part of a dynamic chemical-defense system in which defense chemicals are transported, detoxified, and reallocated and in which investment in defense may be converted into other resources, the entrance of secondary compounds into metabolism is seen as an *expected* feature of chemical-defense substances. In fact, examination of the kinds of metabolic pathways entered into by secondary compounds gives interesting support to the view that these pathways are related to the manipulation of chemical defenses, rather than to the performance of essential internal functions. Some outstanding examples will be documented in later sections of this paper.

The "compound-in-isolation" approach now seems to be giving way to investigation of secondary metabolites in the context of the metabolic matrix in which they are enmeshed. The intent of this paper is to take full advantage of recent advances in the enzymology and regulation of secondary metabolism. However, most of the information in this chapter about distribution of secondary compounds within plants deals with compounds in isolation. The investigator who would attempt to draw generalizations from existing information on intraplant distribution must face several of its major shortcomings. Most importantly, almost none of it was gathered with ecological hypotheses in mind. As a result, in almost all cases the information is grossly incomplete for the purposes of the ecologist. The economic botanist may study secondary compounds only in the commercially useful parts of the plant. The pharmacognosist may present detailed information on the histological distribution of alkaloids in a particular plant, but there may be no information on the toxicity of those particular alkaloids. No one plant species provides the information necessary to make the statement, "Toxic alkaloids tend to have such-and-such a pattern of distribution within the plant," let alone providing the information allowing a generalization across many species. Organic chemists may have documented changes in secondary-compound distribution with development of the plant, but it will probably not be known whether translocation—or catabolism and re-synthesis—are the mechanisms behind these ontogenetic changes in distribution.

A second shortcoming of existing information is its unevenness. The body of literature on secondary-compound distribution within plants emerges from many sources, from investigators with many different sets of aims, and is therefore unsystematic and idiosyncratic in its coverage. Even more scattered and uneven than information about the distribution of secondary compounds within plants is information on the *processes* that have produced these distributions. When information is available, it often concerns individual compounds that are atypical. For example, most information concerning the physiology of alkaloid catabolism deals with ricinine, nicotine, tomatine, and coniine; all of these are highly unusual structures among the alkaloids. Properties of these compounds important in defensive functions, such as the ability to cross membranes, or the ability to be translocated within the plant, may be very different from those of "typical" alkaloids. For the host of isoquinoline and indole alkaloids, and others that predominate in nature, there exists virtually no information beyond their occurrence in specific parts of specific species. Nothing is known of the occurrence or importance of metabolism and translocation, or regulatory mechanisms operative in producing these distribution patterns. Similarly among the phenolics, there is relatively extensive information on the metabolism and regulation of simple aromatic compounds and phenolic acids, but almost none for polyphenolics such as the tannins, which are of widespread importance in plant defense systems (Feeny, 1976; Rhoades and Cates, 1976; Fox and Macauley, 1977).

Faced with this massive but inadequate set of information, postulating particular patterns of distribution as being of general occurrence have been avoided. Though some patterns seem to be consistent with the information reviewed (and these will be duly noted), the unsystematic nature of this information renders exceedingly fragile generalizations drawn from it about patterns of distribution. Instead I have chosen to emphasize the physiological processes related to within-plant distribution and its regulation. These processes should be somewhat more general than are the actual patterns of distribution they generate. Furthermore, they provide the key to understanding the great diversity of patterns that I believe exist.

I hope to demonstrate in this review that the physiology of secondary metabolism in plants exhibits general features to a degree that is remarkable, considering the diverse metabolic origins of these compounds. I also hope to argue convincingly that these general features may be understood as a consequence of their shared ecological function. The spirit of this review is to produce tentative generalizations that may stimulate further research, guided by a unified approach to the study of the physiology of plant chemical-defense systems.

II. GENERAL HYPOTHESES

A. Total Magnitude of Defense Investment

From the time of the first review articles on plant chemical defenses (Whittaker and Feeny, 1971), workers in this field have emphasized that there is a sensitive balance between the adaptive advantage conferred by a herbivore-deterrent chemical and the metabolic cost its production imposes on the plant. Whittaker and Feeny (1971) postulated that patterns of variation in the importance of (*a*) herbivore pressure and (*b*) metabolic costs of chemical defense, would be reflected by variation in the amounts of defensive substances produced. This same kind of balance has been visualized by Janzen (1969), Feeny (1970), Rehr *et al.* (1973a,b,c), Levin (1976), Jones (1972), and virtually every other worker in this field. The concept of costs and benefits of defense has been central to hypotheses that postulate variations in defense investment associated with successional status (Cates and Orians, 1975), soil quality (Janzen, 1974), plant "apparency" (Feeny, 1976), leaf lifespan (Janzen, 1974; Stanton, 1975), environmental variations facing a single species (Cates, 1975), and intraplant distribution (Orians and Janzen, 1974; Rhoades and Cates, 1976).

While the concept of costs and benefits of defense has stimulated the formulation of useful hypotheses that receive support from such empirical studies as have been carried out (e.g., Cates and Orians, 1975; but, see Otte, 1975; McKey *et al.*, 1979), readers of literature in this field will search in vain for any careful quantitative model of costs, benefits, and the outcome of conflicts between the two. We have only qualitative statements about the patterns of variation of costs and of benefits. There has been no assessment, for example, of the ratio of increasing benefit to increasing cost when concentration of a toxic compound is increased. How many resources are saved from herbivory when *x* amount of resources are used to synthesize defense chemicals? Answers to this kind of question are crucial to the refinement of ideas such as that of Janzen (1974) relating the proportion of resources invested in defense to soil quality. If resources are expensive for plants growing on poor soils, why should they be spent on defense? The answer must be that the cost of defense is relatively low compared to the cost of herbivory were the plant not defended. What determines the point when further investment in defense is not rewarded by commensurate benefits? For plants growing on rich soils, if defenses are cheap, why not possess them in abundance? The point is that the predictive power of current formulations is greatly limited by their distinctly qualitative character. Prediction of gross differences between light-gap and understory species, for example, is about the limit of precision allowed by existing models.

Quantification of ideas about defense investment requires two kinds of investigations: (*a*) Simple graphical models, incorporating a simple set of assumptions, that model costs and benefits of chemical defenses, and make predictions about what the equilibrium defense investment should be given various combinations of parameters. (*b*) Extensive studies of the physiology of defense compounds within the plant, and of plant–herbivore interactions in the field. Such studies will lead to refinement and validation of assumptions incorporated in the models. For example, Siegler and Price (1976) bring into question the often implicit assumption of ecologists that the major benefit conferred by secondary compounds is defense. They argue that these compounds might also perform internal metabolic functions. They also point out that defense investment might be "recycled" within the plant by means such as conversion of secondary compounds to nutritive molecules. Only detailed chemical investigations will permit us to refine our assumptions, for example, those about the costs and benefits of "recycling" defense compounds (Swain, this volume, Chapter 19).

Cates (1975) employed a very direct way of assessing relative investment in chemical defenses: measuring the difference in growth rate and in reproductive output (which, after all, are the currency to which investment must be related) between different morphs in a plant population polymorphic for chemical defenses. This example (to be discussed later) is unique in the literature. Similar studies should be pursued with other plants, for example, *Trifolium* and *Lotus* polymorphic for production of cyanogenic glycosides (Jones, 1972). These studies would be most useful if coordinated with detailed investigation of the actual secondary compounds involved (e.g., Mooney and Chu, 1974): their identity, the concentrations encountered in the plant; their physiology within the plant (interrelations with primary metabolic pathways, recycling from senescent tissues, etc.). HCN, for example, is toxic at low concentrations, and the glycosides are in fact present in low concentrations. Furthermore, HCN can be metabolized by plants into asparagine (Nartey, 1969, 1970), so defense investment in this form can be retrieved and converted to common metabolites. Given these considerations, the cost of having cyanogenic glycosides, in terms of lowered growth rate and reproductive output, might be very low. Such studies should be carried out with a variety of kinds of chemical-defense systems: compounds present in high concentrations and in low concentrations; compounds containing nitrogen and those containing only less "valuable" materials such as carbon, hydrogen, and oxygen.

Though it is difficult to quantify, there is much support from empirical studies for the idea that chemical defense imposes a cost, that production of secondary compounds is a significant part of the resource budget of many plants. Cates (1975) studied *Asarum caudatum* (Aristolochiaceae), a species polymorphic for palatability to slugs. This polymorphism is related to presence or absence of unidentified chemical deterrents. He found that in the absence of herbivory, the palatable morph that lacked the chemical deterrent had higher growth rates and produced more seeds than did the unpalatable morph. Rehr *et al.* (1973b) showed that cyanogenic glycosides and other, unidentified deterrent chemicals present in non-ant-acacias (Leguminosae), were absent from *Acacia* species protected by obligate mutualist *Pseudomyrmex* ants. They interpreted this difference to mean that in ant-acacias, metabolic costs of the now-redundant chemical defenses provided selective pressure for their loss. The very existence of polymorphisms for the production of secondary compounds, such as those observed for *Lotus* and *Trifolium* HCN production, suggests that in some environments, their production might be selected against. Circumstantial evidence that defense compounds exert a metabolic cost is provided by the fact that their production is often induced by damage or by invasion by pathogens. Such regulatory systems would not have been evolved had they not been cheaper metabolically than constitutive (Levin, 1971) production of these compounds.

Insight on another aspect of defense investment is provided by a recent study of the secondary compounds of pitcher plants, *Sarracenia* spp. (Sarraceniaceae) (Romeo *et al.*, 1977), that grow in nitrogen-poor bog soils. These plants were found to be virtually devoid of nitrogen-containing defense chemicals such as alkaloids and nonprotein amino acids. This study opens the possibility that not only the amount of defense investment, but also the kinds of nutrients used in construction of defense compounds, may be an important component of the metabolic cost of defense. It is interesting that legumes contain a great diversity and often high concentrations of nitrogen-containing secondary compounds, such as alkaloids, uncommon amino acids, protease inhibitors, and lectins. In this group, the lower metabolic cost of nitrogen, due to symbiotic association with nitrogen-fixing bacteria, may have permitted increased involvement of nitrogen in secondary pathways not essential to internal metabolic functions (cf. Feeny, 1976).

The importance of defense investment as a component of evolutionary–ecological hypotheses depends on its being a trait sufficiently variable and heritable for natural selection to act upon. Numerous studies can be cited in which artificial selection has resulted in greatly increased or decreased concentrations of secondary compounds. These include alkaloids such as nicotine in tobacco (Smith, 1965), quinolizidine alkaloids in *Lupinus* (Leguminosae) (Gustafsson and Gadd, 1965), ergoline alkaloids in the fungus *Claviceps* (Mothes, 1976), phenolics such as condensed tannins in *Sorghum* (Gramineae) (Stitt, 1943, Harris *et al.*, 1970) and *Lespedeza* (Leguminosae) (Cope, 1962; Donnelly and Anthony, 1969), toxic proteins such as chymotrypsin inhibitors in barley (Weiel and Hapner, 1976), and toxic fatty acids such as erucic acid in rapeseed (Mothes, 1976). The strong response to selection in these cases supports the notion that if natural selection is of sufficient intensity, variation in and heritability of secondary compound concentration is sufficiently high that a rapid response to selection will be produced.

The compounds that ecologists consider functionally related as defense compounds emerge from a great diversity of biosynthetic backgrounds. Presence of a compound will be related to underlying genotypes very directly in some cases, very indirectly in others. Thus, we should expect great variation in the heritability of various classes of secondary compounds, and confusing variety in the strength of response to selection, patterns of secondary compounds encountered with hybridization, and so on. Toxic proteins such as proteinase inhibitors and lectins, for example, have a very direct relation to genotype: as proteins (or glycoproteins) they are direct products of transcription and translation of one or more genes (and of the activity of glycosyl transferases, in the case of glycoproteins). Alkaloids, on the other hand, are likely to have very complex relations to the underlying genotypes. An alkaloid is the product of several enzymatic steps. Its production can be influenced by the presence, dosage, and activity levels of any one of several gene products (e.g., Soloman and Crane, 1970).

For few kinds of secondary compounds do we have very much information on the mode of inheritance of the ability to produce and accumulate the compound and on genetic factors influencing its concentration. Review of these studies is beyond the scope of this chapter. The interested reader should consult studies such as Goplen *et al.* (1957), Gustafsson and Gadd (1965), Haskins and Gorz (1965), Haskins and Kosuge (1965), and M. A. Hughes (1973).

Though it is clear that secondary metabolism, like primary metabolism, is mediated by enzymatic activity and thus has a strong element of genetic control, we know very little about the genetic basis for variation in secondary-compound production observed in populations in nature. In most cases, for example, we do not know whether this variation is due to variation in structural genes controlling production of the enzymes involved in synthesis or to variation in regulatory elements.

It is occasionally suggested that later steps in biosynthesis of various secondary compounds are not under enzymatic control. Leete (1967) points out that this is unlikely in the case of alkaloids, since the majority of them are optically active. If their synthesis were not under the control of fairly specific enzymes, racemic mixtures would be expected instead. Scott (1975) points to the existence of such racemic, enantiomeric, and diastereomeric sets in the alkaloids of the Apocynaceae. He considers this may reflect either lack of substrate specificity on the part of enzymes involved in the last steps of synthesis, or the complete absence of such enzymes. Recently Haslam *et al.* (1977) have proposed that the final steps in biosynthesis of condensed tannins—polymerization—may not be under enzymatic control. It should be noted that even though the final reactions may not be under direct enzymatic control, the ability to produce the final product can still be considered to be under genetic control, and hence sensitive to natural selection, since it is the nature of the genetically controlled precursors that determines their ability to polymerize nonenzymatically. There is evidence that in at least some cases even the polymerization of condensed tannins is under strict enzymatic control (Roux, 1972). In another group of compounds, the coumarins of Rutaceae and Umbelliferae, it appears that many of the later synthetic steps, particularly those giving rise to the huge range of prenyl side-chains, are not strictly controlled enzymatically. Even the formation of a furan ring system as opposed to a pyran system, may in these compounds depend only on pH at the site of ring closure (Peter G. Waterman, personal communication).

In summary, there is evidence that production of secondary compounds imposes a metabolic cost to the plant and that the amount produced is a variable and heritable trait. Thus the inclusion of the concept of defense investment in evolutionary hypotheses about defense chemistry is valid.

B. Allocation of Defense Investment

Chemical defenses impose a metabolic cost, so that defense investment is not solely determined by the level of secondary compound production required to confer absolute protection. It follows that there is selection for allocation of the plant's limited supply of defense compounds in ways that most increase its fitness.

What factors influence the intraplant allocation of a defense chemical? Four kinds of factors have been discussed by previous authors: (1) The relative costs to the plant of herbivore-inflicted damage or loss of the parts in question (Rockwood, 1973; McKey, 1974); (2) the relative probability of discovery and successful herbivore attack of different plant parts in the absence of chemical defense (Rockwood, 1973; McKey, 1974; Feeny, 1976; Rhoades and Cates, 1976); (3) physiological constraints on synthesis and/or storage of toxic compounds in different kinds of tissues (Orians and Janzen, 1974; Rhoades and Cates, 1976; Feeny, 1976); and (4) the distribution of other chemical defenses within the plant (Janzen, 1973a). Each of these factors will be discussed briefly here, with more detailed discussion reserved for later sections.

1. *Costs of Herbivory to Different Plant Parts*

On a per-unit weight basis, different parts of the plant do not contribute equally to its fitness. While it would be difficult to attempt to assess the relative contribution of the trunk and the leaves to a tree's fitness, the concept of relative value of different tissues (as reflected in relative costs imposed by their loss), if applied judiciously, can be useful in understanding the intraplant distribution of secondary compounds. Certainly old leaves, for example, are likely to make a much smaller further contribution to plant fitness than are leaves that have just reached maturity.

Most of the available information pertinent to this topic in fact concerns the relative contribution to fitness of leaves of different age. Studies represented fall into three types: (*a*) Studies that measure photosynthetic rate, carbon export, and other measures of production by leaves of different ages (Freeland, 1952; Clark, 1956; Dickman and Kozlowski, 1958; Shiroya *et al.*, 1961; Wardlaw, 1968; Larson and Gordon,

1969; Bidwell, 1974; Aslam *et al.*, 1977); (*b*) studies that compare the effects of simulated herbivory of leaves of different age (see Kulman, 1971); (*c*) Studies that compare the effects of herbivore species with different leaf-age preference patterns (Balch, 1939; Reeks and Barter, 1951; Mazanec, 1966; L. F. Wilson, 1966). In only a few cases (Rees, 1969; Feeny, 1970; Lawton, 1976; Rhoades and Cates, 1976; Ikeda *et al.*, 1977) has leaf-age preference of an herbivore been demonstrated to be due to age-dependent differences in secondary-compound concentration. While all these studies will be dealt with later in detail, it may be said at this point that there is ample evidence that the effect of damage inflicted by a herbivore is strongly dependent on what part of the plant it eats.

2. *Probability of Discovery and Successful Attack in the Absence of Chemical Defenses*

Each part of the plant will have a part-specific rate of discovery by herbivores, depending on factors such as its phenology, density on the plant, and the array of herbivores likely to come into contact with it (Feeny, 1976). The probability that a part will be attacked once discovered will be influenced by many factors other than chemical defenses. These factors include mechanical defenses such as silica, calcium carbonate, or "toughness," nutritional content, and intensity of predation on herbivores of the plant part in question.

3. *Physiological Constraints*

The mechanisms of toxicity of many kinds of plant secondary constituents are so fundamental as to render them at least potentially toxic to the plants that produce them. Orians and Janzen (1974) discussed the possibility that immature tissues may face particularly severe constraints in production and storage of allelochemicals. They considered that immature plant tissues may not possess sufficient cellular substructure for the segregation of toxic materials from cellular contents. They also proposed that the adaptive emphasis in developing leaves and other tissues is on rapid differentiation and development. Production and storage of toxic compounds would compete with rapid development and might occasionally result in release of toxic substances into cells that are especially active biosynthetically and particularly sensitive to toxins.

This is only one of the many kinds of constraints plants face in their handling of toxic chemicals. The paradox inherent in the production by living organisms of substances whose *raison d'etre* is their toxicity to living organisms is a pervasive feature in the organization of plant chemical-defense systems. Its influence in the allocation of secondary compounds within the plant will be a recurring theme.

4. *Distribution of Other Chemical Defenses within the Plant*

Janzen (1973a) has discussed the possibility that the variety of secondary compounds elaborated by a single plant are organized into a system of chemical defense. How several secondary compounds might evolutionarily exert mutual influences on their relative distributions within a plant will be discussed in section IV.

Though it is too early to assess the roles of the factors outlined above, it is clear that, for whatever reasons, secondary compounds are heterogeneously distributed within the plant. This fact becomes tremendously exciting when coupled with the realization that we have no clear idea of the underlying physiological basis of this heterogeneity. Though we have robust hypotheses to explain *why* oak trees exhibit a particular distribution of tannins within the plant, for example, we have no idea *how* this distribution is produced. Evolutionary ecologists who may be tempted to consider this aspect of the problem merely a question of "plumbing", lose sight of the fact that only by understanding the mechanisms leading to a phenotypic trait can we assess its responsivensss to selection. We are not even close to understanding the regulation of intraplant distribution of secondary compounds. Investigation of this area (e.g., Luckner *et al.*, 1977) is vital to a holistic view of plant chemical-defense systems.

If information presently available is any indication, regulation of plant secondary metabolism is likely to be just as complex as metabolic regulation of primary pathways. Heterogenous distribution of secondary compounds within plants may be due to differing rates of synthesis and degradation, or to translocation from one part to another (Mooney and Chu, 1974). In cases where tissue-specific rates of synthetic or catabolic enzymatic activities are known to be involved in producing an heterogeneous distribution, there is little known about the mechanisms by which these enzymatic activities are in turn regulated. There exists only circumstantial evidence, for example, of repression–induction processes in alkaloid synthesis and degradation (Floss *et al.*, 1974).

In addition to the active processes of metabolism and translocation, there are passive processes under, at most, limited control by the plant that result in intraplant differences in concentration. Decreased secondary-compound concentration during leaf maturation, for example, might be due only to dilution effects as other components are synthesized by the leaf or imported into it at higher rates. Similarly, decrease in secondary compound concentration in aging leaves may be due to volatilization, leaching by rainfall (Flück, 1963), or autoxidative polymerization (Goldstein and Swain, 1963). Gentic control of these events may be more complex than control of the activities of specific secondary pathways.

Allocation of chemical defenses within the plant has a temporal component as well: relative defense needs of different parts of the plant may change with its development and aging. The processes involved in producing developmental changes in the amounts and distribution of secondary compounds are the same as those involved in producing the distribution observed at any one point in time. While the consistency of ontogenetic patterns of secondary metabolism typical of most plants implies strong genetic control, there are as yet few examples of heritable variation in behavior of secondary metabolism during development (e.g., alkaloids in *Lupinus angustifolius:* Gustafsson and Gadd, 1965; alkaloids in *Papaver bracteatum:* Luckner *et al.*, 1977).

C. Chinks in the Armor

All living organisms share the same fundamental biochemistry (Swain, 1974). The plant engaged in chemical defense is therefore faced with the problem of having to poison systems that are very similar to its own. Plants are probably precluded by this constraint from the production of kinds of toxins that, while they are effective poisons against other organisms, may be "too hot to handle." The fact that the molecules of defense can be tolerated by plants leads us to suspect that there will never be any absolute barrier to their being tolerated by the fundamentally similar systems of pathogens and herbivores.

This source of compromise interacts in diverse ways with another source of compromise to the effectiveness of plant defense systems that has already been mentioned, the metabolic costs imposed by investment in defense. (*a*) Selection for defense systems that minimize these costs will favor defense substances that can be manipulated in various ways by the plant. Substances may be favored, for example, that can be metabolized to water-soluble derivatives capable of being mobilized from senescent tissues, translocated from seed to seedling, and otherwise shuttled about to achieve maximum protection at minimum cost. If the plant can manipulate its toxic compounds in this way, there should be no absolute barrier to the evolution of such manipulatory ability among herbivores and pathogens. (*b*) Selection for "cheap" defense systems will result in the evolution of secondary compounds that are toxic at very low concentrations. As emphasized by O'Brien (1967), acute toxicity at very low concentrations is achieved by structural features, that enable the toxic compound to interact *specifically* with a crucial component in the body of the target organism. Nicotine is highly toxic in small amounts because it does not interact with many components in the body, but rather interacts specifically with one crucial component (acetylcholine receptor; O'Brien, 1967). The stringency of the set of stereochemical features that the nicotine molecule must possess in order to exert this toxicity means that a very slight change in the molecule might render it much less toxic. Hence "cheap" defense systems based on acute toxins may be automatically susceptible to easy counteradaptation.

Thus the defense systems of plants are the outcome of conflicting demands emerging from internal physiological constraints on the one hand and requirements for effective defense on the other. I present arguments in following sections that some kinds of secondary compounds better satisfy the first of these selective pressures, other kinds better satisfy the second. Which of these two conflicting selective pressures (and, it follows, what sort of defense system) will be most important in forging the plant's response should depend on factors such as the part of the plant concerned (Rhoades and Cates, 1976), the plant's successional status in the community (Feeny, 1976), and the cost of herbivory to the plant (Janzen, 1974).

The first two major hypotheses presented above concern the allocation of "defense investment." Before returning to these hypotheses in Section V, I wish to examine in detail the functional diversity of secondary compounds, in the context of different solutions to the adaptive dilemmas presented in the discussion of the third major hypothesis.

III. WITHIN-PLANT DISTRIBUTION OF DIFFERENT CLASSES OF SECONDARY COMPOUNDS

A. Introduction

Secondary compounds emerge from a tremendous diversity of biochemical backgrounds (Swain, 1974). They exhibit great diversity in their physical and chemical properties, in the relations of the pathways that produce them to fundamental metabolic pathways, and in the ways in which they exert toxic effects on biological systems. Given the diversity in chemical properties alone, it would be very surprising if there did not exist some sort of partitioning of the function of defense between the various classes of secondary compounds. One objective of this section is to elucidate the patterns of partitioning that may exist and to relate these patterns to the properties of different kinds of compounds. In this chapter, the functional diversity of secondary compounds will be discussed mainly in relation to patterns of distribution within plants.

One fundamental property of a secondary compound is its solubility characteristics. These will largely determine its ability to be translocated within the plant, to be absorbed from the gut of a herbivore, and to enter "target" areas such as lipid-rich nervous tissue. Solubility characteristics will also determine the ease with which a compound may be excreted by an animal. Terpenoid compounds, for example, are not very soluble in water and cannot be translocated in the xylem and phloem. Many plants that produce large quantities of terpenoid compounds possess specialized resin or latex systems in which these lipophilic substances can be transported within the plant independent of the vascular system. The ability to cross cell membranes and to enter lipid-rich nervous tissue depends highly on the fat-solubility of a substance. The existence of lipophilic secondary compounds in the form of water-soluble glycosides within the plant may be one outcome of the conflicting selective pressures facing the plant's use of the secondary compound. It may also be no coincidence that many kinds of secondary compounds have intermediate solubility characteristics (e.g., phenolics, Singleton and Kratzer, 1973), or have one moiety that is water-soluble and another that is fat-soluble, or have basic (e.g., nitrogen in a ring system) and acidic (e.g., phenolic hydroxyl) groups in the same molecule.

Another intrinsic property of the secondary compound that may affect its use by the plant is its elemental composition. Different chemical elements are not reserved by the plant for production of secondary metabolites at the same cost. Incorporation of nitrogen into secondary compounds may be subject to more severe constraints than is, for example, the use of carbon. How important such considerations are may depend, however, on other properties of the plant. Incorporation of nitrogen into secondary compounds may be prohibitively expensive for pitcher plants living on nitrogen-poor soils (Romeo *et al.*, 1977), but not for legumes associated with nitrogen-fixing bacteria (Feeny, 1976). Similarly, the allocation of reduced carbon compounds to defense may impose a more severe cost on shaded understory plants than on insolated canopy trees.

Molecular size may also affect the use of a secondary compound by the plant. Very large molecules are, as pointed out by Rhoades and Cates (1976), mostly restricted to the guts of herbivores or the extracellular enzymes of pathogens. Any toxic effects they may exert must affect extracellular processes. This means that such compounds never reach sensitive systems, e.g., nervous tissue, that can be disrupted by low concentrations of a very specific toxin. Such compounds must be present in relatively high concentrations to exert toxicity, unless they inhibit specific digestive enzymes (Rhoades and Cates, 1976) or, like the toxic proteins abrin and ricin, attach themselves to cell membranes and induce their transport into the cytoplasm (Olsnes *et al.*, 1974, 1976).

Other properties not intrinsic to the secondary compound, but completely dependent on extrinsic features of the plant or of the herbivores that confront it, will also lead to functional differences between secondary compounds. For example, we may consider the metabolic background in which each type of secondary compounds is embedded. Some kinds of toxic uncommon amino acids probably arose through relaxed substrate specificity of enzymes catalyzing the biosynthesis of common amino acids (Bell, 1971). These uncommon amino acids, if similar enough to a common amino acid, might serve as substrates for enzymes involved in *catabolism* of the common amino acid. Thus for this type of compound the plant may have a built-in system for converting defense investment into common metabolites. Such a compound might be expected to predominate in parts of the plant in which there is a premium on reclamation of defense investment, for example in germinating seeds. Canavanine is an example of such a compound: Once produced, it is subject to metabolism by the same urea-cycle enzymes that accept arginine and its derivatives. In fact, it is metabolized by this pathway during germination of *Canavalia* seeds (Rosenthal, 1977). In contrast, secondary compounds that originated as something very different from anything the plant's enzymes had ever dealt with (e.g., complex indole–monoterpene alkaloids?) probably do not, at least initially, have enzymes that can degrade them to common metabolites. Until such enzymatic activities are acquired, reclamation of defense investment in the form of such compounds may be difficult.

Another property of a secondary compound not intrinsic to it, is its mode of action. Depending on their specificity and the nature of the systems they attack, some secondary compounds will be effective deterrents at low concentrations, others only at high concentrations. If a plant tissue has only a limited storage capacity for toxic compounds, there may be a premium on allocation to that part of a secondary compound deterrent at low concentration.

In this section, the diversity of secondary compounds will be considered in the context of different solutions to the partially conflicting selective forces facing plant defense systems and the allocation of defense investment within the plant.

B. Avoidance of Autotoxicity

1. *Selection for the Ability to Affect Fundamental Processes*

If a secondary compound is to be effective against the largest possible collection of herbivores, it must affect fundamental biological processes. While some kinds of chemical defenses may exploit biochemical differences between animals and plants, it is to be expected that many secondary compounds will be at least potentially toxic to the plants that produce them, as well as to herbivores and pathogens. This potential for autotoxicity introduces constraints on the plant's physiological manipulation of defense chemicals. While the realization that plants may be susceptible to poisoning by their own secondary compounds is by no means new (McKey, 1974; Orians and Janzen, 1974; Rhoades and Cates, 1976), it may not be widely recognized that this problem potentially exists for representatives of virtually all major classes of secondary compounds, including some considered to exert toxicity specifically in animals. The following examples illustrate the pervasive character of this problem in the functioning of plant defense.

Many phenolic compounds are broad-spectrum, if often weak, toxins to plants, microorganisms, and animals (Singleton and Kratzer, 1969). They are capable of damaging membranes, uncoupling oxidative phosphorylation, and other acts of disruption of general biochemical processes. Tannins (Loomis and Battaile, 1966; Van Sumere *et al.*, 1975; Rhoades and Cates, 1976), some phenolic resins (Rhoades and Cates, 1976), and quinones (Van Sumere *et al.*, 1975) are general protein poisons. Some kinds of phenolic compounds are acutely cytotoxic; podophyllotoxin, for example, is a potent cytotoxin that is particularly effective on dividing cells (Singleton and Kratzer, 1969).

Terpenoids also include many compounds that are toxic to plants as well as animals. Mono- and sesquiterpenes figure prominently as allelopathic agents (Müller, 1966; Müller and Hauge, 1967; Gant and Clebsch, 1975). Sesquiterpene lactones owe their general biological activity to the ability to alkylate thiol groups of proteins; some of these inhibit plant growth (Rodriguez *et al.*, 1976). Saponins alter the permeability of cell membranes (Bondi *et al.*, 1973) and may interact nonspecifically with proteins (Ishaaga and Birk, 1965), thus exerting general toxicity. However, the toxic activity of some kinds of saponins against cells may depend on the presence of cholesterol in their membranes (Shany *et al.*, 1970). The steroidal glycosides known as cardiac glycosides are a classic example of a type of compound that is often considered toxic specifically to animals (Rhoades and Cates, 1976). But Heftmann (1975) notes that ATPases similar to those affected by this group of compounds in animals are present in plants as well and are sensitive to various cardiac glycosides. He states that "it is not easy to understand how plants can contain such substances without being adversely affected by them." Jonas's (1969) study highlights this point. He found that irrigation with cardenolide-containing aqueous extracts or pure cardenolides from *Digitalis purpurea* (Scrophulariaceae) caused chlorosis and necrosis of leaves not only in a number of other plant species, but in *D. purpurea* itself as well. Thus the producer plant is sensitive to its own metabolites if they are present in excess or if they are not properly sequestered.

Many nonprotein amino acids are analogs of common amino acids (Fowden, 1974). Since plants possess the same basic protein-synthesizing machinery as do animals, they are just as susceptible to the toxic action of these compounds. Many of these are in fact highly toxic to plants that do not produce them. In *Phaseolus aureus* (Leguminosae) glutamyl-tRNA^{glu} synthetase activates γ-substituted analogs of glutamic acid (Lea and Fowden, 1972). When azetidine-2-carboxylic acid is supplied to mung bean seedlings, the degree of proline replacement may be so high that insufficient normal enzyme molecules remain to sustain metabolism (Fowden, 1963). Mimosine is highly growth inhibitory to seedlings of nonproducer plants (Smith and Fowden, 1966).

Other plant secondary compounds affect biological processes that are equally fundamental. The A-chains of the toxic proteins abrin and ricin are enzymes acting directly on the 60S subunits of eucaryotic ribosomes, and they are toxic to nonproducer plants such as *Tradescantia* and *Valisnera* (Olsnes and Pihl, 1977), as well as to animals. Cyanogenic glycosides release free HCN, a potent inhibitor of the terminal cytochrome oxidase systems of plants, animals, and microorganisms (Nartey, 1973). Glucosinolates exert physiological effects on plants as well as in animals (Miller, 1973). Other secondary compounds affect fundamental processes, but exploit differences between plant and animal enzymes involved in these processes. Many plant-produced proteinase inhibitors, for example, are strictly specific towards serine proteases of animal origin (Belew *et al.*, 1975; Ryan, 1973).

The mixed-function-oxidase (MFO) detoxifying enzymes of animals represent another plant–animal biochemical difference that can be exploited, as it has been by *Senecio* (Compositae) pyrrolizidine alkaloids (Mattocks, 1972) and by methylenedioxyphenyl synergists (Rhoades and Cates, 1976).

Alkaloids are perhaps the best example of a class of compounds considered to be toxic specifically to animal systems, and thus presenting minimal autotoxicity problems (Rhoades and Cates, 1976). This is not always true, however. Many alkaloids affect systems common to animals and plants. The steroidal alkaloids of *Veratrum* (Liliaceae) inhibit the growth of oats and rye, apparently through a specific effect on DNA stability (Olney, 1968). It is difficult to see how these alkaloids could fail to affect cell division in *Veratrum* (Heftmann, 1975). Steroidal alkaloids of Apocynaceae (Waring, 1970; Waring and Chisholm, 1972), quinine (Estensen *et al.*, 1968), and berberine (Frey and Hahn, 1969) likewise interact with DNA. A variety of alkaloids are anti-mitotic (Piozzi *et al.*, 1968; Malawista *et al.*, 1968), including morphine (Malheiros-Garde, 1950; Fairbairn, 1965). Selection for the avoidance of morphine interference with abundant meiotic and mitotic activity may explain the absence of laticifers, and hence of morphine, from reproductive parts of *Papaver somniferum* (Fairbairn, 1965). Colchicine, another antimitotic alkaloid, inhibits spindle formation by its action on microtubules (Olmsted and Borisy, 1973; Wilson *et al.*, 1974), and induces polyploidy in plants (Kingsbury, 1964). Lycorine, an Amaryllidaceae alkaloid, is an inhibitor of ascorbic acid biosynthesis and powerfully inhibits the growth of higher plants (Arrigoni *et al.*, 1975). Even alkaloids that are toxic to animals due to effects on nervous tissue (of course, absent from plants) can also be toxic to plants. The simple indole pseudoalkaloid gramine is toxic to some plants (Overland, 1966). Cytisine inhibits plant growth (Pöhm, 1957, 1959). Nicotine is toxic to tobacco plants if applied to the surface of the leaf blade or through the petiole (Mothes, 1976).

2. *Mechanisms for Avoiding Autoxicity*

Thus it appears to be a somewhat general pattern that plant secondary compounds are at least potentially toxic to the plants that produce them. In some cases, toxicity to plant cells may be a functionally important part of the defense mechanism. Hypersensitive responses seal off invading pathogens in a ring of dead cells. These cells kill themselves by producing high concentrations of phenolics (Singleton and Kratzer, 1969). But for the most part, potential autotoxicity presents constraints to plant defense systems. This potential has probably greatly restricted the scope of plant defense chemistry. That plants do not produce organophosphorus insecticides, for example, may be due to the great reactivity of such compounds, which would make them difficult to contain within living tissue. This potential has had more conspicuous results in the organization of existing plant chemical-defense systems. A universal component of these defense systems must be mechanisms to avoid autotoxicity, but in only a few cases do we understand what these mechanisms are and how they operate. Several kinds of such mechanisms have been postulated and will be described here.

1. The most frequently cited mechanism is the segregation of toxic compounds in cell vacuoles and other organelles, or even in specialized cells (e.g., tannin cells, Swain, 1965; Beckman and Mueller, 1970), or tissues (latex and resin systems). Tannins (Esau, 1963; Swain, 1965; Rhoades and Cates, 1976), anthocyanins (Barz and Hösel, 1975), flavonoids (Harborne, 1973), and other phenolics are located mostly in cell vacuoles. In callus cultures of *Pinus elliottii* (Pinaceae) tannin accumulations occur in the smooth endoplasmic reticulum, in vesicles, and within vacuoles. Large tannin deposits within the cytoplasm are always surrounded by at least one unit membrane (Baur and Walkinshaw, 1974). Alkaloids often occur in cell vacuoles (Hughes and Genest, 1973; McKey, 1974; Floss *et al.,* 1974). Mothes (1976) observes that segregation of nicotine from cellular contents allows the tobacco plant to tolerate concentrations of this alkaloid that are toxic when applied to an excised leaf blade through the petiole, or even when applied to the surface of the blade. In poppies, alkaloids are stored in specialized vesicles occurring in the latex (Fairbairn *et al.,* 1974). These "alkaloidal vesicles" derive from a dilation of the endoplasmic reticulum of the developing laticifer cell (Nessler and Mahlberg, 1977).

   Low-molecular-weight terpenoids and other volatile oils are often compartmentalized in glandular cells or glandular hairs (Bonner, 1965). Terpenoid oleoresins occur in resin ducts and pockets (Stark, 1965; Werker and Fahn, 1969; Thomas, 1970; Langenheim, 1973).

   In germinating *Ricinus* seeds, the highly toxic protein ricin is localized in the protein bodies, which will become the cell vacuoles (Youle and Huang, 1976), and this segregation may explain its lack of effect on protein synthesis during germination. Though some uncertainty exists over the exact subcellular localization of plant proteinase inhibitors (Richardson, 1977), there is good evidence that they are compartmentalized in vacuoles of cells of tomato leaves (Shumway *et al.,* 1970, 1976; Walker-Simmons and Ryan, 1977) and in other solanaceous plants (Shumway *et al.,* 1972). The role of compartmentalization in avoidance of autotoxicity is in this case questionable, since the products concerned are often active only against serine proteases of animals (Ryan, 1973). Walker-Simmons and Ryan (1977) suggest that storage of these proteins within vacuoles may protect them from degradation by endogenous proteases and thus ensure their long tenure as a defense against herbivores.

2. Secondary compounds as they occur in the plant may be inactive precursors rendered toxic by some event during cell disturbance. Flavonoids usually occur as glycosides. Flavonoid aglycones, like other phenolics, are toxic to living cells and are normally only isolated as aglycones from dead tissues (e.g., heartwood) (Harborne, 1973). Hopkinson (1969) views detoxification of phytotoxic aglycones as an important function of phenol glycosylation. Glycosidases active against phenolic glycosides are widespread in plants, and probably function to release toxic aglycones when cells are damaged or diseased (Pridham, 1963). Cyanogenic glycosides likewise co-occur in plants with specific glycosidases that come into play resulting in release of free HCN only when the cellular structure is disturbed (Butler *et al.*, 1973). Glucosinolates (mustard oil glycosides) are normally accompanied by the hydrolyzing enzyme myrosinase, which is contained in particular cells (idioblasts) and liberated on disintegration of plant tissues (Miller, 1973). In these cases, there need only be segregation of the compound from the enzymes that activate it.
3. For some kinds of secondary compounds there may exist backup detoxification systems that come into play when toxins escape into sensitive tissue. The only documented case concerns cyanide metabolism in *Manihot esculenta* (Euphorbiaceae) (Chew and Boey, 1972; Nartey, 1973). In this plant, activities of two HCN-detoxifying enzymes, rhodanese (also the principal cyanide-detoxifying enzyme in animals, Oke, 1973), and β-cyanoalanine synthase are localized in mitochondrial fractions of tissue homogenates (Nartey, 1973). Here they are well placed to protect the terminal cytochrome oxidase from inhibition by HCN. While the presence of rhodanese in a cyanogenic plant is remarkable, the significance as an evolved adaptation of the ability of cyanogenic plants to assimilate HCN via conversion to asparagine (the first step of which is mediated by β-cyanoalanine synthase) (Abrol and Conn, 1966; Abrol *et al.*, 1966; Nartey, 1969) is problematical, since many noncyanogenic plants possess this ability as well (Blumenthal-Goldschmidt *et al.*, 1963; Nartey, 1970).
4. The effects of some potentially phytotoxic secondary compounds may be avoided in ways that do not require segregation of the

compounds from cellular contents. Modifications of the plant's system corresponding to the "target" system in the herbivore may render the plant immune to the effects of an erstwhile toxin. Plants that produce analogs of protein amino acids have in many cases been demonstrated to possess aminoacyl-tRNA synthetases that have higher specificity than those of nonproducer plants, discriminating between the protein amino acid and its analog (Fowden *et al.*, 1968; Fowden and Frankton, 1968; Lea and Fowden, 1972, 1973; Norris and Fowden, 1972, 1973; Fowden, 1974; Lea and Norris, 1976). Because the analogs may affect the functioning of many other enzymes in the plant (Fowden, 1974), it would be premature to suggest that this greater specificity of amino acid-activated enzymes in analog-producing plants completely obviates the requirement for compartmentalizing these compounds.

While in these cases immunity of the plant's equivalent of the "target" system is achieved by changes in protein primary structure, many other types of modification might serve the same function. The nonprotein amino acid minosine inhibits a variety of animal enzymes by chelating pyridoxal phosphate, which they require as a co-factor. The corresponding plant enzymes, however, are not inhibited, since this co-factor is firmly bound to the enzyme surface (Lea and Norris, 1976).

In other cases, alternative pathways may operate when vital processes are inhibited by endogenously produced toxins. Some plants possess an alternative flavoprotein-mediated respiratory system that is insensitive to cyanide (Solomos, 1977). As is the case with the ability of higher plants to assimilate HCN, the significance of this as an adaptation associated with cyanogenesis is problematical for many noncyanogenic plants possess cyanide-insensitive respiratory systems (Solomos, 1977).

Secondary compounds may be stored in the plant as inactive compounds that are converted to toxic products only during metabolism by the animal. *Dichapetalum* spp. (Dichapetalaceae) contain fluoroacetate that is metabolized by animals to fluorocitrate, a potent inhibitor of Krebs cycle reactions (Peters, 1954). Pyrrolizidine alkaloids are converted to more toxic metabolites by animals (Mattocks, 1972). However, even in these cases, we cannot rule out the possibility that the compound must be segregated from enzymes in the plant that would perform similar conversions.

3. *Conflict and Outcomes*

In most cases it would seem that tolerance of potentially autotoxic compounds by plants requires segregation of these compounds from cellular contents as a whole or from specific components that are sensitive to these compounds or that would activate them to phytotoxic substances. The recognition that such compartmentalization of secondary compounds requires a degree of cellular differentiation has led to the formulation of a family of concepts that relate to their distribution within plants, which will be examined at this point.

1. The costs of storing some kinds of secondary compounds are higher than those of storing others. Rhoades and Cates (1976) draw a distinction between digestion-reducing substances such as tannins that are general protein-complexing agents exerting their harmful effects in the gut of the herbivore and toxins, such as alkaloids and cyanogenic glycosides, that are absorbed from the gut and are active against specific metabolic processes. They argue that tannins and other generalized digestion-inhibitors will impose sequestration costs higher than those associated with toxins, supporting their argument with two kinds of considerations: (*a*) Activity of toxins is often restricted to metabolic systems found in animals but not in plants. Tannins and other digestion-inhibitors affect processes found in plants as well and thus must be completely segregated from cellular contents. (*b*) While toxins are present and active at low concentrations, tannins and other generalized digestion-inhibitors are effective only at relatively high concentrations. Even if toxins affect plant processes and must be segregated from cellular contents, their sequestration requires less storage space than does sequestration of substances such as tannins.
2. Because dividing cells do not have large well-organized vacuoles, rapidly growing and differentiating tissues may have limited storage space for autotoxic substances (Rhoades and Cates, 1976).

These considerations, among others, led Rhoades and Cates (1976) to postulate that tannins and other generalized digestion-reducing substances may be allocated primarily to mature tissues of plants and that immature tissues, when chemically defended, contain primarily toxins or specialized digestion-reducing substances such as protease inhibitors that present minimal autotoxicity problems. A critical evaluation of their hypothesis requires examination of these considerations.

Although I believe that these authors underestimate the degree to which toxins such as alkaloids and cardiac glycosides are potentially autotoxic to plants, I feel this does not seriously undermine their conclusion that sequestration costs in a chemical defense system based on toxins such as alkaloids would be less than those in a defense system based on tannins. This difference, however, may be due primarily to the difference in the amount of toxic substance that must be stored.

The notion that rapidly growing and differentiating tissues may have limited capacity for storage of autotoxic substances is subject to various qualifications. First, in plants that possess latex systems, laticifers may continuously penetrate meristematic tissues (Esau, 1965) and provide chemical defense to tissues chiefly composed of undifferentiated cells. Second, vacuoles are not the only organelle in which toxic compounds may be sequestered. In young cells of *Eucalyptus elaeophora* (Myrtaceae), for example, phenolics arise in vesicles resembling amyloplasts within the cytoplasm. These vesicles coalesce around the periphery of the vacuole as the cell matures (Wardrop and Cronshaw, 1962). Poppy alkaloids are contained in alkaloidal vesicles (Nessler and Mahlberg, 1977). Third, even though meristematic cells do not contain a single large well organized vacuole, they do contain numerous small vacuoles (Esau, 1965). Tannins often appear in these vacuoles at their inception (Esau, 1963; Chafe and Durzan, 1973). Until further studies clarify this point, it seems premature to conclude that tissues comprised of immature cells must contain proportionally lower volumes of space in which secondary compounds may be accumulated than do mature tissues. It is conceivable that selection could favor the formation of allelochemic-containing compartments relatively early in the development of the cell. While the accumulation of high concentrations of substances such as tannins in immature tissues may conflict with their rapid development and thus lower the escape component in their defense (Rhoades and Cates, 1976), there is no absolute physiological barrier to such an event (cf. Orians and Janzen, 1974).

Synthesis of secondary compounds in young leaves (and of the cellular substructure required for their storage) may slow leaf development by competing for materials and machinery available for growth of the cell (Mooney and Chu, 1974; Orians and Janzen, 1974). With this constraint on the chemical defense of young leaves we may expect the following outcomes:

1. Young leaves might often be defended by secondary compounds translocated into the developing leaf, rather than by secondary compounds synthesized *in situ*. As noted by Rhoades and Cates (1976), tannins and other digestion-reducing polymers cannot cross cell membranes, because of their high molecular weight, and must be synthesized at the site of deposition. Their immediate precursors, however (phenolic acids and flavonoids in the

case of tannins), could be translocated as glycosides into young leaves. Water-insoluble compounds are also precluded from translocation into young leaves, but a variety of lipophilic secondary compounds may be rendered water-soluble by glycosylation (Roberts *et al.*, 1956; Roberts, 1960; Harborne, 1964; MacLeod and Pridham, 1966), *N*-oxidation (Bickel, 1969; Phillipson and Handa, 1975), and other reactions. Others, which may not enter such reactions, seem not to undergo translocation in the vascular system of the plant (e.g., some mono- and sesquiterpenes, Flück, 1963).

The judgment that digestion-reducing substances such as tannins may not be capable of translocation into young leaves from elsewhere (Rhoades and Cates, 1976) seems to be supported by the studies of Hillis and Carle (1960, 1962), Hillis and Hasegawa (1963), and S. A. Brown (1964), who subscribe to the view that tannins and other polyphenols are formed at the site of deposition from translocated carbohydrates. However, MacLeod and Pridham (1966), basing their arguments on translocation rates of labelled phenolics through phloem, concluded that there is no reason to suppose that all complex phenolic compounds are synthesized *in situ* from nonaromatic precursors. Their study points up the possibility that tannin-containing young leaves might not be required to synthesize tannins from nonaromatic precursors but only to perform polymerizations of translocated phenolic glycosides.

The relative importance of importation and in situ synthesis in providing young leaves with chemical defense is unclear in the case of other types of secondary compounds as well. Alkaloids seem often to be translocated into young leaves and other tissues remote from sites of synthesis (Floss *et al.*, 1974), though exceptions occur (hordenine: Mann *et al.*, 1963). In some cases, the final step in alkaloid synthesis is performed at the site of deposition from a translocated precursor. For example, in *Duboisia* (Solanaceae) (Hills *et al.*, 1946) tropine and scopine are formed in the root and esterified with tropic acid in the leaf to form atropine and scopolamine. Demethylricinine is transported to the growing apex of *Ricinus communis* (Euphorbiaceae) and methylated there (Skursky and Waller, 1972). Crombie and Crombie (1975) employed an interesting experimental approach to the study of the relative importance of translocation and *in situ* synthesis in determining cannabinoid profile in *Cannabis sativa* (Cannabinaceae). They found that scions of *C. sativa* grafted to other strains of *C. sativa*, or to *Humulus lupulus* (Cannabinaceae), continued to produce their own characteristic composition and quantity of cannabinoids whatever part of the graft system they formed.

2. Young leaves may be invested with secondary compounds that are effective at low concentrations, so that costs of synthesis and of sequestration are minimized (Rhoades and Cates, 1976). Such a response would commit the plant to providing young leaves with phenological escape from specialized herbivores able to tolerate these compounds in the concentrations encountered in the plant (Feeny, 1976).
3. In other cases, young leaves may be invested with high concentrations of tannins and other secondary compounds with high phytotoxic potential. This response requires a completely different pattern of leaf development, characterized by the maintenance of cellular substructure adequate for segregation of these compounds long before full cell expansion, and by slower development of the leaf to maturity.

A basic dichotomy in patterns of leaf morphogenesis and mechanisms of young-leaf defense may explain the existence of the two kinds of young leaves such as that noted by Richards (1952) and familiar to other tropical biologists. Young leaves of many tropical plants expand very rapidly and hang limp for some days. Richards (1952) suggests that in these young leaves expansion precedes differentiation, while in young leaves that are erect and turgid from the outset, differentiation proceeds in concert with expansion. In the context of this discussion, it may be noted that young leaves in which differentiation is delayed until after expansion should be unable to tolerate high concentrations of phytotoxic compounds. Young leaves that differentiate while expanding, in contrast, should be equipped to tolerate such substances throughout their development. My observations in the Douala-Edea Reserve in Cameroon indicate that the drooping young leaves of plants such as *Berlinia bracteosa* and *Anthonotha* sp. (both Caesalpiniaceae) are palatable to herbivores as diverse as *Colobus* monkeys and microlepidoptera, while erect young leaves such as those of *Protomegabaria stapfiana* (Euphorbiaceae), *Coula edulis* (Olacaceae), and *Diospyros* spp. (Ebenaceae) are seldom attacked. Studies of leaf development that approach chemical defense, patterns of differentiation, and development time from a unified conceptual framework, would be most enlightening in the resolution of our thinking about tactics of chemical defense in young leaves.

Insufficient information exists to make conclusions about the frequency of digestion-reducing substances as young-leaf defenses. Ball (1950) noted the usual absence of tannins from actively dividing cells of shoot apices. In *Quercus robur* (Fagaceae) (Feeny, 1970), *Pteridium aquilinum* (Pteridaceae) (Lawton, 1976), and *Heteromeles arbutifolia* (Rosaceae) (Dement and Mooney, 1974), tannins occur in much lower concentrations in young leaves than in mature leaves. In the latter two species, a toxin effective at much lower concentrations (cyanogenic glycosides in both cases) shows the reverse relation of concentration to leaf age. In contrast, in two species of *Larrea* young leaves contain greater concentrations of phenolic resins than do mature leaves. In these cases, the resin coats the leaves and is not sequestered within cells (Rhoades and Cates, 1976). In *Eucalyptus sieberana* (Myrtaceae), concentrations of condensed tannins are higher in young than in mature leaves (Hillis, 1956). Young leaves of several species of *Eucalyptus* contain quite high concentrations of condensed tannins (Fox and Macauley, 1977). A community-wide survey of secondary-compound distribution in a rainforest in Cameroon is revealing that young leaves consistently contain greater concentrations of extractable phenolics than mature leaves of the same species (Waterman *et al.*, 1979). In herbaceous legumes that contain condensed tannins, there is little difference between total tannin content of young and mature leaves (Bate-Smith, 1973).

On the other hand, alkaloids and other toxins (*sensu* Rhoades and Cates, 1976) seem consistently to occur in greater concentrations in young than in mature leaves (Table 1). It is interesting that an apparent exception among the alkaloids is berberine, a compound known to interact with DNA (Krey and Hahn, 1969), so that its presence might disrupt active cell division. However, the antimitotic alkaloid colchicine occurs in higher concentration in young than in mature leaves of *Colchicum autumnale* (Amaryllidaceae) (Cromwell, 1955; James, 1950), demonstrating again that there is no absolute barrier to the presence of cytotoxic compounds in young tissues.

C. Resistance to Detoxification versus Susceptibility to Metabolic Manipulation by the Plant

1. *Selection for Molecules Resistant to Detoxification*

Toxic substances would most deter feeding and would thus confer greatest selective advantage, if they are not easily metabolized by herbivores and pathogens to nontoxic derivatives. In view of the digestive capabilities of herbivores, we might expect selection to have produced

Table 1. Relative concentrations of secondary compounds in young and mature leaves

Class of compound	Plant	Young leaves	Mature leaves	Reference
1. Phenolics				
Tannins	*Quercus robur*	L[a]	H	Feeny (1970)
	Heteromeles arbutifolia	L	H	Dement and Mooney (1974)
	Pteridium aquilinum	L	H	Lawton (1976)
	Eucalyptus sieberana	H	L	Hillis (1956)
	Gossypium hirsutum	H	L	Howell *et al*. (1976)
	herbaceous Leguminosae	Approximately equal		Bate-Smith (1973)
Phenolic resins	*Larrea tridentata*	H	L	Rhoades and Cates (1976)
	L. cuneifolia	H	L	Rhoades and Cates (1976)
Total polyphenols	*Nicotiana tabacum*	H	L	Sheen (1969)
Cholorogenic acid and related	*Helianthus annuus*	L	H	Koeppe *et al*. (1970)
compounds	*Rhamnus purshiana*	H	L	Betts and Fairbairn (1963)
Anthraquinones	*Rheum palmatum*	H	L	Schratz and Niewöhner (1959)
	Hypericum hirsutum	H	L	Rees (1969)
Hypericin				
2. Alkaloids				
Quinolizidines	*Sophora secundiflora*	L	H	Boughton and Hardy (1935)
	Lupinus spp.	H	L	Nowacki (1963); Keeler (1975)
Tropanes	*Atropa belladonna*	H	L	James (1950)
Steroidal alkaloids	*Solanum tuberosum*	H	L	Sinden *et al*. (1973); Schoonhoven (1972); Wolf and Duggar (1946)
	Veratrum spp.	H	L	Flück (1963); Olney (1968)
Colchicine	*Colchicum* spp.	H	L	James (1950); Cromwell (1955)
Tryptamine alkaloids	*Phalaris* spp.	H	L	Simons and Marten (1971)
Isoquinolines and related alkaloids	*Berberis darwinii*	L	H	Cromwell (1933)
	Mahonia trifoliolata	Absent	Traces	Greathouse and Watkins (1938)
	Lophocereus schottii	No consistent differences		Kirchner (1969)

3. Nonprotein amino acids				
N-Methyl-L-serine	*Dichapetalum cymosum*	H	L	Eloff and Grobbelaar (1969)
Stizolobic acid	*Mucuna deeringiana*	H	L	Ellis (1976)
L-Dopa	*Mucuna deeringiana*	L	H	Ellis (1976)
4. Cyanogenic glycosides	Rosaceae	H	L	Robinson (1930)
	Phaseolus lunatus	H	L	Robinson (1930)
	Pangium edule	H	L	Robinson (1930)
	Hydrangea sp.	H	L	Pammel (1911)
	Acacia spp.	H	L	Rehr (1972)
	ferns spp.	H	L	Greshoff (1909)
	Platanus spp.	H	L	Greshoff (1909)
	Pteridium aquilinum	H	L	Bennett (1968)
	Heteromeles arbutifolia	H	L	Dement and Mooney (1974)
	Manihot esculenta	H	L	de Bruijn (1973)
	Nandina domestica	H	L	Abrol *et al*. (1966)
	Alocasia macrorrhiza	H	L	Nahrstedt (1975)
5. Proteinase inhibitors	Solanaceae	H	L	Shumway *et al*. (1972); Ryan and Huisman (1967)
	Species from several dicot families	H	L	Ambe and Sohonie (1956)
6. Glucosinolates	*Brassica* spp.	H	L	Van Emden and Bashford (1969)
7. Steroidal compounds (nonalkaloidal)				
Steroidal saponins	*Agave sisalana*	L	H	Dawidar and Fayez (1961)
Cardiac glycosides	*Asclepias* spp.	H	L	Roeske *et al*. (1976)
	Digitalis purpurea	H	L	Evans and Cowley (1972b)

[a] L = low; H = high

dramatic examples among plant allelochemicals of substances refractory to metabolic degradation. While this notion is not new (Janzen, 1974), it has received little discussion. With the examples presented below, I wish to support my view that this factor has been a pervasive influence in the evolution of the molecules of plant defense.

a. Plant-Produced Inhibitors of Animal Proteases. Many flowering plants produce proteinaceous inhibitors of protein-digesting enzymes (Richardson, 1977). While some of these inhibit endogenous plant proteases and may be involved in regulation of protein degradation within the plant (Richardson, 1977; Ryan, 1973), many are strictly specific against enzymes of microbial or animal origin such as the serine proteases (e.g., trypsin and chymotrypsin, Ryan, 1973; Belew *et al.*, 1975), acidic proteases such as pepsin (Sakato *et al.*, 1975), and microbial alkaline proteases (Kirsi, 1974), and these inhibitors are widely considered to be specific defense substances (Birk, 1968; Ryan, 1973).

Many of these inhibitors are extremely resistant to proteolysis, even by proteases whose activities they do not affect, a property required of a protein molecule whose selective advantage is its ability to inhibit specific proteolytic enzymes in a milieu containing a diversity of them. Both activities of trypsin and chymotrypsin inhibitor AA from soybeans, for example, are extremely resistant to the action of pepsin (Birk, 1961), carboxypeptidase B, and trypsin (Tur-Sinai *et al.*, 1972); incubation with chymotrypsin decreases its ability to inhibit this enzyme but does not affect its activity against trypsin (Tur-Sinai *et al.*, 1972). Similar resistance of various inhibitors to proteolysis is noted by Birk (1974), Ryan (1974), Iwasaki *et al.* (1974), Kassell and Laskowski (1956), and Xavier Filho and Ainouz (1977). Many of these inhibitors are also resistant to denaturation by heat. Richardson (1977) states that several procedures for isolation of plant-produced protease inhibitors include a preliminary step in which the extract is heated for several minutes, during which various contaminatory proteins become precipitated and can be removed. "The inhibitors themselves are left essentially unaltered by this process" (Richardson, 1977).

Investigations of the amino acid compositions of these inhibitors have provided a major clue to the structural basis of their marked stability. In a remarkable number of cases, investigators have reported a high proportion of half-cystine residues and the absence of free sulfhydryl groups, i.e., these proteins are cross-linked by a large number of disulfide bonds. Most legume seed proteinase inhibitors, for example, contain about 20% half-cystine, all of which are involved in disulfide bonds (Birk, 1974). Carboxypeptidase inhibitor from potatoes contains about 15% half-cystine, all of it involved in disulfide bonds (Ryan, 1974). The protease inhibitors of *Cicer arietinum* (Leguminosae) likewise contain about 20% half-cystine, all involved in disulfide bonds (Belew *et al.*, 1975). Similar results are noted by Frattali (1969), Dechary (1970), and Pusztai (1972). The conformational stability provided by large numbers of disulfide bridges (e.g., 7 in the 71-amino-acid-Bowman-Birk inhibitor in soybeans, Ikenaka *et al.*, 1974) is believed to be largely responsible for the stability of these inhibitors to heat, acid, and proteolysis (Birk, 1974; Ryan, 1974; Belew *et al.*, 1975).

b. Lectins. Lectins, or "phytohaemagglutinins," are another class of plant proteins that includes many toxic representatives (Liener, 1976). As with plant-produced protease inhibitors, many plant-produced lectins are resistant to the action of proteolytic enzymes (Jaffé, 1973). The resistance of bean lectin to digestion by animals is shown by the haemagglutinating activity of extracts from feces of rats after eating a raw bean diet (Jaffé and Vega Lette, 1968). Mitogenic activity (due to lectins) of *Phytolacca esculenta* (Phytolaccaceae) is resistant to deproteinizing procedures and treatment with trypsin or chymotrypsin, but is destroyed by digestion with pronase E and Nagarse (Tokuyama, 1973). Lectin in wheat germ is stable over a wide pH and temperature range (Nagatu and Burger, 1974).

It is interesting that a number of plant lectins owe their remarkable stability to digestion to the same structural feature that confers resistance to proteolysis upon protease inhibitors. Very high contents of half-cystine (18–22 percent), and large numbers of disulfide bridges have been reported in lectins from kidney beans (Harms-Ringdahl *et al.*, 1973), wheat germ (Nagatu and Burger, 1974) and two species of *Phytolacca* (Tokuyama, 1973; Reisfeld *et al.*, 1967; Waxdal, 1974). Nagatu and Burger (1974) emphasize that such high contents of half-cystine are highly unusual among proteins and note that "the presence of such a large number of disulfide bridges in the protein molecule must make it stable under various conditions." The contents of half-cystine in the aforementioned lectins and protease inhibitors *are* highly unusual, being exceeded only by the high-sulfur proteins of mammalian hair (Lis and Sharon, 1973). While high half-cystine content in lectins has been considered the exception rather than the rule (Lis and Sharon, 1973), its occurrence has recently been reported in a number of cases.

Extensive cross-linking by disulfide bridges is not the sole mechanism by which toxic proteins have achieved resistance to proteolysis. The lectins of *Abrus* and *Ricinus* possess unexceptional contents of half-cystine yet are extremely resistant to proteolysis (Olsnes and Pihl, 1977). The structural basis for this resistance is unknown.

c. Condensed Tannins. Condensed tannins appear to be resistant to enzymatic degradation (Goldstein and Swain, 1963; S. A. Brown, 1964; □□ Brown *et al.*, 1966; Janzen, 1974). Two kinds of condensed tannins, and catechin, a precursor of some kinds of condensed tannins, were placed in forest soil for 60 days by Lewis and Starkey (1968). Of these, only catechin underwent appreciable decomposition. The generalized ability of tannins to form complexes with proteins, the property that renders them very generally toxic, makes them resistant to attack by enzymes as well (Feeny, 1976).

What are the structural features of condensed tannins that render them stable under enzymatic attack? The presence of numerous phenolic hydroxyl groups that react with functional groups of proteins (Loomis and Battaile, 1966; Van Sumere *et al.*, 1975) must go a long way toward making them truly formidable substrates for enzymes. However, it must be recognized that this structural feature alone cannot explain the refractory nature of condensed tannins, since hydrolysable tannins share this general protein-complexing ability but can be readily degraded enzymically unlike condensed tannins (■■ Brown *et al.*, 1966; S. A. Brown 1964; Haslam and Tanner, 1970; Ribereau-Gayon, 1972). The structural feature of condensed tannins that makes them less susceptible to enzymic degradation is the nature of the linkages between the "monomers" of these heterogeneous polymers. While the ester linkages of hydrolysable tannins are readily hydrolyzed by enzymes from plants (Swain, 1965; Haslam, 1966) and microbes (Haslam, 1966; Haslam and Tanner, 1970), the linkages of condensed tannins (usually carbon—carbon or benzyl ether linkages, Haslam, 1966) are more resistant to chemical and enzymic degradation (S. A. Brown, 1964; Ribereau-Gayon, 1972).

To these structural features of condensed tannins must be added a third that may contribute to their resistance to enzymic attack. The benzene nucleus is a very stable configuration (Dagley, 1967; Ribereau-Gayon, 1972), and while enzymic cleavage of aromatic rings occurs in microbes and in plants, it appears to be a difficult and involved metabolic process (Dagley, 1967; Ellis, 1971, 1974). The condensed tannins, composed of a large number of aromatic rings, linked in ways not easily susceptible to enzymic attack, and bristling with phenolic hydroxyl groups, may be products of the same selective pressures that have resulted in high-cystine, cross-linked proteinaceous defense compounds. It is indeed remarkable that recently an organism (*Penicillium adametzi* Zaleski) has been isolated that is capable of using condensed tannins as its sole carbon source (Grant, 1976). The author notes that "although the biochemical pathway by which this is accomplished is unknown, it is likely that at least the first enzyme involved would possess some unique properties, since enzyme–tannin interactions are usually strong, apparently nonspecific, and result in at least partial inactivation of the enzyme concerned."

In this context it must be noted that benzenoid and other aromatic ring systems are of widespread occurrence in alkaloids, as well as in phenolic compounds. It may be more than coincidental that the aromatic amino acids are so widely involved in alkaloid biosynthesis. It may also be noted that polycyclic structures generally are of widespread occurrence among terpenoids, phenolics, and alkaloids. However, judgment on the significance of these types of structures in providing resistance to animal metabolism seems premature.

Secondary compounds may possess other structural features that enhance their toxic effects by rendering them difficult for animals to detoxify. Singleton and Kratzer (1969) note that gossypol is a difficult compound to detoxify or eliminate, possibly because of the absence of "handles" where substituents can be placed to produce a less toxic compound. Many acutely toxic phenols are substituted in unusual ways that may affect their interaction with detoxifying enzymes (Singleton and Kratzer, 1973). Swain (1974) emphasizes that some types of alkaloids might be degraded by animals by reactions having clear analogies with those already being employed in the catabolism of common metabolites. Degradation of other types of alkaloids presents no analogies to existing catabolic pathways, and evolution of detoxification mechanisms for these types would be more difficult. In the context of this discussion it may be noted that such alkaloids might appear novel to the plant's degradative pathways as well.

2. *Selection for Molecules Metabolically Manipulable by the Plant*

Many instances can be conceived when adaptive advantages would accrue to plants that possess the ability to metabolize their defense compounds. For example, translocation of lipophilic secondary compounds may require their metabolic conversion to water-soluble substances or to substances more reactive with components of transport systems. Translocation of defense compounds would be of value in mobilizing seed defenses into the seedling during germination, in reclaiming defense investment from senescing leaves, and in reallocating defense investment within the plant during development. Secondly, selection may favor defense systems in which secondary compounds that escape into sensitive regions can be detoxified. Finally, selection may favor mechanisms by which resources invested in defense can be converted to other forms of investment.

We may envision various scenarios for the evolution of defense systems in which defense compounds are subject to further metabolism. Specialized pathways for metabolic conversions of defense compounds may evolve some time after the compound itself has been evolved. In other cases, however, the pathways for metabolic conversion of a defense compound may already exist when the compound first appears as a new defense chemical. Two possible examples will be noted here.

1. As noted earlier (p. 21), some toxic non-protein amino acids (e.g., canavanine, Rosenthal, 1972, 1977) may have arisen as a result of altered substrate specificity in enzyme systems catalyzing the biosynthesis of common amino acids (Bell, 1971). If enzyme systems involved in catabolism of the protein amino acid also accept the analog thus produced, the plant has a ready-made system by which the new defense compound can be metabolized.
2. A "new" defense compound may be produced by regulatory changes in pre-existing pathways that result in the accumulation of an intermediary metabolite previously present in trace quantities. A possible example is provided by the toxic amino acid β-cyanoalanine in plants of the genus *Vicia* (Leguminosae). These plants possess the ability to metabolize HCN by the pathway

   $$\text{HCN} \rightarrow \beta\text{-cyanoalanine} \rightarrow \text{asparagine}$$

   (Fowden and Bell, 1965), an ability possibly related to the presence of the cyanogenic glycoside vicianin in many members of this genus (Bell, 1971). Only some species of this genus, however, accumulate β-cyanoalanine in high concentrations (Fowden and Bell, 1965). The first appearance of this compound at toxic concentrations conceivably was due to regulatory changes in pre-existing pathways. The point to be made here is that if β-cyanoalanine did arise as a defense compound in this way, there was already machinery existing for its catabolism.

The origins of defense systems in which secondary compounds are metabolized are undoubtedly diverse and must remain speculative at this time. The point of importance here is that such systems may under certain circumstances confer advantages not conferred by systems in which defense compounds once made are resistant to further metabolism. The possible advantages accompanying metabolic changes of defense compounds in plants have not been the subject of systematic investigation, and only meager circumstantial evidence of their importance is available. The following examples are presented as testimony that studies of secondary-compound catabolism as a facet of chemical-defense systems might be repaid with robust results.

Many toxic plant secondary compounds are lipophilic, a property that is likely to enhance their toxicity to animals (Rhoades and Cates, 1976). Their translocation in the plant may require that they be converted into more water-soluble forms. Metabolic changes may also prepare secondary compounds for subsequent interaction with components of transport systems (Skursky *et al.*, 1969). Many of the kinds of metabolic changes undergone by secondary compounds within plants result in changes in their solubility characteristics. For example, glycosylation of phenolic compounds increases their sap solubility (Harborne, 1964) and results in higher rates of translocation through phloem (MacLeod and Pridham, 1966). Glycosylation may also facilitate translocation by preventing the adsorption of flavonoids by cellulose (Roberts *et al.*, 1956; Roberts, 1960).

Alkaloids may undergo *N*-oxidation (Phillipson, 1971). *N*-oxides are much more polar and water-soluble than the corresponding amines (Bickel, 1969; Phillipson, 1971). Though studies are by no means conclusive, it has been suggested (Phillipson, 1971; Phillipson and Handa, 1975b) that alkaloid *N*-oxides may play a role in the transport of alkaloids in the plant. This is an interesting suggestion in view of the facile interconversion of *N*-oxides and the parent bases (Phillipson and Handa, 1975b) and the large difference in solubility characteristics created by such a simple change. The patterns of appearance and disappearance of pyrrolizidine alkaloid *N*-oxides in *Senecio platyphyllus* (Compositae) (Sokolov, 1959) are consistent with the interpretation that they are involved in alkaloid translocation between root and shoot.

Alkaloids may undergo other reactions that result in increased water-solubility. In *Papaver somniferum* (Papaveraceae), once morphine enters the capsule latex, it is rapidly metabolized into a series of more polar and water-soluble compounds, some of which are transported into the developing seeds (Fairbairn and El-Masry, 1967, 1968). Similar conversions of alkaloids and more polar derivatives have been noted for the alkaloids of *Conium* (Umbelliferae) (Fairbairn and Ali, 1968).

Studies of the biology of alkaloids in the castor plant *Ricinus communis* (Euphorbiaceae) have opened exciting possibilities. Lee and Waller (1972) have demonstrated a cyclic process of metabolism and translocation of ricinine, which results in the virtually complete removal of this alkaloid from senescing leaves and its redeposition in the growing apex. In senescent leaves, but not in green leaves, *O*- and *N*-demethylricinine are produced from ricinine (Lee and Waller, 1972). Demethylation is usually believed to make a compound more active metabolically (Lee and Waller, 1972) and has in this case been postulated to be a preparatory step for translocation (Skursky *et al.*, 1969): "There is no direct evidence that *N*-demethylricinine could be the 'transport form' of ricinine (although experiments performed with whole plants seem to support this possibility)." Extending their studies of alkaloid metabolism and translocation to other plants, Nowacki and Waller (1973) suggested that alkaloids may be present in the plant in two forms, one easily transportable, the other not. In most cases, they found that a plant could export from aging leaves only alkaloids it normally contains, possibly indicating the presence of somewhat specific metabolic systems preparing alkaloids for translocation. However, our knowledge of the physiology of alkaloid transport must still be considered meager (Floss *et al.*, 1974).

Little information exists on the possible role of metabolism in preparing other classes of secondary compounds for translocation. Banthorpe *et al.* (1977) have isolated from three species of Compositae "salvage" enzyme systems that they suggest may convert monoterpenes into water-soluble products for transport. Seasonal variation in the activity of these enzymes (Banthorpe *et al.*, 1977) suggests that they might be of significance in ontogenetic redistribution of monoterpenes. De Bruijn (1973) suggests that cyanogenic glycosides may be translocated in *Manihot esculenta* (Euphorbiaceae), but was not able to rule out synthesis from translocated precursors. Mialonier *et al.* (1973) present evidence for transport of lectins from leaves to seeds. In *Canavalia* (Leguminosae), canavanine is transported from the pod to the developing seeds, and perhaps from leaves to fruits as well (Rosenthal, 1971). Preparatory metabolic changes may not be required for translocation of these water-soluble compounds.

Advantages may also accrue to plants that possess secondary compounds they can detoxify. To some extent, detoxification may be achieved by the same conversions as those implicated in secondary-compound translocation. For example, metabolic conversions of secondary compounds that influence solubility characteristics, such as *N*-oxidation of alkaloids, influence the retention or exclusion of these compounds from cells or cell organelles (Phillipson and Handa, 1975b) and may enhance their segregation from sensitive regions. Glycosylation of phenols probably accomplishes the same function (Harborne, 1964; Hopkinson, 1969). The ability to glycosylate simple phenolic compounds is extremely widespread in terrestrial vascular plants (Pridham, 1964; Glass and Bohm, 1970).

Detoxification of alkaloids by plants has received little attention, and the means by which producer plants avoid the effects of potent phytotoxic alkaloids such as the steroidal alkaloids of *Veratrum* are completely unknown (Heftmann, 1977). *N*-oxidation (Phillipson and Handa, 1975a,b) and *N*-methylation may result in derivatives less toxic to plants. The transformation of cytisine to *N*-methylcytisine during its migration through the cambium into the wood of *Cytisus laburnum* (Leguminosae) is viewed as a detoxification mechanism, since cytisine inhibits plant growth, whereas *N*-methylcytisine does not (Pöhm, 1957, 1959). *N*-methylation and -demethylation are common alkaloid conversion processes in plants, and in some cases (e.g., *Conium maculatum,* Roberts, 1974) it appears the enzymes involved may be specific for alkaloids as substrates. However, the physiological significance of these reactions remains unclear.

The best documented case of detoxification pathways of a plant for its secondary compounds has already been mentioned: the possession by the cyanogenic plant *Manihot esculenta* of two cyanide-metabolizing enzyme systems, rhodanese and β-cyanoalanine synthase, and their localization in the mitochondria of this plant (Nartey, 1973).

Selection may also favor defense system in which accumulated defense compounds may at times be converted to common mainstream metabolites. Such metabolic conversions may be viewed as processes by which defense investment may be reallocated within the plant (by contributing to the pool of metabolites from which secondary compounds may be synthesized elsewhere) or by which defense investment can be converted into other resources. Although pathways leading from secondary compounds to common metabolites may be of little quantitative significance in comparison with other pathways producing these common metabolites (Robinson, 1974), this does not lessen their importance as mechanisms of retrieval of resources that might otherwise have been lost.

Many plants are capable of metabolizing secondary compounds they contain to molecules of recognized worth in primary metabolic pathways, when these compounds are supplied exogenously. Our interpretation of the functional significance of these capabilities must take into account several factors: (*a*) Utilization by plant enzymes of exogenously supplied substrates may not be indicative of pathways or rates of utilization of the same substrate produced endogenously. This is because feeding experiments can seldom duplicate the precise patterns of compartmentalization of substrates and enzymes that exist in the plant (Haslam, 1966). Some of the reactions of phenolic catabolism are evidently tightly compartmentalized (Ellis, 1974). (*b*) Turnover rates reported for various secondary compounds vary from low (Fowden and Bryant, 1959; Dunnill and Fowden, 1965; Hegarty and Peterson, 1973) to high (Burbott and Loomis, 1969; Abbondanza *et al.*, 1965; Breecia and Badiello, 1967), and conflicting reports occur for some compounds (e.g., nicotine: Robinson, 1974; Leete and Chedekal, 1974). (*c*) In many cases, metabolic transformations of secondary compounds may occur only in specific tissues and only at certain stages of development (Fairbairn, 1965; Willuhn, 1966, 1967; Mothes, 1959; Jindra *et al.*, 1964; Fairbairn and Ali, 1968; Fairbairn and Suwal, 1961; Rosenthal, 1970; Heftmann and Schwimmer, 1972; Floss *et al.*, 1974; Shumway *et al.*, 1976; Walker-Simmons and Ryan, 1977). (*d*) In many cases, the nature of the products is unclear.

With these qualifications in mind, we may approach the results of studies documenting metabolism of labelled secondary compounds fed to producer plants. Labelled nicotine can be metabolized by tobacco plants to amino acids, pigments, sugars, and organic acids (Tso and Jeffrey, 1961), and may serve as a methyl donor to choline (Leete and Bell, 1959). Labelled caffeine fed to *Coffea arabica* (Rubiaceae) is degraded to aliphatic compounds (Luckner, 1972). Nitrogen stored in canavanine is reutilized by pathways involving arginase and urease in *Canavalia ensiformis* and *Caragana spinosa* (both Leguminosae) (Rosenthal, 1970, 1972, 1974; Whiteside and Thurman, 1971). Plants that produce γ-methyleneglutamic acid degrade it to common amino acids and to sugars and sugar phosphates (Fowden, 1960). Azetidine-2-carboxylic acid is slowly degraded by producer plants to threonine, glutamine, and unidentified substances (Fowden and Bryant, 1959; Dunnill and Fowden, 1965). Mimosine in *Leucaena leucocephala* (Leguminosae) can be degraded to dihydroxypyridine, pyruvate, and ammonia (Smith and Fowden, 1966).

Numerous studies demonstrate the efficient assimilation of HCN by many plants into asparagine and related metabolites (Blumenthal-Goldschmidt *et al.*, 1963; Abrol *et al.*, 1966; Abrol and Conn, 1966; Nartey, 1969, 1973). Burbott and Loomis (1969) suggest that monoterpenes in peppermint may serve as substrates for energy metabolism in secretory cells after other substances are depleted. Evidence for turnover of a triterpene (β-amyrin, Aexel *et al.*, 1967) and steroids (cardenolides, Evans and Cowley, 1972a) has also been presented, though the nature of the conversion products is unclear.

Studies of metabolism of proteinaceous secondary compounds during seed germination have yielded limited evidence that they may be hydrolyzed to form free amino acids (Kirsi, 1974; Rougé, 1974a,b; Howard *et al.*, 1972), but there are exceptions (Youle and Huang, 1976). Ryan and Huisman (1970) and Shumway *et al.* (1976) postulate a storage role for proteinase inhibitor in potatoes and tomatoes.

Recent studies have even demonstrated that plants can extensively catabolize phenolic compounds, the reactions in many cases involving cleavage of aromatic rings. The A-ring of various flavonoids can be degraded by plants to CO_2 (Ellis, 1974; Barz and Hösel, 1975). Whether significant catabolism of phenolic polymers such as condensed tannins and lignin takes place seems still questionable (Barz and Hösel, 1975).

3. *Conflict*

Thus there are advantages to the plant in possessing defense compounds that are difficult for herbivores to metabolize to water-soluble nontoxic molecules, and there are also advantages associated with having defense compounds that it can metabolize to derivatives that are water-soluble, nontoxic, or useful in primary pathways. When we recognize the fundamental biochemical similarity of plant and animal metabolism, it is not difficult to predict that these two postulated selective forces will often act on the evolution of a particular molecule of defense in partially or completely opposing directions and that in each compound one of these advantages may be gained fully only by some sacrifice of the other. This conflict has two aspects:

1. Any compound that can be metabolized by the plant is potentially subject to the same kind of metabolism by herbivores. There are in fact many convergences in the means by which plants and animals metabolize the same secondary compound. Formation of glycosides, by which plants render non-toxic and water-soluble a great diversity of secondary compounds, is a principal means by which insects convert lipophilic secondary compounds to more water-soluble derivatives that are more easily excreted (J. N. Smith, 1955). Mammals utilize glucuronic acid rather than glucose, and excrete glucuronides of many secondary compounds (Williams, 1964; R. L. Smith and Williams, 1966; Eberhard *et al.*, 1975). *N*-oxidation, a metabolic conversion by which plants produce more water-soluble derivatives of alkaloids, is known to occur in animal metabolism of ingested alkaloids as well. Morphine *N*-oxide is known as a metabolite in animals (Phillipson, 1971). *N*-oxidation of pyrrolizidine alkaloids occurs in rats and sheep (Mattocks, 1972). *N*-oxides are generally less toxic than the parent alkaloids (Bickel, 1969).

   Numerous examples can be cited of similarities in plant and animal metabolism of a particular secondary compound. In sheep, mimosine is degraded by rumen microorganisms to dihydroxypyridine so fast that no depilatory effect results (Hegarty and Peterson, 1973). Metabolism of mimosine in the producer plant *Leucaena glauca* (Leguminosae) results in formation of the same product (I. K. Smith and Fowden, 1966). Atropine is detoxified via the enzyme atropinesterase both by soil microbes that use tropane alkaloids as food (Niemer *et al.*, 1959) and by rabbits eating *Atropa belladonna* (Solanaceae) (Sawin and Glick, 1943); atropine is metabolized by an enzyme catalyzing the same reaction in *Datura stramonium* (Solanaceae) (Jindra *et al.*, 1964).

The most dramatic example of convergence in the means by which a plant and its specialized herbivore avoid the toxic effects of a plant-produced secondary compound emerges from studies of the bruchid beetle *Caryedes brasiliensis,* which feeds on the canavanine-rich seeds of *Dioclea megacarpa* (Leguminosae). This beetle avoids incorporation of canavanine into its protein by having a highly specific arginyl-tRNA^{arg} synthetase that, unlike the corresponding enzyme in other insects, discriminates between arginine and canavanine (Rosenthal *et al.*, 1976). Furthermore, this insect detoxifies and utilizes canavanine via pathways catalyzed by arginase and urease (Rosenthal *et al.*, 1977). Both these means of dealing with canavanine are exactly the same as those found in canavanine-producing plants that have been investigated (Rosenthal, 1970, 1972, 1974) (the physiology of canavanine in *Dioclea* has not been studied). Just as enzymes for the catabolism of canavanine may already have been present in plants when canavanine first evolved, insects as well may have possessed a surprising degree of "preadaptation" for canavanine utilization: insect fat body arginase will hydrolyze canavanine, but more slowly than it hydrolyzes arginine (Reddy and Campbell, 1969).

A major puzzle is the absence of reports that herbivores utilize HCN by the very efficient pathway to asparagine so widespread in plants (Hegarty and Peterson, 1973). This pathway, which results in the recovery of cyanide nitrogen in a useful form, would seem to be superior to the rhodanese pathway most often reported in animals, which results in the loss of this nitrogen as well as an atom of sulfur and the formation of thiocyanate (Oke, 1973), which can itself be toxic (Hill, 1973). The absence of reports may be due to a lack of studies. In insects, for example, there is no evidence for high levels of rhodanese (J. N. Smith, 1964; but see Parsons and Rothschild, 1964), and the report that *Sitophilus* beetles incorporate cyanide carbon into aspartic acid (Bond, 1961) is reminiscent of plant pathways. Cyanide tolerance in some insects (Jones, 1972) and in millipedes (Hall *et al.*, 1971), as in some plants, is due to modifications of the cytochrome system.

2. The second aspect of these two conflicting selective forces is that a secondary compound highly resistant to metabolism by herbivores is likely to be refractory to plant enzymes as well, and present great difficulties (high costs) to the evolution of systems for its translocation, detoxification, and conversion to common metabolites. For example, the remarkable resistance of plant-produced proteinase inhibitors to proteolysis, the feature that makes them effective defenses against animals, may have been achieved only at some sacrifice of the plant's ability to reclaim the amino acids invested in these proteins. Proteinase inhibitor I in tomatoes, for instance, is known not to be easily degraded by plant endopeptidases (Shumway *et al.*, 1976). In like manner, plants can manufacture flavonoids such as tangeretin which cannot be easily glycosylated by animals, are more difficult to convert to water-soluble forms, and are more toxic than other flavonoids (Singleton and Kratzer, 1969), but the biology of tangeretin in the plant may be subject to constraints not affecting the plant's use of other, easily glycosylated flavonoids.

An excellent example of the possible effects of the interplay of these conflicts in shaping the defense systems of plants can be drawn from the literature on hydrolysable and condensed tannins. Both are generalized protein-precipitators with essentially similar modes of action. The linkages joining monomers of hydrolysable tannins are ester linkages, and there appear to be no lack of esterases in producer plants which could presumably hydrolyze them (Swain, 1965; Haslam, 1966). This ability to produce simple phenolic acids (which could then be detoxified by glycosylation) from large phytotoxic polymers could confer significant advantages to the plant, facilitating, for example, the mobilization of defense compounds from seed to seedling during germination, or the translocation of tannins from senescent leaves. But the ability of the plant to metabolize its defense compounds in this way is only achieved by compromising their effectiveness against herbivores, for the same reaction by which plants manipulate hydrolysable tannins can be employed by herbivores to detoxify them (S. A. Brown, 1964; Brown *et al.*, 1966; Singleton and Kratzer, 1969; Haslam and Tanner, 1970). Condensed tannins, on the other hand, appear to be much more resistant to degradation by herbivores to nontoxic compounds (Singleton and Kratzer, 1973).

While no studies have been directed to this point, it seems reasonable to suggest that the resistance of condensed tannins to degradation by herbivores has been achieved only with the sacrifice of the plant's ability to metabolize these defense compounds readily and with a minimum of metabolic cost. While the ability of plants to degrade even such complex aromatic compounds as flavonoids has recently been demonstrated (Ellis, 1974; Barz and Hösel, 1975), it is still considered unlikely that condensed tannins (and lignins) undergo catabolism in the plant to any extent (Barz and Hösel, 1975). The point here is not to deny the prediction of Ellis (1974) that plants may ". . . eventually prove to be capable of degrading virtually any aromatic structure which they have elaborated," but simply to argue that in the case of compounds such as condensed tannins, the difficulties or metabolic costs associated with catabolism may have precluded its extensive occurrence. The situation may be analogous to that of cellulose: though flowering plants contain cellulases (Osborne, 1973), they still lose large quantities of cellulose in shed parts. The cost associated with their mobilization may make condensed tannins less suitable defense compounds under certain circumstances.

4. *Outcomes*

The structural features of secondary compounds such as condensed tannins that make them resistant to enzymic degradation by animals, may also increase the cost to the plant of preparing derivatives that are translocatable, or that are convertible to primary metabolites. The cost to the plant of mobilizing other types of secondary compounds, such as alkaloids or non-protein amino acids, may be much lower. This difference has several implications for the plant's use of different kinds of compounds, which will be discussed at this point.

a. Chemical Defenses of Seeds. Resources invested in seeds, including defense chemicals, are subject to strong selection for the ability to be mobilized and translocated to the developing seedling (Janzen, 1976a). The types of secondary compounds found in seeds (with the exception of such tissues as the seed coat that are not subject to resorption during germination) should possess one or more of the following properties: (*a*) Ability to be translocated to the growing seedling, which requires a degree of water-solubility and the absence of autotoxic effects during translocation. (*b*) Ability to be readily metabolized to compounds having property (*a*). (*c*) Convertibility to nutrient substances. A likely consequence of this is that secondary compounds which are difficult to translocate and metabolize, such as many condensed tannins, might be replaced as seed defenses over evolutionary time by compounds that make a greater contribution to the survival and growth of the seedling.

Information on the behavior of seed secondary compounds during germination is not extensive. Nevertheless, representatives of many classes of secondary compounds appear to be translocated to growing parts of the seedling, either in intact or modified form, or to be extensively degraded to nutrient substances. Canavanine is translocated to growing regions of *Canavalia* seedlings during the early phase of germination (Rosenthal, 1970). Translocation of hydroquinone from seed to seedling is reported to occur during germination of *Xanthium* (Compositae) (Kingsbury, 1964). The results of most existing studies probably do not permit distinction between translocation of the intact molecule as found in the seeds and translocation of a derivative followed by resynthesis in the site of deposition. Wallebroek (1940) reported that in *Lupinus luteus* there is a large decrease in alkaloid content of the cotyledons of the germinating seed concurrent with an increase in the growing regions of the seedling. Perhaps reflective of selection for translocatable seed defenses is the high proportion of *N*-oxides in the seeds of *Atropa belladonna* where they comprise a higher proportion of total alkaloids than in any other part of the plant (Phillipson and Handa, 1975a). It would be interesting to determine whether in other plants producing *N*-oxides and other water-soluble quaternized alkaloids, seeds are often a rich source of such compounds. It must be remembered, however, that there are many other tissues in which the presence of water-soluble alkaloids might be of advantage.

Not uncommonly among the alkaloids, compounds present in the seed are modified during deposition in the seedling. While the significance of these modifications is obscure, it is perhaps not too speculative to suggest that some cases may be related to requirements imposed by translocation. Cytisine, present in *Laburnum anagyroides* (Leguminosae) seeds, is rapidly methylated in young seedlings (Pöhm, 1966). This is interesting in view of the finding by the same author (Pöhm, 1957, 1959) that cytisine is phytotoxic but *N*-methylcytisine is not. Perhaps methylation of the alkaloid serves to avoid autotoxicity during translocation or deposition. In poppies, the water-soluble "bound" forms of alkaloids stored in the seeds are degraded to form substances like the parent alkaloids during germination (Fairbairn and El-Masry, 1968).

In other cases, secondary compounds undergo extensive degradation to nutrient compounds. Perhaps the best documented case is that of canavanine, which in several legumes accumulates in high concentrations in seeds and disappears during germination (Bell, 1971), when it is efficiently converted to primary metabolites via arginase and urease (Rosenthal, 1970; Whiteside and Thurman, 1971). There are suggestions that lectins (Rougé, 1974a,b; Howard *et al.*, 1972) and proteinase inhibitors (Kirsi, 1973, 1974; Horiguchi and Kitagishi, 1971) may in some cases be hydrolyzed to free amino acids that are used by the seedling. It would be interesting to know whether those lectins and protease inhibitors reported to be resistant to denaturation and proteolysis are in fact capable of being hydrolyzed by the producer plants. In at least one plant, *Ricinus communis,* neither the toxic protein ricin nor a lectin (both known to be resistant to degradation) undergo substantial degradation during germination (Youle and Huang, 1976). However, proteinase inhibitor I in potatoes and tomatoes, known to be resistant to plant endopeptidases, still appears to be degraded at certain times (Shumway *et al.,* 1976).

The seed oils of many plant species contain unusual, toxic fatty acids. Little is known about their utilization during germination, but it is difficult to believe that *Ginkgo* seed lipids would comprise 75% antibiotic anacardic acids (Gellerman *et al.,* 1976) or *Sterculia foetida* seed fats would comprise up to 70% or more of the highly toxic (Nixon *et al.,* 1974; Ferguson *et al.,* 1976) cyclopropenoid fatty acid, sterculic acid (Hilditch and Williams, 1964), were these energy-rich molecules not capable of being utilized by the seedling. Seeds of *Ungnadia speciosa* (Sapindaceae) contain 15% dry weight cyanolipids, all of which disappears within 3 days of germination (Siegler and Price, 1976). The authors suggest that these substances are used as energy reserves by the seedling.

These examples illustrate the outcome of selection for seed defenses that can also serve as exceptional sources of crucial nutrients. Nitrogen-rich molecules such as canavanine, and energy-rich molecules such as toxic fatty acids, are interesting solutions to the seed's problem of maximizing both protection and stored reserves in a limited amount of space (Janzen, 1976a). It is interesting that the defenses of seeds so often include toxic lipids and nitrogen-containing secondary compounds such as nonprotein amino acids, alkaloids, protease inhibitors, and lectins.

Large polymeric polyphenols such as condensed tannins would seem not to be suitable defense compounds in seeds, since they should be difficult or costly to degrade to forms that can be translocated to seedlings, as either secondary compounds or nutrients. (The occurrence of hydrolysable tannins in seeds, on the other hand, might be expected more frequently.) Though the information available is far from systematic in its coverage, I have uncovered few reports of large concentrations of condensed tannins in seeds. When present, they are often reported to be localized in the seed husk (sorghum: Harris *et al.,* 1970) or seed coat (*Phaseolus vulgaris:* Swain, 1965) (Bate-Smith and Ribereau-Gayon, 1959). Oligomeric condensed tannins might present less severe problems in this regard than would large polymers. Cotyledons of cacao seeds contain 8.0% dry weight polyphenols. Of this, three-quarters is comprised of catechins, anthocyanins, leucocyanidin, and dimeric leucocyanidins; only one-quarter is polymeric leucocyanidins (Forsyth and Quesnel, 1963). Nevertheless, there are exceptions to the proposed pattern that cannot be explained at present, e.g., the presence of highly polymerized polyphenols in avocado seeds (Geissman and Dittmar, 1965), and the content of 11–26 percent condensed tannins in nuts of *Areca catechu* (Arecaceae) (Singleton and Kratzer, 1973).

b. Reclamation of Secondary Compounds from Senescent Tissues. Plants would be expected to reclaim defense compounds from leaves and other parts before these parts are shed (as other resources are reclaimed), but this will be true only if the cost of reclamation is less than the cost of replacing resources that are not reclaimed. The structural features that confer upon substances such as condensed tannins their resistance to degradation by animal enzymes, may also impose a high cost to the plant of reclaiming them or their derivatives. Other types of secondary compounds such as alkaloids, non-protein amino acids, and cyanogenic glycosides, should impose lower costs of reclamation, for at least two reasons: (*a*) Unlike condensed tannins, which are polymers with linkages resistant to degradation, these compounds are smaller and either water-soluble or convertible to water-soluble forms. (*b*) In contrast to tannins, these compounds do not complex nonspecifically with proteins. It is more likely that the plant will possess, or can evolve, enzymes, membranes, and other components that can interact with these compounds, or their derivatives, without being poisoned. It should be pointed out that these compounds, and many other plant toxins contain nitrogen. The cost of replacing nitrogen lost in shed parts may be relatively greater than the cost of replacing other constituents, especially if the nitrogen is incorporated into complex polycyclic compounds whose production required many biosynthetic steps. These considerations lead me to predict that in the case of condensed tannins, a larger proportion of the content present in mature leaves will be lost when the leaf is shed, than will be the case with many other types of secondary compounds, such as alkaloids and nonprotein amino acids.

Little systematic information is available for evaluating this prediction. Certainly plants release large quantities of phenolic compounds along with shed leaves and other parts (Towers, 1964). Condensed tannins figure prominently among the phenolic compounds contained in leaf litter (Handley, 1954; Davies, 1971). Harrison (1971) collected oak litter in nets as the leaves fell in autumn and found they contained concentrations of tannins inhibitory to most common soil fungi. Tannins, lignins, and other aromatic compounds in litter are chief sources of humic acids in soils (Handley, 1954; Ponomareva, 1969; Janzen, 1974). But while it is well known that condensed tannins and other polyphenols are often abundant components of leaf litter, in no case do we really have a good idea what proportion of tannins present in mature leaves remains at the time of shedding. Leaves of *Populus* hybrids growing in Alberta contain greater dry-weight concentrations of polyphenols and total *o*-dihydroxyphenols in September and October than in mid-season (Dormaar, 1970). In beech and oak, freshly fallen leaves contain significant quantities of polyphenols, but less than those contained in healthy mature leaves (Coulson *et al.*, 1960). Freshly fallen leaves of chestnut also contain appreciable quantities of polyphenols (Anderson, 1973), but in this study the content of polyphenols in mature leaves was not reported. Apparently no information exists concerning the fate of hydrolysable tannins in senescent leaves, but it is reasonable to suggest that they may be subject to greater reclamation than are condensed tannins. Brown *et al.* (1966) noted that unlike *Calluna vulgaris* (Ericaceae), plants in the genus *Chamaenerion* (Onagraceae), which produce hydrolysable tannins, do not generate podosols. They attribute this to the fact that hydrolysable tannin–protein complexes are less resistant to degradation in the soil than are complexes of proteins with the condensed tannins of *Calluna*. It could equally well be explained by the reclamation of hydrolysable tannins by *Chamaenerion* before leaf shed. These authors present no information on the actual tannin content of freshly shed leaves.

Phenolic compounds of other types also persist in appreciable amounts in senescent leaves. In two varieties of tobacco studied by Sheen (1969), senescent leaves contained a greater percentage dry weight of polyphenols, mostly chlorogenic acid, than did mature leaves. Concentrations of chlorogenic acid and related compounds in the oldest leaves of *Helianthus annuus* (Compositae) were about one-quarter to one-third those found in mid-aged leaves (Koeppe *et al.*, 1970). Goodwin (1958) noted that whereas chlorophylls and carotenoids disappear from senescent leaves of *Prunus nigra* (Rosaceae), there was no loss of anthocyanin. Simple phenolic compounds are present in highest concentrations in senescent winter wheat (Berger *et al.*, 1977).

Information available for alkaloids, though not of great extent, presents a very striking contrast to the situation for phenolics. Many statements point to the loss of alkaloids in senescent organs (James, 1950; Mothes, 1955, 1959; Sokolov, 1959; Nowacki and Waller, 1973). In *Atropa belladonna* young mature leaves contain on average 1.96 mg alkaloids. As leaves age, there is a progressive decline in the total alkaloid content per leaf, the oldest leaves containing 0.32 mg alkaloid per leaf (James,

1950). In *Colchicum autumnale* as well, there is a reduction in the actual amount of alkaloid per leaf with aging (James, 1950). Late in the growing season, the root sap of Jimson weed (*Datura stramonium*) contains an enzyme catalyzing the hydrolytic cleavage of atropine to tropine and tropic acid. Activity could not be demonstrated in younger plants, and was related to the decrease in total alkaloid content in the plant near the end of the growing season (Jindra *et al.*, 1964). In the annual plant *Lupinus luteus* alkaloids disappear from roots, leaves, stems and even pods and accumulate in the seeds of senescent plants (Sabalitschka and Jungermann, 1925). In *Senecio platyphyllus,* alkaloids virtually disappear from senescing aerial parts near the end of the growing season, and accumulate in the roots (Sokolov, 1959).

In these cases the metabolic fates of the alkaloids during senescence have not been determined. Substantial information exists, however, on the mobilization of the alkaloid ricinine from senescent leaves of *Ricinus communis*. When labelled ricinine is fed to senescent leaves of this plant, it is rapidly translocated to the raceme of fruiting plants (Skursky and Waller, 1972) or to the growing apex (Lee and Waller, 1972). Yellow senescent leaves contain an enzyme system that demethylates ricinine, a conversion believed to prepare it for translocation. This activity is lacking from green leaves (Lee and Waller, 1972). The demethylated forms of ricinine are translocated and converted to ricinine in their new site of deposition (Skursky *et al.*, 1969). Mobilization of ricinine from senescent leaves is so efficient that its presence (and that of its derivatives) cannot be demonstrated in shed leaves (Lee and Waller, 1972). In plants that have reached the fruiting stage the total content of alkaloid per plant remains almost constant, the alkaloids being translocated from senescent leaves to the growing regions (Skursky and Waller, 1972).

Whether such processes occur in plants containing more typical alkaloids is an interesting question. Nowacki and Waller (1973) demonstrated that senescent leaves of *Lupinus angustifolius* fed lupine alkaloids translocated them so efficiently that the oldest leaves were completely devoid of alkaloids. The results of this study also suggest that transport mechanisms existing in alkaloid-containing plants are specific to their own alkaloids: Plants fed foreign alkaloids usually could not translocate them, whereas they could translocate their own alkaloids. This points to the existence of specific enzyme systems that perhaps represent evolved adaptations for alkaloid transport.

Very little information exists concerning the mobilization of other types of secondary compounds from senescent tissues. Shed conifer needles contain substantial amounts of terpenoid resins (Kononova, 1961). The oldest leaves of *Agave sisalana* (Agavaceae) contain greater dry-weight concentrations of steroidal sapogenins than do younger mature leaves (Dawidar and Fayez, 1961). The behavior of cyanogenic glycosides, on the other hand, seems to resemble that of alkaloids. In Rosaceae, in *Phaseolus lunatus,* and in *Pangium edule* (Flacourtiaceae), cyanide content disappears during senescence (Robinson, 1930). The cyanogenic principle disappears from *Hydrangea* (Hydrangeaceae) leaves in the autumn (Pammel, 1911). Cyanogenic glycosides virtually disappear from older leaves of *Nandina domestica* (Nandinaceae) (Abrol *et al.*, 1966). Dillemann (1958) notes that cyanogenic potential usually falls with leaf age, but cites as exceptions three species in which leaves at the time of leaf fall are reported to be almost as rich in cyanogenic glycoside as are the young leaves. Prakash *et al.* (1977) discuss the possible translocation of a toxic nonprotein amino acid from older tissues to younger ones in *Lathyrus sativus* (Leguminosae), but could not rule out other possible explanations for the decreased content of the amino acid in aging leaves. For other classes of secondary compounds, including lectins and protease inhibitors, I have uncovered no information that relates to their fate in senescent leaves.

c. Reclamation Strategies and Tissue Lifespan. If secondary compounds display such great variation in their ease of reclamation as I propose, this would be expected to influence their allocation within the plant. Plants should maximize the benefits conferred by secondary compounds relative to the costs of reclaiming or replacing them. Particularly, in leaves that are short-lived, there may be an advantage in allocating easily reclaimable chemical defenses; in this way the plant maximizes benefits by re-using the compounds when the leaf is shed. Defense substances that are difficult or expensive to reclaim should be allocated to long-lived leaves, or other long-lived organs. In this way the plant maximizes benefits conferred by an investment that cannot be liquidated. It should be noted that there is no known internal-economic constraint that precludes the allocation of easily reclaimable defenses, as well, to long-lived organs. An external constraint may arise, however, from the requirements of providing effective defense to a plant part that is exceptionally "apparent" (Feeny, 1976) to herbivores owing to its long lifespan. If the tractability of secondary compounds to animal metabolism is at all parallel to their tractability to plant reclamation systems, as I have proposed, then the "quantitative" (Feeny, 1976) defenses expected in long-lived organs may be defenses that are also difficult or expensive to reclaim. In this case selection pressures imposed by herbivores and those emerging from internal-economic constraints may act in the same direction [but, see Fox and Macauley (1977), whose data call into question the assumption that condensed tannins are always effective "quantitative" defenses].

Feeny (1976) argues that in plants of early successional stages, selection pressures favoring fast growth, success in vegetative competition, and production of many propagules, will conflict with a large metabolic allocation for defensive compounds. Selection has in these plants favored the production of small amounts of potent toxins such as alkaloids, cyanogenic glycosides, glucosinolates, and similar compounds (Feeny, 1976). It should be pointed out that these compounds are more likely to be mobilizable than are generalized protein-complexing agents such as condensed tannins. The reclamation of defense investment from senescent tissues, and the ability to reallocate a limited amount of defense chemicals within the plant in response to changing values of different plant parts, are additional features of the defense systems of early successional plants that confer maximum selective advantage at minimum cost. The alkaloid defense system of *Ricinus,* in which a small amount of potent toxin is efficiently shuttled from place to place within the plant, may be considered the embodiment of my expectations about the chemical defense systems of early successional herbs. Whether these features will prove to be general among such plants is a subject for further studies.

D. Other Aspects of Comparative Distribution of Different Secondary Compounds

1. *The Chemical Defenses of Wood*

Owing to its extremely long life relative to other parts of a tree, wood achieves no escape in time. This fact may explain the high concentrations of tannins, flavonoids, stilbenes, tropolones, and other "quantitative" (Feeny, 1976), difficult-to-degrade aromatic secondary compounds often found in wood (Scheffer and Cowling, 1966). The particular susceptibilities of the organisms most likely to invade wood, microbes, have probably also been a major determinant in the kinds of defenses found in wood. Tannins and flavonoids are generally bacteriostatic, fungitoxic, and antibiotic (Swain, 1977). The ability of tannins to inhibit microbial extracellular oxidases may be more important to their ability to inhibit heartwood decay than is their toxicity *per se* (Lyr, 1962). My strong impression is that toxins such as cyanogenic glycosides and alkaloids are not of widespread significance in the chemical defense of wood. Though alkaloids do occur in wood, it is usually in much lower concentrations than in the adjacent bark (Pectu, 1964; Court *et al.*, 1958, 1967; Cromwell, 1955; Kerharo, 1970; Sainsbury and Webb, 1972). For example, Abe (1971) reports the occurrence of the quinolone alkaloid oricine in the wood of *Oricia suaveolens* (Rutaceae) at concentrations of 0.01% dry weight. The stem bark of Nigerian material of this species contains more than 0.1% oricine (Peter G. Waterman, personal communication). Nevertheless, *Ocotea* (Lauraceae) wood has been reported to be rich in alkaloids (Menon, 1956), and antimicrobial alkaloids have been isolated from the heartwood of *Liriodendron tulipifera* (Magnoliaceae) (Hufford *et al.*, 1975). I believe the information justifies considering these cases as exceptional. Three reasons may account for the general absence of alkaloids as wood defenses: (*a*) Resistance of wood to decay organisms is enhanced by its high carbon-to-nitrogen ratio (Scheffer and Cowling, 1966). Concentrations of alkaloids might provide a nitrogen source for a specialist that could detoxify the alkaloids. (*b*) Many kinds of alkaloids seem quite susceptible to counteradaptation by pathogens and herbivores, which would make them less suitable for defense of highly "apparent" tissues such as wood. (*c*) Economic constraints might militate against the consignment of large quantities of nitrogen to metabolic oblivion in the heartwood.

The tannins and other phenolic compounds deposited in the heartwood are synthesized at the sapwood-heartwood boundary (Hillis, 1972; Roux, 1972). The amount of extractable phenolics is often reported to be lower in the inner heartwood than in the outer heartwood (Cartwright, 1941; Scheffer, 1957; Swain, 1965; Hillis, 1972), a difference that seems to be due to autoxidative polymerization of phenolics over time, resulting in highly polymerized molecules that are extremely difficult to extract (Lyr, 1962). In addition to being difficult to extract, these highly polymerized phenols of old heartwood are also much less effective as protein-precipitating antibiotic substances than are the smaller molecules found in recently deposited heartwood, a fact that may account for the lower resistance of inner heartwood to decay (Cartwright, 1941; Scheffer, 1957; Lyr, 1962). This is an interesting concept in that it points to a distinct upper time limit of effectiveness for the class of secondary compounds most suitable for the defense of long-lived tissues. However, the susceptibility of the oldest part of the tree to rotting is not necessarily detrimental (Janzen, 1976b).

2. *The Chemical Defenses of Roots*

There exists a large body of information on the chemical constituents of roots and their role in mediating interactions with soil-inhabiting pathogens, that cannot be adequately reviewed here. The reader is referred to the article by Rohde (1972) and to books edited by Baker and Snyder (1965), Carson (1974), and Marshall (1973). Roots often contain the same secondary compounds as are found in aerial parts of the plant, not infrequently in higher concentrations. Leifertova and Buckova (1969) studied 20 species of *Geranium* and found that underground organs contained higher concentrations of tannins (16–30%) than did aerial parts (11–20%). Total polyphenol content of tobacco roots was found by Sheen (1969) to be more than twice that of the stem. In alkaloid-containing plants, roots and root bark often seem to possess greater alkaloid concentrations than stems and stem bark (Tomczyk, 1964; Court *et al.*, 1967; Kerharo, 1970; Evans *et al.*, 1972; Fish and Waterman 1971a,b, 1972; Phillipson *et al.*, 1974; Fish *et al.*, 1976; Waterman, 1976). These reports concern a limited number of families, Rutaceae, Asclepiadaceae, Apocynaceae, and Solanaceae. In other cases, alkaloids occur in lower concentrations in roots than in aerial parts (Evans and Treagust, 1973; Taber *et al.*, 1963a; White and Spencer, 1964; Gentry *et al.*, 1969), so that no generalizations can be made about the root-versus-shoot allocation of this diverse group of secondary compounds. Very little information exists on this aspect of the intra-plant distribution of other classes of secondary compounds. Highest concentrations of the non-protein amino acid *N*-methyl-L-serine in *Dichapetalum cymosum* occur in the rhizome tips (Eloff and Grobbelaar, 1969). Proteinase inhibitor concentration in tomato roots is extremely low, in both damaged and control plants (Ryan and Green, 1974). Concentrations of steroidal saponins in *Dioscorea* spp. (Dioscoreaceae) are often higher in underground than in the aerial parts (Takeda, 1972).

An interesting aspect of the chemical defense of roots concerns the interaction of roots with mycorrhizal fungi. Many of the secondary compounds produced and/or exuded by roots are fungistatic or fungitoxic (Greathouse, 1939; Gäuman *et al.*, 1960; Hillis and Ishikura, 1969; Melin and Krupa, 1971). Melin and Krupa (1971) suggest that terpenes produced by pines play an important role in restricting the growth of the mycorrhizal fungi within the host root. The problem of allowing limited infection by mycorrhizae while repelling other, pathogenic fungi would not appear to be a simple one.

In some cases the mycorrhizal fungus produces antibiotic substances that inhibit the growth of pathogenic fungi and thus might function in defense of the tree-mycorrhizae association (Marx, 1969a,b, 1970; Marx and Davey, 1969; Park, 1970). It is worth investigating the possibility that in some cases fungal metabolites may largely have replaced host-produced secondary compounds as important chemical defenses of mycorrhizal associations.

IV. COMMUNITY STRUCTURE OF SECONDARY COMPOUNDS WITHIN THE PLANT

Summarizing the arguments of the preceding section, each of the types of secondary compounds at a plant's disposal may possess a set of attributes that suits it for particular roles in defense and thereby influences its allocation within the plant. A question that immediately arises is, to what extent has selection molded the manifold metabolic pathways and the numerous kinds of secondary compounds present in a plant into a *system* of chemical defenses? Does each of a plant's chemical defense modes exert selective pressures influencing the behavior of the others? To what extent do the secondary compounds within an individual plant exhibit a "community structure" (Janzen, 1973a)? Even though secondary compounds emerge from disparate physiological backgrounds, we would expect their evolutionary fates to be tightly linked via the plant's resource budget (Janzen, 1973b). Just as co-evolution with acacia–ants has had profound effects on the chemical defenses of ant–acacias (Rehr *et al.*, 1973b), so would we expect the evolution of a new chemical defense to have dramatic effects on the plant's preexisting defense chemistry (Janzen, 1973a,b).

An individual plant is likely to contain a great diversity of secondary metabolites. For the most part, we have no clear idea how this diversity is organized (or whether it is). We do not know, for example, what proportion of the compounds found in a plant are present in each of its parts. Does each secondary compound tend to occur throughout the plant, or is the plant body differentiated into regions protected by different constellations of defense chemicals? We cannot even confidently compare the diversity of defense chemistry between different species with information presently available.

Few authors have discussed the evolution of the defense system of an entire plant in general terms. As Feeny (1976) notes, selection will often favor a plant's having several lines of chemical defenses. One of the presumed advantages of high intraspecific diversity of defense chemicals is protection against a broad spectrum of herbivores and pathogens. Another advantage is suggested by the experiments of Pimentel and Bellotti (1976): It is easier for a herbivore to evolve resistance to a single defense than to several defenses simultaneously. A third advantage, discussed by Feeny (1976) and by Rhoades and Cates (1976) and gaining emphasis and support from the considerations presented in this study, is that different types of secondary compounds are differentially suited to defense of different plant parts.

A fourth advantage to plants that contain a diversity of secondary compounds is that two co-occurring chemical defenses might achieve the same protective effect at lower cost than either defense occurring alone. The possibilities for synergistic effects among plant chemical defenses are immense (Janzen, 1973a). Though the mechanisms that might result in synergism are undoubtedly diverse, three may be mentioned here. (*a*) Synergism by methylenedioxyphenyl (MDP) compounds. Many compounds containing the MDP group are potent inhibitors of mixed function oxidases (detoxifying enzymes) (Metcalf, 1967), and thus synergists of a wide variety of toxic compounds, apparently by virtue of their conversion to dihydroxyphenols at the reaction site (Kuwatsuka, 1970). The mechanism of inhibition that Kuwatsuka (1970) proposed is in effect a way to smuggle a polyphenol (whose solubility characteristics would normally exclude it from the detoxification site) into the herbivore's detoxification system. Rhoades and Cates (1976) discuss the potential importance of MDP synergists in plant defense chemistry. (*b*) Analog synergism. Slight changes in the structure of a drug may change it to an inhibitor of the enzymes which normally metabolize the drug (Ariëns, 1964). Though to my knowledge it has never been proposed, lowering of the dose-response curve by analog synergism may be one possible outcome of the presence of series of structurally related alkaloids, cardiac glycosides, and other compounds, often found in a single organ of an individual plant. Speculation on this point seems idle until more is known about detoxification of these compounds in a variety of herbivores. (*c*) Synergism in which one compound alters the permeability of cells to other compounds. Surface-active compounds such as saponins sometimes enhance drug absorption from the intestine (Basu and Rastogi, 1967). Kerharo and Adam (1974) note that the seeds of *Erythrophleum guineense* (Caesalpiniaceae) are more toxic than the bark, even though they contain lower concentrations of cardiotonic alkaloids, because of the presence of a saponin that acts synergistically with the alkaloids. It would be interesting to know whether such interactions with saponins, or with other compounds such as lectins that can also influence cell permeability (Liener, 1976), are of general significance in the toxicity of plant secondary compounds. Though the mechanisms are unclear, synergistic effects are commonly observed in toxicity studies of naturally occurring phenolics, according to Singleton and Kratzer (1973), who also point out that it is not uncommon to find a considerable percentage of usual phenolics occurring in the same plant part with an unusual and highly toxic phenolic.

Balanced against these advantages of possessing a diversity of secondary compounds are "competitive" interactions among the secondary compounds within a plant (Janzen, 1973a). Competition might be said to occur when advantages gained by having two or more kinds of defenses are more than offset by disadvantages associated with the consequent lowering of the plant's maximum investment in each. Coexistence would be favored by the following conditions: (*a*) Total defense investment is high. (*b*) Low concentrations of each compound are effective in defense. (*c*) Compounds are suited for allocation to different parts. (*d*) Compounds place different demands on the plant's resource budget. (*e*) Synergisms occur.

Incorporation of a greater diversity of chemical defenses may be accomodated by partial displacement, within the plant, of one mode of chemical defense by another. Thus in some "apparent" plants, tannins and toxins such as cyanogenic glycosides display distributions that tend to be mutually exclusive (Feeny, 1976; Rhoades and Cates, 1976), tannins occurring in long-lived tissues (mature leaves), potent toxins occurring in lower concentrations in short-lived tissues (young leaves) more susceptible to autotoxicity. Similarly, many "unapparent" herbaceous plants that lack condensed tannins in leaves (possibly because of the high metabolic cost associated with a tannin-based defense system) still possess them in seed coats (Bate-Smith and Ribereau-Gayon, 1959), where their fungistatic and bacteriostatic properties may continue to confer a net selective advantage. The point of importance here is that these plants have not lost the ability to synthesize these compounds; they have only excluded them from leaves, where selective pressures have led to their replacement by "cheap" defense systems based on small amounts of reclaimable toxins. (It may be added that the coat of a dormant seed of an ephemeral plant is likely to be much more apparent to herbivores in time than is the adult plant.) Similar kinds of selective pressures may have resulted in the pattern of distribution of cyanogenic glycoside in *Barteria fistulosa* (Passifloraceae). The root bark contains much greater concentrations than the stem bark (Paris *et al.*, 1969). The aerial parts of this plant (but not, of course, the roots) are patrolled by obligate mutualist ants (Janzen, 1972).

Considerations presented in this review outline a number of factors exerting possibly general effects on allocation of different kinds of chemical defenses. However, the selective factors that drive the evolution of intraplant distribution of chemical defenses may be expected to be diverse, and sometimes so idiosyncratic as to defy attempts to predict them. Allocation of a toxic compound to a given part might often be explained in terms of the properties of the herbivores that affect that part. For example, water-soluble benzophenanthridine alkaloids (chelerythrine, nitidine, and related compounds) tend to occur in especially high concentration in the cork of the outer root bark in a number of Rutaceae (Fish and Waterman, 1971a,b; Fish *et al.*, 1975a,b), with much lower concentrations being found in the inner root bark and root wood, and in the stem bark. These alkaloids are extremely fungitoxic (chelerythrine and another benzophenanthridine, sanguinarine, completely inhibit the growth of *Phymatotrichum* root-rot fungus at 0.0001 M: Greathouse and Rigler, 1940) and water-soluble. Their allocation to the outermost layer of the root bark may reflect advantages conferred by the formation of a fungitoxic zone around the roots of these plants (Fish *et al.*, 1975a).

In other cases (e.g., when a new highly effective broad-spectrum defense appears), selection may favor its allocation to all parts of the plant. If the new defense mode and pre-existing ones place similar and conflicting demands on the plant's resource budget and the advantages associated with maintaining a diversity of chemical defenses are not compensatory, selection may favor the replacement of pre-existing defense modes by the new one. In species of the genus *Astragalus*, for example, accumulation of canavanine and accumulation of S-substituted nonprotein amino acids tend to be mutually exclusive species-specific traits. Pathways leading to synthesis of both these types of secondary compounds conceivably compete for the limited supply of nitrogen that is invested in defense (cf. Dunnill and Fowden, 1967). Complete replacement of an old defense mode by a new one would be most likely when both require a particular resource that may be in limited supply, as in this case, when total defense investment is low and the concentrations of each defense must be high to achieve toxic effects, when the two modes are not suited for allocation to different parts, or for defense against different potential enemies.

Thus the interactions among the different kinds of secondary compounds within a plant may have a variety of expected outcomes, in terms of their effects on the variety of chemical defenses present in a plant and on their distribution patterns within the plant. And while we may proceed with the expectation that selection has "organized" the secondary metabolism of plants into coherent defense systems, only a few of the features of these systems are apparent at this time. The recent papers by Feeny (1976) and Rhoades and Cates (1976) are important in that they form the robust beginnings of a holistic approach to the investigation of plant defense chemistry. Further similar studies, which transcend the historical and technical barriers that have to a large extent perpetuated the independent investigation of alkaloids, phenolics, etc and emphasize the common functions shared by these chemically di- verse groups, are likely to add much to our understanding of pla herbivore interactions.

V. THE ALLOCATION OF DEFENSE INVESTMENT

Preceding sections have dealt with the comparative apportionment of different types of defense chemicals. This section concerns several aspects of the allocation of "defense investment" to different parts of the plant. Not only will different kinds of compounds be allocated to different plant parts; some parts will contain higher total amounts of defense chemicals than other parts (e.g., Mooney and Chu, 1974). The aim of this section is to examine some patterns of the quantitative distribution of resources invested in defense, in different parts of the plant. In cases where different parts of the plant contain very different types of defense chemicals, it is not obvious how "defense investment" in the different parts can be compared. Nevertheless, there do exist examples of tissues whose high or low content of defense chemicals seems to be general across all types of secondary compounds, and these will be considered here.

A. Allocation of Secondary Compounds to External Tissues

The external tissues of plants and animals are usually the first to come into contact with herbivores and pathogens, and physical and chemical features of external regions constitute the organism's first line of defense. It is not surprising that toxic secondary compounds are often present in relatively high concentrations in external tissues. To the references cited in an earlier study (McKey, 1974) may be added those listed in Table 2.

Another frequent avenue by which pathogens and sometimes herbivores may enter the plant is through the vascular system, and the frequent occurrence of high concentrations of secondary compounds in vascular bundles (e.g., James, 1950) may be analogous to their concentration in external tissues. That sucking insects may evade such chemical barriers is suggested by reports such as that of McCarthy (1968) that one *Aloe* cultivator in South Africa successfully treats whitescale infection on *Aloe* leaves with the juice of the plant being attacked.

Deposition of chemical defenses in restricted cell layers or tissues (e.g., external tissues or vascular bundles) is of course open to exploitation by herbivores such as leaf-miners or stem-borers. Many of the herbivores of oak mature leaves are leaf-miners, which appear to avoid tannin-rich cell layers within the leaf (Feeny, 1970). Similarly, the pine needle miner *Exoteleia pinifoliella* (Lepidoptera: Gelechiidae) utilizes pine species whose needles have few small resin canals, and is excluded

Table 2. Relative concentrations of secondary compounds in internal and external parts of plants.

Class of compound	Plant	Organs examined	Internal	External	Reference
1. Phenolics					
Tannins	*Geranium* spp.	Roots, stems, leaves, seeds	L[a]	H	Kostecka-Madalska *et al*. (1967)
Protocatechuic acid	*Allium* spp.	Bulb	L	H	Link *et al*. (1929)
2. Alkaloids					
Isoquinolines	*Mahonia* spp.	Roots	L	H	Greathouse and Watkins (1938)
	Lophocereus schottii	Stems	L	H	Kircher (1969)
Indoles	*Rauvolfia* spp.	Roots	L	H	Court *et al*. (1967); Kerharo (1970)
Quinolizidines	*Lupinus luteus*	Stems	L	H	White and Spencer (1964)
Steroidal glycoalkaloids	*Solanum tuberosum*	Tubers	L	H	Allen and Kuć (1968)
Various	*Zanthoxylum* spp.	Stems, roots	L	H	P. G. Waterman, personal communication
3. Coumarins	*Zanthoxylum* spp.	Stems, roots	L	H	P. G. Waterman, personal communication
4. Cyanogenic glycosides	*Manihot esculenta*	Roots	L	H	de Bruijn (1973)
5. Steroidal sapogenins	*Dioscorea spiculiflora*	Tubers	L	H	Preston *et al*. (1964)

[a] L = low; H = high

from those species whose needles have numerous resin canals (Bennett, 1954).

B. Fate of Defense Compounds during Maturation of Fleshy Fruits

In fleshy fruits dispersed by animals, chemical defenses allocated to the immature fruit should be neutralized upon maturation so that the fruit is attractive to dispersal agents (McKey, 1974). Limited information on how this is accomplished, in the case of two contrasting kinds of chemical defenses, alkaloids and tannins, shows yet another facet of the different roles of these types of compounds in plant defense systems.

Almost all the information on alkaloids concerns the atypical steroidal glycoalkaloids of Solanaceae. Solasodine in *Solanum incanum* (Zaitschek and Segal, 1972), tomatine in tomatoes (Sander, 1956, 1963; Heftmann and Schwimmer, 1972) and in *Solanum nigrum* (Schreiber *et al.*, 1961), and the alkaloids of *S. dulcamare* (Willuhn, 1966, 1967) are all degraded almost completely during maturation of the fruit flesh (in contrast, mature berries of *S. khasianum* are reported to contain high concentrations of solasodine: Maiti *et al.*, 1965). Degradation of tomatine in tomato has been most thoroughly studied. Green tomatoes are apparently unable to degrade tomatine, the required enzyme system being activated or synthesized during fruit maturation (Heftmann and Schwimmer, 1972).

Information presented in this chapter points to the conclusion that condensed tannins are not so susceptible to degradation by the plant as are secondary compounds such as many alkaloids. Thus it is interesting that in the species that have been studied "disappearance" of tannins during maturation of fleshy fruits is achieved not by their degradation or transport from the fruit but rather by their increased polymerization to compounds of very high molecular weight that are nonastringent (Goldstein and Swain, 1963; Joslyn and Goldstein, 1964; Swain, 1965).

Janzen (1975) has briefly noted that the presence of secondary compounds in mature fruits may "focus" seed dispersal through a subset of the large number of potential dispersal agents. However, there is as yet no indication that ability to tolerate such compounds bears any relation to qualities desirable in dispersal agents.

C. Leaf Demography and Chemical Defense

The normal lifespan of leaves of vascular plants may range from a few weeks in the case of some desert ephemerals, to as long as 30 years in some species of *Araucaria* (Woolhouse, 1967). The complement of leaves present on an evergreen tree at any one time is composed of leaves varying greatly in age (Pease, 1917; Silver, 1962). If leaves of different ages do not contribute equally to the fitness of the plant, one would expect that the plant's limited supply of chemical defenses would not be apportioned equally among them. Unfortunately, there is little information that can be applied to analysis of the relation of defense investment to leaf age. However, there is information on the contribution of leaves of various ages to various components of fitness. Most of this information is from north-temperate gymnosperms, whose growth patterns make possible straightforward determination of leaf age (Pease, 1917; Silver, 1962).

Studies of net photosynthetic rate in relation to leaf age have usually led to the conclusion that maximum photosynthetic rate, and maximum export of carbohydrate from a leaf, occur soon after the leaf attains maximum size, after which the photosynthetic efficiency and rate of carbohydrate export both decline (R. H. Brown *et al.*, 1966; Wardlaw, 1968; Larson and Gordon, 1969; Shiroya *et al.*, 1961; Bidwell, 1974; Aslam *et al.*, 1977). Among conifers, one-year-old needles of *Pinus resinosa* supply more photosynthate to expanding shoots than do two- and three-year-old needles. Current year needles are not major exporters of carbohydrates until fully expanded late in summer (Dickman and Kozlowski, 1958). Similarly in *Abies balsamea* and *Picea glauca* young needles, with the exception of immature current foliage, are photosynthetically more active than old needles (Clark, 1956). Young leaves are also more active in the production of hormones (Kozlowski, 1973). The greater activity of younger age classes of leaves in producing assimilate and hormones, along with their greater *future* contribution to plant fitness (due to their greater expectancy of further life), should mean that they make a greater contribution to the fitness of the plant than do older leaves. It must be noted, however, that a leaf may be the site of *storage* of assimilate long after it has ceased to actively produce it. Old leaves may be an important source of reserves at certain times of the year (Kramer and Wetmore, 1943; Onaka, 1950; Kozlowski and Winget, 1964).

Nevertheless, artificial defoliation studies with conifers have consistently demonstrated that the removal of older age classes of leaves has a much less damaging effect on the plant than does the removal of one-year-old foliage or foliage of the current year (Reeks and Barter, 1951; Dochinger, 1963; Craighead, 1940; Linzon, 1958; O'Neil, 1962; Lanier, 1967; Burger, 1953). Spruce withstand several years of almost complete loss of old needles as long as most of the new needles are intact (Reeks and Barter, 1951). While removal of two- or three-year-old foliage of *Pinus banksiana* has no appreciable effect on growth, removal of current-year foliage induces high bud mortality and reduction in the rate of shoot elongation.

We would expect the allocation of chemical defenses among leaves of different ages to have been affected by this difference in their value. While this prediction cannot be adequately evaluated at present, it is interesting that juvenile foliage of *Pinus banksiana* contains two chemicals that deter feeding of diprionid sawflies and that concentrations of these feeding deterrents decline as the needles mature (Ikeda *et al.*, 1977). Nine species of *Neodiprion* (Ikeda *et al.*, 1977) and other diprionid sawflies (Reeks and Barter, 1951; Balch, 1939; Wilson, 1966) strongly avoid feeding on new growth of conifers. If the allocation pattern of feeding deterrents in *Pinus banksiana* proves to be widespread, it could explain why this important family of conifer herbivores is restricted to those needle age classes of least value to the plant, so that even several successive years of severe defoliation sometimes results in no host mortality (Wilson, 1966).

Preferential allocation of chemical defenses to younger leaves may be achieved by a variety of mechanisms. The simplest, and perhaps the most widespread, would be passive dilution (perhaps coupled with decreased production) of these defenses by other constituents during maturation. In some cases, however, selection may favor actual withdrawal of resources invested in defense for reallocation elsewhere. The extreme case, of course, is when the leaf begins to senesce and its further contribution to fitness falls to a figure near zero. The difference between the export of defense investment at senescence and its reallocation based on different leaf value is one of degree, not kind. As mentioned previously (Section III,C,4,b), there is an added qualification to the predictions about reallocation: Some forms of defense investment may not be reclaimable.

VI. THE REGULATION OF SECONDARY METABOLISM

A. Endogenous Regulation

The arguments about intraplant allocation of chemical defenses presented here presuppose the existence of regulatory controls on secondary metabolism. The information available on the heterogeneous distribution of secondary compounds within plants confirms this supposition. Secondary metabolism exhibits the same degree of within-plant differentiation and specialization as do primary metabolic pathways

(Swain, 1974; Luckner *et al.*, 1977). Understanding the physiological regulation of secondary metabolism would yield insight into the evolutionary processes by which existing patterns of secondary-compound distribution have been produced and into the likely responses of plants to selective pressures favoring evolutionary changes in these patterns. If other evolutionary studies (e.g., King and Wilson, 1975) may serve as a guide, the evolution of new chemical-defense phenotypes may prove to be due as much to regulatory changes in existing pathways as to the ability to produce new metabolites. Two examples may be cited to illustrate the potential importance of regulatory changes as sources of new defense phenotypes. (*a*) Ability of some species of *Vicia* to accumulate the toxic amino acid β-cyanoalanine may have been conferred by regulatory changes in pre-existing pathways that previously elaborated only trace quantities of this compound (Fowden and Bell, 1965), as discussed in Section III,C,2,b, (*b*) Beets produce trace quantities of a number of "uncommon" amino acids, some of them thought to be so restricted in distribution that they have been used as taxonomic markers (Fowden, 1974). The latent ability to produce these compounds might serve as the basis for rapid changes in defense chemistry under appropriate selective regimes.

Unfortunately, existing information on the regulation of secondary metabolism is meager. Luckner *et al.* (1977) note the "very poor experimental state of the molecular biology of secondary metabolism as compared to that of other differentiation processes." We do not know, for example, why synthesis and storage of anthraquinones in intact *Morinda citrifolia* (Rubiaceae) is restricted to the root, while tissues of leaf, stem, and fruits give rise to cell cultures whose anthraquinone spectrum is identical to that of cell cultures derived from root tissues (Luckner *et al.*, 1977). Nor do we know what mechanisms produce the heterogeneous microdistribution of alkaloids observed in *Lupinus luteus* (White and Spencer, 1964), or how the leaves of *Ipomoea violacea* (Convolvulaceae) contain very low concentrations of alkaloids, even though these organs are the principal site of synthesis (Mockaitis *et al.*, 1973). The temporal component of regulation of secondary metabolism is similarly unexplored. It is unknown what mechanisms account for the production of alpiniginene in young plants only of some strains of *Papaver bracteatum*, while other strains produce this alkaloid throughout life (Luckner *et al.*, 1977). Likewise, we cannot account for the appearance in soybeans, during days 4–6 of germination, of trypsin inhibitors different from those present in the dry seed (Orf *et al.*, 1977).

Since studies of the endogenous regulation of secondary metabolism in higher plants have as yet yielded few generalizations, they will not be further dealt with here. The reader should consult recent reviews (Del Moral, 1972; H. Smith, 1973; Floss *et al.*, 1974; Grisebach and Hahlbrock, 1974; Creasy and Zucker, 1974; Butt, 1976; Camm and Towers, 1977; Luckner *et al.*, 1977). The intent here is to discuss endogenous regulation only in terms of its mediation of the responses of plant secondary metabolism to external influences.

B. Effects of External Environment

While in some cases, (e.g., monoterpene production in *Pinus*, Hanover, 1965) secondary metabolism is relatively unaffected by variations in the external environment, in many other cases such variations have a considerable effect on the concentrations and composition of the spectrum of secondary metabolites produced by plants. The most general effect, judging from the information available, is that under stress conditions there is a tendency for expansion or activation of many secondary metabolic pathways (Brian, 1973). This activation is nonspecific in that it generally occurs whether the stress is produced by pathogenic infection, wounding or by physical factors such as drought and cold (e.g., Stoessl *et al.*, 1976). While the increase in activity of secondary metabolism under physiological stress is subject to a variety of interpretations (see Del Moral, 1972, for one that contrasts with the one entertained here), it might be rationalized as an adaptive response to conditions in which (further) herbivore damage to the plant is likely to be unusually costly. According to such an interpretation, the defense system of the plant may incorporate mechanisms by which its production of defensive substances is varied in a manner more or less correlated with the potential costs of herbivory. The increased production of secondary metabolites might be localized to certain areas (e.g., areas actually being invaded by pathogens) or systemic (e.g., in response to a general stress).

Though the secondary metabolites produced by the plant in response to invasion by pathogens are usually not specific to this source of stress, response to infection has historically been studied separately from response to soil quality, etc., and will be considered separately here. Most of the research in this area concerns those secondary metabolites known to phytopathologists as phytoalexins. The literature on this group of compounds is immense and cannot be adequately discussed here. The reader is referred to recent reviews (Ingham, 1972; Kuć, 1972; Byrde and Cutting, 1973; Friend and Threlfall, 1976). The following summary remarks are supported by this literature [these remarks are in fact paraphrased from those of Brian (1973)]: (*a*) Though originally considered to be specific defense mechanisms induced by infection by fungi, bacteria, or viruses, many phytoalexins are produced in response to a wide range of stress conditions (e.g., Stoessl *et al.*, 1976). (*b*) There is no real dichotomy between the production of completely new compounds upon infection (part of the original phytoalexin concept) and the quantitatively greater production of substances already present (e.g., Stoessl, 1970). (*c*) Although phenolics and to a lesser extent terpenoids have been most often implicated as phytoalexins, in fact a wide variety of secondary metabolites are accumulated in response to infection, wounding, and other stresses. In addition to phenolics such as chlorogenic acid (Friend *et al.*, 1973), isoflavonoids (Rathwell, 1973), and coumarins (Scheel *et al.*, 1963), these include tropane alkaloids in *Datura* (Singh, 1970), aporphine alkaloids in *Liriodendron tulipifera* (Chen *et al.*, 1976), steroidal glycoalkaloids in potato and tomato plants (Locci and Kuć, 1967; Drysdale and Langcake, 1973), sesquiterpenoids in Solanaceae (Stoessl *et al.*, 1976; Ishizaka and Tomiyama, 1972; Ward *et al.*, 1975), stilbenes in *Pinus* (Jorgensen, 1961; Shain, 1967), and proteinase inhibitors in potato and tomato (Green and Ryan, 1972; Ryan and Green, 1974) and in tobacco (Wong *et al.*, 1976).

A second aspect of the plant's response to infection by pathogens or damage by herbivores is the possible presence of systems that use the damage produced as information. By this view, defense investment might be locally increased in adjacent tissues in response to an increased likelihood of attack. While there is no question that production of phytoalexins can be induced by a variety of chemical and other stimuli, the frequent reports that substances such as fungal wall glucans are especially good inducers (Albersheim and Anderson-Prouty, 1975; Anderson-Prouty and Albersheim, 1975; Ebel *et al.*, 1976) lends credence to the notion that information, as well as stress, is being responded to by the plant. The induction in potato and tomato of proteinase-inhibitor accumulation in leaves adjacent to damaged leaves, due to an inducing factor exported from the damaged leaf (Ryan and Green, 1974), is a clear example of information transfer that permits a localized response of defense investment to likelihood of attack. It may be pointed out that the message is capable of interception by herbivores, for, as demonstrated by Ryan and Green (1974), export of the inducing factor from the damaged leaf can be prevented by the simple act of excising the leaf at the petiole [chewing off the petioles of damaged leaves after feeding, a behavior noted for *Papilio polyxenes* (Feeny, 1976), would accomplish this.] Such observations must be accorded slight significance, however, in light of our lack of knowledge of the generality of such inducible defense mechanisms in nature.

Stresses may also arise from deficiencies in mineral nutrition related to soil quality. Both in tissue cultures (Westcott and Henshaw, 1976) and in growth experiments with intact plants (Davies *et al.*, 1964; Armstrong *et al.*, 1970), nitrogen deficiency stimulates an increase in biosynthesis of phenolic compounds, including chlorogenic acid, coumarins, and tannins. Response to stress might be the basis for the finding by Davies *et al.* (1964) that for all eight tree species examined, leaves of individuals growing on base-poor sites contained higher levels of condensed tannins than did leaves of individuals growing on base-rich sites. Subsequent growth experiments with four of these species established that nitrogen- and phosphorus-deficient plants consistently contained higher levels of polyphenols than did the controls. Stitt *et al.* (1946) and Wilson (1955) present partially conflicting results concerning the effect of soil quality on tannin content of *Lespedeza cuneata* (Leguminosae). The outcome of this phenoplastic mechanism of varying phenolic-compound concentration in response to soil quality is similar to the postulated evolutionary outcome (Janzen, 1974).

Stress associated with drought may lead to increased levels of phenolics in sunflower (Del Moral, 1972), and increased levels of cyanogenic glycosides in *Manihot esculenta* (de Bruijn, 1973; Butler *et al.*, 1973) and in sorghum (Nelson, 1953).

In contrast to the results obtained for phenolics, concentrations of alkaloids tend to increase with increasing nitrogen availability from the soil (Sokolov, 1959; Flück, 1963; Nowacki *et al.*, 1976). Present evidence does not seem to allow distinction between a preferential activation of alkaloid synthetic pathways on the one hand, and consequences of higher levels of soluble nitrogen within the plant, on the other (Nowacki *et al.*, 1976).

Diverse effects on secondary metabolism of factors such as light, soil, moisture, and so on have been reported. Those desiring extensive information should consult Sokolov (1959) and Flück (1963).

C. Properties of the Response System

Plants would be expected to possess the ability to respond to environmental challenges with increased activity of secondary metabolic pathways. The effects of external factors, reviewed in the preceding section, lend support to the idea that many plants do possess this ability. Unfortunately, little is known about the physiological events leading to expression of these effects.

Systems that regulate the response of secondary metabolism to external factors might be expected to have a number of interesting properties. Among them might be the ability to buffer defense investment from the effects of unimportant environmental variations with minimum restriction of the capacity to respond to real challenges. A second, which will be briefly considered here, is the ability to regulate secondary metabolism independently of primary metabolic pathways: the ability to respond by increased production of defensive metabolites with minimum effects on nondefensive pathways.

The regulatory problems posed by the requirement that defense-related pathways be selectively activated, would depend to some extent on the position of these pathways in the metabolic matrix of the plant. Many pathways of secondary metabolism seem to be quite isolated biosynthetically once they branch from the rest of metabolism, and might be relatively easily regulated as a unit. In other cases, the synthesis of "secondary" metabolites seems to be enmeshed in a network of "primary" pathways. For example, ricinine synthesis in *Ricinus communis* is connected to the pyridine nucleotide cycle (Johnson and Waller, 1974). Increasing the production of ricinine, without affecting the activity of this cycle, might pose regulatory problems absent from the regulation of alkaloid pathways that are more isolated from primary pathways. Similarly, if a single enzyme catalyzes the formation of a protein amino acid and that of a similar amino acid that is a toxic analog, regulation of the production of the analog independently of the common of precursors that affect the activity of primary pathways.

VII. OVERVIEW AND CONCLUSIONS

The objective of this chapter has been to combine information about the physiology of secondary metabolism within the plant with knowledge about the ecology of plant-herbivore interactions, to produce an integrated view of chemical defenses within the plant. Several main themes recur in this view.

1. Defense-related pathways must be examined in the context of the metabolic matrix in which they are enmeshed. As noted at various points in the chapter, this perspective yields insight on the origin and functioning of the defense metabolite itself, and of mechanisms for its metabolic manipulation within the plant. The relation of the defense-related pathways to the rest of metabolism may also determine to some extent the mechanisms by which these pathways will be regulated.
2. The fundamental biochemical similarities between plants and the organisms that drive the evolution of their chemical defenses, introduce two major constraints to plant chemical defense systems: a) The requirement of avoiding autotoxicity lim-

its the kinds of defense chemicals that can be produced, and limits the within-plant distribution of those that are produced. b) Resistance of defense chemicals to detoxification by animals may be achieved only by the sacrifice of the plant's ability to manipulate its defense chemicals metabolically.

3. Different types of allelochemics represent different solutions to the conflicting selective pressures imposed by internal physiological constraints on the one hand, and by herbivores on the other. Some solutions work better in some parts of the plant, other solutions in other parts. Thus bark (and to a lesser extent wood) is rich in compounds such as condensed tannins that provide long-term defense against a very broad range of potential enemies, and seeds are often defended by nitrogen-rich or fat-rich toxins that can be re-used by the germinating seedling.
4. Interactions between the diverse types of secondary compounds within the plant, mediated via the plant's resource budget, have the potential to organize the defenses of the plant into what may be termed a defense system. The regulatory mechanisms that accomplish the complex, heterogeneous patterns of secondary-compound distribution observed during differentiation and development, are just coming under study.

In conclusion, it must be emphasized that though this view of the chemical defenses of plants is admittedly speculative, it is offered with the twin hopes that its speculations will not be considered unreasonable and that it will stimulate further research based on a unified approach to plant defense chemistry.

ACKNOWLEDGMENTS

Special thanks are extended to D. H. Janzen, whose constant encouragement, constructive criticism, and tutelage have been a major shaping force in the development of the ideas presented here. M. Huston must also be given special acknowledgment, for extensive service as a critic available at a moment's notice during the weeks when the manuscript was being prepared. Several people read and commented on the manuscript: P. Dillon, B. Hazlett, S. Kleinfeldt, M. Martin, P. Morrow, T. Swain, and P. Waterman. However, responsibility for any errors of fact or interpretation, or for quirks of organization, rests with the author.

REFERENCES

Abbondanza, A., Badiello, R., and Breccia, A. (1965). *Tetrahedron Lett.* pp. 4337–4341.
Abe, M. O. (1971). *Phytochemistry* **10**, 3328–3330.

Abrol, Y. P., and Conn, E. E. (1966). *Phytochemistry* **5,** 237–242.
Abrol, Y. P., Conn, E. E., and Stoker, J. R. (1966). *Phytochemistry* **5,** 1021–1027.
Aexel, R., Evans, S., Kelley, M., and Nicholas, H. J. (1967). *Phytochemistry* **6,** 511–524.
Albersheim, P., and Anderson-Prouty, A. J. (1975). *Annu. Rev. Plant Physiol.* **26,** 31–52.
Allen, E., and Kuć, J. (1968). *Phytopathology* **58,** 776–781.
Ambe, K. S., and Sohonie, K. (1956). *Experientia* **12,** 302–303.
Anderson, J. M. (1973). *Oecologia* **12,** 275–288.
Anderson-Prouty, A., and Albersheim, P. (1975). *Plant Physiol.* **56,** 286–291.
Ariëns, E. J., (1964). "Molecular Pharmacology," Vol. I. Academic Press, New York.
Armstrong, G. M., Rohrbaugh, L. M., Rice, E. L., and Wender, S. H. (1970). *Phytochemistry* **9,** 945–948.
Arrigoni, O., Arrigoni-Liso, R., and Calabrese, G. (1975). *Nature (London)* **256,** 513–514.
Aslam, M., Lowe, S. B., and Hunt, L. A. (1977). *Can. J. Bot.* **55,** 2288–2295.
Baker, K. F., and Snyder, W. C., eds. (1965). "Ecology of Soil-borne Plant Pathogens—Prelude to Biological Control." Univ. of California Press, Berkeley.
Balch, R. E. (1939). *J. Econ. Entomol.* **32,** 412–418.
Ball, E. (1950). *Growth* **14,** 295–323.
Banthorpe, D. V., Bucknall, G. A., Gutowski, J. A., and Rowan, M. G. (1977). *Phytochemistry* **16,** 355–358.
Barz, W., and Hösel, W. (1975). *In* "The Flavonoids" (J. B. Harborne, T. J. Mabry, and H. Mabry, eds.), Part 2, pp. 916–969. Academic Press, New York.
Basu, N., and Rastogi, R. P. (1967). *Phytochemistry* **6,** 1249–1270.
Bate-Smith, E. C. (1973). *Phytochemistry* **12,** 1809–1812.
Bate-Smith, E. C., and Ribereau-Gayon, P. (1959). *Qual. Plant. Mater. Veg.* **5,** 189–198.
Baur, P. S., and Walkinshaw, C. H. (1974). *Can. J. Bot.* **52,** 615–619.
Beckman, C. H., and Mueller, W. C. (1970). *Phytopathology* **60,** 79–82.
Belew, M., Porath, J., and Sundberg, L. (1975). *Eur. J. Biochem.* **60,** 247–258.
Bell, E. A. (1971). *In* "Chemotaxonomy of the Leguminosae" (J. B. Harborne, D. Boulter, and B. L. Turner, eds), pp. 179–206. Academic Press, New York.
Bennett, W. D. (1968). *Phytochemistry* **7,** 151–152.
Bennett, W. H. (1954). *Can. Entomol.* **86,** 49–60.
Berger, P. J., Sanders, E. H., Gardner, P. D., and Negus, N. C. (1977). *Science* **195,** 575–577.
Betts, T. J., and Fairbairn, J. W. (1963). *Planta Med.* **12,** 64–70.
Bickel, M. H. (1969). *Pharmacol. Rev.* **21,** 325–355.
Bidwell, R. G. S. (1974). "Plant Physiology." Macmillan, New York.
Birk, Y. (1961). *Biochim. Biophys. Acta* **54,** 378–381.
Birk, Y. (1968). Chemistry, pharmacology and chemical applications of proteinase inhibitors. *Ann. N. Y. Acad. Sci.* **146,** 361–787.
Birk, Y. (1974). *Proteinase Inhibitors Bayer Symp., 5th, 1973* pp. 355–361.
Blumenthal-Goldschmidt, S., Butler, G. W., and Conn, E. E. (1963). *Nature (London)* **197,** 718–719.
Bond, E. J. (1961). *Can. J. Biochem. Physiol.* **39,** 1793–1802.
Bondi, A., Birk, Y., and Gestetner, B. (1973). *In* "Chemistry and Biochemistry of Herbage" (G. W. Butler and R. W. Bailey, eds.), Vol. 1, pp. 511–528. Academic Press, New York.
Bonner, J. (1965). *In* "Plant Biochemistry" (J. Bonner and J. E. Varner, eds.), 2nd ed., pp. 665–692. Academic Press, New York.
Boughton, I. B., and Hardy, W. T. (1935). *Tex. Agric. Exp. Stn., Bull.* **519.**
Breccia, A., and Badiello, R. (1967). *Z. Naturforsch., Teil B* **22,** 44–49.
Brian, P. W. (1973). *In* "Fungal Pathogenicity and the Plant's Response" (R. J. W. Byrde and C. V. Cutting, eds.), pp. 469–474. Academic Press, New York.
Brown, B. R., Brown, P. E., and Pike, W. T. (1966). *Biochem. J.* **100,** 733–738.

Brown, R. H., Cooper, R. B., and Blaser, R. E. (1966). *Crop Sci.* **6,** 206–209.
Brown, S. A. (1964). *In* "Biochemistry of Phenolic Compounds" (J. B. Harborne, ed.), pp. 361–398. Academic Press, New York.
Burbott, A. J., and Loomis, W. D. (1969). *Plant Physiol.* **44,** 173–179.
Burger, H. (1953). *Abstr. Pap., Int. Bot. Congr., 7th, 1950; For. Abstr.* **13,** No. 156 (1952).
Butler, G. W., Reay, P. F., and Tapper, B. A. (1973). *In* "Chronic Cassava Toxicity" (B. Nestel and R. MacIntyre, Monogr. IDRC-010e, pp. 67–71. Int. Dev. Res. Cent., Ottawa.
Butt, V. S. (1976). *Perspect. Exp. Biol.* **2,** 357–367.
Byrde, R. J. W., and Cutting, C. V., eds. (1973). "Fungal Pathogenicity and the Plant's Response." Academic Press, New York.
Camm, E. L., and Towers, G. H. N. (1977). *Prog. Phytochem.* **4,** 169–188.
Carson, E. W., ed. (1974). "The Plant Root and its Environment." University Press of Virginia, Charlottesville.
Cartwright, K. St. G. (1941). *Forestry* **15,** 65–75.
Cates, R. G. (1975). *Ecology* **56,** 391–400.
Cates, R. G., and Orians, G. H. (1975). *Ecology* **56,** 410–418.
Chafe, S. C., and Durzan, D. J. (1973). *Planta* **113,** 251–262.
Chen, C.-H., Chang, H.-M., Cowling, E. B., HuangHsu, C.-Y., and Gates, R. P. (1976). *Phytochemistry* **15,** 1161–1167.
Chew, M. Y., and Boey, C. G. (1972). *Phytochemistry* **11,** 167–169.
Clark, J. (1956). *Bi-Mon. Prog. Rep., Div. For. Biol., Dep. Agric., Can.* **12** (5).
Clarke, A. E., Knox, R. B., and Jermyn, M. A. (1975). *J. Cell Sci.* **19,** 157–167.
Cope, W. A. (1962). *Crop Sci.* **2,** 10–12.
Corcoran, M. R., Geissman, T. A., and Phinney, B. O. (1972). *Plant Physiol.* **49,** 323–330.
Coulson, C. B., Davis, R. I., and Lewis, D. A. (1960). *J. Soil Sci.* **11,** 20–29.
Court, W. E., Evans, W. C., and Trease, G. E. (1958). *J. Pharm. Pharmacol.* **10,** 380–383.
Court, W. E., Hakim, F. S., and Stewart, A. F. (1967). *Planta Med.* 15, 282–286.
Craighead, F. C. (1940). *J. For.* **38,** 885–888.
Creasy, L. L., and Zucker, M. (1974). *In* "Metabolism and Regulation of Secondary Plant Products" (V. C. Runeckles and E. E. Conn, eds.), pp. 1–19. Academic Press, New York.
Crombie, L., and Crombie, W. M. L. (1975). *Phytochemistry* **14,** 409–412.
Cromwell, B. T. (1933). *Biochem. J.* **27,** 860–872.
Cromwell, B. T. (1955). *In* "Moderne Methoden der Pflanzenanalyse", (K. Paech and M. V. Tracey, eds.), Vol. 4, pp. 367–516. Springer-Verlag, Berlin and New York.
Dagley, S. (1967). *Soil Biochem.* **1,** 287–317.
Davies, R. I. (1971). *Soil Sci.* **111,** 80–85.
Davies, R. I., Coulson, C. B., and Lewis, D. A. (1964). *J. Soil Sci.* **15,** 310–318.
Dawidar, A. A., and Fayez, M. B. E. (1961). *Arch. Biochem. Biophys.* **92,** 420–423.
de-Bruijn, G. H. (1973). *In* "Chronic Cassava Toxicity" (B. Nestel and R. MacIntyre, eds.), Monogr. IDRC-010e, pp. 43–48. Int. Dev. Res. Cent., Ottawa.
Dechary, J. M. (1970). *Econ. Bot.* **24,** 113–122.
Del Moral, R. (1972). *Oecologia* **9,** 289–300.
Dement, W. A., and Mooney, H. A. (1974). *Oecologia* **15,** 65–76.
Dickman, D. I., and Kozlowski, T. T. (1958). *Am. J. Bot.* **55,** 900–906.
Dillemann, G. (1958). *In* "Handbuch der Pflanzenphysiologie" (W. Ruhland, ed.), Vol. 8, pp. 1050–1075. Springer-Verlag, Berlin and New York.
Dochinger, L. S. (1963). *U. S., For. Serv., Res. Note CS* **CS-16.**
Donnelly, E. D., and Anthony, W. B. (1969). *Crop Sci.* **9,** 361–362.

Dormaar, J. F. (1970). *J. Soil Sci.* **21,** 105–110.
Drysdale, R. B., and Langcake, P. (1973). *In* "Fungal Pathogenicity and the Plant's Response" (R. J. W. Byrde and C. V. Cutting, eds.), pp. 423–433, Academic Press, New York.
Dunnill, P. M., and Fowden, L. (1965). *Phytochemistry* **4,** 445–451.
Dunnill, P. M., and Fowden, L. (1967). *Phytochemistry* **6,** 1659–1663.
Ebel, J. A., Ayers, A. R., and Albersheim, P. (1976). *Plant Physiol.* **57,** 775–779.
Eberhard, I. H., McNamara, J., Pearse, R. J., and Southwell, I. A. (1975). *Aust. J. Zool.* **23,** 169–179.
Ellis, B. E. (1971). *FEBS Lett.* **18,** 228–230.
Ellis, B. E. (1974). *Lloydia* **37,** 168–184.
Ellis, B. E. (1976). *Phytochemistry* **15,** 489–491.
Ellis, B. E., and Towers, G. H. N. (1970). *Phytochemistry* **9,** 1457–1461.
Eloff, J. N., and Grobbelaar, N. (1969). *Phytochemistry* **8,** 2201–2204.
Esau, K. (1963). *Am. J. Bot.* **50,** 495–506.
Esau, K. (1965). "Plant Anatomy," 2nd ed. Wiley, New York.
Estensen, R. D., Krey, A. K., and Hahn, F. E. (1968). *Fed. Proc., Fed. Am. Soc. Exp. Biol.* **27,** 713.
Evans, F. J., and Cowley, P. S. (1972a). *Phytochemistry* **11,** 2729–2733.
Evans, F. J., and Cowley, P. S. (1972b). *Phytochemistry* **11,** 2971–2975.
Evans, W. C., and Treagust, P. G. (1973). *Phytochemistry* **12,** 2505–2507.
Evans, W. C., Ghani, A., and Woolley, V. A. (1972). *Phytochemistry* **11,** 470–472.
Fairbairn, J. W. (1965). *In* "Beitrage zur Biochemie und Physiologie von Naturstoffen" (D. Gröger, H.-B. Schröter, and H. R. Schütte, eds.), pp. 113–119. Fischer, Jena.
Fairbairn, J. W., and Ali, A. A. E. R. (1968). *Phytochemistry* **7,** 1593–1597.
Fairbairn, J. W., and El-Masry, S. (1967). *Phytochemistry* **6,** 499–504.
Fairbairn, J. W., and El-Masry, S. (1968). *Phytochemistry* **7,** 181–187.
Fairbairn, J. W., and Suwal, P. N. (1961). *Phytochemistry* **1,** 38–46.
Fairbairn, J. W., Hakim, F., and El Kheir, Y. (1974). *Phytochemistry* **13,** 1133–1139.
Feeny, P. P. (1970). *Ecology* **51,** 565–581.
Feeny, P. P. (1976). *Recent Adv. Phytochem.* **10,** 1–40.
Ferguson, T. L., Wales, J. H., Sinnhuker, R. O., and Lee, D. J. (1976). *Food Cosmet. Toxicol.* **14,** 15–18.
Fish, F., and Waterman, P. G. (1971a). *J. Pharm. Pharmacol.* **23,** 67–68.
Fish, F., and Waterman, P. G. (1971b). *J. Pharm. Pharmacol.* **23,** Suppl., 132S–135S.
Fish, F., and Waterman, P. G. (1972). *Phytochemistry* **11,** 3007–3014.
Fish, F., Gray, A. I., and Waterman, P. G. (1975a). *Phytochemistry* **14,** 310–311.
Fish, F., Gray, A. I., Waterman, P. G., and Donachie, F. (1975b). *Lloydia* **38,** 268–270.
Fish, F., Meshal, I. A., and Waterman, P. G. (1976). *Planta Med.* **29,** 310–317.
Floss, H. G., Robbers, J. E., and Heinstein, P. F. (1974). *In* "Metabolism and Regulation of Secondary Plant Products" (U. C. Runeckles and E. E. Conn, eds.), pp. 141–178. Academic Press, New York.
Flück, H. (1963). *In* "Chemical Plant Taxonomy" (T. Swain, ed.), pp. 167–186. Academic Press, New York.
Forsyth, W. G. C., and Quesnel, V. C. (1963). *Adv. Enzymol.* **25,** 457–492.
Fowden, L. (1960). *Nature (London)* **186,** 897–898.
Fowden, L. (1963). *J. Exp. Bot.* **14,** 387–398.
Fowden, L. (1974). *In* "Metabolism and Regulation of Secondary Plant Products" (V. C. Runeckles and E. E. Conn, eds.), pp. 95–122. Academic Press, New York.
Fowden, L., and Bell, E. A. (1965). *Nature (London)* **206,** 110–112.

Fowden, L., and Bryant, M. (1959). *Biochem. J.* **71,** 210–217.
Fowden, L., and Frankton, J. B. (1968). *Phytochemistry* **7,** 1077–1086.
Fowden, L., Smith, I. K., and Dunnill, P. M. (1968). *In* "Recent Aspects of Nitrogen Metabolism in Plants" (E. J. Hewitt, ed.), pp. 165–177. Academic Press, New York.
Fox, L., and Macauley, B. J. (1977).
Frattali, V. (1969). *J. Biol. Chem.* **244,** 274–280.
Freeland, R. O. (1952). *Plant Physiol.* **27,** 685–690.
Friend, J., and Threlfall, D. R., eds. (1976). "Biochemical Aspects of Plant-Parasite Relationships." Academic Press, New York.
Friend, J., Reynolds, S. B., and Aveyard, M. A. (1973). *Physiol. Plant Pathol.* **3,** 495–507.
Gant, R. E., and Clebsch, E. E. C. (1975). *Ecology* **56,** 604–615.
Gäuman, E., Nuesch, J., and Rimpau, R. H. (1960). *Phytopathol. Z.* **38,** 274–308.
Geissman, T. A., and Dittmar, H. K. F. (1965). *Phytochemistry* **4,** 359–368.
Gellerman, J. L., Anderson, W. H., and Schlenk, H. (1976). *Phytochemistry* **15,** 1959–1961.
Gentry, C. E., Chapman, R. A., Henson, L., and Buckner, R. C. (1969). *Agron. J.* **61,** 313–316.
Glass, A. D. M., and Bohm, B. A. (1970). *Phytochemistry* **9,** 2197–2198.
Goldstein, J. L., and Swain, T. (1963). *Phytochemistry* **2,** 371–383.
Goodwin, T. W. (1958). *Biochem. J.* **68,** 503–511.
Goplen, B. P., Greenshields, J. E. R., and Baenziger, H. (1957). *Can. J. Bot.* **35,** 583–593.
Grant, W. D. (1976). *Science* **193,** 1137–1139.
Greathouse, G. A. (1939). *Plant Physiol.* **14,** 377–380.
Greathouse, G. A., and Rigler, N. E. (1940). *Phytopathology* **30,** 475–485.
Greathouse, G. A., and Watkins, G. M. (1938). *Am. J. Bot.* **25,** 743–748.
Green, T. R., and Ryan, C. A. (1972). *Science* **175,** 776–777.
Greshoff, M. (1909). *Kew Bull.* **23,** 397–418.
Grisebach, H., and Hahlbrock, K. (1974). *In* "Metabolism and Regulation of Secondary Plant Products" (V. C. Runeckles and E. E. Conn, eds.), pp. 21–52.
Gross, D. (1975). *Phytochemistry* **14,** 2105–2112.
Gustafsson, A., and Gadd, I. (1965). *Hereditas* **53,** 15–39.
Hall, F. R., Hollingworth, R. M., and Shankland, D. L. (1971). *Comp. Biochem. Physiol. B* **38,** 723–737.
Handley, W. R. C. (1954). *For. Comm. Bull.* **23.**
Hanover, J. W. (1965). *Phytochemistry* **5,** 713–717.
Harborne, J. B. (1964). *In* "Biochemistry of Phenolic Compounds" (J. B. Harborne, ed.), pp. 129–169. Academic Press, New York.
Harborne, J. B. (1973). *Phytochemistry* **2,** 344–380.
Harms-Ringdahl, M., Fedorcsák, I., and Ehrenberg, L. (1973). *Proc. Natl. Acad. Sci. U.S.A.,* **70,** 569–573.
Harris, H. B., Cummins, D. G., and Burns, R. E. (1970). *Agron. J.* **62,** 633–635.
Harrison, A. F. (1971). *Soil Biol. & Biochem.* **3,** 167–172.
Haskins, F. A., and Gorz, H. J. (1965). *Genetics* **51,** 733–738.
Haskins, F. A., and Kosuge, T. (1965). *Genetics* **52,** 1059–1068.
Haslam, E. (1966). "Chemistry of Vegetable Tannins." Academic Press, New York.
Haslam, E., and Tanner, R. J. N. (1970). *Phytochemistry* **9,** 2305–2309.
Haslam, E., Opie, T., and Porter, L. J. (1977). *Phytochemistry* **16,** 99–102.
Heftmann, E. (1975). *Phytochemistry* **14,** 891–901.
Heftmann, E. (1977). *Prog. Phytochem.* **4,** 257–276.
Heftmann, E., and Schwimmer, S. (1972). *Phytochemistry* **11,** 2783–2787.

Hegarty, M. P., and Peterson, P. J. (1973). *In* "Chemistry and Biochemistry of Herbage" (G. W. Butler and R. W. Bailey, eds.), Vol. 1, pp. 1–62. Academic Press, New York.
Hilditch, T. P., and Williams, P. N. (1964). "The Chemical Constitution of Natural Fats," 4th ed. Spottiswoode, Ballantyne, London.
Hill, D. C. (1973). *In* "Chronic Cassava Toxicity" (B. Nestel and R. MacIntyre, eds.), Monogr. IDRC-010e, pp. 105–111. Int. Dev. Res. Cent., Ottawa.
Hillis, W. E. (1956). *Aust. J. Biol. Sci.* **9,** 263–280.
Hillis, W. E. (1972). *Phytochemistry* **11,** 1207–1218.
Hillis, W. E., and Carle, A. (1960). *Biochem. J.* **74,** 607–615.
Hillis, W. E., and Carle, A. (1962). *Biochem. J.* **82,** 435–439.
Hillis, W. E., and Hasegawa, M. (1963). *Phytochemistry* **2,** 195–199.
Hillis, W. E., and Ishikura, N. (1969). *Aust. J. Biol. Sci.* **22,** 1425–1436.
Hills, K. L., Trautner, E. M., and Radwell, C. N. (1946). *Aust. j. sci.* **9,** 24–25.
Hopkinson, S. M. (1969). *Q. Rev., Chem. Soc.* **23,** 98–124.
Horiguchi, T., and Kitagishi, K. (1971). *Plant Cell Physiol.* **12,** 907–915.
Howard, I. K., Sage, H. J., and Horton, C. B. (1972). *Arch. Biochem. Biophys.* **149,** 323–326.
Howell, C. R., Bell, A. A., and Stipanovic, R. D. (1976). *Physiol. Plant Pathol.* **8,** 181–188.
Hufford, C. D., Funderburk, M. J., Morgan, J. M., and Robertson, L. W. (1975). *J. Pharm. Sci.* **64,** 789–792.
Hughes, D. W., and Genest, K. (1973). *Phytochemistry* **2,** 118–170.
Hughes, M. A. (1973). *In* "Chronic Cassava Toxicity" (B. Nestel and R. MacIntyre, eds.), Monogr. IDRC-010e, pp. 49–54. Int. Dev. Res. Cent., Ottawa.
Ikeda, T., Matsumura, F., and Benjamin, D. M. (1977). *Science* **197,** 497–499.
Ikenaka, T., Odani, S., and Koide, T. (1974). *Proteinase Inhibitors, Bayer Symp., 5th, 1973* pp. 325–343.
Ingham, J. L. (1972). *Bot. Rev.* **38,** 343–424.
Ishaaga, I., and Birk, Y. (1965). *J. Food Sci.* **30,** 118–120.
Ishizaka, N., and Tomiyama, K. (1972). *Plant Cell Physiol.* **13,** 1053–1063.
Iwasaki, T., Iguchi, I., Kiyohara, T., and Yoshikawa, M. (1974). *J. Biochem. (Tokyo)* **75,** 1387–1390.
Jaffé, W. G. (1973). *In* "Toxicants Occurring Naturally in Foods," pp. 106–129. Nat. Acad. Sci., Washington, D. C.
Jaffé, W. G., and Vega Lette, C. L. (1968). *J. Nutr.* **94,** 203–210.
James, W. O. *The Alkaloids* (R. H. Manske, ed.), Vol. 1, pp. 16–90. Academic Press, New York.
Janzen, D. H. (1969). *Evolution* **23,** 1–27.
Janzen, D. H. (1972). *Ecology* **53,** 885–892.
Janzen, D. H. (1973a). *Pure Appl. Chem.* **34,** 529–538.
Janzen, D. H. (1973b). *Am. Nat.* **107,** 786–790.
Janzen, D. H. (1974). *Biotropica* **6,** 69–103.
Janzen, D. H. (1975). "Ecology of Plants in the Tropics." Arnold, London.
Janzen, D. H. (1976a). *Ecology* **57,** 826–828.
Janzen, D. H. (1976b). *Biotropica* **8,** 110.
Jindra, A., Cihák, A., and Kovács, P. (1964). *Czech. Chem. Commun.* **29,** 1059–1064.
Johnson, R. D., and Waller, G. R. (1974). *Phytochemistry* **13,** 1493–1500.
Jonas, H. (1969). *Z. Pflanzenphysiol.* **60,** 359–369.
Jones, D. A. (1972). *In* "Phytochemical Ecology" (J. B. Harborne, ed.), pp. 103–122. Academic Press, New York.
Jorgensen, E. (1961). *Can. J. Bot.* **39,** 1765–1772.
Joslyn, M. A., and Goldstein, J. L. (1964). *J. Agric. Food Chem.* **12,** 511–520.

Kassell, B., and Laskowski, M. (1956). *J. Biol. Chem.* **219,** 203–210.
Keeler, R. F. (1975). *Lloydia* **38,** 56–86.
Kerharo, J. (1970). *J. Agric. Trop. Bot. Appl.* **17,** 353–368.
Kerharo, J., and Adam, J. G. (1974). "La pharmacopée sénégalaise traditionelle." Vigot, Paris.
King, M.-C., and Wilson, A. C. (1975). *Science* **188,** 107–116.
Kingsbury, J. M. (1964). "Poisonous Plants of the United States and Canada." Prentice-Hall, Englewood Cliffs, New Jersey.
Kircher, H. W. (1969). *Phytochemistry* **8,** 1481–1488.
Kirsi, M. (1973). *Physiol. Plant.* **29,** 141–144.
Kirsi, M. (1974). *Physiol. Plant.* **32,** 89–93.
Koeppe, D. E., Rohrbaugh, L. M., Rice, E. L., and Wender, S. H. (1970). *Phytochemistry* **9,** 297–301.
Kononova, M. M. (1961). "Soil Organic Matter" (Engl. ed., transl. by T. Z. Nowakowski and G. A. Greenwood). Pergamon, Oxford.
Kostecka-Madalska, O., Cetnarowska, B., and Bańkowski, Cz. (1967). *Acta Pol. Pharm.* **24,** 193–197.
Kozlowski, T. T. (1973). *In* "Shedding of Plant Parts" (T. T. Kozlowski, ed.), pp. 1–44. Academic Press, New York.
Kozlowski, T. T., and Winget, C. H. (1964). *Am. J. Bot.* **51,** 522–529.
Kramer, P. J., and Wetmore, T. H. (1943). *Am. J. Bot.* **30,** 428–431.
Krey, A. K., and Hahn, F. E. (1969). *Science* **166,** 775–757.
Kúc, J. (1972). *Annu. Rev. Phytopathol.* **10,** 207–232.
Kulman, H. M. (1971). *Annu. Rev. Entomol.* **16,** 289–324.
Kuwatsuka, S. (1970). *In* "Biochemical Toxicology of Insecticides" (R. D. O'Brien and I. Yamamoto, eds.), pp. 131–144. Academic Press, New York.
Langenheim, J. H. (1973). *In* "Tropical Forest Ecosystems in Africa and South America: A Comparative Review" (B. J. Meggers, E. S. Avensu, and W. D. Duckworth, eds.), pp. 89–104. Smithson. Inst. Press, Washington, D. C.
Lanier, L. G. (1967). *Proc. Congr. Int. Union For. Res. Organ., 14th, 19* . Sec. 24, Vol. 5, pp. 501–509.
Larson, P. R., and Gordon, J. C. (1969). *Am. J. Bot.* **56,** 1058–1066.
Lawton, J. H. (1976). *Bot. J. Linn. Soc.* **73,** 187–216.
Lea, P. J., and Fowden, L. (1972). *Phytochemistry* **11,** 2129–2138.
Lea, P. J., and Fowden, L. (1973). *Phytochemistry* **12,** 1903–1916.
Lea, P. J., and Norris, R. D. (1976). *Phytochemistry* **15,** 585–595.
Lee, H. J., and Waller, G. R. (1972). *Phytochemistry* **11,** 965–973.
Leete, E. (1967). *Annu. Rev. Plant Physiol.* **18,** 179–196.
Leete, E., and Bell, V. M. (1959). *J. Am. Chem. Soc.* **81,** 4358–4359.
Leete, E., and Chedekal, M. R. (1974). *Phytochemistry* **13,** 1853–1859.
Leifertova, I., and Buckova, B. (1969). *Cesk. Farm.* **18,** 172–176.
Levin, D. A. (1971). *Am. Nat.* **105,** 157–181.
Levin, D. A. (1976). *Annu. Rev. Ecol. Syst.* **7,** 121–159.
Lewis, A., and Starkey, R. L. (1968). *Soil Sci.* **106,** 241–247.
Liener, I. E. (1976). *Annu. Rev. Plant Physiol.* **27,** 291–319.
Link, K. P., Dickson, A. D., and Walker, J. C. (1929). *J. Biol. Chem.* **84,** 719–725.
Linzon, S. N. (1958). *For. Chron.* **34,** 50–56.
Lis, H., and Sharon, N. (1973). *Annu. Rev. Biochem.* **42,** 541–574.
Locci, R., and Kuć, J. (1967). *Phytopathology* **57,** 1272–1273.
Loomis, W. D., and Battaile, J. (1966). *Phytochemistry* **5,** 423–438.

Luckner, M. (1972). "Secondary Metabolism in Plants and Animals". Chapman & Hall, London.

Luckner, M., Nover, L., and Böhm, H. (1977). "Secondary Metabolism and Cell Differentiation. Springer-Verlag, Berlin and New York.

Lyr, H. (1962). *Nature (London)* **195,** 289–290.

McCarthy, T. J. (1968). *Planta Med.* **16,** 348–356.

McKey, D. (1974). *Am. Nat.* **108,** 305–320.

McKey, D., Waterman, P. J., Mbi, C. N., Gartlan, J. S., Struhsaker, T. T. (1978). *Science* **202,** 61–64.

MacLeod, N. J., and Pridham, J. B. (1966). *Phytochemistry* **5,** 777–781.

Maiti, P. C., Mookerjee, S., and Mathew, R. (1965). *J. Pharm. Sci.* **54,** 1826–1827.

Malawista, S. E., Sato, H., and Bensch, K. G. (1968). *Science* **160,** 770–772.

Malheiros-Garde, N. (1950). *Genet. Iber.* **2,** 29–38.

Mann, J. D., Steinhart, C. E., and Mudd, S. H. (1963). *J. Biol. Chem.* **238,** 676–681.

Marshall, J. K., ed. (1973). "The Below-ground Ecosystem: A Synthesis of Plant Associated Processes," Vol. 2. Dowden, Hutchinson, & Ross, Stroudsburg, Pennsylvania.

Marx, D. H. (1969a). *Phytopathology* **59,** 153–163.

Marx, D. H. (1969b). *Phytopathology* **59,** 411–417.

Marx, D. H. (1970). *Phytopathology* **60,** 1472–1473.

Marx, D. H., and Davey, C. B. (1969). *Phytopathology* **59,** 549–558.

Mattocks, A. R. (1972). *In* "Phytochemical Ecology" (J. B. Harborne, ed.), pp. 179–200. Academic Press, New York.

Mazanec, Z. (1966). *Aust. For.* **30,** 125–130.

Melin, E., and Krupa, S. (1971). *Physiol. Plant.* **25,** 337–340.

Menon, P. K. B. (1956). *Malay. For. Rec.* **19,** 000–000.

Metcalf, R. H. (1967). *Annu. Rev. Entomol.* **12,** 229–256.

Mialonier, G., Privat, J.-P., Monsigny, M., Kahlen, G., and Durand, R. (1973). *Physiol. Veg.* **11,** 519–537.

Miller, L. P. (1973). *Phytochemistry* **1,** 297–375.

Mockaitis, J. M., Kivilaan, A., and Schulze, A. (1973). *Biochem. Physiol. Pflanz.* **164,** 248–257.

Mooney, H. L., and Chu, C. (1974). *Oecologia* **14,** 275–306.

Mothes, K. (1955). *Annu. Rev. Plant Physiol.* **6,** 393–432.

Mothes, K. (1959). *J. Pharm. Pharmacol.* **11,** 193–210.

Mothes, K. (1976). *Recent Adv. Phytochem.* **10,** 385–405.

Múller, C. H. (1966). *Bull. Torrey Bot. Club* **93,** 332–351.

Müller, C. H., and Hauge, R. (1967). *Bull. Torrey Bot. Club* **94,** 182–191.

Nagatu, Y., and Burger, M. M. (1974). *J. Biol. Chem.* **249,** 3116–3122.

Nahrstedt, A. (1975). *Phytochemistry* **14,** 1339–1340.

Nartey, F. (1969). *Physiol. Plant.* **22,** 1085–1096.

Nartey, F. (1970). *Z. Pflanzenphysiol.* **62,** 398–400.

Nartey, F. (1973). *In* "Chronic Cassava Toxicity" (B. Nestel and R. MacIntyre, eds.), Monogr. IDRC-010e, pp. 73–87. Int. Dev. Res. Cent., Ottawa.

Nelson, C. E. (1953). *Agron. J.* **45,** 615–617.

Nessler, C. L., and Mahlberg, P. G. (1977). *Am. J. Bot.* **64,** 541–551.

Niemer, H., Bucherer, H., and Kohler, A. (1959). *Hoppe-Seyler's Z. Physiol. Chem.* **317,** 238–242.

Nixon, J. E., Eisele, T. A., Wales, J. H., and Sinnhuber, R. O. (1974). *Lipids* **9,** 314–321.

Norris, R. D., and Fowden, L. (1972). *Phytochemistry* **11,** 2921–2935.

Norris, R. D., and Fowden, L. (1973). *Phytochemistry* **12,** 2829–2841.

Nowacki, E. (1963). *Genet. Pol.* **4,** 161–202.
Nowacki, E., and Waller, G. R. (1973). *Z. Pflanzenphysiol.* **69,** 228–241.
Nowacki, E., Jurzysta, M., Gorski, P., Nowacka, D., and Waller, G. R. (1976). *Biochem. Physiol. Pflanz.* **169,** 231–240.
O'Brien, R. D. (1967). "Insecticides: Action and Metabolism." Academic Press, New York.
Oke, O. L. (1973). *In* "Chronic Cassava Toxicity" (B. Nestel and R. MacIntyre, eds.), Monogr. IDRC-010e, pp. 97–104. Int. Dev. Res. Cent., Ottawa.
Olmsted, J. B., and Borisy, G. G. (1973). *Annu. Rev. Biochem* **42,** 507–540.
Olney, H. O. (1968). *Plant Physiol.* **43,** 293–302.
Olsnes, S., and Pihl, A. (1977). *In* "The Specificity and Action of Animal, Bacterial and Plant Toxins" (P. Cuatrecasas, ed.), pp. 129–173. Chapman & Hall, London.
Olsnes, S., Refsnes, K., and Pihl, A. (1974). *Nature (London)* **249,** 627–631.
Olsnes, S., Sandvig, K., Refsnes, K., and Pihl, A. (1976). *J. Biol. Chem.* **251,** 3985–3992.
Onaka, F. (1950). *Bull. Kyoto Univ. For.* **18,** 55–95.
O'Neil, L. C. (1962). *Can. J. Bot.* **40,** 273–280.
Orf, J. H., Mies, D. W., and Hymowitz, T. (1977). *Bot. Gaz. (Chicago)* **138,** 255–260.
Orians, G. H., and Janzen, D. H. (1974). *Am. Nat.* **108,** 581–592.
Osborne, D. J. (1973). *In* "Shedding of Plant Parts" (T. T. Kozlowski, ed.), pp. 125–147. Academic Press, New York.
Otte, O. (1975). *Oecologia (Berl.)* **18:** 129–144.
Overland, L. (1966). *Am. J. Bot.* **53,** 423–432.
Pammel, L. H. (1911). "A Manual of Poisonous Plants." Torch Press, Cedar Rapids, Iowa.
Paris, M., Bouquet, A., and Paris, R.-R. (1969). *C. R. Hebd. Seances Acad. Sci., Ser. D* **268,** 2804–2806.
Park, J. Y. (1970). *Can. J. Microbiol.* **16,** 798–800.
Parsons, J., and Rothschild, M. (1964). *Entomol. Gaz.* **15,** 58–59.
Pease, V. A. (1917). *Am. J. Bot.* **4,** 145–160.
Pectu, P. (1964). *Pharmazie* **19,** 53–55.
Peters, R. (1954). *Endeavour* **13,** 147–154.
Phillipson, J. D. (1971). *Xenobiotica* **1,** 419–447.
Phillipson, J. D., and Handa, S. S. (1975a). *Phytochemistry* **14,** 999–1003.
Phillipson, J. D., and Handa, S. S. (1975b). *Phytochemistry* **14,** 2683–2690.
Phillipson, J. D., Tezcan, I., and Hylands, P. J. (1974). *Planta Med.* **25,** 301–309.
Pimentel, D., and Bellotti, A. C. (1976). *Am. Nat.* **110,** 877–888.
Piozzi, F., Fuganti, C., Mondelli, R., and Ceriotti, G. (1968). *Tetrahedron* **24,** 1119–1131.
Pöhm, M. (1957). *Abh. Otsch. Akad. Wiss. Berlin, Kl. Chem., Geol. Biol.* pp. 000–000.
Pöhm, M. (1959). *Monatsch. Chem.* **90,** 58–61.
Pöhm, M. (1966). *Abh. Dsch. Akad. Wiss. Berlin, Kl. Chem., Geol. Biol.* **3,** 251–254.
Ponomareva, V. V. (1969). "Theory of Podzolization." U. S. Dept. of Commerce, Washington, D. C.
Prakash, S., Misra, B. K., Adsulf, R. N., and Barat, G. K. (1977). *Biochem. Physiol. Pflanz.* **171,** 369–374.
Preston, W. H., Jr., Haun, J. R., Garvin, J. W., and Daum, R. J. (1964). *Econ. Bot.* **18,** 323–328.
Pridham, J. B. (1963). *In* "Enzyme Chemistry of Phenolic Compounds" (J. B. Pridham, ed.), pp. 73–80. Macmillan, New York.
Pridham, J. B. (1964). *Phytochemistry* **3,** 493–497.
Pusztai, A. (1972). *Planta* **107,** 121–129.
Rapport, D. J., and Turner, J. E. (1977). *Science* **195,** 367–373.
Rathwell, W. G. (1973). *Physiol. Plant Pathol.* **3,** 255–267.

Reddy, S. R. R., and Campbell, J. W. (1969). *Comp. Biochem. Physiol.* **28,** 515–534.
Reeks, W. A., and Barter, G. W. (1951). *For. Chron.* **27,** 140–156.
Rees, C. J. C. (1969). *Entomol. Exp. Appl.* **12,** 565–583.
Rehr, S. S. (1972). M. S. Thesis, Cornell University, Ithaca, New York.
Rehr, S. S., Bell, E. A., Janzen, D. H., and Feeny, P. P. (1973a). *Biochem. Syst. Ecol.* **1,** 63–67.
Rehr, S. S., Feeny, P. P., and Janzen, D. H. (1973b). *J. Anim. Ecol.* **42,** 405–416.
Rehr, S. S., Janzen, D. H., and Feeny, P. P. (1973c). *Science* **181,** 81–82.
Reisfeld, R. A., Borjeson, J., Chessin, L. N., and Small, P. A., Jr. (1967). *Proc. Natl. Acad. Sci. U.S.A.* **58,** 2020–2027.
Rhoades, D. H., and Cates, R. G. (1976). *Recent Adv. Phytochem.* **10,** 168–213.
Ribereau-Gayon, P. (1972). "Plant Phenolics." Oliver & Boyd, Edinburgh.
Richards, P. W. (1952). "The Tropical Rain Forest." Cambridge Univ. Press, London and New York.
Richardson, M. (1977). *Phytochemistry* **16,** 159–169.
Roberts, E. A. H. (1960). *Nature (London)* **185,** 536–537.
Roberts, E. A. H., Cartwright, R. A., and Wood, D. J. (1956). *J. Sci. Food Agric.* **7,** 637–646.
Roberts, M. F. (1974). *Phytochemistry* **13,** 1847–1851.
Robinson, M. E. (1930). *Biol. Rev. Cambridge Philos. Soc.* **5,** 126–141.
Robinson, T. (1974). *Science* **184,** 430–435.
Rockwood, L. L. (1973). *Ecology* **54,** 1363–1369.
Rodriguez, E., Towers, G. H. N., and Mitchell, J. C. (1976). *Phytochemistry* **15,** 1573–1580.
Roeske, C. N., Seiber, J. N., Brower, L. P., and Moffitt, C. M. (1976). *Recent Adv. Phytochem.* **10,** 92–167.
Rohde, R. A. (1972). *Annu. Rev. Phytopathol.* **10,** 233–252.
Romeo, J. T., Bacon, J. D., and Mabry, T. J. (1977). *Biochem. Syst. Ecol.* **5,** 117–120.
Rosenberg, H., and Stohs, S. J. (1974). *Phytochemistry* **13,** 1861–1863.
Rosenthal, G. A. (1970). *Plant Physiol.* **46,** 273–276.
Rosenthal, G. A. (1971). *Plant Physiol.* **47,** 209–211.
Rosenthal, G. A. (1972). *Plant Physiol.* **50,** 328–331.
Rosenthal, G. A. (1974). *J. Exp. Bot.* **25,** 609–613.
Rosenthal, G. A. (1977). *Q. Rev. Biol.* **52,** 155–178.
Rosenthal, G. A., Dahlman, D. L., and Janzen, D. H. (1976). *Science* **192,** 256–258.
Rosenthal, G. A., Janzen, D. H., and Dahlman, D. L. (1977). *Science* **196,** 658–660.
Rouge, P. (1974a). *C. R. Hebd. Seances Acad. Sci., Ser. D* **278,** 449–452.
Rouge, P. (1974b). *C. R. Hebd. Seances Acad. Sci., Ser. D* **278,** 3083–3087.
Roux, D. G. (1972). *Phytochemistry* **11,** 1219–1230.
Ryan, C. A. (1973). *Annu. Rev. Plant Physiol.* **24,** 173–196.
Ryan, C. A. (1974). *Proteinase Inhibitors, Bayer Symp., 5th, 1973* pp. 565–573.
Ryan, C. A., and Green, T. R. (1974). *In* "Metabolism and Regulation of Secondary Plant Products" (V. C. Runeckles and E. E. Conn, eds.), pp. 123–139. Academic Press, New York.
Ryan, C. A., and Huisman, O. C. (1967). *Nature (London)* **214,** 1047–1049.
Ryan, C. A., and Huisman, W. (1970). *Plant Physiol.* **45,** 484–489.
Sabalitschka, I., and Jungermann, C. (1925). *Biochem. Z.* **163,** 445–456.
Sainsbury, M., and Webb, B. (1972). *Phytochemistry* **11,** 2337–2339.
Sakato, K., Tanaka, H., and Misawa, M. (1975). *Eur. J. Biochem.* **55,** 211–219.
Sander, H. (1956). *Planta* **47,** 374–400.
Sander, H. (1963). *Planta Med.* **11,** 23–36.
Sawin, P. B., and Glick, D. (1943). *Proc. Natl. Acad. Sci. U.S.A.* **29,** 55–59.

Scheel, L. D., Perone, V. B., Larkin, R. L., and Kupel, R. E. (1963). *Biochemistry* **2,** 1127–1131.
Scheffer, T. (1957). *J. For.* **55,** 434–442.
Scheffer, T., and Cowling, E. B. (1966). *Annu. Rev. Phytopathol.* **4,** 147–170.
Schoonhoven, L. M. (1972). *Recent Adv. Phytochem.* **5,** 197–224.
Schratz, E., and Niewöhner, C. (1959). *Planta Med.* **7,** 137–170.
Schreiber, K., Hammer, U., Ithal, E., Ripperger, H., Rudolph, W., and Weissenborn, A. (1961). *Tagungsber., Dtsch. Akad. Landwirtschaftswiss. Berlin* **27,** 47–73.
Scott, A. I. (1975). *Recent Adv. Phytochem.* **9,** 189–241.
Shain, L. (1967). *Phytopathology* **57,** 1034–1045.
Shany, S., Gestetner, B., Birk, Y., and Bond, A. (1970). *J. Sci. Food Agric.* **21,** 508–510.
Sheen, S. J. (1969). *Phytochemistry* **8,** 1839–1847.
Shiroya, T., Lister, G. R., Nelson, C. D., and Krotkov, G. (1961). *Can. J. Bot.* **39,** 855–864.
Shumway, L. K., Rancour, J. M., and Ryan, C. A. (1970). *Planta* **93,** 1–14.
Shumway, L. K., Cheng, V., and Ryan, C. A. (1972). *Planta* **106,** 279–290.
Shumway, L. K., Yang, V. V., and Ryan, C. A. (1976). *Planta* **129,** 161–165.
Siegler, D., and Price, P. W. (1976). *Am. Nat.* **110,** 101–105.
Silver, G. T. (1962). *For. Chron.* **38,** 433–438.
Simons, A. B., and Marten, G. C. (1971). *Agron. J.* **63,** 915–919.
Sinden, S. L., Goth, R. W., and O'Brien, M. J. (1973). *Phytopathology* **63,** 303–307.
Singh, P. (1970). *Indian J. Pharm.* **32,** 94–95.
Singleton, V. L., and Kratzer, F. H. (1969). *J. Agric. Food Chem.* **17,** 497–512.
Singleton, V. L., and Kratzer, F. H. (1973). *In* "Toxicants Occurring Naturally in Foods," pp. 309–345. Natl. Acad. Sci., Washington, D. C.
Skursky, L., and Waller, G. R. (1972). *Biochem. Physiol. Alkaloide, Int. Symp., 4th, 1969* pp. 181–186.
Skursky, L., Burleson, D., and Waller, G. R. (1969). *J. Biol. Chem.* **244,** 3238–3242.
Smith, B. D. (1966). *Nature (London)* **212,** 213–214.
Smith, H. (1973). *In* "Biosynthesis and Its Control in Plants" (B. V. Millborrow, ed.), pp. 303–321. Academic Press, New York.
Smith, H. H. (1965). *Am. Nat.* **99,** 73–79.
Smith, I. K., and Fowden, L. (1966). *J. Exp. Bot.* **17,** 750–761.
Smith, J. N. (1955). *Biol. Rev. Cambridge Philos. Soc.* **30,** 455–475.
Smith, J. N. (1964). *Comp. Biochem.* **6,** 403–457.
Smith, R. L., and Williams, R. T. (1966). *In* "Glucuronic Acid" (G. J. Dutton, ed.), pp. 457–491. Academic Press, New York.
Sokolov, V. S. (1959). *Symp. Soc. Exp. Biol.* **13,** 230–257.
Soloman, M. J., and Crane, F. A. (1970). *J. Pharm. Sci.* **59,** 1670–1672.
Solomos, T. (1977). *Annu. Rev. Plant Physiol.* **28,** 279–297.
Stanton, N. (1975). *Biotropica* **7,** 8–11.
Stark, R. W. (1965). *Annu. Rev. Entomol.* **10,** 303–324.
Stitt, R. E. (1943). *J. Am. Soc. Agron.* **35,** 944–954.
Stitt, R. E., Hyland, H. L., and McKee, R. L. (1946). *J. Am. Soc. Agron.* **38,** 1003–1009.
Stoessl, A. (1970). *Recent Adv. Phytochem.* **3,** 143–180.
Stoessl, A., Stothers, J. B., and Ward, E. W. B. (1976). *Phytochemistry* **15,** 855–872.
Swain, T. (1965). *In* "Plant Biochemistry" (J. Bonner and J. E. Varner, eds.), 2nd ed., pp. 552–580. Academic Press, New York.
Swain, T. (1974). *Compr. Biochem. A* **29,** 125–302.
Swain, T. (1977). *Annu. Rev. Plant Physiol.* **28,** 479–501.
Taber, W. A., Vining, L. C., and Heacock, R. A. (1963a). *Phytochemistry* **2,** 65–70.

Taber, W. A., Heacock, R. A., and Mahon, M. E. (1963b). *Phytochemistry* **2,** 99–101.
Takeda, K. (1972). *Prog. Phytochem.* **3,** 287–333.
Thomas, B. R. (1970). *In* "Phytochemical Phylogeny" (J. B. Harborne, ed.), pp. 59–79. Academic Press, New York.
Tokuyama, H. (1973). *Biochim. Biophys. Acta* **317,** 338–350.
Tomczyk, H. (1964). *Diss. Pharm.* **16,** 297–308.
Towers, G. H. N. (1964). *In* "Biochemistry of Phenolic Compounds" (J. B. Harborne, ed.), pp. 249–294. Academic Press, New York.
Tso, T. C., and Jeffrey, R. N. (1961). *Arch. Biochem. Biophys.* **92,** 253–256.
Tur-Sinai, A., Birk, Y., Gertler, A., and Rigbi, M. (1972). *Biochim. Biophys. Acta* **147,** 402–404.
Van Emden, H. F., and Bashford, M. A. (1969). *Entomol. Exp. Appl.* **12,** 351–364.
Van Sumere, C. F., Albrecht, J., Dedonder, A., de Pooter, H., and Pé, I. (1975). *In* "The Chemistry and Biochemistry of Plant Proteins" (J. B. Harborne and G. F. Van Sumere, eds.), pp. 211–264. Academic Press, New York.
Walker-Simmons, M., and Ryan, C. A. (1977). *Plant Physiol.* **60,** 61–63.
Wallebroek, J. C. J. (1940). *Recl. Trav. Bot. Neerl.* **37,** 78–132.
Ward, E. W. B., Unwin, C. H., Hill, J., and Stoessl, A. (1975). *Phytopathology* **65,** 859–863.
Wardlaw, I. F. (1968). *Bot. Rev.* **34,** 79–105.
Wardrop, A. B., and Cronshaw, J. (1962). *Nature (London)* **193,** 90–92.
Waring, M. J. (1970). *J. Mol. Biol.* **54,** 247–279.
Waring, M. J., and Chisholm, J. W. (1972). *Biochim. Biophys. Acta* **262,** 18–23.
Waterman, P. G. (1976). *Phytochemistry* **15,** 578–579.
Waterman, P. G., McKey, D., and Gartlan, J. S. (1979). In preparation.
Waxdal, M. (1974). *Biochemistry* **13,** 3671–3677.
Weiel, J., and Hapner, K. D. (1976). *Phytochemistry* **15,** 1885–1887.
Werker, E., and Fahn, A. (1969). *Bot. J. Linn. Soc.* **62,** 379–411.
Westcott, R. J., and Henshaw, G. G. (1976). *Planta* **131,** 67–73.
White, H. A., and Spencer, M. (1964). *Can. J. Bot.* **42,** 1481–1484.
Whiteside, J. A., and Thurman, D. A. (1971). *Planta* **98,** 279–284.
Whittaker, R. H., and Feeny, P. P. (1971). *Science* **171,** 757–770.
Williams, R. T. (1964). *In* "Biochemistry of Phenolic Compounds" (J. B. Harborne, ed.), pp. 205–248. Academic Press, New York.
Willuhn, G. (1966). *Biochem. Physiol. Alkaloide, Int. Symp., 3rd, 1965* No. 3, pp. 97–103.
Willuhn, G. (1967). *Planta Med.* **15,** 58–73.
Wilson, C. M. (1955). *Agron. J.* **47,** 83–86.
Wilson, L., Bamburg, J. B., Mizel, S. B., Grisham, L. M., and Creswell, K. M. (1974). *Fed. Proc., Fed. Am. Soc. Exp. Biol.* **33,** 158–166.
Wilson, L. F. (1966). *J. Econ. Entomol.* **59,** 1043–1049.
Wolf, M. J., and Duggar, B. M. (1946). *J. Agric. Res.* **73,** 1–32.
Wong, P. P., Kuo, T., Ryan, C. A., and Kado, C. I. (1976). *Plant Physiol.* **57,** 214–217.
Woolhouse, H. W. (1967). *Symp. Soc. Exp. Biol.* **21,** 179–213.
Xavier Filho, J., and Ainouz, I. L. (1977). *Biol. Plant.* **19,** 183–189.
Youle, R. J., and Huang, A. H. C. (1976). *Plant Physiol.* **58,** 703–709.
Zaitschek, D. V., and Segal, R. (1972). *Lloydia* **35,** 192.

Chapter **3**

Mechanism of Plant Avoidance of Autotoxicity by Secondary Metabolites, Especially by Nonprotein Amino Acids

L. FOWDEN *And* **P. J. LEA**

I. INTRODUCTION AND GENERAL CONSIDERATION OF AVOIDANCE OF AUTOTOXICITY

Other authors contributing to this treatise have provided detailed accounts of the secondary products encountered in plants. For the present, it is sufficient to record that these include phenolic compounds such as the flavonoids and related compounds; a diverse range of other aromatic compounds derived by either the shikimate or the polyacetate biogenetic pathway; the many types of terpenoid products arising by isoprene-type condensations, including the steroid group; the alkaloids,

ISBN 0-12-597180-X

now numbering approximately 10,000 known compounds; the cardiac glycosides and the cyanogenetic glucosides; and the group of nonprotein amino and imino acids. Many of these secondary products exhibit physiological activity against organisms other than the producer plant. The most striking behavioral responses are observed in animals, and accounts of the toxicity of certain alkaloids and glycosides can be found in the oldest written records of mankind. Chemical antagonisms, however, can involve other forms of living organisms, and there are many demonstrations of toxic or inhibitory effects produced by higher plant secondary compounds on microorganisms or green plants. Yet against this background it would seem that inherently toxic products synthesized by particular plants do not inhibit the growth, or otherwise interfere with metabolic processes, in the producer plants. The possible reasons for such avoidance of autotoxicity form the basis of this chapter.

Clearly, plants must have adjusted their physiology and biochemistry to accommodate the presence of such potentially toxic substances within their cells and tissues. Some compounds, including many of the alkaloids, exert their physiological action in animals by interfering with brain and nerve function and, since plants do not have counterparts governing their behavior, there is no *a priori* reason to expect that such compounds should antagonize plant functions. More generally, however, plants have adapted to the presence of toxic compounds in more subtle ways and have used a variety of physical, chemical, and biochemical mechanisms to achieve this purpose. An obvious mechanism for avoiding autotoxicity is the physical removal by the producer plant of noxious products from cellular sites and enzyme systems where antagonism might otherwise occur. This can be achieved by rendering toxic substances insoluble, as is the case in a number of plants producing large amounts of oxalic acid; in those instances the material is present largely as crystals (raphides) of calcium oxalate, which are identifiable microscopically within the plant cells. By the polymerization of small molecules to give macromolecular tannins, lignins, rubber, and similar compounds, considerable quantitites of material can be stored innocuously, partly because the giant molecules are sparingly soluble and so have little osmotic or metabolic activity and also because they frequently are deposited outside the living cells of the plant. Less commonly, significant quantities of secondary products can be discharged by the plant. Volatilization of certain essential oils (terpenes) from plant surfaces is well established, whereas other materials may be leached from leaf laminae by rain or may either be actively exuded from living roots or enter the soil following the decay of cells sloughed from root surfaces. The method for physically isolating problematic compounds depends on the subcellular compartmentalization of plant cells. The organellar structure of cells is a dominant feature of their physiology and biochemistry, and a wealth of information now exists concerning the metabolic activities and interdependence of the various organelles. In many mature plant cells, the vacuole occupies the largest fraction of the cell volume and is known to be the subcellular site to which many secondary compounds are transported and in which they are stored in an almost inert metabolic state, enzymes generally being located elsewhere in the cell. Occasionally, the amount of a particular secondary compound produced by plants is so large and the percentage of cell volume occupied by the vacuole is so high, that there is a clear physical necessity for major proportions of the compound to be accumulated in the vacuole. Direct evidence in support of vacuolar accumulation of plant products is of two principal types. Certain types of compound, such as the alkaloids, can be located within cells in tissue sections by specific histochemical staining techniques, and microscopic examination clearly reveals a preponderance of material in the vacuole. An adaptation of this technique, applicable in some situations, uses radioisotopically labeled compounds in association with microautoradiographic location techniques. More recently, new techniques for handling cells in ways that permit the isolation of intact subcellular organelles, including vacuoles, have been developed; separation of the different types of organelles released from cells involves their differential sedimentation in density gradients. For example, chemical assay for cyanogenetic glucosides indicates the organelle in which the material is principally stored. An example of this approach is the recent experiment of Saunders and Conn (1977) showing unequivocally that the cyanogenetic glucoside of *Sorghum bicolor* is present predominantly in cell vacuoles.

The toxic effect resulting from the ingestion of some secondary products by animals is indirect; the animal modifies the original compound by the action of its own enzymes and so generates the toxin indirectly. Examples include the cyanogenetic glucosides, which are hydrolyzed rapidly by animal β-glucosidases to products including cyanide, and also the amino acid hypoglycin A, which causes hypoglycemia in human beings and animals only after oxidative deamination and decarboxylation (see later).

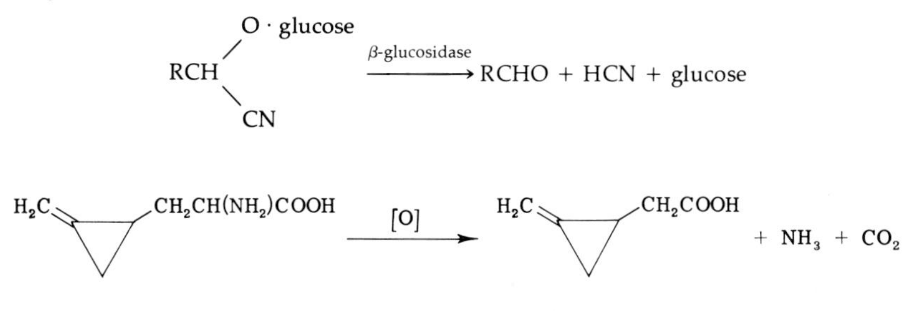

The producer plant may resist its own toxic products by failing to effect such lethal metabolic conversions. With the cyanogenetic glucosides, hydrolysis seems not to occur in intact plant tissues (or occurs only at a slow rate compatible with other processes operating to maintain nontoxic intracellular concentrations of cyanide), but cyanide production can be demonstrated clearly when tissues are crushed. This indicates strongly that in intact cells the separation of cyanogenetic glucosides and the hydrolytic β-glucosidases is strictly regulated by subcellular fine structural elements. More generally, cyanogenetic glucosides present an example of a process of glycosylation encountered more widely among secondary products, for conjugation with sugar moieties seems to be a further means adopted by plants for the innocuous storage of materials that could otherwise antagonize aspects of cell metabolism.

Most forms of toxic action involve some type of metabolic antagonism normally involving enzyme systems. In the simplest cases of antimetabolic action, the toxic compound acts either as an alternative substrate or as an inhibitor of enzymes catalyzing basic cellular processes. As described above, separation at the subcellular level provides one means of avoiding interference with enzyme action, but such physical mechanisms may be "leaky" under certain physiological conditions or environmental stresses. Therefore, it is interesting to find situations in which plants have developed more sophisticated protective mechanisms involving altered amino acid sequences in proteins to produce enzymes with altered affinities (i.e., little or none) for the toxic materials as substrates or inhibitors. This concept of avoidance of autotoxicity is developed below and is illustrated by examples drawn from the group of toxic nonprotein amino acids.

There is little current information to indicate the extent to which plants regulate the synthesis and expression of activity of enzymes involved in the metabolism of many secondary products. These features of enzyme behavior are subject to delicate regulatory control at many points in basic metabolic processes, and the activity of sequential enzymes in a pathway frequently appears to be coordinated. This ensures that no excessive buildup of intermediates that might otherwise present problems of toxicity occurs in cells. An example is found in the stepwise process of assimilation of nitrate nitrogen into protein nitrogen, in which the activity of the first enzyme, nitrate reductase, seems to set the pace of the assimilatory process. The functional activities of later enzymes, especially nitrite reductase and glutamine synthetase, are normally sufficient to maintain cellular concentrations of nitrite and ammonia at nontoxic levels. Many secondary products are end products of biosynthetic pathways, and therefore identical considerations do not apply, but their accumulation in vacuoles has a counterpart in the deposition of protein as discrete bodies and of lipids as globules in storage tissues of plants.

II. MECHANISMS FOR AVOIDANCE OF AUTOTOXICITY WITH SPECIAL REFERENCE TO TOXIC AMINO ACIDS

A. Nonprotein Amino Acids

Plants synthesize the 20 amino acids required as components of proteins; in addition, some produce other types of amino acids that occur unbound or combined only as γ-glutamyl peptides. The numbers of such amino acids isolated and identified from members of the plant kingdom increase year by year as more species are analyzed systematically, and at present there are about 300 known examples of nonprotein amino acids. Many novel chemical structures are found among this group, and some of the compounds exhibit strong metabolic interaction (toxicity) when applied to other living species. Such toxicities often appear to be caused by a close similarity in chemical structure (molecular shape and size) with a particular protein amino acid: The compound is isosteric with the essential metabolite. In these circumstances, it is not unreasonable to expect that the nonprotein amino acid shows toxic behavior because it antagonizes certain basic intermediary reactions by mimicking the substrate action of the normal amino acid reactant or by acting as a competitive or end product inhibitor; it is then termed an amino acid analog.

The simplest type of analog involves compounds homologous with protein amino acids. The addition or loss of CH_2 from a molecule, although altering its dimensions slightly, often produces only a very small change in general shape and chemical reactivity, especially within larger molecules. Substitution of a particular atom or group of atoms by others of similar size may also lead to analog molecules. Such substitutions are seen in canavanine (O for CH_2 in arginine), albizziine (NH for CH_2 in glutamine), and *S*-aminoethylcysteine (S for CH_2 in lysine). The structural distribution and toxic effects of naturally occurring nonprotein amino acids are discussed by Rosenthal and Bell (Chapter 9).

Often the noxious action of a particular plant species was recorded long before the underlying chemical nature of the toxic compounds was established. For example, the poisoning of cattle and horses after grazing herbage containing certain *Astragalus* species was recorded a century ago in the American Midwest, and similar old records tell how ingestion of the foliage of the tropical legume *Leucaena leucocephala* by horses led to a loss of the tail hairs. Poisoning of human beings, especially children, can result when unripe fruits of akee (*Blighia sapida*) are eaten. Akee features importantly in the diet of the Jamaican people, and the poisoning is often referred to as Jamaican vomiting sickness. Now an amino acid is known to be responsible for each of these toxic situations, but in all cases the underlying antagonism is more complex than direct interference with the metabolism of a related protein amino acid.

1. *Indirect Metabolic Antagonisms*

Mimosine **(1)**, a toxic constituent of *Leucaena*, is degraded to dihydroxypyridine **(2)** by the microflora present in the digestive tract of animals, and there are differences among species in the ease with which this process occurs. Over a period of months, dihydroxypyridine production by animals ingesting mimosine produces symptoms of goiter, but the microfloral degradation appears to protect the herbivore against the direct and seemingly more damaging toxic effects of mimosine itself (Vietmeyer and Cotton, 1977). Since *Leucaena* seedlings possess an enzyme system that degrades mimosine to dihydroxypyridine, pyruvate, and ammonia (Smith and Fowden, 1966), the assumption must be made that **2** antagonizes processes of animal metabolism that have no counterpart in plants.

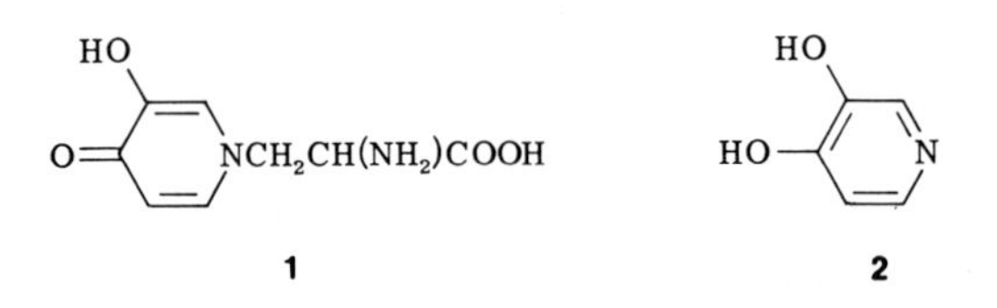

The hazardous component of unripe akee fruits was shown to be β-(methylenecyclopropyl)alanine, hypoglycin A **(3)** (Ellington *et al.*, 1959). The structure of hypoglycin A **(3)** shows some relationship that of leucine **(4)** and/or isoleucine **(5)**, but antagonism of amino acid or protein biosynthetic systems does not seem to provide an explanation of the toxicity of the amino acid to mammals. The metabolic action is more indirect because, after ingestion, hypoglycin A undergoes oxidative deamination and decarboxylation to give rise to α-(methylenecyclopropyl)acetic acid, which strongly inhibits the β-oxidation steps of fatty acid utilization (see Corredor, 1975). The animal is then compelled to rely heavily on available carbohydrate as an immediate respiratory substrate, with the result that blood glucose levels fall dramatically, producing vomiting, convulsions, or death in severe cases. Respiration and energy supply in maturing fruits are not likely to depend on the oxidative degradation of fatty acids, for lipids generally are synthesized and stored, rather than catabolized, during the development of these plant organs.

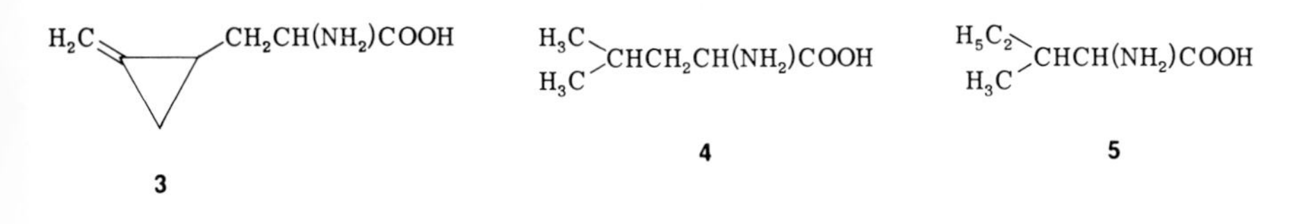

2. *Direct Analog Actions*

There are now several well-documented examples of nonprotein amino acids that interfere with the growth and development of foreign organisms by reason of their direct antagonism of aspects of basic amino acid and protein metabolism. In this section, only compounds used later to illustrate the metabolic basis of antagonism are described.

The adverse effects produced by an antagonistic molecule frequently can be reduced or completely reversed by providing, simultaneously with the analog, enhanced concentrations of the structurally related protein amino acid. This situation is clearly demonstrated in the case of azetidine-2-carboxylic acid **(6)** inhibition of the growth of *Phaseolus aureus* seedlings, which is mitigated progressively when increasing amounts of proline **(7)** are supplied together with the analog (Fowden, 1963).

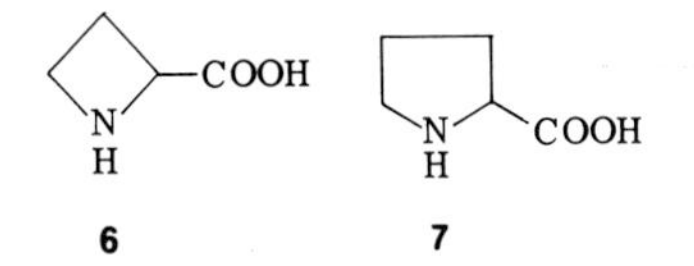

A large number of nonprotein amino acids possessing antimetabolic activity have been derived from legume species. Of particular interest as analog molecules are canavanine **(8)** and indospicine **(9)**, which are isosteres of arginine **(10)**. Albizziine **(11)**, a lower homolog of citrulline, displays a structural resemblance to glutamine **(12)** as well as similar metabolic antagonism. It occurs as a constituent of species from the *Acacia* and *Albizzia* genera (Seneviratne and Fowden, 1968).

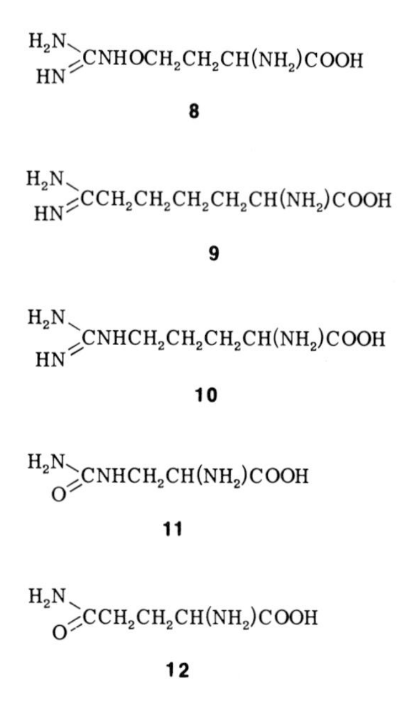

The families Sapindaceae (which includes *Blighia sapida*), Hippocastanaceae, and Aceraceae produce a related group of amino acids containing either a branched carbon skeleton or a cyclopropyl ring system. Hypoglycin A is a member of this group of compounds; it is a component of selected species from each of the three families. A different C_7 amino acid, 2-amino-4-methylhex-4-enoic acid **(13)** is the major soluble nitrogenous constituent of seeds of *Aesculus californica* (Fowden and Smith, 1968) and acts in an interesting way as an analog of phenylalanine (Anderson and Fowden, 1970). The spatial conformation about C-4 is planar, so that the terminal hydrophobic part of the molecule resembles an incomplete phenyl ring.

The final illustrative analogs are *S*-aminoethylcysteine (AEC) **(14)** and a group of γ-substituted glutamic acids. The former was mentioned previously as being an isostere of lysine **(15)** and is sometimes given the trivial name thialysine. The amino acid is synthesized by the fungus *Rozites caperata* (Warin *et al.*, 1969), although it has never been isolated from higher plants, which are known to produce a wide range of S-substituted cysteines. γ-Substituted glutamic acids are found principally in species of the Leguminosae and Liliaceae, but some ferns also produce these types of amino acid. Two compounds that are interesting as analog molecules are *erythro*-γ-methyl-L-glutamic acid **(16a)** and *threo*-γ-hydroxy-L-glutamic acid **(16b)** (Lea and Fowden, 1972).

$$\begin{array}{l} H_3CCH \\ \quad\quad \searrow CCH_2CH(NH_2)COOH \\ H_3C \nearrow \end{array}$$

13

$H_2NCH_2CH_2SCH_2CH(NH_2)COOH$

14

$H_2NCH_2CH_2CH_2CH_2CH(NH_2)COOH$

15

$HOOCCHRCH_2CH(NH_2)COOH$

16a R = CH_3
16b R = OH

B. Metabolic Targets of Direct Amino Acid Antagonists

Studies in the past 10–15 years have revealed the principal metabolic processes that are subject to direct interference by amino acid analogs and, in consequence, those that must be protected from the adverse effects of analogs in "producer" species. Three main activities are involved:

1. Amino acid uptake and transport processes at the intra- and intercellular levels. Permease enzyme systems are known to facilitate the uptake of amino acids across cell membranes. Detailed descriptions exist of the properties of certain microbial permeases, but those present in higher plants are less well defined, especially in terms of substrate specificity.

2. Amino acid biosynthesis. The stepwise pathways of synthesis of the majority of the 20 protein amino acids are now adequately characterized for higher plant systems, and information is increasing rapidly concerning the detailed ways in which the activities of the enzymes catalyzing each step may be influenced by substrate, product, and allied (analog) molecules.

3. Activation and incorporation of amino acids into protein molecules. Protein synthesis is initiated by an activation of amino acids involving their transfer to specific transfer RNA molecules catalyzed by 20 specific aminoacyl-tRNA synthetase enzymes. Although messenger RNA and chain initiation and termination factors determine the nature of the polypeptide/protein product, antagonism caused by amino acid analogs during the activation reaction can diminish the rate of protein synthesis or lead to the formation of anomalous protein molecules.

The metabolic options available to plants to overcome these types of antagonism form the basis of the following sections.

1. *Altered Permease Specificity*

Many analog amino acids permeate cell membranes readily and, where detailed studies have been made, this movement seems to be effected by the permease enzyme responsible for conveying the normal structurally related protein amino acid. Differences in the specificities of permeases then might represent one contributory mechanism whereby "producer" plants can prevent, or greatly reduce, the access of analog molecules to metabolically sensitive areas of cells. Altered permease specificities have been encountered in some mutant lines of bacterial cells showing resistance to the usual growth inhibitory effects produced by analog molecules. Tristram and Neale (1968) provided an example of this using strains of *Escherichia coli* C4, whose proline-specific transport mechanism is well documented. The permease facilitates the entry of a number of proline analog, including azetidine-2-carboxylic acid, into the cells. Several classes of mutants selected for resistance to the analog failed to incorporate azetidine-2-carboxylate into cells and thus avoided adverse effects resulting from its ultimate introduction into protein molecules. At least one of these resistant strains possessed an altered permease showing little affinity for azetidine-2-carboxylate but still capable of transporting proline normally.

There are few comparable studies of the permeases of higher plants, but those that exist generally are confined to the cellular uptake of the protein amino acids, often by young root tips or storage tissue discs. Characterizations of the permeases involved tend to be less precise, and perhaps uptake processes are less specific in terms of the range of amino acids acceptable as permease substrates. Even with these limitations, there is evidence suggesting that competition occurs between analog and normal amino acids for permease sites and that the antagonistic effects of analogs may be less in "producer" species than in others. An example is found in the study of amino acid uptake by seedling root tips of *Cucumis melo* and *Caesalpinia tinctoria*. The latter legume species produces two unusual amino acids, 3-hydroxymethylphenylalanine **(17a)** and 4-hydroxy-3-hydroxymethylphenylalanine **(17b)**. These two aromatic amino acids inhibited phenylalanine uptake more strongly in root tips of *Cucumis* (a "nonproducer" of the analogs) than in those of *Caesalpinia* (Watson and Fowden, 1975).

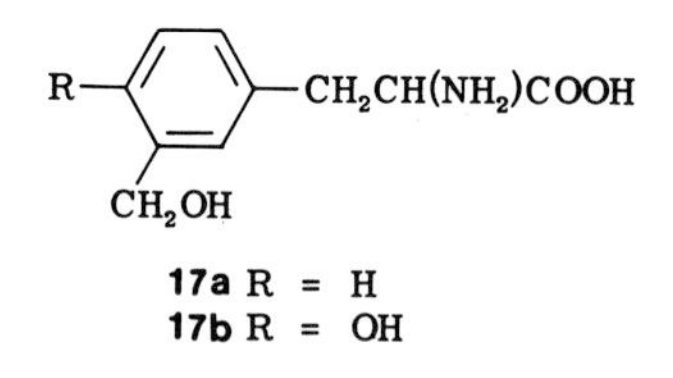

17a R = H
17b R = OH

2. *Amino Acid Biosynthesis*

a. Lysine Biosynthesis. It is now well established that the only mechanism of regulation of amino acid biosynthesis in higher plants is via end product inhibition (Miflin, 1977; Miflin and Lea, 1977). The end product amino acid is able to inhibit the first enzyme involved in its biosynthesis. If an amino acid is formed by a branched-chain pathway, the amino acid is also able to inhibit the first enzyme unique to its own biosynthesis. A good example can be seen in lysine biosynthesis, in which lysine is able to inhibit aspartate kinase and dihydrodipicolinate synthase. A complete discussion of the regulation of the amino acids formed from aspartate can be found in Bright *et al.* (1978a) and Miflin *et al.* (1978).

If an amino acid analog is able to bind to the regulatory site of an enzyme in place of the end product amino acid, the synthesis of the protein amino acid will be reduced. Thus, ultimately the cell will be starved of a particular amino acid, and the possibility that the analog will be incorporated into protein is thereby increased.

Lysine is the nutritionally limiting amino acid in the protein of cereal seeds eaten by monogastric animals, and a program of work at Rothamsted seeks an understanding of the regulation of lysine biosynthesis. In these studies the fungal product AEC **(14)** has been employed since it behaves as a close analog of lysine in viral (Scioscia-Santoro *et al.*, 1975), bacterial (Stern and Mehler, 1965), and mammalian systems (DeMarco *et al.*, 1976). *S*-Aminoethylcysteine (2 m*M*) kills normal rice callus cell cultures, but, if mutagen-treated cultures are exposed to it, a very small number of cells are able to grow slowly (Chaleff and Carlson, 1975). Analysis of these resistant cells showed that the levels of lysine, isoleucine, leucine, and valine in the free amino acid pool and in protein had increased. Unfortunately, the mechanism of resistance has not been further studied, and it has not been possible to regenerate the cultured cells to whole plants.

The growth inhibitory action of AEC on barley plantlets is shown in Fig. 1. Lysine is able to reverse this effect (Fig. 2), but at concentrations above 1 m*M* lysine itself becomes inhibitory (Bright *et al.*, 1978a; S. W. J. Bright, unpublished results). Mutant barley lines that are resistant to AEC have been selected, the plants have been grown to maturity, and the characteristics of the progeny (which retain their resistance) are now being studied.

The mechanism of resistance to AEC in the fungus has not been investigated but may well be different from that in barley plants, since lysine is synthesized via 2-aminoadipic acid and saccharopine (Broquist, 1971). Although higher plant aspartate kinase is inhibited by AEC, concentrations between 2 and 6 m*M* are required for 50% inhibition (Shewry and Miflin, 1977; Davies, 1977; Bright *et al.*, 1978b). Dihydrodipicolinate synthase from wheat germ, however, is far more sensitive to AEC, 0.07 m*M* being required to give 50% inhibition (M. Mazelis,

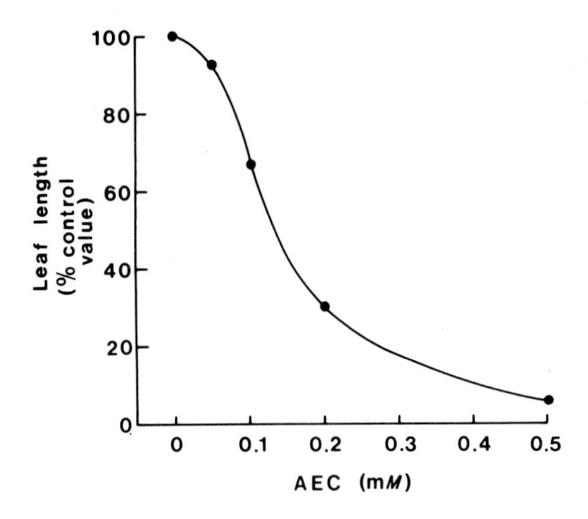

Figure 1. Growth inhibition of barley plantlets by *S*-(2-aminoethyl)cysteine.

unpublished results; see Mazelis *et al.*, 1977). Studies on these two enzymes in the AEC-resistant barley strains will be carried out to determine whether there are any alterations in the regulatory properties.

It is also hoped that AEC-resistant mutants will have higher levels of lysine, which may improve the nutritional quality of cereals. The possi-

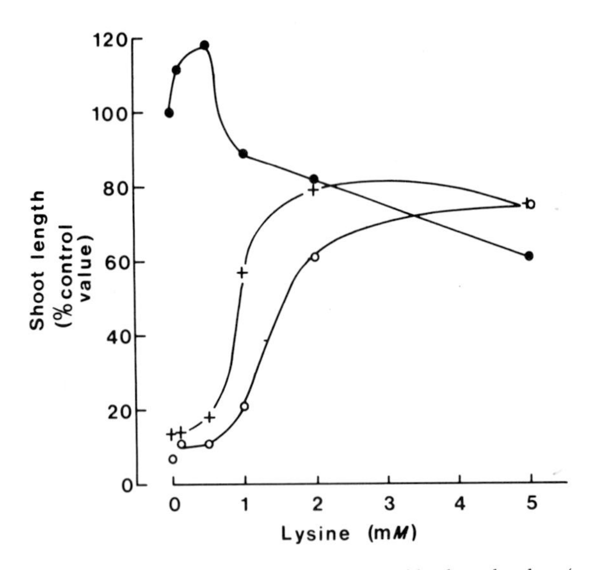

Figure 2. Reversal by lysine of the growth inhibition of barley plantlets (+—+, 0.3 m*M* AEC; ○—○, 0.5 m*M* AEC; ●—●, lysine alone).

bility that AEC may also interfere with the incorporation of lysine into protein cannot be excluded (Davies, 1977), and mutants having lysyl-tRNA synthetase with altered active sites (similar to the situations described in Section II,B,3) may also exist. Mutants that could discriminate against AEC at the level of protein synthesis or by possession of an altered permease (Section II,B,1) would not necessarily have increased levels of lysine in the cell.

b. Tryptophan Biosynthesis. The biosynthesis of tryptophan in higher plants is apparently controlled by feedback inhibition of the enzyme anthranilate synthetase (see Miflin and Lea, 1977). The growth of carrot tissue cells can be completely inhibited by 44 μM DL-5-methyltryptophan **(18)**.

H_3C — $CCH_2CH(NH_2)COOH$ — N — H

18

Widholm (1972), however, showed that 1 in 3.6×10^6 cells was resistant to the analog. These cells could be cultured and were shown to be resistant to levels of 5-methyltryptophan 50 times higher than levels that normal cells could survive (Fig. 3). Although 5-methyltryptophan is not

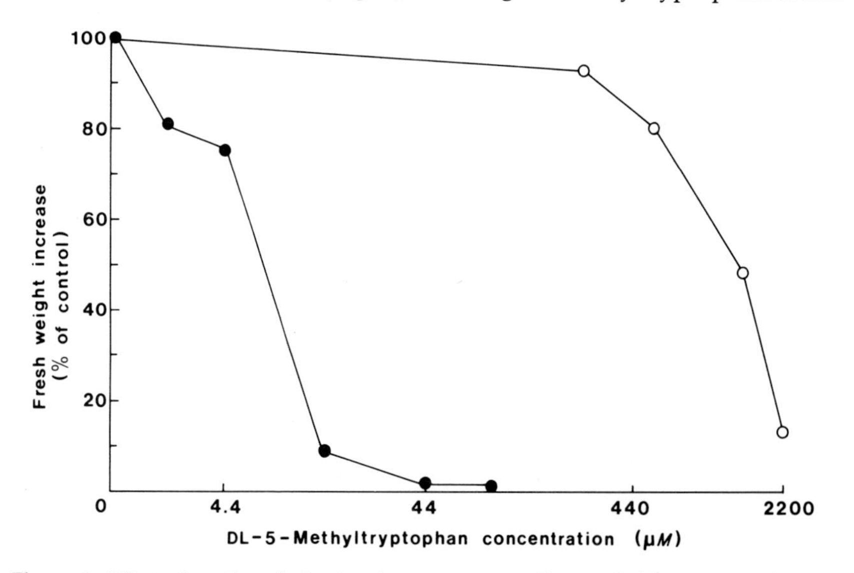

Figure 3. Effect of DL-5-methyltryptophan on carrot cell growth (●—●, normal cells; ○—○, resistant cells). Reproduced from Widholm (1972).

a natural product, similar substituted tryptophans, e.g., 5-hydroxytryptophan, can be major constituents of mature legume seeds (Bell, 1976). Investigations relating to the mechanism of resistance of tissue culture cells may provide a "model system" to explain the resistance shown by producer plants toward their own toxic amino acid constituents.

The concentration of free tryptophan in normal carrot cells was 81 μM, whereas that in the resistant cells was 2170 μM (Widholm, 1972). Anthranilate synthetase from normal cells was inhibited by 50% in the presence of 3.3 μM tryptophan; the corresponding concentration for the enzyme from resistant cells was 17 μM (Fig. 4). The concentrations of 5-methyltryptophan required to produce 50% of enzyme activity showed similar differences (Fig. 5).

This modification of the regulatory site of anthranilate synthetase has two consequent advantages for the resistant cell lines: (a) 5-methyltryptophan has a lower inhibitory effect on anthranilate synthetase, and therefore the cell is less likely to be starved of tryptophan. (b) The decreased end product inhibitory action of tryptophan permits higher concentrations of free tryptophan to accumulate, which in turn competes more effectively with 5-methyltryptophan at other possible sites of growth inhibition (e.g., attachment to tRNA in protein synthesis).

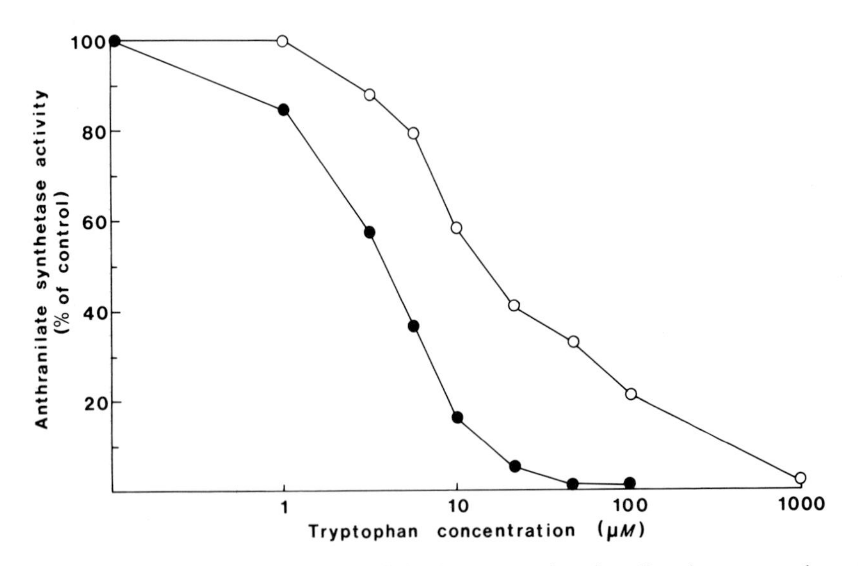

Figure 4. Inhibition by tryptophan of the formation of anthranilate by extracts from normal (●—●) and resistant (○—○) carrot cells. Reproduced from Widholm (1972).

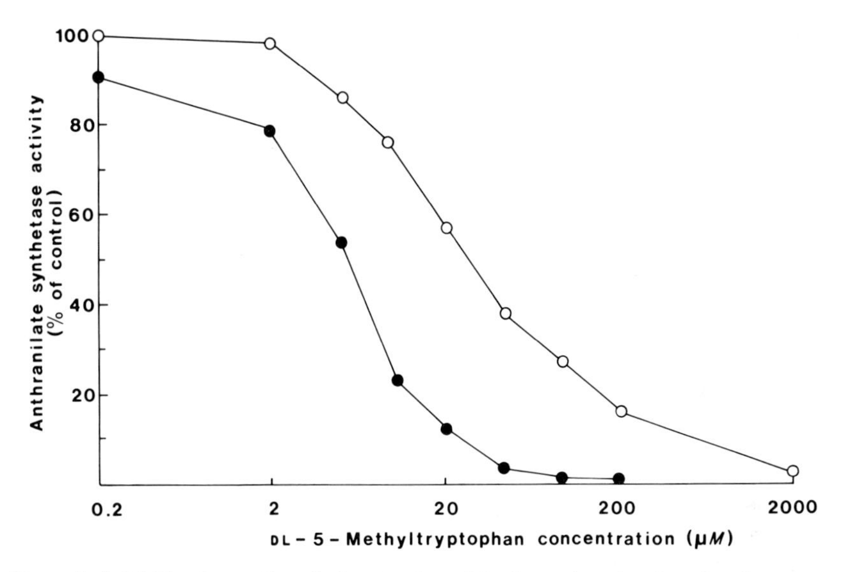

Figure 5. Inhibition by DL-5-methyltryptophan of the formation of anthranilate by extracts from normal (●—●) and resistant (○—○) carrot cells. Reproduced from Widholm (1972).

The thesis that, in tissue culture cells, resistance to an analog may be due to overproduction of the normal protein amino acids has been extended by Widholm, who obtained cell cultures overproducing phenylalanine (Palmer and Widholm, 1975) and lysine, methionine, and proline (Widholm, 1976). Such selection procedures may have future practical application in attempts to produce crop plants with more desirable levels of essential amino acids.

c. Glutamine Amide Transfer Reactions. Albizziine **(11)** is an inhibitor of glutamine amide transfer reactions in bacteria and animals (Buchanan, 1973). Recent studies at Rothamsted have shown that the analog is able to inhibit similar enzyme reactions in higher plants (Table I).

Asparagine synthetase catalyzes the following reaction in which the amide NH_2 group of glutamine is transferred to aspartate to form asparagine:

$$\text{Aspartate} + \text{glutamine} \xrightarrow{\text{ATP/Mg}^{2+}} \text{asparagine} + \text{glutamate}$$

Albizziine acts as a competitive inhibitor of the reaction and, if the enzyme is isolated from *Lupinus albus* (Lea and Fowden, 1975), asparagine synthesis is still strongly inhibited (71.3%) when equal concentrations of glutamine and albizziine are present. In albizziine-producing species, *Albizzia julibrissin* and *Acacia armata*, there is no significant inhibition when equal concentrations of analog and sub-

Table I. Action of albizziine as an inhibitor of glutamine-dependent amide transferases isolated from plant sources

Enzyme	Source	Concentration (m*M*)		Inhibition (%)	Reference
		Glutamine	Albizziine		
Asparagine synthetase	*Lupinus* seedlings	1.0	5.0	85.2	Lea and Fowden (1975)
Glutamate synthase (NADH)	*Pisum* roots	5.0	10.0	89	Miflin and Lea (1977)
Glutamate synthase (ferredoxin)	*Vicia* leaves	2.5	10.0	45	Wallsgrove *et al.* (1978)
Glutamate synthase (NADH)	*Phaseolus* root nodules	5.0	20.0	83.4	K. O. Awonaike, P. J. Lea and B. J. Miflin (unpublished)

strate are present. Thus, there is apparent variability within the active site of asparagine synthetase, the particular conformation associated with the producer species ensuring that albizziine is unable to compete efficiently with glutamine at the binding site. Albizziine has no inhibitory action on glutaminyl-tRNA synthetases, and no differences could be detected between the enzymes of producer and nonproducer species regarding their affinity for albizziine (Lea and Fowden, 1973).

3. *Incorporation of Amino Acid Analogs into Protein*

Probably the major mechanism by which an amino acid analog exerts its toxic effect is by incorporation into proteins at the sites normally occupied by the amino acid that it mimicks. The first step in protein synthesis is carried out by aminoacyl-tRNA synthetases, which catalyze a two-step reaction:

$$\text{Amino acid} + \text{ATP} \rightleftharpoons \text{aminoacyl-AMP} + \text{PP}_i \quad (1)$$

$$\text{Aminoacyl-AMP} + \text{tRNA} \rightleftharpoons \text{aminoacyl-tRNA} + \text{AMP} \quad (2)$$

Once this reaction has taken place, it is the tRNA moiety of the complex and not the attached amino acid that is recognized by mRNA (Chapeville *et al.*, 1962). Therefore, the substrate specificity of aminoacyl-tRNA synthetases has an overriding importance in the selection of the amino acid finally incorporated into protein (see Lea and Norris, 1977, for further discussion of this subject).

a. Azetidine-2-carboxylic Acid. Azetidine-2-carboxylic acid (A2C, **6**) is synthesized in large quantities by some members of the Liliaceae and by a limited number of legumes, of which *Delonix regia* has been studied extensively (Fowden, 1955; Sung and Fowden, 1971). The imino acid also occurs in trace quantities in sugar beet (Fowden, 1972), probably at concentrations that have no effect on the normal metabolism of proline **(7)**. Azetidine-2-carboxylic acid is incorporated into the proteins of the nonproducer plant *Phaseolus aureus* (Fowden *et al.*, 1967) and into the hemoglobin of blood erythrocytes (Trasko *et al.*, 1976).

Early studies on the properties of prolyl-tRNA synthetase showed that the enzyme isolated from *Convallaria majalis* (an abundant producer of A2C) was unable to activate A2C, but the enzyme from a nonproducer species, e.g., *P. aureus,* was able to use A2C as a substrate (Peterson and Fowden, 1965).

Studies on prolyl-tRNA synthetase were extended by Norris and Fowden (1972) to include a wider range of producer and nonproducer plants and a larger number of analogs. A comparison of the affinity of the enzyme for A2C and an analog larger than proline, *exo*(*cis*)-3,4-methano-L-proline **(19),** is shown in Table II.

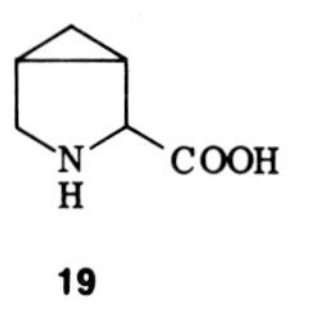

19

In general prolyl-tRNA synthetases from plants producing A2C showed no affinity for this imino acid, whereas they were able to activate the larger molecule *cis*-3,4-methano-L-proline. Conversely, the enzymes from the nonproducer species were able to activate A2C but showed no affinity for the larger analog. Thus, it would appear that plants producing A2C have evolved an active site that can accommodate substrates larger than that of the nonproducer species. In consequence, analogs smaller than proline, e.g., A2C, fit too loosely at the active site of the producer species to allow the correct binding of the 2-carboxyl group to the phosphate of AMP, and no activation and transfer of A2C to tRNA occur. However, the increased size of the active site means that molecules larger than proline are now substrates for the enzyme, but such larger analogs have not been isolated with A2C in the producer species investigated.

Prolyl-tRNA synthetases from *Delonix regia* and *Phaseolus aureus,* a producer and nonproducer species, respectively, have been studied fur-

Table II. Kinetic parameters of prolyl-tRNA synthetases from azetidine-2-carboxylic acid producer and nonproducer plants[a]

Species	Production of large amounts of A2C	L-Proline, $K_m (\times 10^{-4} M)$	L-Azetidine-2-carboxylic acid		*cis*-3,4-Methano-L-proline	
			$K_m (\times 10^{-3} M)$	V_{max}[b]	$K_m (\times 10^{-3} M)$	V_{max}[b]
Parkinsonia aculeata	Yes	4.35	∞	0–5	7.1	42
Delonix regia	Yes	1.82	∞	0–5	4.6	22
Convallaria majalis	Yes	4.5	∞	0–5	2.5	36
Beta vulgaris	No	4.5	2.2	73	n.d.[c]	<3
Hemerocallis fulva	No	6.25	5.3	75	n.d.[c]	<3
Phaseolus aureus	No	1.37	1.43	55	n.d.[c]	<2
Ranunculus bulbosa	No	2.9	2.0	66	∞	0

[a] All data derived from measurement of the ATP–PP_i exchange reaction as described by Norris and Fowden (1972).
[b] The V_{max} values are expressed as a percentage of that obtained for proline.
[c] n.d., not determined.

ther by Norris and Fowden (1973a,b). The ability of analogs to protect the enzymes against inactivation by heat, *p*-chloromercuribenzoate, and dye-dependent light reactions suggests that cysteinyl and histidinyl residues are present at the active sites of both enzymes.

b. γ-Substituted Glutamic Acids. A number of γ-substituted glutamic acids accumulate in high concentrations in some plant species. *erythro*-γ-Methylglutamic acid **(20)** is distributed among the Liliaceae and Leguminosae and accumulates in *Phyllitis scolopendrium* (Blake and Fowden, 1964) and *Caesalpinia bonduc* (Watson and Fowden, 1973). *threo*-γ-Methylglutamic acid **(21)** does not occur naturally. *threo*-γ-Hydroxyglutamic acid **(22)** occurs sporadically among species of Lilaceae including *Hemerocallis fulva* (Fowden and Steward, 1957); it also has been isolated from various ferns (Virtanen and Hietala, 1955). The *erythro* isomer **(23)** is not known as a higher plant constituent. γ-Hydroxy-γ-methylglutamic acid exists as the 2(*S*),4(*S*) isomer **(24)** in certain legumes (Sung and Fowden, 1971) and ferns (Blake and Fowden, 1964). γ-Methyleneglutamic acid **(25)** also occurs in some legumes (Done and Fowden, 1952) and members of the Liliaceae (Fowden and Steward, 1957).

$$\begin{array}{c} COOH \\ | \\ H-C-CH_3 \\ | \\ H-C-H \\ | \\ H-C-NH_2 \\ | \\ COOH \\ \mathbf{20} \end{array} \qquad \begin{array}{c} COOH \\ | \\ H_3C-C-H \\ | \\ H-C-H \\ | \\ H-C-NH_2 \\ | \\ COOH \\ \mathbf{21} \end{array}$$

$$\begin{array}{c} COOH \\ | \\ HO-C-H \\ | \\ H-C-H \\ | \\ H-C-NH_2 \\ | \\ COOH \\ \mathbf{22} \end{array} \qquad \begin{array}{c} COOH \\ | \\ H-C-OH \\ | \\ H-C-H \\ | \\ H-C-NH_2 \\ | \\ COOH \\ \mathbf{23} \end{array} \qquad \begin{array}{c} COOH \\ | \\ HO-C-CH_3 \\ | \\ H-C-H \\ | \\ H-C-NH_2 \\ | \\ COOH \\ \mathbf{24} \end{array} \qquad \begin{array}{c} COOH \\ | \\ C=CH_2 \\ | \\ H-C-H \\ | \\ H-C-NH_2 \\ | \\ COOH \\ \mathbf{25} \end{array}$$

Data relating to the amino acid specificity of the glutamyl-tRNA synthetase for three plant species, as obtained by Lea and Fowden (1972), are given in Table III. The plants studied were (a) *Phaseolus aureus*, in which there is no record of occurrence of substituted glutamic acids, (b) *Hemerocallis fulva*, a liliaceous species producing *threo*-γ-hydroxyglutamic acid, and (c) *Caesalpinia bonduc*, a legume whose seed contains large amounts of *erythro*-γ-methylglutamic acid. The results show that enzymes from species producing particular glutamate analogs

Table III. Kinetic parameters determined for glutamic acid and several analogs using glutamyl-tRNA synthetase preparations from higher plants[a]

Species	L-Glutamic acid		*erthro*-γ-Methyl-L-glutamic acid		*threo*-γ-Methyl-DL-glutamic acid		*threo*-γ-Hydroxy-L-glutamic acid		*erythro*-γ-Hydroxy-DL-glutamic acid		2(*S*),4(*S*)-γ-Hydroxy-γ-methylglutamic acid	
	K_m	V_{max}	K_m	V_{max}	K_m	V_{max}	K_m	V_{max}	K_m	V_{max}	K_m	V_{max}
Phaseolus aureus seed	7.2×10^{-3}	100	1.55×10^{-2}	68.1	—	55.2 (75 m*M*)	2.11×10^{-2}	54.7	—	58.2 (75 m*M*)	3.43×10^{-2}	42.2
Hemerocallis fulva leaf	5.24×10^{-3}	100	2.81×10^{-2}	40.2	∞	0	∞	0	—	34.2 (75 m*M*)	1.25×10^{-1}	Calculated as 10.2
Caesalpinia bonduc seed	9.3×10^{-3}	100	∞	0	—	20.1 (75 m*M*)	5.21×10^{-2}	23.6	∞	0	∞	0

[a] The K_m values are expressed as molar concentrations with respect to the L form. The V_{max} values are expressed as percentages of the values determined for glutamic acid.

fail to activate the plants' own products. The discriminatory behavior appears to be stereospecific; the *H. fulva* enzyme can activate *erythro*-substituted compounds but not *threo*-substituted glutamic acids, whereas the converse is seen for the *C. bonduc* enzyme. (The *P. aureus* enzyme appears to be unable to discriminate between either the type of substituent group or its stereochemical position on the glutamic acid skeleton.) Thus, both producer species have evolved enzymes with similar but stereospecifically distinct changes at their active sites to prevent incorporation of their own products into protein molecules.

c. 2-Amino-4-methylhex-4-enoic Acid. Species of *Aesculus* belonging to the subgeneric group Calothyrsus synthesize large quantities of 2-amino-4-methylhex-4-enoic acid (AMHA, **13**) (Fowden *et al.*, 1970). This amino acid was activated at higher rates than was phenylalanine by the phenylalanyl-tRNA synthetase from *Phaseolus aureus* (Smith and Fowden, 1968). Studies with the enzyme from *Aesculus hippocastanum* (a nonproducer species) showed that a number of related compounds, all coplanar about C-4, could also mimic phenylalanine (Table IV) (Anderson and Fowden, 1970).

The phenylalanyl-tRNA synthetases were purified from five species of *Aesculus* (a representative of each subgeneric group). The affinities of these enzymes toward phenylalanine and AMHA are shown in Table V. Enzyme from *A. californica* (an AMHA-producing species) activated AMHA only 30% as efficiently as phenylalanine, whereas the other four *Aesculus* enzymes activated AMHA and phenylalanine at about identi-

Table IV. Structure–activity relationships of compounds used as substrates for the phenylalanyl-tRNA synthetase from *Aesculus hippocastanum*[a]

$$\mathrm{Y{-}C(Z){=}C(X)CH_2CH(NH_2)COOH}$$

X	Y	Z	Compound	V_{max}[b]
H	H	H	Allylglycine (5 m*M*)	0
H	CH_2		2-Aminohex-4,5-dienoic acid (5 m*M*)	23
H	CH_3	H	Crotylglycine (2.5 m*M*)	29
CH_3	H	H	Methallylglycine (5 m*M*)	35
C_2H_5	H	H	Ethallylglycine (5 m*M*)	40
H	CH_3	CH_3	2-Amino-5-methylhex-4-enoic acid (10 m*M*)	48
CH_3	CH_3	H	2-Amino-4-methylhex-4-enoic acid (AMHA) (10 m*M*)	100

[a] After Anderson and Fowden (1970).
[b] Expressed as a percentage of phenylalanine activity.

Table V. Kinetic data for phenylalanyl-tRNA synthetases from five subgeneric groups of *Aesculus*

Subgeneric group and species	K_m ATP (M)	K_m Phe (M)	K_m AMHA[a] (M)	K_m AMHA/K_m Phe	V_{max} AMHA/V_{max} Phe × 100
Calothyrus (*A. californica*)	3.3×10^{-4}	1.05×10^{-4}	3.2×10^{-3}	30	30
Pavia (*A. glabra*)	3.6×10^{-4}	4.9×10^{-5}	1.78×10^{-3}	36	90
Aesculus (*A. hippocastanum*)	2.8×10^{-4}	3.1×10^{-5}	1.18×10^{-3}	38	100
Parryanae (*A. parryi*)	3.8×10^{-4}	1.9×10^{-5}	8.9×10^{-4}	47	100
Macrothyrsus (*A. parviflora*)	2.8×10^{-4}	1.6×10^{-5}	6.8×10^{-4}	43	100

[a] AMHA, 2-Amino-4-methylhex-4-enoic acid.

cal rates (Anderson and Fowden, 1970). Thus, there is only partial discrimination by the *A. californica* enzyme against AMHA. The K_m values determined for this enzyme indicate that AMHA would have to be present at 30-fold higher concentrations than phenylalanine in order to saturate the enzyme, and it is therefore possible that the partial discrimination observed is sufficient to prevent incorporation of AMHA into protein in *A. californica*. In addition, further discrimination may occur against the analog during its transfer from the enzyme–aminoacyl-AMP complex to $tRNA^{Phe}$.

d. Selenoamino Acids. The toxicity of selenium (see Chapter 9 by Rosenthal and Bell) is thought to be caused by the incorporation of the two selenium analogs selenocysteine **(26)** and selenomethionine **(27)** into

$$CH_2CH(NH_2)COOH \qquad CH_3SeCH_2CH_2CH(NH_2)COOH$$

26 **27**

protein in place of the natural sulfur amino acids. The altered properties of the polypeptide formed are sufficient to cause a malfunctioning of enzymes and other important proteins. The replacement of sulfur by selenium in parsley (*Petroselinum crispum*) ferredoxin causes changes in spectral and oxidation–reduction properties and a 20% reduction in its ability to reduce cytochrome *c* (Fee *et al.*, 1971).

In accumulator species selenium is apparently stored in a number of nonprotein amino acids, *Se*-methylselenocysteine **(28)** (Virupaksha and Shrift, 1965), selenocystathionine **(29)** (Peterson and Butler, 1967), and

$$Se{-}CH_2{-}CH(NH_2)COOH \qquad HOOCCH(NH_2)CH_2SeCH_2CH_2CH(NH_2)COOH$$

28 **29**

their γ-glutamyl peptides (Nigam and McConnell, 1969, 1976a). The suggestion has been made that, by introducing selenium into these nonprotein amino acids, the plants prevent incorporation into protein. However, there is some evidence that selenomethionine is a precursor of *Se*-methylselenocysteine (Chow *et al.*, 1973), and cystathionine is a known precursor of methionine (Datko *et al.*, 1974). Thus, unless there is a strict compartmentation of the metabolism of selenoamino acids, selenomethionine is likely to be synthesized at sites where it may be incorporated into the protein of accumulator species.

Evidence presented by Nigam and McConnell (1976b) suggests that 0.5% of the total [^{75}Se]selenate fed to *Astragalus bisulcatus* may be incorporated into protein. However, the possibility of the nonenzymatic binding of selenate to protein cannot be ruled out, since no evidence for the occurrence of protein-bound selenocysteine or selenomethionine was presented.

Cysteinyl-tRNA synthetase was purified 300-fold from *Phaseolus aureus* (Burnell and Shrift, 1977). Selenocysteine was able to act as a substrate for the enzyme at 70% the rate of cysteine, and the K_m values for the two substrates were very similar. Unpublished data of J. N. Burnell and A. Shrift has shown that the cysteinyl-tRNA synthetase from *A. bisulcatus* is unable to activate selenocysteine, although the enzyme from other accumulator species may be able to use the analog as a substrate for the pyrophosphate exchange reaction.

Thus, the mechanism of avoidance of toxicity of selenium amino acids in accumulator species is still not clearly resolved. There is considerable evidence that the production of nonprotein amino acids may represent tolerance mechanisms. There is probably a second line of defense as well, depending on differential substrate specificities of the cysteinyl- and methionyl-tRNA synthetases that reduce the extent to which selenium analogs of cysteine and methionine become bound to tRNA.

REFERENCES

Anderson, J. W., and Fowden, L. (1970). *Biochem. J.* **119,** 677–690.
Bell, E. A. (1976). *FEBS Lett.* **64,** 29–38.
Blake, J., and Fowden, L. (1964). *Biochem. J.* **92,** 136–142.
Bright, S. W. J., Wood, E., and Miflin, B. J. (1978a). *Planta* **139,** 113–117.
Bright, S. W. J., Shewry, P. R., and Miflin, B. J. (1978b). *Planta* **139,** 119–125.
Broquist, H. P. (1971). *In* "Methods in Enzymology" (H. Tabor and C. W. Tabor, eds.), **17,** (Part B), 112–129. Academic Press, New York.
Buchanan, J. M. (1973). *In* "The Enzymes of Glutamine Metabolism" (S. Prusiner and E. R. Stadtman, eds.), pp. 387–408. Academic Press, New York.
Burnell, J. N., and Shift, A. (1977). *Plant Physiol.* **60,** 670–674.
Chaleff, R. S., and Carlson, P. S. (1975). *In* "Modification of the Information Content of Plant Cells" (R. Markham *et al.,* eds.), pp. 197–214. North-Holland Publ., Amsterdam.
Chapeville, F., Lipmann, F., von Ehrenstein, G., Weisblum, B., Ray, W. J., and Benzer, S. (1962) *Proc. Natl. Acad. Sci. U.S.A.* **48,** 1086–1092.
Chow, C. M., Nigam, S. N., and McConnell, W. B. (1973). *Can. J. Biochem.* **51,** 489–490.
Corredor, C. P. (1975). *In* "Hypoglycin" (E. A. Kean, ed.), pp. 93–107. Academic Press, New York.
Datko, A. H., Giovanelli, J., and Mudd, S. H. (1974). *J. Biol. Chem.* **249,** 1139–1145.
Davies, H. M. (1977). Ph. D. Thesis, University of London.
De Marco, C., Busiello, V., Di Girolamo, M., and Cavallini, D. (1976). *Biochim. Biophys. Acta* **454,** 298–308.
Done, J., and Fowden, L. (1952). *Biochem. J.* **51,** 451–458.

Ellington, E. V., Hassall, C. H., Plimmer, J. R., and Seaforth, C. E. (1959). *J. Chem. Soc.* pp. 80–85.
Fee, J. A., Mayhew, S. G., and Palmer, G. (1971). *Biochim. Biophys. Acta* **245,** 196–200.
Fowden, L. (1955). *Nature (London)* **176,** 347–348.
Fowden, L. (1963). *J. Exp. Bot.* **14,** 387–398.
Fowden, L. (1972). *Phytochemistry* **11,** 2271–2276.
Fowden, L., and Smith, A. (1968). *Phytochemistry* **7,** 809–819.
Fowden, L., and Steward, F. C. (1957). *Ann. Bot. (London)* [N.S.] **21,** 53–67.
Fowden, L., Lewis, D., and Tristram, H. (1967). *Adv. Enzymol.* **29,** 89–163.
Fowden, L., Anderson, J. W., and Smith, A. (1970). *Phytochemistry* **9,** 2349–2357.
Lea, P. J., and Fowden, L. (1972). *Phytochemistry* **11,** 2129–2138.
Lea, P. J., and Fowden, L. (1973). *Phytochemistry* **12,** 1903–1916.
Lea, P. J., and Fowden, L. (1975). *Proc. R. Soc. London, Ser. B* **192,** 13–26.
Lea, P. J., and Norris, R. D. (1977). *Prog. Phytochem.* **4,** 121–167.
Mazelis, M., Whatley, F. R., and Whatley, J. (1977). *FEBS Lett.* **84,** 236–240.
Miflin, B. J. (1977). *In* "Regulation of Enzyme Synthesis and Activity in Higher Plants" (H. Smith, ed.), pp. 23–40. Academic Press, New York.
Miflin, B. J., and Lea, P. J. (1977). *Annu. Rev. Plant Physiol.* **28,** 299—329.
Miflin, B. J., Bright, S. W. J., Davies, H. M., Shewry, P. R., and Lea, P. J. (1978). "Proceedings of the Second Long Ashton Symposium on Nitrogen Assimilation." Academic Press, London (in press).
Nigam, S. N., and McConnell, W. B. (1969). *Biochim. Biophys. Acta* **192,** 185–190.
Nigam, S. N., and McConnell, W. B. (1976a). *Biochim. Biophys. Acta* **437,** 116–121.
Nigam, S. N., and McConnell, W. B. (1976b). *J. Exp. Bot.* **27,** 565–571.
Norris, R. D., and Fowden, L. (1972). *Phytochemistry* **11,** 2921–2935.
Norris, R. D., and Fowden, L. (1973a). *Phytochemistry* **12,** 2109–2122.
Norris, R. D., and Fowden, L. (1973b). *Phytochemistry* **12,** 2829–2841.
Palmer, J. E., and Widholm, J. M. (1975). *Plant Physiol.* **56,** 233–238.
Peterson, P. J., and Butler, G. W. (1967). *Nature (London)* **213,** 599–600.
Peterson, P. J., and Fowden, L. (1965). *Biochem. J.* **97,** 112–124.
Rosenfeld, I., and Beath, O. A. (1964). "Selenium," 2nd ed., pp. 91–140. Academic Press, New York.
Saunders, J. A., and Conn, E. E. (1977). Abstracts of 1st Joint Symposium of Phytochemical Society of Europe and Phytochemical Society of North America, held at Gent (Belgium) Aug 29–Sept 2, 1977.
Scioscia-Santoro, A., Cavallini, D., Degener, A. M., Perze-Bercoff, R., and Rita, G. *Congr. Soc. Ital. Microbiol., 17th., 1975 p. 17.*
Seneviratne, A. S., and Fowden, L. (1968). *Phytochemistry* **7,** 1039–1045.
Shewry, P. R., and Miflin, B. J. (1977). *Plant Physiol.* **59,** 69–73.
Smith, I. K., and Fowden, L. (1966). *J. Exp. Bot.* **17,** 750–761.
Smith, I. K., and Fowden, L. (1968). *Phytochemistry* **7,** 1065–1069.
Stern, R., and Mehler, A. H. (1965). *Biochem. Z.* **342,** 400–409.
Sung, M.-L., and Fowden, L. (1971). *Phytochemistry* **10,** 1523–1528.
Trasko, C. S., Franzblau, C., and Troxler, R. I. (1976). *Biochim. Biophys. Acta* **447,** 425–455.
Tristram, H., and Neale, S. M. (1968). *J. Gen. Microbiol.* **50,** 121–137.
Vietmeyer, N., and Cotton, B. (1977). *In* "Leucaena: Promising Forage and Tree Legume Crop for the Tropics" (F. R. Ruskin, ed.), pp. 32–39. Nat. Acad. Sci., Washington, D.C.
Virtanen, A. I., and Hietala, P. K. (1955). *Acta Chem. Scand.* **9,** 175–176.

Virupaksha, T. K., and Shrift, A. (1965). *Biochim. Biophys. Acta* **107,** 69–80.
Wallsgrove, R. M., Harel, E., Lea, P. J., and Miflin, B. J. (1978). *J. Exp. Bot.* **28,** 588–596.
Warin, R., Jadot, J., and Casimir, J. (1969). *Bull. Soc. R. Sci. Liege* **38,** 280–287.
Watson, R., and Fowden, L. (1973). *Phytochemistry* **12,** 617–622.
Watson, R., and Fowden, L. (1975). *Phytochemistry* **14,** 1181–1186.
Widholm, J. M. (1972). *Biochim. Biophys. Acta* **279,** 48–57.
Widholm, J. M. (1976). *Can. J. Bot.* **54,** 1523–1529.

Chapter **4**

How Animals Perceive Secondary Compounds

R. F. CHAPMAN *And* **W. M. BLANEY**

I. INTRODUCTION

A. Need to Perceive Secondary Plant Chemicals

Secondary plant chemicals often have adverse effects on the physiology, development, and survival of animals (see, e.g., Beck and Reese, 1976; Bell, 1972, 1976; Feeny, 1975; Jones, 1974; Mattocks, 1972; Rhoades and Cates, 1976; Schoonhoven, 1972; and various chapters in this book), and it is probable that all herbivores are vulnerable to some extent even though they may be adapted to tolerate or even to utilize certain groups of these chemicals. The effects may be sudden and dramatic, as in the

ISBN 0-12-597180-X

death of cattle following the ingestion of highly cyanogenic plants, but commonly they are much more subtle and may become apparent only when the organism is subjected to additional stress. In any case it is not surprising that the host-selection behavior of many herbivorous species is such that plants containing potentially harmful chemicals are rejected (Bernays and Chapman, 1977; Jones, 1972).

In contrast, some insects are so adapted that they can sequester certain secondary plant chemicals and so acquire toxic properties that provide a defense against predators (Brower, 1969; Fishelson, 1960; Marsh and Rothschild, 1974; Rothschild and Kellett, 1972). When reared on other plants lacking the chemicals, such insects may be eaten readily by predators so that selection favors individuals that eat plants containing the appropriate secondary plant chemicals.

A quite different type of dependence of some animals on secondary plant chemicals is their use as sex pheromones. Males of *Danaus chrysippus* derive the main component of their aphrodisiac pheromone from a pyrrolizidine alkaloid precursor in *Heliotropium*. If they do not have access to this chemical they are deficient in the pheromone. Such dependence is probably widespread in the Danaidae (Edgar and Culvenor, 1974; Schneider, 1975, 1977) and some other families of Lepidoptera (Pliske, 1975).

It is apparent that there is an important, and in some cases vital, need for animals to be able to perceive and recognize secondary plant substances, and selection will have favored the development of this capacity to various degrees.

B. Mechanisms of Recognition

It is to be expected that secondary plant chemicals are commonly perceived by the chemical senses of olfaction and gustation, and there is ample evidence that this is the case. There are, however, no chemosensory structures associated specifically with the perception of secondary plant substances, and in the following account a general description of chemosensory systems and their functioning is given with emphasis on aspects concerned with the perception of secondary plant chemicals where appropriate.

Not all secondary plant chemicals are perceived in this way, however. Some, commonly called urticating compounds, cause pain and irritation of the skin of vertebrates, which does not involve the chemical senses. The perception of these compounds is considered briefly.

The response of the sense cells to secondary plant chemicals is innate, although it may be modified by experience (Schoonhoven, 1969; Städler and Hanson, 1976). The overt behavioral response, resulting in avoidance or acceptance of the compound, is also often innate, but there are examples of animals learning to avoid ingesting secondary plant chemicals as a result of an earlier "unpleasant" experience. This behavior is discussed briefly in Section V.

Where information is available this review concentrates on terrestrial herbivores. However, the shortage of well-documented data is such that it has often proved necessary to consider other animals in an attempt to make the review more useful.

II. STRUCTURE AND ARRANGEMENT OF CHEMORECEPTORS

The perception of chemicals in the environment requires that sensitive nerve endings be exposed to the environment, and this raises the problem of desiccation. The problem is solved in different ways in different animal groups, resulting in the development of very different types of receptor organs. For this reason, the form of chemoreceptors in terrestrial arthropods, vertebrates, nematodes, and terrestrial mollusks will be considered separately.

A. Terrestrial Arthropods

Nearly all the studies on chemoreceptors of terrestrial arthropods have been concerned with insects, and this account is restricted to this group. Some work on mites, although not on herbivorous species, indicates that the basic structures are similar to those of insects (Axtell *et al.*, 1973; Foelix and Axtell, 1971).

1. General Structure

The structure of insect chemoreceptors has been reviewed by Lewis (1970), Slifer (1970), and Stürckow (1970). There are three basic elements: one or more bipolar neurons, two or three accessory cells, and the cuticular apparatus. The neurons are primary sense cells connected directly to the central nervous system (CNS) without any intervening synapses (see, e.g., Blaney and Chapman, 1969; Steinbrecht, 1969; Stürckow *et al.*, 1967).

The innermost accessory cell, the thecogen cell of Steinbrecht and Müller (1976), ensheathes the neurons. It is wrapped around by the trichogen and tormogen cells in concentric array (Fig. 1), and adjacent epidermal cells may also have a sheathing function. The trichogen and tormogen cells enclose an extracellular cavity, the receptor lymph cavity, through which the scolopale and dendrites pass. These cells have nu-

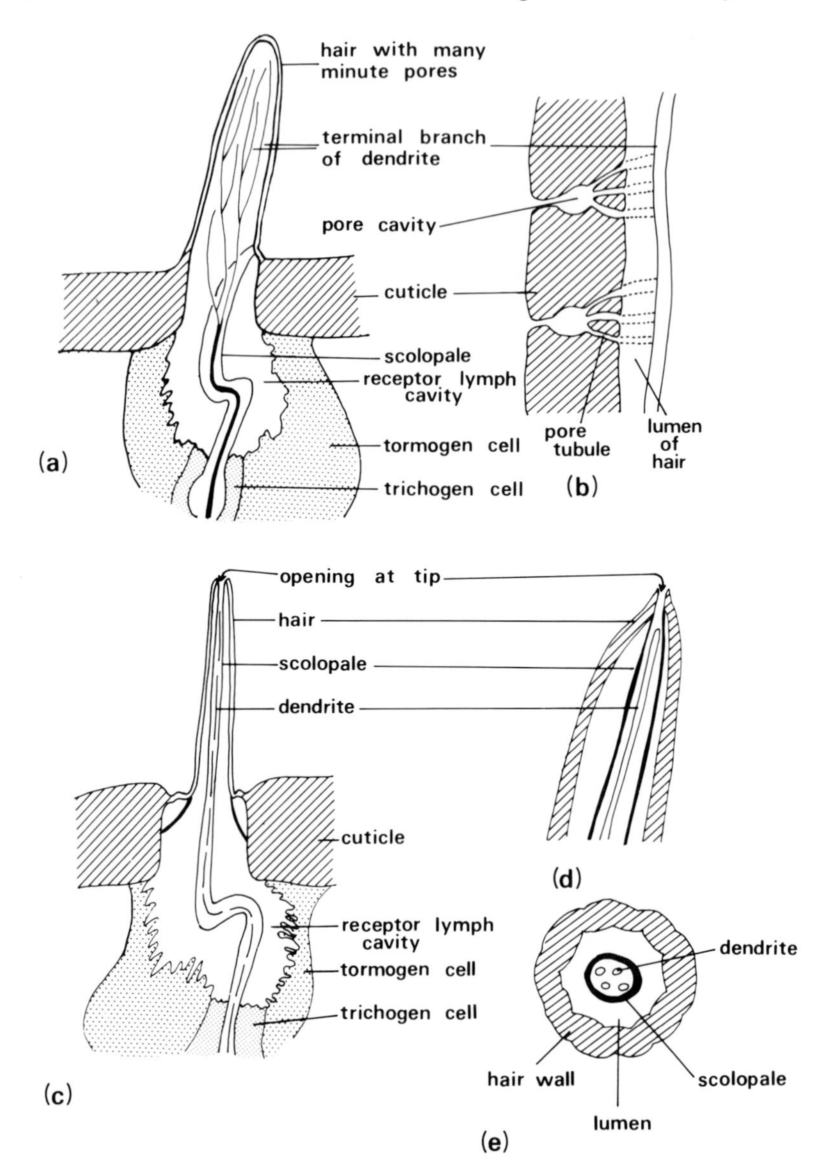

Figure 1. Chemosensory sensilla of insects. (a) Diagram of an olfactory sensillum; (b) part of the wall of an olfactory sensillum showing the pores through which stimulating molecules reach the dendrite; (c) diagram of a contact chemoreceptor; (d) longitudinal section through the tip of a chemoreceptor hair; (e) transverse section through a chemoreceptor hair.

merous microvilli where they bound the cavity, and they appear to be metabolically active, possibly secreting fluid into the cavity. Phillips and Vande Berg (1976) showed that protein from the hemolymph is taken up by endocytosis in nonneural cells of the sensillum and transported into the cavity. In olfactory sensilla the fluid may replace material lost through the pores in the cuticle (see below) or it may have a trophic function for the distal dendrites. Thurm (1972) suggested that secretion into this cavity may be coupled with an electrogenic ion pump providing an additional energy source for the receptor potential. He showed that the cavity has a positive potential of approximately 60 to 80 mV with respect to the hemolymph.

Stimulating chemicals reach the dendrites of chemoreceptors through minute perforations in the cuticle. The cuticular structure immediately associated with the dendrite may be a flat, perforated plate but is commonly a hollow hair measuring from a few microns to several hundred microns long. The organ so formed is called a sensillum.

2. *Olfactory Sensilla*

In the majority of olfactory sensilla, the cuticular structure is a thin-walled hair perforated by several thousand pores, which may be as little as 10–15 nm in diameter (Schneider and Steinbrecht, 1968) or over 200 nm (Slifer, 1972). Each pore gives access to a pore canal leading to a cavity from which several pore tubules, with diameters of 10 to 20 nm, penetrate the remainder of the hair wall. These tubules appear to continue for about 150 nm into the hair lumen, where they may end in contact with the receptor cell dendrite (Steinbrecht, 1974) (Fig. 1b). The number of neurons in olfactory sensilla is very variable, ranging from 1 to over 50 (Slifer, 1970). Distally the dendrites leave the scolopale and are free in the receptor lymph (Fig. 1a). In many cases, the dendrites of olfactory sensilla branch within the lumen of the hair, sometimes very extensively. In *Glossina austeni* one form of olfactory sensillum contains about 120 branches which derive from only two or three receptor cells (Lewis, 1970), and in a sensillum of *Locusta* dendrites show a variable degree of branching that results in dendrite elements having a wide range of diameters and therefore of input impedance. It is suggested that this may equip the sensillum to respond efficiently to a wide range of levels of stimulation (Blaney, 1977). The dendrites in the long olfactory sensilla of *Bombyx* do not branch at all (Steinbrecht, 1973).

3. *Gustatory Sensilla*

In gustatory sensilla the dendrites are exposed through a terminal pore in the cuticular hair or cone. The precise structure of the hair tip is incompletely understood, chiefly because of the extreme difficulty of cutting thin sections of this region for study with the electron microscope. A subject of continuing debate is the presence or absence of a connection between the inner and outer lumina at the tip of the hair, although the penetration of ^{14}C-labeled glucose into both cavities provides strong evidence for the presence of such a connection (Hanamori, 1976). The distal regions of the dendrites are totally enclosed by the scolopale (Fig. 1d), and presumably the material in which they are bathed and which protects them at the opening is secreted by the neurilemma cell. This material is a mucopolysaccharide (Bernays *et al.*, 1975).

4. *Distribution of Sensilla*

Olfactory sensilla are concentrated on the antennae, commonly with the majority toward the distal end (Harbach and Larsen, 1977; Hatfield *et al.*, 1976; Masson and Friggi, 1971). When the sense of smell is of particular importance to the species, the structure of the antennae may be elaborated so that there is space for more sensilla. For instance, the surface area of the pectinate antenna of the male silkmoth, *Bombyx mori*, is 29.0 mm^2, compared with 4.8 mm^2 in the simple antenna of the female. Correspondingly, there are about 17,000 long olfactory sensilla on the male antenna compared with only 6000 on the female (Steinbrecht, 1973). It is common for a large number of olfactory sensilla to be present on the antennae, and there are, for instance, more than 26,000 in *Leucophaea* (Schafer, 1971) and 2800 in adult *Tenebrio* (Harbach and Larsen, 1977), but sometimes the number is much lower. Small numbers of olfactory sensilla are present on other parts of the body.

The main concentrations of gustatory receptors are on the mouthparts. In general, the orthopteroid orders possess large numbers of gustatory sensilla; *Locusta* has a total of 3000 gustatory sensilla on the mouthparts incorporating 15,000 sensory neurons (Chapman and Thomas, 1978). In contrast, hemipteroid insects and endopterygote insects have relatively few, with a total of just over 100 neurons in caterpillars (Schoonhoven, 1973; Schoonhoven and Henstra, 1972). Relatively large numbers of gustatory sensilla are also present on the tarsi (Dethier, 1976; Kendall, 1970), and smaller numbers are often present on the ovipositor and elsewhere.

5. *Central Nervous Connections*

The chemoreceptor cells of insects are connected directly to the CNS by their own axons, and all integration occurs within the CNS. The region of the CNS principally involved depends on the location of the receptor cells since in general their axons pass to the ganglion of the segment from which they are primitively derived. Axons of antennal receptor cells synapse in the deutocerebrum, many of those from the mouthparts in the subesophageal ganglion and those from the legs in the thoracic ganglia. Most information is available for the blowfly (Dethier, 1976; Geisert and Altner, 1974), the cockroach, and the locust (Aubele and Klemm, 1977; Boeckh *et al.*, 1975; Ernst *et al.*, 1977).

B. Vertebrates

In terrestrial vertebrates, the problem of desiccation is solved by restricting all the chemoreceptors to mucus-covered sensory epithelia in cavities in which the humidity normally remains high. The sensations of smell and taste are generally mediated by different receptor systems, the former located in the nasal cavity and the latter in the oral cavity. In addition, amphibians, reptiles, and mammals have developed, to a variable degree, a third system in the vomeronasal, or Jacobson's organ. This organ is situated close to the olfactory system and has a similar cellular arrangement. There is still doubt, however, about the exact nature of the vomeronasal sensation. The three organ systems have different nervous pathways carrying information to different parts of the brain (Scalia *et al.*, 1968; Heimer, 1969; Winans and Scalia, 1970; Scalia, 1972).

The organization of the sensory epithelium differs between organs of taste and organs of smell in a number of ways. In organs of smell, the receptor cells are true neurons occurring singly in the epithelium and sending their axons directly to the CNS (Fig. 2). In contrast, the receptor cells in taste organs are secondary sensory cells grouped together with

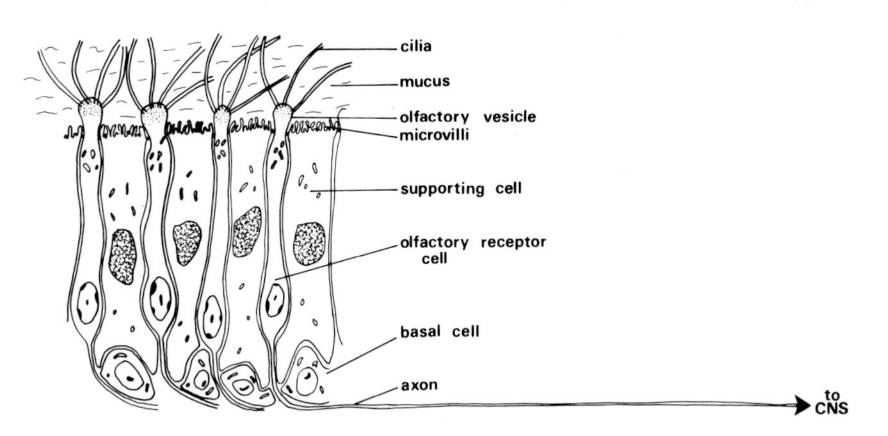

Figure 2. Diagram of part of the olfactory epithelium of a vertebrate.

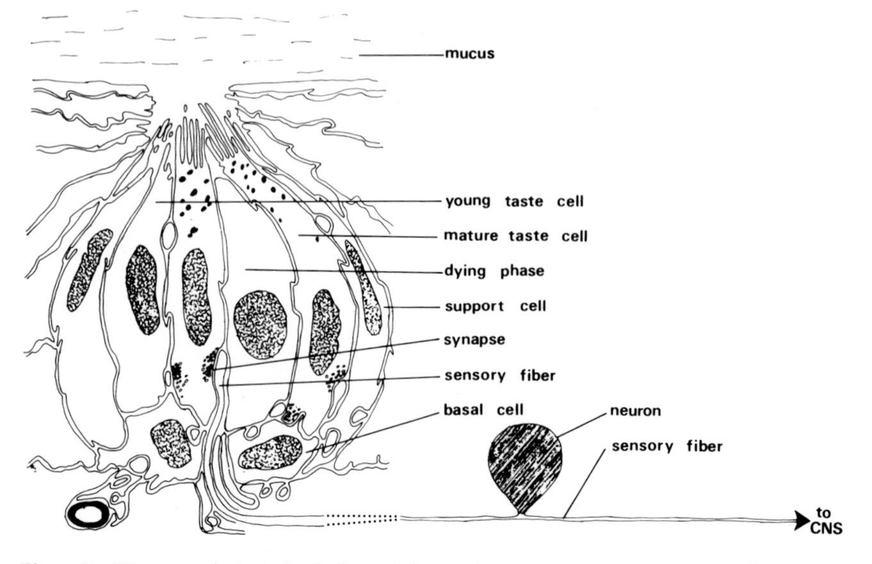

Figure 3. Diagram of a taste bud of a vertebrate. A sensory neuron connecting the receptor cells to the CNS is shown diagrammatically.

supporting and basal cells to form clusters called taste buds, which are surrounded by normal epithelial cells. The taste receptor cells form synapses with the fibers of neurons whose perikarya lie in a sensory ganglion of a cranial nerve and whose axons synapse in the CNS (Fig. 3).

1. *Olfactory Organs*

The nasal chamber, lying between the internal and external nares, is more or less divided into two chambers: an outer or anterior respiratory chamber and an inner or posterior olfactory chamber. Some variations in structure of nasal organs occurring in amphibians, reptiles, birds, and mammals are described by Allison (1953) and Parsons (1971).

The olfactory epithelium conforms to a common plan and is composed of receptor cells, support cells, and basal cells (Fig. 2) [see reviews by Allison (1953) and Le Gros Clark (1957)]. Distally the receptor cell commonly bears a number of long cilia that protrude a short distance from the surface of the epithelium and then bend to lie parallel along the surface in the mucous covering. Alternatively, or in addition, microvilli may be present. These structures increase the exposed surface of the receptor, in some cases by as much as 1000 or more times (Graziadei, 1971). They have been implicated as the sites of interaction with odorous stimulants, but there is no clear evidence of the site of energy transduction. Little is known of the sensitivities of individual receptor cells or how cells with differing sensitivities may be disposed in the epithelium as a whole (Adrian, 1950, 1951). It is suggested by Mozell (1971) that spatial separation of odorants may result from differential rates of migration across the mucosa and that this may be related to the distribution of receptors with differing sensitivities. Moulton (1976) reviews the evidence for a spatial component to odor quality coding.

Most receptor cells are surrounded, and separated from each other, by columnar epithelial cells that usually have a series of long, irregular microvilli on the apical surface in contact with the mucous layer. The support cells contact each other and the receptor cells in a variety of ways, but the significance of the various intercellular contacts is largely unresolved (see de Lorenzo, 1970; Graziadei, 1971; Reese and Brightman, 1970; Kolnberger and Altner, 1972). Secretory granules are prominent in these cells, and this secretory material, of unknown composition and function, is released during strong odor stimulation (Graziadei, 1971). In addition, the support cells contain much of the pigment that gives olfactory epithelium its color, ranging in different species from pale yellow to black. Evidence implicating pigments in the olfactory transduction process is inconclusive, and Moulton (1971) points out that pigment is lacking in the highly chemosensitive vomeronasal organ.

The basal cells undergo frequent mitoses (Seifert and Uie, 1967), and studies with tritiated thymidine have shown that they replace the other cell types in the epithelium (Moulton and Fink, 1972). Similar autoradiographic studies by Graziadei (1974) have shown that the axons from newly formed peripheral neurons grow along the existing olfactory nerve to the olfactory bulb, where they form synapses with second-order neurons. Behavioral and electrophysiological studies on pigeons have shown that these connections are "normal" and functional (Graziadei, 1974).

In all vertebrates, the axons of the olfactory receptor cells collectively form the olfactory nerve, which makes contact with the CNS in the olfactory bulb lying before the main forebrain mass. On the surface of the olfactory bulb, the receptor axons end in spherical masses of dense nervous tissue, the olfactory glomeruli. Here they synapse with a variety of second-order cells, principally the so-called mitral and tufted cells. Adrian (1950) and others showed some topographical localization of the receptor endings on the olfactory bulb, but the primary axons are interwoven in a complicated fashion. The synaptic connections between cells in glomeruli are also complex, and it has been suggested that the interaction between second-order cells could provide for "lateral" inhibition of mitral cells and thus sharpening of discrimination between odors

(Reese and Brightman, 1970). The degree of convergence in this system is considerable; about 25,000 axons enter each glomerulus in the rabbit, whereas from each glomerulus about 25 mitral cells send axons to the forebrain (Allison and Warwick, 1949).

2. *Vomeronasal Organ*

In mammals, the vomeronasal organ is usually in the form of a long, thin tube, partly lined by sensory epithelium, that opens anteriorly into the nasal cavity. In reptiles, it opens into the oral cavity. It is absent or vestigial in adult higher primates and in birds.

The sensory epithelium resembles the olfactory epithelium, the chief difference being that the receptor cells of the vomeronasal organ lack cilia, although they bear branched and somewhat irregular microvilli (Graziadei, 1971). In the rabbit, there are about 1,650,000 vomeronasal receptor cells on each side, that is, about one-thirtieth of the number of olfactory receptor cells (Allison, 1953). The axons from these cells form synapses in a distinct olfactory bulb.

3. *Gustatory Organs*

The taste buds of terrestrial vertebrates are situated principally in the oral cavity and on the surface of the tongue. The morphology of the tongue varies in relation to its role as an organ for gustation, mastication, grooming, or a combination of all of these. The dorsal surface of the tongue is rough and bears small papillae, which are divided into four categories on the basis of their morphology and distribution. Filiform or conical papillae are nongustatory and have a mechanical function. Fungiform papillae are buttonlike structures raised above the surrounding epithelium and bearing a few taste buds in their surface epithelium. They usually occur on the oral part of the tongue. The circumvalate papillae are similar but are sunk beneath the general surface of the tongue. They bear numerous taste buds. One to three circumvalate papillae commonly occur toward the back of the tongue. Foliate papillae lie on the lateral border of the tongue and are often called lateral organs. They consist of a series of clefts, each cleft bearing taste buds. Circumvalate and foliate papillae are restricted to mammals, and most physiological work has been done on foliate papillae of rats. The forms of vertebrate tongues and the occurrence of papillae on them are reviewed by Bradley (1971). Taste buds are not restricted to the tongue but have also been found in various vertebrates on the palate, the epiglottis, the larynx, the pharynx, and the esophagus (Bradley, 1971). The "bitter" sense is commonly associated with the taste buds at the back of the tongue.

Taste buds consist of more or less spherical assemblies of fusiform cells distributed as discrete units amidst the epithelial cells. They contain at least three different cell types: receptor, support, and basal cells (Fig. 3). Despite numerous histochemical and transmission electron microscope studies (see, e.g., Murray *et al.*, 1972; Graziadei, 1974), there is no general agreement on the exact role of all the cell types. The basal cells are undifferentiated and are believed to replace the other two cell types (Murray and Murray, 1970). The rate of turnover can be high. For example, in rats the life span of receptor cells is 10–12 days (Beidler and Smallman, 1965) so that, in one taste bud, cells of a given type may appear in a range of developmental forms (Fig. 3). In general, receptor cells have microvilli projecting into the apical pore of the taste bud, where they might encounter stimulant molecules. These cells have no axons but make synaptic contacts with sensory fibers of true neurons. The innervation is complex, and fiber branches from a single neuron may contact more than one receptor cell. Both "afferent" and "efferent" synapses occur between taste receptor cells and sensory fibers (Graziadei, 1974), the latter possibly indicating some sort of feedback mechanism or a trophic function of the nerve on the sensory cell. Certainly, cutting the sensory fiber results in the degeneration of receptor cells, which is detectable in 12–24 hours (Graziadei, 1971), and regrowth of gustatory nerves is followed by the reappearance of apparently normal taste buds (Guth, 1971). The life span of receptor cells is brief compared with that of the innervating neurons. Thus, the synaptic connections must continually be switched from dying receptor cells to their replacements.

The axons of neurons innervating gustatory receptors reach the brain via the fifth (trigeminal), seventh (facial), and ninth (glossopharyngeal) cranial nerves. In birds, the sense of taste is served only by the glossopharyngeal nerve (Wenzel, 1973), presumably because the relatively few taste buds in birds are situated in the posterior part of the oral cavity.

C. Nematodes

In no case as yet has the role of supposed chemoreceptors of nematodes been confirmed by electrophysiological recording, but nematodes have been shown to respond to chemicals (Croll and Matthews, 1977), and there is much information on the food and feeding habits of plant-parasitic species (reviewed in Doncaster, 1971).

Ward *et al.* (1975) reconstructed the anterior sensory anatomy of *Caenorhabditis elegans* from serial-section electron micrographs. Of the

58 anterior neurons, 52 are found in sensilla; these include 6 inner labial sensilla, 6 outer labial sensilla, 4 cephalic sensilla, and 2 anterior–lateral pits, the amphids. Each sensillum consists of ciliated sensory neurons ensheathed by nonnervous cells. The perikarya of the neurons lie near the nerve ring, their axons projecting into the ring or into ventral ganglia. The cilia of the neurons of the inner labial sensilla and of the amphids are exposed to the exterior by channels in the cuticle and may be chemoreceptors. Ward (1973) showed that the chemical sense of *C. elegans* includes detection of cyclic nucleotides, anions, cations, and hydroxyl ions, but if the amphids are malformed, as in the "blister mutants" of *C. elegans,* the response to chemical gradients is lost. Behavioral studies (Croll, 1970) have shown that both free-living and parasitic nematodes are attracted by various odorants and, on the basis of ultrastructural evidence, Burr and Burr (1975) suggested that the amphid of *Oncholaimus resicarius* contains olfactory cilia. Endo and Wergin (1977) studied the labial and cephalic organs of *Meloidogyne incognita* and found the ciliary endings identified in other nematodes. The cuticular pores associated with the labial organs indicate a chemoreceptive or secretory function.

D. Terrestrial Mollusks

The body wall of mollusks has sensory endings scattered evenly among epidermal cells in many regions, but there are clusters of sense cells in places and even special sense organs consisting of areas of specialized sensory epithelium, over which the cuticular covering may be reduced.

The receptor cells are primary, with the perikarya near the surface epithelium and axons proceeding to various parts of the nervous system. Receptor neurons have been ascribed to a number of categories, partly on the basis of their morphology and partly in an attempted correlation with the predominant sensitivity occurring in particular areas of epithelium. Schultz (1938), studying *Helix pomatia,* found chemoreceptors on both pairs of tentacles, on the front of the head and the oral lobes, on the upper surface of the foot, and on the anterior and posterior margins of the foot. Salánki and Van Bay (1975) found both tactile receptors and chemoreceptors on the surface of the lip of *Helix.*

In terrestrial pulmonates, the distal tip of each of the large tentacles bears a knob covered with sensory epithelium. The large number of receptor cells involved suggests that this organ has an important sensory role, and there is evidence that it is a chemoreceptor used in detecting smell stimuli from potential food material (Kittel, 1956). The fine structure of this organ was studied in *Helix pomatia* by Wondrak (1975). The chemoreceptor is covered with columnar epithelial cells, between which lie the more or less evenly spaced dendrites of sensory cells. The dendrites pass through the subepithelial layer of muscle cells and, converging in groups of 10 to 30, form bundles, each of which is enveloped by a glial cell. From the perikarya, lying beneath the muscle cell layer, bundles of axons extend to the digital ganglion in the tentacle, in which synaptic contacts are made with interneurons. Most of the sensory cells are bipolar neurons, and their dendritic endings typically bear cilia and long, tortuous microvilli. Wondrak identified six classes of neurons based on the occurrence and form of these organelles. The epithelial support cells surrounding the dendrites also bear apical microvilli, and membrane junctional complexes unite and encircle all the cells subapically.

In a similar study of the sensory epithelium at the tips of the large tentacles of the slug *Arion ater*, Wright (1974a,b) identified four cell types. In this mollusk, 80% of the dendrites had microvilli but lacked cilia. Sensory endings without cilia are rare in both vertebrates and invertebrates. It is suggested that in *A. ater* they are chemoreceptors since they are the most numerous type and chemoreception is the principal function of the organ. Wright suggests that the ciliated dendrites are mechanoreceptors since they are almost identical in structure with those of the free nerve endings occurring in the epithelium of the tentacle wall. Direct evidence for the role of these receptors is lacking, and their identification must be made with care because of the similar appearance of the apical region of support cells and the difficulty of tracing dendritic endings back to their cell bodies.

III. FUNCTIONING OF CHEMORECEPTORS

The perception of stimulating molecules involves similar processes in chemoreceptor cells irrespective of whether they are concerned with olfaction or gustation, and some cells respond to both types of stimulation. For instance, soluble, odorless chemicals stimulate the olfactory cells of the frog if the olfactory epithelium is perfused with a solution containing them. The electrical responses of the cells cannot be distinguished from those produced by odors (Gesteland, 1971). Conversely, contact chemoreceptors of the blowfly respond to high concentrations of odors (Dethier, 1972). For these reasons olfaction and gustation are considered together in this chapter.

A. Receptor Mechanisms

In vertebrates the close association of the olfactory epithelium with the respiratory system ensures that a flow of air carries odor molecules to the sensory epithelium, where they are trapped in the mucous covering. The anterior position of antennae in insects and the tentacles of terrestrial gastropods similarly increases the exposure of olfactory sense cells, and in insects the form and size of the sensilla favor the physical attraction of molecules. Experiments with ^{14}C- or ^{3}H-labeled odorants have shown that cuticular surfaces are good adsorbents and show negligible subsequent desorption (Kasang and Kaissling, 1972). Furthermore, the factors governing diffusion around cylindrical objects with very small diameters, such as olfactory hairs on a moth antenna, cause the hairs to act as very effective molecular traps. Thus, in *Bombyx*, the hairs account for less than 13% of the surface area of the antenna, but when ^{3}H-labeled bombykol is taken up by the antenna more than 80% of it is adsorbed by the hairs (Steinbrecht and Kasang, 1972). The measured rate of subsequent conveyance of molecules to the receptor sites indicates that they do not rely on hitting a pore to be effective (Steinbrecht, 1974). Instead, it is hypothesized, molecules move along the cuticular surface by a process of two-dimensional diffusion and through the pore tubules by one-dimensional diffusion (see Steinbrecht, 1974, for references). Most authors consider this process of capture of odor molecules to be nonselective, but Kasang *et al.* (1974) believe that in the case of insect pheromones there may be some degree of selectivity in the capture mechanism.

Gustation in vertebrates involves a process that is essentially similar to olfaction, except that the stimulating chemical is in higher concentration in solution. This is also often true for insect sensilla, but these also have the capacity to perceive chemicals on dry surfaces (see Chapman, 1977). It is conceivable that the mucopolysaccharide clothing the tips of the dentrites of insect gustatory sensilla to some extent determines the nature of chemicals reaching the dendrite endings (Bernays *et al.*, 1975; Stürckow, 1970), although there is no firm information on this point. It was shown by Bernays *et al.* (1972) that in *Locusta migratoria* a centrally controlled mechanism regulates peripheral sensory input in relation to the degree of satiation by causing the pore at the tip of some sensilla to be closed when the insect has fed, but this mechanism has not yet been found in other insects.

After capture on the outer surface of the chemoreceptor, the stimulating molecule diffuses in and reaches the tip of the dendrite. The response of a receptor cell to stimulation by a chemical is to develop a generator potential. This is induced by changes in the permeability of the cell membrane to ions as a result of the interaction of stimulating molecules with molecules in the cell membrane. It is at this stage that cell specificity to particular compounds or classes of compounds is likely to occur.

Molecular configuration of the stimulating molecule is critically important in determining specificity. The most intensive studies have been concerned with insect olfactory systems, and Kafka (1970), for instance, found that the coeloconic sensilla on the antenna of *Locusta* were highly sensitive to only 3 closely related compounds (hexanoic acids) out of 370 tested. The activity of the cell varied as the form of the molecule changed so that, whereas 2-oxohexanoic acid is maximally stimulating, 2-hydroxyhexanoic acid does not stimulate at all. Small configurational changes also profoundly reduce the stimulating power of insect pheromones (Kaissling, 1974).

Some data exist on nonvolatile secondary plant compounds (see Schoonhoven, 1972). For instance, the glucosinolate receptor of larval *Pieris brassicae* responds to only seven glucosides with the formula

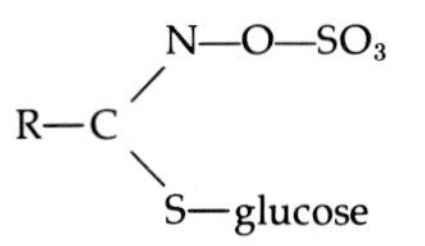

Other compounds are not stimulating. However, there is no obvious correlation between the structure of the R group and the relative stimulating power of the glucoside. In contrast, the deterrent cell (see Section III,B,2) of larval *Bombyx mori* is stimulated by a range of compounds with no obvious structural relationships.

Kubota and Kubo (1969) consider that the bitterness of some diterpenes for man is dependent on the molecular structure and involves a "bitter unit" consisting of a proton-donor group and a proton-acceptor group close enough together to be hydrogen bonded. Examples of proton-donor groups are OH, CH, $COOCH_3$, and CHCO; examples of proton-acceptor groups are CHO, CO, and COOH. Diterpenes without this arrangement do not taste bitter, but the presence of the "bitter unit" alone may not produce a bitter sensation if the whole molecule is not also of an appropriate size and shape.

In general, it is presumed that the precise molecular form determines the readiness with which a stimulating molecule combines with an acceptor molecule on the receptor cell membrane. The number of compounds capable of "fitting" the acceptor molecule may be limited. Structurally different compounds all of which stimulate a single receptor cell may involve a range of acceptor sites with different characteristics unless, as with the bitter-tasting compounds described by Kubota and Kubo (1969), all the stimulants have a common characteristic. They assume that at the receptor cell membrane a proton donor of the protein acceptor molecule severs the hydrogen bond in the "bitter unit" of the stimulating molecule, and its proton donor and acceptor groups then bond with groups on the acceptor molecule. There are, however, other possibilities (see below).

There is a good deal of evidence that specific acceptor molecules do exist (Dastoli, 1974; Price, 1974; Schoonhoven, 1974) and some evidence that different taste modalities have distinct acceptor molecules.

The most extensive studies have been concerned with "sweet" taste in mammals. A protein that forms complexes with chemicals that taste sweet to man has been isolated from the tongues of cows (Dastoli and Price, 1966), rats (Hiji *et al.*, 1971), and dogs. Although the protein occurs in other tissues, as well as in taste buds, it complexes with sugars, saccharin, and sweet-tasting amino acids, and it is probable that it is the receptor for sweet taste stimuli (Price, 1974). The recent immunological studies of Hough and Edwardson (1978) provide a new approach to this study. In insects (Hansen, 1974), it is considered certain that the enzyme α-glucosidase is involved in the reception of sugars, and numerous studies of flies seem to confirm this (see Kijima *et al.*, 1977, for references).

The information on the reception of bitter-tasting substances by mammals has been reviewed briefly by Dastoli (1974) and Price (1974). Dastoli *et al.* (1968) isolated a protein from the papillae at the back of pigs' tongues that formed complexes with quinine hydrochloride, brucine hydrochloride, and caffeine, but their conclusion that this protein represents a receptor protein has been questioned (Price, 1969).

Norris and associates (see Singer *et al.*, 1975) showed that 1,4-naphthoquinones, which are feeding deterrents for *Scolytus multistriatus* and *Periplaneta*, bind with proteins in the receptor cell membranes from the antennae of *Periplaneta*. It is not certain that this mechanism represents a receptor mechanism *sensu strictu* for 1,4-naphthoquinones (see Section III,B,2).

Price (1974) suggested that bitter substances may react directly with the enzyme phosphodiesterase or affect it indirectly through a phosphodiesterase inhibitor. He found (1973) that the enzyme was activated by most bitter substances including strychnine and nicotine. Increased phosphodiesterase activity would reduce the level of cyclic AMP in the cell and so reduce the permeability of the cell membrane, in keeping with the observations of Ozeki (1971). Kurihara (1972) showed that the activity of the enzyme phosphodiesterase is higher in bovine sensory papillae than in the surrounding tissue but found that its activity was *inhibited* by 11 bitter substances, including a number of plant secondary compounds. In a histochemical study of taste buds on the tongue of the rhesus monkey, Trefz (1972) found that papillae from different parts of the tongue mostly contained the same enzymes although in different amounts, but β-hydroxybutyrate dehydrogenase was found only in presumed bitter receptors at the back of the tongue.

A quite different mechanism of reception of bitter, salt, and acid stimuli was proposed by Kurihara (Kurihara *et al.*, 1975). He showed (Kurihara, 1975) that there is more phospholipid in the sensory papillae on the bovine tongue than in the epithelium in between them and suggested (Kurihara, 1973) that there is some interaction between bitter substances and the phospholipids because more lipid is extracted from the excised papillae following incubation with brucine or strychnine. He suggested that the initial event in stimulation of a receptor cell by a bitter substance is the interaction of the substance with the phospholipid of the cell membrane. The development of a receptor potential stems from changes in electrical potential at the membrane–solution interface (Kurihara *et al.*, 1975). Koyama and Kurihara (1972) suggested that different classes of stimulating chemicals interact with membrane lipid in different ways. Salts may associate with polar regions of the lipid, whereas bitter substances associate with the nonpolar parts. They did not postulate the presence of receptor proteins in the cell membrane. This is supported by Hiji and Ito (1977), who found that treatment of the rat tongue with proteases, which presumably interact with acceptor proteins, reversibly eliminated the electrophysiological response to sucrose but had no effect on the responses to quinine, hydrochloric acid, and sodium chloride.

In general, it is assumed that the association of the stimulating molecule with the acceptor molecule leads to an increase in conductance across the receptor cell membrane, followed by a depolarization of the cell (Kaissling, 1974). In the taste receptors of frogs and rats, which are not primary neurons (see Section II,B,3), depolarization is similarly associated with an increase in conductance when the cells are stimulated by sodium chloride, sucrose, or hydrochloric acid, but a *decrease* in conductance (increased membrane resistance) occurs when they are stimulated by quinine (Akaike and Sato, 1976; Akaike *et al.*, 1976; Ozeki, 1971). It is suggested that in the first case the membrane increases in permeability to sodium and chloride ions, but in the latter the permeability to potassium ions is reduced.

At present there is no clear picture of how secondary plant compounds are received. It may be that wherever some degree of specificity occurs, as in the glucosinolate receptor of *Pieris,* a specific acceptor molecule exists. Elsewhere, with cells responding to a wide range of chemicals, a more generalized interaction with the receptor cell membrane may occur. Following the production of a receptor potential it is necessary that inactivation of the stimulating molecules occurs rapidly so that further stimulation is possible. With insect pheromones this inactivation involves their degradation by enzymes (see Mayer, 1975).

B. Neurophysiology

It is important to appreciate some of the problems of neurophysiological recording in order that the possible limitations in interpreting results be understood. For this reason we give a brief account of methodology before describing the work carried out.

1. *Methodology*

The essential feature of neurophysiology is that electrical activity emanating from a receptor organ or part of an organ is detected as a variation in potential difference between two metal electrodes. Commonly, one electrode, the recording electrode, is placed close to the nerve cells and the other, the indifferent electrode, some distance away in the body of the animal. It is often desirable to position the recording electrode very precisely in order to record from only one cell or a few cells. With tungsten this can be achieved by electrolytically etching a piece of tungsten wire to a pointed tip of less than 1 μm diameter and insulating it except at the extreme tip. Alternatively a glass capillary can be drawn to form a microcapillary pipette of similar tip diameter. The pipette is then filled with an electrolyte, and a chlorided silver wire is placed inside it. Variations of these techniques have been developed to suit particular recording situations in different animals and to detect different types of electrical activity.

When an insect antenna is excised and a glass microcapillary electrode inserted into each end, the summed receptor potentials from many olfactory sensilla can be recorded when the antenna is suitably stimulated. This slow, graded potential is called an electroantennogram and was first demonstrated by Schneider (1957). It is useful in studying antennae having large numbers of receptor cells responding identically but otherwise gives a confused picture because many cells may react to the same stimulus in different ways.

Recordings can be made from individual olfactory sensilla in insects by penetrating the cuticle adjacent to the sensillum with a tungsten or glass electrode. Action potentials are recorded, and if there are few cells in the sensillum it may be possible to identify the output from individual cells on the basis of the amplitude, shape, and temporal occurrence of the spikes (see e.g., Dethier and Schoonhoven, 1969). Such identifications should always be made with extreme caution.

An alternative method for gustatory hairs is to place a glass capillary electrode with a tip diameter of about 10 μm over the tip of the hair. The electrode contains both an electrolyte, to ensure electrical conductivity, and the stimulating chemical. The solution gains access to the dendrites through the apical pore in the sensillum, and action potentials are recorded antidromically through the dendrites, which act as an extension of the capillary electrode right down to the site of spike initiation (Hodgson *et al.*, 1955). This technique, called tip recording, is not without handicaps. The effect of the electrolyte cannot be properly assessed, and only rarely has an amplifier been developed that allows recording with distilled water in the stimulating capillary (see, e.g., van der Starre, 1972). Only substances miscible with water can be tested. This constraint was overcome by Blaney (1975), who sonicated leaf surface waxes to disperse them in the stimulus solution.

The disadvantage of the presence of electrolyte in the stimulating solution has been overcome in some studies by piercing the side of the hair shaft with the recording electrode so that electrical continuity is achieved between the recording capillary and the dendrites (Morita, 1959). Stimulants can then be applied by an independent capillary over the hair tip. In addition, the unstimulated response of the hair can be recorded. Unfortunately, changes of pressure applied to the side of the hair throughout recording affect the firing threshold of the receptor cells so that the results can be rather variable (Blaney, 1974; Dethier, 1974). This side-wall method and the tip recording method share a disadvantage common to almost all chemoreceptor recording, namely, the impossibility of stimulating and recording from only one receptor cell at a time and the consequent difficulty of identifying the responses of individual receptors as soon as four or more cells are involved. This problem was discussed more fully by Blaney (1974) together with the problem of variability of responses, which was also discussed by Dethier (1974).

Recordings from the CNS of insects, mollusks, and vertebrates are made by penetrating the nerve tissue with a very fine tipped metal electrode or glass microcapillary. Responses are recorded when the capillary tip is fairly closely apposed to or is within an axon. Even in extracellular recording in the CNS, individual cells can be marked by cobalt chloride leached from the recording capillary and subsequently converted to cobalt sulfide (Pitman *et al.*, 1972).

It is possible to record summed receptor potentials from many cells in vertebrate olfactory epithelium, giving a response similar to the electroantennogram of insects. This electroolfactogram is recorded by placing a glass capillary recording electrode into the mucous coat and in contact with the epithelial layer. Finer microcapillaries may be used to record from individual receptor cells in the olfactory epithelium or from their axons in the olfactory bulb (Gesteland *et al.*, 1963; Shibuya and Tonosaki, 1972).

Neurophysiological studies of taste in vertebrates have focused on axons in the chorda tympani, a branch of the seventh cranial nerve, because of the accessibility of these axons to exposure by superficial surgery on a tranquilized animal and the ease of stimulating their receptor fields on the tongue. Typically, the nerve is cut and placed on fine recording electrodes and teased apart so that only a few fibers remain in effective contact with the electrodes, and single fiber responses, judged by the appearance of spikes of constant height, are obtained (e.g., Pfaffman *et al.*, 1976).

Depolarizations can be recorded from the taste receptors on the tongue with a glass microcapillary electrode manipulated through the pore of a taste bud and into one of the receptor cells. Penetration is recognized by detection at the microelectrode tip of a sudden negativity of about 20 to 80 mV (Sato and Ozeki, 1972). Support cells, however, show a similar potential change on penetration, and identification of gustatory cells is made on the basis of depolarization in response to chemical stimulation, to which the support cells are unresponsive.

2. *Neurophysiological Responses*

The most extensive studies on the neurophysiology of chemoreceptors have been on insects, and most work involving phytophagous insects has been done on caterpillars by Dethier (1973), Ma (1972, 1977), and Schoonhoven (1973, 1974) and on locusts by Blaney (1974, 1975), Haskell and Schoonhoven (1969), and Kafka (1970). The extensive work on flies is drawn together by Dethier (1976), and Kaissling (1971) reviews insect olfaction. In most of these cases recordings have been made from sensilla rather than individual receptor cells (see Section III,B,1), but it is often possible to distinguish among different cells from the different characteristics of their action potentials. The vertebrate studies have been done mainly on rats (Frank, 1975; Ozeki and Sato, 1972) and monkeys (Pfaffmann *et al.*, 1976; Sato, 1975) and have commonly involved single-cell recording from the receptor cells or the neurons of the chorda tympani.

Inorganic salts and sugars have been most widely used as stimuli in experiments aimed at achieving an understanding of the functioning of chemoreceptors, but relatively extensive ranges of secondary plant substances have been tested against some insects [notably by Schoonhoven (1973) and Wieczorek (1976)]. In vertebrates, the approach has been colored by a consideration of the four taste modalities: sweet, salt, sour, and bitter. Quinine, either as the sulfate or the hydrochloride, has been used as a representative bitter substance in almost every instance, and consequently it is the only secondary plant chemical to have been extensively studied.

Olfactory and gustatory receptor cells vary in the range of chemicals to which they respond. At one extreme are specialized cells, which are highly sensitive to one specific chemical and show little, if any, sensitivity to other chemicals; at the other extreme are "generalist" cells, which respond to a wide range of chemicals of many different classes. There is no fundamental difference between specialist and generalist cells, and between these extremes are cells showing various degrees of specificity to certain chemicals or chemical classes.

Highly specific receptors are to be expected only when an animal regularly encounters a chemical that is of special significance to it at some stage in its life. The best known examples are the pheromone receptors of male moths, in which the receptor cells are hundreds or thousands of times more sensitive to the pheromone than to any other compounds (Kaissling, 1971). Cells with similar properties might also be expected to occur in herbivorous animals that feed only on plant species characterized by a specific chemical or group of chemicals.

Some insects are restricted to feeding on Cruciferae and in two cases are known to possess receptor cells that are particularly sensitive to the glucosinolates present in the host plants. These are the cabbage white butterfly (*Pieris brassicae*) and the cabbage moth (*Mamestra brassicae*). The larva of the former has one cell in each of the maxillary medial and lateral sensilla that responds only to the glucosinolates (Ma, 1972; Schoonhoven, 1973). Extensive studies with *Mamestra* by Wieczorek (1976) showed a similar cell in the lateral sensillum of the larval maxilla. This cell does not respond to sugars, salts, most amino acids, or phenolic compounds, but it is stimulated by a range of glucosinolates and nonheterocyclic arylglycosides. The threshold for sinigrin and 1-naphthyl-β-glucoside is about $10^{-3}\,M$. This is not a particularly high level of sensitivity but is appropriate to the concentrations found in host plants. Specific glucosinolate receptors may also occur on the tarsi of adult females of *Pieris,* which lay their eggs on Cruciferae (Ma and Schoonhoven, 1973).

The chrysomelid beetle *Chrysolina brunsvicensis* provides another example of a specific receptor. This insect feeds on species of *Hypericum* containing hypericin (hexahydroxydimethylnaphthodianthrone), and the tarsal receptors contain a cell that is highly sensitive to this chemical, with a threshold response at 10^{-5} *M* reaching a plateau at 10^{-2} *M*. Homologous sensilla on the tarsi of other species of *Chrysolina* have no response to hypericin at 5×10^{-3} *M* (Rees, 1969).

Comparable specificity of feeding is uncommon in vertebrates, and no electrophysiological studies have been carried out on relevant species. It is to be expected, however, that wherever such specificity occurs, as in the dependence of the koala bear on *Eucalyptus*, highly specific receptor cells will occur.

Such a high degree of specificity of a receptor to a secondary compound that is deterrent is not known, nor is it to be expected since the occurrence of any one deterrent chemical in the life of an animal is not likely to be consistently important. Hence, selection for this feature is unlikely to have occurred. It is known, however, that food selection by many phytophagous insects is determined by the distribution of different deterrent chemicals in plants that are not normally eaten (see, e.g., Bernays and Chapman, 1977; Thorsteinson, 1960). Schoonhoven (1973) records the presence of a "deterrent receptor" in the caterpillars of 8 of 18 species of Lepidoptera that have been studied. In each insect, a cell in one of the maxillary sensilla respond to a range of secondary compounds from different chemical classes, although the spectrum of response varies with the insect species. Commonly, it is the only cell in a sensillum to respond to these chemicals.

Comparable data with a range of secondary chemicals are not available for vertebrates, but some cases have been recorded of cells being particularly sensitive to quinine. For instance, 19 of 52 single fibers in the glossopharyngeal nerve of the rat responded more strongly to stimulation of the tongue with 0.001 *M* quinine hydrochloride than they did to 0.3 *M* sucrose, 0.3 *M* sodium chloride, or 0.003 *M* hydrochloric acid. The response to quinine was correlated with the response to caffeine, but not with the other substances tested (Frank, 1975). Eleven of 63 fibers in the chorda tympani of the macaque showed a similar high sensitivity to quinine (Sato, 1975). The output of these neurons in vertebrates is derived from a number of receptor cells so that its significance, in terms of receptor cell function, is difficult to determine. However, Ozeki and Sato (1972) found a correspondence in the proportion of receptor cells on the rat tongue and neurons in the chorda tympani responding to particular stimuli.

Quinine has no particular significance in the normal lives of these animals, and the range and concentrations of other chemicals tested are very limited. Nevertheless, it appears reasonable to conclude that the receptor cells do show marked differences in their responsiveness to chemicals and that some cells are very sensitive to quinine. If quinine can be regarded as a representative secondary compound, as it is regarded as a typical "bitter" compound, it may be that these cells are comparable with the "deterrent" cells of caterpillars.

In most cases, however, the degree of specificity to individual compounds or classes of compounds is relatively slight. Most receptor cells respond to a range of chemicals, although with differing sensitivities. Dethier (1973) investigated the responses of the maxillary sensilla of the caterpillars of seven species of Lepidoptera to a range of glycosides, salts, sugar, and plant extracts. In all the species, he found that cells in any one sensillum responded to a wide range of different chemicals, and he concluded that there was no consistent difference between the sensory patterns associated with the acceptance of a material and those associated with rejection.

Dethier and Schoonhoven (1969) reached a similar conclusion with regard to olfaction in the caterpillar of the tobacco hornworm *Manduca sexta*. Using extracellular recording from receptor cells in the antenna, they found that each cell responded to a range of odors, including those of secondary plant chemicals such as limonene and citral, and that the temporal patterns of spike activity produced in different neurons by different chemicals could provide a code by which each substance could be recognized (Fig. 4).

The terminal sensilla on the palps of the migratory locust *Locusta migratoria* generally contain six neurons, which show a marked lack of specificity (Blaney, 1974, 1975). From one to three cells in each sensillum respond to sodium chloride, and each sensillum responds to a wide range of chemicals, sometimes with a decrease in the overall firing rate (Table I). Considerable variation occurs within and among sensilla in the response to any one chemical.

The same pattern emerges from studies on gustation and olfaction in vertebrates, in which most neural elements are multisensitive (Bernard, 1975; Gesteland, 1971; Macrides, 1976). Ozeki and Sato (1972), for instance, measured the receptor potentials in cells of the fungiform papillae on the tongue of the rat and found that most cells responded to all the chemicals and each one had a characteristic profile of responses that distinguished it from other cells. Pfaffmann *et al.* (1976) used quinine and sodium chloride in addition to a range of sugars and ammonium chloride in examining the responses of receptors in the tongue of the squirrel monkey. They recorded from fibers in the chorda tympani. Of 43 fibers, 13 responded similarly to all the substances tested, and the rest

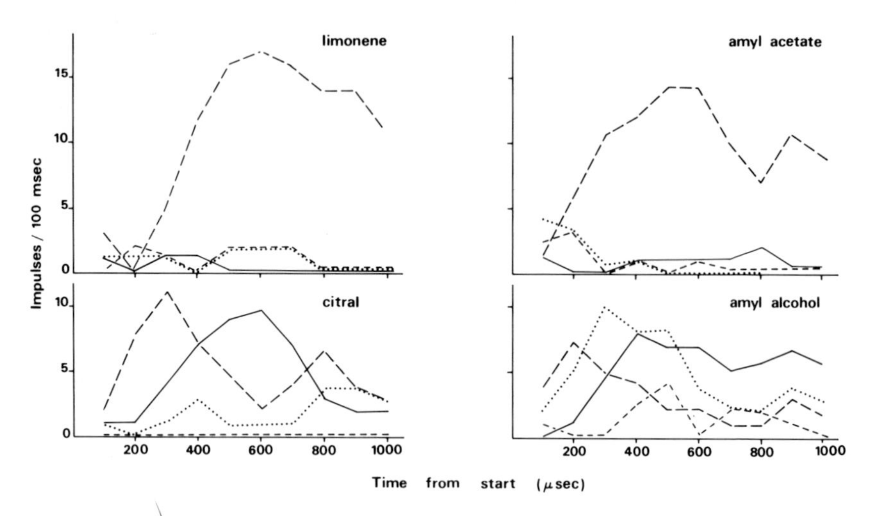

Figure 4. Temporal patterns of the frequency with which nerve impulses are produced in a sensillum of the caterpillar of *Manduca sexta* in response to different chemicals. Each symbol indicates the frequency of occurrence of spikes of a particular amplitude. After Dethier and Schoonhoven (1969).

ranged from some giving no response to quinine to others responding to quinine only, but nearly all of them responded to more than one chemical.

These examples show that in both insects and mammals a majority of chemosensory cells are "generalists" responding to a range of chemicals. Most chemicals that an animal smells or tastes are perceived by these cells rather than by specialist cells adapted to the reception of a particular compound.

Table I. Responses of ten sensilla on the maxillary palp of *Locusta* to various compounds

	Sensillum[a]									
Stimulus solution	1	2	3	4	5	6	7	8	9	10
Sodium chloride	+	+	+	+	0	+	+	−	0	+
Glucose	0	+	0	+	0	+	+	+	0	+
Serine	−	0	+	0	−	+	0	0	−	+
Quinine hydrochloride	0	0	+	+	0	+	0	0	0	0
Neem[b]	0	+	+	0	−	0	−	−	0	0
Anthroquinone	−	−	0	−	−	−	0	−	−	−

[a] Key: 0, no change from spontaneous firing rate; +, increase in firing rate; −, decrease in firing rate.

[b] Extract of seeds from the neem tree *Azadirachta indica*.

In most of the instances so far cited, the characteristics of the responses of the cells to secondary plant compounds are no different from those to any other chemical. The firing rate increases sharply immediately after stimulation and then rapidly adapts to a low tonic level. Both the peak firing rate and the tonic level are proportional to the concentration of the stimulating chemical (Ma and Schoonhoven, 1973; Rees, 1969).

In other cases, however, secondary plant chemicals and other chemicals producing some kind of aversive behavior result in the production of action potentials in irregular sequences or bursts. For example, high concentrations of monocarboxylic acids cause the fly *Phormia regina* to extend the proboscis partially and wipe it with the forelegs; fluid may be regurgitated. This behavior is associated with irregular firing of the sensilla receiving the stimulus, the spikes often occurring in bursts or volleys in several cells of each sensillum (McCutchan, 1969). Stürckow (1959) recorded similar activity in receptors of the Colorado potato beetle *Leptinotarsa decemlineata* during stimulation by the alkaloids tomatine, solanine, and demissine, although solanine is a normal constituent of the potato plant. High concentrations of some odors, including citral and limonene, produce similar irregular activity in the labellar and tarsal receptors of *Phormia* (Dethier, 1972). These substances also cause evasive behavior by the insect, but Haskell and Schoonhoven (1969) recorded bursting activity of some mouthpart sensilla of locusts following stimulation by extracts of grass, which is readily eaten. The last instance cannot be regarded as an abnormal response but, in some cases of irregular bursting, the neurons may continue to produce action potentials after the stimulus has been removed, perhaps indicating that the cell has been damaged by the stimulus.

A total cessation of sensory output is known to occur after stimulation by some substances. A well-documented example is the effect of warburganol on the maxillary sensilla of the caterpillar of the African armyworm *Spodoptera exempta* (Ma, 1977). Warburganol is a sesquiterpene dialdehyde occurring in *Warburgia.* It produces a normal response in the receptor cells when first applied at a concentration of 0.001 *M*, but after 1 minute bursting activity begins in the neuron, and subsequently neural activity declines. A 3-minute application of warburganol to a sensillum subsequently reduces the response to sugar, and two such applications cause a very marked reduction (Fig. 5), from which recovery is slow.

Receptor activity in *Spodoptera* is also depressed by *p*-chloromercuribenzoate (PCMB), a sulfhydryl-blocking reagent (Ma, 1977) that also affects receptors in other animals. The activity of salt and sugar

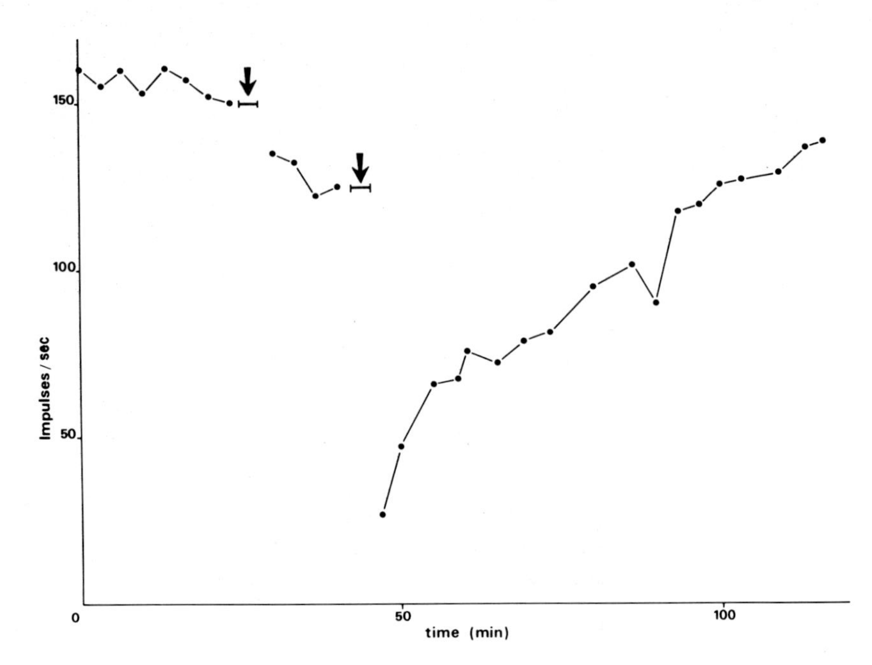

Figure 5. Effect of 1 m*M* warburganol on the response of a sensillum of the caterpillar of *Spodoptera exempta* to 10 m*M* sucrose. Each horizontal bar represents a 3-minute treatment with warburganol. After Ma (1977).

receptors of the flesh fly *Boettcherisca peregrina* is reduced by 40–60% following a 3-minute exposure to 0.005 *M* PCMB. After a 10-minute exposure the cells do not recover (Shimada *et al.*, 1972). The same concentration of PCMB perfused over the tongue for 20 minutes reduces the total response to sugar from the chorda tympani of the rat. At higher concentrations the responses to salt and quinine are also reduced (Noma and Hiji, 1970).

Norris and associates (Norris *et al.*, 1970, 1971; Rozental and Norris, 1973; Singer *et al.*, 1975) showed that 5-hydroxy-1,4-naphthoquinone and other 1,4-naphthoquinones are feeding deterrents for the bark beetle *Scolytus multistriatus* and the cockroach *Periplaneta americana.* They showed that these compounds form complexes with sulfhydryl groups of the receptor proteins and consider this to be a mechanism by which deterrent compounds are detected. Warburganol and PCMB probably act in a similar way (Ma, 1977; Shimada *et al.*, 1972). However, all the experiments involved exposure of the receptors for relatively long periods and, although it is certain that the blocking of sulfhydryl groups at or near receptor sites does occur, it is questionable whether this is the normal mechanism of reception.

Cyanide also acts by inhibiting normal sensory responses. Levinson *et al.* (1973) showed that a 30-second exposure of the antenna of the male silk moth *Bombyx mori* eliminated the development of receptor potentials when the antenna was subsequently exposed to the female pheromone, and a 1-minute exposure of the antenna of the burnet moth *Zygaena filipendulae* reduced the response to geraniol. Recovery did not occur in the first example and took some time in the second.

The effects of prolonged exposure of chemoreceptor cells to quinine are known to be similar to these in vertebrates. Akaike and Sato (1976) record that prolonged exposure of the fungiform papillae on the tongue of the bullfrog *Rana catesbeiana* reduced the magnitude of the depolarization produced by various salts and observed a parallel reduction in neural activity in the glossopharyngeal nerve in response to a range of stimulating chemicals (Fig. 6). Similar effects were produced by other bitter-tasting chemicals and by the local anesthetics procaine hydrochloride and lidocaine hydrochloride (Akaike and Sato, 1975). Gymnemic acid, from *Gymnema sylvestre,* reduces the sensory response to

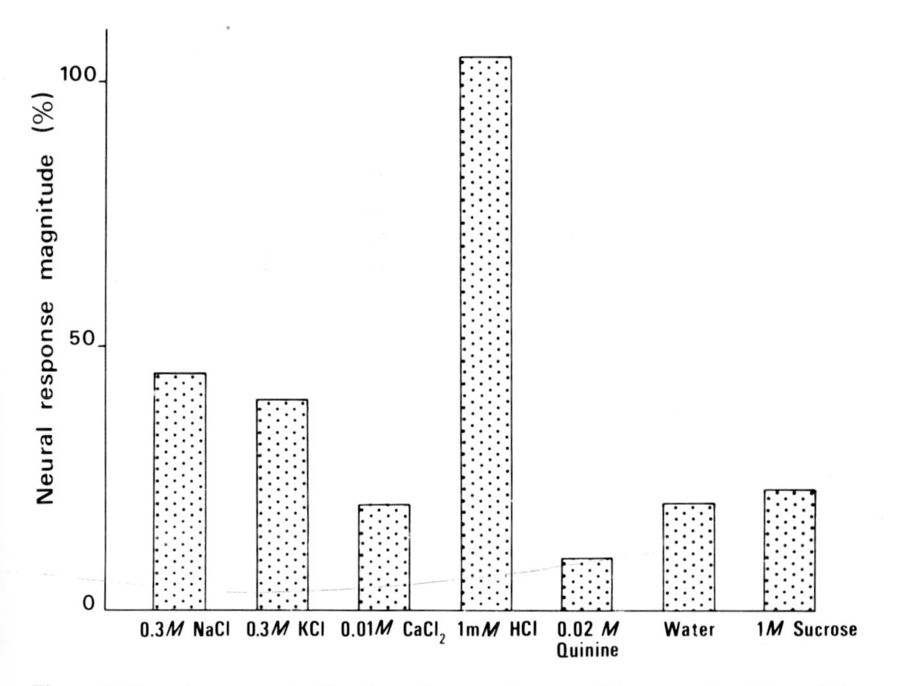

Figure 6. Neural response in the glossopharyngeal nerve of frogs to stimulation of the tongue with various chemicals following stimulation with 1 m*M* quinine. The magnitude of the response is expressed as a percentage of the response occurring in the absence of quinine. After Akaike and Sato (1976).

sucrose in dogs (Andersson *et al.*, 1950) and hamsters (references in Faull and Halpern, 1971) but not in monkeys (Hellekant *et al.*, 1974).

In all these cases the effect of prolonged exposure of the chemoreceptor to the secondary plant compound is to reduce the response of the receptor cells to stimulation by other classes of compounds. It is not known whether this effect is produced by a wide range of secondary plant compounds, but it could be an important mechanism by which these compounds influence the behavior of animals. Perception of the secondary compounds themselves probably involves normal electrophysiological responses, as in the case of quinine and warburganol, and it is important to distinguish between the effects of short-term and long-term exposure to these substances.

C. Information Processing

The information reaching the CNS from peripheral sensilla partly determines the subsequent overt behavior of the animal. In the case of a secondary plant chemical, the animal may make some kind of aversive behavior, or it may respond to the stimulus by feeding. The organism is clearly able to categorize the information it receives as indicating acceptability or otherwise; in other words, it is capable of some degree of recognition.

Wherever specialist cells exist, the potential for recognition of a particular chemical exists peripherally. The information transmitted by these cells to the CNS indicates the presence of one specific chemical, and the behavior of the animal is programmed genetically so that it responds appropriately. It is presumed that no central analysis of the information is required, although the response will depend on other conditions being appropriate. The recognition of the qualities of most chemicals, however, poses a different problem. They stimulate many different cells, and the only information reaching the CNS from each receptor is a train of action potentials. These trains may differ in the frequency with which action potentials occur, the rate of increase in frequency at the beginning of stimulation, and the subsequent rate of adaptation. In addition, the ratio of the peak frequency of spike production to the adapted frequency may vary. Bernard (1975) considers that these different features in the input from a single cell can indicate the general character, or modality, of the stimulating chemical. Thus, in the case of man they might indicate whether it is sweet, salt, sour, or bitter. In other organisms the range of modalities might be greater. However, these features would not be sufficient to permit more precise recognition of stimulating chemicals, although it is certain from behavioral studies

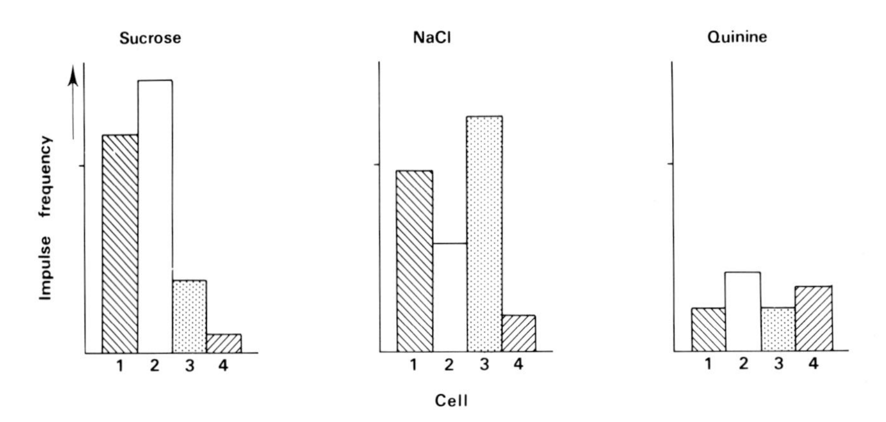

Figure 7. Diagram showing the manner in which across-fiber patterning among four different cells would enable an animal to distinguish different chemicals.

that this is possible. Bernard considers that this level of recognition depends on across-fiber patterning. Most authors are partial to the view that recognition at all levels is dependent on this process (e.g., Blaney, 1975; Dethier, 1976; Pfaffmann *et al.*, 1976; van der Starre, 1972; Yamada, 1971).

Since generalist receptor cells differ in their responses to a chemical, the overall pattern of sensory output generated by any one chemical in all the receptors or afferent fibers is unique. Recognition of this chemical then depends on the interpretation of this total pattern within the CNS. This is the phenomenon of across-fiber patterning. Figure 7 gives a very simple example of the basis of such discrimination. All the cells respond to sucrose and to salt, but the two can be differentiated by the different levels of response in cells 1, 2, and 3. Quinine is distinguished from both these substances by the relatively high firing rate in cell 4 compared with that in the other cells, even though the absolute rate of firing is low.

IV. URTICATION

Contact dermatitis, or urtication, is due to pharmacologically active chemicals associated with plants having stinging hairs. These plants occur in four families: Urticaceae, Euphorbiaceae, Loasaceae, and Hydrophyllaceae. On contact, the tip of the hair breaks off, and its contents are injected into the skin of the animal. In some plant species, the contents include acetylcholine, histamine, and 5-hydroxytryptamine, but these are absent from other species and are probably not the main toxic agents (Fastier and Laws, 1975; Thurston and Lersten, 1969).

The effect of the sting in most instances is to cause a local flare on the skin surface and a local increase in skin temperature due to the dilation of arterioles. Subsequently, the skin erupts in a rash of small papules. These effects cause pain and itching of the affected area. In more extreme cases, the reaction may be more general, resulting in vomiting and sometimes even death (Thurston and Lersten, 1969).

V. LEARNING

The avoidance or acceptance by a herbivore of a plant containing particular secondary plant chemicals is often innate. This is obviously true of many insect responses and of the vertebrate response to bitter substances. If perception of the chemical is associated with some other stimulus, it is possible that subsequently the behavior of the animal is affected by that stimulus alone, without direct contact with the chemical. Such a learning process has been shown to occur in snails (*Helix aspersa*), which, after climbing up a vertical rod and coming into contact with a quinine-soaked thread, subsequently climbed more slowly than control snails, which had not encountered quinine (Siegel and Jarvik, 1974; Stephens and McGaugh, 1972). Gelperin (1975) showed that the slug *Limax maximus* learned to avoid a palatable food if its presentation was associated with an aversive stimulus, in this case CO_2 poisoning.

An example of such behavior in a field situation is described by van Lawick-Goodall (1968). She saw a chimpanzee pick unripe figs and drop them after one bite. Subsequently, the ape selected fruit from the feel, squeezing each one gently with his fingers. Unsuitable fruits were left attached to the branch. It is a reasonable inference that this animal had learned to associate hardness of fruit with an unpleasant taste. This type of association is probably widespread in herbivorous vertebrates, although there is no adequate documentation. Man, having once been stung, learns to avoid urticating plants by their visual form, and presumably some grazing animals do the same thing.

The most extensive studies on such learned aversion have been concerned with vertebrate predators feeding on insects that sequester toxic secondary plant chemicals. Brower (1969) showed that the monarch butterfly *Danaus plexippus* sequesters cardiac glycosides from its host plant *Asclepias currassavica*. Blue jays accepted butterflies reared on host plants lacking the glycosides but vomited after eating a butterfly that contained the glycosides and subsequently rejected any adults of *D.*

plexippus that were offered. Similar examples involving the storage of cardiac glycosides are quoted by Rothschild and Kellet (1972), and that of pyrrolizidine alkaloids by Rothschild *et al.* (1977) and Bernays *et al.* (1977). This learned aversion has also been extensively studied in rats and in carnivores (Garcia and Hankins, 1975), where commonly the toxicosis has been induced by lithium chloride.

The mechanism of perception of the toxic effect varies with the chemical, its concentration, and the way it acts. Much of the published work describes vomiting. This might be induced by the direct stimulation of esophageal chemoreceptors but, when there is a delay between feeding and vomiting, it may result from absorption of the chemical, which then acts directly on brainstem cells. Respiratory distress and cardiac irregularities may result from the ingestion of some secondary plant chemicals and may be registered by the animal as indicators of toxicosis. Urticating substances are perceived by their irritant effect on the skin.

The subsequent avoidance of the food source need not involve the toxic chemical itself, but qualities that the animal associates with it. With carnivores the taste of meat that caused sickness was of major importance, but visual or olfactory stimuli may also have been involved (Garcia and Hankins, 1975). Man and presumably other animals recognize urticating plants by their visual form. With predators of distasteful insects, vision is of primary importance in subsequent recognition and avoidance. Insect species that sequester plant toxins are aposematically colored so that they are readily recognized. This association of a toxic principle with recognizable characteristics that are not themselves harmful provides a basis for mimicry, in which other individuals of the same or other species avoid predation, even though they are not unpalatable, by adopting the characteristics of individuals that are unpalatable (see, e.g., Brower, 1969).

VI. EVOLUTION OF PERCEPTION

On the basis of the current distribution of secondary plant chemicals, it is probable that the plants in existence when the first herbivorous insects evolved were already well endowed with condensed tannins as a defense against microorganisms and plant pathogens (Swain, 1976). It seems likely that these insects already possessed the sensory capacity to perceive these compounds since sense organs capable of perceiving potentially toxic chemicals are found even in Coelenterata (Garcia and Hankins, 1975). Of present-day herbivorous insects, the grasshoppers

(Acridoidea) are among the most ancient, and they have very large numbers of sensilla on the mouthparts, the generalist function of which enables them to perceive many classes of stimulating chemicals. There is, however, a suggestion that grasshoppers that feed on a very restricted range of plants have fewer sensilla than more broad-spectrum feeders (Chapman and Thomas, 1978). This tendency to reduction becomes obvious when orthopteroids are compared with hemipteroid and endopterygote insects. All the available information indicates that the latter groups possess much smaller numbers of contact chemoreceptors on the mouthparts, and presumably this reduction reflects the high cost to the organism of producing, maintaining, and integrating large numbers of sensilla. A reduction in the number of receptor cells may be facilitated by a tendency to select host plants containing specific chemicals rather than to avoid all those with distasteful chemicals, although many endopterygote insects appear still to depend on the latter strategy.

In the course of evolution, adaptation to previously unacceptable plants will often have occurred so that a stimulus that initially indicated "unacceptable" came to mean "acceptable." With a receptor system dependent on the pattern of signals generated by an array of generalist cells, this implies that the overall pattern produced by the chemical is interpreted in a different sense. The difficulty of achieving this might partly account for the general lack of host-plant specificity in Acridoidea (Bernays and Chapman, 1978).

An alternative possibility would involve the specialization of a new or an existing receptor cell so that it would perceive only the appropriate stimulus. This perhaps is the case with the sinigrin receptor of *Pieris*. In this case, the modifications within the CNS would not necessarily be extensive. It is interesting, and perhaps significant, that relative specificity to particular species of plants is much more widespread among the endopterygotes and hemipteroids, which have only a few gustatory receptors, than it is in orthopteroids, which have many.

Specificity appears to demand a receptor mechanism of the "lock and key" type, with acceptor molecules capable of combining only with stimulating molecules having a certain molecular configuration. But it is difficult to visualize how such a mechanism could operate in the perception of a wide range of chemically different secondary plant substances since this would demand an extensive array of different types of acceptor molecules. A more feasible possibility seems to be a mechanism depending on a general physicochemical interaction of the stimulating molecule with the cell membrane, as Kurihara (1975) suggests. But even in this case some degree of cell specialization occurs because not all chemoreceptor cells respond to such substances.

In mollusks and vertebrates, the inherent response to secondary plant chemicals is associated with an ability to learn so that learning may play a dominant role in plant recognition. This change of emphasis perhaps provides for greater versatility within the individual.

The ability to perceive and recognize secondary plant chemicals has influenced the evolution of both the herbivores and the plants. Without this ability, adaptation to a chemically new host would not be possible, and the chemical diversity of plants would be pointless if herbivores could not distinguish between them.

VII. CONCLUSIONS

The striking feature that has emerged from this chapter is the paucity of information dealing with the perception of secondary plant compounds. Although it has been possible to interpret some of the existing work in terms of this topic, very few studies have been undertaken with this aim in view.

Electrophysiological studies have, until recently, been dominated by the idea that different receptor cells are specialized to perceive different chemical modalities. Insects have been most fully studied because of their ease of manipulation, and studies with secondary plant compounds have primarily emphasized their specialized role as phagostimulants rather than their more general role as deterrents. This partly reflects the problems of interpreting data derived from studies of generalist cells. We are totally ignorant of how the information they transmit is coded and utilized. Only with much more work in neuroanatomy and neurophysiology with refined techniques will we increase our knowledge of this aspect.

In the learning and behavioral field, most of the information available remains at the anecdotal level. To what extent is the avoidance of certain plants by grazing herbivores innate, and how much is learned? There seem to be no adequate studies, although the literature abounds with stories of cattle dying after eating strange plants or of certain plant species remaining untouched in grazed areas. Here, there is a need for critical observation and experiment.

In vertebrates in particular, but to a lesser extent also in insects, the range of secondary plant chemicals investigated in any type of experiment, physiological or behavioral, has been very limited, and we would stress the need for comparative studies with compounds of different chemical classes and, if possible, related organisms with different feeding habits.

ACKNOWLEDGMENTS

We are indebted to the Information Service at the Centre for Overseas Pest Research and especially to Miss P. Schofield, Miss F. Sherwood, and Mr. P. Shannon for their assistance with the literature survey. Dr. E. A. Bernays and Mrs. C. Winstanley critically read the manuscript, and Miss E. Leather prepared the illustrations.

REFERENCES

Adrian, E. D. (1950). *Br. Med. Bull.* **6,** 330–332.
Adrian, E. D. (1951). *J. Physiol. (London)* **115,** 42.
Akaike, N., and Sato, M. (1975). *Jpn. J. Physiol.* **25,** 585–597.
Akaike, N., and Sato, M. (1976). *Jpn. J. Physiol.* **26,** 29–40.
Akaike, N., Noma, A., and Sato, M. (1976). *J. Physiol.* **254,** 87–107.
Allison, A. C. (1953). *Biol. Rev. Cambridge Philos. Soc.* **28,** 195–244.
Allison, A. C., and Warwick, R. T. T. (1949). *Brain* **72,** 186–197.
Andersson, B., Landgren, S., Olsson, L., and Zotterman, Y. (1950). *Acta Physiol. Scand.* **21,** 105–119.
Aubele, E., and Klemm, N. (1977). *Cell Tissue Res.* **178,** 199–219.
Axtell, R. C., Foelix, R. F., Coons, L. B., and Roshdy, M. A. (1973). *Proc. Int. Congr. Acarol., 3rd, 1971* pp. 35–39.
Beck, S. D., and Reese, J. C. (1976). *In* "Biochemical Interactions Between Plants and Insects" (J. W. Wallace and R. L. Mansell, eds.), pp. 41–92. Plenum, New York.
Beidler, L. H., and Smallman, R. L. (1965). *J. Cell Biol.* **27,** 263–272.
Bell, E. A. (1972). *In* "Phytochemical Ecology" (J. A. Harborne, ed.), pp. 163–177. Academic Press, New York.
Bell, E. A. (1976). *FEBS Lett.* **64,** 29–35.
Bernard, R. (1975). *In* "Olfaction and Taste V" (D. A. Denton and J. P. Coghlan, eds.), pp. 11–14. Academic Press, New York.
Bernays, E., Edgar, J. A., and Rothschild, M. (1977). *J. Zool.* **182,** 85–87.
Bernays, E. A., and Chapman, R. F. (1977). *Ecol. Entomol.* **2,** 1–18.
Bernays, E. A., and Chapman, R. F. (1978). *In* "Biochemical Aspects of Plant and Animal Co-evolution" (J. A. Harborne, ed.), pp 99–141. Academic Press, New York.
Bernays, E. A., Blaney, W. M., and Chapman, R. F. (1972). *J. Exp. Biol.* **57,** 745–753.
Bernays, E. A., Blaney, W. M., and Chapman, R. F. (1975). *In* "Olfaction and Taste V" (D. A. Denton and J. P. Coghlan, eds.), pp. 227–230. Academic Press, New York.
Blaney, W. M. (1974). *J. Exp. Biol.* **60,** 275–293.
Blaney, W. M. (1975). *J. Exp. Biol.* **62,** 555–569.
Blaney, W. M. (1977). *Cell Tissue Res.* **184,** 397–409.
Blaney, W. M., and Chapman, R. F. (1969). *Z. Zellforsch. Mikrosk. Anat.* **99,** 74–97.
Boeckh, J., Ernst, K.-D., Sass, H., and Waldow, U. (1975). *In* "Olfaction and Taste V" (D. A. Denton and J. P. Coghlan, eds.), pp. 239–245. Academic Press, New York.
Bradley, R. M. (1971). *Hand. Sens. Physiol.* **4,** 1–30.
Brower, L. P. (1969). *Sci. Am.* **220,** 22–29.
Burr, A. H., and Burr, C. (1975). *J. Ultrastruct. Res.* **51,** 1–15.
Chapman, R. F. (1977). *Colloq. Int. C. N. R. S.* **265,** 133–149.
Chapman, R. F., and Thomas, J. (1978). *Acrida* **7,** 115–148.
Croll, N. A. (1970). "The Behaviour of Nematodes," Arnold, London.

Croll, N. A., and Matthews, B. E. (1977). "Biology of Nematodes," Blackie, Glasgow and London.

Dastoli, F. R. (1974). *Life Sci.* **14,** 1417–1426.

Dastoli, F. R., and Price, S. (1966). *Science* **154,** 905–907.

Dastoli, F. R., Lopiekes, D. V., and Doig, A. R. (1968). *Nature (London)* **218,** 884–885.

de Lorenzo, A. J. D. (1970). *Taste Smell Vertebr., Ciba Found. Symp., 1969* pp. 151–176.

Dethier, V. G. (1972). *Proc. Natl. Acad. Sci. U.S.A.* **69,** 2189–2192.

Dethier, V. G. (1973). *J. Comp. Physiol.* **82,** 103–134.

Dethier, V. G. (1974). *In* "Experimental Analysis of Insect Behaviour" (L. Barton Browne, ed.), pp. 21–31. Springer-Verlag, Berlin and New York.

Dethier, V. G. (1976). "The Hungry Fly." Harvard Univ. Press, Cambridge, Massachusetts.

Dethier, V. G., and Schoonhoven, L. M. (1969). *Entomol. Exp. Appl.* **12,** 535–543.

Doncaster, C. C. (1971). *In* "Plant Parasitic Nematodes" (B. M. Zuckerman, W. F. Mai, and R. A. Rohde, eds.), Vol. 2, pp. 137–157. Academic Press, New York.

Edgar, J. A., and Culvenor, C. C. J. (1974). *Nature (London)* **248,** 614–615.

Endo, B. Y., and Wergin, W. P. (1977). *J. Ultrastruct. Res.* **59,** 231–249.

Ernst, K.-D., Boeckh, J., and Boeckh, V. (1977). *Cell Tissue Res.* **176,** 285–308.

Fastier, F. N., and Laws, G. F. (1975). *Search* **6,** 117–120.

Faull, J. R., and Halpern, B. P. (1971). *Physiol. Behav.* **7,** 903–907.

Feeny, P. (1975). *In* "Coevolution of Animals and Plants" (L. E. Gilbert and P. H. Raven, eds.), pp. 3–19. Univ. of Texas Press, Austin.

Fishelson, L. (1960). *Eos* **36,** 41–62.

Foelix, R. F., and Axtell, R. C. (1971). *Z. Zellforsch. Mikrosk. Anat.* **114,** 22–37.

Frank, M. (1975). *In* "Olfaction and Taste V" (D. A. Denton and J. P. Coghlan, eds.), pp. 59–64. Academic Press, New York.

Garcia, J., and Hankins, W. G. (1975). *In* "Olfaction and Taste V" (D. A. Denton and J. P. Coghlan, eds.), pp. 39–45. Academic Press, New York.

Geisert, B., and Altner, H. (1974). *Cell Tissue Res.* **150,** 249–259.

Gelperin, A. (1975). *Science* **189,** 567–570.

Gesteland, R. C. (1971). *Handb. Sens. Physiol.* **4,** Part 1, pp. 132–150.

Gesteland, R. C., Lettvin, J. Y., Pitts, W. H., and Rojas, A. (1963). *In* "Olfaction and Taste I" (Y. Zotterman, ed.), pp. 19–34. Pergamon, Oxford.

Graziadei, P. P. C. (1971). *Handb. Sens. Physiol.* **4,** Part 1, 27–58.

Graziadei, P. P. C. (1974). *In* "Transduction Mechanisms in Chemoreception" (T. M. Poynder, ed.), pp. 3–14. Information Retrieval, London.

Guth, L. (1971). *Handb. Sens. Physiol.* **4,** Part 2, 63–74.

Hanamori, T. (1976). *Chem. Senses Flavor* **2,** 229–239.

Hansen, K. (1974). *In* "Biochemistry of Sensory Functions" (L. Jaenicke, ed.), pp. 207–233. Springer-Verlag, Berlin and New York.

Harbach, R. E., and Larsen, J. R. (1977). *Int. J. Insect Morphol. Embryol.* **6,** 41–60.

Haskell, P. T., and Schoonhoven, L. M. (1969). *Entomol. Exp. Appl.* **12,** 423–440.

Hatfield, L. D., Frazier, J. L., and Coons, L. B. (1976). *Int. J. Insect Morphol. Embryol.* **5,** 279–287.

Heimer, L. (1969). *Ann. N.Y. Acad. Sci.* **167,** 129–147.

Hellekant, G., Hagström, E. C., Kasahara, Y., and Zotterman, Y. (1974). *Chem Senses Flavor* **1,** 137–145.

Hiji, Y., and Ito, H. (1977). *Comp. Biochem. Physiol. A* **58,** 109–113.

Hiji, Y., Kobayashi, H., and Sato, M. (1971). *Comp. Biochem. Physiol. B* **39,** 367–375.

Hodgson, E. S., Lettvin, J. Y., and Roeder, K. D. (1955). *Science* **122,** 417–418.

Hough, C. A. M., and Edwardson, J. A. (1978). *Nature (London)* **271,** 381–383.
Jones, D. A. (1972). *In* "Phytochemical ecology" (J. A. Harborne, ed.), pp. 103–124. Academic Press, New York.
Jones, D. A. (1974). *In* "Taxonomy and Ecology" (V. H. Heywood, ed.), pp. 213–242. Academic Press, New York.
Kafka, W. A. (1970). *Z. Vergl. Physiol.* **70,** 105–143.
Kaissling, K.-E. (1971). *Handb. Sens. Physiol.* **4,** Part 1, 351–431.
Kaissling, K.-E. (1974). *In* "Biochemistry of Sensory Functions" (L. Jaenicke, ed.), pp. 243–273. Springer-Verlag, Berlin and New York.
Kasang, G., and Kaissling, K.-E. (1972). *In* "Olfaction and Taste IV" (D. Schneider, ed.), pp. 207–213. Wiss. Verlagsges., Stuttgart.
Kasang, G., Knauder, B., and Beroza, M. (1974). *Experientia* **30,** 147.
Kendall, M. D. (1970). *Z. Zellforsch. Mikrosk. Anat.* **109,** 112–137.
Kijima, H., Amakawa, T., Nakashima, M., and Morita, H. (1977). *J. Insect Physiol.* **23,** 469–479.
Kittel, R. (1956). *Zool. Anz.* **157,** 185–195.
Kolnberger, I., and Altner, H. (1972). *In* "Olfaction and Taste IV" (D. Schneider, ed.), pp. 34–39. Wiss. Verlagsges., Stuttgart.
Koyama, N., and Kurihara, K. (1972). *Biochim. Biophys. Acta* **288,** 22–26.
Kubota, T., and Kubo, I. (1969). *Nature (London)* **223,** 97–99.
Kurihara, K. (1972). *F.E.B.S. Lett.* **27,** 279–281.
Kurihara, K., Kamo, N., Ueda, T., and Kobatake, Y. (1975). *In* "Olfaction and Taste V" (D. A. Denton and J. P. Coghlan, eds.), pp. 77–82. Academic Press, New York.
Kurihara, Y. (1973). *Biochim. Biophys. Acta* **306,** 478–482.
Kurihara, Y. (1975). *Chem. Senses Flavor* **1,** 251–255.
Le Gross Clark, W. E. (1957). *Proc. R. Soc. London Ser. B* **146,** 299–319.
Levinson, H. Z., Kaissling, K.-E., and Levinson, A. R. (1973). *J. Comp. Physiol.* **86,** 209–214.
Lewis, C. T. (1970). *Symp. R. Entomol. Soc. London* **5,** 59–76.
Ma, W.-C. (1972). *Meded. Landbouwhogesch. Wageningen* **72,** No. 11, 1–162.
Ma, W.-C. (1977). *Physiol. Entomol.* **2,** 199–207.
Ma, W.-C., and Schoonhoven, L. M. (1973). *Entomol. Exp. Appl.* **16,** 343–357.
McCutchan, M. C. (1969). *Z. Vergl. Physiol.* **65,** 131–152.
Macrides, F. (1976). *In* "Mammalian Olfaction, Reproductive Processes, and Behaviour" (R. L. Doty, ed.), pp. 29–65. Academic Press, New York.
Marsh, H., and Rothschild, M. (1974). *J. Zool.* **174,** 89–122.
Masson, C., and Friggi, A. (1971). *C.R. Hebd. Seances Acad. Sci., Ser. D.* **272,** 618–621.
Mattocks, A. R. (1972). *In* "Phytochemical Ecology" (J. A. Harborne, ed.), pp. 179–200. Academic Press, New York.
Mayer, M. S. (1975). *Experientia* **31,** 452–454.
Morita, H. (1959). *J. Cell. Comp. Physiol.* **54,** 189–204.
Moulton, D. G. (1971). *Handb. Sens. Physiol.* **5,** Part 1, 59–74.
Moulton, D. G. (1976). *Physiol. Rev.* **56,** 578–593.
Moulton, D. G., and Fink, R. P. (1972). *In* "Olfaction and Taste IV" (D. Schneider, ed.), pp. 20–26. Wiss. Verlagsges., Stuttgart.
Mozell, M. M. (1971). *Handb. Sens. Physiol.* **4,** Part 1, 205–214.
Murray, R. G., and Murray, A. (1970). *Taste Smell Vertebr., Ciba Found. Symp., 1969* pp. 3–30.
Murray, R. G., Murray, A., and Hellekant, G. (1972). *In* "Olfaction and Taste IV" (D. Schneider, ed.), pp. 56–62. Wiss. Verlagsges., Stuttgart.

Noma, A. and Hiji, Y. (1970). *Kumamoto Med. J.* **23,** 114–116.

Norris, D. M., Baker, J. E., Borg, T. K., Ferkovich, S. M., and Rozental, J. M. (1970). *Contrib. Boyce Thompson Inst.* **24,** 263–274.

Norris, D. M., Ferkovich, S. M., Baker, J. E., Rozental, J. M., and Borg, T. K. (1971). *J. Insect Physiol.* **17,** 85–97.

Ozeki, M. (1971). *J. Gen. Physiol.* **58,** 688–699.

Ozeki, M., and Sato, M. (1972). *Comp. Biochem. Physiol.* **41,** 391–407.

Parsons, T. S. (1971). *Hand. Sens. Physiol.* **4,** Part 1, 1–26.

Pfaffmann, C., Frank, M., Bartoshuk, L. M., and Snell, T. C. (1976). *Prog. Psychobiol. Physiol.* **6,** 1–27.

Phillips, C. E., and Vande Berg, J. S. (1976). *Int. J. Insect Morphol. Embryol.* **5,** 423–431.

Pitman, R. M., Tweedle, C. D., and Cohen, M. J. (1972). *Science* **176,** 412–414.

Pliske, T. E. (1975). *Environ. Entomol.* **4,** 455–473.

Price, S. (1969). *Nature (London)* **221,** 779.

Price, S. (1973). *Nature (London)* **241,** 54–55.

Price, S. (1974). *In* "Transduction Mechanisms in Chemoreception" (T. M. Poynder, ed.), pp. 177–184. Information Retrieval, London.

Rees, C. J. C. (1969). *Entomol. Exp. Appl.* **12,** 565–583.

Reese, T. S., and Brightman, M. W. (1970). *Taste Smell Verteb., Ciba Found. Symp., 1969* pp. 115–149.

Rhoades, D. F., and Cates, R. G. (1976). *In* "Biochemical Interactions Between Plants and Insects" (J. W. Wallace and R. L. Mansell, eds.), pp. 168–213. Plenum, New York.

Rothschild, M., and Kellett, D. N. (1972). *J. Entomol., Ser. A* **46,** 103–110.

Rothschild, M., Rowan, M. G., and Fairbairn, J. W. (1977). *Nature (London)* **266,** 650–651.

Rozental, J. M., and Norris, D. M. (1973). *Nature (London)* **244,** 370–371.

Salánki, J., and Van Bay, T. (1975). *Annls Inst. biol., Tihany* **42,** 115–128.

Sato, M. (1975). *In* "Olfaction and Taste V" (D. A. Denton and J. P. Coghlan, eds.), pp. 23–26. Academic Press, New York.

Sato, M., and Ozeki, M. (1972). *In* "Olfaction and Taste IV" (D. Schneider, ed.), pp. 252–258. Wiss. Verlagsges., Stuttgart.

Scalia, F. (1972). *Brain Res.* **36,** 409–411.

Scalia, F., Halpern, M., Knapp, H., and Riss, W. (1968). *J. Anat.* **103,** 245–262.

Schafer, R. (1971). *J. Morphol.* **134,** 91–104.

Schneider, D. (1957). *Z. Vergl. Physiol.* **40,** 8–41.

Schneider, D. (1975). *In* "Olfaction and Taste V" (D. A. Denton and J. P. Coghlan, eds.), pp. 327–328. Academic Press, New York.

Schneider, D. (1977). *Colloq. Int. C.N.R.S.* **265,** 353–356.

Schneider, D., and Steinbrecht, R. A. (1968). *Symp. Zool. Soc. London* **23,** 279–297.

Schoonhoven, L. M. (1969). *Proc. K. Ned. Akad. Wet., Ser. C* **72,** 491–498.

Schoonhoven, L. M. (1972). *Recent Adv. Phytochem.* **4,** 197–224.

Schoonhoven, L. M. (1973). *Symp. R. Entomol. Soc. London* **6,** 87–99.

Schoonhoven, L. M. (1974). *In* "Transduction Mechanisms in Chemoreception" (T. M. Poynder, ed.), pp. 189–201. Information Retrieval, London.

Schoonhoven, L. M., and Henstra, S. (1972). *Neth. J. Zool.* **22,** 343–346.

Schultz, F. (1938). *Z. Morphol. Oekol. Tiere* **33,** 553–581.

Seifert, K., and Uie, G. (1967). *Z. Zellforsch. Mikrosk. Anat.* **76,** 147–169.

Shibuya, T., and Tonosaki, K. (1972). *In* "Olfaction and Taste IV" (D. Schneider, ed.), pp. 102–108. Wiss. Verlagsges., Stuttgart.

Shimada, I., Shirasishi, A., Kijima, H., and Morita, H. (1972). *J. Insect Physiol.* **18,** 1845–1855.

Siegel, R. K., and Jarvik, M. E. (1974). *Bull. Psychom. Soc.* **4,** 476–478.
Singer, G., Rozental, J. M., and Norris, D. M. (1975). *Nature (London)* **256,** 222–223.
Slifer, E. H. (1970). *Annu. Rev. Entomol.* **15,** 121–142.
Slifer, E. H. (1972). *Acrida* **1,** 1–5.
Städler, E., and Hanson, F. E. (1976). *Symp. Biol. Hung.* **16,** 267–273.
Steinbrecht, R. A. (1969). *J. Cell Sci.* **4,** 39–53.
Steinbrecht, R. A. (1973). *Z. Zellforsch. Mikrosk. Anat.* **139,** 533–565.
Steinbrecht, R. A. (1974). *In* "Transduction Mechanisms in Chemoreception" (T. M. Poynder, ed.), pp. 15–21. Information Retrieval, London.
Steinbrecht, R. A., and Kasang, G. (1972). *In* "Olfaction and Taste IV" (D. Schneider, ed.), pp. 193–199. Wiss. Verlagsges., Stuttgart.
Steinbrecht, R. A., and Müller, B. (1976). *Tissue & Cell* **8,** 615–636.
Stephens, G. J., and McGaugh, J. L. (1972). *Anim. Behav.* **20,** 309–315.
Stürckow, B. (1959). *Z. Vergl. Physiol.* **42,** 255–302.
Stürckow, B. (1970). *Adv. Chemoreception* **13,** 107–159.
Stürckow, B., Holbert, P. E., and Adams, J. R. (1967). *Experientia* **23,** 780–782.
Swain, T. (1976). *In* "Secondary Metabolism and Coevolution" (M. Luckner, K. Mothes, and L. Nover, eds.), pp. 551–561. Dtsch. Akad. Naturforsch., Leopoldina (Halle).
Thorsteinson, A. J. (1960). *Annu. Rev. Entomol.* **5,** 193–218.
Thurm, U. (1972). *In* "Olfaction and Taste IV" (D. Schneider, ed.), pp. 95–101. Wiss. Verlagseges., Stuttgart.
Thurston, E. L., and Lersten, N. R. (1969). *Bot. Rev.* **35,** 393–412.
Trefz, B. (1972). *J. Dent. Res.* **51,** 1203–1211.
van der Starre, H. (1972). *Neth. J. Zool.* **22,** 227–282.
van Lawick-Goodall, J. (1968). *Anim. Behav. Monogr.* **1,** 161–311.
Ward, S. (1973). *Proc. Natl. Acad. Sci. U.S.A.* **70,** 817–821.
Ward, S., Thomson, N., White, J. G., and Brenner, S. (1975). *J. Comp. Neurol.* **160,** 313–338.
Wenzel, B. M. (1973). *In* "Avian Biology" (D. S. Farner and J. R. King, eds.), Vol. 3, pp. 389–415. Academic Press, New York.
Wieczorek, H. (1976). *J. Comp. Physiol.* **106,** 153–176.
Winans, S., and Scalia, F. (1970). *Science* **170,** 330–332.
Wondrak, G. (1975). *Cell Tissue Res.* **159,** 121–140.
Wright, B. R. (1974a). *Cell Tissue Res.* **151,** 229–244.
Wright, B. R. (1974b). *Cell Tissue Res.* **151,** 245–257.
Yamada, M. (1971). *J. Physiol. (London)* **214,** 127–143.

Chapter 5

Biochemical Defense Mechanisms in Herbivores against Plant Allelochemicals

LENA B. BRATTSTEN

ISBN 0-12-597180-X

I. INTRODUCTION

A. The Poisoned Platter

Herbivores expose themselves to the hazard of being poisoned by every meal. Feeding is a necessity, and the fact that herbivores are able to survive their meals and thrive and reproduce is witness to their successful adaptations to the elaborate equipment of toxic allelochemicals in plants.

Once a toxic compound has found its way into an animal, a sufficiently large fraction of it has to get through to the critical target site to exert its toxic effect. There are several severe obstacles en route. The compound may undergo a multitude of rapid metabolic degradations by an arsenal of enzymes. It may be rapidly excreted. It may be stopped by special barriers or channeled away for storage into specialized tissues or organelles. Or it may be unable to interact with the target organ due to some subtle modification in that organ.

There are large numbers of cases in which herbivores are adapted to feed with impunity on plants containing highly toxic chemicals. Insect herbivores are notorious for feeding on poisonous plants. An outstanding example is the polyphagous larva of a tiger moth, *Arctia caja,* which freely ingests foxglove, lily-of-the-valley, potato, aconitum, and several species of *Senecio,* to mention only a few (Rothschild *et al.,* 1970). The black cherry, rich in cyanogenic glycosides, is the favorite food plant of the cecropia larvae, although these big silkworms feed on a great variety of trees and shrubs. Many other insects such as zygaenid larvae also ingest cyanogenic plants (Rothschild *et al.,* 1970). An entire specialized insect fauna including pierid larvae and flea beetles is associated with crucifers, despite the high levels of glucosinolates found in these plants (Thorsteinson, 1953; Feeny *et al.,* 1970). Another guild of insects including monarch larvae, milkweed bugs, and milkweed beetles are well-known specialist feeders on cardenolide-containing milkweeds. Even though the main use of nicotine is as an insecticide, many insects, e.g., the tobacco hornworm, the tobacco budworm, and the green peach aphid feed on tobacco plants. The Colorado potato beetle feeds on potato plants with high enough glycoalkaloid content to make them poisonous to vertebrates. Cotton growers are among the heaviest users of synthetic insecticides to protect their crops, even though cotton plants contain the highly toxic chemical gossypol.

One of the most successful cases of biological weed control was the introduction of the *Chrysolina* beetle to the northwestern United States. This beetle reduced the rampantly spreading Klamath weed, which had also been introduced into the area, to insignificant occurrence by larval feeding, although the plant contains high concentrations of the phototoxin hypericin (Robinson, 1967). Black swallowtail butterfly larvae are specialized to feed on umbellifers (Feeny, 1975), which contain a variety of toxic coumarins and polyacetylenes (Hegnauer, 1971). A bruchid beetle larva, *Caryedes brasilensis,* feeds exclusively on seeds of *Dioclea megacarpa* with very high concentrations of the toxic amino acid L-canavanine (Rosenthal *et al.,* 1977). Several insect species including honeybees, danaid butterflies, larvae of arctiid moths, and grasshoppers are known to ingest the hepatotoxic pyrrolizidine alkaloids that occur in a large number of plant species, mainly in the families Compositae, Leguminosae, and Boraginaceae. Cycads containing the poisonous glycoside cycacin are fed on by at least one lepidopteran (Teas, 1967). Particularly interesting is the association of a desert fruit fly, *Drosophila pachea,* with the senita cactus *Lophocereus schotti,* which contains the alkaloids lophocereine and its trimer pilocereine. The cactus is lethal to eight other desert fruit fly species due to the alkaloids but is the sole breeding site of *D. pachea*. The cactus also contains a sterol, schottenol, which is essential to *D. pachea* but not to any of the other fruit flies (Kircher *et al.,* 1967).

It is very unusual to find specialist feeders of poisonous plants among the vertebrate herbivores. Many mammals, including people, have little trouble surviving a low frequency of toxic meals, provided that a diversity of food materials is available. Some mammals appear to be more tolerant than others; it is well known that the white-tailed deer frequently feeds on yew, rhododendron, and moutain laurel, meals that would be poisonous to cattle. The gray squirrel intersperses its diet with *Amanita* mushrooms containing the highly toxic polypeptide α-amanitin. The mountain viscacha *Lagidium peruvianum* lives at high altitudes in the Peruvian Andes, where it has access to a very meager fare consisting of grasses, lichens, *Opuntia* cacti, and several species of hepatotoxic *Senecio* (Pearson, 1948). The koala specializes in eating a variety of *Eucalyptus* foliage rich in terpenoids and phenolic compounds, and camels in Australia also eat *Eucalyptus* (Whittaker, 1970). Several more examples of mammals feeding on toxic plants are given in Freeland and Janzen (1974). Among birds, the carrion-feeding turkey vulture has developed immunity to one of the most potent toxins known, botulinus toxin.

All these cases in which vertebrates or invertebrates avoid poisoning by toxic components in their meals are testimony to the abundance in nature of highly effective mechanisms for dealing with toxic chemicals. The explanation for each of these interactions is in a few cases known but is in many cases unexplored. Metabolic detoxification mechanisms are usually the most important factor in cases of sustained tolerance. This chapter presents an overview of the biochemical detoxification mechanisms available to organisms.

Chemicals that do not appear to be involved in or contribute to the basic structural–nutritional metabolism in organisms are referred to as foreign compounds or xenobiotics. Those occurring naturally in plants have been assigned the unfortunate label "secondary plant substances." In view of the great ecological importance and the infinitely sophisticated interactions of these chemicals, a fact that is now being fully recognized and explored, a new, less downgrading term is in order. An appropriate name would be "allelochemical," in line with the implications of Whittaker (1970) and Whittaker and Feeny (1971) and also the discussions by Thorsteinson (1960) and Beck (1965). Allelochemicals do indeed play a primary role in the ecologically interrelated systems of which all organisms are part.

B. Metabolic Fate of Foreign Compounds

Eventually, a foreign compound that has reached the internal environment of an organism is degraded metabolically to an excretable end product. Metabolic degradation is conveniently divided into primary and secondary processes, as shown in Fig. 1. Many enzymes or enzyme systems are ready to attack the foreign chemical. The resulting product

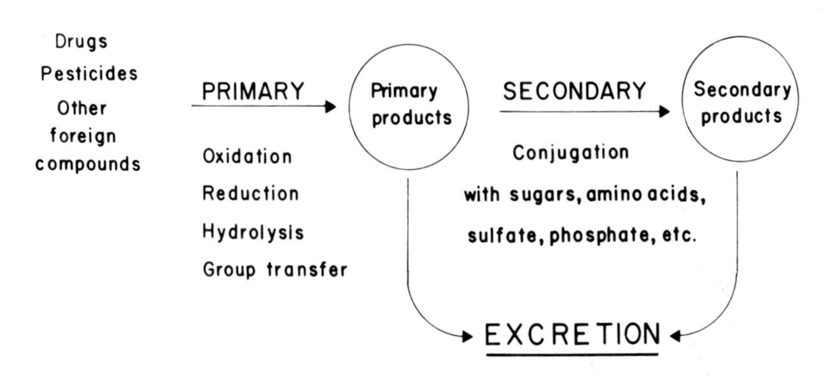

Figure 1. Schematic representation of xenobiotic metabolism.

is more water soluble (hydrophilic) than the parent compound. The primary product is sometimes polar enough to be excreted directly but is usually converted by secondary processes to an easily excretable, conjugated material. Here again, many enzymes participate in the secondary metabolism.

All organisms with a water-based excretory system have difficulties in eliminating lipophilic foreign compounds from their tissues. Lipophiles tend to accumulate in lipid-containing membranes and storage material in tissues, and with time they reach sufficiently high concentrations to interfere with fundamental biochemical processes in cells. As a consequence, lipophilic compounds are potentially toxic in small doses.

Mixed-function oxidase (MFO) enzymes possess three major characteristics that make them ideally suited as a biochemical waste disposal mechanism. They catalyze numerous oxidative reactions, resulting in more polar, excretable products. They accept a large variety of chemicals as substrates with a remarkable nonspecificity. And they adjust very rapidly to the presence of external chemicals by the process of induction. A variety of esterases, reductases, and a few special group transfer enzymes are also involved in the primary metabolism.

II. MIXED-FUNCTION OXIDASES

The MFO's are by far the most important enzymes involved in the primary metabolism of lipophilic foreign compounds. Several excellent reviews on the MFO's have been published recently (e.g., Hutson, 1977; Nakatsugawa and Morelli, 1976; Testa and Jenner, 1976; Agosin and Perry, 1974; Mannering, 1972). In fact, this enzyme system is so important in chemical therapeutics and insect pest control programs that the information available on it with very few exceptions is to be found in the context of drug and pesticide metabolism.

The microsomal enzymes involved in oxidations of lipophilic foreign compounds are often called mixed-function oxidases (according to Mason, 1957) due to the double role of the oxygen molecule as both oxidizing (to form water) and oxygenating agent. They are also called mono oxygenases (according to Hayaishi, 1962) to distinguish them from the dioxygenases, in which case both oxygen atoms oxygenate the substrate molecule. The Enzyme Commission labels them hydroxylases with the assigned numbers 1.14. Only the microsomal monooxygenases or MFO's involved in xenobiotic metabolism will be described in some detail in this chapter.

A. MFO-Catalyzed Reactions

Due to the large number of molecules that are acceptable as MFO substrates, there can be no assignment of any one normal or natural substrate for this enzyme system. Rather, it is becoming clear that the MFO enzymes are specialized to metabolize the great array of foreign lipophilic compounds that organisms unavoidably encounter in the course of their normal life processes (Brodie *et al.*, 1958; Nakatsugawa and Morelli, 1976; Brattsten *et al.*, 1977). The allelochemicals of plants are first-rate examples of such compounds. However, more is known about chemicals of commercial importance.

All MFO-catalyzed reactions can be viewed as hydroxylations (Brodie *et al.*, 1958). Nakatsugawa and Morelli (1976) recognize three categories of oxidative metabolic transformations that lipophilic foreign compounds undergo in organisms, namely, CH hydroxylations, π-bond oxygenations, and oxygenation at an unshared electron pair.

The first group of reactions, CH hydroxylations, in which a hydroxyl group replaces a hydrogen, results in aliphatic hydroxylations. The conversion of *p,p'*-DDT, 1,1,1-trichloro-2,2-bis(p-chlorophenyl)ethane, to kelthane is an example of an aliphatic hydroxylation (Fig. 2a). There was considerable uncertainty about the identity of the ethanol-oxidizing enzyme of liver microsomes. This enzyme has been identified as a typical cytochrome P-450-containing MFO and converts ethanol to acetaldehyde (Teschke *et al.*, 1974) by a CH hydroxylation.

Nitrogen, oxygen, and sulfur dealkylations can, by analogy, as pointed out by Nakatsugawa and Morelli (1976), be regarded as CH hydroxylations. They all start with an attack on the carbon atom α to the heteroatom and proceed via at least one highly unstable α-hydroxy intermediate. N-Dealkylations are extremely important in the metabolism of drugs and alkaloids. An alkyl group of any size, e.g., methyl, ethyl, benzyl, isopropyl, cyanoethyl, and hydroxyethyl (Testa and Jenner, 1976), can be the leaving substituent. The conversion of morphine to the inactive normorphine exemplifies an N-demethylation (Fig. 2b). Oxidative O-dealkylations occur frequently. They do not involve ester bonds, which instead undergo hydrolytic cleavage by esterases. The most famous example of an O-demethylation is the conversion of codeine to morphine (Fig. 2c). Another important target for O-demethylation is rotenone (Fig. 2d). Very few cases of S-dealkylation are known. Houseflies produce an S-demethylated metabolite of the insecticide aldicarb *in vivo* (Metcalf *et al.*, 1967). Thiocarbamate herbicides are converted *in vivo* and *in vitro* in mammals to unstable sulfoxide metabolites (Casida *et al.*, 1975). Characteristically, the nitrogen,

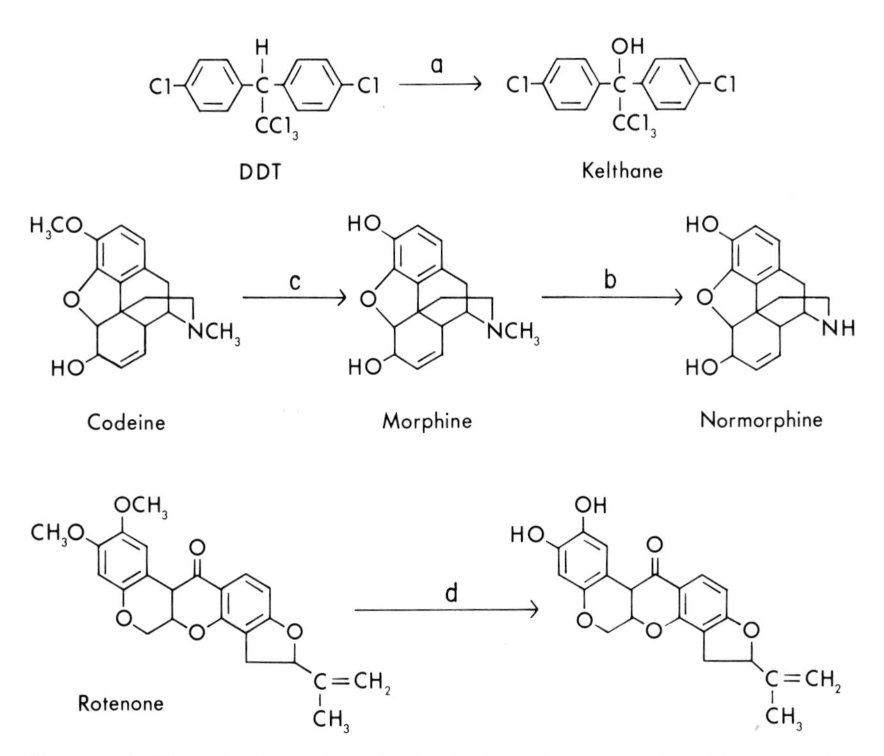

Figure 2. MFO-catalyzed reactions. (a) Aliphatic hydroxylation; (b) N-demethylation; (c and d) O-demethylation.

oxygen, sulfur, or carbon atom involved in these reactions does not undergo any redox change, but the leaving group is found to be oxidized. The participation of cytochrome P-450 in the dealkylations of heteroatoms occurs only to a limited extent. Cytochrome P-450-containing MFO's are probably not involved in S-dealkylations, which may be catalyzed by other monooxygenases (Hutson, 1977). There is evidence that a microsomal mixed-function amine oxidase without cytochrome P-450 is responsible for part of the N-dealkylations (Prough and Ziegler, 1977). Little is known about the relative participation in O-dealkylations of MFO's that are not mediated by cytochrome P-450 (Hutson, 1977).

The second category of reactions, π-bond oxygenations, includes epoxidations, hydroxylations, and phospho ester oxidations. The initial formation of an epoxide across a carbon–carbon double bond has been suggested as the initial step in oxidative hydroxylation reactions (Jerina *et al.*, 1970). The epoxides formed are highly unstable and undergo enzymatic hydration to phenols, dihydrodiols, or glutathione conjugates (Testa and Jenner, 1976) quite rapidly. The unstable epoxide intermediate also explains the migration of substituents (NIH shift) on aromatic rings (Guroff *et al.*, 1967). A grasshopper (*Romalea microptera*) has 2,5-dichlorophenol as a component in its defensive froth (Eisner *et al.*, 1971). Clorinated chemicals are not very common in nature, and the authors hypothesize that the grasshopper might have ingested the herbicide 2,4-D (2,4-dichlorophenoxyacetic acid), which during its metabolic processing might have undergone an NIH shift. An exception is the stable epoxides formed from the cyclodiene insecticides (Brooks, 1974a), for instance, dieldrin, the 6,7-epoxide of aldrin (Fig. 3a), although these epoxides, too, undergo hydration to dihydrodiols with time (see Nakatsugawa and Morelli, 1976). Both aliphatic and aromatic carbon–carbon double bonds are hydroxylated by this mechanism. The hydroxylations of pyrethrin I and naphthalene are illustrated in Fig. 3b.

Nakatsugawa and Morelli (1976) point out the analogy between the oxidative desulfuration of the P—S bond as it occurs in a large number of organophosphorous insecticides and the hydroxylation and rear-

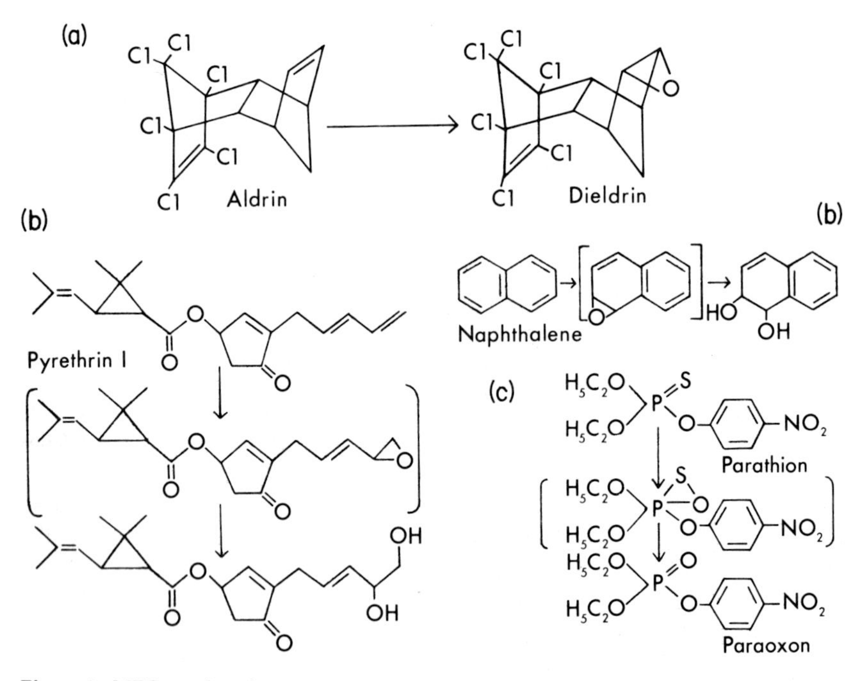

Figure 3. MFO-catalyzed reactions. (a) Cyclodiene epoxidation; (b) aromatic and aliphatic epoxidation; (c) oxidative desulfuration. (Adapted from Nakatsugawa and Morelli, 1976).

rangements of the C—C bond in a cyclic oxide. The basic assumption that unstable cyclic

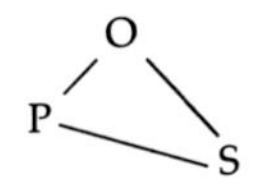

oxide intermediates are formed with the π bonds of the P→S structure is in perfect analogy with the unstable carbon–carbon epoxide

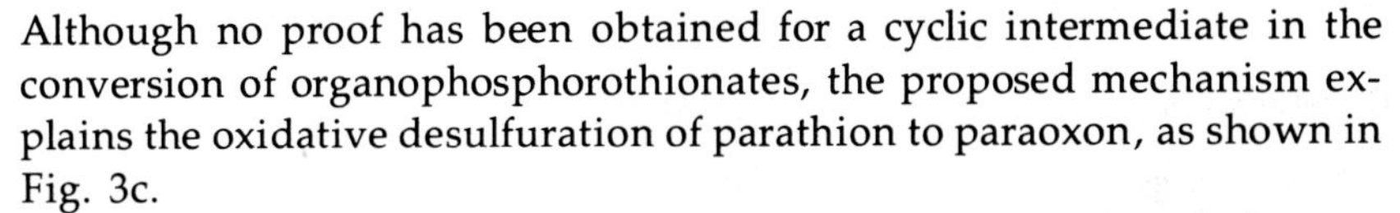

Although no proof has been obtained for a cyclic intermediate in the conversion of organophosphorothionates, the proposed mechanism explains the oxidative desulfuration of parathion to paraoxon, as shown in Fig. 3c.

The third group of reactions, oxygenation at an unshared electron pair, results in oxidation of thio ethers to sulfoxides and sulfones and in oxidation of tertiary amines to N-oxides. Thio ether oxidation is illustrated by the oxidation of the carbamate insecticide aldicarb (Fig. 4a), which results in the sulfoxide as the major metabolite (Oonnithan and Casida, 1968). Sulfoxide formation is catalyzed by a P-450-independent MFO system in pig liver microsomes (Poulsen *et al.*, 1974). The N-oxide is one of the major metabolites of nicotine (Fig. 4b) and other alkaloids, e.g., the pyrrolizidine alkaloids and morphine, as well as of a large number of drug molecules. Reactions in the third group are catalyzed by MFO's with and without cytochrome P-450. Gorrod (1973) divided

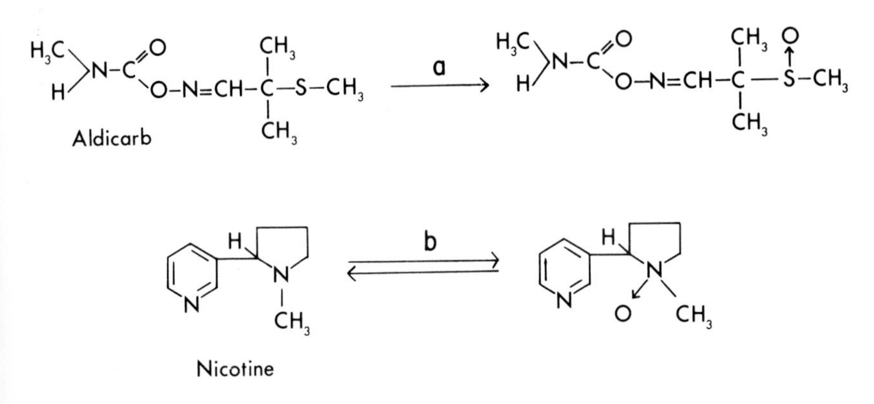

Figure 4. MFO-catalyzed reactions. (a) Thio ether oxidation; (b) tertiary amine N-oxidation.

nitrogen-containing compounds into three groups: basic amines with a pK_a of 8 to 11, intermediary amines that are weak bases with pK_a values between 1 and 7, and compounds containing nonbasic nitrogen. Those in the last group undergo predominantly cytochrome P-450-mediated mixed-function oxidation, resulting in N-dealkylation. The middle group of compounds are sometimes N-dealkylated by P-450 depending on external factors and the polarity of the compound. The strong bases in the first group, the tertiary amines, are N-dealkylated by cytochrome P-450-dependent oxidases but N-oxidized by another mixed-function amine oxidase, which is a flavoprotein (Prough and Ziegler, 1977). Tertiary amines usually undergo N-oxidation in preference to N-dealkylation (Testa and Jenner, 1976).

B. Methods for Measuring MFO Activity

It would be charitable to call the MFO enzyme system a joy to work with. Problems arise from several factors, the first of which is the membrane-bound nature of the system; it loses activity if the supporting membrane is disrupted. Second, the enzyme accepts only fairly lipophilic chemicals as substrates, which sometimes must be added in special vehicles. Third, the enzyme consists of at least three parts: cytochrome P-450, a flavoprotein reductase, and a phospholipid. It requires atmospheric oxygen and a lasting supply of the cofactor NADPH or a system that generates this cofactor. Mammalian MFO enzymes also need $MgCl_2$. This complicated constitution of the enzyme, which makes purification of the components and reconstitution of the purified components a major effort, leads to the routine utilization of a microsomal preparation. The microsomal suspension contains many things besides the active enzyme, some of which may interfere with the measurements. This relative crudeness of the preparation in combination with the relative inaccessibility of the lipophilic substrate makes kinetic studies questionable and difficult to interpret with any certainty. Fourth, the liver is usually the best source of enzyme activity in vertebrates, but a choice of a suitable tissue must be made with invertebrates. This usually involves screening tissues for activity initially, whenever a new species is being investigated. This is particularly necessary with insects.

Another complication is the lability of the MFO system. All operations must be performed at around zero-degree temperatures to prevent the conversion of cytochrome P-450 to its inactive form, cytochrome P-420. This conversion is triggered mainly by lipid peroxidation activity in vertebrate liver tissue or by the action of several so-called endogenous inhibitors in invertebrate tissues.

In addition to the small size of insects and the scarcity of active protein unless large numbers are dissected, another problem with insect tissues is the occurrence of endogenous inhibitors. The proteolytic enzymes in the guts of caterpillars and crickets (Krieger and Wilkinson, 1970; Orrenius *et al.*, 1971; Brattsten and Wilkinson, 1973a) were found to solubilize the NADPH–cytochrome P-450 reductase away from the membrane. It remained in the supernatant of the high-speed centrifugation, and the microsomal pellet contained an incomplete MFO system without activity. An unusual endogenous MFO inhibitor was found in honeybee tissues (Gilbert and Wilkinson, 1974) and was characterized as an intracellular RNA. Ribonucleic acids from bakers' yeast and *Torula* yeast were also found to be strongly inhibitory of MFO activities in armyworm midgut preparations.

There are two early reports (Matthews and Hodgson, 1966; Chakraborty *et al.*, 1967) about endogenous MFO inhibitors in houseflies, which were homogenized whole and used for MFO measurements. The eye pigment xanthommatin was found to be one of the culprits (Schonbrod and Terriere, 1971; Wilson and Hodgson, 1972; references in Wilkinson and Brattsten, 1972). Xanthommatin is thought to attract electrons carried by the reductase on their way between NADPH and cytochrome P-450 (Wilkinson and Brattsten, 1972) to form reduced dihydroxanthommatin, which can undergo autoxidation or can be oxidized by other cell substituents, e.g., cytochrome c or tyrosinase. The oxidized form is thus perpetually regenerated and continues to deplete cytochrome P-450 of electrons.

In addition, Crankshaw *et al.* (1977) reported that tyrosinase is an MFO inhibitor by virtue of its ability to cross-link microsomal proteins with its substrate, catechol. The tyrosinase complex (a soluble monooxygenase) is prominently involved in the darkening and hardening of the insect cuticle after molting and is therefore present in insect tissues.

Hansen and Hodgson (1971) reported that, of a number of media used during the process of isolation, cleaning, homogenization, and centrifugation of the active tissue, 0.20 *M* potassium phosphate buffer, pH 7.8, gives the most active microsomal preparation from housefly abdomina. The ionic strength (0.57) was found to be the most important parameter of the homogenization medium. Rat liver microsomes are routinely prepared in phosphate buffer. Isotonic 1.15% potassium chloride and 0.25 *M* sucrose are also widely used for the preparation of microsomes. The latter two can be somewhat improved if buffered with potassium phosphate to a pH suitable for the tissue [pH 7.8 for armyworm midgut, pH 7.5 for rat liver (L. B. Brattsten, personal observation)] provided that the ionic strength of the resulting medium is also adjusted.

The process of homogenization must be as gentle, cool, and quick as possible. For rat liver a chilled, smooth-walled glass tube with a motor-driven, fitted Teflon pestle is probably the most widely used type of homogenizer. For lepidopterous larval guts a rough-walled glass tube allows a shorter time for homogenization than does a smooth tube. The combination of chilled glass tube and motor-driven Teflon pestle gives microsomes of superior quality with most species of insects and mammals. Very little if any cytochrome P-420 is seen in homogenates prepared this way. Morello *et al.* (1971) report that they are able to obtain a very active microsomal preparation from housefly abdomina by crushing these in a mortar with a pestle, followed by centrifugation. This method is gentle enough that the endogenous inhibitors (see above) normally found in flies are not released. For a detailed discussion of the preparation of insect microsomes, see Wilkinson and Brattsten (1972) and Agosin and Perry (1974).

Centrifugal fractionation according to the method of Siekevitz (1963) is employed to obtain a microsomal pellet. After an initial spin of about 10,000 *g* for 10–15 minutes to precipitate mitochondria and all heavier cell constituents and fragments, the supernatant is recentrifuged at a high speed (100,000 *g*) for 1 hour. The pellet (microsome) is suspended, sometimes after washing, in a suitable buffer and assayed for activity. An extensive recipe for the preparation of rat liver microsomes is given by Mazel (1971). These are many *in vitro* MFO assays that utilize model molecules as substrates (Table I). Ideally the product should be convenient to unambiguously extract and measure. The substrate should be able to undergo only one MFO-catalyzed reaction, and the enzyme should have a high enough turnover number with respect to the substrate to minimize statistical errors and consumption of active protein. However, actual drugs and insecticides are often incubated with a microsomal preparation, and product formation is monitored by thin-layer chromatography techniques or, in recent years, high-pressure liquid chromatography techniques. Table I is a summary of some of the most frequently used *in vitro* MFO assays.

The epoxidation of the cyclodienes aldrin and heptachlor is perhaps the most reliable assay because of the exceptional stability of the cyclodiene epoxides. This assay may provide a "frozen image," reflecting what is considered to be normally an extremely short-lived intermediate step in all aromatic hydroxylation reactions (Jerina *et al.*, 1970). It is probably safe to say that the cyclodiene epoxidase reaction is catalyzed exclusively by the cytochrome P-450-mediated MFO system, without any interference from any other monooxygenase that may be present in the incubation mixture. The cyclodiene epoxidase assay is easy to monitor by means of a gas–liquid chromatograph equipped with an electron-capture detector. Normally only one product, the stable epoxide, is formed. The product and remaining substrate can be extracted virtually completely from the reaction mixture. The enzyme has a high enough turnover number for the cyclodiene substrates that the reaction proceeds at an easily measurable rate even with a crude homogenate of an active tissue. A drawback is that kinetic data must be interpreted with great caution when obtained with the highly lipophilic cyclodiene substrates.

In addition to measuring the catalytic activity of the MFO system, it is possible to measure the concentration of cytochrome P-450. The cytochrome forms a complex with carbon monoxide, which in its reduced form absorbs light at 450 nm (Fig. 5a). The original method for measur-

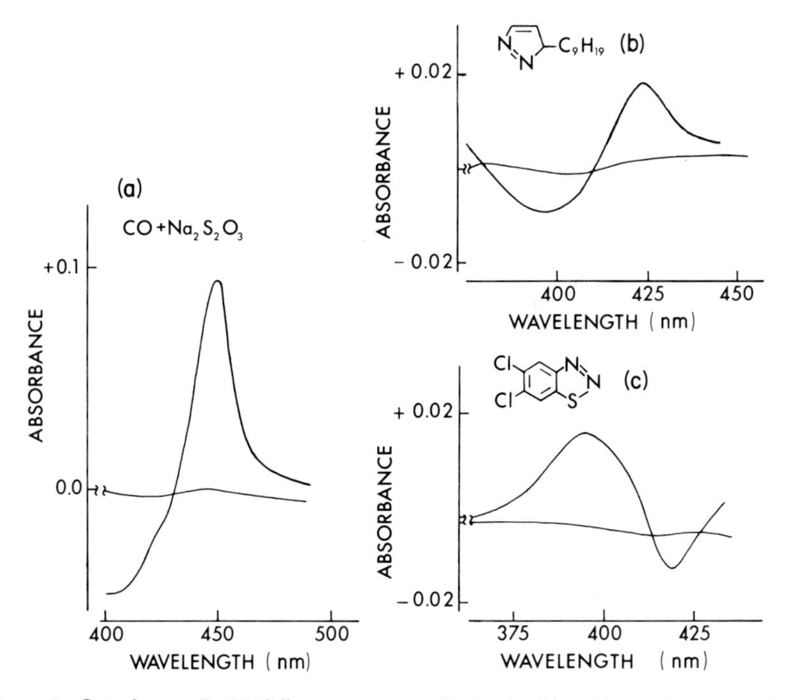

Figure 5. Cytochrome P-450 difference spectra obtained with midgut microsomes from last instar larvae of the southern armyworm *Spodoptera eridania.* (a) Reduced cytochrome P-450–CO spectrum; (b) oxidized cytochrome P-450–1-nonylimidazole difference spectrum, type II; (c) oxidized cytochrome P-450–5,6-dichlorobenzothiadiazole difference spectrum, type I.

Table I. Assays for *in vitro* measurement of MFO activity

Substrate	Product	Method of detection	Reference
		Epoxidation	
Aldrin	Dieldrin	Gas–liquid chromatography	Lewis *et al*. (1967)
Heptachlor	Heptachlor epoxide	Gas–liquid chromatography	Yu *et al*. (1971)
		Aliphatic hydroxylation	
p-Nitrotoluene	*p*-Nitrobenzoic acid	Spectrophotometric	Bratton and Marshall (1939), Chakraborty and Smith (1964), Reddy and Krishnakumaran (1974)
Dihydroisodrin	6-Hydroxy-6,7-dihydroisodrin	Gas–liquid chromatography	Krieger and Wilkinson (1971)
		Aromatic hydroxylation	
Benzo[*a*]pyrene	3-Hydroxybenzo[*a*]pyrene and other hydroxylated products	Spectroflurometric	Nebert and Gelboin (1968)
Aniline	*p*-Aminophenol	Spectrophotometric	Kato and Gilette (1965)
Naphthalene	1-Naphthol and other hydroxylated products	Spectrophotometric	Powis *et al*. (1976)
		Scintillation counting	Yu and Terriere (1972)
Biphenyl	2-Hydroxybiphenyl, 4-hydroxybiphenyl, 4,4′-dihydroxybiphenyl	Spectrofluorometric	Creaven *et al*. (1965b)

Table I. (*Continued*)

Substrate	Product	Method of detection	Reference
Coumarin	7-Hydroxycoumarin, 3-hydroxycoumarin	Spectrofluorometric	Creaven *et al*. (1965a)
		N-Demethylation	
p-Chloro-*N*-methylaniline	*p*-Chloroaniline, formaldehyde	Spectrophotometric	Kupfer and Bruggeman (1966), Nash (1953)
N,N-Dimethyl *p*-nitrophenol carbamate	*p*-Nitrophenol	Spectrophotometric	Hodgson and Casida (1961)
Aminopyrine	Monomethyl-4-aminoantipyrine, 4-aminoantipyrine, formaldehyde	Spectrophotometric	Brodie and Axelrod (1950), Nash (1953)
Ethylmorphine	Norethylmorphine, formaldehyde	Spectrophotometric	Anders and Mannering (1966)
		O-Dealkylation	
Methoxyresorufin	Resorufin	Spectrofluorometric	Mayer *et al*. (1977)
p-Nitroanisole	*p*-Nitrophenol	Spectrophotometric	Netter and Seidel (1964)
7-Ethoxycoumarin	7-Hydroxycoumarin	Spectrophotometric	Addison *et al*. (1977)
		Thio ether oxidation	
p-Chlorothioanisole	*p*-Chlorophenyl methylsulfinyl ether	Gas–liquid chromatography	Nigg *et al*. (1972)

ing the concentration of the complex was described by Omura and Sato (1964). The oxidized MFO system also forms ligand complexes that yield characteristic absorption spectra (Schenkman *et al.*, 1967). The spectra are stable and can be measured as long as there are no electrons (NADPH) available to start the oxidation of the ligand. Difference spectra are of two basic types (Fig. 5).

It is also possible to measure the activity of the NADPH-dependent microsomal reductase due to its ability to reduce cytochrome c. The assay is conveniently performed in spectrophotometric cuvettes by monitoring the increase in absorbance at 550 nm (the α band of reduced cytochrome c) with time (Williams and Kamin, 1962; Wilson and Hodgson, 1971).

It has become common practice to also measure the concentration of cytochrome b_5 in the microsomal suspension used for cytochrome P-450 measurements. Cytochrome b_5 can be measured simply as the difference spectrum between the reduced and the oxidized cytochrome (Omura and Sato, 1964).

In addition to the *in vitro* enzyme assays and the analysis of metabolic products formed after radiolabeled drug or insecticide administration *in vivo* (for review, see Ahmad, 1975), there are also at least two reliable bioassays for MFO activity. The one used with mammals measures "hexobarbital sleeping time" (Mannering, 1968) and is based on the return of the "righting reflex" after hexobarbital administration. This is an accurate and nondestructive assay that yields information about any changes in MFO activity with reference to a control resulting from induction or inhibition.

The standard bioassay used for screening insecticides involves measurement of mortality resulting from a series of dosages after a certain time period (Busvine, 1971). The measurements result in values of LD_{50} (the dose that kills 50% of the population), usually within 24 hours. The percent mortality data are converted to probit units and plotted against the logarithm of the dosage (Bliss, 1935), and the LD_{50} value can be obtained from the resulting straight line (Fig. 6). The data can also be programmed for computer analysis (Rosiello *et al.*, 1977). The most commonly used animal for this bioassay is the housefly, although it is possible to use any animal that can be killed in sufficiently large numbers. The administration of an insecticide in conjunction with a synergist yields the SR (synergistic ratio), which is an indirect indicator of MFO activity based on the fact that the effectiveness of the insecticide is inversely dependent on its metabolic degradation by MFO's (Metcalf, 1967; Brattsten and Metcalf, 1970; Benke and Wilkinson, 1971a).

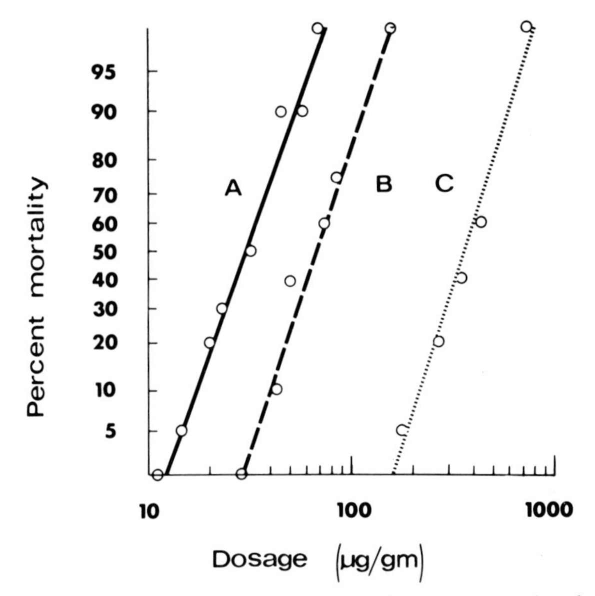

Figure 6. Regression lines resulting from probit mortality versus log-dosage plots of 24-hour toxicity of carbaryl (1-naphthyl N-methylcarbamate) to the last instar larvae of the southern armyworm *Spodoptera eridania* fed diets containing (B) 0.2% hexamethylbenzene, (C) 0.2% pentamethylbenzene, or (A) control diet for 3 days. The LD_{50} values are (A) 30 µg/gm of body weight (control), (B) 60 µg/gm of body weight (hexamethylbenzene), (C) 330 µg/gm of body weight (pentamethylbenzene). From Brattsten and Wilkinson (1973b).

C. MFO Enzymes

1. *Organization of MFO Enzymes*

The MFO system is firmly attached to the endoplasmic reticulum membranes of cells. High activity is, in particular, associated with the smooth endoplasmic reticulum membranes. When rat liver tissue is mechanically disrupted by homogenization and the resulting homogenate is subjected to a fractional centrifugation schedule, the vesiculated endoplasmic reticulum fragments can be obtained as a pellet from a high-speed spin of the postmitochondrial supernatant (Palade and Siekevitz, 1956). This pellet is termed the microsomal pellet and is routinely used in toxicological studies. The enzymes contained in this pellet are habitually referred to as microsomal enzymes.

The MFO system consists of several components. As shown schematically in Fig. 7, the catalytic center is occupied by cytochrome P-450. The reaction depends on reducing equivalents being transported from

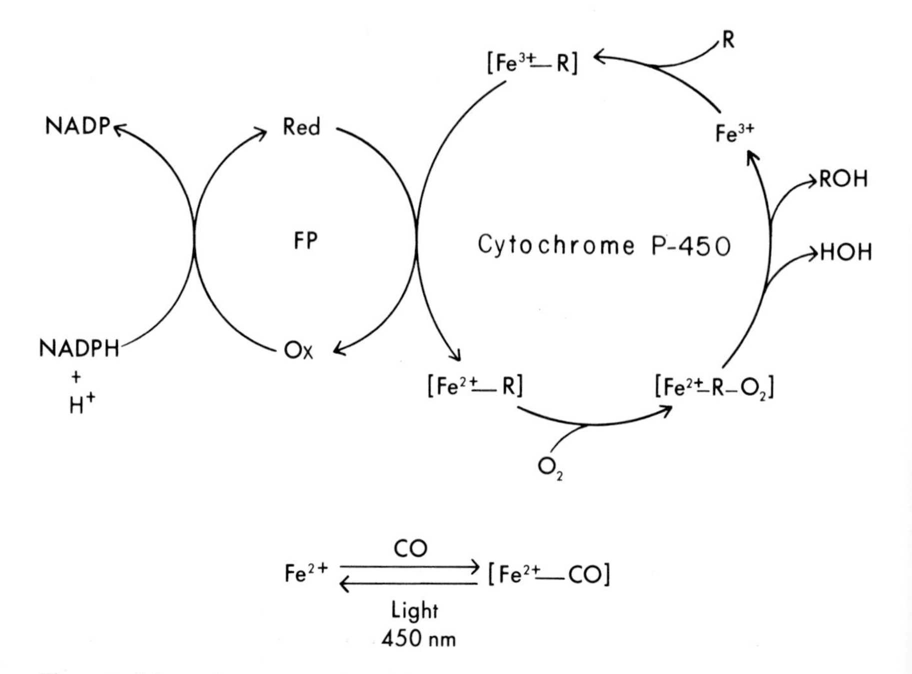

Figure 7. Schematic representation of the cytochrome P-450-dependent microsomal MFO system. Also shown is the complex formation of ferrocytochrome P-450 with carbon monoxide and its reversal by light of 450 nm.

NADPH to the cytochrome by a special flavoprotein (FP) enzyme, NADPH–cytochrome P-450 reductase. A third component, phosphatidylcholine, is also necessary for activity (Lu and Coon, 1968). In xenobiotic-metabolizing microsomal monooxygenases containing cytochrome P-450, these are probably the only three essential components. Another very similar form of the system was found in adrenal cortex mitochondria. The mitochondrial system appears to also depend on a nonheme iron containing protein operating between the reductase and the terminal cytochrome for activity (Omura *et al.*, 1966). This system is involved mainly in the metabolism of steroid hormones (Kupfer, 1970).

There is a second cytochrome attached to the endoplasmic reticulum membranes, cytochrome b_5, and a second flavoprotein enzyme associated with it that transports electrons preferentially from NADH. This sytem is responsible for fatty acid desaturation (Sato *et al.*, 1969). It is not known whether there is any cooperation or exchange of electrons between the two adjacent systems, although there are indications of interactions (Lu *et al.*, 1974).

2. *Cytochrome P-450*

The terminal oxidase of the MFO system is a b-type cytochrome that binds carbon monoxide. The reduced cytochrome–carbon monoxide complex shows a difference spectrum with an absorption peak at 450 nm. Omura and Sato (1964) first described this binding spectrum and named the cytochrome P-450 after pigment absorbing at 450 nm. The exact location of the peak consistently depends on the source of the cytochrome (Fig. 5a).

The cytochrome–substrate complex shows a characteristic absorption spectrum depending on the nature of the substrate. Compounds containing basic nitrogen, e.g., lipophilic aromatic and aliphatic amines, usually yield a spectrum with a peak at 430 nm and a trough at 395 nm (Fig. 5c). This kind of spectrum is called a type II spectrum. Most drugs, insecticides, and other acceptable substrates form complexes that show a type I spectrum with a peak at 385 nm and a trough at 420 nm (Fig. 5b) (Schenkman *et al.*, 1967). Carbamates containing nonbasic nitrogen give rise to type I spectra with liver cytochrome P-450 from several mammalian species (Kulkarni *et al.*, 1975).

The spectral interactions are taken to reflect the existence of two binding sites on the cytochrome. Type II spectra may be due to binding only to the sixth ligand of the iron (Schenkman *et al.*, 1967), whereas type I spectra are thought to arise from interactions with the protein–phospholipid complex immediately adjacent to the active site (Leibman and Estabrook, 1971). The interaction with lipophilic areas close to the heme group is also of critical importance in type II interactions (Wilkinson *et al.*, 1974). In fact, the cytochrome changes to a catalytically inactive form, cytochrome P-420, when separated from the membrane. Cytochrome P-420 shows an absorbance peak at 420 nm in the reduced CO complex. This dependence on a lipophilic surrounding has rendered the solubilization and purification of cytochrome P-450 extremely difficult.

Recently, the cytochrome has been obtained in pure form from mammals by solubilization with detergents followed by purification by chromatographic methods. Purified forms of P-450 differ only very slightly from each other and can be separated electrophoretically. Molecular weights are from 44,000 to 54,000 daltons. The forms have slightly different spectral properties and show differences in substrate preferences (Welton and Aust, 1974; Coon *et al.*, 1975). By the use of a sample of partially purified cytochrome P-420, the molar extinction coefficient of cytochrome P-450 from rabbit liver was estimated to be 91 mM^{-1} cm^{-1} (Omura and Sato, 1964). This value is still used for expressing concentrations of cytochrome P-450 from widely different sources.

3. *NADPH–Cytochrome P-450 Reductase*

In contrast to cytochrome P-450, the reductase associated with it (NADPH oxidase, EC 1.6.2.3) has been successfully solubilized and purified. The reductase is a flavoprotein with a molecular weight of about 80,000 daltons and contains one molecule of FAD and one molecule of FMN per molecule of protein. Cytochrome P-450 reductase from southern armyworm midgut microsomes is virtually identical to that in rat liver (D. L. Crankshaw and C. F. Wilkinson, personal communication). Antibodies to the reductase from rat liver microsomes inhibit *in vitro* MFO activity in a variety of mammalian tissue homogenates (Masters *et al.*, 1971), supporting the contention that the reductase is extremely similar in organisms and not responsible for any differences in catalytic specificity (Welton and Aust, 1974).

4. *Reaction Mechanism*

The details of the catalytic events are not fully understood. Initially, the substrate molecule combines with the oxidized cytochrome (Fig. 7). The complex then undergoes reduction and subsequently interacts with molecular oxygen in such a way that the hydroxylated substrate and a molecule of water leave the now reoxidized cytochrome (Estabrook *et al.*, 1971). Two electrons from NADPH transported by the flavoprotein interact in the hydroxylation of one substrate molecule. The two one-step reductions of the cytochrome substrate and oxygen complex produce the activated oxygen that is capable of hydroxylating the lipophilic substrate. The precise nature of the activated oxygen species that effects the hydroxylation is unknown, but the most likely candidate is the superoxide anion (O_2^-) free radical (Ullrich, 1971). Early experiments with the heavy oxygen isotope $^{18}O_2$ established that molecular oxygen, rather than oxygen atoms from the aqueous medium, participates in the reaction and that one oxygen is reduced to water, and the other hydroxylates the substrate molecule (Mason *et al.*, 1955).

D. MFO Enzymes in Animals

1. *Occurrence*

MFO-Catalyzed degradation of foreign compounds has been observed in organisms representing virtually all phyla of the animal kingdom as well as in plants and aerobic microorganisms. The MFO enzymes have been intensively investigated in standard laboratory animals (the rat, mouse, guinea pig, golden hamster, and rabbit), and these studies have supplied the backbone of current knowledge about the enzymes. There is also considerable information about MFO enzymes from insect tissues. Detailed studies have been done with the housefly *Musca domestica*, the larva of the southern armyworm *Spodoptera eridania*, and less than a dozen other insect species (Wilkinson and Brattsten, 1972).

Table II is a summary of most of the animal species in which MFO activity has been observed. Frequently, activity was measured by more than one method (for commonly used MFO assays, see Section II,B), although only one is mentioned.

The activity levels of the MFO enzymes have been found to be extremely different, even in closely related species. The activity levels are not included in Table II because of the difficulties in comparing results from many different laboratories. Some of the included studies were carried out at quite an early date, when techniques were less sophisticated. Some were undertaken for the sole purpose of establishing the presence or absence of activity and may consequently not show activity at optimal conditions. The O-demethylation of *p*-methoxyphenol in several detritus-feeding invertebrate species was measured in an *in vivo* experimental design (Neuhauser and Hartenstein, 1976) not comparable to the *in vitro* assays of, for instance, aldrin epoxidation in several of the insect species.

Attempts have been made to correlate levels of MFO activity in organisms with selected features in their life styles. The only successful study was done with 35 species of herbivorous lepidopterous larvae (Krieger *et al.*, 1971) and shows that polyphagous larvae have significantly higher aldrin epoxidase activity than oligo- or monophagous larvae. Dewaide (1971) measured aniline hydroxylation and N-demethylation of 4-dimethylaminopyrine in fishes caught in the Rhine River. The sample included perch, eel, salmon, houting, loach, twait, rudd, golden orfe, ide, dace, bream, tench, carp, and roach. Dewaide pointed out that the roach *Leuciscus rutilus* is a highly mobile species that feeds on a great variety of small aquatic plants and animals near the surface of the water, whereas fishes such as the bream, carp, and tench are all slow-swimming bottom dwellers. In analogy with the insect larvae studied by Krieger *et al.* (1971), liver preparations from the latter fishes were found to have lower MFO activities than those from the active and exposed roach.

In fact, as we shall see later (Section II,F,2), MFO activity levels in an organism are highly variable and so directly correlated to the meal the animal recently ingested and to the immediate environmental conditions as to make comparisons with other studies of questionable value. Table II lists about 60 species of invertebrates, and about as many verte-

Table II. Occurrence of mixed-function oxidase activity in animals

Animal (life stage; tissue)	Observed activity	Reference
Insects		
House cricket, *Acheta domesticus* (adult; Malpighian tubules)	Aldrin epoxidation	Benke and Wilkinson (1971b)
Tampa cockroach, *Nauphoeta cinerea* (adult; fat body)	Aldrin epoxidation	Benke and Wilkinson (1971a)
Madagascar cockroach, *Gromphadorhina portentosa* (adult; fat body)	Aldrin epoxidation	Benke *et al*. (1972)
American cockroach, *Periplaneta americana* (adult; fat body)	Heptachlor epoxidation	Nakatsugawa *et al*. (1965)
German cockroach, *Blattella germanica* (adult)	DDT hydroxylation	Agosin *et al*. (1961)
Caddisfly, *Limnephilus* sp. (larva)	Aldrin epoxidation	Krieger and Lee (1973)
Sawfly, *Macremphytus varianus* (larva; gut)	Aldrin epoxidation	Krieger *et al*. (1970)
Housefly, *Musca domestica* (adult; abdomen)	Cytochrome P-450	Hodgson *et al*. (1974)
Fleshfly, *Sarcophaga bullata* (adult)	Aldrin epoxidation	Terriere and Yu (1976b)
Blowfly, *Phormia regina* (adult)	Aldrin epoxidation	Terriere and Yu (1976b)
Mosquito, *Culex pipiens quinquefasciatus* (larva)	*p*-Nitroanisole O-demethylation	Hansen *et al*. (1972)
Honeybee, *Apis mellifera* (adult, larva; gut)	Aldrin epoxidation	Gilbert and Wilkinson (1974)
Grassgrub, *Costelytra zealandica* (larva)	Phenyl N-methylcarbamate N-demethylation	Hook and Smith (1967)
Tobacco budworm, *Heliothis virescens* (gut, fat body)	Aldrin epoxidation	Bull and Whitten (1972)
Pink bollworm, *Pectinophora gossypiella* (larva)	Aldrin epoxidation	Williamson and Schechter (1970)
Black cutworm, *Agrotis ypsilon* (larva)	Dihydroisodrin hydroxylation	Thongsinthusak and Krieger (1976)
Silkworm, *Telea polyphemus* (larva), and 5 other saturniids	Aldrin epoxidation	Krieger *et al*. (1976)
Cabbage looper, *Trichoplusia ni* (larva; fat body)	Carbaryl hydroxylation	Kuhr (1971)
Greater wax moth, *Galleria mellonella* (larva; gut)	*p*-Nitroanisole O-demethylation	Ahmad and Brindley (1971)
Southern armyworm, *Spodoptera eridania*, and 34 other lepidopterous larvae (gut)	Aldrin epoxidation	Krieger *et al*. (1971)

Other arthropods		
Wool-handed crab, *Eriocheir sinensis* (gill)	Aniline hydroxylation	Dewaide (1971)
Spot shrimp, *Pandalus platyceros* (larva)	Naphthalene hydroxylation	Sanborn and Malins (1977)
Lobster, *Homarus americanus* (hepatopancreas)	Aniline hydroxylation	Elmamlouk and Gessner (1976)
Sourbug, *Oniscus asellus*	*p*-Methoxyphenol O-demethylation	Neuhauser and Hartenstein (1976)
Centipede, *Lithobius forficatus*	*p*-Methoxyphenol O-demethylation	Neuhauser and Hartenstein (1976)
Millipede, *Pseudopolydesmus serratus*	*p*-Methoxyphenol O-demethylation	Neuhauser and Hartenstein (1976)
Ostracod, *Chlamydotheca arcuata*	Aldrin epoxidation	Kawatski and Schmulbach (1971)
Other invertebrates		
Earthworm, *Lumbricus terrestris* (gut, typhlosole)	Aldrin epoxidation	Nelson *et al*. (1976)
Slug, *Arion hortensis*	*p*-Methoxyphenol O-demethylation	Neuhauser and Hartenstein (1976)
Snail, *Physa elliptica*	*p*-Nitroanisole O-demethylation	Hansen *et al*. (1972)
Snail, *Lymnea pallustris*	*p*-Nitroanisole O-demethylation	Hansen *et al*. (1972)
Fishes		
Sea lamprey, *Petromyzon marinus* (liver)	Aniline hydroxylation	Dewaide (1971)
Pike, *Esox lucius,* and 16 other species of fish (liver)	Aniline hydroxylation	Dewaide (1971)
Fathead minnow, *Pimephales promelas* (female)	*p*-Nitroanisole O-demethylation	Hansen *et al*. (1972)
Coho salmon, *Oncorhynchus kisutsch,* and 5 other species of fish (liver)	Aniline hydroxylation	Buhler and Rasmusson (1968)
Mosquito fish, *Gambusia affinis,* and 4 other species of fish (liver)	Aldrin epoxidation	Ludke *et al*. (1972)
Amphibians		
Frog, *Rana temporaria* (adult; liver)	Biphenyl hydroxylation	Creaven *et al*. (1967)
Bullfrog, *Rana catesbiana* (tadpole)	Cytochrome P-450	Machinist *et al*. (1968)
Reptiles		
Lizard, *Lacerta viridis* (liver)	Aniline hydroxylation	Dewaide (1971)
Alligator snapping turtle, *Chelydra serpentina* (liver)	Cytochrome P-450	Machinist *et al*. (1968)
Birds		
Puffin, *Fratercula artica* (liver)	Aniline hydroxylation	Bend *et al*. (1977)
Mallard, Anas platyrhynchos (liver)	Aniline hydroxylation	Davidson and Sell (1972)
Chicken, *Gallus domesticus* (liver)	Aniline hydroxylation	Parke (1960)
Pigeon, *Columba livia* (liver)	Coumarin hydroxylation	Creaven *et al*. (1965a)

(*Continued*)

Table II. *(Continued)*

Animal (life stage; tissue)	Observed activity	Reference
Starling, *Sturnus vulgaris*, and 3 other species of birds (liver)	N,N-Dimethylaniline N-demethylation	Pan *et al*. (1975)
Grackle, *Quiscalus quiscala* (liver)	N,N-Dimethylaniline N-demethylation	Pan *et al*. (1975a)
Red-winged blackbird, *Agelaius phoeniceus* (liver)	Aniline hydroxylation	Pan *et al*. (1975b)
Brown-headed cowbird, *Molothrus ater* (liver)	N,N-Dimethylaniline N-demethylation	Pan *et al*. (1975a)
English sparrow, *Passer domesticus* (liver)	Cytochrome P-450	Machinist *et al*. (1968)
Zebra finch, *Taeniopygia guttata* (liver)	Aniline hydroxylation	Dewaide (1971)
Japanese quail, *Cornutix cornutix* (liver)	Cytochrome P-450	Hinderer and Menzer (1976)
Blackbird, *Turdus merula* (liver)	Cytochrome P-450	Yawetz *et al*. (1978)
Rock partridge, *Alectoris graeca* (liver)	Cytochrome P-450	Yawetz *et al*. (1978)
Barn owl, *Tyto alba* (liver)	Cytochrome P-450	Yawetz *et al*. (1978)
African bulbul, *Pychnonotus capensis* (liver)	Cytochrome P-450	Yawetz *et al*. (1978)
Marsupials		
Opossum, *Didelphis virginiana* (liver)	Cytochrome P-450	Machinist *et al*. (1968)
Quokka, *Setonix brachyurus* (liver)	Aniline hydroxylation	McManus and Ilett (1976)
Mammals		
Dog, *Canis familiaris* (liver)	Aniline hydroxylation	Parke (1960)
Mouse, *Mus musculus* (liver)	Aniline hydroxylation	Parke (1960)
Fox, *Vulpes vulpes* (liver)	Biphenyl hydroxylation	Creaven *et al*. (1965b)
Cat, *Felis domesticus* (liver)	Biphenyl hydroxylation	Creaven *et al*. (1965b)
Coypu, *Myocastor coypus* (liver)	Biphenyl hydroxylation	Creaven *et al*. (1965b)
Hamster, *Cricetus cricetus* (liver)	Biphenyl hydroxylation	Creaven *et al*. (1965b)
Guinea pig, *Cavia porcellus* (liver)	Biphenyl hydroxylation	Creaven *et al*. (1965b)
Bat, *Myotis velixer* (liver)	Cytochrome P-450	Machinist *et al*. (1968)
Ground squirrel, *Citellus tridecemlineatus* (liver)	Cytochrome P-450	Machinist *et al*. (1968)
Armadillo, *Tolypeutes tricinctus* (liver)	Cytochrome P-450	Machinist *et al*. (1968)
Racoon, *Procyon lotor* (liver)	Cytochrome P-450	Machinist *et al*. (1968)
Pig, *Sus scrofa* (liver)	Cytochrome P-450	Machinist *et al*. (1968)

brate species have been studied. It is probably reasonable to assume that *all* aerobic organisms have the capacity for mixed-function oxidation, even though it has been demonstrated in a comparatively small number of organisms.

2. *Tissue Distribution of MFO Activity*

The liver is the major site of foreign compound metabolism in mammals and other vertebrates. Not only has a greater variety of reactions been observed with liver tissue than with any other (Testa and Jenner, 1976), but also the activities are often more than 10-fold higher in liver tissue than in other organs (Machinist *et al.*, 1968; Booth and Boyland, 1971). Mammalian MFO activity is also high in kidney, lung, small intestine, and placenta, and low levels of activity are found in skin and brain.

A study of aminopyrine *N*-demethylase in the domestic chicken showed that duodenum has as high activity as liver, followed by lower levels in the midintestine and kidney. Crop and rectum have very low activity levels. A similar distribution was seen in the domestic turkey and goose, whereas in the domestic duck the activity levels in liver and kidney were comparable (Bartlet and Kirinya, 1976).

In insects, high MFO activity is observed in the fat body and in the tissues associated with the alimentary tract. As pointed out by Wilkinson and Brattsten (1972), there is not any one insect organ that consistently has higher activity than all others in all species as does the vertebrate liver. Honeybee midgut has 10 times higher aldrin epoxidase activity than other honeybee tissues (Gilbert and Wilkinson, 1974). Like the bee, most lepidopterous larvae have highest activity in their midgut tissues. Krieger and Wilkinson (1969) reported dramatically higher aldrin epoxidase activity in southern armyworm midgut tissue than in Malpighian tubules, fat body, larval head, or other tissues. The midgut is also the most active tissue in larvae of the silk moth *Antheraea pernyi* (Krieger *et al.*, 1976) and in the corn earworm *Heliothis zeae* (Khan, 1969). Bull and Whitten (1972) found high aldrin epoxidase activity in both gut and fat body of tobacco budworm larvae, whereas the major site of microsomal oxidases in the cabbage looper (Kuhr, 1970, 1971) and wax moth larvae (Reddy and Krishnakumaran, 1974) is the fat body.

The midgut is also an important location for MFO activity in cockroaches. Benke *et al.* (1972) found the highest aldrin epoxidase and dihydroisodrin hydroxylase activities in crude homogenates from midgut, Malpighian tubules, and gastric coeca of the Madagascar cockroach. Turnquist and Brindley (1975) reported the highest detoxification activity of EPN, O-ethyl-O-*p*-nitrophenyl phenylphosphorothioate, in fat body and midgut microsomes of the American cockroach. The Malpighian tubules (followed by the gut) are also the site of maximal MFO activity in the house cricket (Benke and Wilkinson, 1971b) as well as in the lubber grasshopper *Romalea microptera* (L. B. Brattsten, unpublished).

The highest aldrin epoxidase activity of the earthworm was found to be in the intestine with the mucous-cell-lined and infolded typhlosole (Nelson *et al.*, 1976).

Clearly, the distribution of MFO activity in both vertebrates and insects has a strategic aspect. In vertebrates the liver is the major screen through which all metabolites are passed by the blood from the portal vein. The intestine and lungs are important barriers that foreign compounds must penetrate before reaching the interior environment of the animal and are thus ideal locations for a biochemical detoxification system. The mobile cockroaches, crickets, and grasshoppers have one line of defense against ingested xenobiotics in their gut MFO enzymes and another one against possible cuticular or tracheal contamination in their fat bodies. This pattern persists in the lepidopterous larvae, with predominating emphasis on the gut MFO's (Wilkinson and Brattsten, 1972), which reflects feeding as their major activity.

E. Physiological Factors Influencing MFO Activity

During the life span of an animal the MFO enzymes are constantly under the influence of physiological factors including age, sex, and reproductive and nutritional condition.

1. *Age–Activity Variation*

It is well known that fetal and newborn mammals essentially lack MFO activity. Jondorf *et al.* (1959) showed that fetal and 1-day-old guinea pigs and mice lack the ability to metabolize several drugs. This ability develops during the first week of life to about one-third of the adult capacity, which is attained after about 8 weeks. The low or lacking perinatal MFO activity has been studied in detail in rats. Iba *et al.* (1975) suggested that the level of unsaturation of the fetal phospholipids might be involved in the suppression of fetal MFO activity. They showed that a fetal lipid extract, containing 28% unsaturated fatty acids compared to 20% for adult male rats, inhibited *in vitro* ethylmorphine N-demethylation but not aniline hydroxylation. Dallner *et al.* (1966) reported that the endoplasmic reticulum in liver of fetal rats and rabbits appears to largely lack the smooth-surfaced portions with which the highest MFO activity is associated. The deficient MFO activity in fetal and neonatal mammals not only may be due to a deficiency in the tissue per se, as indicated by the absence of smooth endoplasmic reticulum, but may also result from the hormonal status of the immature organism (Testa and Jenner, 1976).

The age-related variation in MFO activity in birds differs somewhat from that in mammals. Fetal chickens lack several MFO activities, but these attain a maximal level 1 day after hatching and subsequently decline toward the lower adult levels, which are reached after about 1 week (Powis *et al.*, 1976). Bartlet and Kirinya (1976) made a similar observation with the chicken and also with the domestic goose.

It is generally true that in insects high MFO activity levels coincide with an active life stage. As early as 1958, Fenwick concluded that MFO activity is high only between molts in the desert locust. This observation has since been confirmed every time a new insect species has been carefully studied. Insects are prevented from continuous growth by their chitinous exoskeleton and instead grow by molts. During the complicated process of molting (see Wigglesworth, 1972), insects shed their old skin after a new, bigger integument has formed underneath. The newly molted insect inflates the new, soft integument to allow for additional increase in size. Upon exposure to air the new integument undergoes pigmentation and hardens to serve as a new exoskeleton. During molts, which take several hours, insects seek out secluded places and cease to feed. Hemimetabolous insects such as the cricket, cockroach, and grasshopper grow to adult size by a series of molts in which size increase and development of sexual maturity are the major features. In holometabolous insects, e.g., flies, moths, and butterflies, the life stages are highly specialized. The immature larvae are feeders, and the adults disperse and reproduce the species, often without any further feeding. The two life stages are morphologically adapted to their functions and look entirely different from each other. In between the two is an immobile, enclosed pupal stage during which there is a complete reorganization of the body form and function.

Fig. 8 shows the variation with age of MFO activity in the southern armyworm (Krieger and Wilkinson, 1969), the house cricket (Benke and Wilkinson, 1971b), and the Madagascar cockroach (Benke *et al.*, 1972). In the case of the Madagascar cockroach the active life stages of high MFO activity and the molt periods with low activity are dramatically obvious. The house cricket shows a similar low activity during the molts and higher activity between the molts. The southern armyworm displays a generally increasing MFO activity with age, although there is, during every larval molt, a temporary drop in activity. When the larvae cease to feed, void their guts, and prepare to pupate there is a total and almost

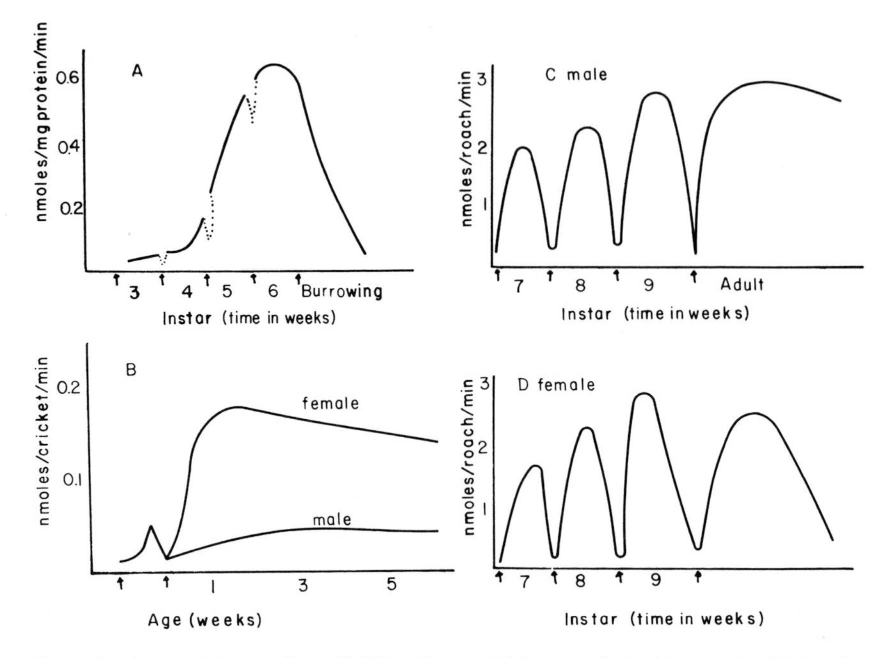

Figure 8. Age–activity profiles of MFO activity (aldrin epoxidation) in insects. (A) Southern armyworm, *Spodoptera eridania;* (B) house cricket, *Achaeta domesticus;* (C and D) Madagascar cockroach, *Gromphadorhina portentosa.* Redrawn from Wilkinson and Brattsten (1972).

instantaneous disappearance of activity. A very similar age–activity profile of oxidase activity was observed in larvae of the silk moth *Saturnia pyri* (Krieger *et al.*, 1976), in larvae of the black swallowtail *Papilio polyxenes* (L. B. Brattsten, unpublished), in the cabbage looper (Kuhr, 1970, 1971), and in the black cutworm (Thongsinthusak and Krieger, 1976).

Krieger and Wilkinson (1969) did not find any oxidase activity in adult armyworm moths, even though these are known to feed on nectar after emergence. There are no reports of microsomal oxidase measurements in any other adult lepidopteran. Adult flies, on the other hand, have higher oxidase activity levels than any other fly life stage. Eggs and pupae of the housefly (Yu and Terriere, 1971a), the fleshfly *Sarcophaga bullata,* and the blowfly *Phormia regina* (Terriere and Yu, 1976b) were found to lack heptachlor and aldrin epoxidase activity totally or almost totally. During most of the larval development a low level of activity was observed, and in all three species there was a sharp peak of activity 12–15 hours before pupation. The activity dropped to a very low level as abruptly as it appeared. Larval activity levels were in all cases lower than those of the adult flies.

Nelson *et al.* (1976) reported a steady increase in aldrin epoxidase activity in the gut of the earthworm with age. There is a sixfold higher activity in adult earthworms than in infants.

2. *Sex-Related Variation*

It is well known that male rats have higher MFO activity levels than female rats (Giri, 1973; Kato, 1974). However, this clear-cut sex-dependent difference in drug-metabolizing capacity has not been observed in other vertebrates. Testa and Jenner (1976) noted the absence of observed sex-related differences in mice, hamsters, guinea pigs, rabbits, dogs, pigs, and monkeys.

Bartlet and Kirinya (1976) reported higher aminopyrine N-demethylase activity in cocks than in hens and in male domestic geese than in females. Female domestic ducks had higher activity than males but were also of a different strain of duck, possibly reflecting a genetic difference as well.

A careful investigation with numerous MFO substrates would undoubtedly reveal sex-dependent differences in mammals besides the rat. For instance, Yaffe *et al.* (1968) observed a more active hexobarbital metabolism in female mice than in males and could correlate the rate of metabolism with the reproductive state of the female. Testa and Jenner (1976) discuss at length the influence of sex hormones on the observed sex-dependent MFO differences in rats. They conclude that the balance between male and female hormones is an important factor in the regulation of MFO enzymes.

Sex-related differences in MFO activity have also been reported for some insect species. There are no differences in aldrin epoxidase activity in adult male or female fleshflies or blowflies (Terriere and Yu, 1976b), but male honeybees (drones) have higher oxidase activities than worker bees (Gilbert and Wilkinson, 1974). As can be seen in Fig. 8, the aldrin epoxidase activity increases rapidly after the molt in adult Madagascar cockroaches. In males the activity appears to stay at a high level or to only slowly decline with age. In females there is a drastic and rapid decrease in activity about 6 weeks after the molt. This decrease coincides with the ovoviviparous development of the young, which requires approximately 6 weeks (Benke *et al.*, 1972). During pregnancy oxidase activities in Dutch rabbits were found to be as much as seven times lower (Gut *et al.*, 1976). Turnquist and Brindley (1975) reported a very similar rise in cytochrome P-450 content and *p*-nitroanisole O-demethylase activity in the female American cockroach after the adult molt. The activities reach a peak after about 90 days and then decline rapidly.

In the adult cricket the activity distribution appears to be reversed (Fig. 8), with higher oxidase activity in female crickets than in males.

In analogy with the lack of sex-related differences in fetal or neonatal rats, there are no differences in activity in immature (nymphal) cockroaches (Fig. 8).

3. *Genetic Control*

Cram *et al.* (1965) showed an impressive variation in several MFO-catalyzed reactions in six strains of rabbits. The rate of N-demethylation of aminopyrine was 15 times higher in English rabbits than in cottontails. English rabbits had almost 4 times higher aniline hydroxylase activity than Dutch rabbits. But there were no differences in chlorpromazine hydroxylation among the six strains. Much more moderate (two- to threefold) differences in drug-metabolizing ability were reported in genetically defined strains of rats and mice (Mitoma *et al.*, 1967; Vesell, 1968).

Insects notoriously develop strains that are resistant to insecticides. There is usually a strong positive correlation between high metabolic capacity and resistance to insecticides, even though several factors other than metabolism are involved in resistance. In a survey of 14 strains of the housefly, Schonbrod *et al.* (1968) found a more than 10-fold difference in both aldrin epoxidase and naphthalene hydroxylase activities but not a corresponding difference in resistance to naphthalene vapors. High oxidase activity levels in houseflies were found to be associated with semidominant genes on chromosomes II and V (Plapp and Casida, 1969; Hodgson and Plapp, 1970). Backcross experiments with visible marker genes showed that in one resistant strain, R-Baygon, the gene(s) on chromosome II was responsible, and in another resistant strain, R-Fc, the gene(s) on chromosome V was linked to the resistance and high oxidase activity. In another strain, R-Isolan, a semidominant gene (*Ox*) on chromosome II was linked to high aldrin epoxidase activity (Khan, 1970). A gene on chromosome III contributes to the observed resistance by coding for decreased cuticular penetration of contact insecticides (Schafer and Terriere, 1970).

Wilkinson and Brattsten (1972) and Agosin and Perry (1974) discuss quantitative and qualitative differences in microsomal oxidation in susceptible and resistant housefly strains. Hodgson and Tate (1976) summarize the qualitative differences as follows: (a) There is a shift in the cytochrome P-450–carbon monoxide complex difference spectrum from 451–452 nm in susceptible flies to 448–449 nm in resistant flies; (b) microsomes from susceptible flies do not form a type I difference spectrum, whereas microsomes from resistant flies do; and (c) microsomes from susceptible flies do not form a single-trough type II n-octylamine spectrum, whereas microsomes from resistant flies do.

4. *Influence of Nutrition*

Rats fed a diet containing 5% casein as the protein source, compared to 20% casein in the control diet, were found to have significantly reduced body weight, microsomal protein content, and cytochrome P-450 and cytochrome P-450 reductase levels but not lower phosphatidylcholine levels. The ability of the rats fed 5% casein to bind and metabolize ethylmorphine and aniline was reduced (Hayes *et al.*, 1973; Hayes and Campbell, 1974). Even deficiencies in single essential amino acids, e.g., tryptophan or valine, can cause reduced levels of cytochrome P-450 in rats (Truex *et al.*, 1977). Drug-metabolizing capacity and cytochrome P-450 levels are lower in rats as a result of deficiencies in any of a number of nutritional factors besides protein, e.g., lipids (Marshall and McLean, 1971), vitamin C (Zannoni and Lynch, 1973), vitamin E (Carpenter, 1972), and several metals.

Experiments of this kind, which contribute information about the molecular biology of the enzymes, have been performed only with standard laboratory mammals.

F. External Factors Influencing MFO Activity

Superimposed on the fluctuation in MFO levels due to physiological factors are the changes caused by factors in the environment of the organism. The presence of chemicals of almost any kind, excluding nutrients, has the most profound influence on the MFO enzymes. Many chemicals are attacked and metabolized by the MFO enzymes; many chemicals induce higher activity levels; and some chemicals are very potent inhibitors of MFO activity.

1. *MFO Inhibitors*

Several kinds of small molecules, illustrated in Fig. 9, are known inhibitors of the MFO enzymes. The active molecules are aryloxyalkylamines such as SKF 525-A, a therapeutically used drug extender, aryl 2-propynyl ethers and oximes, organothiocyanates, 1,2,3-benzothiadiazoles, and substituted imidazoles. A commercially used synergist is the phthalimide MGK-264. Some compounds, notably piperonyl butoxide, Tropital, sulfoxide, and propyl isome (Fig. 10), are also used commercially as insecticide synergists. Synergism denotes the

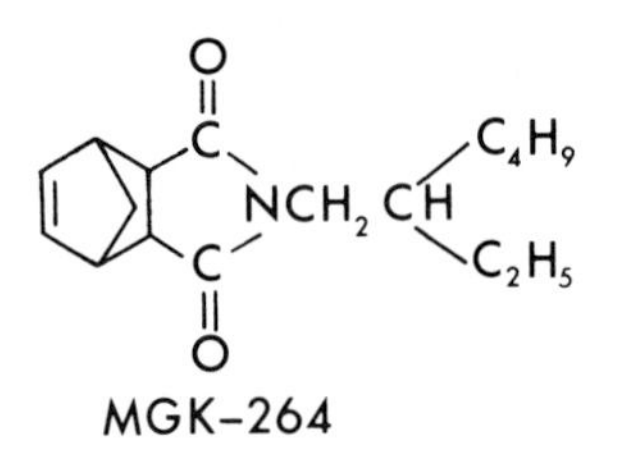

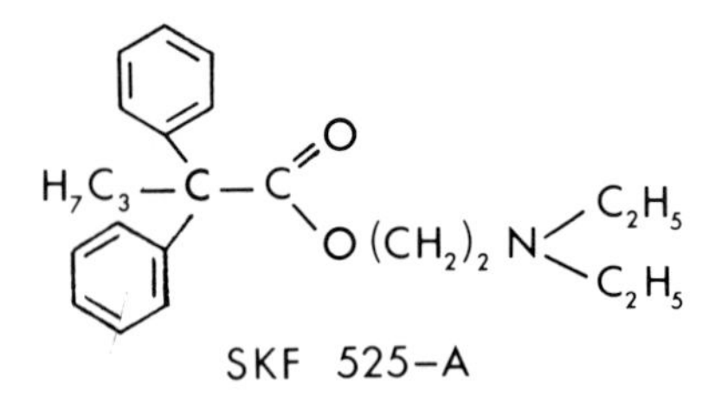

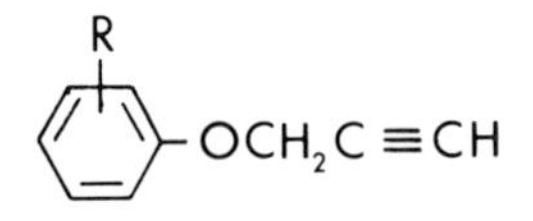

Aryl 2-propynyl ether

Aryl 2-propynyl ether

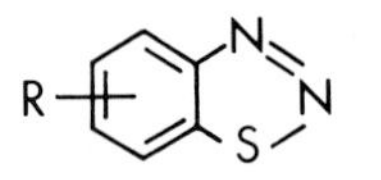

1,2,3 — Benzothiadiazole

Imidazole

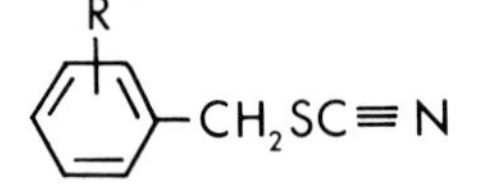

Benzyl thiocyanate

Figure 9. Cytochrome P450 inhibitors.

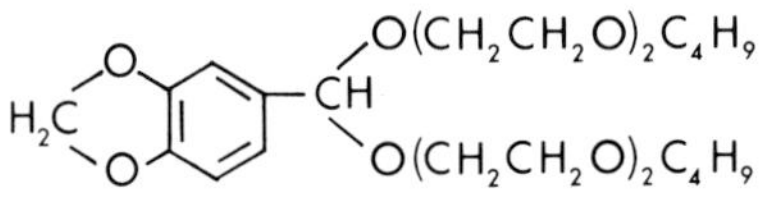

Tropital

Piperonyl butoxide

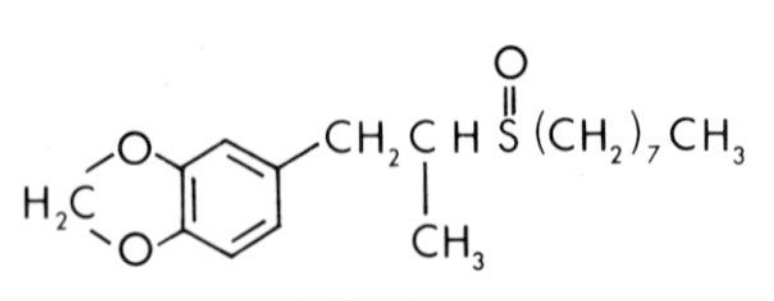

Propyl isome

Figure 10. 1,3-Benzodioxole inhibitors of cytochrome P-450.

significantly increased biological activity (toxicity) of one chemical as a result of addition of another chemical, the synergist, which is without acute toxicity at the concentration used. Inhibition by synergists is one of the most intensively studied aspects of MFO biochemistry due to the resulting, practically important changes in the toxicity of insecticides and the effectiveness of drugs.

It has been known for almost 40 years (see references in Hodgson and Tate, 1976) that sesame oil improves the insecticidal action of pyrethrum. Sesame oil contains two compounds, sesamin and sesamolin (Fig. 11), which both have a 1,3-benzodioxole (methylenedioxyphenyl) group and are MFO inhibitors. Within this group of compounds the benzodioxole structure is essential to the inhibitory action. Even minor changes such as replacing one of the oxygen atoms with sulfur or (worse) a methyl group drastically reduces the inhibitory activity (Wilkinson *et al.*, 1966). Figure 12 shows the synergistic ratio (LD_{50} of insecticide alone/LD_{50} of insecticide plus synergist) of carbaryl in a 1:5 mixture with several structural analogs of the 1,3-benzodioxole nucleus. The rigid planarity of the ring system and the integrity of the methylene group are essential features. Replacing the methylenic hydrogens with deuterium atoms decreased the inhibitory activity (Metcalf *et al.*, 1966), indicating homolytic free-radical interaction as a possible inhibitory mechanism (Hansch, 1968). Although it appears that only the 1,3-benzodioxole nucleus interacts with the MFO catalytic site, the side chain (R) is important in giving a suitable lipophilicity to the molecule so that it can penetrate to the site of action (Wilkinson *et al.*, 1966). Indications were obtained (Graham *et al.*, 1970) that 1,3-benzodioxole

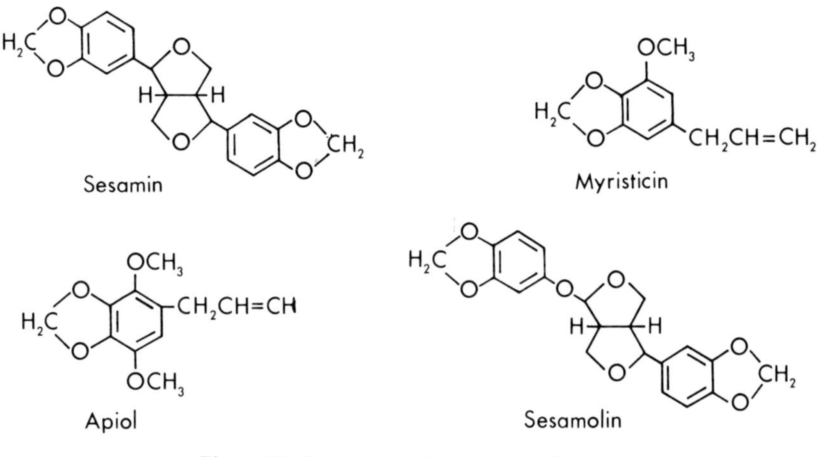

Figure 11. Some naturally occurring lignans.

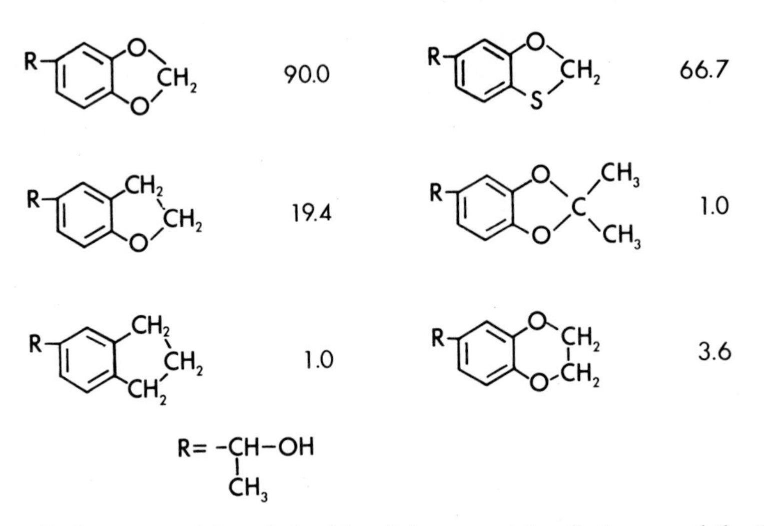

Figure 12. Structure–activity relationship of the synergistic effectiveness of the 1,3-benzodioxole nucleus as indicated by synergistic ratios (5 : 1, w/w, with insecticide) of carbaryl toxicity to adult houseflies. Redrawn from Wilkinson (1968).

compounds themselves undergo metabolism, and subsequent work (Wilkinson and Hicks, 1969; Kamiensky and Casida, 1970) has shown that they are, in fact, acceptable MFO substrates. Excellent reviews of 1,3-benzodioxole synergist interactions with insecticides are available (Wilkinson, 1968; Casida, 1970).

Compounds containing a 1,3-benzodioxole nucleus are frequently encountered in nature. It is well known that the source of pyrethrum insecticides, the flowers of the composite plant *Chrysanthemum cinerariaefolium,* also contain 1,3-benzodioxole compounds, e.g., sesamin. Known as lignans, these chemicals are found in leaves, heartwood, resins, and other plant parts. About three dozen naturally occurring lignans are known (Robinson, 1967). Some compounds are shown in Fig. 11. Myristicin, found in parsnip, is a good synergist for carbaryl (Casida, 1970). Apiol occurs in the leaves of parsley (Robinson, 1967).

More recently, the spectral interactions of cytochrome P-450 and 1,3-benzodioxoles have been investigated (Hodgson and Philpot, 1974; Hodgson and Tate, 1976). Piperonyl butoxide was found to produce a type I difference spectrum with oxidized microsomes. Reduction of the microsome–synergist complex by NADPH produced a spectrum with two peaks, one at 455 nm and one at 427 nm. This spectrum has been seen with microsomes from both mammals and insects. It is called a type III spectrum, and it cannot be displaced by carbon monoxide, unlike most cytochrome P-450–substrate complex spectra. The presence of the type III spectrum prevents the formation of type I or type II substrate spectra. A direct correlation has been established between the type III interaction and synergism of insecticides by MFO inhibition (Franklin, 1972; see Hodgson and Tate, 1976, for detailed discussion).

There is less detailed information available about the interactions of the MFO system and other kinds of inhibitors. A study with 56 substituted 1,2,3-benzothiadiazoles showed that the synergistic activity again correlates directly with the ability of the compounds to form homolytic free radicals and with their lipophilic properties (Gil and Wilkinson, 1976, 1977). Studies with 1-alkyl-substituted imidazoles also showed the importance of the lipophilic nature of the molecule (Wilkinson *et al.*, 1974). A chain length of 8 to 12 carbon atoms most effectively inhibited aldrin epoxidation in rat liver and armyworm gut microsomes, increased pentobarbital sleeping time in mice, and synergized carbaryl toxicity to houseflies. Some of these substituted 1-alkylimidazoles are the most potent inhibitors of MFO activity reported so far. The I_{50} value (molar concentration that inhibits 50% of the enzyme activity *in vitro*) of 1-nonylimidazole to aldrin epoxidation in armyworm midgut was 8×10^{-8} M (Wilkinson *et al.*, 1974) compared to an I_{50} value of 1.8×10^{-5} M for piperonyl butoxide under identical conditions (Wilkinson and Brattsten, 1972). Additional work with 1-,2-, and 4(5)-aryl-substituted imidazoles showed that at least one of the imidazole nitrogens must be sterically unhindered to interact with the cytochrome. This work led to the proposition of a double interaction of the imidazole (and other) inhibitors with the cytochrome. The inhibitors bind to the fifth or sixth ligand of the porphyrin heme with the unbonded imidazole nitrogen electrons and in addition bind to at least one sterically defined spot in the lipophilic membrane adjacent to the heme group.

2. *Induction of MFO Activity*

Since it was first discovered (Brown *et al.*, 1954) that certain chemicals stimulate the activity of MFO enzymes, the phenomenon of induction has been intensively studied. Induction, *sensu strictu*, refers to the production of enzyme protein in addition to the amount that is normally present by a temporarily accelerated process of *de novo* protein synthesis. Most work with induced enzyme systems do not include the actual demonstration of *de novo* protein synthesis, and the term "induction" is used broadly to denote an enzyme with a higher than usual specific activity.

A vast number of chemicals are known to induce the MFO system in mammals. Conney (1967) lists about 60 compounds ranging from hypnotics and sedatives to the carcinogenic polycyclic aromatic hydrocarbons. Sher (1971, a literature tabulation) and Remmer (1972) each list at least 70 compounds. These three extensive and basic review articles are mainly concerned with the pharmacological implications of MFO induction. In addition to hypnotics, sedatives, and polycyclic hydrocarbons, inducing chemicals are found among anticonvulsants, tranquilizers, antipsychotics, antidepressives, analgesics, antihistamines, antibiotics, steroids, insecticides, alkaloids, and more. Plant allelochemicals are also important MFO inducers. Hutson (1977) summarizes the molecular properties that inducing chemicals appear to share: (a) suitable lipophilicity, (b) suitable biological stability, and (c) ability to interact for a prolonged period (form a fairly stable complex) with cytochrome P-450. Inducing chemicals fall more or less into two categories depending on their specific effect on cytochrome P-450. Compounds that have a general inducing effect, stimulating a large number of reactions, effect an increase in cytochrome P-450 with a carbon monoxide complex spectrum at 450 nm. This category is exemplified by phenobarbital and comprises most of the active inducers such as drugs and insecticides. The other group of inducers is exemplified by 3-methylcholanthrene and contains the polycyclic aromatic hydrocarbons. This group has a much more limited effect and stimulates only a few MFO-catalyzed reactions. The cytochrome resulting from 3-methylcholanthrene induction has a carbon monoxide complex absorption spectrum with the peak shifted to 448 nm. This cytochrome is called P-448 or P_1-450. Inducing compounds of both categories tend to give rise to a type I substrate complex spectrum with cytochrome P-450, supporting the hypothesis that inducers must be able to interact with the hemoprotein part of the cytochrome (Remmer, 1972).

General inducers of the phenobarbital type effect several changes in mammals. There is an increase of up to 40% in the liver weight in rats and rabbits (Fouts and Rogers, 1965), and a proliferation of the endoplasmic reticulum membranes can be seen. The smooth portions of the endoplasmic reticulum account for most of the increase (Remmer and Merker, 1963). An increase in the rate of turnover of phospholipids (Orrenius and Ericson, 1966) was shown by increased incorporation of [^{32}P] phosphate. Davidson and Wills (1974) also reported that any increase in phospholipid content is transient but that the turnover rate is increased. *De novo* protein synthesis was demonstrated by increased incorporation of [^{14}C]leucine into microsomal proteins (Kato and Gillette, 1965; Agosin *et al.*, 1965), whereas there was almost zero radiolabeling of proteins in other subcellular fractions, e.g., mitochondria and nuclei. Experiments with protein synthesis inhibitors such as ethionine, cycloheximide, actinomycin D, and puromycin also indicate the prerequisite of *de novo* protein synthesis for induction (Conney, 1967; Testa and Jenner, 1976; Gelboin, 1971).

Cytochrome P-450 depends on an iron-containing porphyrin nucleus, the so-called heme group, for activity. The biosynthesis of heme proceeds via approximately six enzyme-catalyzed reactions beginning with the condensation of succinyl coenzyme A and glycine to δ-aminolevulinic acid. The first enzyme, δ-aminolevulinate synthetase (EC 2.3.1.37), has a short half-life and is inhibited by the end product of the pathway, heme. It is therefore seen as a regulatory point in heme biosynthesis (Baron and Tephly, 1969). This enzyme was shown to be stimulated by MFO inducers within very short time periods, 3–5 hours, after exposure to an inducer (Marver, 1969), thus making more heme available. It was also shown by the use of radiolabeled δ-aminolevulinic acid that subsequently labeled heme was preferentially incorporated into cytochrome P-450 (Remmer, 1972).

Induction involves an increase in DNA-dependent RNA polymerase activity (Bresnick, 1966; Balazs and Agosin, 1968; Gelboin *et al.*, 1967; Elshourbagy and Wilkinson, 1978), as well as an increase in RNA template activity (Litvak and Agosin, 1968; Loeb and Gelboin, 1964).

Induction by MFO enzyme inducers is a fairly specific process. Even the closely associated microsomal enzymes cytochrome b_5 and NADH–cytochrome b_5 reductase are much less influenced by induction than cytochrome P-450. Mitochondrial enzymes other than δ-aminolevulinate synthetase appear to be largely unaffected (Orrenius *et al.*, 1969). Glucose-6-phosphate dehydrogenase and isocitrate dehydrogenase, both enzymes associated with the production of NADPH, a necessary cofactor for MFO activity, showed higher activities following phenobarbital treatment of rats. Higher activity of the enzymes involved in the pentose phosphate pathway was observed in houseflies after DDT treatment (Agosin *et al.*, 1966). The activity of NAD kinase was also stimulated by DDT in *Triatoma infestans* (Agosin *et al.*, 1967).

The incorporation of [^{14}C]leucine into NADPH–cytochrome c reductase in mice (Jick and Shuster, 1966) and the rate of its subsequent disappearance from the purified enzyme showed that, although *de novo* protein synthesis is a major factor in the mechanism of induction, stabilization of the enzyme protein is also important.

Stabilization is probably involved to some extent in the observed increased levels of all components of the MFO system and supporting membrane (Orrenius, 1968) and seems to contribute in particular to the residual enhanced activity observed in some cases after the inducer has been removed.

Induction of microsomal oxygenases appears to result from a local interaction with the nuclei in the exposed cells. Hormonal mediation is not involved, as shown in experiments in which induction by benzopyrene was observed in perfused rat liver (Juchau *et al.*, 1965). Additional support for a local effect is the induction of MFO activity (aryl hydrocarbon hydroxylase) in cell cultures (Nebert and Gelboin, 1968; Whitlock and Gelboin, 1974).

Much more is known about induction in mammals than in any other group of animals, even though only very few species have been studied. Those that have include the rat, rabbit, mouse, guinea pig, and man (Hutson, 1977). It is naturally difficult to justify terminal experiments in most species of vertebrates other than small rodents that can easily be mass-reared. Studies with human beings utilize, whenever possible, liver biopsy samples, but more often, the *in vivo* rate of production and excretion of a drug metabolite that is easily isolated and measured is monitored under controlled conditions. Most of these studies must employ hospital patients on multiple drug therapy regimens. To the extent that data obtained with subjects in extraordinary conditions can be applied to the human species at large, they show that human liver MFO activities are in a perpetual state of change in response to external chemical exposure and that the changes are short-lived and often as dramatic as those seen in laboratory animals (Remmer, 1971).

Induction was reported in one marsupial, the quokka, in which ethylmorphine N-demethylation and cytochrome P-450 levels were higher after intraperitoneal injections on each of 7 days of phenobarbital (McManus and Ilett, 1976).

The reproductive failure of raptorial and marine birds has been linked to the accumulation of potential inducers, e.g., DDT, its major metabolite DDE, and other "hard" chlorinated hydrocarbon insecticides, in sufficiently high levels to disturb the MFO-mediated endocrine regulation of calcium deposition in the eggshells. Liver microsomal MFO activities in the puffin (*Fratercula arctica*) are sensitive to induction by low (50 ppm) dietary levels of DDE (Bend *et al.*, 1974). No other study has yet been done with seabirds; raptors are not feasible objects for terminal studies. Chan *et al.* (1967) reported the induction of epoxidase activity in Japanese quail by dieldrin and DDT analogs. In one study, the domestic chicken was found to have increased cytochrome P-450 levels and N-demethylase activities following treatment with phenobarbital (Strittmatter and Umberger, 1969) or DDT (Aboudonia and Menzel 1968). Stephen *et al.* (1971) found consistent induction by phenobarbital of cytochrome P-450 and *N*- and *O*-demethylase activities in domestic chicken. Young chickens were more induced than mature hens. The chicken MFO system was also induced by DDT, but inconsistently, and in this case there was no obvious effect on eggshell thickness attributable to the induction. Undoubtedly, birds have species-characteristic responses to potential inducers and also characteristic levels of sensitivity to the endocrine consequences of MFO induction.

The recent large-scale oil spills in marine coastal areas have led to concern about the effects of crude petroleum products on marine life forms. For many years it was uncertain whether fishes and aquatic animals were able to metabolize foreign compounds. Many species of fish and marine invertebrates are now known to have an MFO enzyme system (see Section II,D) which resembles that in mammals. Fish MFO activities are also inducible. Burns (1976) found an increase in cytochrome P-450 content and aldrin epoxidase activity in an estuarine fish (*Fundulus heteroclitus*) following its exposure in the water to the antiinflammatory drug phenylbutazone. Bend *et al.* (1977) reported up to 10 times higher benzo[*a*]pyrene hydroxylase activity in the sheepshead (*Archosargus probatocephalus*) and increased cytochrome P-450 levels after intraperitoneal injections of 3-methylcholanthrene. They also mentioned increased benzo[*a*]pyrene hydroxylase activity in the little skate (*Raja erinacea*) following treatment with 3-methylcholanthrene, tetrachlorodibenzo-*p*-dioxin, and dibenzanthracene. On the other hand, Addison *et al.* (1977) did not detect any inductive effect of DDT or DDE on trout (*Salvelinus fontinalis*) liver cytochrome P-450, aniline hydroxylase, or 7-ethoxycoumarin *O*-deethylation activities. And in another study Bend *et al.* (1973) were unable to induce MFO activities in the liver of the little skate with phenobarbital by either oral, intravenous, or intraperitoneal administration. Payne and Penrose (1975) reported a 14-fold increase in benzo(a)pyrene hydroxylase activity in liver from trout (*Salmo trutta*) caught in an "oil-polluted pond" in comparison to trout caught in a control pond. Payne (1975) also showed induction by effluents from an oil refinery in liver and gill tissue of the cunner (*Tautogolabrus adspersus*) and suggested that field evaluation of MFO activities in fish might serve as a monitor for marine petroleum pollution. However, species variation in response to the presence of potential inducers is obvious, and field evaluations always suffer from problems in establishing stringent control situations. Analytical chemical methods are available for direct monitoring. Payne (1975) also mentioned that DDT and other pesticides or the chlorinated biphenyl Arochlor 1016 did not induce benzopyrene hydroxylase in the trout at the concentrations used.

Since the duration and effectiveness of an insecticide are almost exclusively dictated by the rate with which it undergoes metabolic degradation primarily by MFO enzymes, the phenomenon of induction has received a fair amount of attention in insects. Detailed studies have been done with adult houseflies and southern armyworm larvae. Induction has been observed in several species. Extended exposure of wax moth (*Galleria mellonella*) larvae to dietary chlorcyclizine and phenobarbital was reported (Ahmad and Brindley, 1971) to induce EPN detoxication, *p*-nitroanisole *O*-demethylation, and NADPH–neotetrazolium (cytochrome *c*) reductase activity in gut homogenates. Ahmad and Brindley (1969) also showed that exposure to the inducers changed the toxicity of parathion (*O*,*O*-diethyl *O*-4-nitrophenyl phosphorothioate) to the wax moth larvae, and Youssef and Brindley (1972) showed ultrastructural changes in larval cells after exposure to the same compounds. Khan and Matsumura (1972) reported an increase in benzopyrene hydroxylation, incorporation of [^{14}C]leucine, and incorporation of bases into RNA following treatment of American and German cockroaches with DDT or dieldrin. Gil *et al.* (1974) demonstrated increased oxidase and conjugating activities in the Madagascar cockroach following injection of phenobarbital. The increase in activities in fat body and Malpighian tubule tissues was greater than in those in the gut even though control (uninduced) activity levels were lowest in the fat body. Piperonyl butoxide, one of the most widely used insecticide synergists, was shown by Thongsinthusak and Krieger (1974) to have a biphasic effect on MFO-catalyzed dihydroisodrin hydroxylation when applied topically to black cutworm (*Agrotis ypsilon*) larvae. At first, for up to 2 hours there was an increasing degree of inhibition followed by an equally long period of recovery. About 5 hours after treatment, induction began to be noticeable and continued for an additional 15 hours before enzyme activity leveled off at an approximately 1.5 times higher level than in controls. In this case, too, the increase was highest in fat body and Malpighian tubules despite the higher control activity in the gut tissue.

Induction in insects was first demonstrated by Morello (1964), who found that *Triatoma infestans* nymphs were more tolerant (12%) to DDT after injection of 3-methylcholanthrene and that treated insects had a higher rate of production of polar metabolites (Kelthane) than controls. This experiment also demonstrated for the first time a direct correlation between resistance to an insecticide and the rate of its metabolic degradation. Morello (1965) also showed that DDT is capable of inducing its own metabolism in rat liver, a process that could be reversed by puromycin, indicating the involvement of *de novo* protein synthesis.

The development of insecticide-resistant strains of insects is a genetic phenomenon in which insects with preadapted, inheritable high metabolic capacity (high tolerance level) are selected for survival by a discriminating agent (the insecticide) during many generations. Induction of high enzyme activity, on the other hand, is a noninheritable, temporary adaptation that follows from exposure to chemical stress. Work with the genetically very well known housefly indicates that, in this species, individuals with high original MFO activity are more inducible than individuals with low original activity (Terriere *et al.*, 1971). In this case the phenomenon of induction leads to the more rapid evolution of genetically pure resistant strains. In emergency situations of heavy pesticide exposure, species might be forced to abandon gene diversity in return for immediate survival and rely on subsequent mutations to reestablish a more flexible heterozygous condition.

Dieldrin and DDT were shown to be inducers of MFO enzymes in the housefly by Plapp and Casida (1970), who found increased metabolism of several insecticides. Walker and Terriere (1970) showed enhanced naphthalene hydroxylation and heptachlor epoxidation in flies following exposure to dieldrin and also showed a reversal of cabaryl toxicity to dieldrin-exposed flies. Both these groups used several genetically defined strains of flies and found the response to inducers characteristic of each strain. Walker and Terriere (1970) hypothesize that a resistant strain, R-Isolan, has a tripled capacity to be induced compared to a relatively sensitive strain, dield : cyw, whereas an intermediary strain, Orlando, responds at twice the rate, reflecting a condition of gene duplication or triplication in the more resistant strains. This is in agreement with the suggestion by Gil *et al.* (1968) that induction occurs only in certain resistant strains of flies that have the genetic capacity to respond. Terriere *et al.* (1971) also demonstrated that the degree of induction in a hybrid strain obtained by crossing a high MFO and a low MFO strain showed an intermediary level of induction as well as of original heptachlor epoxidase activity. These experiments clearly demonstrate that induction of MFO enzyme activity is possible only if the basic genetic equipment for MFO activity is present originally. The inducer thus interacts with the genetic material, a fact that is also indicated by the demonstrated *de novo* protein synthesis. Even though the action of the inducer is at the gene level, there is no mediation by the endocrine control system as indicated by studies of induction in cell cultures (see Section II,F). Yu and Terriere (1971b) demonstrated a similar direct and local action on the cell exposed to the inducer by using isolated fly abdomina, and flies ligated between the thorax and the abdomen to stop the circulation of hormones in the hemolymph. They found increased heptachlor epoxidation in the isolated abdomina. But in another experiment they found that flies treated topically with dieldrin on the heads showed higher levels of induction in their abdomina than when dieldrin was applied to other body parts, indicating that an indirect effect possibly contributes to the observed induction.

Yu and Terriere (1972) showed that a number of cyclodiene insecticidal chemicals are inducers of MFO activity at exposure levels high enough to allow the inductive effect to occur but low enough so as not to kill the insects. They also showed that fly enzymes are induced by phenobarbital (1973) and by insect juvenile and molting hormones (1971a) and that the 1,3-benzodioxole synergist piperonylbutoxide has a biphasic effect in at least one strain of flies (1974b), being first inhibitory and subsequently inducing. Perry *et al.* (1971) showed induction by phenobarbital, butylated hydroxytoluene, pentobarbital, and several trisubstituted phosphoric acid esters. They, too, noticed a relatively higher increase in cytochrome P-450 levels in resistant fly strains than in susceptible strains. They also reported the occurrence of a cytochrome with a CO complex absorption peak at 448 nm in the resistant strains, whereas the corresponding peak was at 452 nm in the susceptible strains. They did not observe any change in absorbance peak position due to induction. However, feeding 1,3-benzodioxole compounds and substituted propynyl ethers to Diazinon-resistant flies shifted the location of their cytochrome–CO complex absorbance from 448 to 450 or 452 nm. The cytochrome with the absorbance peak at 448 nm has been observed in several resistant strains (Philpot and Hodgson, 1971; Tate *et al.*, 1973; Perry and Buckner, 1970) but appears to be absent in susceptible strains, which have cytochromes with the absorbance peak at 450–452 nm. Detailed knowledge of the relative contribution to the resulting MFO profile in a strain of houseflies selected to resistance by any agent from the genetic predisposition, the chemical nature of the agent, and the possibly superimposed state of induction in addition to physiological factors is needed to explain the occurrence of a distinctly different cytochrome in resistant flies. Interestingly, the ligand binding that gives rise to a type I difference spectrum is associated with cytochrome P-448 in the resistant strains and is lacking in susceptible strains (Plapp *et al.*, 1976). This observation may explain the absence or relatively lower degree of induction (see Remmer, 1972) in susceptible strains. Hodgson and Tate (1976) describe other differences in the cytochrome in resistant and susceptible flies and discuss in depth recent progress in the genetics of fly MFO enzymes.

The midgut microsomal MFO system in the southern armyworm responds with increased activities to a large number of foreign chemicals. A series of methylated benzenes effected up to 3-fold higher activities, increasing with each additional methyl group to a maximal effect with five. Hexamethylbenzene is a less effective inducer than pentamethylbenzene. As in mammals and adult houseflies, the gut tissue in the armyworm responded to an increase in dietary dose up to a maximal concentration with increasing MFO activities. Increasing exposure times also effected corresponding increases in MFO activities (Brattsten and Wilkinson, 1973b). The armyworm MFO system is also sensitive to induction by phenobarbital and 3-methylcholanthrene, although both of these inducers are less potent with both aniline hydroxylation and *p*-chloro N-methylaniline N-demethylation (Brattsten *et al.*, 1976) than is pentamethylbenzene. There was a dramatic increase in tolerance to the insecticide carbaryl (1-naphthyl N-methylcarbamate) with increasing MFO activity. Pentamethylbenzene-induced armyworms are 11-fold more tolerant than controls (Fig. 6), and armyworms induced by commercial insecticide spray solvents (Brattsten and Wilkinson, 1977) show an equally increased resistance to carbaryl in some cases. A similar increase in *in vivo* tolerance to an insecticide following treatment by drug inducers was reported in the wax moth larva (Ahmad and Brindley, 1969). Most interesting was the observation that a variety of plant allelochemicals have the same effect on armyworm larvae. Terpenoids such as the pinenes, the steroids stigmasterol and sitosterol, and several other kinds of plant compounds were moderate to extremely potent inducers of the armyworm midgut MFO enzymes. Induction effected by α-pinene also rendered the larvae more tolerant to the insecticide nicotine. Leaf-feeding insects may encounter both of these compounds in their normal feeding habits, sometimes at quite high concentrations. Induction by α-pinene or sinigrin occurred very rapidly. Within 30 minutes of ingesting a "realistic" dose, the gut MFO system had significantly higher activity (Brattsten *et al.*, 1977). It is likely that plant allelochemicals in nature, although restricting herbivore feeding rates to tolerable levels, nevertheless help herbivore survival by effecting small but important changes in their metabolic equipment.

G. Natural Function of MFO Enzymes

The ubiquitous and highly diversified microsomal MFO enzymes with their documented ability to metabolize a large number of lipophilic foreign compounds certainly did not arise in response to the rather recently acquired ability of human beings to synthesize organic chemicals. The natural role or function of the MFO's is not by any means obvious. They are energetically expensive to operate and to synthesize. The source of the reducing equivalents, NADPH, are the NADP-linked isocitrate and glucose-6-phosphate dehydrogenases in the tricarboxylic acid cycle and the phosphogluconate (pentose phosphate) pathway, respectively (Lehninger, 1976). The NADPH represents in each case an energy content of about 5 kcal/mole. Until the catalytic mechanism of the MFO reaction has been established it is only possible to say that the hydroxylation of a substrate costs at least 5 kcal/mole. In the overall energy budget of an organism (of fair size) this is a minor expense. Energy is also used to maintain a baseline level of MFO enzyme proteins. The energy cost in protein synthesis is approximately 29 kcal per peptide bond (Lehninger, 1976). To make a protein of about 50,000 daltons (cytochrome P-450) and another one of about 80,000 daltons (cytochrome P-450 reductase) costs approximately 14,500 and 23,200 kcal/mole each if one assumes that the proteins contain about 500 and 800 peptide bonds, respectively. In one case (Remmer, 1972) a content of 0.7 nmole of cytochrome P-450 and 0.03 nmole of NADPH–cytochrome P-450 reductase per milligram of microsomal protein were estimated, or a total of about 122.5 nmoles cytochrome P-450 and 5.25 nmoles of the reductase in the liver of a 150-gm rat. Cytochrome P-450 has a half-life of about 22 hours (Remmer, 1972) and that of the reductase is of similar magnitude. A rat thus must spend less than 1 cal (0.95) every day to maintain basic MFO activities. This figure does not include some extra energy for the fabrication of phospholipids, for the production of the heme group, and for assembling MFO components on the endoplasmic reticulum membrane. It is nevertheless a trivial part of the total energy turnover of an animal the size of a rodent or bigger. The sudden need for a greater detoxifying capacity that may arise when a herbivore starts feeding on a new plant may prove to be energetically costly for smaller animals such as insect herbivores.

Freeland and Janzen (1974) mention that the rat typically samples each new food item very carefully and does not devour any novel food. A similar feeding behavior was seen in a Costa Rica rain forest rat (*Tylomys*). The typical feeding behavior of southern armyworm larvae is several minutes of feeding alternating with periods of rest lasting up to 20 minutes (Crowell, 1943). A typical sampling behavior was also observed in howler monkeys (*Alouatta palliata*) in a Costa Rica rain forest (Glander, 1977). Both insect and vertebrate herbivores thus give an opportunity to their MFO enzymes to become induced in anticipation of a larger meal containing a potential toxicant. It has been shown with the southern armyworm larva (Brattsten and Wilkinson, 1973b; Brattsten *et al.*, 1977) that the induction indeed takes place with enough rapidity to afford the larva a significantly higher detoxifying capacity within as short a time as half an hour. Adult houseflies are also capable of being induced within minutes of exposure (Terriere and Yu, 1976a). A similar speed of induction has not been observed in vertebrates.

Protein synthesis is a fast process. In eukaryotic cells such as rabbit reticulocytes polypeptides can be formed at the rate of 1 new bond per second. Provided that the nutritional status of the animal is adequate, the rate of peptide bond formation is thus certainly high enough to double the amount of relatively small enzymes within minutes only.

It is well established that the MFO enzymes are involved in the metabolism of naturally occurring foreign compounds (plant allelochemicals) such as pyrethrins, rotenoids, nicotine, lignans, and several opium alkaloids. Krieger *et al.* reported in 1971 that a group of polyphagous lepidopterous larvae have significantly higher aldrin epoxidase activity than comparable groups of oligo- and monophagous larvae. Gordon (1961) suggested that polyphagous pest insects are generally more resistant to insecticides than are monophagous insects because they have been exposed to a wider variety of potential toxins in their food plants to which they have been forced to develop detoxification mechanisms. Indeed, as early as 1939 Swingle suggested that the feeding antecedents of an insect have a bearing on its subsequent ability to tolerate poisons. After establishing that the MFO system in the polyphagous southern armyworm does in fact respond with enough speed to realistically low dietary doses of a variety of plant allelochemicals to protect larvae against chemical exposure, Brattsten *et al.* (1977) conclude that MFO enzymes play a crucial role in allowing herbivores their feeding strategies. They showed that larvae respond with increased MFO activities only in a graded fashion so as to establish an equilibrium between their detoxifying capacity and the chemical stress. These authors also point out that even though the MFO enzymes of herbivores may operate primarily in response to the pressure of potentially toxic plant allelochemicals this does not preclude other functions, possibly more original, of the MFO enzymes.

Plant allelochemicals have also been shown to induce MFO activities in vertebrates. Aboudonia and Dieckert (1971) reported increased *p*-chloro N-methylaniline O-demethylase activity and carbamate metabolism in rats exposed to gossypol. Pyrethrum was reported to increase several MFO activities at high dose levels and after several days in rats (Springfield *et al.*, 1973). Cinti *et al.* (1976) found that rats, but not rabbits, had induced MFOs after being exposed to a disinfectant containing the volatile terpenes α-terpineol and isobornyl acetate. In this case, too, the effect appeared only after 1 or more days. Isosafrole induces *p*-nitroanisole O-demethylation and several other activities in rats (Vainio and Parkki, 1976), and α-pinene effects proliferation of the smooth endoplasmic reticulum membranes in rat liver cells (Pap and Szarvas, 1974). The essential oil eucalyptol, but not α-pinene, induces pentobarbital metabolism and several MFO activities in rats (Jori *et al.*,

1969). Mice are induced by spironolactone (Feller and Gerald, 1971), and caffeine has an inducing effect on rat liver MFO activities, although in this case, too, the treatment had to be maintained for several days (Lombrozo and Mitoma, 1970). In all of these cases a relatively high dose of the inducer was necessary. Interestingly, Babish and Stoewsand (1975a,b) reported that freeze-dried cauliflower leaves or tea solids included in diets for rats or rabbits, respectively, increased their MFO activities. Again rather lengthy feeding periods of several days or even weeks were needed to achieve an effect.

It thus appears that an omnivorous and relatively long lived vertebrate, the rat, is much less sensitive to the inducing effects of plant allelochemicals than is the herbivorous southern armyworm, which lives in each larval instar for only 3–6 days. Whatever aspect of the widely different natural histories of the rat and the armyworm accounts for the difference in their sensitivity to external inducers, it points to the probability that the major natural or normal role of the MFO's need not be identical in all modern life forms.

It is well known that a cytochrome P-450-dependent monooxygenase is involved in steroidogenesis in mammals (Conney, 1967; Estabrook *et al.*, 1975). This cytochrome is located in mitochondria, and electron transport from NADPH is mediated by a flavoprotein reductase and a ferrodoxin proximally to the cytochrome. This MFO system is quite refractory to induction by external agents. Pig testes (Mason and Boyd, 1975) and bovine adrenal cortex appear to be particularly rich in this form of the MFO system. A similar special form of the MFO enzymes appears to operate in insects, as well. Bollenbacher *et al.* (1977) found a mitochondrial MFO enzyme system in the fat body of the tobacco hornworm *Manduca sexta* that catalyzes the C-20 hydroxylation of the insect steroid molting prehormone, α-ecdysone, to the active form, β-ecdysone (ecdysterone). Yu and Terriere (1974a) reported that induction by phenobarbital or piperonyl butoxide of the larval MFO system in houseflies led to subsequent inhibition of pupation and adult emergence, indicating MFO involvement also in the inactivation of the molting hormone. Not enough is known about either microsomal or mitochondrial MFO enzymes in insects to predict the extent of typical microsomal MFO involvement in insect hormonal regulation.

The fact that the MFO enzymes sometimes activate originally rather inert molecules to more toxic metabolites also clearly shows that their primary function is not detoxication, but rather, as suggested by Brodie *et al.* (1958) and later by Wilkinson and Brattsten (1972) and Nakatsugawa and Morelli (1976), their primary role is in the conversion of lipophilic compounds to water-soluble, excretable metabolites. MFO's bioactivate the polycyclic aromatic hydrocarbons to carcinogenic epoxide intermediates (Jerina and Daly, 1974). Not only synthetic chemicals, such as the carcinogen precursors and the organophosphorothionate insecticides parathion, diazinon, etc., which are converted to the potent acetylcholinesterase inhibitors paraoxon, diazoxon, etc., but also naturally occurring plant allelochemicals undergo MFO-catalyzed bioactivation. Notable compounds are the pyrrolizidine alkaloids, which are converted to hepatotoxic pyrrole derivatives (Mattocks and White, 1971), and the aflatoxins, which are rendered highly carcinogenic (Tilak *et al.*, 1975).

III. EPOXIDE HYDRASES

The epoxide metabolites resulting from cytochrome P-450-mediated hydroxylation of aromatic and aliphatic foreign compounds are more polar and also more reactive than the parent compounds. In fact, some of these epoxide intermediates, known as arene oxides, form covalent bonds with cellular macromolecules such as DNA, RNA, and proteins very rapidly. They are therefore strongly implicated as the ultimate carcinogens arising by metabolic activation from the polycyclic aromatic hydrocarbons such as benzo[a]pyrene (Jerina and Daly, 1974).

Arene oxides are short-lived molecules and undergo transformation to diols, phenols, or glutathione conjugates very rapidly. The most important inactivating agent is epoxide hydrase (EC 4.2.1.63), an enzyme that catalyzes the conversion of epoxides to *trans*-diols by stereospecific addition of water.

Epoxide hydrase has been studied in relatively few species. Oesch *et al.* (1974) reported activity in liver tissue of rhesus monkey, man, rabbit, guinea pig, rat, and mouse. Several insects also have an active epoxide hydrase, e.g., the housefly (Brooks *et al.*, 1970), a blowfly (*Calliphora erythrocephala*), the mealworm (*Tenebrio molitor*) (Brooks, 1973), the southern armyworm, and the Madagascar cockroach (Slade *et al.*, 1975). In mammals the liver is the major source of the enzyme (Hook and Bend, 1976), although low-level activity is also found in lung tissue.

Bend *et al.* (1977) reported epoxide hydrase activity comparable to that in rat liver in the liver of several marine fishes, e.g., the sheepshead, black drum, Atlantic stingray, and dogfish shark, toward styrene 7,8-oxide, octene 1,2-oxide, and benzo[a]pyrene 4,5-oxide. They also observed high activity in several marine invertebrates including the clam, blue crab, and rock crab. And they reported extremely high activity from spiny lobster hepatopancreas, although this tissue is apparently almost devoid of cytochrome P-450-mediated oxidation.

Epoxide hydrase is bound to the smooth endoplasmic reticulum membranes of cells, in very close association with cytochrome P-450 (Oesch and Daly, 1971), but does not require molecular oxygen or NADPH for activity. The enzyme was solubilized and purified from guinea pig liver microsomes and weighs about 50,000 daltons (Oesch and Daly, 1971). It accepts a wide variety of arene and alkene oxides as substrates. The three most commonly used substrates are styrene 7,8-oxide, 3-methylcholanthrene 11,12-oxide, and the cyclodiene epoxide 1,2,3,4,9,9-6,7-epoxy-1,4,4a,5,6,7,8,8a-octahydro-1,4-methanonaphthalene (HEOM) (Oesch *et al.*, 1971; Brooks *et al.*, 1970). The enzyme prefers an unusually high pH for optimal activity. A pH value of 8.2 to 9 was reported to give optimal activity of mammalian epoxide hydrase toward 11 different substrates (Jerina *et al.*, 1977). With HEOM as substrate the optimal activity of blowfly and armyworm enzyme was at pH 9.0, whereas the Madagascar cockroach enzyme showed an optimum at pH 8.1 (Slade *et al.*, 1975).

The enzyme can be induced in mammals (Oesch, 1973; Jerina *et al.*, 1977) by phenobarbital and 3-methylcholanthrene. Low dietary doses of dieldrin induced rat liver epoxide hydrase without stimulating simultaneously the rat liver cytochrome P-450 activity (Bellward *et al.*, 1975), but injection of aldrin, dieldrin, or isosafrole in rats for 6 days enhanced both hydrase and P-450 systems (Vainio and Parkki, 1976).

The existence of isoenzymes or slightly different forms of epoxide hydrase is uncertain. Jerina *et al.* (1977) found a wide variation in catalytic effectiveness toward 11 different substrates, reflected as a 20-fold variation in optimal substrate concentration, a 15-fold difference in optimal incubation time, and a 100-fold variation in protein concentration in their *in vitro* assay mixtures. They concluded, however, that in rat liver there is probably only one epoxide hydrase with a low level of specificity because the inducers 3-methylcholanthrene and phenobarbital both produced a fairly even increase in activity with respect to all the substrates. On the contrary, the interaction of inhibitors with epoxide hydrases from different species indicates the existence of different forms of the enzyme. Oesch (1973) observed no inhibition of rat liver enzyme by piperonyl butoxide or SKF 525-A with styrene oxide as substrate, whereas these P-450 inhibitors effectively inhibited HEOM epoxide hydrase from the blowfly (Brooks, 1974b). Slade *et al.* (1975) likewise found cytochrome P-450 inhibitors effective against insect epoxide hydrase with HEOM as substrate. They also reported inhibition of blowfly and armyworm HEOM epoxide hydrase by about 20 other P-450 inhibitors, numerous epoxides and glycidyl ethers, and other compounds. However, the P-450 inhibitors metyrapone and 1-(2′-iso-propylphenyl)-imidazole, which strongly stimulate *in vitro* styrene oxide and cyclodiene epoxide hydration by mammalian enzyme (Oesch *et al.*, 1973), do not stimulate insect epoxide hydration (Slade *et al.*, 1975).

Epoxide hydrase thus is an important enzyme that is closely associated with the microsomal cytochrome P-450 oxidase system and acts in concert with it in the inactivation of numerous foreign compounds. Epoxide hydrase in insects is also involved in the regulation of juvenile hormone, a highly important factor in insect growth and development. Slade and Wilkinson (1973) reported the conversion of the cecropia juvenile hormone epoxide to the corresponding *trans*-diol by tissue homogenates of southern armyworm midguts. The juvenile hormone hydration proceeded optimally at pH 7.9, slightly lower than the typical range of pH 8.2–9 found with most substrates. This could indicate a modified form of the enzyme, possibly specialized for juvenile hormone regulation.

IV. REDUCTASES

Although oxidative reactions are of major importance in the degradation of foreign compounds, reductions are also of frequent occurrence. Enzymes that catalyze reductions of nitro and azo compounds have been found in liver microsomes of mammals (Mitchard, 1971) and fishes (Adamson *et al.*, 1965). Azo and nitro reductases have also been demonstrated in lobster hepatopancreas (Elmamlouk and Gessner, 1976) and in the intestinal epithelium of the nematode *Ascaris lumbricoides* (Douch, 1975). Rose and Young (1973) found an enzyme in the Madagascar cockroach fat body cells that reduces flavins (FAD and FMN). The reduced flavin in turn is capable of nonenzymatic reduction of nitrobenzene. Azo and nitro reductase activity is also localized in the soluble fraction of the cell. Activity requires NADPH but can also be supported by NADH and is stimulated by flavin nucleotides. Azo and nitro reductases can be measured *in vitro* only under anaerobic assay conditions and in the presence of high levels of cofactors. It is unlikely that activities measured *in vitro* under such conditions reflect reductive capacities of intact organisms.

Rat liver cells also contain enzymes that reduce tertiary amine N-oxides. Cytochrome P-450 is implied as a crucial participant in these reductions. Sugiura *et al.* (1977) reported that the enzyme that reduces tiaramide N-oxide and N-oxides of several other tertiary amines is recovered in the microsomal fraction and requires NADPH as cofactor. The enzyme can be induced to some extent by phenobarbital and

3-methylcholanthrene and is stimulated in *in vitro* assays by flavin nucleotides. Sugiura *et al.* (1976) suggested that the N-oxide group seems to bind directly to the heme iron. The inhibitory effect of atmospheric oxygen could thus result from competition with the N-oxide for binding as the sixth ligand to the reduced heme. Sugiura and Kato (1977) also reported the localization of tiaramide N-oxide reductase on the inner membranes of mitochondria. This activity cannot be induced by MFO inducers and appears to depend on isocitrate and other tricarboxylic acid cycle intermediates for activity.

Soluble cytosolic aldehyde and ketone (AK) reductases are of widespread occurrence in the tissues of mammals and birds (Bachur, 1976). Cytosolic AK reductases appear to comprise a class of enzymes characterized by the common utilization of NADPH as reducing agent and the ability to attack a broad range of aldehydes and ketones. They are known to reduce naturally occurring compounds such as benzaldehyde (Culp and McMahon, 1968) and daunorubicin (Bachur and Gee, 1971) as well as synthetic substrates. Some of the AK reductases have been purified and have a molecular weight of 30,000 to 40,000 daltons. They depend on endogenous sulfhydryl groups for activity. They are not inducible.

V. HYDROLYTIC ENZYMES

The action of hydrolases together with that of transferases constitutes the most important alternative to oxidative degradation of foreign compounds. The delineation of the hydrolases from other groups of enzymes is not entirely clear-cut since some of the glutathione *S*-transferases perform what can be regarded as hydrolytic cleavage of organophosphorus insecticides.

The A and B esterases and the carboxylamidases included in this category are known as a result of their involvement in organophosphorous insecticide degradation. With one or two exceptions, no natural substrate or normal role has been assigned to these esterases. Although some of them have been purified and are known in considerable biochemical detail, there are still a multitude of questions associated with them. The hydrolases were reviewed recently by Ahmad and Forgash (1976) and Dauterman (1976). They will be described very briefly only, since their involvement in the metabolism of plant allelochemicals is virtually unknown. This is clearly an area ripe for further experimentation.

Aldridge (1953) classified the esterases active toward organophosphorus insecticides into A esterases and B esterases. The A esterases hydrolyze *p*-nitrophenyl acetate faster than the butyrate derivative and are not inhibited by paraoxon, whereas the B esterases hydrolyze the *p*-nitrophenyl butyrate faster than the acetate and are very sensitive to inhibition by paraoxon. Mammalian A esterases are soluble and occur in a multitude of tissues, with rabbit serum as a particularly rich source. Insect A esterases were found in the microsomal fraction of a cockroach preparation and can hydrolyze diazoxon (Shishido and Fukami, 1972). The A esterases fall into two groups. Group I consists of A esterases that are activated by Ca^{2+}. They occur in mammals but not in insects, and they hydrolyze organophosphates but are inactive toward organophosphorothionates. Group II consists of A esterases that are activated by Mn^{2+} or Co^{2+}. They are present in both mammals and insects. In mammals they appear to be active against organophosphorothionates, but in insects they hydrolyze organophosphates (Dauterman, 1976).

The B esterases are also called aliesterases (EC 3.1.1.1, carboxylic ester hydrolase). They hydrolyze water-soluble aliphatic and aromatic esters and appear to be important in the degradation of malathion (O'Brien, 1960) and in synthetic pyrethroid metabolism (Abernathy and Casida, 1973). Jao and Casida (1974) found esterases in milkweed bugs and cockroaches that are involved in the hydrolysis of synthetic pyrethroid insecticides. Oxidative degradation of pyrethroids is, however, more effective with the naturally occurring compounds. The B esterases are also found in the microsomal fraction of mammalian liver cells.

Carboxylamidases are a third kind of esterases. Amidases are known to hydrolyze organophosphorous insecticides containing an amide group, e.g., dimethoate. Carboxylamidases were purified from sheep liver microsomes by Chen and Dauterman (1971). The enzyme showed increasing affinity for N-methylalkylamides with increasing size of the alkyl group. Maximal affinity occurred with N-phenylcaproamide, which was therefore considered the most likely natural substrate of the enzyme. Carboxylamidases appear to be inhibited by phosphates and may only be able to hydrolyze phosphorothionates (Dauterman, 1976).

VI. GROUP TRANSFER ENZYMES

Most group transfer enzymes, with the exception of rhodanese (see Section VI,A) and certain glutathione *S*-transferases, are typically involved in the secondary metabolism of foreign compounds and result in so-called conjugations. A conjugated compound is sometimes a product of primary metabolic processes. Endogenous substances as well as foreign compounds undergo conjugation. Conjugations invariably result in more polar and therefore more excretable products, but detoxification is by no means the sole purpose of conjugation reactions. Conjugations are implicated in transport across membranes and in the production of inactive storage forms of bioactive substances, e.g., cyanogenic glycosides in plants. Conjugations result in biologically inactive products, with the exception of conjugated N-hydroxy compounds. Glucuronic acid, sulfate, or phosphate conjugates of these, as exemplified by 2-acetylaminofluorene (Irving, 1971), are known to be highly reactive toward nucleophilic centers in cellular macromolecules and are therefore strongly implicated in chemical carcinogenesis (Yang, 1976).

With the exception of the conversion of cyanide to thiocyanate by the enzyme rhodanese, group transfer enzymes merely put together two unchanged and perfectly "recognizable" parts. However, this process is never a simple addition but proceeds via a high-energy intermediate. The intermediate can include the conjugating agent, as in the case of sulfate, phosphate, glucuronide, or glucoside formation. It can also be an activated form of the molecule to be conjugated, as when conjugations to amino acids or glutathione occur.

Several outstanding reviews have been published on conjugating mechanisms by Yang (1976), Testa and Jenner (1976), Ahmad and Forgash (1976), and Williams (1974). Westley (1973) reviewed the literature about rhodanese.

A. Rhodanese

The enzyme rhodanese [thiosulfate:cyanide sulfur transferase (EC 2.8.1.1)] is a group transfer enzyme that effects the conversion of cyanide (CN^-) to the metabolite thiocyanate (SCN^-), which is 200 times less toxic. The enzyme was first studied and described by Lang (1933), who gave it the name rhodanese. The enzyme is strictly a mitochondrial enzyme (Sörbo, 1951) and transfers the outer sulfur of a sulfenyl sulfur compound to a suitable nucleophilic acceptor (Westley, 1973). The sulfur donor does not have to be thiosulfate, and the acceptor can be nucleophiles other than cyanide. The natural role or function of the enzyme is not known, and its involvement in cyanide detoxification may be utterly fortuitous. Sörbo (1957) showed that addition of rhodanese can reactivate cyanide-inhibited cytochrome aa_3.

The enzyme activity can be measured *in vitro* by spectrophotometric estimation of the complex between the product thiocyanate, formerly called rhodanate, and ferric ions in an acidic solution. The *in vitro* activity depends on the availability of thiosulfate. This may be the case in the intact organism as well. Lang (1933) reported enzyme activity in several organisms and tissues. He found the highest activity in frogs, followed by rabbits and cattle. He also reported activity in human beings, chickens, cats, dogs, and *E. coli*. The liver and kidney were the major sources of activity, but activity was widespread and occurred in virtually all tissues examined except blood and muscle. In a later study Himwich and Saunders (1948) reported very similar species and tissue distribution of rhodanese.

Rhodanese has also been observed in a few insect species including the horse botfly larvae living in the stomach of horses (Bertran, 1952), in the pupae of the common blue butterfly *Polyommatus icarus* (Parsons and Rothschild, 1964), in the pupae, larvae, and adults of the blowfly *Calliphora vomitoria* (Parsons and Rothschild, 1962), and in southern armyworm larvae (L. B. Brattsten, unpublished).

Rhodanese has been obtained in pure crystalline form from bovine liver (Sörbo, 1953) and shown to consist of two equal dimers. The molecular weight of the enzyme is 37,500 daltons. Rhodanese is inhibited by thiol reagents. It is also inhibited by pyridoxal 5′-phosphate and other aromatic aldehydes but not by aliphatic ones. Addition of thiosulfate to the inhibited enzyme restores activity (Canella *et al.*, 1975). The tricarboxylic acid cycle intermediates, in particular α-ketoglutarate, inhibit the enzyme (Oi, 1975).

Schievelbein *et al.* (1969) hypothesized that rhodanese might be a biochemical relic surviving from times when cyanide was prevalent in the atmosphere. They did not find any clear-cut support for this idea in the species distribution of the enzyme activity. However, they found that, whereas the highest activity of terrestrial organisms is in the liver and kidneys, the highest activity of marine animals is in the gills.

B. Glutathione *S*-Transferases

Conjugation to the tripeptide glutathione results in the formation of hippuric acid by a four-step process:

$$\text{Glutathione} + \text{ROH} \xrightarrow{\text{GSH-}S\text{-transferase}} \text{RSG} + H_2O$$

$$\text{RSG} \xrightarrow{\text{glutathionase}} \text{R-Cys-Gly} + \text{glutamate}$$

$$\text{R-Cys-Gly} \xrightarrow{\text{peptidase}} \text{R-Cys} + \text{glycine}$$

$$\text{R-Cys} + \text{acetyl-CoA} \xrightarrow{\text{N-acetyltransferase}} \text{R-hippuric acid} + \text{CoA}$$

Of the four enzymes involved, the first, glutathione *S*-transferase, has received most attention. There are at least 10 different GSH-transferases, which can be distinguished on the basis of their substrate preference.

They are located in the cytosol of cells, and several have been obtained in pure form. Appleton and Nakatsugawa (1972) demonstrated the presence of a GSH-alkyltransferase in rat liver that is involved in the deethylation of the insecticide paraoxon. This enzyme exemplifies the involvement of GSH-transferases in primary xenobiotic metabolism. Another identified enzyme is a GSH-aryltransferase, which is involved in the detoxification of diazinon (Shishido *et al.*, 1971) in rat liver and American cockroach fat body.

Arene oxides are further metabolized by GSH-epoxide transferases in addition to epoxide hydrases (see Section III). One of these transferases was isolated from rat liver (Fjellstedt *et al.*, 1973). In addition, six different forms of a GSH-alkenetransferase are known (Testa and Jenner, 1976). A GSH-aralkyltransferase active against aralkyl halides such as benzyl chloride has been observed.

The capacity to form glutathione conjugates is quite general and has been demonstrated in mammals, amphibians, reptiles, birds, fishes, and insects (Smith, 1968, Yang, 1976). The liver is the most active tissue in mammals. The capacity to form a glutathione conjugate with naphthalene 1,2-oxide was 10 times greater in sheep liver than in sheep lung (Hayakawa *et al.*, 1974). Hook and Bend (1976) reported much higher activities in the liver of rat, rabbit, and guinea pig than in the lung tissues of these species. The noted inability of the guinea pig to form mercapturic acid metabolites has been explained as an inability to acetylate the cysteine derivative (see Testa and Jenner, 1976) resulting from glutathione conjugation.

C. Amino Acid Conjugations

In amino acid conjugation the active intermediate is formed with the foreign compound in a two-step reaction that requires ATP:

$$\text{RCOOH} + \text{ATP} \xrightarrow[\text{synthase}]{\text{acyl}} \text{RCO–AMP} + \text{PP}_i$$

$$\text{RCO–AMP} + \text{CoASH} \xrightarrow[\text{thiokinase}]{\text{acyl}} \text{RCO–S–CoA} + \text{ATP}$$

In a third reaction the conjugate with glycine is formed:

$$\text{RCO–S–CoA} + \text{glycine} \xrightarrow[\text{glycine:N-acyltransferase}]{\text{acyl-CoA}} \text{RCO–Gly} + \text{CoASH}$$

All three enzymes occur in the mitochondria of liver and kidney cells, and species specificity resides in the transferases. Although glycine is the most widely used amino acid for conjugations in most animal species, other amino acids are also used, e.g., glutamine, glycylglycine, and taurine. For instance, the cat is known to conjugate quinaldic acid with the dipeptides glycyltaurine and glycylglycine (Kaihara and Price, 1965). Hens, turkeys, ducks, and geese produce ornithine conjugates with benzoic acid. The hen also conjugates benzoic acid to glycine. Williams (1974) extensively reviewed the species variation in amino acid conjugations. Testa and Jenner (1976) pointed out that many of the peculiarities in amino acid and other conjugations may be a result of limited information. With all conjugations there are probably a number of different transferases with different substrate specificities. A limited choice of substrates could thus artificially limit the detectable conjugating capabilities in a species. Amino acid conjugates are formed as part of the secondary metabolism.

D. Acyltransferases

Methylations and acetylations are performed on intermediates from glutathione S-transferase reactions or amino acid transferase reactions by special N-acyltransferases. Acylations do not always result in biologically less active metabolites and are consequently not always detoxifying. Acylations are catalyzed by soluble and mitochondrial enzymes with limited specificity. N-Acyltransferases are found in liver and kidney cells of mammals and have not been studied in invertebrates. There is considerable variation in the ability of mammals to perform acetylations. The dog is unable to acetylate aromatic amines and hydrazines but can acetylate aliphatic amines perfectly well (Williams, 1974).

E. UDP-Hexosyltransferases

UDP-Hexosyltransferases utilize a carbohydrate for conjugation to phenols, aromatic carboxylic acids and amines, sulfhydryls, and alcohols. Conjugation to glucuronic acid is probably the most widespread reaction among secondary metabolism mechanisms. It seems to be replaced in insects and plants by conjugation to glucose. The active intermediate in glucuronide formation is UDP-glucuronic acid, which is made from glucose 1-phosphate:

$$\text{Glucose 1-phosphate} + \text{UTP} \xrightarrow[\text{pyrophosphorylase}]{\text{UDPG-}} \text{UDP-}\alpha\text{-glucose} + \text{PP}_i$$

$$\text{UDP-}\alpha\text{-glucose} + 2\text{NAD}^+ + \text{H}_2\text{O} \xrightarrow[\text{dehydrogenase}]{\text{UDPG-}} \text{UDP-}\alpha\text{-glucuronic acid} + 2\text{NADH} + 2\text{H}^+$$

UDP-Glucuronate transferase (EC 2.4.1.14) then catalyzes the formation of a β-glucuronide with the active intermediate and the foreign compound:

$$\text{ROH} + \text{UDP-}\alpha\text{-glucuronic acid} \xrightarrow[\text{transferase}]{\text{UDPG-glucuronate}} \text{RO-}\beta\text{-glucuronide} + \text{UDP} + H_2O$$

The formation of β-glucosides in plants and insects may be simplified by the use of UDP-α-glucose as the active intermediate (Yang, 1976):

$$\text{Glucose 1-phosphate} + \text{UTP} \xrightarrow[\text{phosphorylase}]{\text{UDPG-pyro}} \text{UDP-}\alpha\text{-glucose} + PP_i$$

$$\text{UDP-}\alpha\text{-glucose} + \text{ROH} \xrightarrow[\text{transferase}]{\text{UDP-glucosyl}} \text{RO-}\beta\text{-glycoside} + \text{UDP} + H_2O$$

Glucose conjugation occurs in vertebrates, although glucuronide formation predominates. Gessner *et al.* (1973) found a glucosyltransferase (EC 2.4.1.35) in mouse liver microsomes that conjugates a variety of phenolic compounds. Mehendale and Dorough (1972) demonstrated the presence of a glucosyltransferase in the soluble fraction of tobacco hornworm midgut and fat body cells with activity toward 1-naphthol.

The UDP-glucuronyltransferases in mammals are strictly bound to the endoplasmic reticulum membranes. In fact, treatment of the microsomal fraction with detergents or other membrane-disrupting agents produces higher glucuronidation activity than observable with intact membranes. The enzyme is either present in a latent form or removed by compartmentalization from maximal *in vitro* activity measurements (Berry *et al.*, 1975). UDP-Glucuronyltransferase is found mainly in liver tissue but also in the intestine, lung, and kidney.

The domestic cat is unable to make glucuronic acid conjugates of certain xenobiotics due to the lack of certain UDP-glucuronyltransferases even though the cat can make the active intermediate. The cat possesses the transferases that are active toward phenolphthalein and bilirubin. UDP-Glucuronyltransferases occur in several different forms with different activity spectra in different species (see Williams, 1974). Other cats, e.g., the civet and the genet, are also unable to make glucuronides out of phenol, 1-naphthylacetic acid, or sulfadimethoxine (Caldwell *et al.*, 1975).

The UDP-glucuronyltransferase in rat liver could be induced to higher activity after several days of exposure to isosafrole but was not induced by aldrin or dieldrin (Vainio and Parkki, 1976).

In a study with domesticated birds, Bartlet and Kirinya (1976) observed very low UDP-glucuronyltransferase activity toward *o*-aminophenyl in geese, ducks, and chickens, although activity was high in one strain of ducks. Activity was also high in turkeys. The activities were higher in neonatal and immature birds than in adults. Activity was highest in the liver but also occurred in kidney and duodenum.

F. Sulfotransferases

Sulfate conjugations are probably the energetically most costly form of conjugations that organisms perform. They are probably also the most susceptible to a shortage of the conjugating agent—sulfate.

The active intermediate, 3′-phosphoadenosine 5′-phosphosulfate (PAPS), is synthesized in a two-step reaction via adenosine 5′-phosphosulfate (APS) involving two enzymes and two molecules of ATP per molecule of sulfate:

$$ATP + SO_4^{2-} \underset{\text{adenylyltransferase}}{\overset{\text{ATP-sulfate}}{\rightleftharpoons}} APS + PP_i$$

$$ATP + APS \underset{\text{3'-phosphotransferase}}{\overset{\text{ATP-adenylylsulfate}}{\longrightarrow}} PAPS + ADP$$

$$PAPS + ROH \xrightarrow{\text{sulfotransferase}} ROSO_3H + PAP$$

The first two enzymatic steps are common to all sulfate conjugations and are catalyzed by soluble enzymes with high specificity for their substrates. Cytosolic sulfotransferases catalyze the sulfation of endogenous or foreign molecules, leaving 3′-phosphoadenosine 5′-phosphate (PAP) behind. The sulfotransferases are found mainly in liver cells of vertebrates and also in kidney, intestinal mucosa, and placenta. At least 12 rather specific sulfotransferases are known which catalyze the sulfation of important endogenous substances such as cerebrosides, mucopolysaccharides, and steroids (Testa and Jenner, 1976).

Organisms are capable of sulfation of a wide variety of endogenous and foreign phenols. This capacity may reflect the existence either of several similar phenol sulfotransferases or of one enzyme with a low substrate specificity for phenolic compounds.

Sulfotransferases that conjugate foreign compounds have been found in mammals, birds, amphibians, and invertebrates (Smith, 1968; Yang, 1976). Sulfate conjugation of phenol was reported in the dog, ferret, hyena, cat, civet, and genet (Caldwell *et al.*, 1975). Yang and Wilkinson (1972) reported a very active sulfotransferase in the midgut of the southern armyworm, which catalyzes the sulfate conjugation of *p*-nitrophenol and several plant and animal sterols. The armyworm enzyme is more active than the corresponding enzyme preparation from rat liver, but exogenous sulfate must be added to the *in vitro* incubations to attain maximal activity. Yang and Wilkinson (1973) also found a very active sulfotransferase in the gut tissues of seven other species of insects. They suggested that insect sulfotransferases may be involved in the regulation of insect molting hormone titers by acting in concert with arylsulfatases, which they (Yang *et al.*, 1973) also found in southern armyworm gut tissues. Mammalian steroid hormones are known to exist in a sulfate-conjugated storage form. If indeed sulfotransferases are very active and of widespread occurrence in herbivorous insects, they might serve to inactivate α-ecdysone (the most immediate and most important insect molting hormone precursor) ingested with the food. Ecdysone is a common plant allelochemical that sometimes occurs at very high concentrations.

Bartlet and Kirinya (1976) detected sulfotransferase activity toward *p*-nitrophenol in the liver and kidney preparations of chicken, turkey, duck, and goose. Activity was highest in goose liver. The liver activity was significantly increased by added sulfate, but that in the kidney preparations of the birds was not stimulated by exogenous sulfate. Bartlet and Kirinya (1976) also observed that there is in general an inverse relationship between sulfate conjugating activity and glucuronide formation in the birds, e.g., the goose liver was practically devoid of UDP-glucuronyltransferase activity.

G. Phosphotransferases

Conjugation of foreign compounds or their metabolites to phosphate groups is known to occur in mammals (Boyland *et al.*, 1961) and insects (Yang, 1976). The mechanism appears to have received relatively little attention and is of minor importance in comparison to glucuronidation or any other forms of conjugation. Yang and Wilkinson (1973) reported the existence of a phosphotransferase in the gut of the Madagascar cockroach and the tobacco hornworm. The enzyme requires Mg^{2+} and ATP in the *in vitro* incubation system for activity toward *p*-aminophenol. In both species sulfotransferase activity was higher than that of phosphotransferase. Similar results were observed with preparations from houseflies. The active intermediate of this system is not known. The relative importance of the phosphotransferase system and its natural function remain obscure. Several phosphotransferases (kinases) involved in intermediary metabolism have been investigated in detail.

VII. CASE HISTORIES

A few selected interactions between potentially toxic plant allelochemicals and insect herbivores will be described. The selection is based mainly on availability of information. The mechanisms behind these interactions have been studied in detail only in a very few instances. The knowledge acquired so far shows that herbivores maintain several different lines of defense against both chronic exposure and acute intoxication. The chronic exposure of an organism to a toxic chemical necessitates the development of adaptations. Some of the mechanisms of adaptation may be unique; others may involve normally occurring biochemical features of the organism. The mechanisms that protect organisms from chronic exposure may or may not be identical to those capable of rescuing the organism from an acute dose of a toxicant.

A. Cyanide Toxicology

The occurrence of cyanide in the form of cyanogenic glycosides is very widespread in higher plants. Cyanogenic glycosides were found to have a defensive value in South American non-ant–acacias (Rehr *et al.*, 1973b) and also in clovers (Jones, 1972). Cyanide is also used as a defense in arthropods. Polydesmoid millipedes (Blum and Woodring, 1962; Eisner *et al.*, 1963, 1967; Towers *et al.*, 1972) and a geophilid centipede (Schildknecht *et al.*, 1968) fabricate cyanogenic compounds in a specially designed glandular apparatus. Some insects feeding on cyanogenic plant material apparently sequester cyanogenic glycosides into their body tissues, which release cyanide when crushed (Jones *et al.*, 1962).

Like all poisons, cyanide has acute and chronic effects. The acute effect of cyanide is death, indirectly resulting from the inhibition of the terminal oxidase, cytochrome aa_3, in the mitochondrial respiratory pathway. Inhibition of cytochrome aa_3 is the major and most devastating effect of cyanide.

Cyanide interacts in numerous ways with subcellular body constituents. Cyanide is metabolized by the enzyme rhodanese (see Section VI,A) to a less toxic metabolite. Another enzyme, EC 2.8.1.2, is of uncertain identity but may detoxify cyanide by combination with β-mercaptopyruvate (Baumeister *et al.*, 1975). Cyanide also enters the intermediary metabolism of organisms. Labeled cyanide was recovered in several amino acids and amines in the granary weevil (Bond, 1961) and the rat (Boxer and Richards, 1952). The rat was also found to make the metabolically inert 2-imino-4-thiazolidine carboxylic acid with labeled cyanide (Wood and Cooley, 1956). Cyanide also enters into an equilibrium with vitamin B_{12}, and excess cyanide can form an inactive vitamin B_{12}–cyanide complex (see Baumeister *et al.*, 1975).

Even though entomologists go insect collecting equipped with cyanide jars to kill their booty, cyanide is, in fact, vastly more toxic to the collector than to the insect. It is not known with certainty how arthropods can survive cyanide exposure. Hall *et al.* (1971) investigated the mitochondrial respiratory system of insects and millipedes. They concluded that millipedes may have a cytochrome aa_3 that is unusually resistant to cyanide inhibition. An alternative, cyanide-insensitive respiratory pathway has been postulated in plants (Henry and Nyns, 1975). Yust and Shelden (1952) reported that cyanide-resistant respiration in the California red scale *Aeonidiella aurantii* possibly relies on metal-free autoxidizable respiratory enzymes, e.g., flavoproteins, rather than cytochromes.

The ability of insects to feed for extended periods on cyanogenic plant material may reflect a combination of biochemical defenses including a modified target system, adaptable detoxification enzymes, or ability to sequester the toxic material into special organs or tissues or even cells. Detailed investigations of the biological characteristics and significance of rhodanese have not been done in insects. There are indications that rhodanese undergoes induction in bacteria exposed to cyanide (Bowen *et al.*, 1965) or mercaptosuccinate (Tabita *et al.*, 1969).

The release of cyanide from crushed insect tissues does not reveal the existence of any special compartmentalization, cellular or subcellular. Sequestration of ingested plant chemicals into special organs has, however, been shown. An Australian sawfly stores eucalyptus oils in a pair of foregut diverticula (Morrow *et al.*, 1976), and a grasshopper stores ingested cardenolides in a special gland (von Euw *et al.*, 1967). The larva of a lepidopteran (*Tortrix viridana*) feeds on oak leaves and stores the ingested tannins in special cells in the gut wall (Hollande, 1923).

B. Ouabain Nontoxicity

Cardenolides are responsible for the distastefulness of monarch butterflies (*Danaus plexippus*) to starlings, blue jays, and other birds (Brower and van Zandt Brower, 1964; Brower, 1969) due to their emetic properties. The major mode of toxic action of cardenolides is their inhibitory effect on the Na^+–K^+-ATPase (EC 3.6.1.4) responsible for transporting alkali ions across membranes. Cardenolides have historically been used in very low doses as heart tonics. They are poisonous to vertebrates in still relatively low doses. Phytophagous insects have potassium as the major cation in their body fluids, with sodium playing a very minor role (Jungreis, 1977). In vertebrates the sodium ion concentration is usually about 10 times higher than the potassium ion concentration. The reversed distribution of ions in herbivorous insects presumably reflects the prevalence of ion species in the plant tissues. In insect tissues there is thus transport of potassium ions into cells without the need for the accompanying countertransport of sodium ions out of cells. All vertebrates have in their epithelia Na^+-K^+-ATPases that are sensitive to inhibition by cardenolides, including ouabain. Insects possess ATPases that are insensitive to high ouabain concentrations even in the absence of high concentrations of potassium ions (Vaughan and Jungreis, 1977). The presence of high potassium ion concentrations in insect hemolymph in itself antagonizes the binding of the cardenolide to the enzyme active site (Vaughan and Jungreis, 1977). An additional protection for insect epithelia such as midgut, cloaca, rectum, and Malpighian tubules is the apparent lack of a conventional potassium-stimulated ATPase. Ion distribution across the membranes in these tissues is maintained by other kinds of transport systems (Jungreis and Vaughan, 1977). Insects have ouabain-sensitive Na^+–K^+-ATPase in their nerve tissues. However, these tissues are encapsulated within a neural sheath, which is impermeable to cardenolides (Jungreis and Vaughan, 1977).

In this case of tolerance to a toxic plant allelochemical, there are thus at least four different mechanisms operating in concert. The target interaction of the chemical or its failure is more important in the resulting tolerance to ouabain than any other interaction. In fact, it appears that ouabain is not attacked by any metabolizing enzymes. Rather, most of a given dose was excreted within 48 hours (Vaughan and Jungreis, 1977), presumably after a fairly high tissue concentration of the alkaloid had been attained.

C. Noninsecticidal Nicotine

Forty percent of Black Leaf 40 consists of nicotine sulfate. It is the major marketing form for the insecticide nicotine and is recommended for use in gardens. Nicotine is extremely toxic to mammals and fish and fairly toxic upon contact to hordes of small, soft-bodied insects such as aphids. Nicotine is biosynthesized in the roots of tobacco plants and translocated via the xylem to the vegetative parts. The concentrations of nicotine in leaves is highly variable and can be as much as 18% in *Nicotiana rustica* (see Schmeltz, 1971). More or less nicotine is present in all species of *Nicotiana* and several other plants. Nevertheless, insects are seen feeding on them.

The green peach aphid *Myzus persicae* has a behavioral mechanism for avoiding nicotine poisoning. It restricts its feeding to the phloem of the plants, which contains no nicotine (Guthrie *et al.*, 1962). Several other biochemical physiological defense mechanisms put nicotine out of business with many species of tobacco-feeding insects.

The tobacco hornworm (*Manduca sexta*), tobacco budworm (*Heliothis virescens*), and cabbage looper (*Trichoplusia ni*) have an extremely fast and efficient excretory system. These insects pass the ingested meal through the gut so fast that a damaging dose of nicotine (or other poison) cannot accumulate (Self *et al.*, 1964). The explanation of this mechanism is found in a special alkaloid transport system in the Malphigian tubules of the larvae of *M. sexta, Pieris brassicae,* and *Rhodnius prolixus* (Maddrell and Gardiner, 1976). The system is absent in adult hornworm moths, which are nectar feeders and presumably do not need to excrete alkaloids. This transport system, which also transports basic but not acidic dyes, is not universally occurring in insects.

Nicotine is a nerve poison. It binds to the acetycholine receptor in the synapses (Yamamoto, 1965). At physiological pH values most of the nicotine exists in ionized form (Yang and Guthrie, 1969) and can only very poorly penetrate the neural sheath, which is impermeable to ions (O'Brien, 1957). This is an additional form of protection for the hornworm (Yang and Guthrie, 1969). The nicotine-susceptible silkworm *Bombyx mori,* on the other hand, has a less efficient neural sheath.

Another major defense mechanism against nicotine is metabolism. Self *et al.* (1964) found very effective metabolism of nicotine in several tobacco insects including the tobacco wireworm (*Conoderus vespertinus*), larvae of the cigarette beetle (*Lasioderma serricorne*), and the differential grasshopper (*Melanoplus differentialis*). The major metabolite was in all cases cotinine, but two or three additional metabolites were found. These insects consequently must have a highly active MFO system in their alimentary tract.

In the case of nicotine tolerance there are thus several different defense mechanisms that operate independently or to some degree complement each other.

D. Toxic Amino Acids as Nitrogen Source

Specialized seed predators among insects are more than likely to be exposed to high concentrations of nonprotein amino acids. These amino acids are normally not incorporated as structural elements into proteins. When they are incorporated, it is usually a metabolic mistake with detrimental consequences. Therefore, they are toxic amino acids. It appears that leguminous plants in particular store toxic amino acids in their seeds (Dahlman and Rosenthal, 1975). L-Canavanine, the naturally occurring oxyguanidine analog of L-arginine, is highly toxic to the tobacco hornworm (Dahlman and Rosenthal, 1975), the larva of the cowpea weevil *Callosobruchus maculatus* (Janzen *et al.*, 1977), the southern armyworm (Rehr *et al.*, 1973a), and the boll weevil *Anthonomus grandis* (Vanderzant and Chremos, 1971). However, Rosenthal *et al.* (1976) found that the larvae of another bruchid seed specialist are resistant to very high concentrations of L-canavanine. This beetle larva not only avoids poisoning by L-canavanine, but is in fact able to take advantage of it as a valuable nitrogen source.

The aminoacyl-tRNA synthetases are enzymes that activate amino acids for incorporation into polypeptides by attaching them to specific tRNAs. Studies with ^{14}C-labeled L-canavanine revealed that this amino acid is not incorporated into any proteins. Since aminoacyl-tRNA synthetases are extremely specific enzymes, this could result from the absence of an aminoacyl-tRNA synthetase specific for canavanine, or it might result from the absence of a tRNA with a specific binding site for L-canavanine. Parallel experiments with the tobacco hornworm showed that hornworm proteins contain L-canavanine after exposure, possibly explaining the toxicity of the compound to the hornworm (Rosenthal *et al.*, 1976).

Rosenthal *et al.* (1977) also found that beetle larvae have appreciable levels of arginase and an exceedingly active urease. They concluded that the larvae hydrolyze, by arginase action, a significant portion of the L-canavanine to canaline and urea. Subsequently, the urea is metabolized to CO_2 and ammonia, which can be utilized as a nitrogen source for primary metabolic pathways. The means by which the larvae conserve ammonia, which is also toxic at high concentrations, are not known.

This rather specialized case shows that enzymes that are normally engaged in basic subcellular processes can become available for detoxification in circumstances in which an organism has adapted to the presence of a toxic substance in its environment.

VIII. CONCLUSIONS

In the light of the natural interactions just described (Section VII), it is highly conceivable that the biochemical defense mechanisms described earlier (Sections II–VI) are designed for normal, although in some cases obscure, functions. Special attention has been directed toward these metabolic mechanisms due to their great importance in practical insect pest control and human drug therapy programs. It seems safe to say that an extended and intensified study of naturally occurring interactions will eventually reveal a normal mechanism that can be slightly adapted to deal with almost any kind of toxic chemical. This, of course, is inherent in the statement of Paracelsus (1540) that the toxicity of any chemical depends on the dosage.

REFERENCES

Abernathy, C. O., and Casida, J. E. (1973). *Science* **179,** 1235.
Aboudonia, M. B., and Dieckert, J. (1971). *Toxicol. Appl. Pharmacol.* **18,** 507.
Aboudonia, M. B., and Menzel, D. B. (1968). *Biochemistry* **7,** 3788.
Adamson, R. H., Dixon, R. L., Francis, F. L., and Rall, D. P. (1965). *Proc. Natl. Acad. Sci. U.S.A.* **54,** 1386.
Addison, R. F., Zinck, M. E., and Willis, D. E. (1977). *Comp. Biochem. Physiol. C* **57,** 39.
Agosin, M., and Perry, A. S. (1974). *In* "The Physiology of Insecta" (M. Rockstein, ed.), Vol. 5, p. 538. Academic Press, New York.
Agosin, M., Michaelis, D., Miskus, R., Nakasawa, S., and Hoskins, W. M. (1961). *J. Econ. Entomol.* **54,** 340.
Agosin, M., Aravena, L., and Neghme, A. (1965). *Exp. Parasitol.* **16,** 318.
Agosin, M., Fine, B. C., Scaramelli, N., Ilivicky, J., and Aravena, L. (1966). *Comp. Biochem. Physiol.* **19,** 339.
Agosin, M., Ilivicky, J., and Litvak, S. (1967). *Can. J. Biochem.* **45,** 619.
Ahmad, N., and Brindley, W. A. (1969). *Toxicol Appl. Pharmacol.* **15,** 433.
Ahmad, N., and Brindley, W. A. (1971). *Toxicol. Appl. Pharmacol.* **18,** 124.
Ahmad, S. (1975). *Drug Metab. Rev.* **4,** 177.
Ahmad, S., and Forgash, A. J. (1976). *Drug Metab. Rev.* **5,** 141.
Aldridge, W. N. (1953). *Biochem. J.* **53,** 110.
Anders, M. W., and Mannering, G. J. (1966). *Mol. Pharmacol.* **2,** 319.
Appleton, H. T., and Nakatsugawa, T. (1972). *Pestic. Biochem. Physiol.* **2,** 286.
Babish, J. G., and Stoewsand, G. S. (1975a). *Nutr. Rep. Int.* **12,** 109.
Babish, J. G., and Stoewsand, G. S. (1975b). *J. Nutr.* **105,** 1592.
Bachur, N. R. (1976). *Science* **193,** 595.
Bachur, N. R., and Gee, M. (1971). *J. Pharmacol. Exp. Ther.* **177,** 567.
Balazs, I., and Agosin, M. (1968). *Biochim. Biophys. Acta* **123,** 142.
Bartlet, A. L., and Kirinya, L. M. (1976). *Q. J. Exp. Physiol. Congr. Med. Sci.* **61,** 105.
Baron, J., and Tephly, T. R. (1969). *Biochem. Biophys. Res. Commun.* **36,** 526.
Baumeister, R. G. H., Schievelbein, H., and Zickgraf-Rüdel, G. (1975). *Drug Res.* **25,** 1056.
Beck, S. D. (1965). *Annu. Rev. Entomol.* **10,** 207.
Bellward, G. D., Dawson, R., and Otten, M. (1975). *Res. Commun. Chem. Pathol. Pharmacol.* **12,** 669.
Bend, J. R., Pohl, R. J., and Fouts, J. R. (1973). *Bull. Mt. Desert Island Biol. Lab.* **13,** 9.
Bend, J. R., Miller, D. S., Kinter, W. B., and Peakall, D. B. (1974). *Biochem. Pharmacol.* **26,** 1000.
Bend, J. R., James, M. O., and Dansette, P. M. (1977). *Ann. N.Y. Acad. Sci.* **298,** 505.
Benke, G. M., and Wilkinson, C. F. (1971a). *J. Econ. Entomol.* **64,** 1032.
Benke, G. M., and Wilkinson, C. F. (1971b). *Pestic. Biochem. Physiol.* **1,** 19.
Benke, G. M., Wilkinson, C. F., and Telford, J. N. (1972). *J. Econ. Entomol.* **65,** 1221.
Berry, C., Stellon, A., and Hallinan, T. (1975). *Biochim. Biophys. Acta* **403,** 335.
Bertran, E. C. (1952). *An. Fac. Vet. Univ. Madrid Inst. Invest. Vet.* [2] **4,** 334.
Bliss, C. I. (1935). *Ann. Appl. Biol.* **22,** 134.
Blum, M. S., and Woodring, J. P. (1962). *Science* **138,** 512.
Bollenbacher, W. E., Smith, S. L., Wielgus, J. J., and Gilbert L. I. (1977). *Nature* (*London*) **268,** 660.
Bond, E. J. (1961). *Can. J. Biochem. Physiol.* **39,** 1793.
Booth, J., and Boyland, E. (1971). *Biochem. Pharmacol.* **20,** 407.
Bowen, T. J., Butler, P. J., and Happold, F. C. (1965). *Biochem. J.* **95,** 5p.

Boxer, G. E., and Richards, J. C. (1952). *Arch. Biochem. Biophys.* **39,** 7.
Boyland, E., Kinder, C. H., and Manson, D. (1961). *Biochem. J.* **78,** 175.
Bratton, A. C., and Marshall, E. K. (1939). *J. Biol. Chem.* **128,** 537.
Brattsten, L. B., and Metcalf, R. L. (1970). *J. Econ. Entomol.* **63,** 101.
Brattsten, L. B., and Wilkinson, C. F. (1973a). *Comp. Biochem. Physiol. B* **45,** 59.
Brattsten, L. B., and Wilkinson, C. F. (1973b). *Pestic. Biochem. Physiol.* **3,** 393.
Brattsten, L. B., and Wilkinson, C. F. (1977). *Science* **196,** 1211.
Brattsten, L. B., Wilkinson, C. F., and Root, M. M. (1976). *Insect Biochem.* **6,** 615.
Brattsten, L. B., Wilkinson, C. F., and Eisner, T. (1977). *Science* **196,** 1349.
Bresnick, E. (1966). *Mol. Pharmacol.* **2,** 406.
Brodie, B. B., and Axelrod, J. (1950). *J. Pharmacol. Exp. Ther.* **99,** 171.
Brodie, B. B., Gillette, J. R., and LaDu, B. N. (1958). *Annu. Rev. Biochem.* **27,** 427.
Brooks, G. T. (1973). *Nature (London)* **245,** 382.
Brooks, G. T. (1974a). "Chlorinated Insecticides." Chem. Rubber Publ. Co., Cleveland, Ohio.
Brooks, G. T. (1974b). *Pestic. Sci.* **5,** 177.
Brooks, G. T., Harrison, A., and Lewis, S. E. (1970). *Biochem. Pharmacol.* **19,** 255.
Brower, L. P. (1969). *Sci. Am.* **220,** 22.
Brower, L. P., and van Zandt Brower, J. (1964). *Zoologica (N.Y.)* **49,** 137.
Brown, R. R., Miller, J. A., and Miller, E. C. (1954). *J. Biol. Chem.* **209,** 211.
Buhler, D. R., and Rasmusson, M. E. (1968). *Comp. Biochem. Physiol.* **25,** 223.
Bull, D. L., and Whitten, C. J. (1972). *J. Agric. Food Chem.* **20,** 561.
Burns, K. A. (1976). *Comp. Biochem. Physiol. B* **53,** 443.
Busvine, J. R. (1971). "A Critical Review of the Techniques for Testing Insecticides." Commonw. Agric. Bur., Slough, SL2 3BN England.
Caldwell, J., French, M. R., Idle, J. R., Renwick, A. G., Bassir, O., and Williams, R. T. (1975). *FEBS Lett.* **60,** 391.
Cannella, C., Pecci, L., Costa, M., Pensa, B., and Cavallini, D. (1975). *Eur. J. Biochem.* **56,** 283.
Carpenter, M. P. (1972). *Ann. N.Y. Acad. Sci.* **203,** 81.
Casida, J. E. (1970). *J. Agric. Food Chem.* **18,** 753.
Casida, J. E., Kimmel, E. C., Ohkawa, H., and Ohkawa, R. (1975). *Pestic. Biochem. Physiol.* **5,** 1.
Chakraborty, J., and Smith, J. N. (1964). *Biochem. J.* **93,** 389.
Chakraborty, J., Sissons, C. H., and Smith, J. N. (1967). *Biochem. J.* **102,** 492.
Chan, T. M., Gillett, J. W., and Terriere, L. C. (1967). *Comp. Biochem. Physiol.* **20,** 731.
Chen, P. R. S., and Dauterman, W. C. (1971). *Biochim. Biophys. Acta* **250,** 216.
Cinti, D. L., Lemelin, M. A., and Christian, J. (1976). *Biochem. Pharmacol.* **25,** 100.
Conney, A. H. (1967). *Pharmacol. Rev.* **19,** 317.
Coon, M. J., Nordblom, G. D., White, R. E., and Haugen, D. A. (1975). *Biochem. Soc. Trans.* **3,** 813.
Cram, R. L., Juchau, M. R., and Fouts, J. R. (1965). *Proc. Soc. Exp. Biol. Med.* **118,** 872.
Crankshaw, D. L., Zabik, M., and Aust, S. D. (1977). *Pestic. Biochem. Physiol.* **7,** 564.
Creaven, P. J., Parke, D. V., and Williams, R. T. (1965a). *Biochem. J.* **96,** 390.
Creaven, P. J., Parke, D. V., and Williams, R. T. (1965b). *Biochem. J.* **96,** 879.
Creaven, P. J., Davies, W. H., and Williams, R. T. (1967). *Life Sci.* **6,** 105.
Crowell, H. H. (1943). *Ann. Entomol. Soc. Am.* **36,** 243.
Culp, H. W., and McMahon, R. E. (1968). *J. Biol. Chem.* **243,** 848.
Dahlman, D. L., and Rosenthal, G. A. (1975). *Comp. Biochem. Physiol. A* **51,** 33.
Dallner, G., Siekevitz, P., and Palade, G. E. (1966). *J. Cell Biol.* **30,** 73.

Dauterman, W. C. (1976). *In* "Pesticide Biochemistry and Physiology" (C. F. Wilkinson, ed.), p. 149. Plenum, New York.
Davidson, K. L., and Sell, J. L. (1972). *J. Agric. Food Chem.* **20,** 1198.
Davidson, S. C., and Wills, E. D. (1974). *Biochem. J.* **142,** 19.
Dewaide, J. H. (1971). "Metabolism of Xenobiotics." Drukkerij Leijn, Nijmegen.
Douch, P. G. C. (1975). *Xenobiotica* **5,** 293.
Eisner, H. E., Eisner, T., and Hurst, J. J. (1963). *Chem. Ind. (London)* p. 124.
Eisner, H. E., Alsop, D. W., and Eisner, T. (1967). *Psyche* **71,** 107.
Eisner, T., Hendry, L. B., Peakall, D. B., and Meinwald, J. (1971). *Science* **172,** 277.
Elmamlouk, T. H., and Gessner, T. (1976). *Comp. Biochem. Physiol. C* **53,** 57.
Elshourbagy, N. A., and Wilkinson, C. F. (1978). *Insect Biochem.* **8,** 425.
Estabrook, R. W., Hildebrandt, A. G., Baron, J., Netter, K. J., and Leibman, K. (1971). *Biochem. Biophys. Res. Commun.* **42,** 132.
Estabrook, R. W., Martinez-Zedillo, G., Young, S., Peterson, J. A., and McCarthy, J. (1975). *J. Steroid Biochem.* **6,** 419.
Feeny, P. P. (1975). *In* "Coevolution of Animals and Plants" (L. E. Gilbert and P. H. Raven, eds.), p. 3. Univ. of Texas Press, Austin.
Feeny, P. P., Paauwe, K. L., and Demong, N. J. (1970). *Ann. Entomol. Soc. Am.* **63,** 832.
Feller, D. R., and Gerald, M. C. (1971). *Biochem. Pharmacol.* **20,** 1991.
Fenwick, M. L. (1958). *Biochem. J.* **70,** 373.
Fjellstedt, T. A., Allen, R. H., Duncan, B. K., and Jacoby, W. B. (1973). *J. Biol. Chem.* **248,** 3702.
Fouts, J. R., and Rogers, L. A. (1965). *J. Pharmacol. Exp. Ther.* **147,** 112.
Franklin, M. R. (1972). *Biochem. Pharmacol.* **21,** 3287.
Freeland, W. J., and Janzen, D. H. (1974). *Amer. Naturalist,* **108,** 269.
Gelboin, H. V. (1972). *In* "Fundamentals in Drug Metabolism and Drug Disposition" (B. N. LaDu, H. G. Mandel, and E. L. Way, eds.), p. 279. Williams & Wilkins, Baltimore, Maryland.
Gelboin, H. V., Wortham, J. S., and Wilson, R. G. (1967). *Nature (London)* **241,** 281.
Gessner, T., Jacknowitz, A., and Vollmer, C. A. (1973). *Biochem. J.* **132,** 249.
Gil, D. L., and Wilkinson, C. F. (1976). *Pestic. Biochem. Physiol.* **6,** 338.
Gil, D. L., and Wilkinson, C. F. (1977). *Pestic. Biochem. Physiol.* **7,** 183.
Gil, D. L., Fine, B. C., Dinamarca, M. L., Balaz, I., Busvine, J. R., and Agosin, M. (1968). *Entomol. Exp. Appl.* **11,** 15.
Gil, D. L., Rose, H. A., Yang, R. S. H., Young, R. G., and Wilkinson, C. F. (1974). *Comp. Biochem. Physiol. B* **47,** 657.
Gilbert, M. D., and Wilkinson, C. F. (1974). *Pestic. Biochem. Physiol.* **4,** 56.
Giri, S. N. (1973). *Toxicol. Appl. Pharmacol.* **24,** 513.
Glander, K. E. (1977). *Nat. Hist. (N.Y.)* **86,** 35.
Gordon, H. T. (1961). *Annu. Rev. Entomol.* **6,** 27.
Gorrod, J. W. (1973). *Chem.-Biol. Interact.* **7,** 289.
Graham, P. S., Hellyer, R. O., and Ryan, A. J. (1970). *Biochem. Pharmacol.* **19,** 769.
Guroff, G., Daly, J., Jerina, D., Renson, J., Udenfriend, S., and Witkop, B. (1967). *Science* **157,** 1524.
Gut, I., Becker, B. A., and Gutová, M. (1976). *Arch. Toxicol.* **35,** 41.
Guthrie, F. E., Campbell, W. B., and Baron, R. L. (1962). *Ann. Entomol. Soc. Am.* **55,** 42.
Hall, F. R., Hollingworth, R. M., and Shankland, D. L. (1971). *Comp. Biochem. Physiol. B* **38,** 723.
Hansch, C. (1968). *J. Med. Chem.* **11,** 920.
Hansen, L. G., and Hodgson, E. (1971). *Biochem. Pharmacol.* **20,** 1569.

Hansen, L. G., Kapoor, I. P., and Metcalf, R. L. (1972). *Comp. Gen. Pharmacol.* **3,** 339.
Hayaishi, O., ed. (1962). "Oxygenases." Academic Press, New York.
Hayakawa, J., Lemahieu, R. A., and Udenfriend, S. (1974). *Arch. Biochem. Biophys.* **162,** 223.
Hayes, J. R., and Campbell, T. C. (1974). *Biochem. Pharmacol.* **23,** 1721.
Hayes, J. R., Mgbodile, M. U. K., and Campbell, T. C. (1973). *Biochem. Pharmacol.* **22,** 1005.
Hegnauer, R. (1971). *In* "The Biology and Chemistry of the Umbelliferae" (V. H. Heywood, ed.), p. 267. Academic Press, New York.
Henry, M. F., and Nyns, E. J. (1975). *Sub-Cell. Biochem.* **4,** 1.
Himwich, W. A., and Saunders, J. P. (1948). *Am. J. Physiol.* **153,** 348.
Hinderer, R. K., and Menzer, R. E. (1976). *Pestic. Biochem. Physiol.* **6,** 161.
Hodgson, E., and Casida, J. E. (1961). *Biochem. Pharmacol.* **8,** 179.
Hodgson, E., and Philpot, R. M. (1974). *Drug Metab. Rev.* **3,** 231.
Hodgson, E., and Plapp, F. W. (1970). *J. Agric. Food Chem.* **18,** 1048.
Hodgson, E., and Tate, L. G. (1976). *In* "Insecticide Biochemistry and Physiology" (C. F. Wilkinson, ed.), p. 115. Plenum, New York.
Hodgson, E., Tate, L. G., Kulkarni, A. P., and Plapp, F. W. (1974). *J. Agric. Food Chem.* **22,** 360.
Hollande, A. C. (1923). *Arch. Anat. Microsc. Morphol. Exp.* **19,** 349.
Hook, G. E. R., and Bend, J. R. (1976). *Life Sci.* **18,** 279.
Hook, G. E. R., and Smith, J. N. (1967). *Biochem. J.* **102,** 504.
Hutson, D. H. (1977). *Foreign Compd. Metab. Mamm.* **4,** 259.
Iba, M. M., Soyka, L. F., and Schulman, M. P. (1975). *Biochem. Biophys. Res. Commun.* **65,** 870.
Irving, C. C. (1971). *Xenobiotica* **1,** 387.
Janzen, D. H., Juster, H. B., and Bell, E. A. (1977). *Phytochemistry* **16,** 223.
Jao, L. T., and Casida, J. E. (1977). *Pestic. Biochem. Physiol.,* **4,** 465.
Jerina, D. M., and Daly, J. W. (1974). *Science* **185,** 573.
Jerina, D. M., Daly, J. W., Witkop, B., Zaltzman-Nirenberg, P., and Udenfriend, S. (1970). *Biochemistry* **9,** 147.
Jerina, D. M., Dansette, P. M., Lu, A. Y. H., and Levin, W. (1977). *Molec. Pharmacol.,* **13,** 342.
Jick, H., and Shuster, L. (1966). *J. Biol. Chem.* **241,** 5366.
Jondorf, W. R., Maickel, R. P., and Brodie, B. B. (1959). *Biochem. Pharmacol.* **1,** 352.
Jones, D. A. (1972). *In* "Phytochemical Ecology" (J. B. Harborne, ed.), p. 103. Academic Press, New York.
Jones, D. A., Parsons, J., and Rothschild, M. (1962). *Nature (London)* **193,** 52.
Jori, A., Bianchetti, A., and Prestini, P. E. (1969). *Biochem. Pharmacol.* **18,** 2081.
Juchau, M. R., Cram, R. L., Plaa, G. R., and Fouts, J. R. (1965). *Biochem. Pharmacol.* **14,** 423.
Jungreis, A. M. (1977). *In* "Water Relationships in Membrane Transport in Plants and Animals" (A. M. Jungreis *et al.,* eds.), p. 89. Academic Press, New York.
Jungreis, A. M., and Vaughan, G. L. (1977). *J. Insect Physiol.* **23,** 503.
Kaihara, M., and Price, J. M. (1965). *J. Biol. Chem.* **340,** 454.
Kamienski, F. X., and Casida, J. E. (1970). *Biochem. Pharmacol.* **19,** 91.
Kato, R. (1974). *Drug Metab. Rev.* **3,** 1.
Kato, R., and Gillette, J. R. (1965). *J. Pharmacol. Exp. Ther.* **150,** 279.
Kawatski, J. A., and Schmulbach, J. C. (1971). *J. Econ. Entomol.* **64,** 316.
Khan, M. A. Q. (1969). *J. Econ. Entomol.* **62,** 723.

Khan, M. A. Q. (1970). *Biochem. Pharmacol.* **19,** 903.

Khan, M. A. Q., and Matsumura, F. (1972). *Pestic. Biochem. Physiol.* **2,** 236.

Kircher, H. W., Heed, W. B., Russell, J. S., and Groove, J. (1967). *J. Insect Physiol.* **13,** 1869.

Krieger, R. I., and Lee, P. W. (1973). *J. Econ. Entomol.* **66,** 1.

Krieger, R. I., and Wilkinson, C. F. (1969). *Biochem. Pharmacol.* **18,** 1403.

Krieger, R. I., and Wilkinson, C. F. (1970). *Biochem. J.* **116,** 781.

Krieger, R. I., and Wilkinson, C. F. (1971). *Pestic. Biochem. Physiol.* **1,** 92.

Krieger, R. I., Gilbert, M. D., and Wilkinson, C. F. (1970). *J. Econ. Entomol.* **63,** 1322.

Krieger, R. I., Feeny, P. P., and Wilkinson, C. F. (1971). *Science* **172,** 579.

Krieger, R. I., Wilkinson, C. F., Hicks, L. J., and Taschenberg, E. F. (1976). *J. Econ. Entomol.* **69,** 1.

Kuhr, R. J. (1970). *J. Agric. Food Chem.* **18,** 1023.

Kuhr, R. J. (1971). *J. Econ. Entomol.* **64,** 1373.

Kulkarni, A., Mailman, R. B., and Hodgson, E. (1975). *J. Agric. Food. Chem.* **23,** 177.

Kupfer, D. (1970). *BioScience* **20,** 705.

Kupfer, D., and Bruggeman, L. L. (1966). *Anal. Biochem.* **17,** 502.

Lang, K. (1933). *Biochem. Z.* **263,** 262.

Lehninger, A. L. (1976). "Biochemistry." Worth Publ., New York.

Leibman, K. C., and Estabrook, R. W. (1971). *Mol. Pharmacol.* **7,** 26.

Lewis, S. E., Wilkinson, C. F., and Ray, J. W. (1967). *Biochem. Pharmacol.* **16,** 1195.

Litvak, S., and Agosin, M. (1968). *Biochemistry* **7,** 1560.

Loeb, L. A., and Gelboin, H. V. (1964). *Proc. Natl. Acad. Sci. U.S.A.* **52,** 1219.

Lombrozo, L., and Mitoma, C. (1970). *Biochem. Pharmacol.* **19,** 2317.

Lu, A. Y. H., and Coon, M. J. (1968). *J. Biol. Chem.* **243,** 1331.

Lu, A. Y. H., Levin, W., Lander, H., and Deiring, D. (1974). *Biochem. Biophys. Res. Commun.* **61,** 1348.

Ludke, J. L., Gibson, J. R., and Lusk, C. I. (1972). *Toxicol. Appl. Pharmacol.* **21,** 89.

Machinist, J. M., Dehner, E. W., and Ziegler, D. M. (1968). *Arch. Biochem. Biophys.* **125,** 858.

Maddrell, S. H. P., and Gardiner, B. O. C. (1976). *J. Exp. Biol.* **64,** 267.

Mannering, G. J. (1968). *In* "Selected Pharmacological Testing Methods" (A. Burger, ed.), p. 51. Dekker, New York.

Mannering, G. J. (1972). *In* "Fundamentals in Drug Metabolism and Drug Disposition" (B. N. LaDu, H. G. Mandel, and E. O. Way, eds.), p. 206, Williams & Wilkins, Baltimore, Maryland.

Marshall, W. J., and McLean, A. E. M. (1971). *Biochem. J.* **122,** 569.

Marver, H. S. (1969). *In* "Microsomes and Drug Oxidations" (J. R. Gillette *et al.*, eds.), p. 495. Academic Press, New York.

Mason, H. S. (1957). *Adv. Enzymol.* **19,** 79.

Mason, H. S., Fowlks, W., and Peterson, E. (1955). *J. Am. Chem. Soc.* **77,** 2914.

Mason, J. I., and Boyd, G. S. (1975). *Biochem. Soc. Trans.* **3,** 832.

Masters, B. S. S., Baron, J., Taylor, W. E., Isaacson, E. L., and Spalluto, J. L. (1971). *J. Biol. Chem.* **246,** 4143.

Matthews, H. B., and Hodgson, E. (1966). *J. Econ. Entomol.* **59,** 1286.

Mattocks, A. R., and White, I. N. H. (1971). *Chem.-Biol. Interact.* **3,** 383.

Mayer, R. T., Jermyn, J. W., Burke, M. D., and Prough, R. A. (1977). *Pestic. Biochem. Physiol.* **7,** 349.

Mazel, P. (1972). *In* "Fundamentals in Drug Metabolism and Drug Disposition" (B. N. LaDu, H. G. Mandel, and E. L. Way, eds.), p. 546. Williams & Wilkins, Baltimore, Maryland.

McManus, M. E., and Ilett, K. F. (1976). *Drug Metab. Dispos.* **4,** 199.
Mehendale, H. M., and Dorough, H. W. (1972). *J. Insect Physiol.* **18,** 981.
Metcalf, R. L. (1967). *Annu. Rev. Entomol.* **12,** 229.
Metcalf, R. L., Fukuto, T. R., Wilkinson, C. F., Fahmy, M. H., Elaziz, S. A., and Metcalf, E. R. (1966). *J. Agric. Food Chem.* **14,** 555.
Metcalf, R. L., Osman, M. F., and Fukuto, T. R. (1967). *J. Econ. Entomol.* **60,** 445.
Mitchard, M. (1971). *Xenobiotica* **1,** 469.
Mitoma, C., Neubauer, S. E., Badger, N. L., and Sorich, T. J. (1967). *Proc. Soc. Exp. Biol. Med.* **125,** 284.
Morello, A. (1964). *Nature (London)* **203,** 785.
Morello, A. (1965). *Can. J. Biochem.* **43,** 1289.
Morello, A., Bleecker, W., and Agosin, M. (1971). *Biochem. J.* **124,** 199.
Morrow, P. A., Bellas, T. E., and Eisner, T. (1976). *Oecologia* **24,** 193.
Nakatsugawa, T., and Morelli, M. A. (1976). *In* "Insecticide Biochemistry and Physiology" (C. F. Wilkinson, ed.), p. 61. Plenum, New York.
Nakatsugawa, T., Ishida, J., and Dahm, P. A., (1965). *Biochem. Pharmacol.* **14,** 1853.
Nash, T. (1953). *J. Biol. Chem.* **55,** 416.
Nebert, D. W., and Gelboin, H. V. (1968). *J. Biol. Chem.* **243,** 6250.
Nelson, P. A., Stewart, R. R., Morelli, M. A., and Nakatsugawa, T. (1976). *Pestic. Biochem. Physiol.* **6,** 243.
Netter, K. J., and Seidel, G. (1964). *J. Pharmacol. Exp. Ther.* **146,** 61.
Neuhauser, E., and Hartenstein, R. (1976). *Comp. Biochem. Physiol. C* **53,** 37.
Nigg, H. N., Kapoor, I. P., Metcalf, R. L., and Coats, J. R. (1972). *J. Agric. Food Chem.* **20,** 446.
O'Brien, R. D. (1957). *Ann. Entomol. Soc. Am.* **50,** 223.
O'Brien, R. D. (1960). "Toxic Phosphorus Esters." Academic Press, New York.
Oesch, F. (1973). *Xenobiotica* **3,** 305.
Oesch, F., and Daly, J. (1971). *Biochim. Biophys. Acta* **227,** 692.
Oesch, F., Jerina, D. M., and Daly, J. W. (1971). *Biochim. Biophys. Acta* **227,** 685.
Oesch, F., Jerina, D. M., Daly, J. W., and Rice, J. M. (1973). *Chem.-Biol. Interact.* **6,** 189.
Oesch, F., Thoenen, H., Fahrländer, H., and Suda, K. (1974). *Biochem. Pharmacol.* **23,** 1307.
Oi, S. (1975). *J. Biochem. (Tokyo)* **78,** 825.
Omura, T., and Sato, R. (1964). *J. Biol. Chem.* **239,** 2379.
Omura, T., Sanders, E., Estabrook, R. W., Cooper, D. Y., and Rosenthal, O. (1966). *Arch. Biochem. Biophys.* **117,** 660.
Oonnithan, E. S., and Casida, J. E. (1968). *J. Agric. Food Chem.* **16,** 2.
Orrenius, S. (1968). *In* "The Interaction of Drugs and Subcellular Components in Animal Cells" (P. N. Campbell, ed.), p. 97. Churchill, London.
Orrenius, S., and Ericsson, J. L. E. (1966). *J. Cell Biol.* **28,** 181.
Orrenius, S., Das, M., and Gnosspelius, Y. (1969). *In* "Microsomes and Drug Oxidations" (J. R. Gillette *et al.,* eds.), p. 251. Academic Press, New York.
Orrenius, S., Berggren, M., Moldeus, P., and Krieger, R. I. (1971). *Biochem. J.* **124,** 427.
Palade, G. E., and Siekevitz, P. (1956). *J. Biophys. Biochem. Cytol.* **2,** 171.
Pan, H. P., Fouts, J. R., and Devereux, T. R. (1975a). *Life Sci.* **17,** 819.
Pan, H. P., Hook, G. E. R., and Fouts, J. R. (1975b). *Xenobiotica* **5,** 17.
Pap, Á., and Szarvas, F. (1974). *Acta Morphol. Acad. Sci. Hung.* **22,** 187.
Parke, D. V. (1960). *Biochem. J.* **77,** 493.
Parsons, J., and Rothschild, M. (1962). *J. Insect Physiol.* **8,** 285.
Parsons, J., and Rothschild, M. (1964). *Entomol. Gaz.* **15,** 58.

Payne, J. F. (1975). *Science* **191,** 945.
Payne, J. F., and Penrose, W. R. (1975). *Bull. Environ. Contam. Toxicol.* **14,** 112.
Pearson, O. P. (1948). *J. Mammal.* **29,** 345.
Perry, A. S., and Buckner, A. J. (1970). *Life Sci.* **9,** 335.
Perry, A. S., Dale, W. E., and Buckner, A. J. (1971). *Pestic. Biochem. Physiol.* **1,** 131.
Philpot, R. M., and Hodgson, E. (1971). *Chem.-Biol. Interact.* **4,** 399.
Plapp, F. W., and Casida, J. E. (1969). *J. Econ. Entomol.* **62,** 1174.
Plapp, F. W., and Casida, J. E. (1970). *J. Econ. Entomol.* **63,** 1091.
Plapp, F. W., Tate, L. G., and Hodgson, E. (1976). *Pestic. Biochem. Physiol.* **6,** 175.
Paulsen, L. L., Hyslop, R. M., and Ziegler, D. M. (1974). *Biochem. Pharmacol.* **23,** 3431.
Powis, G., Drummond, A. H., MacIntyre, D. E., and Jondorf, W. R. (1976). *Xenobiotica* **6,** 69.
Prough, R. A., and Ziegler, D. M. (1977). *Arch. Biochem. Biophys.* **180,** 363.
Reddy, G., and Krishnakumaran, A. (1974). *Insect Biochem.* **4,** 355.
Rehr, S. S., Bell, E. A., Janzen, D. H., and Feeny, P. P. (1973a). *Biochem. Syst.* **1,** 63.
Rehr, S. S., Feeny, P. P., and Janzen, D. H. (1973b). *J. Anim. Ecol.* **42,** 405.
Remmer, H. (1971). *Pestic. Chem., Proc. Int. IUPAC Congr. Pestic. Chem., 2nd, 1971* Vol. 2, p. 167.
Remmer, H. (1972). *Eur. J. Clin. Pharmacol.* **5,** 116.
Remmer, H., and Merker, H. J. (1963). *Science* **142,** 1657.
Robinson, T. (1967). "The Organic Substituents of Higher Plants." Burgess, Minneapolis, Minnesota.
Rose, H. A., and Young, R. G. (1973). *Pestic. Biochem. Physiol.* **3,** 243.
Rosenthal, G. A., Dahlman, D. L., and Janzen, D. H. (1976). *Science* **192,** 256.
Rosenthal, G. A., Janzen, D. H., and Dahlman, D. L. (1977). *Science* **196,** 658.
Rosiello, A. P., Essigman, J. M., and Wogan, G. N. (1977). *J. Toxicol. Environ. Health* **3,** 797.
Rothschild, M., Reichstein, T., von Euw, J., Aplin, R., and Harman, R. R. M. (1970). *Toxicon* **8,** 293.
Sanborn, H. R., and Malins, D. C. (1977). *Proc. Soc. Exp. Biol. Med.* **154,** 151.
Sato, R., Nishibayashi, H., and Ito, A. (1969). *In* "Microsomes and Drug Oxidations" (J.R. Gillette *et al.,* eds.), p. 111. Academic Press, New York.
Schafer, J. A., and Terriere, L. C. (1970). *J. Econ. Entomol.* **63,** 787.
Schenkman, J. B., Remmer, H., and Estabrook, R. W. (1967). *Mol. Pharmacol.* **3,** 113.
Schievelbein, H., Baumeister, R., and Vogel, R. (1969). *Naturwissenschaften* **56,** 416.
Schildknecht, H., Maswitz, U., and Krauss, D. (1968). *Naturwissenschaften* **55,** 230.
Schmeltz, I. (1971). *In* "Naturally Occurring Insecticides" (M. Jacobson and D. G. Crosby, eds.), p. 99. Dekker, New York.
Schonbrod, R. D., and Terriere, L. C. (1971). *J. Econ. Entomol.* **64,** 44.
Schonbrod, R. D., Khan, M. A. Q., Terriere, L. C., and Plapp, F. W. (1968). *Life Sci.* **7,** 681.
Self, L. S., Guthrie, F. E., and Hodgson, E. (1964). *Nature (London)* **204,** 200.
Sher, S. P. (1971). *Toxicol. Appl. Pharmacol.* **18,** 780.
Shishido, T., and Fukami, J. (1972). *Pestic. Biochem. Physiol.* **2,** 39.
Shishido, T., Usui, K., Sato, M., and Fukami, J. (1972). *Pestic. Biochem. Physiol.* **2,** 51.
Siekevitz, P. (1963). *Annu. Rev. Physiol.* **25,** 15.
Slade, M., and Wilkinson, C. F., (1973). *Science,* **181,** 672.
Slade, M., Brooks, G. T., Hetnarski, H. K., and Wilkinson, C. F. (1975). *Pestic. Biochem. Physiol.* **5,** 35.
Smith, J. N. (1968). *Adv. Comp. Physiol. Biochem.* **3,** 173.
Sörbo, B. H. (1951). *Acta Chem. Scand.* **5,** 724.
Sörbo, B. H. (1953). *Acta Chem. Scand.* **7,** 1129.

Sörbo, B. H. (1957). *Acta Chem. Scand.* **11,** 628.
Springfield, A. C., Carlson, G. P., and DeFeo, J. J. (1973). *Toxicol. Appl. Pharmacol.* **24,** 298.
Stephen, B. J., Garlich, J. D., and Guthrie, F. E. (1971). *Bull. Environ. Contam. Toxicol.* **5,** 569.
Strittmatter, C. F., and Umberger, F. T. (1969). *Biochim. Biophys. Acta* **180,** 18.
Sugiura, M., and Kato, R. (1977). *J. Pharmacol. Exp. Ther.* **200,** 25.
Sugiura, M., Iwasaki, K., and Kato, R. (1976). *Mol. Pharmacol.* **12,** 322.
Sugiura, M., Iwasaki, K., and Kato, R. (1977). *Biochem. Pharmacol.* **26,** 489.
Swingle, M. C. (1939). *J. Econ. Entomol.* **32,** 884.
Tabita, R., Silver, M., and Lundgren, D. G. (1969). *Can. J. Biochem.* **47,** 1141.
Tate, L. G., Plapp, F. W., and Hodgson, E. (1973). *Chem.-Biol. Interact.* **6,** 237.
Teas, H. J. (1967). *Biochem. Biophys. Res. Commun.* **26,** 686.
Terriere, L. C., and Yu, S. J. (1976a). *Insect Biochem.* **6,** 109.
Terriere, L. C., and Yu, S. J. (1976b). *Pestic. Biochem. Physiol.* **6,** 223.
Terriere, L. C., Yu, S. J., and Hoyer, R. F. (1971). *Science* **171,** 581.
Teschke, R., Hasumura, Y., and Lieber, C. S. (1974). *Arch. Biochem. Biophys.* **163,** 404.
Testa, B., and Jenner, P. (1976). "Drug Metabolism, Chemical and Biochemical Aspects." Dekker, New York.
Thongsinthusak, T., and Krieger, R. I. (1974). *Life Sci.* **14,** 2131.
Thongsinthusak, T., and Krieger, R. I. (1976). *Comp. Biochem. Physiol. C* **54,** 7.
Thorsteinson, A. J. (1953). *Can. J. Zool.* **31,** 52.
Thorsteinson, A. J. (1960). *Annu. Rev. Entomol.* **5,** 193.
Tilak, T. G., Nagarajan, V., and Tupule, P. G. (1975). *Experientia* **31,** 953.
Towers, G. H. N., Duffey, S. S., and Siegel, S. M. (1972). *Can. J. Zool.* **50,** 1047.
Truex, C. R., Brattsten, L. B., and Visek, W. J. (1977). *Biochem. Pharmacol.* **26,** 667.
Turnquist, R. L., and Brindley, W. A. (1975). *Pestic. Biochem. Physiol.* **5,** 211.
Ullrich, V. (1971). *Angew. Chem., Int. Ed. Engl.* **11,** 701.
Vainio, H., and Parkki, M. G. (1976). *Toxicology* **5,** 279.
Vanderzant, E. S., and Chremos, J. H. (1971). *Ann. Entomol. Soc. Am.* **64,** 780.
Vaughan, G. L., and Jungreis, A. M. (1977). *J. Insect Physiol.* **23,** 585.
Vesell, E. S. (1968). *Ann. N.Y. Acad. Sci.* **151,** 900.
von Euw, J., Fishelson, L., Parsons, J. A., Reichstein, T., and Rothschild, M. (1967). *Nature (London)* **214,** 35.
Walker, C. R., and Terriere, L. C. (1970). *Entomol. Exp. Appl.* **13,** 260.
Welton, A. F., and Aust, S. D. (1974). *Biochem. Biophys. Res. Commun.* **56,** 898.
Westley, J. (1973). *Adv. Enzymol.,* **39,** 327.
Whitlock, J. P., and Gelboin, H. V. (1974). *J. Biol. Chem.* **249,** 2616.
Whittaker, R. H. (1970). *In* "Chemical Ecology" (E. Sondheimer and J. B. Simeone, eds.), p. 43. Academic Press, New York.
Whittaker, R. H., and Feeny, P. P. (1971). *Science* **171,** 757.
Wigglesworth, V. B. (1972). "The Principles of Insect Physiology." Methuen, London.
Wilkinson, C. F. (1968). *In* "Enzymatic Oxidation of Toxicants" (E. Hodgson, ed.), p. 113. North Carolina State Univ. Press, Raleigh.
Wilkinson, C. F., and Brattsten, L. B. (1972). *Drug Metab. Rev.* **1,** 153.
Wilkinson, C. F., and Hicks, L. J. (1969). *J. Agric. Food Chem.* **17,** 829.
Wilkinson, C. F., Metcalf, R. L., and Fukuto, T. R. (1966). *J. Agric. Food Chem.* **14,** 73.
Wilkinson, C. F., Hetnarski, K., Cantwell, C. P., and DiCarlo, F. J. (1974). *Biochem. Pharmacol.* **23,** 2377.
Williams, C. H., and Kamin, H. (1962). *J. Biol. Chem.* **237,** 587.
Williams, R. T. (1974). *Biochem. Soc. Trans.* **2,** 359.

Williamson, R. L., and Schechter, M. S. (1970). *Biochem. Pharmacol.* **19,** 1719.
Wilson, T. G., and Hodgson, E. (1971). *Insect Biochem.* **1,** 19.
Wilson, T. G., and Hodgson, E. (1972). *Pestic. Biochem. Physiol.* **2,** 64.
Wood, J. L., and Cooley, S. L. (1956). *J. Biol. Chem.* **218,** 449.
Yaffe, S. J., Krasner, J., and Catz, C. S. (1968). *Ann. N.Y. Acad Sci.* **151,** 887.
Yamamoto, I. (1965). *Adv. Pest Control Res.* **6,** 231.
Yang, R. S. H. (1976). *In* "Pesticide Biochemistry and Physiology" (C. F. Wilkinson, ed.), p. 177. Plenum, New York.
Yang, R. S. H., and Guthrie, F. E. (1969). *Ann. Entomol. Soc. Am.* **62,** 141.
Yang, R. S. H., and Wilkinson, C. F. (1972). *Biochem. J.* **130,** 487.
Yang, R. S. H., and Wilkinson, C. F. (1973). *Comp. Biochem. Physiol. B* **46,** 717.
Yang, R. S. H., Pellicia, J. G., and Wilkinson, C. F. (1973). *Biochem. J.* **136,** 817.
Yawetz, A., Agosin, M., and Perry, A. S. (1978). *Pestic. Biochem. Physiol.* **8,** 44.
Youssef, N. N., and Brindley, W. A. (1972). *Cytobios* **6,** 143.
Yu, S. J., and Terriere, L. C. (1971a). *Life Sci.* **10,** 1173.
Yu, S. J., and Terriere, L. C. (1971b). *Pestic. Biochem. Physiol.* **1,** 173.
Yu, S. J., and Terriere, L. C. (1972). *Pestic. Biochem. Physiol.* **2,** 184.
Yu, S. J., and Terriere, L. C. (1973). *Pestic. Biochem. Physiol.* **3,** 141.
Yu, S. J., and Terriere, L. C. (1974a). *J. Insect Physiol.* **20,** 1901.
Yu, S. J., and Terriere, L. C. (1974b). *Pestic. Biochem. Physiol.* **4,** 160.
Yu, S. J., Kiigemagi, U., and Terriere, L. C. (1971). *J. Agric. Food Chem.* **19,** 5.
Yust, H. R., and Shelden, F. F. (1952). *Ann. Entomol. Soc. Am.* **45,** 220.
Zannoni, V. G., and Lynch, M. M. (1973). *Drug Metab. Rev.* **2,** 57.

Chapter **6**

Plant Resources for Chemical Defense

FRANCES S. CHEW *And* **JAMES E. RODMAN**

I. INTRODUCTION

Terrestrial plants produce an extremely diverse array of organic molecules that have potent effects on other organisms. These compounds are often termed "secondary" because explicit physiological ("primary") functions are rarely known (e.g., Sachs, 1875, pp. 628ff.; Fraenkel, 1959; Seigler and Price, 1976; Jones, 1979), even though many of the compounds are actively metabolized within the plants that produce them (e.g., Pfeffer, 1900, pp. 129ff.; Robinson, 1974; Scott, 1974; Heftmann, 1977). Numerous functions have been suggested for various members of this complex array. Among these are (1) regulators of plant growth or biosynthetic activities (e.g., Del Moral, 1972; Fowden, 1973; Nakanishi *et al.*, 1974; Robinson, 1974), (2) storage forms of plant growth regulators (e.g., Kutáček and Kefeli, 1968; Elliot and Stowe, 1971), (3) energy reserves (e.g., Blum and Ebercon, 1976; Jager and Meyer, 1977), (4) trans-

ISBN 0-12-597180-X

port facilitators (e.g., Liener, 1976; Harborne, 1977), (5) waste products (e.g., Sachs, 1887, p. 329; Muller, 1969; Luckner, 1972, pp. 3ff.), (6) detoxification products of environmental poisons (e.g., Harborne, 1977; Kjaer, 1977), (7) shields against excessive radiation (e.g., McClure, 1975; Rhoades, 1977), and (8) effectors of allelochemical interactions between plants and their plant competitors (e.g., Muller, 1970; Rice, 1974, 1977) and between plants and heterotrophic organisms (e.g., Stahl, 1888, cited by Fraenkel, 1959; Whittaker, 1970; Whittaker and Feeny, 1971). Thus, these compounds are components of both physiological and external environments of plants. For heterotrophic organisms that exploit plants as food, these compounds are variables of the external environment encountered during ingestion or infection of plant tissue. Investigations of heterotroph adaptation to plant environments have often revealed a clear dependence on the chemical profile of the host plant (e.g., Verschaffelt, 1911; Atabekov, 1975; Ikeda *et al.*, 1977); however, reciprocal investigations of plant responses, especially of a chemical nature, to herbivore or pathogen attack have yielded no such insight of comparable generality.

The variety of functions that secondary compounds may perform in plants renders examination of the "costs" and "benefits" of producing these compounds worthwhile but complex. We are particularly interested in allelochemical interactions of plants and their herbivores and pathogens because the specificity of many such interactions is mediated by secondary compounds. When demonstrable, this causality permits us to identify both specific heterotroph agents whose activities may reduce plant fitness and specific plant products that may increase plant resistance to heterotrophs. In principle at least, we can then measure the effects of heterotroph damage to the plant and the effects of specific plant products on the damaging activities of specific herbivores and pathogens. These measurements then permit us to calculate the direction and intensity of natural selection on each interacting organism and thus permit us in principle to examine the potential for evolutionary change. "Benefits" need not relate to allelochemical interactions with heterotrophic organisms. We suggest examining plant–heterotroph interactions primarily because the entities may be identified, observed, and manipulated.

In this chapter, we examine the costs and benefits of plant production of secondary compounds and consider the significance of qualitative and quantitative variations in physiological, ecological, and genetic contexts. As products of plant metabolic pathways, these compounds reflect the presence of appropriate genes whose expression is regulated by the genetic and physiological environments of the plant. As products of living organisms, these compounds reflect the evolutionary histories and present ecological situations of plants. With these considerations in mind, we focus on three problems in particular: (1) What is the nature of plant metabolic response to herbivore or pathogen attack? (2) How might the biosynthetic costs and ecological benefits of producing secondary compounds be calculated and compared? (3) What physiological and genetic constraints influence the evolution of plant chemical defense against herbivores and pathogens? Many authors have argued that the diversity and complexity of secondary compounds must be evolutionarily advantageous (e.g., Fraenkel, 1959; Ehrlich and Raven, 1965; Levin, 1976; Swain, 1977). We agree. However, our present knowledge suggests that the magnitude of such advantage, as well as the cost of producing particular compounds, may vary considerably.

II. PLANT VULNERABILITY TO HETEROTROPH ATTACK

Because plant growth is usually indeterminate, plant capacity for biosynthesis varies not only with availability of light energy for photosynthesis, appropriate inorganic molecules, and appropriate environmental conditions (e.g., Ehleringer *et al.*, 1976; Wareing and Allen, 1977), but also with the developmental stage of the plant (e.g., Loomis *et al.*, 1976). Thus, factors that limit the rate of plant biosynthesis from photosynthetic assimilates change during development; very probably the proportion of biosynthetic effort devoted to any particular plant part or product also changes (e.g., Mooney and Chu, 1974; Ho, 1976). The phenotypic plasticity of terrestrial plants (e.g., Wallace *et al.*, 1976; Bracken and Bucher, 1977) and our incomplete knowledge of biosynthetic pathways and transport patterns (e.g., Penning de Vries *et al.*, 1974; Penning de Vries, 1975a,b; McCree, 1976) make compiling and interpreting information on plant budgets difficult. We would like to know how herbivore or pathogen attack affects overall plant biosynthetic capacity for secondary compounds in general, as well as for specific classes of such compounds. Published work addresses three questions that bear on the problem (Fig. 1). First, how is the concentration (or absolute amount) of secondary compounds in plant tissue related to plant productivity (biomass yield)? Second, how do shortages of specific assimilates (e.g., carbon, mineral nutrients) affect production of plant secondary compounds and/or biomass? Third, how does defoliation due to herbivore or pathogen attack affect biomass and/or production of secondary compounds? As Fig. 1 suggests, a particular perturbation

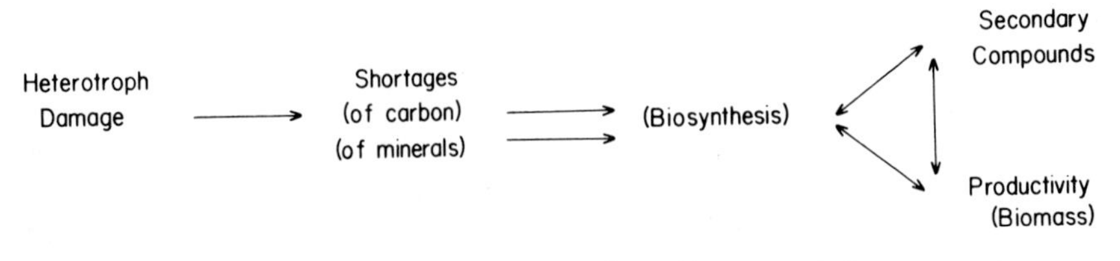

Figure 1. Interactive factors that influence the effect of heterotroph damage on plant productivity and secondary-compound production. Damage to plant tissue may cause loss of photosynthetic capacity and/or mineral nutrients, which in turn affects biosynthetic capacity and plant resource allocations.

may be several steps removed from observable effects. For example, by affecting photosynthetic capacity, defoliation can affect the availability of carbon for biosynthetic processes, which in turn affects overall plant productivity; this in turn affects the proportion of biosynthetic resources allocated to secondary compounds. Because experimental work to examine correlations between secondary-compound production and other plant biosynthetic processes has only begun to reveal the nature of underlying physiological mechanisms, we cannot satisfactorily answer our question. However, in examining work done in these areas, we shall explore the nature of physiological constraints on plant ability to compensate for heterotroph damage.

A. Plant Resources for Biosynthesis

1. *Plasticity in Allocation*

Some plants, perhaps most, do not usually photosynthesize at their maximal rates (e.g., Maggs, 1964, 1965; Sweet and Wareing, 1966; Wareing and Allen, 1977), just as most animals do not continuously breathe or eat at their maximal rates. Variation in the rate of carbon fixation may result from different fixation routes (e.g., Mooney, 1972; Mooney *et al.*, 1976; Bassham, 1977). Some plants may use more than one photosynthetic pathway (e.g., Ting, 1976; Lange and Zuber, 1977). Although C_4 plants may commit a greater proportion of photosynthetic assimilates to biomass production than do C_3 plants, Caldwell *et al.* (1977) suggest that variation in the proportion of photosynthate allocated to biomass (and presumably to other plant products as well) reflects physiological differences other than those related to a particular route of carbon fixation. Caswell *et al.* (1973) hypothesized, on the basis of a preliminary survey of the literature on animal feeding, that C_4 plants are preferentially avoided as food sources compared to C_3 associates in a community, perhaps because of lower levels of utilizable nitrogen in C_4 plants. A test of their prediction (Boutton *et al.*, 1978) found no statistically significant differences in herbivore utilization of C_3 and C_4 grasses, but the researchers did postulate a "trend" toward greater use of C_3 plants.

Some wild plants such as the annual herb *Polygonum cascadense* (Polygonaceae) show plasticity in both total biomass production and the proportion of photosynthates allocated to various plant parts. Individuals from populations growing in different habitats respond similarly to greenhouse culture but show significant differences in the proportion of biomass allocated to reproductive organs when grown under field conditions (Hickman, 1975). Plants growing in harsh environments (here defined as those in which plants were water-stressed, as measured by minimal xylem sap potentials) committed a greater proportion of biomass to reproductive tissue than those growing in more mesic habitats. Larger plants (generally those growing in more mesic areas) committed a smaller proportion of biomass to reproductive parts, even though the absolute weight of reproductive organs in these plants was greater than in smaller plants, a relationship also observed in several *Lupinus* species (Leguminosae) (Hickman and Pitelka, 1975; Pitelka, 1977). Resource allocation to seed reproduction may remain constant under different environmental conditions, whereas allocation to vegetative reproduction (i.e., clonal spread) varies considerably (Holler and Abrahamson, 1977). Work with crops suggests that both the photosynthetic capacity of a plant and the demands of a sink organ [a region of net import (Evans, 1975)] affect the biomass of specific plant organs. Reciprocal grafting experiments in which photosynthetic parts (sources) are grafted onto sinks (e.g., tubers) show that, in the absence of environmental limitations, a larger sink can stimulate greater photosynthetic capacity in a source. Conversely, a source with greater photosynthetic capacity results in greater yield in the sink organ (e.g., Thorne and Evans, 1964; Maggs, 1964, 1965; Allen and Morgan, 1975; Hahn, 1977). High-yielding crop varieties, i.e., those that commit a large proportion of biomass to an economically desirable plant part, may be those in which photosynthetic rate is stimulated by high sink demand (e.g., Wareing and Patrick, 1975; Wallace *et al.*, 1976).

The correlation of biomass yield with levels of secondary compounds varies. Chaplin and Weeks (1976), for example, reported a reduction in yield associated with several "low-alkaloid" lines of tobacco, but Matzinger *et al.* (1972), investigating progeny of crosses of two tobacco cultivars, observed that increased nicotine (the major alkaloid in tobacco) correlated with decreased yield. Studies of natural plant populations reveal similar variability (e.g., Rice, 1974; Hare, 1977). In *Xanthium strumarium* (Compositae) a negative correlation between thickness of seed coat and its concentration of secondary compounds may be due not only to physiological constraints but also to ecological constraints exerted by seed predators (Hare, 1977).

Negative correlation between compounds that require the same minerals has also been observed. Romeo and Bell (1974) examined seeds of *Erythrina* (Leguminosae) for nonprotein amino acids and alkaloids, both of which require nitrogen. They found that concentrations of the two groups of secondary compounds were usually negatively correlated, which suggests that the nitrogen allocated to secondary compounds is concentrated in one group (Romeo and Bell, 1974). Kuč *et al.* (1976) and Russell and Berryman (1976) observed similar negative correlations between biosynthetically related phytoalexins. If the biological, especially allelochemical, effects of secondary compounds are concentration or dose dependent, such allocation patterns could result in the maintenance of one line of effective chemical defense rather than the maintenance of ineffective concentrations of two or more groups of compounds.

Seasonal changes in biomass allocation to various tissues and in concentrations of secondary compounds in the evergreen chaparral shrub *Heteromeles arbutifolia* (Rosaceae) are closely correlated with photosynthetic capacity, which is limited primarily by water availability (Mooney *et al.*, 1975). Dement and Mooney (1974) suggest that the high concentrations of tannins and cyanogenic glycosides in very young leaves may be related to the ready availability of both carbon and nitrogen before leaf initiation. Despite the year-round photosynthetic capacity of *H. arbutifolia*, not all demands for photosynthetic assimilates can be met simultaneously (Mooney and Chu, 1974), a conclusion reached by other workers for other plants (e.g., Jameson, 1963; Kulman, 1971; Wallace *et al.*, 1976; Mooney and Bartholomew, 1974). It would be interesting to know whether levels of secondary compounds are correlated with ecological as well as physiological requirements, i.e., whether high levels of leaf tannins and cyanogenic glycosides are correlated with herbivore activity (and presumed need for chemical defense) as well as with availability of biosynthetic substrates. The fruit–seed dynamics of tannins and cyanogenic glycosides in *H. arbutifolia* suggest that ecological constraints are important. Immature fruits, which are rarely eaten by animals, contain high concentrations of tannins and cyanogenic glycosides in the pulp; on maturation, the pulp concentration of these secondary compounds declines, cyanogenic glycosides are concentrated in the seeds, and the fruits are rapidly removed by birds (Dement and Mooney, 1974).

2. *Response to Mineral Shortage*

Interpreting plant responses to shortages of inorganic nutrients is often difficult because we do not yet understand how specific shortages affect plant metabolism. For example, work on the effects of boron shows that the relationship between a specific mineral deficiency and plant production of secondary compounds can be complex. Boron is an essential mineral nutrient for all higher plants (Rains, 1976). Boron-deficient plants accumulate significant amounts of phenolic compounds, 10- to 20-fold normal concentrations (Wender, 1970). Lee and Aronoff (1967) suggest that, by regulating pathways of carbon utilization, boron influences the rate of synthesis of phenolics. Boron as borate normally combines with 6-phosphogluconate, the initial substrate in the 6-phosphogluconate pathway (pentose phosphate shunt), to form a complex that inhibits 6-phosphogluconate dehydrogenase and further activity along this pathway. The pathway ultimately produces erythrose-4-phosphate, an important intermediate in the formation of phenolic acids. Boron-deficient plants may thus produce elevated concentrations of phenolic compounds because the pathway regulated by boron furnishes more erythrose-4-phosphate than would normally be the case (Lee and Aronoff, 1967). A similar negative correlation has been observed between the availability of other mineral nutrients to plants and concentrations of phenolic acids. Lehman and Rice (1972) and Del Moral (1972) observed that plants deficient in nitrogen, potassium, or phosphorus may produce 10- to 20-fold increases in various phenolic compounds compared to control plants. These observations suggest that plants growing in nutrient-poor soils may produce effective concentrations of allelochemicals.

Positive correlation between availability of mineral nutrients and concentrations of secondary compounds that contain these minerals has also been observed. Romeo *et al.* (1977) found that *Sarracenia* plants (Sarraceniaceae), which grow in nitrogen-deficient bogs, lack nonprotein amino acids and other nitrogenous secondary compounds and contain few free amino acids. Freeman and Mossadeghi (1970, 1971) found a positive correlation between sulfate supply and sulfur-containing compounds in garlic and onion (Liliaceae). For several cruciferous crops (Cruciferae) they found that supplying plants with increasing levels of sulfate increased the concentrations of sulfur-containing isothiocyanates (presumably arising from parent glucosinolates) in these plants (cf. Josefsson, 1970a) but that these increases reached a plateau when plants were fertilized with very high sulfate levels (Freeman and Mossadeghi, 1972a,b,c).

In some cases, a positive correlation between levels of inorganic minerals and secondary compounds may reflect plant responses to toxic levels of minerals. Kjaer (1977) suggests that the elaborate array of sulfur-containing compounds in mangroves (Rhizophoraceae), including cyclic disulfides, dithiolanes, and alkaloid-type molecules, may be related to the presence of high (potentially toxic) levels of sulfur in the habitat. In some aquatic and marine habitats, hydrogen sulfide may accumulate; high levels inhibit metabolism in rice seedlings (Joshi and Hollis, 1977). Harborne (1977) suggests that flavonoid sulfates may assist in the detoxification of exogenous phenolic compounds or in the incorporation and transfer of sulfur in aquatic environments (cf. Nissen and Benson, 1964). Mathys (1977) suggests a similar hypothesis to explain the greater concentrations of "mustard oil glycoside [*sic*]" in natural populations of *Thlaspi alpestre* (Cruciferae) which are tolerant of high soil levels of zinc. Thus, in these cases, secondary compounds may serve dual functions (at least); moreover, the origin of these secondary compounds as products of detoxification does not preclude their serving allelochemical functions (Muller, 1969; Whittaker, 1970).

B. Plant Response to Heterotroph Damage

Herbivores and pathogens damage plants in a variety of ways. In some cases, ovule or seed destruction provides a direct measure of damage and, by implication, fitness reduction (e.g., Breedlove and Ehrlich, 1968, 1972). In other cases, however, herbivore damage to seeds has varying effects on germination and seedling success (e.g., Janzen, 1976b). More generally, heterotroph damage may not translate directly into observable changes in plant performance. For example, loss of mineral nutrients in damaged tissue may impose as important a limitation on productivity of a defoliated plant as the more obvious loss of photosynthetic tissue (Harper, 1977, p. 385). We have chosen to examine the effects of experimental defoliation on plant productivity and secondary-compound content in an effort to achieve a more general understanding of the effects of heterotroph damage.

The timing of defoliation with respect to the developmental stage of the plant is critical. During development, the capacity of the photosynthetic source varies with respect to the demands of specific sink organs or products (e.g., secondary compounds). Wareing and Patrick (1975) suggest that by removing either sources or sinks at appropriate times (and thus changing their ratio), one can obtain the various positive, negative, and neutral effects observed in experimental defoliation of crops and natural vegetation (e.g., Harris, 1972; Mattson and Addy,

1975). Chopping or clipping may increase the production of dry matter, especially early in the growing season (e.g., Mueggler, 1967; Harris, 1972; Mattson and Addy, 1975); it may also have little or no effect (e.g., Begum and Eden, 1965; Turnipseed, 1972; Caldwell, 1973; Singh and Nair, 1975). Defoliation at other times, however, may result in significant reduction in biomass production. In some cases these reductions result in reduced fecundity or survivorship (e.g., Begum and Eden, 1965; Romig and Calpouzos, 1970; James *et al.*, 1971; Maun and Cavers, 1971; Todd and Morgan, 1972; Caldwell, 1973; Rockwood, 1973; Hampton, 1975; Singh and Nair, 1975; Dyer and Bokhari, 1976; Janzen, 1976a; Bracken and Bucher, 1977; Kugler *et al.*, 1977; Sanders *et al.*, 1977; Schultz and Allen, 1977; Moore, 1978; Morrow and LaMarche, 1978).

Although field observations suggest that significant natural defoliation occurs in a variety of woody and herbaceous noncrop plants (e.g., Dethier, 1959; Wilson and Janzen, 1972; Rehm and Humm, 1973; Rockwood, 1973; White, 1974; Stanton, 1975), the quantitative relationship between damage and biomass yield is not clear. Harris (1972), Mattson and Addy (1975), and Dyer and Bokhari (1976) suggest that the relationship is not necessarily linear. For some plants, perhaps especially those that are periodically subjected to damage, the relationship may be ditonic; i.e., if one graphs the degree of damage to the plant on the abscissa and the resulting productivity (biomass yield) on the ordinate, one finds that a modest degree of damage may increase productivity over that of intact plants. However, increasing levels of damage beyond some threshold would significantly reduce plant productivity.

The relationship between damage and biomass yield is influenced by two additional considerations: transport pathways within the plant and the nature of biotic, as opposed to mechanical, damage. First, anatomical restrictions on the transport of assimilates between different plant parts, "canalizing effects of the vascular system" (Wareing and Patrick, 1975), are important determinants of how photosynthates and other biosynthetic products are distributed and, thus, of how plant productivity is buffered against losses of specific organs. Work on both herbaceous and woody plants shows that not all photosynthetic organs contribute equally to biomass accumulation in all plant parts (e.g., Davidson, 1964; Maggs, 1965; Lucas and Asano, 1968). In some cases, there may be extremely restricted transport. For example, in *Phragmites australis* (=*P. communis*; Gramineae), a grass that forms large clones, restriction of lateral transport through rhizomes not only facilitates intraclonal differentiation but also makes it virtually impossible to eradicate the plant by applying systemic herbicides to one part of the clone (Bjork, 1967). In the Kentucky coffee tree *Gymnocladus dioicus* (Leguminosae), productivity of individual branches and their levels of secondary compounds are supported only by the photosynthetic activities of those branches; no translocation between branches is evident (Janzen, 1976a). Anatomical restrictions on product distribution may be extremely important to the functioning of secondary compounds because many of these, for example, alkaloids, are synthesized in one part of the plant but sequestered in another (e.g., Cromwell, 1965; Waller *et al.*, 1965; Robinson, 1974).

Second, Dyer and Bokhari (1976) showed that mechanical clipping of *Bouteloua gracilis* (Gramineae) does not adequately mimic insect defoliation. They found that the regrowth potential of plants damaged by grasshoppers was greater than that of plants whose foliage was clipped and suggested that herbivores with chewing mouthparts not only inflict physical damage but also trigger plant physiological responses to components of insect saliva injected into the wound. These observations suggest that herbivore defoliation may induce a generalized increase in metabolic activity (Dyer and Bokhari, 1976). We find this idea interesting because a generalized increase in metabolic activity is also observed when plants are infected by pathogens (e.g., Schwochau and Hadwiger, 1970). Changes in levels of secondary compounds following herbivore damage (Green and Ryan, 1972; Russell and Berryman, 1976) may thus parallel pathogen-induced phytoalexin production (e.g., Cruickshank, 1963; Pierre and Bateman, 1967; Stoessl, 1970; Schonbeck and Schroder, 1972; Friend, 1976; Kuč *et al.*, 1976; Van Etten and Pueppke, 1976; Dewick, 1977). Although the induction mechanisms of plant response to herbivore defoliation and pathogen infection may differ, we find the similarity of plant metabolic responses intriguing.

Our examination of plant physiological responses to perturbation, although hardly comprehensive or exhaustive, provides substance to the view that plant resources for growth and other biosynthetic processes are limited and thus worth defending by chemical or other means. In view of the complexity of physiological mechanisms underlying observable responses and the possible multiple functions served by secondary compounds, we find it premature to draw conclusions about the involvement of specific classes of compounds in plant metabolic responses to heterotroph damage. However, further knowledge of the function of specific compounds and of physiological mechanisms governing their production (e.g., the hypothesis suggested for production of phenolic compounds in boron-deficient plants) would permit us to predict which compounds ought to show correlated changes in their production when the plant is subjected to a specific perturbation. In the next section, we consider whether physiological allocation of plant resources to defense functions is advantageous to individual plants in an ecological context, and how the biosynthetic cost of chemical defenses might be estimated.

III. HOW MIGHT COST BE CALCULATED AND COMPARED?

Ideally an approach to calculating and comparing biosynthetic cost involves two steps: (1) an estimate of cost to produce, transport, and sequester some number of secondary-compound molecules (or to maintain some concentration of an actively metabolized compound); and (2) an estimate of fitness reduction that results from attack by a specified herbivore or pathogen against which the secondary compound is a presumed defense. This combination of information from metabolic and pharmacological considerations is not without serious problems, some of which we discuss. However, such data would permit us to construct a graph relating cost to effectiveness (and, by implication, benefit). Although the linkage between effectiveness of a compound against a particular heterotroph and the fitness of the plant that synthesizes the compound may involve several steps about which we know little, we are persuaded that this relationship is worth examining in a quantitative manner.

A. Estimating the Cost of Producing Secondary Compounds

In practice, estimating cost is extremely difficult. In bacteria and higher plants, protein synthesis and transport of molecules utilize very large proportions of ATP-derived energy (Lehninger, 1971; Penning de Vries, 1975a). These processes contribute to the currently inestimable cost of maintaining a pathway for synthesis of a particular compound or group of compounds. The cost of maintaining enzyme levels might be calculable in principle from knowledge of the molecular weight of the enzyme and its rate of synthesis and degradation, but such information is only beginning to emerge for enzymes involved in the metabolism of specific secondary compounds (e.g., Björkman, 1976). Furthermore, the cost of maintaining a pathway for secondary-compound synthesis may be spread over the lifetime of a plant, just as a budget for nitrogen or other nutrients might be considered over a plant's lifetime (see Chapter 2 by McKey).

Three approaches seem to promise useful results. The first involves examining a single nutrient such as sulfur, nitrogen, or carbon (e.g.,

Mooney, 1972) and determining what proportion of that nutrient assimilated by the plant is allocated to specific secondary compounds. Such an estimate was obtained by Rosenthal (1977) for nitrogen incorporated into the nonprotein amino acid canavanine, which is found in large quantities in the seeds of plants such as *Dioclea megacarpa* (Leguminosae). Rosenthal's examination of total and free amino acid nitrogen in seven canavanine-producing legumes showed that canavanine represents at least 80% of the free amino acid content of the seeds. At least one-third of total seed nitrogen is incorporated into canavanine, which constitutes 6–13% of seed dry weight. This estimate does not include the cost of assimilating nitrogen, which may vary, for example, with the occurrence of symbiotic nitrogen-fixing bacteria in legumes (Hewitt, 1975). Such data do not necessarily establish the allocation priorities of the plant, but they permit us to compare the proportions of a mineral resource allocated to secondary compounds in different plants.

A second approach involves estimating the cost of maintaining some concentration of a particular compound on the basis of its turnover rate, its molecular formula, and its biosynthetic precursors (Robinson, 1974). From a knowledge of the molecular formula and biosynthetic starting substrate of a compound, one can calculate the number of substrate molecules needed per unit time to maintain a particular concentration of the secondary compound. From a consideration of the rate of carbon fixation or rate of assimilation of some mineral resource, one can calculate the proportion of plant resources needed for these biosynthetic activities at a specific time during development (e.g., Mooney and Chu, 1974; Robinson, 1974). In practice, such an approach will be complicated if radioactively labeled precursors are used to estimate turnover rate. Kirby (1967) and Scott (1974) discuss problems in choosing an appropriately labeled precursor (which presumes some knowledge of the pathways involved in synthesis and catabolism of a particular compound) and in achieving plant incorporation of it. This second approach measures cost in terms of the resources available to the plant but is complicated by the possible mobilization of stored reserves (e.g., Mooney and Bartholomew, 1974). An assimilation rate for carbon dioxide or other nutrients may not accurately represent the resources available for biosynthesis.

A third method for calculating biosynthetic costs has been advanced by Penning de Vries and colleagues (1975a,b; Penning de Vries *et al.*, 1974). He suggests that if one considers an overall equation for the conversion of a substrate (such as glucose) to another compound and assumes that intermediate reactions occur at equilibrium rates, one can estimate a minimal number of substrate molecules required for carbon skeletons and for ATP and reducing power. Penning de Vries *et al.* (1974) have done such calculations for a large number of essential metabolites, e.g., amino acids, organic acids, and simple carbohydrates, from biosynthetic pathways outlined by Dagley and Nicholson (1970). Although we can in principle estimate the number of ATP molecules required for the synthesis of a specific compound, we may not be able to estimate the proportion of the plant's ATP budget that this synthesis represents. Nonetheless, the calculations provide a basis for comparing biosynthetic costs of different groups of compounds. In principle, the methodology could be extended to the calculation and comparison of the costs of physical and biotic (e.g., ant guards) defense systems as well as of diverse chemical ones.

B. An Attempt to Calculate and Compare Costs

In this section, we sketch an attempt to calculate and compare the costs of producing three presumptive allelochemicals: a cyanogenic glucoside, a glucosinolate, and a nonprotein amino acid. Costs are calculated as the summation of substrates required for materials (carbon skeletons, inorganic nutrients) and for energy (ATP, $NADH_2$) from an overall biosynthetic balance sheet (cf., Penning de Vries *et al.*, 1974). Overall costs would include those of cellular transport and compartmentation as well as the costs of "maintenance of the tools for biosynthesis (nucleic acids and enzymes)" (Penning de Vries *et al.*, 1974); all are theoretically reducible to substrate and energy units. In practice, these transport, compartmentation, and maintenance costs are difficult to compute (Penning de Vries, 1975a) yet may be the most critical in cost comparisons of specific anabolic processes.

For convenience, we have chosen three end products corresponding to a common precursor protein amino acid, isoleucine [giving lotaustralin, 2-butylglucosinolate anion, and 2-amino-4-methylhex-4-enoic acid (AMHA), respectively]. We restrict our calculations to the substrates delimited by Penning de Vries *et al.* (1974) (i.e., glucose, $NADH_2$, ATP), and we compute costs of chemical conversions from considerations of known or hypothesized biosynthetic sequences for these compounds. For the difficult calculation of "maintenance" costs, we substitute an estimate of the number of enzymatic steps required as an admittedly crude approximation. In principle, these costs could be computed from knowledge of the molecular formulas of the enzymes and their turnover rates. We further ignore the costs of cellular transport and compartmentation because little is known about these with regard to the compounds under consideration. We arrive, then, at a minimal and speculative estimate of the cost of conversion of material and energy substrates for examples of three classes of presumptive allelochemicals.

We have chosen to focus on examples of cyanogenic glycosides, glucosinolates, and nonprotein amino acids for five reasons. (1) Examples from these three classes have been demonstrated to function as allelochemical defenses (Jones, 1972; Erickson and Feeny, 1974; Bell, 1972; respectively). (2) Representatives of these classes would be found in a wide array of angiosperms (Gibbs, 1974). (3) Information is available on the biosynthesis of particular examples in each class and in as much detail as is available for other secondary plant products (Kjaer and Larsen, 1973, 1976). (4) These three classes of nitrogenous compounds exemplify a limited spectrum of "fundamental metabolic reaction types" (Kjaer and Larsen, 1973), thus simplifying to some extent calculation of the costs of chemical conversions while inviting a comparison of their relative substrate and enzymatic costs. (5) Similar concentrations of cyanogenic glycosides, glucosinolates, and AMHA occur in seeds (Dement and Mooney, 1974; Phelan and Vaughan, 1976; Fowden and Mazelis, 1971; respectively). This equivalency is even more striking if one considers strictly the biologically active aglycone component of the cyanogenic glycoside and the glucosinolate compared to AMHA concentration. We emphasize in particular the cyanogenic glycosides and glucosinolates because these share certain analogous early biosynthetic steps (from amino acids to aldoximes), they are mutually exclusive in botanical distribution (Mahadevan, 1973; Tapper and Reay, 1973), and they appear to be functionally equivalent in part (Tapper and Reay, 1973). The nonprotein amino acids, as a class, are structurally and biosynthetically more diverse and may well be functionally more diverse (Fowden, 1970). Some of them serve as precursors to cyanogenic glycosides and glucosinolates (Kjaer and Larsen, 1976), although AMHA apparently does not.

The biosynthetic steps on which we base our calculations of substrate requirements and number of required enzymes are presented in Figs. 2–4. These schemes are directly based on pathways outlined by Conn (1973) and Kjaer and Larsen (1973) for cyanogenic glycosides; by Kjaer and Larsen (1973, 1976) and Underhill *et al.* (1973) for glucosinolates; and by Boyle and Fowden (1971) and Fowden and Mazelis (1971) for AMHA. Strictly by analogy, we have adapted for lotaustralin and 2-butylglucosinolate the known and hypothesized intermediate steps to the precursor isoleucine. In the several instances in which the true intermediates are not known, we have made arbitrary choices among the possibilities suggested by the above-cited authors. We do not believe that alternatives would change our qualitative conclusions or

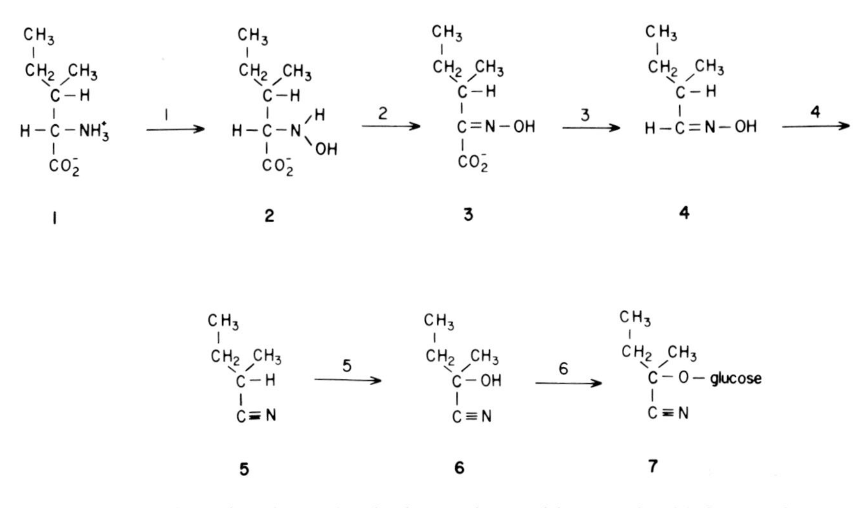

Figure 2. Hypothetical pathway for the biosynthesis of lotaustralin (**7**) from isoleucine (**1**) [adapted from Conn (1973) and Kjaer and Larsen (1973)] with an estimate of substrate cost (ATP, $NADH_2$, and glucose). Compounds not demonstrated to be true intermediates are marked in the tabulation below with an asterisk. Each enzymatic step, hypothetical unless a specific reference is cited, is numbered and formally described as follows:

Enzymatic step	Formal characterization and cost
1	Isoleucine (**1**) to *N*-hydroxyamino acid* (**2**); a monooxygenase reaction presumably requiring 1 $NADH_2$
2	*N*-Hydroxyamino acid (**2**) to 2-oximino acid* (**3**); a dehydrogenase reaction presumably generating 1 $NADH_2$
3	2-Oximino acid (**3**) to aldoxime (**4**); a decarboxylation assumed to be energy neutral
4	Aldoxime (**4**) to the corresponding nitrile (**5**); a dehydratase reaction assumed to be energy neutral
5	Nitrile (**5**) to α-hydroxynitrile* (**6**); catalyzed by a "mixed function oxidase" (Kjaer and Larsen, 1976), presumed to require 1 $NADH_2$
6	α-Hydroxynitrile (**6**) is glucosylated by a UDP-glucose : aldehyde cyanohydrin β-glucosyltransferase (Kjaer and Larsen, 1976), requiring 2 ATP and 1 glucose
	Overall cost: 2 ATP, 1 $NADH_2$, and 1 glucose, in addition to the cost of isoleucine

change the numbers of computed ATP and $NADH_2$ molecules by more than a few. For each hypothesized or known enzymatic step, we compute an ATP and $NADH_2$ cost by analogy with known costs for enzymatic conversions of that class; for example, a dehydrogenase reaction is presumed to involve 1 $NADH_2$. Although anabolic reactions may often

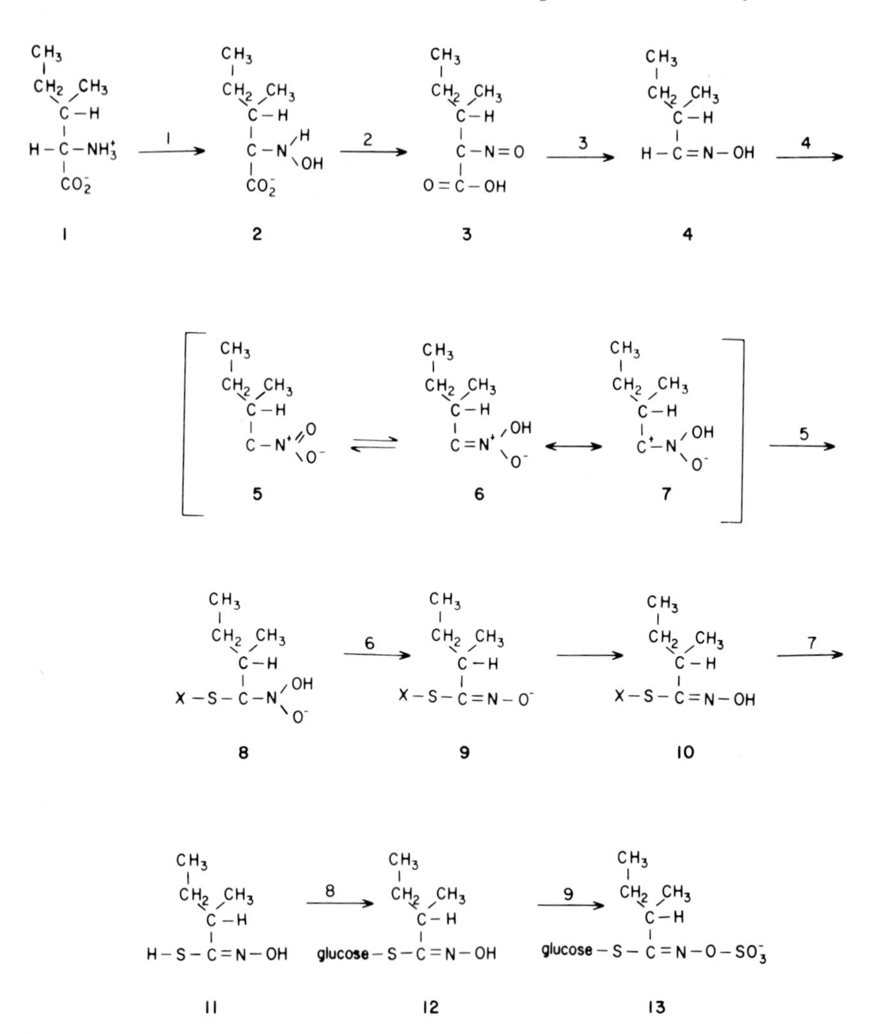

Figure 3. Hypothetical pathway for the biosynthesis of 2-butylglucosinolate anion (**13**) from isoleucine (**1**) [adapted from Kjaer and Larsen (1973, 1976) and Underhill *et al.* (1973)] with an estimate of substrate cost (ATP, $NADH_2$, and glucose). Compounds not demonstrated to be true intermediates are marked in the table at right with an asterisk. Each enzymatic step, hypothetical unless a specific reference is cited, is numbered and formally described as follows:

involve $NADPH_2$ instead of $NADH_2$, we have chosen to calculate the requirement as the latter in the absence of specific information. A simplified alternative would be to adopt the conversion equation of Penning de Vries *et al.* (1974) (1 $NADPH_2$ equals 1 $NADH_2$ plus 1 ATP); such an alternative would not change our qualitative results but would increase the ATP and $NADH_2$ costs. From the reaction balances and estimated cost of nitrate reduction given by Penning de Vries *et al.* (1974), we compute the cost of 1 isoleucine molecule at 7 $NADH_2$, 7 ATP, and 1 glucose molecule; treating the glucose as equivalent to 38 ATP, we arrive at a cost for the isoleucine precursor of 7 $NADH_2$ and 45 ATP. Again following Penning de Vries *et al.* (1974), we figure the cost of reduced sulfur as 4 $NADH_2$ and 8 ATP.

From our calculations (Figs. 2–4) we project the greatest costs, in terms of substrate requirements for carbon skeletons and for chemical conversions, for the glucosinolate, which is primarily a reflection of the additional costs of introducing sulfur into the molecule. For the less costly cyanogenic glucoside, the intermediary substrate costs are slightly

Enzymatic step	Formal characterization and cost
1	Isoleucine (**1**) to *N*-hydroxyamino acid* (**2**); a monooxygenase reaction presumably requiring 1 $NADH_2$
2	*N*-Hydroxyamino acid (**2**) to α-nitroso acid* (**3**); a dehydrogenase reaction presumed to generate 1 $NADH_2$
3	α-Nitroso acid (**3**) to aldoxime (**4**); a decarboxylation assumed to be energy neutral
4	Aldoxime (**4**) to 1-nitro-2-substituted ethane* (**5**) [or its *aci*–tautomer (**6**, **7**)], a postulated mercaptide acceptor; a monooxygenase reaction presumably requiring 1 $NADH_2$. The positive charge on the nitrogen on compound (**6**) shifts to the adjacent carbon in compound 7.
5	This acceptor adds reduced sulfur ($S \cdot X^*$) [presumed to cost 4 $NADH_2$ and 8 ATP (Penning de Vries *et al.*, 1974)]; a sulfotransferase reaction to (**8**) assumed to be energy neutral.
6	Compound **8** to S-substituted hydroximic acid* (**9**); a dehydratase reaction assumed to be energy neutral, followed by a presumably nonenzymatic protonation to (**10**)
7	Compound **10** to thiohydroximic acid (**11**); possibly a hydratase reaction assumed to be energy neutral and regenerating $X \cdot OH$
8	Thiohydroximic acid (**11**) to desulfoglucosinolate (**12**); a UDP-glucose : thiohydroximate β-glucosyltransferase (Kjaer and Larsen, 1973) requiring 2 ATP and 1 glucose
9	Desulfoglucosinolate (**12**) to glucosinolate (**13**); a sulfotransferase reaction with PAPS (Kjaer and Larsen, 1973) requiring 2 ATP.
	Overall cost: 12 ATP, 5 $NADH_2$, and 1 glucose, in addition to the cost of isoleucine

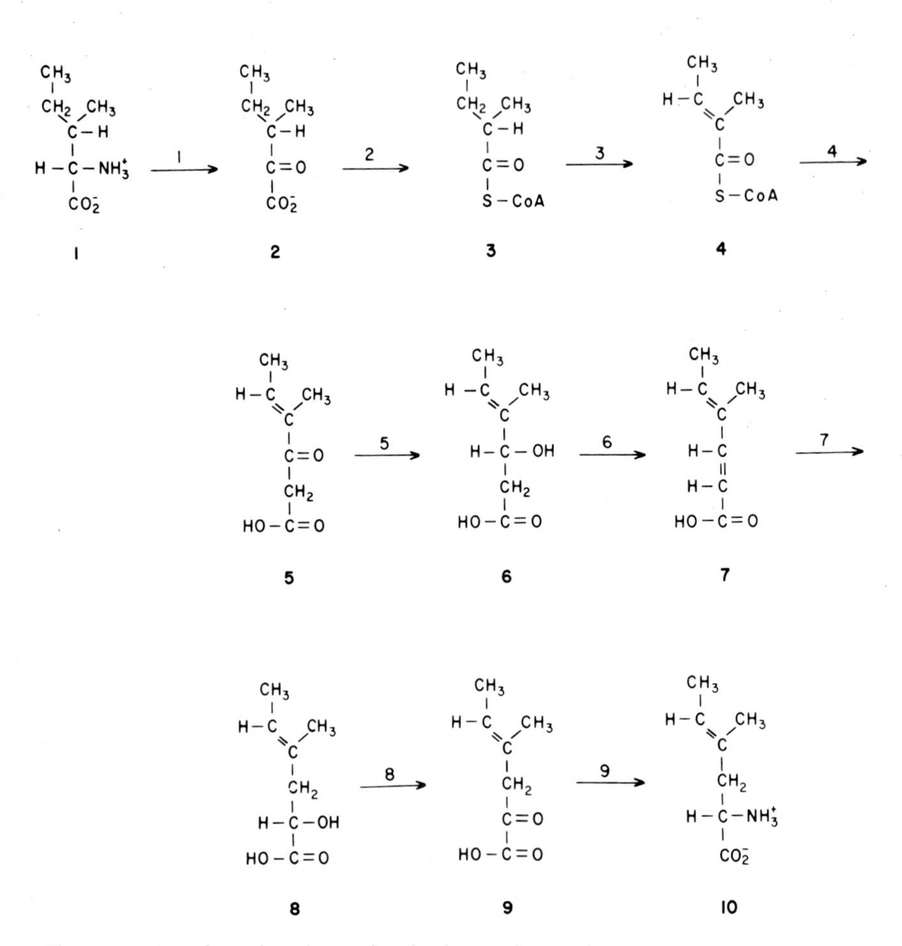

Figure 4. Hypothetical pathway for the biosynthesis of 2-amino-4-methylhex-4-enoic acid (AMHA) (**10**) from isoleucine (**1**) [from Boyle and Fowden (1971) and Fowden and Mazelis (1971)] with an estimate of substrate cost (ATP, $NADH_2$, and glucose). Compounds not demonstrated to be true intermediates are marked in table at right with an asterisk. Each hypothetical enzymatic step is numbered and formally described as follows:

less than the cost of the isoleucine precursor. The intermediary substrate cost of the nonprotein amino acid is less than half that of the glucosinolate, and the intermediary conversions could conceivably generate a few $NADH_2$ (and ATP). Comparing the three presumptive allelochemicals, the differences in calculated substrate cost for one molecule of each are much less than an order of magnitude. In terms of number of enzymatic steps, the cost is equal for the glucosinolate and the nonprotein amino acid and less for the cyanogenic glycoside. Interestingly, this correlates with the known botanical distribution of these compounds. Cyanogenic glycosides biosynthesized from protein amino acids are found in a wide array of vascular plant families (Seigler, 1977); glucosinolates occur only in a few dicotyledonous families (Ettlinger and Kjaer, 1968; Kjaer, 1960, 1974); and AMHA and biosynthetically related compounds are apparently restricted to the Hippocastanaceae (Fowden, 1970, 1973).

Our estimates of the number of enzymatic steps for the three pathways are conservative. For step 2 hypothesized in the AMHA pathway,

Enzymatic step	Formal characterization and cost
1	Isoleucine (**1**) to α-keto-β-methylvaleric acid* (**2**); a transamination assumed to be energy neutral
2	α-Keto-β-methylvaleric acid (**2**) to 2-methylbutyryl-CoA* (**3**); condensation with coenzyme A, generating 1 $FADH_2$ and, hence, possibly 1 $NADH_2$
3	2-Methylbutyryl-CoA (**3**) to tiglyl-CoA* (**4**); a dehydrogenase reaction presumably generating 1 $NADH_2$
4	Tiglyl-CoA adds acetate* (**5**) [acetate calculated to cost one-half glucose molecule from calculations for pyruvate given by Penning de Vries *et al.* (1974) and biosynthesis of acetate from pyruvate given by Dagley and Nicholson (1970; p. 118)]; a condensation regenerating coenzyme A
5	3-Keto-4-methylhex-4-enoic acid* (**5**) to 3-hydroxy-4-methylhex-4-enoic acid* (**6**); a dehydrogenase reaction presumably requiring 1 $NADH_2$
6	3-Hydroxy-4-methylhex-4-enoic acid (**6**) to 4-methylhex-2,4-enoic acid* (**7**); a dehydratase reaction assumed to be energy neutral
7	4-Methylhex-2,4-enoic acid (**7**) to 2-hydroxy-4-methylhex-4-enoic acid* (**8**); a hydratase reaction assumed to be energy neutral
8	2-Hydroxy-4-methylhex-4-enoic acid (**8**) to 2-keto-4-methylhex-4-enoic acid* (**9**); a dehydrogenase reaction presumably generating 1 $NADH_2$
9	2-Keto-4-methylhex-4-enoic acid (**9**) to AMHA (**10**); a transamination assumed to be energy neutral
	Overall cost: ½ glucose, in addition to the cost of isoleucine, while possibly generating 1 or 2 $NADH_2$

for example, the reaction may require a three-enzyme complex analogous to that known for pyruvate dehydrogenase or α-ketoglutarate dehydrogenase (Lehninger, 1975; pp. 450ff.). More fundamentally, these kinds of calculations can represent only crude estimates until more is known about the enzymology of the reactions involved and about the requirements for active transport and compartmentation. Specifically, one must know the ratio of enzyme molecules to end product molecules and the turnover rates for the enzymes. This would make it possible to calculate the maintenance costs for the enzymatic machinery relative to the substrate costs of carbon skeletons and chemical conversions. One can contrast two possible extremes. In the first, long-lived enzymes with high catalytic activity would constitute a relatively small part of overall cost; in the second, short-lived enzymes with lower activity would constitute a more substantial share of total "production" costs. Finally, this method of cost computation requires data on what might be termed "regulatory" costs reflecting the plant's genetically programmed developmental control over the tissue- and organ-specific patterns of biosynthesis, storage, and catabolism of these allelochemicals. The absence of data here precludes a realistic estimate of total substrate costs and hence an evaluation of relative magnitude of the component costs of biosynthesis. However, we find striking the result that the substrate cost of the isoleucine precursor is comparable to the further overall substrate costs of these secondary metabolites. Based on these three products as representative of the complexity of secondary compounds, we speculate that further biochemical elaborations of primary plant metabolites may not be energetically expensive and, furthermore, that the number and nature of the required enzymes are a more critical determinant of such elaborations.

C. Relating Biosynthetic Cost to Ecological Benefit

In this section we are concerned with assessing the possible benefits of biosynthetic allocation to secondary compounds and of scaling those benefits against costs. We would like to determine the benefit of producing secondary compounds in terms of the fitness of various chemical phenotypes in plant populations subject to heterotroph damage. One approach, which we characterize as pharmacological, is to assume that susceptibility to heterotroph attack (or intensity of attack) is a component of plant fitness and then to examine how secondary compounds mediate susceptibility by poisoning or repelling the heterotroph or by masking chemical attractants in the plant. This approach has the advantage of permitting us, in principle, to examine the scaling of benefit against cost among plants from different populations and taxa as well as among plants within a single population. However, estimates of fitness are appropriately made only among individuals within a population. Examination of heterotroph activity in a plant population that is polymorphic for some chemical defense would permit us to make such fitness estimates or other measures of benefit in an ecologically relevant context, that is, under conditions in which selection may occur if individual plants are differentially susceptible to heterotroph damage because of their chemical constitution (Jones, 1971). Such an analysis would then permit formulation of the scaling relationship between benefit (i.e., fitness) and cost (i.e., biosynthetic cost of the chemical defense) for a particular plant–heterotroph interaction. The literature on the dynamics of plant–heterotroph systems does not establish the form of this relationship between benefit and cost. By analogy with the pharmacological approach, we speculate on three possible forms (graphically presented) of this scaling, and review some ecological observations relevant to these, in the concluding part of this section.

The pharmacological approach relates the biosynthetic cost of some chemical defense to some measure of efficacy in conferring resistance to heterotroph attack. A dose–response curve for a secondary compound (or group of compounds) is constructed for a particular heterotroph by observing the behavioral or toxicological effects of various concentrations of the compound on that herbivore or pathogen. Varying concentrations can in principle be translated into varying measures of cost (see Section III,B), which would permit comparison among compounds and classes of compounds. Although proportionately greater quantities of carbon substrates are required for the synthesis of increased concentrations of a particular compound, we cannot yet accurately predict how costs associated with other components of the biosynthetic process (e.g., enzyme turnover) would increase (see Section III,A). Despite this drawback, the pharmacological approach does permit us to make comparisons among plants with very different chemical defenses since it does not depend on comparing individuals within a population.

Dose–response experiments might involve observing herbivore or pathogen activity on plants that contain different concentrations of a compound (e.g., Burnett *et al.*, 1974; Mitchell, 1977), boosting levels of a compound in plant tissue (e.g., Harris and Mohyuddin, 1965; Erickson and Feeny, 1974; Marsh and Rothschild, 1974), or introducing the compound into an artificial diet (e.g., Nayar and Thorsteinson, 1963, David and Gardiner, 1966). Problems involved in distinguishing behavioral from physiological effects are discussed by Beck and Reese (1976) and Slansky and Feeny (1977). Concentrations that naturally occur in plant tissue can then be compared to the dose–response curve to estimate the efficacy of the compound against a particular heterotroph. This kind of pharmacological information is available for a number of secondary compounds, for example, for 2-phenylethyl isothiocyanate (presumably deriving from its parent glucosinolate in the plant), tested on a number of insect pests (Lichtenstein *et al.*, 1962). If such curves for two or more secondary compounds are constructed for a particular heterotroph, concentrations that exert equivalent effects can be compared. The compounds under study may belong to the same biogenetic class or to different classes of secondary metabolites. Nayar and Thorsteinson (1963), for example, tested examples of a single class (glucosinolates) on *Plutella maculipennis* and found that compounds within this class have varying behavioral and toxicological effects. Schoonhoven (1967, 1969) and Wieczorek (1976), also investigating the effects of glucosinolates, similarly observed varying behavioral and electrophysiological responses from *Pieris* and *Mamestra* caterpillars, respectively. Data on compounds in different classes were obtained by Janzen *et al.* (1977) for the seed-eating larvae of the bruchid beetle *Callosobruchus maculatus*, which were reared on diets adulterated with a variety of secondary compounds. Similar data for other insects and for fungal pathogens are available from a variety of sources; for example, a number of different secondary compounds have been tested on the southern armyworm *Spodoptera* (=*Prodenia*) *eridania* (e.g., Soo Hoo and Fraenkel, 1964; Rehr *et al.*, 1973a,c; Burnett *et al.*, 1974, 1977; Berenbaum, 1978). Comparisons of plants or diets containing more than one chemical defense would provide an opportunity to consider synergism between compounds of the same or different classes (e.g., Eisner and Halpern, 1971; Rehr *et al.*, 1973a; Janzen *et al.*, 1977). This pharmacological approach thus permits us to assess the potential efficacy of a broad range of compounds, and different concentrations of these, as chemical defenses against specified herbivores and pathogens. Its primary advantage is that it permits comparison of plants with very different chemical phenotypes. However, to avoid the possibility of erroneous extrapolation from laboratory experiments (see, for example, Burnett *et al.*, 1977), we must assess benefit in an ecological context.

Scaling benefit against cost in an ecologically relevant context requires relating variability of herbivore or pathogen activity within plant populations to variability among chemical phenotypes. Many studies have documented variation in herbivore activity on plants within natural populations (e.g., Dethier, 1959; Ralph, 1977) as well as among plants of different co-occurring species (e.g., Singer, 1971; Chew, 1977). Likewise, quantitative and qualitative variation in secondary compounds is well documented for natural plant populations [e.g., for alkaloids: Alston (1967), Dolinger *et al.* (1973); for cardenolides: Duffey and Scudder (1972); for cyanide production: Daday (1954), Jones (1966, 1977), Cooper-Driver and Swain (1976); for flavonoids: Alston (1967); for glucosinolates: Al-Shehbaz (1973), Rodman (1974); for terpenoids: Irving and Adams (1973), Thorin and Nommik (1974), von Rudloff (1975)]. However, the relationship between variation in heterotroph activity and allelochemical variability among individual plants is not clear. Examination of some of these chemical polymorphisms in field and laboratory situations reveals a correlation with heterotroph attack (e.g., Jones, 1962, 1966; Crawford-Sidebotham, 1972; Angseesing and Angseesing, 1973; Dolinger *et al.*, 1973; Whitman, 1973; Angseesing, 1974; Cates, 1975; Greenhalgh and Mitchell, 1976; Cooper-Driver *et al.*, 1977; Hare, 1977). In well-studied cases, however, such as those involving cyanogenesis in *Lotus* and *Trifolium* (Leguminosae), climatic and edaphic factors may also contribute to maintenance of the polymorphism (Armstrong *et al.*, 1912; Daday, 1965; Bishop and Korn, 1969; Jones, 1970, 1972, 1973; Foulds and Grime, 1972; de Araujo, 1976; Ellis *et al.*, 1976, 1977a,b; Foulds, 1977). Thus far, no clear picture has emerged of how chemical variation in natural plant populations relates to variation in susceptibility to heterotroph attack. At issue here are the problems of determining fitnesses in natural populations (Lewontin, 1974; pp. 19ff.) and, particularly with plant populations, of recognizing the appropriate individual (Harper, 1977) and of distinguishing phenotypic plasticity from genotypic variability.

The widespread occurrence of quantitative and qualitative chemical variation within natural plant populations provides a spectrum of biosynthetic costs comparable to the range of variation that might be used to construct a pharmacological dose–response curve. If we scale the differential effects of chemical phenotypes on specified heterotrophs in units presumed to reflect plant fitness, we can, in principle at least, examine the form of the relationship between allelochemical variation and heterotroph damage. We can visualize three forms of this relationship, shown by the three curves in Fig. 5. These do not exhaust the possible relationships between biosynthetic cost of a defense (or, for example, concentration of a particular secondary compound as a measure of cost) and its efficacy against heterotrophs. Nonetheless, we feel that a quantitative approach to the interactions of plants and heterotrophs should help us to understand the extent to which variances in cost and benefit influence the direction and intensity of natural selection in heterotroph-stressed plant populations.

The first hypothetical relationship (curve A) is a linear one, in which

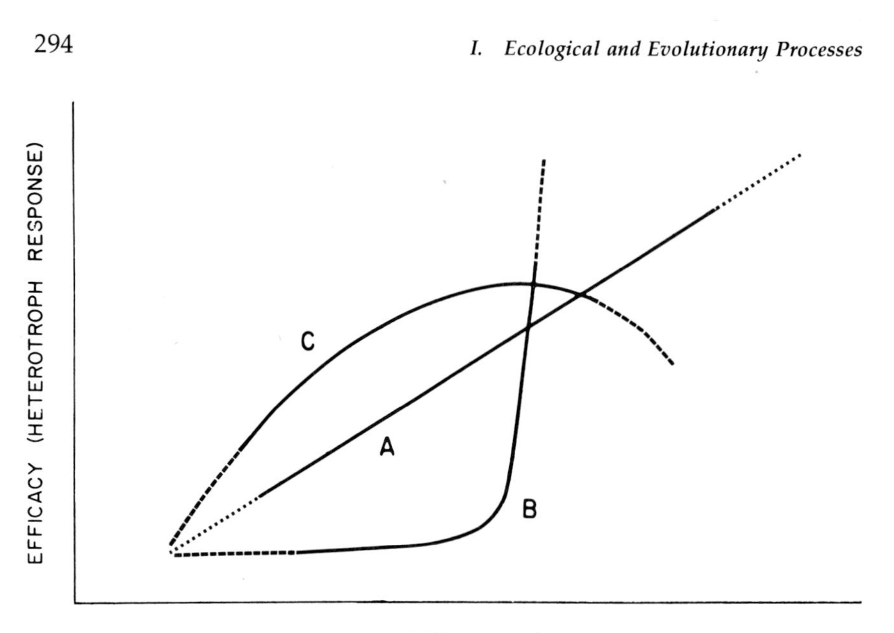

Figure 5. Three hypothetical scaling relationships between the biosynthetic cost of chemical defense and its efficacy (see text). End points of curves are deliberately left ambiguous.

increased plant allocation (increased biosynthetic cost) is associated with a proportional increase in benefit, that is, efficacy against heterotrophs. If cost is expressed in units of concentration of a particular secondary compound, this linear relation is consistent with the known effects of a number of secondary compounds on insect and mammalian herbivores. Williams *et al.* (1971), for example, found a nearly linear relation between alkaloid content and palatability to sheep of reed canary grass; Wilson and Shaver (1973) reported a linear relation between gossypol concentrations in cotton seedlings and damage by tobacco budworm larvae; and Long *et al.* (1977) observed a linear relation between hydroxamic acid concentrations in maize and infestation intensity of the corn leaf aphid.

Curve B represents a second hypothetical relationship between cost and benefit, in which a threshold effect is evident (the nature of the end points of all the curves is deliberately left ambiguous). Our interpretation of such a relationship, in which some plants incur a small cost without any compensatory benefit being realized, would include cases in which individual plants lack part of a biosynthetic pathway or produce suboptimal levels of an allelochemical. Acyanogenic *Lotus* plants may, for example, lack the requisite hydrolyzing enzyme(s) but incur all the biosynthetic costs of producing the cyanophoric molecule (cf. Jones, 1972, 1973). Threshold effects for concentrations of various secondary compounds are known for a number of heterotrophs. Lichtenstein *et al.* (1962), for example, reported such threshold concentrations for the toxic effects of 2-phenylethyl isothiocyanate on the Mexican bean beetle. Erickson and Feeny (1974) similarly observed a threshold for the effects of sinigrin (allylglucosinolate) on *Papiliopolyxenes.*

The third hypothetical relationship we consider, graphed as the ditonic curve C, is perhaps the most speculative. Here, increasing cost confers increasing benefit up to some optimal level, and further increases (e.g., greater concentrations of a secondary compound) result in a decline in resistance to heterotroph attack. Drug or spice plants or other crops that have increased levels of secondary compounds due to artificial selection (e.g., Tetenyi, 1970; Hegnauer, 1975) may provide examples for such a ditonic relationship. One mechanism by which such crop plants might incur heavier heterotroph loads is stimulation of increased activity by specialist herbivores due to higher levels of attractant compounds (e.g., Smith, 1966; Nault and Styer, 1972; Mitchell, 1977).

It is presently unclear whether natural plant populations, in their interactions with particular heterotrophs, conform to any of these relationships. One example that we have already suggested for relationship B might be *Lotus corniculatus* populations polymorphic for cyanogenesis, in which acyanogenic phenotypes would appear to require some form of balancing selection to compensate for their susceptibility to herbivore attack (cf. Jones, 1972, 1973). The problem depends in part on whether particular classes of secondary compounds are heterotroph dependent or independent in their effects and thus whether particular classes are likely to mediate particular scaling relationships. Some presumed allelochemicals may have similar effects on a wide variety of heterotrophs. Feeny (1975, 1976) and Rhoades and Cates (1976), for example, argue that tannins function as "digestibility-reducing" agents (Rhoades and Cates, 1976) against numerous heterotrophs. Assuming that tannin effects are so general, it remains to be demonstrated what selective mechanisms, if any, account for the chemical variability and taxonomic diversity of tannins (Swain, 1965; Sarkar *et al.*, 1976). In contrast, the effects of many allelochemicals are likely to be heterotroph dependent. Certain classes such as glucosinolates and their hydrolysis products (Gmelin and Virtanen, 1960; Benn, 1977) may exhibit both "digestibility-reducing" and "toxic" effects (Rhoades and Cates, 1976). For the glucosinolates, data of a number of workers demonstrate that many of the various behavioral, electrophysiological, and toxicological effects of these secondary compounds are heterotroph specific (Lichtenstein *et al.*, 1962; Nayar and Thorsteinson, 1963; David and Gardiner, 1966; Schoonhoven, 1967, 1969; Van Emden and Bashford, 1969; Nair *et al.*, 1976; Wieczorek, 1976). Furthermore, the data of Lichtenstein *et al.* (1962) show that whether linear or threshold effects are observed in dose–response experiments depends on the particular heterotroph. The heterotroph specificity of the effects of these and possibly many other secondary compounds must condition attempts to predict changes in a plant's chemical arsenal. Thus, an estimate of benefit for a specific plant population must include field studies to determine what heterotrophs attack the plant population and how allelochemical variation is related to patterns of heterotroph activity. These data, together with pharmacological data from laboratory tests, would permit us to assess the efficacy of a particular compound or suite of compounds as defense(s) and to predict how heterotrophs that presently attack the plant population might respond to specific quantitative and qualitative changes in the plant's allelochemical array. In principle, these considerations would permit us to predict the kinds of changes that would be selectively advantageous to members of the plant population.

IV. GENETIC BACKGROUND TO ALLELOCHEMICAL DIVERSIFICATION

We recognize that plant defenses against herbivores and pathogens may involve physical and biotic components as well as allelochemical agents (e.g., Janzen, 1966; Gilbert, 1971; Levin, 1973; Johnson, 1975; Hocking, 1975; Rathcke and Poole, 1975; Pillemer and Tingey, 1976; Bentley, 1977). Many plant–heterotroph interactions, perhaps most, involve a combination of morphological, chemical, and phenological factors (e.g., Breedlove and Ehrlich, 1968; Tahvanainen, 1972; Johnson and Beard, 1977). However, by mediating the specificity of many plant–heterotroph interactions (Dethier, 1954, 1970), secondary compounds influence the direction and intensity of natural selection in plant populations. An understanding of the dynamics of these interactions would allow us to predict the probabilities of various changes in these systems. We are far from that ideal, although a number of workers have speculated on parameters in addition to chemical aspects of defense that influence these coevolutionary interactions. For example, the relationship of life history and successional status to various strategies of heterotroph avoidance has been explored (Cates and Orians, 1975; Feeny, 1975, 1976; Futuyma, 1976; Rhoades and Cates, 1976).

Community-level interactions among plant species with similar chemical defenses have been discussed by Tahvanainen and Root (1972), Janzen (1973c), Atsatt and O'Dowd (1976), and Cates and Rhoades (1977). Whatever the pattern of heterotroph exploitation of co-occurring plant populations (e.g., Singer, 1971; Chew, 1977), the probability of change in plant populations will depend at least in part on genetic mechanisms for generating phenotypic variability. In this concluding section, we attempt a limited consideration of the genetic background of changes in chemical defense and the relation of these mechanisms to plant resources for the production of secondary compounds.

We envision three general ways in which the chemical defenses of a plant might be modified: (1) quantitative change in levels of a compound already present; (2) qualitative change involving addition or substitution of a compound belonging to a biogenetic class already present; (3) qualitative change involving addition or substitution of a compound belonging to a novel class.

Quantitative variation for presumptive allelochemicals is widespread among individuals in a population [e.g., for glucosinolates, Josefsson (1970b); Al-Shehbaz (1973); Van Etten *et al.* (1976)] and among populations within species [e.g., for glucosinolates, Al-Shehbaz (1973); Rodman (1974, 1976)]. Although much variation may well result from phenotypic response to environmental influences (e.g., edaphic conditions), quantitative variation can reflect genetic differences. For example, Bazzaz *et al.* (1975) documented ecotypic variation in cannabinoid content for tropical and temperate populations of *Cannabis sativa* (Cannabaceae). The number of genetic differences involved in such quantitative variation may be small. For a different class of secondary compounds, for example, Wilson and Shaver (1973) showed that single allele substitutions alter quantities of gossypol in cotton plants (Malvaceae).

Individuals within natural and cultivated plant populations often contain arrays of biosynthetically related secondary compounds. Such qualitative variation has been observed for alkaloids (e.g., Dolinger *et al.*, 1973), certain biosynthetically related nonprotein amino acids (Fowden, 1970, 1973), cyanogenic glycosides (e.g., Butler, 1965), glucosinolates (Joseffson, 1970b; Rodman, 1976), monoterpenes (e.g., Irving and Adams, 1973; Thorin and Nommik, 1974) and other terpenoids (e.g., Zavarin and Cobb, 1970; von Rudloff, 1975), and sesquiterpene lactones (e.g., Mabry, 1970). Where such variation has been examined statistically (e.g., Zavarin and Cobb, 1970) or genetically, the results suggest that the number of loci involved in producing such diversity within a biogenetic class is small (e.g., Walker and Stahmann, 1955; Flor, 1956, 1971; Schwochau and Hadwiger, 1970; Irving and Adams, 1973; Barker and Hovin, 1974; Hegnauer, 1975; Harlan, 1976b; Dewick, 1977; cf., Da Costa and Jones, 1971; Hughes and Conn, 1976). Thus, diversification within a biogenetic class is likely to be a primary response of plant populations to heterotroph attack (cf., Harper, 1977, p. 414), particularly attack by specialized feeders, which exert persistent and consistent selective effects (Root, 1973).

Many plants characteristically produce a variety of secondary compounds in several biogenetic classes (Hegnauer, 1962–1973; Gibbs, 1974). This diversity may well have evolved under the influence of a succession of heterotroph species (Janzen, 1968, 1973a). Within plant families that exhibit a characteristic spectrum of secondary metabolites (cf. Ehrlich and Raven, 1965), one may find species or genera that contain compounds not typical of the family. For example, a few genera of the family Cruciferae, which characteristically produce glucosinolates (Ettlinger and Kjaer, 1968), also produce cardenolides [e.g., *Erysimum* (Gmelin and Bredenberg, 1966)] or cucurbitacins [e.g., *Iberis* (Curtis and Meade, 1971)]. Such genera often escape attack by heterotrophs that specialize on plants in this family (for insect herbivores of Cruciferae, see, e.g., Feeny *et al.*, 1970; Hicks and Tahvanainen, 1974; Chew, 1975; Feeny, 1977; Nielsen *et al.*, 1977; Nielsen, 1978; but see Thorsteinson, 1953). Unfortunately, we do not know the magnitude of genetic differences between these taxa and others whose chemistry is more typical of the Cruciferae.

Several lines of evidence suggest that relatively simple genetic changes can produce qualitative changes in plant secondary compounds. First, the appearance of novel compounds may result from the accumulation of biosynthetic intermediates. Certain novel flavonoids in *Phlox* allotetraploids have been shown to be accumulations of biosynthetic intermediates of end products found in the parental species (Levy and Levin, 1971, 1974). Second, novel compounds may arise by enzyme complementation in hybrids, resulting in a new or extended biosynthetic pathway (Levy and Levin, 1975). Third, novel compounds may already be present in the plant in trace amounts, as documented for nonprotein amino acids by Fowden (1972) in commercial-scale preparations of beets (Chenopodiaceae). These compounds may represent unaccumulated biosynthetic intermediates [as in the case of novel flavonoids found by Levy and Levin (1971, 1974)] or rapidly degraded compounds [Mothes, 1965, cited by Robinson (1968, 1974)]. More interestingly, these novel compounds may result from low substrate specificity of enzymes involved in anabolic processes (Hegeman and Rosenberg, 1970; Jensen, 1976). Hegeman and Rosenberg (1970) provide numerous examples of this phenomenon in bacteria, in which such "substrate ambiguity" (Jensen, 1976) generates small quantities of novel compounds. If small amounts of plant metabolites produced in this way were fortuitously to increase plant fitness, selection might result in increased production of these compounds. Genetic changes underlying such increases might include (1) specific gene derepression, as may occur in phytoalexin production (Keen, 1975; Dewick, 1977); (2) regulatory changes including position effects (Wilson, 1976); and (3) modifications of enzymes that increase substrate specificity (e.g., Stanier, 1968; Ornston, 1971).

If mechanisms to generate variability in plant secondary compounds are relatively simple, why then do individual plants and plant species contain only a fraction of the extremely diverse array of secondary compounds found in nature? Very probably, some forms of defense have replaced others, as has been argued for chemical defenses in flowering plants (e.g., Gardner, 1977) and for biotic defenses (e.g., Rehr *et al.*, 1973b). The existence of defense cost constraints on plant populations is consonant with the observation that plants may rapidly lose resistance to their fungal pathogens when grown in environments free from them (Harlan, 1976a; cf., Janzen, 1973b, 1975), just as resistant pest insects may lose resistance when no longer exposed to a pesticide (e.g., Georghiou, 1972). Furthermore, as a number of authors have suggested, plant resistance to heterotrophs may involve population or community-level patterning of secondary compounds rather than the continual addition of new compounds in a chemical arms race (Jones, 1968; Dolinger *et al.*, 1973; Atsatt and O'Dowd, 1976; Cates and Rhoades, 1977; J. E. Rodman and F. S. Chew, unpublished data).

We hope to have demonstrated that the first-order response of plant populations under heterotroph stress is likely to be quantitative and/or qualitative diversification of existing allelochemicals. But plant secondary compounds cannot be considered in isolation from other defense mechanisms with which they interact in physiological and ecological contexts. We suggest that future work to calculate and compare the costs of plant defense against heterotrophs might well focus on three areas: (1) examination of the impact of plant secondary compounds on heterotroph activity in a quantitative manner; (2) examination of the significance of allelochemical variation in populations or communities of cultivated and wild plants under field conditions; and (3) examination of the costs associated with multiple defense functions in a plant, including physical defenses and maintenance of biotic defenders (e.g., Hocking, 1975). We are far from predicting changes in plant–heterotroph systems with any accuracy; however, we trust that data from work in these areas will make it possible to assess the magnitude of allelochemical differences among genotypes in plant populations (Jones, 1971; Harlan, 1976a) and of genetic differences among conspecific populations subjected to different degrees of heterotroph-related stress (e.g., Janzen, 1973b, 1975). That knowledge would then permit us to identify factors that influence the intensity, direction, and consistency of natural selection.

ACKNOWLEDGMENTS

We are grateful to L. N. Ornston, R. S. Feldberg, and S. Long for advice on biochemical matters, to M. Glenn, J. D. Hare, and R. Nakamura for helpful discussion and comments on an early draft, and to J. A. Chamberlain for drawing the figures. We thank the Andrew W. Mellon Foundation (to FSC through Tufts) and the National Science Foundation (BMS75-03311 and DEB78-11124 to JER) for support.

REFERENCES

Allen, E. J., and Morgan, D. G. (1975). *J. Agric. Sci.* **85,** 159–174.
Al-Shehbaz, I. A. (1973). *Contrib. Gray Herb., Harv. Univ.* **204,** 3–148.
Alston, R. E. (1967). *Evol. Biol.* **1,** 197–305.
Angseesing, J. P. A. (1974). *Heredity* **32,** 73–83.
Angseesing, J. P. A., and Angseesing, W. J. (1973). *Heredity* **31,** 276–282.
Armstrong, H. E., Armstrong, E. F., and Horton, E. (1912). *Proc. R. Soc. London, Ser. B* **84,** 471–484.
Atabekov, J. G. (1975). *Annu. Rev. Phytopathol.* **13,** 127–145.
Atsatt, P. R., and O'Dowd, D. J. (1976). *Science* **193,** 24–29.
Barker, R. E., and Hovin, A. W. (1974). *Crop Sci.* **14,** 50–53.
Bassham, J. A. (1977). *Science* **197,** 630–638.
Bazzaz, F. A., Dusek, D., Seigler, D. S., and Haney, A. W. (1975). *Biochem. Syst. Ecol.* **3,** 15–18.
Beck, S. D., and Reese, J. C. (1976). *Recent Adv. Phytochem.* **10,** 41–92.
Begum, A., and Eden, W. G. (1965). *J. Econ. Entomol.* **58,** 591–592.
Bell, E. A. (1972). *In* "Phytochemical Ecology" (J. B. Harborne, ed.), pp. 163–167. Academic Press, New York.
Benn, M. (1977). *Pure Appl. Chem.* **49,** 197–210.
Bentley, B. L. (1977). *Annu. Rev. Ecol. Syst.* **8,** 407–427.
Berenbaum, M. (1978). *Science* **201,** 532–534.
Bishop, J. A., and Korn, M. E. (1969). *Heredity* **24,** 423–430.
Bjork, S. (1967). *Folia Limnol. Scand.* **14,** 1–248.
Björkman, R. (1976). *In* "The Biology and Chemistry of the Cruciferae" (J. G. Vaughan, A. J. MacLeod, and B. M. G. Jones, eds), pp. 191–205. Academic Press, New York.
Blum, A., and Ebercon, A. (1976). *Crop Sci.* **16,** 428–431.
Boutton, T. W., Cameron, G. N., and Smith, B. N. (1978). *Oecologia* **36,** 21–32.
Boyle, J. E., and Fowden, L. (1971). *Phytochemistry* **10,** 2671–2678.
Bracken, G. K., and Bucher, G. E. (1977). *J. Econ. Entomol.* **70,** 701–705.

Breedlove, D. E., and Ehrlich, P. R. (1968). *Science* **162,** 671–672.
Breedlove, D. E., and Ehrlich, P. R. (1972). *Oecologia* **10,** 99–104.
Burnett, W. C., Jones, S. B., Mabry, T., and Padolina, W. G. (1974). *Biochem. Syst. Ecol.* **2,** 25–29.
Burnett, W. C., Jones, S. B., and Mabry, T. J. (1977). *Plant Syst. Evol.* **128,** 277–286.
Butler, G. W. (1965). *Phytochemistry* **4,** 127–131.
Caldwell, B. E., ed. (1973). "Soybeans: Improvement, Production, and Uses," Agronomy, No. 16. Am. Soc. Agron., Madison, Wisconsin.
Caldwell, M. M., White, R. S., Moore, R. T., and Camp, L. B. (1977). *Oecologia* **29,** 275–300.
Caswell, H., Reed, F., Stephenson, S. N., and Werner, P. A. (1973). *Am. Nat.* **107,** 465–480.
Cates, R. G. (1975). *Ecology* **56,** 391–400.
Cates, R. G., and Orians, G. H. (1975). *Ecology* **56,** 410–418.
Cates, R. G., and Rhoades, D. F. (1977). *Biochem. Syst. Ecol.* **5,** 185–193.
Chaplin, J. F., and Weeks, W. W. (1976). *Crop Sci.* **16,** 416–418.
Chew, F. S. (1975). *Oecologia* **20,** 117–127.
Chew, F. S. (1977). *Evolution* **31,** 568–579.
Conn, E. E. (1973). *Biochem. Soc. Symp.* **38,** 277–302.
Cooper-Driver, G. A., and Swain, T. (1976). *Nature (London)* **260,** 604.
Cooper-Driver, G. A., Finch, S., and Swain, T. (1977). *Biochem. Syst. Ecol.* **5,** 177–183.
Crawford-Sidebotham, T. J. (1972). *Heredity* **28,** 405–411.
Cromwell, B. T. (1965). *In* "Biosynthetic Pathways in Higher Plants" (J. B. Pridham and T. Swain, eds.), pp. 147–157. Academic Press, New York.
Cruickshank, I. A. M. (1963). *Annu. Rev. Phytopathol.* **1,** 351–374.
Curtis, P. J., and Meade, P. M. (1971). *Phytochemistry* **10,** 3081–3083.
Da Costa, C. P., and Jones, C. M. (1971). *Science* **172,** 1145–1146.
Daday, H. (1954). *Heredity* **8,** 61–78.
Daday, H. (1965). *Heredity* **20,** 355–366.
Dagley, S., and Nicholson, D. E. (1970). "An Introduction to Metabolic Pathways." Wiley, New York.
David, W. A. L. and Gardiner, B. O. (1966). *Entomol. Exp. Appl.* **9,** 247–255.
Davidson, J. L. (1964). *Aust. J. Agric. Res.* **16,** 721–731.
de Araujo, A. M. (1976). *Heredity* **37,** 291–293.
Del Moral, R. (1972). *Oecologia* **9,** 289–300.
Dement, W. A., and Mooney, H. A. (1974). *Oecologia* **15,** 65–76.
Dethier, V. G. (1954). *Evolution* **8,** 33–54.
Dethier, V. G. (1959). *Can. Entomol.* **91,** 581–596.
Dethier, V. G. (1970). *In* "Chemical Ecology" (E. Sondheimer and J. B. Simeone, eds.), pp. 83–102. Academic Press, New York.
Dewick, P. M. (1977). *Phytochemistry* **16,** 93–97.
Dolinger, P. M., Ehrlich, P. R., Fitch, W. L., and Breedlove, D. E. (1973). *Oecologia* **13,** 191–204.
Duffey, S. S., and Scudder, G. G. E. (1972). *J. Insect Physiol.* **18,** 63–78.
Dyer, M. I., and Bokhari, U. G. (1976). *Ecology* **57,** 762–772.
Ehleringer, J., Bjorkman, O., and Mooney, H. A. (1976). *Science* **192,** 376–377.
Ehrlich, P. R., and Raven, P. H. (1965). *Evolution* **18,** 586–608.
Eisner, T., and Halpern, B. P. (1971). *Science* **172,** 1362.
Elliott, M. C., and Stowe, B. B. (1971). *Plant Physiol.* **48,** 498–503.
Ellis, W. M., Keymer, R. J., and Jones, D. A. (1976). *Heredity* **36,** 245–252.
Ellis, W. M., Keymer, R. J., and Jones, D. A. (1977a). *Heredity* **38,** 339–347.

Ellis, W. M., Keymer, R. J., and Jones, D. A. (1977b). *Heredity* **39,** 45–66.
Erickson, J. M., and Feeny, P. (1974). *Ecology* **55,** 103–111.
Ettlinger, M., and Kjaer, A. (1968). *Recent Adv. Phytochem.* **1,** 60–144.
Evans, L. T. (1975). *In* "Photosynthesis and Productivity in Different Environments" (J. P. Cooper, ed.), IBP Synth. Vol. 3, pp. 501–507. Cambridge Univ. Press, London and New York.
Feeny, P. (1975). *In* "Coevolution of Animals Plants" (L. E. Gilbert and P. H. Raven, eds.), pp. 3–19. Univ. of Texas Press, Austin.
Feeny, P. (1976). *Recent Adv. Phytochem.* **10,** 1–40.
Feeny, P. (1977). *Ann. Mo. Bot. Gard.* **64,** 221–234.
Feeny, P., Paauwe, K. L., and Demong, N. J. (1970). *Ann. Entomol. Soc. Am.* **63,** 832–841.
Flor, H. H. (1956). *Adv. Genet.* **8,** 29–54.
Flor, H. H. (1971). *Annu. Rev. Phytopathol.* **9,** 275–296.
Foulds, W. (1977). *Heredity* **39,** 219–234.
Foulds, W., and Grime, J. P. (1972). *Heredity* **28,** 143–146.
Fowden, L. (1970). *Prog. Phytochem.* **2,** 203–266.
Fowden, L. (1972). *Phytochemistry* **11,** 2271–2276.
Fowden, L. (1973). *In* "Biosynthesis and Its Control in Plants" (B. V. Milborrow, ed.), pp. 323–339. Academic Press, New York.
Fowden, L., and Mazelis, M. (1971). *Phytochemistry* **10,** 359–365.
Fraenkel, G. (1959). *Science* **129,** 1466–1470.
Freeman, G. G., and Mossadeghi, N. (1970). *J. Sci. Food Agric.* **21,** 610–615.
Freeman, G. G., and Mossadeghi, N. (1971). *J. Sci. Food Agric.* **22,** 330–334.
Freeman, G. G., and Mossadeghi, N. (1972a). *J. Sci. Food Agric.* **23,** 387–402.
Freeman, G. G., and Mossadeghi, N. (1972b). *J. Sci. Food Agric.* **23,** 1335–1345.
Freeman, G. G., and Mossadeghi, N. (1972c). *J. Hortic. Sci.* **47,** 375–387.
Friend, J. (1976). *In* "Biochemical Aspects of Plant-Parasite Relationships" (J. Friend and D. R. Threlfall, eds.), pp. 291–303. Academic Press, New York.
Futuyma, D. J. (1976). *Am. Nat.* **110,** 285–292.
Gardner, R. Y. (1977). *Biochem. Syst. Ecol.* **5,** 29–35.
Georghiou, G. P. (1972). *Annu. Rev. Ecol. Syst.* **3,** 133–168.
Gibbs, R. D. (1974). "Chemotaxonomy of Flowering Plants," 4 vols. McGill-Queen's Univ. Press, Montreal.
Gilbert, L. E. (1971). *Science* **172,** 585–586.
Gmelin, V. R., and Bredenberg, J. B. (1966). *Arzneim.-Forsch.* **16,** 123–127.
Gmelin, V. R., and Virtanen, A. I. (1960). *Acta Chem. Scand.* **14,** 507–510.
Green, T. R., and Ryan, C. A. (1972). *Science* **175,** 776–777.
Greenhalgh, J. R., and Mitchell, N. D. (1976). *New Phytol.* **77,** 391–398.
Hahn, S. K. (1977). *Crop Sci.* **17,** 559–562.
Hampton, R. O. (1975). *Phytopathology* **65,** 1342–1346.
Harborne, J. B. (1977). *Prog. Phytochem.* **4,** 189–208.
Hare, J. D. (1977). Ph.D. Dissertation, State University of New York, Stony Brook.
Harlan, J. R. (1976a). *Annu. Rev. Phytopathol.* **14,** 31–51.
Harlan, J. R. (1976b). *Crop Sci.* **16,** 330–333.
Harper, J. L. (1977). "Population Biology of Plants." Academic Press, New York.
Harris, P. (1972). *Symp. R. Entomol. Soc. London,* **6,** 201–209.
Harris, P., and Mohyuddin, A. I. (1965). *Can. Entomol.* **97,** 830–833.
Heftmann, E. (1977). *Prog. Phytochem.* **4,** 257–276.
Hegeman, G. D., and Rosenberg, S. L. (1970). *Annu. Rev. Microbiol.* **24,** 429–462.
Hegnauer, R. (1962–1973). "Chemotaxonomie der Pflanzen," 6 vols. Birkhaeuser, Basel.

Hegnauer, R. (1975). *In* "Crop Genetic Resources for Today and Tomorrow" (O. H. Frankel and J. G. Hawkes, eds.), pp. 249–265. Cambridge Univ. Press, London and New York.

Hewitt, E. J. (1975). *Annu. Rev. Plant Physiol.* **26,** 73–100.

Hickman, J. C. (1975). *J. Ecol.* **63,** 689–701.

Hickman, J. C., and Pitelka, L. F. (1975). *Oecologia* **21,** 117–121.

Hicks, K. L., and Tahvanainen, J. O. (1974). *Am. Midl. Nat.* **91,** 406–423.

Ho, L. C. (1976). *Ann. Bot. (London)* [N.S.] **40,** 163–165.

Hocking, B. (1975). *In* "Coevolution of Animals and Plants" (L. E. Gilbert and P. H. Raven, eds.), pp. 78–90. Univ. of Texas Press, Austin.

Holler, L. C., and Abrahamson, W. G. (1977). *Am. J. Bot.* **64,** 1003–1007.

Hughes, M. A., and Conn, E. E. (1976). *Phytochemistry* **15,** 697–701.

Ikeda, T., Matsumura, F., and Benjamin, D. M. (1977). *Science* **197,** 497–499.

Irving, R. S., and Adams, R. P. (1973). *Recent Adv. Phytochem.* **6,** 187–214.

Jager, H. J., and Meyer, H. R. (1977). *Oecologia* **30,** 83–96.

James, W. C., Callbeck, L. C., Hodgson, W. A., and Shih, C. S. (1971). *Phytopathology* **61,** 1471–1476.

Jameson, D. A. (1963). *Bot. Rev.* **29,** 532–594.

Janzen, D. H. (1966). *Evolution* **20,** 249–275.

Janzen, D. H. (1968). *Am. Nat.* **102,** 592–595.

Janzen, D. H. (1973a). *Am. Nat.* **107,** 786–790.

Janzen, D. H. (1973b). *Biotropica* **5,** 15–28.

Janzen, D. H. (1973c). *Pure Appl. Chem.* **34,** 529–538.

Janzen, D. H. (1975). *Science* **189,** 145–147.

Janzen, D. H. (1976a). *Am. Midl. Nat.* **95,** 474–478.

Janzen, D. H. (1976b). *Ecology* **57,** 826–828.

Janzen, D. H., Juster, H. B., and Bell, E. A. (1977). *Phytochemistry* **16,** 223–237.

Jensen, R. A. (1976). *Annu. Rev. Microbiol.* **30,** 409–425.

Johnson, A. L., and Beard, B. H. (1977). *Crop Science* **17,** 369–372.

Johnson, H. B. (1975). *Bot. Rev.* **41,** 233–258.

Jones, D. A. (1962). *Nature (London)* **193,** 1109–1110.

Jones, D. A. (1966). *Can. J. Genet. Cytol.* **8,** 556–567.

Jones, D. A. (1968). *Heredity* **23,** 453–455.

Jones, D. A. (1970). *Heredity* **25,** 633–641.

Jones, D. A. (1971). *Science* **173,** 945.

Jones, D. A. (1972). *In* "Phytochemical Ecology" (J. B. Harborne, ed.), pp. 103–124. Academic Press, New York.

Jones, D. A. (1973). *In* "Taxonomy and Ecology" (V. H. Heywood, ed.), pp. 213–242. Academic Press, New York.

Jones, D. A. (1977). *Heredity* **39,** 27–44.

Jones, D. A. (1979). *Am. Nat.* **113,** 445–451.

Josefsson, E. (1970a). *J. Sci. Food Agric.* **21,** 98–103.

Josefsson, E. (1970b). "Pattern, Content, and Biosynthesis of Glucosinolates in Some Cultivated Cruciferae." Swedish Seed Assoc., Lund.

Joshi, M. M., and Hollis, J. P. (1977). *Science* **195,** 179–180.

Keen, N. T. (1975). *Science* **187,** 74–75.

Kirby, G. W. (1967). *Science* **155,** 170–173.

Kjaer, A. (1960). *Prog. Chem. Org. Nat. Prod.* **18,** 122–176.

Kjaer, A. (1974). *In* "Chemistry in Botanical Classification" (G. Bendz and J. Santesson, eds.), pp. 229–234. Academic Press, New York.

Kjaer, A. (1977). *Pure Appl. Chem.* **49,** 137–152.
Kjaer, A., and Larsen, P. O. (1973). *Biosynthesis* **2,** 71–105.
Kjaer, A., and Larsen, P. O. (1976). *Biosynthesis* **4,** 179–203.
Kuć, J., Currier, W. W., and Shih, M. J. (1976). *In* "Biochemical Aspects of Plant-Parasite Relationships" (J. Friend and D. R. Threlfall, eds.), pp. 225–237. Academic Press, New York.
Kugler, J. L., Kehr, W. R., and Ogden, R. L. (1977). *Crop Sci.* **17,** 621–624.
Kulman, H. M. (1971). *Annu. Rev. Entomol.* **16,** 289–324.
Kutáček, M., and Kefeli, V. I. (1968). *In* "Biochemistry and Physiology of Plant Growth Substances" (F. Wightman and G. Setterfield, eds), pp. 127–152. Runge Press, Ottawa.
Lange, O. L., and Zuber, M. (1977). *Oecologia* **31,** 67–72.
Lee, S., and Aronoff, S. (1967). *Science* **158,** 789–799.
Lehman, R. H., and Rice, E. L. (1972). *Am. Midl. Nat.* **87,** 71–80.
Lehninger, A. (1971). "Bioenergetics," 2nd ed. Benjamin, Menlo Park, California.
Lehninger, A. (1975). "Biochemistry," 2nd ed. Worth, New York.
Levin, D. A. (1973). *Q. Rev. Biol.* **48,** 3–15.
Levin, D. A. (1976). *Annu. Rev. Ecol. Syst.* **7,** 121–159.
Levy, M., and Levin, D. A. (1971). *Proc. Natl. Acad. Sci. U.S.A.* **68,** 1627–1630.
Levy, M., and Levin, D. A. (1974). *Am. J. Bot.* **61,** 156–167.
Levy, M., and Levin, D. A. (1975). *Evolution* **29,** 487–499.
Lewontin, R. C. (1974). "The Genetic Basis of Evolutionary Change." Columbia Univ. Press, New York.
Lichtenstein, E. P., Strong, F. M., and Morgan, D. G. (1962). *J. Agric. Food Chem.* **10,** 30–33.
Liener, I. E. (1976). *Annu. Rev. Plant Physiol.* **27,** 291–319.
Long, B. J., Dunn, G. M., Bowman, J. S., and Routley, D. G. (1977). *Crop Sci.* **17,** 55–58.
Loomis, R. S., Ng, E., and Hunt, W. F. (1976). *In* "CO_2 Metabolism and Plant Productivity" (R. H. Burris and C. C. Black, eds.), pp. 269–287. Univ. Park Press, Baltimore, Maryland.
Lucas, D., and Asano, R. D. (1968). *Physiol. Plant.* **21,** 1217–1223.
Luckner, M. (1972). "Secondary Metabolism in Plants and Animals." Chapman & Hall, London.
Mabry, T. J. (1970). *In* "Phytochemical Phylogeny" (J. B. Harborne, ed.), pp. 269–300. Academic Press, New York.
McClure, J. W. (1975). *In* "The Flavonoids" (J. B. Harborne, T. J. Mabry, and H. Mabry, eds.), Part 2, pp. 970–1055. Academic Press, New York.
McCree, K. J. (1976). *In* "CO_2 Metabolism and Plant Productivity" (R. H. Burris and C. C. Black, eds.), pp. 177–184. Univ. Park Press, Baltimore, Maryland.
Maggs, D. H. (1964). *J. Exp. Bot.* **15,** 574–584.
Maggs, D. H. (1965). *J. Exp. Bot.* **16,** 387–404.
Mahadevan, S. (1973). *Annu. Rev. Plant Physiol.* **24,** 69–88.
Marsh, N., and Rothschild, M. (1974). *J. Zool.* **174,** 89–122.
Mathys, W. (1977). *Physiol. Plant.* **40,** 130–136.
Mattson, W. J., and Addy, N. D. (1975). *Science* **190,** 515–522.
Matzinger, D. F., Wernsman, E. A., and Cockerham, C. C. (1972). *Crop Sci.* **12,** 40–43.
Maun, M. A., and Cavers, P. B. (1971). *Can. J. Bot.* **49,** 1123–1130.
Mitchell, N. D. (1977). *Entomol. Exp. Appl.* **22,** 208–219.
Mooney, H. A. (1972). *Annu. Rev. Ecol. Syst.* **3,** 315–346.
Mooney, H. A., and Bartholomew, B. (1974). *Bot. Gaz. (Chicago)* **135,** 306–313.
Mooney, H. A., and Chu, C. (1974). *Oecologia* **14,** 295–306.

Mooney, H. A., Harrison, A. T., and Morrow, P. A. (1975). *Oecologia* **19,** 293–302.
Mooney, H. A., Ehleringer, J., and Berry, J. A. (1976). *Science* **194,** 323–324.
Moore, L. R. (1978). *Oecologia* **34,** 185–202.
Morrow, P. A., and La Marche, V. C., Jr. (1978). *Science* **201,** 1244–1246.
Mueggler, W. F. (1967). *Ecology* **48,** 942–494.
Muller, C. H. (1969). *Science* **165,** 415–416.
Muller, C. H. (1970). *In* "Biochemical Coevolution" (K. Chambers, ed.), pp. 13–31. Oregon State Univ. Press, Corvallis.
Nair, K. S. S., McEwen, F. L., and Snieckus, V. (1976). *Can. Entomol.* **108,** 1031–1036.
Nakanishi, K., Goto, T., Ito, S., Natori, S., and Nozoe, S., eds. (1974). "Natural Products Chemistry," Vol. 1. Academic Press, New York.
Nault, L. R., and Styer, W. E. (1972). *Entomol. Exp. Appl.* **15,** 423–437.
Nayar, J. K., and Thorsteinson, A. J. (1963). *Can. J. Zool.* **41,** 923–929.
Nielsen, J. K. (1978). *Entomol. Exp. Appl.* **24,** 41–54.
Nielsen, J. K., Larsen, L. M., and Sorensen, H. (1977). *Phytochemistry* **16,** 1519–1522.
Nissen, P., and Benson, A. A. (1964). *Biochim. Biophys. Acta* **82,** 400–402.
Ornston, L. N. (1971). *Bacteriol. Rev.* **35,** 87–116.
Penning de Vries, F. W. T. (1975a). *Ann. Bot. (London)* [N.S.] **39,** 77–92.
Penning de Vries, F. W. T. (1975b). *In* "Photosynthesis and Productivity in Different Environments" (J. P. Cooper, ed.), IBP Synth. Vol. 3, pp. 459–480. Cambridge Univ. Press, London and New York.
Penning de Vries, F. W. T., Brunsting, A. H. M., and van Laar, H. H. (1974). *J. Theor. Biol.* **45,** 339–377.
Pfeffer, W. (1900). "The Physiology of Plants" (transl. by A. J. Ewart), 2nd ed., Vol. 1. Oxford Univ. Press (Clarendon), London and New York.
Phelan, J. R., and Vaughan, J. G. (1976). *Biochem. Syst. Ecol.* **4,** 173–178.
Pierre, R. E., and Bateman, D. F. (1967). *Phytopathology* **57,** 1154–1160.
Pillemer, E. A., and Tingey, W. M. (1976). *Science* **193,** 482–484.
Pitelka, L. F. (1977). *Ecology* **58,** 1055–1065.
Rains, D. W. (1976). *In* "Plant Biochemistry" (J. Bonner and J. E. Varner, eds.), 3rd ed., pp. 561–597. Academic Press, New York.
Ralph, C. P. (1977). *Ecology* **58,** 799–809.
Rathcke, B. J., and Poole, R. W. (1975). *Science* **187,** 175–176.
Rehm, A., and Humm, H. J. (1973). *Science* **182,** 173–174.
Rehr, S. S., Bell, E. A., Janzen, D. H., and Feeny, P. P. (1973a). *Biochem. Syst.* **1,** 63–67.
Rehr, S. S., Feeny, P. P., and Janzen, D. H. (1973b). *J. Anim. Ecol.* **42,** 405–416.
Rehr, S. S., Janzen, D. H., and Feeny, P. P. (1973c). *Science* **181,** 81–82.
Rhoades, D. F. (1977). *Biochem. Syst. Ecol.* **5,** 281–290.
Rhoades, D. F., and Cates, R. G. (1976). *Recent Adv. Phytochem.* **10,** 168–213.
Rice, E. L. (1974). "Allelopathy." Academic Press, New York.
Rice, E. L. (1977). *Biochem. Syst. Ecol.* **5,** 201–206.
Robinson, T. (1968). "The Biochemistry of Alkaloids." Springer-Verlag, Berlin and New York.
Robinson, T. (1974). *Science* **184,** 430–435.
Rockwood, L. L. (1973). *Ecology* **54,** 1363–1369.
Rodman, J. E. (1974). *Contrib. Gray Herb., Harv. Univ.* **205,** 3–146.
Rodman, J. E. (1976). *Syst. Bot.* **1,** 137–148.
Romeo, J. T., and Bell, E. A. (1974). *Lloydia* **37,** 543–568.
Romeo, J. T., Bacon, J. D., and Mabry, T. J. (1977). *Biochem. Syst. Ecol.* **5,** 117–120.
Romig, R. W., and Calpouzos, L. (1970). *Phytopathology* **60,** 1801–1805.

Root, R. B. (1973). *Ecol. Monogr.* **43,** 95–124.
Rosenthal, G. A. (1977). *Biochem. Syst. Ecol.* **5,** 219–220.
Russell, C. E., and Berryman, A. A. (1976). *Can. J. Bot.* **54,** 14–18.
Sachs, J. (1875). "Text-book of Botany, Morphological and Physiological" (transl. by A. W. Bennett and W. T. Thiselton Dyer). Oxford Univ. Press (Clarendon), London and New York.
Sachs, J. (1887). "Lectures on the Physiology of Plants" (transl. by H. M. Ward). Oxford Univ. Press (Clarendon), London and New York.
Sanders, T. H., Ashley, D. A., and Brown, R. H. (1977). *Crop. Sci.* **17,** 548–550.
Sarkar, S. K., Howarth, R. E., and Goplen, B. P. (1976). *Crop Sci.* **16,** 543–546.
Schonbeck, F., and Schroder, C. (1972). *Physiol. Plant Pathol.* **2,** 91–99.
Schoonhoven, L. M. (1967). *Proc. K. Ned. Akad. Wet., Ser. C* **70,** 556–568.
Schoonhoven, L. M. (1969). *Entomol. Exp. Appl.* **12,** 555–564.
Schultz, D. E., and Allen, D. C. (1977). *Environ. Entomol.* **6,** 276–283.
Schwochau, M. E., and Hadwiger, L. E. (1970). *Recent Adv. Phytochem.* **3,** 181–189.
Scott, A. I. (1974). *Science* **184,** 760–764.
Seigler, D. S. (1977). *Prog. Phytochem.* **4,** 83–120.
Seigler, D. S., and Price, P. W. (1976). *Am Nat.* **110,** 101–105.
Singer, M. C. (1971). *Evolution* **25,** 383–389.
Singh, R. P., and Nair, K. P. P. (1975). *J. Agric. Sci.* **85,** 241–245.
Slansky, F., Jr., and Feeny, P. (1977). *Ecol. Monogr.* **47,** 209–228.
Smith, B. D. (1966). *Nature (London)* **212,** 213–214.
Soo Hoo, C. F., and Fraenkel, G. (1964). *Ann. Entomol. Soc. Am.* **57,** 788–790.
Stanier, R. Y. (1968). *In* "Chemotaxonomy and Serotaxonomy" (J. G. Hawkes, ed.), pp. 201–225. Academic Press, New York.
Stanton, N. (1975). *Biotropica* **7,** 8–11.
Stoessl, A. (1970). *Recent Adv. Phytochem.* **3,** 143–180.
Swain, T. (1965). *In* "Plant Biochemistry" (J. Bonner and J. E. Varner, eds.), 2nd ed., pp. 552–580. Academic Press, New York.
Swain, T. (1977). *Annu. Rev. Plant Physiol.* **28,** 479–501.
Sweet, G. B., and Wareing, P. F. (1966). *Nature (London),* **210,** 77–79.
Tahvanainen, J. O. (1972). *Entomol. Scand.* **3,** 120–138.
Tahvanainen, J. O., and Root, R. B. (1972). *Oecologia* **10,** 321–346.
Tapper, B. A., and Reay, P. F. (1973). *In* "Chemistry and Biochemistry of Herbage" (G. W. Butler and R. W. Bailey, eds.), Vol. 1, pp. 447–476. Academic Press, New York.
Tetenyi, P. (1970). "Infraspecific Chemical Taxa of Medicinal Plants." Chem. Publ. Co., New York.
Thorin, J., and Nommik, H. (1974). *Phytochemistry* **13,** 1879–1881.
Thorne, G. N., and Evans, A. F. (1964). *Ann. Bot. (London)* [N.S.] **28,** 499–508.
Thorsteinson, A. J. (1953). *Can. J. Zool.* **31,** 52–72.
Ting, I. P. (1976). *In* "CO_2 Metabolism and Plant Productivity" (R. H. Burris and C. C. Black, eds.), pp. 251–268, University Park Press, Baltimore.
Todd, J. W., and Morgan, L. W. (1972). *J. Econ. Entomol.* **65,** 567–570.
Turnipseed, S. G. (1972). *J. Econ. Entomol.* **65,** 224–229.
Underhill, E. W., Wetter, L. R., and Chisholm, M. D. (1973). *Biochem. Soc. Symp.* **38,** 303–326.
Van Emden, H. F., and Bashford, M. A. (1969). *Entomol. Exp. Appl.* **12,** 351–364.
Van Etten, C. H., Waxenbichler, M. E., Williams, P. H., and Kwolek, W. F. (1976). *J. Agric. Food Chem.* **24,** 452–455.

Van Etten, H. D., and Pueppke, S. G. (1976). *In* "Biochemical Aspects of Plant-Parasite Relationships" (J. Friend and D. R. Threlfall, eds.), pp. 238–289. Academic Press, New York.
Verschaffelt, E. (1911). *Proc. K. Ned. Akad. Wet.* **13,** 536–542.
von Rudloff, E. (1975). *Phytochemistry* **14,** 1319–1329.
Walker, J. C., and Stahmann, M. A. (1955). *Annu. Rev. Plant Physiol.* **6,** 351–366.
Wallace, D. H., Peet, M. M., and Ozbun, J. L. (1976). *In* "CO_2 Metabolism and Plant Productivity" (R. H. Burris and C. C. Black, eds.), pp. 43–58. Univ. Park Press, Baltimore, Maryland.
Waller, G. R., Tang, M. S., Scott, M. R., Goldberg, F. J., Mayes, J. S., and Auda, H. (1965). *Plant Physiol.* **40,** 803–807.
Wareing, P. F., and Allen, E. J. (1977). *Philos. Trans. R. Soc. London, Ser. B* **281,** 107–119.
Wareing, P. F., and Patrick, J. (1975). *In* "Photosynthesis and Productivity in Different Environments" (J. P. Cooper, ed.), IBP Synth. Vol. 3, pp. 481–499. Cambridge Univ. Press, London.
Wender, S. H. (1970). *Recent Adv. Phytochem.* **3,** 1–29.
White, R. R. (1974). *Oecologia* **14,** 307–315.
Whitman, J. (1973). *Heredity* **30,** 241–245.
Whittaker, R. H. (1970). *In* "Chemical Ecology" (E. Sondheimer and J. B. Simeone, eds.), pp. 43–70. Academic Press, New York.
Whittaker, R. H., and Feeny, P. P. (1971). *Science* **171,** 757–770.
Wieczorek, H. (1976). *J. Comp. Physiol.* **106,** 153–176.
Williams, M., Barnes, R. F., and Cassady, J. M. (1971). *Crop Sci.* **11,** 213–217.
Wilson, A. C. (1976). *In* "Molecular Evolution" (F. J. Ayala, ed.), pp. 225–262. Sinauer, Sunderland, Massachusetts.
Wilson, D. E., and Janzen, D. H. (1972). *Ecology* **53,** 954–959.
Wilson, F. D., and Shaver, T. N. (1973). *Crop Sci.* **13,** 107–110.
Zavarin, E., and Cobb, F. W. (1970). *Phytochemistry* **9,** 2509–2515.

Chapter **7**

Interactions of Allelochemicals with Nutrients in Herbivore Food

JOHN C. REESE

I. INTRODUCTION

An animal in its environment faces a chemical milieu of almost unbelievable complexity [over 12,000 described molecules from plants alone (Scott, 1974)]. On a chemosensory level, it must make some sense of this

ISBN 0-12-597180-X

complexity to locate food, avoid danger, etc., and it must integrate these chemosensory signals with visual and auditory signals. Once an animal locates a suitable food source, it may ingest a portion of it. Then the "postingestive" phase begins. It is this phase that will be considered in this chapter. What happens to that vast mixture of compounds when an animal eats a plant? More specifically, how do particular chemicals in the plant interact with others?

The term "herbivore" here refers not only to classic herbivores that feed largely on plants in nature, but also to various omnivores that eat plant materials. A large portion of the research in this area has been conducted on such "herbivores" as chickens, rats, rabbits, cattle, pigs, and man and has utilized such "plants" as soybean meal, cottonseed meal, pepper, and nutmeg. Through an examination of such studies, perhaps we can better understand some of the more usual plant–herbivore interactions and how interactions among specific chemical compounds (and their concentrations) influence these ecological relationships.

Those readers interested in interactions between nutrients and nonnutrients of nonplant as well as plant origin are referred to the extensive (896 references cited) review by Oltersdorf *et al.* (1977).

Two terms should be defined at the outset. A nutrient is a compound required for the normal growth, development, and maintenance of an organism's functions. Thus, a nutrient for man might not be one for a grasshopper. A nutrient for a growing, immature organism may not be a nutrient for the same organism as an adult. Similarly, a nutrient for an aphid that has lost its symbiotes may be nonessential for an aphid that has its symbiotes. Furthermore, a compound that is a requirement at one concentration may be highly toxic at a 10-fold greater concentration. Thus, the definition of the word "nutrient" depends on the situation. Being a nutrient is not an intrinsic characteristic of a molecule but rather depends on the biochemistry of the ingesting organism, the concentration of the compound, etc. However, it is clear that certain compounds are more likely to be treated as nutrients than are others.

A second key term is allelochemicals. Allelochemicals are nonnutritional chemicals produced by one organism that affect the growth, health, behavior, or population biology of members of other species. Again, this term does not describe an intrinsic characteristic of a particular compound but depends on the situation.

Another vague area of nutritional concepts centers around host finding. If a particular herbivore cannot find its host unless that host emits a particular attractant or cueing compound, the herbivore might starve in areas where mutants of its usual host no longer produce this attractant. Is such a substance, which is in fact required for normal growth but which interacts primarily with chemoreceptors, a nutrient? Perhaps in such a case it would be useful to distinguish between physiological nutrients and ecological nutrients. To circumvent some of these terminology problems, the concept of dietetics or dietetic requirements may be more useful than the concept of classic nutrition (Beck, 1972, 1974; Reese, 1977). The organism must find and ingest food "that not only meets its nutritional requirements, but is also capable of being assimilated and converted into the energy and structural substances required for normal activity and development" (Beck and Reese, 1976). Thus, animal dietetics encompasses not only nutrition in the more classic sense, but also any allelochemical effects on feeding behavior (attractants, feeding stimulants, repellents, feeding inhibitors), any allelochemical effects on survival (toxicants), and any chronic effects of certain compounds (antivitamins, digestibility-reducing factors, chelators of essential minerals) on growth, development, or reproduction. Understanding all of these aspects is crucial to understanding a given plant–herbivore interaction. An adverse effect in any one of these areas could cause a given plant to become more resistant to the attack of herbivores.

The different aspects of animal dietetics interact with each other. An essential nutrient may be bound by an allelochemical, or a nutrient at one concentration may act as a deleterious allelochemical at another concentration. A very toxic allelochemical may be used as a source of nitrogen by a highly coevolved herbivore. In this chapter I shall discuss some of these interactions.

II. KAIROMONES

Kairomones are allelochemicals that are presumed to be useful to the receiving organism. Attractants and feeding stimulants are good examples. Another example would be a compound that is sequestered by an herbivore and then utilized in some type of defense mechanism. For example, monarch butterfly larvae apparently sequester cardenolides from their host plant (see review by Roeske *et al.*, 1976); these compounds are extremely toxic to avian predators. Certain compounds from the cotton plant that are not necessary for growth and development are nevertheless essential for production of pheromones for bringing male and female boll weevils together for mating. Some sawfly larvae have a pouch of the foregut in which they apparently store a defensive fluid that includes compounds sequestered from *Eucalyptus* leaves, on which they feed (Morrow *et al.*, 1976). Eisner *et al.* (1974) clearly demonstrated this with a sawfly species feeding on pine, and Rothschild *et al.* (1977) showed that the tiger moth and a species of grasshopper can even store cannabinoids.

Attractants are obviously necessary to the herbivore in locating host plants. In this way they interact ecologically with essential nutrients; they bring the herbivore to its source of nutrients. Since this nutrient–allelochemical interaction is more ecological than physiological and since vast amounts have already been written on the subject [see recent reviews by Beck and Schoonhoven (1979) and Shorey and McKelvey (1977) and discussion by Atsatt (1977)], it is sufficient here to simply point out the existence of this type of interaction.

Compounds that are normally toxic can be growth stimulants. Luckey (1968) demonstrated that insecticides at very low concentrations could actually stimulate growth. He termed this phenomenon "hormoligosis" (from the Greek: *hormo* = excite and *oligo* = small quantities). Similarly, Dittrich *et al.* (1974) showed that exposure to low levels of insecticides stimulated increased egg production by mites. As will often be pointed out in this chapter, similar concentration-dependent changes in biological activity occur with many compounds. Intensive study could probably demonstrate such a dual action for a very large number of both harmful and beneficial compounds.

III. ALLOMONES

Allomones are allelochemicals that are deleterious to the receiving organism. Obviously, the distinction between allomones and kairomones is similar to the distinction between nutrients and nonnutrients in that whether a particular compound is an allomone or a kairomone may depend on its concentration, the species being exposed to it, etc. Allomones include repellents (unless they repel the herbivore from something that would be highly toxic if the herbivore did ingest it), feeding inhibitors, digestibility-reducing factors, and toxicants (see review on alkaloids by Levin, 1976), antihormones, and antivitamins. Allomones can also be defined as allelochemicals that are beneficial to the emitter and so would include attractants when the animal being attracted is a pollinator or a spreader of seeds. Allomone interactions with nutrients may be direct, as in the case of antivitamins and antihormones, or they may be more indirect. An example of an indirect interaction would be a feeding inhibitor that causes a herbivore to refuse to ingest a perfectly nutritious plant (for an extensive review of insect feeding inhibitors, see Chapman, 1974) or a feeding stimulant causing the herbivore to eat more. Such ideas have been utilized in developing successful diets for rearing insects, and man has certainly flavored and spiced his nutrients for many centuries.

Classic nutrients can themselves influence behavior, in which case the same compound is both a nutrient and an allelochemical. Srivastava and Auclair (1975) found that several amino acids acted as feeding stimulants for a species of aphid. Cabbage looper (*Trichoplusia ni*) larval feeding was shown to be stimulated by proteins, sugars, wheat germ oil, and inorganic salts (Gothilf and Beck, 1967). European corn borers have a strong predilection for feeding on diets or tissues containing high sugar concentrations. This has been called "saccharotrophism" (Beck, 1956, 1957). Chlorogenic acid has been termed a vitamin for silkworm (*Bombyx mori*) larvae, but in this case it is more likely that the compound is simply a feeding stimulant (Kato and Yamada, 1963–1964, 1966; Yamada and Kato, 1966).

A. Analogs of Nutrients

Any given nutrient is metabolized by an animal through often complex pathways such as glycolysis and the Krebs cycle. If a nonnutrient that closely resembles a nutrient is ingested, it may interfere with the normal metabolism of the herbivore. Although theoretically this could occur with analogs of any group of nutrients, the analogs of essential amino acids and of vitamins have received by far the most attention.

Mattson *et al.* (1977) found that cholesterol absorption decreased when various plant sterols were added to the diets of rats. They proposed that the plant sterol esters are hydrolyzed in the intestine. The resulting free sterol may decrease the solubility of cholesterol with a consequent decrease in cholesterol absorption. Mattson *et al.* (1977) also discuss the co-precitation mechanism. The physico-chemical interactions of dietary sterols are not clearly understood yet.

A number of plant amino acids are not ordinarily required by herbivores and are not usually incorporated into normal proteins (for recent reviews of toxic amino acids in legumes, see Bell, 1972, 1976). For example, the structure of 3,4-dihydroxyphenylalanine (L-dopa) is similar to that of tyrosine. L-Dopa has been found in relatively high concentrations in a number of legumes. It may play a role in favism (Bell, 1973) and inhibits growth, assimilation of food, efficiency of conversion of food to biomass, and normal development in some insects (Reese and Beck, 1976b; Rehr *et al.*, 1973; Janzen *et al.*, 1977). How many of these effects are actually due to its similarity to tyrosine is not known, but this is an area that warrants further investigation.

Mimosine is another analog of tyrosine and phenylalanine. It inhibits growth in rats; this inhibition can be partially reversed by supplementing the diet with phenylalanine and completely reversed with tyrosine (Liener, 1969).

L-Canavanine is a structural analog of arginine that is probably readily incorporated into the proteins of most insects (for detailed review of the subject, see Chapter 9, Section I, and review by Rosenthal, 1977), acts as a competitive inhibitor of arginine metabolism (Vanderzant and Chremos, 1971; Dahlman and Rosenthal, 1976), and can inhibit insect survival (Isogai *et al.*, 1973a; Harry *et al.*, 1976; Janzen *et al.*, 1977), reproduction (Hegdekar, 1970), and metamorphosis (Isogai *et al.*, 1973b). In an elegant example of a counter to a plant's defenses, it has been shown that one of the few species that can attack a plant containing high concentrations (up to 13% by dry weight) of L-canavanine possesses an arginyl-tRNA synthetase that discriminates between arginine and canavanine (Rosenthal *et al.*, 1976). Thus, in this species the interaction between the nutrient (arginine) and the allelochemical (L-canavanine) is not directly deleterious but must have a certain energetic or nitrogenous cost for the extra enzymatic machinery (for an excellent discussion of nitrogen–energy relationships, see Slansky and Feeny, 1977). Furthermore, this species has the enzymes necessary to degrade canavanine and utilize it as a source of nitrogen (Rosenthal *et al.*, 1977). This again would have a certain cost. L-Canavanine, then, depending on the herbivore, is ordinarily a highly toxic allelochemical, but in certain rare cases can be utilized as a nutrient, for a price. Presumably the cost of utilizing an unusual compound is paid for by the advantages of having fewer competitors for a food source.

In addition to forming defective proteins, amino acid analogs may also inhibit protein synthesis, block enzymatic reactions, and compete with protein amino acids for transport sites. Any extensive discussion of molecular mechanisms, however, quickly tends to get into the realm of speculation. There is little direct evidence for specific mechanisms; when they are elucidated, however, the toxicity of any one compound will probably depend on a variety of mechanisms working together (Harper *et al.*, 1970).

B. Excessive Amounts of Essential Amino Acids

Not only structural analogs of amino acids, but also essential amino acids themselves can be deleterious, if they are ingested in excessive quantities or if they are not in balance with other amino acids (Strong, 1973; Janzen *et al.*, 1977). Too much of almost any amino acid can be harmful. Ironically, the indispensable amino acids are generally less well tolerated in excessive amounts than are the dispensable ones (Harper, 1973), but nonprotein amino acids are certainly much more toxic than protein amino acids (Janzen *et al.*, 1977). Protein amino acid concentrations of 3 to 10 times those for maximal growth will inhibit rat growth (Harper, 1973). Excessive amounts of methionine are particularly poorly tolerated; only about 3 times the required amount can cause growth retardation (Harper, 1973), if the protein content of the diet is low. Glycine or serine supplementation will partially alleviate the condition (Benevenga *et al.*, 1976), apparently by bringing the dietary amino acid concentrations back into a favorable balance (for discussions of nutrient proportions in insects, see House, 1969, 1971). Recent work by Rotruck and Boggs (1977) with the rat confirmed the toxicity of methionine as well as a potential methionine substitute, *N*-acetyl-L-methionine. The latter compound was slightly less toxic than the former at very high concentrations. Vitamin B_6, riboflavin, and niacin deficiency also reduced the rat's tolerance for excessive amounts of methionine (Harper, 1973). On the other hand, methionine can alleviate tyrosine toxicity (Yamamoto *et al.*, 1976). Interestingly, this characteristic of methionine toxicity being most striking in association with low protein diets seems to be true for amino acids in general (Harper, 1973). In the case of excessive amino acid–protein interaction, apparently excessive amounts of individual amino acids, except for tryptophan and tyrosine, do not stimulate sufficient increases of amino acid catabolic enzymes in contemporary time. Animals inclined to feed on higher-protein diets, on the other hand, have relatively high levels of these enzymes, and so can better handle excessive amounts of specific amino acids.

Other aspects of imbalanced amino acid diets were extensively reviewed by Harper *et al.* (1970). Adverse effects include reduced food intake, growth inhibition, low survival rates, and pathological lesions. Ingestion of excessive amounts of tyrosine by rats fed a low-protein diet causes severe eye and paw lesions, depressed ingestion, and retarded growth.

C. Rapeseed Oils

The deleterious effects of rapeseed oils on livestock growth have been known for many years (Mattson, 1973). The oils containing exceptionally high levels of erucic acid are especially active biologically. Hornstra (1972) proposed that one effect on the molecular level might be the uncoupling of oxidative phosphorylation. Closely related are recent studies on thyroid metabolism (Summers and Leeson, 1977) and the studies by Clement and Renner (1977) on the effects of rapeseed oils containing high and low concentrations of erucic acid on energy utilization in chicks. The latter researchers found that the oils with high erucic acid concentrations did indeed suppress efficient energy utilization, as well as growth, in the chicks. A caveat discussed later in Section G must be mentioned here. That is, in order for conclusive statements to be made about the effects of a particular compound, that compound should be added to an artificial diet in pure form. Testing plants or plant materials believed to differ from one another in the concentration of only one compound is naive unless the plants have actually been analyzed for the 12,000 or so natural products known to occur in plants.

D. Antivitamins

Vitamins, like amino acids, have structural analogs. In fact, an early definition of an antivitamin included only structural analogs that competed with the vitamin in question. That definition has now been expanded, however, to include any compound that "diminishes or abolishes the effect of a vitamin in a specific way" (Somogyi, 1973). Examples include antithiamin compounds in the bracken fern (*Pteridium equilinum*), ragi (*Eleusine cora*), a bean species (*Phaseolus radiatus*), mustard seed, cottonseed, flaxseed, blackberries, black currants, red cicerone, and brussels sprouts. The active compound in fern was identified as an enzyme (thiamase I) that breaks down thiamin (Somogyi 1973). Bracken fern also contains caffeic acid, which has antithiamin activity. Catechol has antithiamin activity, which may account for its growth-inhibiting and toxic effects on insects (Reese and Beck, 1976a,b Todd *et al.*, 1971; Desmarchelier and Fukuto, 1974). Apparently the antithiamin activity resides in the *o*-hydroxy portions of these molecules (Somogyi, 1973). Linatine (γ-glutamyl-1-amino-D-proline) from flax is an antagonist of vitamin B_6. A particularly well known vitamin antagonist is dicoumarol in sweet clover. This compound blocks vitamin K synthesis and so blocks the biochemistry of the clotting reaction of mammals. Adding vitamin K to the diet decreases the problem (Stephenson, 1973). Pea seedlings contain an antagonist of pantothenic acid. Phytic acid in several plants is an antivitamin D compound. Raw soybean contains lipoxidase, which oxidizes (and thus destroys) carotene, from which vitamin A is derived (Liener, 1969). In corn, complexes are formed with niacin, rendering it unavailable (Gontzea and Sutzescu, 1968). The list could go on, but the point is that a number of plants contain compounds that in some herbivores antagonize, destroy or block utilization of vitamins.

E. Vitamin Toxicity

Just as the deficiency of a vitamin can be deleterious, so too can excessive amounts. Thus, we have another example of a group of nutrients becoming deleterious allomones due to a change in concentration. Nicotinamide can inhibit growth when fed at 1% of a diet high in fat and low in choline (Hayes and Hegsted, 1973). Vitamin A is toxic in high concentrations (Strong, 1973). Vitamin A may be teratogenic in several animals and may cause liver damage and hypercalcemia (DiPalma and Ritchie, 1977).

Vitamin D toxicity usually relates to calcium metabolism. Increasingly, vitamin D is being thought of more as a hormone than as a vitamin, partly because it is not a dietary requirement under normal conditions (DiPalma and Ritchie, 1977). Niacin excesses may interfere with liver function and the formation of diphosphopyridine nucleotide and triphosphopyridine nucleotide. Pyridoxine can cause convulsive disorders in both excessive and deficient amounts. It also can interfere with the therapeutic effect of L-dopa. This could point the way toward some interesting experiments on the relationship between increased dietary pyridoxine and toxic effects of L-dopa from various legumes. For further information on vitamin toxicity, see the recent review by DiPalma and Ritchie (1977).

The important point here, just as for amino acids, is that excessive amounts of even required nutrients can be quite deleterious. The possibility of plants utilizing excess vitamin concentrations as defense mechanisms is probably remote; vitamin toxicity has only rarely resulted from the ingestion of natural foodstuffs (Hayes and Hegsted, 1973).

F. Protease Inhibitors

Certain plant compounds known to inhibit trypsin and chymotrypsin activity were first isolated many years ago. It is certainly an interesting notion that plants might defend themselves against excessive herbivore attack by producing substances that would block the action of proteolytic enzymes, making the plant a less digestible food source. Although still basically reasonable, like so many neat, logical pictures, this one has become a good deal more complex than was first thought.

A trypsin inhibitor was isolated from soybeans (Kunitz, 1945, 1946, 1947). Since then, soybeans have been intensively studied (see Chapter 17, Section I), but inhibitors of proteolytic enzymes have been found in plants other than the soybean. Inhibitors have been found in potatoes, tomatoes (Green and Ryan, 1972), lima beans (Feeney *et al.*, 1969), peas, alfalfa (Whitaker and Feeney, 1973), corn (Mitchell *et al.*, 1976), barley (Weiel and Hapner, 1976), and a very wide variety of other plants (Ryan, 1973; McFarland and Ryan, 1974; Walker-Simmons and Ryan, 1977).

The growth-inhibiting effects of soybeans and other plants containing proteolytic enzyme inhibitors are well known (Liener and Kakade, 1969). Birk and Applebaum (1960) studied the effect of soybean trypsin inhibitors on the development of *Tribolium castaneum* larvae. Soybean trypsin inhibitor reduced beetle production by a bruchid beetle (Janzen *et al.*, 1977). Khayambashi and Lyman (1966) found that, even though food intakes were controlled, soybean trypsin inhibitor reduced growth. Other effects, such as mortality and decreased reproduction, have also been observed (Su *et al.* 1974). Finlay *et al.* (1973) investigated the mechanism of gossypol (a phenolic pigment of cotton) inactivation of pepsinogen.

The idea that the growth-depressing effects of proteolytic enzyme inhibitors is due to more than simply inhibiting the breakdown of protein in the gut tract of an animal has existed for some time. Liener *et al.* (1949) postulated a twofold mechanism. The first part involves a reduction in the availability of methionine for growth due to the anti-proteolytic activity; addition of supplemental methionine largely counteracts this effect. The second part seems to have little to do directly with proteolytic activity. The observation that these inhibitors caused stimulation of the pancreas and thus excessive loss of nitrogen into the gut tract helped explain the effects (Lyman and Lepkovksy, 1957; Khayambashi and Lyman, 1966). A number of papers have now been published on proteolytic enzyme inhibition and pancreatic activity (Lyman *et al.*, 1974; Yen *et al.*, 1977; Mallory and Travis, 1975; Schingoethe *et al.*, 1974; Madar *et al.*, 1976). Apparently soybean trypsin inhibitor stimulates pancreatic secretion by binding trypsin inhibitor so tightly that it cannot fully activate chymotrypsinogen. This, in turn (since both trypsin and chymotrypsinogen are in effect removed), prevents the normal negative feedback systems from operating, so that pancreatic enlargement and excessive pancreatic secretion both occur (Lyman *et al.*, 1974). This is complicated by the fact that cattle and swine do not develop pancreatic hypertrophy (Liener and Kakade, 1969). Thus, in contrast to such animals as the rat and chick, the pig may indeed suffer from the direct inhibition of intestinal proteolysis when fed raw soybean meal, but the evidence suggests that some inhibitor other than the trypsin inhibitor may be responsible (Yen *et al.*, 1977). These interactions are discussed in more detail by Liener and Kakade (1969) and in Chapter 17, Section I, of this volume.

A particularly exciting area of protease inhibitor research is the recent work of Green and Ryan (1972, 1973; Ryan and Green, 1974) on inhibitors in potatoes and tomatoes. At least some plants can apparently be induced to produce much higher levels of enzyme inhibitors. Thus, the cost of defense should be less, since the compound is produced by the plant only at the time it is most needed.

Despite the interest in protease inhibitors in legumes, it is important to keep in mind that these are by no means the only growth inhibitors present. Liener (1976a) pointed out that characteristics of undenatured proteins and phytohemagglutinins probably also contribute to the growth-inhibiting properties of certain legumes. Ward *et al.* (1977) reported studies on a thermolabile growth inhibitor from faba beans that has no trypsin inhibitor or hemagglutinin activity and may in fact be a condensed tannin (Marquardt *et al.*, 1977).

G. Compounds That Block Utilization of Nutrients

No matter how nutritious a food source is in terms of essential nutrients present, it may be dietetically unacceptable if any essential nutrient is bound in an unavailable form or if its utilization is blocked in some way. Condensed and hydrolyzable tannins are both powerful precipitants of proteins. They can cause growth depression of various experimental animals. Although several mechanisms of growth inhibition apparently occur, they all more or less relate to their ability to bind to proteins. Foods may be so astringent that animals starve rather than eat them. The indigestible complexes may lower the efficiency of conversion of food to animal biomass and raise the nitrogen excretion level. They may bind to and therefore inactivate digestive enzymes and thus mimic the action of the protease inhibitors discussed above (Singleton and Kratzer, 1973).

As mentioned briefly above, phytohemagglutinins are important to the nutritional value of some legumes. Although their name refers to their ability to agglutinate red blood cells, they strongly bind to a variety of saccharide and saccharide-containing proteins. The phytohemagglutinin concanavalin A from jack beans has been intensively studied. Many other phytohemagglutinins have also been isolated (Jaffé, 1969), including abrin and ricin (Liener, 1974, 1976b). Various phytohemagglutinins have been shown to inhibit growth (Janzen *et al.*, 1976). The toxic effects of these compounds may very well be due to their ability to bind to specific receptor sites on the epithelial cells of the intestine (Liener, 1976b).

Feeny (1968a,b, 1970) showed that oak leaf tannins are capable of binding to protein and probably account for the poor growth of winter moth (*Operophtera brumata*) larvae on mature oak leaves. Bernays (1978) on the other hand, found no reduction in digestion by the addition of condensed tannin to grasshopper diets. Condensed tannins of sainfoin (*Onobrychis viciifolia*) form complexes with both dietary proteins and salivary mucoproteins of cattle (Jones and Mangan, 1977). Perhaps the habit of many people of putting milk or cream into such drinks as coffee and tea, both of which are high in tannins, has the effect of binding some of the tannin before it binds to mouth and intestinal proteins. Tannins can apparently also reduce the availability of smaller molecular weight compounds, such as lysine (Delort-Laval and Viroben, 1969).

Similarly, breakdown products of tannins (phenols) and many related or free phenols can bind to various nutrients, including proteins (especially sulfhydryl groups) and metal ions. Elliger *et al.* (1976) demonstrated an apparent interaction between cholesterol and levopimaric acid, a compound with a close structural and biogenetic relationship to various acids found in sunflowers. Adding excessive amounts of cholesterol greatly decreased the deleterious effects of levopimaric acid. Although Norris and colleagues were working with nerve membrane proteins when they developed their energy transduction theory (Norris *et al.*, 1971; Rozental and Norris, 1973; Rozental *et al.*, 1975), there is no reason why a similar mechanism of shifting the sulfhydryl–disulfide ratio could not be occurring with membrane carrier proteins of the gut tract, enzymes of the gut tract, etc. Just as these shifts change the characteristics of chemosensory nerve proteins, so too they may change the characteristics of many other proteins. There are numerous other examples of phenolics acting as astringents (Joslyn and Goldstein, 1964), in browning reactions (Kefford and Chandler, 1970; Mathew and Parpia, 1971; Reynolds, 1963), and as hypocholesteremic agents (Singleton and Esau, 1969).

Binding of materials may also be utilized to the advantage of the ingesting organism, as in the example cited above in which cream is added to tea or coffee. High-gossypol diets can be made less toxic by adding iron salts (Schaible *et al.*, 1934; Schaible and Bandemer, 1946).

There are interesting interactions between sterols and saponins (Shany *et al.*, 1970). Some saponins can form complexes with cholesterol, which could be helpful in the treatment of atherosclerosis but which can apparently inhibit the growth of animals in which the available dietary cholesterol is below the required level (Birk, 1969).

How do diluting substances of a diet interact with nutrients? In certain cases, a more diluted diet has apparently put a stress on an organism such that other stresses, such as an allelochemical, may have a more pronounced effect. For example, the expression of resistance to the

European corn borer is partly dependent on the sugar concentration of the diet used; thus, a useful bioassay diet had a lower dextrose (D-glucose) content than the basal diet (Beck, 1957; Beck and Smissman, 1960). Dilution of nutrients can also increase the efficiency with which dietary matter is converted to biomass (Reese and Beck, 1978).

Allelochemicals may act as antioxidants of nutrients and so essentially prevent certain nutrients from being degraded in diets (Cheeke, 1972).

Phytate, the hexaphosphoric ester of inositol, is a common plant compound that is found in high concentrations in cottonseed meal (Stephenson, 1973). It also occurs in corn, rice, wheat, potatoes, sweet potatoes, artichokes, nuts, legumes, blackberries, strawberries, and figs and can form a wide variety of insoluble salts with several heavy metals (Oberleas, 1973). In the case of decreasing calcium absorption, there is also an interaction with vitamin D. For example, if the vitamin D concentration of the diet is low, the phytate effect on calcium is much more striking (Oberleas, 1973). Phytate seems to be best known for complexing zinc, making it unavailable to monogastric animals (Nelson, 1967). Other compounds such as oxalic acid do not complex with zinc (Welch *et al.*, 1977). Phytate also ties up phosphorus in a less available form (Rojas and Scott, 1969). Perhaps phytate complexes with proteins, too. The availability of proteins was increased by treating cottonseed meal with phytase (Rojas and Scott, 1969). Gossypol toxicity was also reduced, suggesting that increasing the nutrient availability of a diet may allow the metabolism of an organism to appropriate more energy for detoxification. In pigs gossypol may form a complex with iron in the liver (Skutches *et al.*, 1974). This complex may then be secreted from the liver via the bile, thus lowering the liver iron concentration.

In the literature on insects there is considerably less evidence for the complexing of specific allelochemicals with specific nutrients. There is, however, a rapidly growing body of evidence that many different plant allelochemicals clearly affect such nutritional parameters of insect physiology as assimilation of material across the gut wall (approximate digestibility, AD), efficiency of conversion of assimilated material to insect biomass (ECD), and overall efficiency of conversion of all ingested material to insect biomass (ECI).

$$\mathrm{AD} = \frac{\text{weight of food ingested} - \text{weight of feces}}{\text{weight of food ingested}} \times 100$$

$$\mathrm{ECD} = \frac{\text{weight gained}}{\text{weight of food ingested} - \text{weight of feces}} \times 100$$

$$\mathrm{ECI} = \frac{\text{weight gained}}{\text{weight of food ingested}} \times 100$$

The use of these indices in insect dietetics was discussed extensively by Waldbauer (1968). Certainly, we have known for a number of years that different host plants are converted to insect biomass at different rates of efficiency, and yet it is only relatively recently that specific plant compounds have been tested as potential inhibitors of assimilation or efficiency. Shaver *et al.* (1970) studied the effects of gossypol on two species of lepidopterous larvae. In bollworm (*Heliothis zea*) larvae, gossypol decreased assimilation. On the other hand, gossypol had no measurable effect on tobacco budworm (*H. virescens*) larvae. Erickson and Feeny (1974) tested the hypothesis that the lepidopterous larval host plant range of *Papilio polyxenes asterius* is partially determined by compounds that are not required for perception of host or nonhost plants. They demonstrated that sinigrin (or one of its breakdown products) reduced assimilation but did not reduce the efficiency with which assimilated food was converted to insect biomass.

Beck and Reese (Beck and Reese, 1976; Reese and Beck, 1976a,b,c,d; Reese, 1977) have examined the effects of a number of plant allelochemicals on the growth and development of another lepidopterous insect, the black cutworm (*Agrotis ipsilon*). Growth inhibition may be due to the inhibition of assimilation or lowering the efficiency of conversion of assimilated food or a combination of both. Inhibition of either of these processes would, of course, reduce the overall efficiency of conversion of ingested food. Simultaneous inhibition of both of these processes would have a multiplying effect and might reduce the overall efficiency of conversion of ingested food more than if either assimilation or efficiency of conversion of assimilated food were reduced.

The use of nutritional indices as a basis for comparing dietary utilization by a herbivore has some limitations (Waldbauer, 1968). Variations in water and fiber contents may result in variations in the indices (Soo Hoo and Fraenkel, 1966a,b; Waldbauer, 1964; Kogan and Cope, 1974). A relatively low AD might be the result of high fiber content of the diet of a nonruminant. In this case, could cellulose fiber be considered an allelochemical, simply because it is a long-chain polymer that is difficult for nonruminants to break down? To answer yes to this question might be stretching a point, but it is certainly an interesting notion. A low ECD could result from antinutrients present in the food, unfavorable amino acid ratios, etc.

Nutritional indices are useful in a preliminary assessment of host plant or dietary utilization but are not sufficient to identify the specific factors influencing the efficiency of the utilization. The effects of chemical plant factors, including both allelochemicals and nutritional factors, on the efficiency of dietary utilization can best be investigated by incorporating these factors into a standardized artificial diet and then determining nutritional indices. In this way chemical factors can be studied with less equivocation. Also, by using the technique of perfusion of known amounts of a compound into plant material, most of the limitations discussed above can be avoided. Erickson and Feeny (1974) used this technique quite successfully in their work with sinigrin, as discussed earlier. The obvious advantage of using an artificial diet and the perfusion technique rather than using different plants is that one can vary one factor at a time. Even with such an intensively analyzed plant as cotton (Hedin *et al.*, 1976), to say that two lines differ only in the level of one compound is speculation at best. That compound must be synthesized by the plant, and so surely the two lines vary in terms of concentrations of compounds closely related biochemically to the one in question. In addition, a plant comprises an exceedingly complex milieu of chemicals. To say that two cultivars are isogenic except for compound *x* is really saying that they are isogenic for the compounds we have thought to analyze for, except for compound *x*.

H. Hormones

Substances that mimic herbivore hormones or that are structurally similar to herbivore hormones have been found in plants for many years [see reviews by Heftmann (1970, 1975a,b) and Leopold *et al.* (1976)]. Other than thyroxine interactions, most of the examples are relatively peripheral to allelochemical–nutrient interactions.

The only sterols that can enter an insect's system are those that can pass through the metabolic and regulatory mechanisms of the midgut epithelium. Thus, unless a hormone acts like some of the other compounds previously discussed and interacts with a nutrient while it is still external to the insect or while it is in the gut tract, it could pass harmlessly through the digestive system. In several cases, this is an extremely important point to keep in mind. A hormone might also remain in the digestive system but have an effect on the microflora of the gut tract.

α-Ecdysone (molting hormone) is synthesized from cholesterol in insects. Ecdysone may be metabolized to form the more active hormones, and these conversions can occur in various parts of the body (King, 1972; Nakanishi *et al.*, 1972). Plants synthesize nutritionally utilizable sterols, ecdysones, and structural analogs of sterols. Nearly 30 different phytoecdysones have been found (Beck and Reese, 1976). When injected, some of these compounds show a great deal of biological activity. Herbivores, however, are not injected by their host plants. Of all the interesting compounds in plants, only ponasterone A and inokosterone have been shown to influence insect development when ingested (Beck and Reese, 1976; Shigematsu *et al.*, 1974; Williams, 1970). A similar situation appears to exist for juvenile hormone-active compounds, which can block the normal functions of normal juvenile hormone (Bowers *et al.*, 1976; Pratt and Bowers, 1977; Ohta *et al.*, 1977).

As suggested above, apparently most phytoecdysones that are ingested by insects are degraded by the midgut epithelium. Thus, they do not ordinarily reach other tissues of the insect in a hormonally active form. For example, desert locusts that were fed bracken fern (high in both ecdysone and 20-hydroxyecdysone) showed no particular abnormalities (Carlisle and Ellis, 1968). It was suggested that the phytoecdysones were dehydroxylated by the midgut and then excreted.

I. Hypoglycemic Agents

Substances that lower sugar reserves have been known for several years. Hypoglycine A [β-(methylenecyclopropyl)alanine] is found in the fruit and seed of *Blighia sapida*, a plant that grows in Jamaica and Nigeria and is consumed by man. Through a rather complicated metabolic pathway, this compound depletes the sugar reserves of the body (Liener, 1969). Potassium atractylate from *Atractylis gummifera* can also produce severe hypoglycemic convulsions and even death. It apparently acts by a different mechanism. It inhibits oxidative phosphorylation and energy transfer reactions in the liver (Liener, 1969).

J. Interactions of Pesticides with Nutrients

Certainly the nutritional status of an herbivore can influence how it responds to pesticides (Gordon, 1961). Since these are not allelochemical–nutrient interactions in the strict sense, they are not considered in much detail here. However, it is possible that some of the same hypotheses in this area of research could be applied to allelochemical–nutrient dietetics. Certain allelochemicals are insecticides (e.g., nicotine and pyrethrins), and others may have modes of action that are similar to those of certain insecticides. For example, Krieger *et al.* (1971) found that lepidopterous larvae with wider host plant ranges tended to have higher mixed-function oxidase (enzymes associated with insecticide detoxification systems) levels. Also, some allelochemicals interact with insecticides (Abou-Donia *et al.*, 1974; Shaver and Wolfenbarger, 1976). There may also be certain underlying principles for the relationship between any toxicant and the nutritional status of the organism. For example, excess nutrients might allow detoxification systems such as the mixed-function oxidase system to operate at a higher level of efficiency. At least the nutritional status of the host plant can have an effect on susceptibility to insecticides (Henneberry, 1964). Morcos (1970) suggested that low protein content of the diets of test animals can accentuate the effects of toxins in general.

IV. DISCUSSION

From the above examples, it is obvious that allelochemicals can interact with nutrients in a number of ways and that in many cases these interactions can explain the deleterious effects of a given allelochemical. Although there are several examples of allelochemical–nutrient interactions cited here, when one thinks about the number of nutrients essential to herbivore health and the vast number of allelochemicals in plants, one realizes that the possibilities for studying interactions between the two groups are nearly limitless. Obviously, only a tiny fraction of the possibilities have been investigated so far. When one adds to this such variables as concentration, the number of possibilities becomes even more staggering. And yet such a variable is vitally important, as illustrated by the toxic effects of high levels of amino acids and vitamins, the stimulating effects of low levels of insecticides, and other effects of certain allelochemicals.

Can one array the secondary compounds into two groups, with one group including those that are often actually used by some animals, either for defense or for metabolic purposes, and a group including those that are always treated as something to be gotten rid of, avoided, or degraded? For example, is cellulose a nutrient to a ruminant but an allelochemical to other herbivores in the sense that it is inert stuffing in the food and therefore lowers the efficiency of conversion of that food to biomass? Or are certain fibers nutrients in the sense of lowering potentially deleterious plasma concentrations of such compounds as cholesterol, as shown by Mathé *et al.* (1977)? As shown in recent studies of high-fiber diets, there is certainly no simple answer to this question. Can the placing of nutrients in an unavailable form be regarded as a type of allelochemical production? If not a type of allelochemical production, it certainly could be an important consideration in areas of herbivore dietetics.

When is an allelochemical an allelochemical, and when is it a nutrient? The fascinating story of the bruchid beetle apparently utilizing L-canavanine, as discussed earlier, is an excellent illustration of the fact that the definitions of both "nutrient" and "allelochemical" are situation dependent and are not characteristics of the molecule as such. Similarly, ascorbic acid is required by most insects but can inhibit reproduction in some (Bridges and Norris, 1977).

Perhaps some of these questions for which there is no clear-cut answer will point the way toward increased awareness of the tremendous complexity of herbivore dietetics.

ACKNOWLEDGMENTS

I wish to thank D. H. Janzen for his advice and encouragement during the preparation of this chapter. For valuable suggestions I thank S. D. Beck, A. E. Harper, R. Lindsay, H. R. Bird, D. Ryan, F. Gould, M. Scriber, F. Slansky, and D. McKey.

REFERENCES

Abou-Donia, M. B., Taman, F., Bakery, N. M., and El-Sebae, A. H. (1974). *Experientia* **30,** 1151–1152.

Atsatt, P. R. (1977). *Am. Nat.* **111,** 579–586.

Beck, S. D. (1956). *Ann. Entomol. Soc. Am.* **49,** 552–558.

Beck, S. D. (1957). *Ann. Entomol. Soc. Am.* **50,** 247–250.

Beck, S. D. (1972). *In* "Insect and Mite Nutrition" (J. G. Rodriguez, ed.), pp. 1–6. North-Holland Publ., Amsterdam.

Beck, S. D. (1974). *In* "Proceedings of the Summer Institute on Biological Control of Plant Insects and Diseases" (F. G. Maxwell and F. A. Harris, eds.), pp. 290–311. University Press of Mississippi, Jackson.

Beck, S. D., and Reese, J. C. (1976). *Recent Adv. Phytochem.* **10,** 41–92.

Beck, S. D., and Schoonhoven, L. M. (1979). *In* "Breeding Plants Resistant to Insects" (F. G. Maxwell and P. R. Jennings, eds.). Wiley, New York (in press).

Beck, S. D., and Smissman, E. E. (1960). *Ann. Entomol. Soc. Am.* **53,** 755–762.

Bell, E. A. (1972). *In* "Phytochemical Ecology" (J. B. Harborne, ed.), pp. 163–177. Academic Press, New York.

Bell, E. A. (1973). *In* "Toxicants Occurring Naturally in Foods" 2d Ed. pp. 153–169. Nat. Acad. Sci., Washington, D. C.

Bell, E. A. (1976). *FEBS Lett.* **64,** 29–35.

Benevenga, N. J., Yeh, M.-H., and Lalich, J. J. (1976). *J. Nutr.* **106,** 1714–1720.

Bernays, E. A. (1978). *Entomol. Exp. Appl.* **24,** 244–253.

Birk, Y. (1969). *In* "Toxic Constituents of Plant Foodstuffs" (I. E. Liener ed.), pp. 169–210. Academic Press, New York.

Birk, Y., and Applebaum, S. W. (1960). *Enzymologia* **22,** 318–326.

Bowers, W. S., Ohta, T., Cleere, J. S., and Marsella, P. A. (1976). *Science* **193,** 542–547.

Bridges, J. R., and Norris, D. M. (1977). *J. Insect Physiol.* **23,** 497–501.

Carlisle, D. B., and Ellis, P. E. (1968). *Science* **159,** 1472–1474.

Chapman, R. F. (1974). *Bull. Entomol. Res.* **64,** 339–363.

Cheeke, P. R. (1972). *Nutr. Rep. Int.* **5,** 159–170.

Clement, H., and Renner, R. (1977). *J. Nutr.* **107,** 251–260.
Dahlman, D. L., and Rosenthal, G. A. (1976). *J. Insect Physiol.* **22,** 265–271.
Delort-Laval, J., and Viroben, G. (1969). *C. R. Hebd. Seances Acad. Sci.* **269,** 1558.
Desmarchelier, J. M., and Fukuto, T. R. (1974). *J. Econ. Entomol.* **67,** 153–158.
DiPalma, J. R., and Ritchie, D. M. (1977). *Annu. Rev. Pharmacol. Toxicol.* **17,** 133–148.
Dittrich, V., Streibert, P., and Bathe, P. A. (1974). *Environ. Entomol.* **3,** 534–540.
Eisner, T., Johnessee, J. S., Carrel, J., Hendry, L. B., and Meinwald, J. (1974). *Science* **184,** 996–999.
Elliger, C. A., Zinkel, D. F., Chan, B. G., and Waiss, A. C., Jr. (1976). *Experientia* **32,** 1364–1365.
Erickson, J. M., and Feeny, P. (1974). *Ecology* **55,** 103–111.
Feeney, R. E., Means, G. E., and Bigler, J. C. (1969). *J. Biol. Chem.* **244,** 1957–1960.
Feeny, P. P. (1968a). *J. Insect Physiol.* **14,** 805–817.
Feeny, P. P. (1968b). *Phytochemistry* **7,** 871–880.
Feeny, P. P. (1970). *Ecology* **51,** 565–581.
Finlay, T. H., Dharmgrongartama, E. D., and Perlman, G. E. (1973). *J. Biol. Chem.* **248,** 4827–4833.
Gontzea, I., and Sutzescu, P. (1968). "Natural Antinutritive Substances in Foodstuffs and Forages." Karger, Basel.
Gordon, H. T. (1961). *Annu. Rev. Entomol.* **6,** 27–54.
Gothilf, S., and Beck, S. D. (1967). *J. Insect Physiol.* **13,** 1039–1053.
Green, T. R., and Ryan, C. A. (1972). *Science* **175,** 776–777.
Green, T. R., and Ryan, C. A. (1973). *Plant Physiol.* **51,** 19–21.
Harper, A. E. (1973). *In* "Toxicants Occurring Naturally in Foods" 2d Ed. pp. 130–152. Nat. Acad. Sci., Washington, D.C.
Harper, A. E., Benevenga, N. J., and Wohlheuter, R. M. (1970). *Physiol. Rev.* **50,** 428–558.
Harry, P., Dror, Y., and Applebaum, S. W. (1976). *Insect Biochem.* **6,** 273–279.
Hayes, K. C., and Hegsted, D. M. (1973). *In* "Toxicants Occurring Naturally in Foods," (2d Ed.) pp. 235–253. Nat. Acad. Sci., Washington, D.C.
Hedin, P. A., Thompson, A. C., and Gueldner, R. C. (1976). *Recent Adv. Phytochem.* **10,** 271–350.
Heftmann, E. (1970). *Recent Adv. Phytochem.* **3,** 211–277.
Heftmann, E. (1975a). *Phytochemistry* **14,** 891–901.
Heftmann, E. (1975b). *Lloydia* **38,** 195–209.
Hegdekar, B. M. (1970). *J. Econ. Entomol.* **63,** 1950–1956.
Henneberry, T. J. (1964). *J. Econ. Entomol.* **57,** 674–676.
Hornstra, G. (1972). *Nutr. Metab.* **14,** 282–297.
House, H. L. (1969). *Entomol. Exp. Appl.* **12,** 651–669.
House, H. L. (1971). *J. Insect Physiol.* **17,** 1225–1238.
Isogai, A., Chang, C.-F., Murakoshi, S., Suzuki, A., and Tamura, S. (1973a). *J. Agric. Chem. Soc. Jpn.* **47,** 443–447.
Isogai, A., Murokoshi, S., Suzuki, A., and Tamura, S. (1973b). *J. Agric. Chem. Soc. Jpn.* **47,** 449–453.
Jaffé, W. G. (1969). *In* "Toxic Constituents of Plant Foodstuffs" (I. E. Liener, ed.), pp. 69–101. Academic Press, New York.
Janzen, D. H., Juster, H. B., and Liener, I. E. (1976). *Science* **192,** 795–796.
Janzen, D. H., Juster, H. B., and Bell, E. A. (1977). *Phytochemistry* **16,** 223–227.
Jones, W. T., and Mangan, J. L. (1977). *J. Sci. Food Agric.* **28,** 126–136.
Joslyn, M. A., and Goldstein, J. I. (1964). *Adv. Food Res.* **13,** 179–217.
Kato, M., and Yamada, H. (1963–1964). *Rev. Ver Soie* **15–16,** 85–92.

Kato, M., and Yamada, H. (1966). *Life Sci.* **5,** 717–722.
Kefford, J. F., and Chandler, B. V. (1970). "The Chemical Constituents of Citrus Fruits." Academic Press, New York.
Khayambashi, H., and Lyman, R. L. (1966). *J. Nutr.* **89,** 455–464.
King, D. S. (1972). *Am. Zool.* **12,** 343–345.
Kogan, M., and Cope, D. (1974). *Ann. Entomol. Soc. Am.* **74,** 66–72.
Krieger, R. I., Feeny, P. P., and Wilkinson, C. F. (1971). *Science* **172,** 579–581.
Kunitz, M. (1945). *Science* **101,** 668–669.
Kunitz, M. (1946). *J. Gen. Physiol.* **29,** 149–154.
Kunitz, M. (1947). *J. Gen. Physiol.* **30,** 291–310.
Leopold, A. S., Erwin, M., Oh, J., and Browning, B. (1976). *Science* **191,** 98–100.
Levin, D. A. (1976). *Am. Nat.* **110,** 261–284.
Liener, I. E. (1969). *In* "Toxic Constituents of Plant Foodstuffs" (I. E. Liener, ed.), pp. 409–448. Academic Press, New York.
Liener, I. E. (1974). *J. Agric. Food Chem.* **22,** 17–22.
Liener, I. E. (1976a). *J. Food Sci.* **41,** 1076–1081.
Liener, I. E. (1976b). *Annu. Rev. Plant Physiol.* **27,** 291–319.
Liener, I. E., and Kakade, M. L. (1969). *In* "Toxic Constituents of Plant Foodstuffs" (I. E. Liener, ed.), pp. 8–68. Academic Press, New York.
Liener, I. E., Deuel, H. J., Jr., and Fredd, H. L. (1949). *J. Nutr.* **39,** 325–339.
Luckey, T. D. (1968). *J. Econ. Entomol.* **61,** 7–12.
Lyman, R. L., and Lepkovsky, S. (1957). *J. Nutr.* **62,** 269–284.
Lyman, R. L., Olds, B. A., and Green, G. M. (1974). *J. Nutr.* **104,** 105–110.
McFarland, D., and Ryan, C. A. (1974). *Plant Physiol.* **54,** 706–708.
Madar, Z., Tencer, Y., Gertler, A., and Birk, Y. (1976). *Nutr. Metab.* **20,** 234–242.
Mallory, P. A., and Travis, J. (1975). *Am. J. Clin. Nutr.* **28,** 823–830.
Marquardt, R. R., Ward, A. T., Campbell, L. D., and Cansfield, P. E. (1977). *J. Nutr.* **107,** 1313–1324.
Mathé, D., Lutton, C., Rautureau, J., Coste, T., Gouffier, E., Sulpice, J. C., and Chevallier, F. (1977). *J. Nutr.* **107,** 466–474.
Mathew, A. G., and Parpia, H. A. B. (1971). *Adv. Food Res.* **19,** 75–145.
Mattson, F. H. (1973). *In* "Toxicants Occurring Naturally in Foods," 2d Ed. pp. 189–209. Nat. Acad. Sci. Washington, D.C.
Mattson, F. H., Volpenhein, R. A., and Erickson, B. A. (1977). *J. Nutr.* **107,** 1139–1146.
Mitchell, H. L., Parrish, D. B., Cormey, M., and Wassom, C. E. (1976). *J. Agric. Food Chem.* **24,** 1254–1255.
Morcos, S. R. (1970). *J. Nutr.* **100,** 1165–1172.
Morrow, P. A., Bellas, T. E., and Eisner, T. (1976). *Oecologia* **24,** 193–206.
Nakanishi, K., Moriyama, H., Okauchi, T., Fujioka, S., and Koreeda, M. (1972). *Science* **176,** 51–52.
Nelson, T. S. (1967). *Poult. Sci.* **46,** 862–871.
Norris, D. M., Ferkovich, S. M., Baker, J. E., Rozental, J. M., Borg, T. K. (1971). *J. Insect Physiol.* **17,** 85–97.
Oberleas, D. (1973). *In* "Toxicants Occurring Naturally in Foods," 2d Ed. pp. 363–371. Nat. Acad. Sci., Washington, D.C.
Ohta, T., Kuhr, R. J., and Bowers, W. S. (1977). *Agric. Food Chem.* **25,** 478–481.
Oltersdorf, U., Miltenberger, R., and Cremer, H.-D. (1977). *World Rev. Nutr. Diet.* **26,** 41–134.
Pratt, G. E., and Bowers, W. S. (1977). *Nature (London)* **265,** 548–550.
Reese, J. C. (1977). *ACS Symp. Ser.* **62,** 129–152.

Reese, J. C., and Beck, S. D. (1976a). *Ann. Entomol. Soc. Am.* **69,** 59–67.
Reese, J. C., and Beck, S. D. (1976b). *Ann. Entomol. Soc. Am.* **69,** 68–72.
Reese, J. C., and Beck, S. D. (1976c). *Ann. Entomol. Soc. Am.* **69,** 999–1003.
Reese, J. C., and Beck, S. D. (1976d). *In* "The Host-Plant in Relation to Insect Behaviour and Reproduction" (T. Jermy, ed.), pp. 217–221. Plenum, New York.
Reese, J. C., and Beck, S. D. (1978). *J. Insect Physiol.* **24,** 473–479.
Rehr, S. S., Janzen, D. H., and Feeny, P. P. (1973). *Science* **181,** 81–82.
Reynolds, T. M. (1963). *Adv. Food Res.* **12,** 1–52.
Roeske, C. N., Seiber, J. N., Brower, L. P., and Moffitt, C. M. (1976). *Recent Adv. Phytochem.* **10,** 93–167.
Rojas, S. W., and Scott, M. L. (1969). *Poult. Sci.* **48,** 819–835.
Rosenthal, G. A. (1977). *Q. Rev. Biol.* **52,** 155–178.
Rosenthal, G. A., Dahlman, D. L., and Janzen, D. H. (1976). *Science* **192,** 256–258.
Rosenthal, G. A., Janzen, D. H., and Dahlman, D. L. (1977). *Science* **196,** 658–660.
Rothschild, M., Rowan, M. G., and Fairbairn, J. W. (1977). *Nature (London)* **266,** 650–651.
Rotruck, J. T., and Boggs, R. W. (1977). *J. Nutr.* **107,** 357–362.
Rozental, J. M., and Norris, D. M. (1973). *Nature (London)* **244,** 370–371.
Rozental, J. M., Singer, G., and Norris, D. M. (1975). *Biochem. Biophys. Res. Commun.* **65,** 1040–1046.
Ryan, C. A. (1973). *Annu. Rev. Plant Physiol.* **24,** 173–196.
Ryan, C. A., and Green, T. R. (1974). *Recent Adv. Phytochem.* **8,** 123–140.
Schaible, P. J., and Bandemer, S. (1946). *Poult. Sci.* **25,** 456–459.
Schaible, P. J., Moore, L. A., and Moore, J. M. (1934). *Science* **79,** 372.
Schingoethe, D. J., Tideman, L. J., and Uckert, J. R. (1974). *J. Nutr.* **104,** 1304–1312.
Scott, A. I. (1974). *Science* **184,** 760–764.
Shany, S., Gestener, B., Birk, Y., and Bondi, A. (1970). *J. Sci. Food Agric.* **21,** 508–510.
Shaver, T. N., and Wolfenbarger, D. A. (1976). *Environ. Entomol.* **5,** 192–194.
Shaver, T. N., Lukefahr, M. J., and Garcia, J. A. (1970). *J. Econ. Entomol.* **63,** 1544–1546.
Shigematsu, H., Moriyama, H., and Arai, N. (1974). *J. Insect Physiol.* **20,** 867–875.
Shorey, H. H., and McKelvey, J. J., Jr. (1977). "Chemical Control of Insect Behavior." Wiley, New York.
Singleton, V. L., and Esau, P. (1969). "Phenolic Substances in Grapes and Their Wine and Their Significance." Academic Press, New York.
Singleton, V. L., and Kratzer, F. H. (1973). *In* "Toxicants Occurring Naturally in Foods," 2d Ed. pp. 309–345. Nat. Acad. Sci., Washington, D.C.
Skutches, C. L., Herman, D. L., and Smith, F. H. (1974). *J. Nutr.* **104,** 415–422.
Slansky, F., Jr., and Feeny, P. (1977). *Ecol. Monogr.* **47,** 209–228.
Somogyi, J. C. (1973). *In* "Toxicants Occurring Naturally in Foods," 2d Ed. pp. 254–275. Natl. Acad. Sci., Washington, D.C.
Soo Hoo, C. F., and Fraenkel, G. (1966a). *J. Insect Physiol.* **12,** 693–709.
Soo Hoo, C. F., and Fraenkel, G. (1966b). *J. Insect Physiol.* **12,** 711–730.
Srivastava, P. N., and Auclair, J. L. (1975). *J. Insect Physiol.* **21,** 1865–1871.
Stephenson, E. L. (1973). *In* "Effect of Processing on the Nutritional Value of Feeds," pp. 67–71. Nat. Acad. Sci., Washington, D.C.
Strong, F. M. (1973). *In* "Toxicants Occurring Naturally in Foods," 2d Ed. pp. 1–5. Natl. Acad. Sci., Washington, D.C.
Su, H. C. F., Speirs, R. D., and Mahany, P. G. (1974). *J. Ga. Entomol. Soc.* **9,** 86–87.
Summers, J. D., and Leeson, S. (1977). *Poult. Sci.* **56,** 25–35.
Todd, G. W., Getahun, A., and Cress, D. C. (1971). *Ann. Entomol. Soc. Am.* **64,** 718–722.
Vanderzant, E. S., and Chremos, J. H. (1971). *Ann. Entomol. Soc. Am.* **64,** 480–485.

Waldbauer, G. P. (1964). *Entomol. Exp. Appl.* **7,** 253–269.
Waldbauer, G. P. (1968). *Adv. Insect Physiol.* **5,** 229–288.
Walker-Simmons, M., and Ryan, C. A. (1977). *Plant Physiol.* **59,** 437–439.
Ward, A. T., Marquardt, R. R., and Campbell, L. D. (1977). *J. Nutr.* **107,** 1325–1334.
Weiel, J., and Hapner, K. D. (1976). *Phytochemistry* **15,** 1885–1887.
Welch, R. M., House, W. A., and Van Campen, D. (1977). *J. Nutr.* **107,** 929–933.
Whitaker, J. R., and Feeney, R. E. (1973). *In* "Toxicants Occurring Naturally in Foods," pp. 276–298. Nat. Acad. Sci., Washington, D.C.
Williams, C. M. (1970). *In* "Chemical Ecology" (E. Sondheimer and J. B. Simeone, eds.), pp. 103–132. Academic Press, New York.
Yamada, H., and Kato, M. (1966). *Proc. Jpn. Acad.* **42,** 399–403.
Yamamoto, Y., Katayama, H., and Muramatsu, K. (1976). *J. Nutr. Sci. Vitaminol.* **22,** 467–475.
Yen, J. T., Jensen, A. H., and Simon, J. (1977). *J. Nutr.* **107,** 156–165.

Chapter **8**

New Horizons in the Biology of Plant Defenses

DANIEL H. JANZEN

I. INTRODUCTION

The answer to Why do "all the good things which an animal likes have the wrong sort of swallow or too many spikes"? (Milne, 1928) is that the herbivores selected the plants to be that way. Given the acceptance of this answer, there appears to be no more intellectual content to be discovered in the study of the biology of secondary compounds. There appears to be only the working out of the detailed mechanics of how secondary compounds are made, what they cost, how they affect an herbivore, how they are avoided, how they are genetically programmed, etc. However, once Darwinian selection became linked with genetics, the same could be said for all areas of evolutionary biology. So do we pack up and go home? No, I vote for absorption in the challenge of figuring out the details of how systems work, systems that are by and large invisible to us because we are too large, because we cannot go back in time, or because our presence stops the system. I cannot watch a cell construct a morphine molecule, I cannot see how a mastodon responded to a *Simaba cedron* fruit 15,000 years ago, and I cannot watch an

ISBN 0-12-597180-X

agouti eating wild seeds since it refuses to eat when I approach it in the rain forest. There are new horizons in the biology of secondary compounds that are of great importance in understanding human feeding and medical biology and involve multiple intricate puzzles about how animals and plants interact. After all, herbivores are responsible for the caffeine in your morning coffee, the tannin to make leather shoes, and synthetic pesticides in the environment.

It is my intent in this chapter to underline some areas of research in plant defense biology where major questions are seemingly being ignored or the major questions do not seem to correspond to observations in the field.

II. FACULTATIVE DEFENSES

The defense systems of all organisms contain facultative as well as standing components. The slower, the more heterogeneous, flexible, and unpredictable the challenges, the more important become the facultative components. It is conspicuous that understanding the defenses of plants against fungi, nematodes, and microbial attack has always involved the study of facultative defenses (e.g., Stoessl, 1970). Secondary deposition of tannin around an area of wood or leaf invaded by fungi has been known since the turn of the century (and probably much earlier), and many phytopathologists have gone so far as to describe an ecological group of secondary compounds as "phytoalexins" (Cruickshank, 1963). Phytoalexins are narrowly defined as those defensive compounds produced in direct response to microbial or fungal invasion of cells (e.g., Stoessl, 1970; Strong, 1977; Harborne, 1978). It is equally conspicuous that researchers working with herbivores larger than fungi were extremely slow to become aware of either the fact of facultative defenses against small beasts or the generalization that there ought to be facultative defenses against all sorts of herbivores. Until Ryan and associates studied the induction of protease inhibitors in foliage following herbivory (for a review, see Ryan, 1978), about the only example was that of African acacias, which make longer spines on shorter internodes following browsing of shoot tips by big mammals. Incidentally, this phenomenon is readily observable in other species of African arid-land shrubs and presumably reflects the facts that spininess is expensive, a substantial number of the members of the plant population are inaccessible to the animals in any given generation, and a plant grows through a susceptible low stage to an unsusceptible tall stage.

Given that we recognize environmental challenges of a type that should produce facultative defenses in plants, where are the new horizons? Ryan (1978) sees the new horizons through ever finer dissections of the system to determine how the plant hormonally controls biosynthesis and storage of defense compounds on the spot at the time of damage—a biochemist's approach. An ecologist, however, tends to go in the other direction. When a leaf-cutter ant colony (*Atta* spp.), a population of moth larvae, or a howler monkey (*Alouatta* spp.) eats the new leaf crop off a tropical deciduous forest tree at the beginning of the rainy season but does not do the same to a replacement leaf crop produced a few weeks later, is it because these animals are prudent harvesters of their resources, is it because they have to mix their diets through time (Freeland and Janzen, 1974), or is it because the replacement leaf crop is different from the original? Development of the last idea is at best embryonic (Janzen, 1978a). When Ryan (1978) tells us that proteinase inhibitors accumulate within a few hours to the highest levels in a leaf that has just had bites taken out of it, the field ecologist's mind turns immediately to the view up through the crown of a tree, the leaves of which have recently been severely damaged by a generation of caterpillars or beetles. Ryan certainly has given us a testable hypothesis as to why these insect larvae often seem to choose an undamaged leaf blade to eat when starting a feeding bout, with the result that the feeding holes are quite evenly distributed over the leaf surface of the tree crown.

This observation, and the postulated cause, bring to mind a totally unexplored area of leaf defense biology. In agricultural entomology it should be a standing principle that the crop pest is reduced in density only to the point where its effects are just slightly greater than the cost of getting rid of it. One doesn't pay $21 an acre to get rid of a pest that damages $3 worth of crops an acre, but one might pay $2.95 an acre to remove the pest. The same applies to the budget of a wild plant. At the time of the evolution of facultative proteinase inhibitor defenses, natural selection should have driven up the speed and quantity of production until the return to the plant in lowered damage about equaled or slightly surpassed the cost of production. But note that "lowered damage" potentially comes in more ways than just lowering the amount of leaf tissue eaten. There are at least two other ways that facultative defenses could be selected for other than in lowering the amount of leaf tissue eaten. One could be in forcing the caterpillars of highly host-specific insects to run their development cycles at maximal rates, thereby minimizing both the duration of the period before the tree can put out a replacement leaf crop and the number of leaves sufficiently damaged to make retention of the remains uneconomical. The aborted partly damaged leaves are replaced, but all that was aborted is a loss caused by the herbivore yet not consumed by the herbivore. The other is that there is undoubtedly some pattern by which, say, 40% loss of leaf surface area can be distributed over the surface of a tree for minimal reduction of the fitness of the tree. It would be most interesting if it turned out that the dispersed pattern of feeding generated by facultative chemical defenses was just this pattern. Here, then, I am postulating that facultative chemical defenses may have been strongly selected for without actually reducing the amount of leaf tissue removed by chewing but rather by influencing the pattern of chewing. This may even be one of those magical cases in which a large return is gained by the plant at a small loss to the herbivore. In fact, if the facultative response by the plant were to raise the levels of proteinaceous protease inhibitor, for example, in all leaves as high as in the leaf recently chewed on (surely this is a physiological possibility, even if an expensive one), the specialist herbivores might simply respond evolutionarily by becoming good at harvesting all that protein.

Although beyond even the wildest of a Darwinist's dreams, it is just possible that some plants can evaluate the kind of leaf-clipping herbivore to which they have been subject and generate an appropriate facultative response. It has been shown that when grass blades are clipped off by a grasshopper or cow, a prairie grass plant responds differently than when the blades are cut with clippers; apparently the saliva makes the difference (Mel Dyer, personal communication). This is an area of plant defense biology that must be approached with extreme caution. Adding various primary metabolites to bacterial systems that are synthesizing secondary compounds can strongly perturb the production status quo (e.g., Drew and Demain, 1977). Damage to living plants can easily mimic this process, and the products may be nothing more than the nonsensical results of a complex machine running out of control. Furthermore, a change in secondary-compound chemistry following foliage damage may be the result of a rearrangement of internal priorities rather than an explicit attempt to produce facultative defenses. For example, it is unlikely that the lowering of alkaloid production by clipping of larkspur (*Delphinium occidentale*) (Laycock, 1975) is a facultative defense against the herbivores that clipping mimics.

Agriculturalists growing plants for commercial extraction of secondary compounds are well aware that the absolute quantity per unit plant tissue harvested changes with the physical environmental conditions and the health of the plant. Of a slightly more esoteric nature, several studies have shown that the concentrations of phenolics in tobacco and *Helianthus annuus* plants vary with "stress" and nutrient conditions for the plant (Wender, 1970; del Moral, 1972). It is obvious that many of these changes in secondary-compound chemistry may occur simply because the "stress" mimics the changes in plant physiology generated by herbivory or microbial attack in nature. On the other hand, it may well be that plants are so finely tuned to the challenges of herbivory and to the degree of fitness-lowering generated by a given act of herbivory in the context of the plant at the moment that they alter their standing defenses according to their individual circumstances. This is roughly like different householders buying different amounts of fire insurance depending on the neighborhood they live in, the details of their house, and the value of the items they have in the house at different times of the year. In nature it is obvious that different individual plants of similar age and stage are subject to different degrees of seed predation and herbivory (also see Moore, 1978a,b; Janzen 1975a, 1977a, 1978b). It is not obvious to what degree this is due to the capricious settling behavior of the herbivores, herbivore response to invisible interindividual differences in nutrient levels, and herbivore response to interindividual differences in secondary-compound defenses. Clearly all three are operative, but the last is habitually ignored by field biologists because of their frustration at being unable to instantaneously assay concentrations of the multitude of secondary-compound defenses possessed by a given plant. Only when the entire suite of defenses of a few species is worked out can those species be studied in nature with respect to the questions raised at the beginning of this paragraph. On the other hand, it is easy to predict that, once these defenses are determined we may find plants to be amazingly finely tuned in their combination of standing and facultative defenses. It might even be that the only reason why herbivores get through at all is that the time of their appearance at a given plant is extremely unpredictable, whether a given plant is to be attacked at all is very unpredictable, and the impact of that attack on the plant's fitness is very unpredictable.

III. ONE BEAST'S DRINK IS ANOTHER BEAST'S POISON

Toxicity is not an intrinsic property of any naturally occurring molecule, and secondary compounds are no exception. The allyl glucosinolate in cruciferous plants is very toxic to black swallowtail larvae (*Papilio polyxenes*) moderately toxic to southern armyworm larvae (*Spodoptera eridania*) and harmless to cabbage butterfly larvae (*Pieris rapae*) even when the concentrations in food are considerably higher than those occurring in the plants in nature (Blau *et al.*, 1978). The canavanine in seeds of *Dioclea megacarpa* is used as a nitrogen source by the larvae of the bruchid beetle *Caryedes brasiliensis* (Rosenthal *et al.*, 1978; Rosenthal, 1977). It is, however, extremely lethal at the same concentrations to the larvae of the bruchid beetle *Callosobruchus maculatus* (Janzen *et al.*, 1977) and toxic to a wide variety of other organisms (Rosenthal, 1977). Even three protein amino acids (tryptophan, cystine, and methionine) were lethal when added to the diets of *C. maculatus* larvae (Janzen *et al.*, 1977). These three compounds occur in exceptionally low concentrations in southern cowpea seeds (*Vigna unguiculata*), the normal food of *C. maculatus*. In short, it has become obvious that with respect to a particular organism a toxin or other disrupter of development is to be defined by its effect rather than by some intrinsic characteristic of the molecule. Of course, certain molecular structures have a higher probability of being metabolically disruptive than others [e.g., generally alkaloids kill *C. maculatus* larvae at concentrations of 0.1 to 1% in the diet, whereas uncommon amino acids generally require 1–5% concentrations for lethal effects (Janzen *et al.*, 1977)], but it still holds that a particular compound has not been shown to lack a protective function if it is harmless to one or even many herbivores, or vice versa.

A major frontier in the biology of secondary compounds involves relaxing the phytochemist's mind to the point where a given study derives its value not from traits of the molecules at hand but from how they influence the interactions of animals and plants (and, as well, other kinds of interactions). It is clear, for example, that lectins (phytohemagglutinins) have many potential and likely biological roles both inside and outside the organism (e.g., Liener, 1976). That many different functions of lectins have recently been reported does not mean that the state of the art is very primitive or that no "real function" has appeared. There is nothing biologically improbable about lectins functioning, for example, as site-specific binding agents at low concentrations within the organism or on its surface (Liener, 1976), as effective antiherbivore compounds when in high concentrations in dormant seeds (Janzen *et al.*, 1976), and as nutrient storage proteins for the developing seedling. Biochemical biology has grown up with a tradition of "one molecule, one function"; ecology operates on a tradition of "one unit, many functions," or at least "one unit, many responses."

IV. HERBIVORES DO NOT EAT LATIN BINOMIALS

Animals generally do not feed on all parts of a plant; they usually consume quite specific parts. Red colobus monkeys (Struhsaker, 1975), black colobus monkeys (McKey, 1978), and howler monkeys (Glander,

1975, 1979) are all conspicuous in their choice of, for example, leaf petioles over leaf blades for some species, and vice versa for others. Leaf-cutter ants (Attini) feed on many Latin binomials but in fact eat only new foliage from one, only old from another, only flowers from a third species, and only new green fruit from a fourth (G. Stevens and S. Hubbell, personal communication). These preferences may even change within the year. Bruchid beetles eat only the contents of seeds, not the seed coats as well (Janzen, 1977c). Costa Rican tapirs consume ripe fruits and grind up the seeds of the Costa Rican tree *Mastichodendron capiri* but emphatically reject the foliage. The adults of *Tetraopes* beetles eat flowers and green pods but only the tips of the leaves of common milkweed (*Asclepias syriaca*) in Michigan.

Likewise, Latin binomials do not eat plants. Juvenile black rats (*Rattus rattus*) in the Galapagos peel the small fruits of *Miconia* before eating them, whereas adults eat the peel and all (Clark, 1979). Young *Ctenosaura similis* lizards are insectivorous; the adults feed almost entirely on leaves and fruits in lowland deciduous forests of Costa Rica. Adult bruchid beetles eat pollen and nectar from flowers; larvae eat the contents of ripening seeds (usually of quite different species of plants). Most adult lepidopterans take nectar from flowers or rotting fruits of species of plants totally different from the plants whose leaves are consumed by their caterpillars. A thirsty howler monkey may be unable to deal with toxins that can be flushed out of the system by a monkey satiated with water (Glander, 1978).

In short, a list of Latin binomials feeding on other Latin binomials carries almost no information when it is remembered that the secondary-compound chemistry of two different plant parts on the same plant is much more likely to be different than the same. For example, the caffeine content (percent of dry weight) of *Cola nitida* fruit is a trace, that of the seed coat is 0.44%, and that of the seed contents is 1.58% (Ogutuga, 1975); 0.1% is lethal to the seed-eating bruchid *Callosobruchus maculatus* (Janzen *et al.*, 1976). The moral becomes even clearer when it is remembered that two different life forms of the same species of animal are very likely to have different abilities to deal with secondary compounds and different physiologies to be affected by them.

By now it should also be apparent that Latin binomials do not contain secondary compounds, but rather plant parts do. There are two huge offenders in the contemporary literature. First, it is common to report results for a plant and not specify leaves, stems, or other kinds of vegetative foliage. Since concentrations are usually not given for wet or dry weights, one cannot even use such information to understand an herbivore large enough to eat entire tops off of plants. Second, this ecological grievance applies even to apparently carefully divided samples. A phytochemist might be quite proud to report that the fruit pulp and seeds were separated in the analysis (after all, the secondary compounds in a fruit are usually evolutionarily designed to get a fruit eaten and digested by some animal; the secondary compounds in a seed are usually evolutionarily designed to keep the seed from being digested by many animals). However, it is generally forgotten that many seeds are consumed by animals that eat only the contents and discard or avoid the seed coat. The seed coat may constitute as much as 70% of the seed and contains vastly different secondary compounds than do the seed contents (Janzen, 1977c). Therefore, reports of concentrations of secondary compounds in seeds are very commonly off by a factor of 20 to 100% from the viewpoint of the animal who might try to digest one. To make analysis even more difficult, there are additional facts that must be taken into consideration. For example the lectins in bean seeds are not there until just before the seed is mature (J. Hamblin, personal communication); this means that insects and vertebrates that eat the immature seeds, as they commonly do, may be avoiding this major form of defense.

The new horizon is in understanding a plant as an enormous suite of secondary-compound defenses and working with ecologists to figure out how herbivores get past a defense or complex of defenses. We do not need another random screen of qualitative alkaloid content of 5000 species of central African plants. We need a team, or a person who thinks like a team, to try to develop defense profiles through the year and development stages for a given population of plants. At present, if one shows that three of the seven species of *Acacia* in lowland Costa Rica contain cyanogenic glycosides in their foliage (E. E. Conn and D. S. Seigler, personal communication), one has not shown that the four acyanogenic species are any less well protected chemically. Likewise, the demonstration that there is a 20-fold interindividual variation in cyanogenic glycoside content in the foliage within a Costa Rican population of *Acacia farnesiana* occupying a few hectares, whereas the flavonoid content of these same leaves stays constant in kind and quantity (D. S. Seigler and E. E. Conn, personal communication), should cause one to be very wary of correlating herbivore preferences with secondary-compound foliage analyses until all of the defensive repertoire has been reported.

In short, there seem to be two productive directions to take in working out the preferences displayed by animals. One is to develop realistic artificial diets and then tip in solitary and combined secondary compounds and nutrients to define the limits of tolerances and pick out those compounds that invoke exceptionally strong reactions. The other is to focus on a few key species of plants, work out their defense repertoires in detail, and then focus on the specialist herbivores that get around these defenses and the generalists that are deterred by them. The recent work on Cruciferae and Umbelliferae by Feeny, Root, and associates (e.g., Blau *et al.*, 1978; Berenbaum, 1978), Passifloraceae by Gilbert (e.g., Smiley, 1978), canavanine-containing plants by Rosenthal, and legume seeds by Janzen has evolved in these two directions, but it is a miniscule fraction of what should be, considering the importance of secondary compounds in pharmacology, agriculture, and more esoteric ecological studies.

V. PLANTS ARE ANACHRONISMS

Evolutionary biologists are very fond either of pretending that the plant traits we see are selected for and maintained by current interactions or, at the least, of choosing to work on those systems that seem to match this assumption. However, we all know perfectly well that a plant (and its herbivores) is a collection of anachronistic traits that at any given time have caught up with contemporary selective pressures to a highly variable degree. One reason why evolutionary biologists like to sweep this fact under the rug is that for a long time it was a standard loophole for dealing with some conspicuous trait, the natural history of which was not known (the selective pressures for secondary-compound defenses have often been ignored on these grounds). To invoke currently extinct selective pressures to explain the presence of a trait, however, was to mask an incomplete study of natural history. Another reason for avoiding the anachronistic aspects of evolutionary biology is that one is quickly caught in a morass of untestable hypotheses. How does one field-test the assertion, "The single most important aspect of browsing pressure is height [of dinosaurs]" (Bakker, 1978)?

However, there is a huge body of information on the natural history of plant defenses that deserves examination in the light of a strong historical perspective. Lignin, clearly a secondary-compound defense when attached to cellulose, has been identified from Triassic age fossil wood (Sigleo, 1978). Most coal beds are fossilized, polyphenol-rich, peatlike deposits from the beds of swamps that were probably very similar in natural history to contemporary sites of peat deposition. As such they were probably populated by very few herbivores and the plants in them were very well defended with chemical traits (especially polyphenolics) selected for by the herbivores (Janzen, 1974). Regal (1977) stressed that bird-dispersed fruit biology may have been very important in the evolution of angiosperms, which brings to mind an enormous number of questions about the interactions of fruit secondary-compound chemistry and vertebrates. Do fruit flavors come from millennia of adaptation by animals to those secondary compounds placed in fruits by plants to attract the right dispersal agents and repel the wrong ones? Alternatively, do fruit flavors result from millennia of adaptation by plants to the age-old sensory receptors possessed by animals long before they ate fruits? Some combination of the two is, of course, what we are dealing with. T. Swain's examination of the taste perceptiveness of tortoises (they have trouble detecting bitter compounds) is a step in the right direction. Pollination biology studies have long assumed a large historical element to floral morphology (e.g., Sussman and Raven, 1978), but for some reason herbivore relationships do not carry this tradition. It is obvious that contemporary browsing pressures did not produce and maintain the extreme spininess still present in southwestern United States desert plants, but I know of no study that has taken up this subject in detail and related it to the large herbivore faunas that roamed this terrain until a few tens of thousands of years ago. The question of how long defense traits will persist once the herbivore is removed has simply been ignored in the ecological literature. It can be stated with certainty that the rate of appearance of a trait will be in large part a function of the intensity of selection for it, whereas its disappearance rate may depend in great part on its cost of maintenance. For example, spines (dead tissue of low initial cost and no maintenance cost) may appear very rapidly and extensively when a flora is subjected to heavy browsing but disappear very slowly because of their small drain on the resource budget.

There is one largely undeveloped approach to the extinct interactions between large vertebrate herbivores and plant defenses. Field ecologists have long recognized that when herbivorous vertebrates were (are) introduced onto oceanic islands, these herbivores grossly altered the plant species composition and interrelationships (e.g., Pickard, 1976). This is presumed to have occurred because the plant defenses were only weakly developed owing to millennia of evolution without this herbivory. However, island plants give every evidence of having come from mainland floras originally, floras presumably subject to the kinds of herbivory experienced on mainlands today. This means that the island floras have presumably lost many of their defensive traits and not that they never had them in the first place. Oddly, I cannot locate a single study of the defense repertoires of island plants as contrasted with mainland congeners, except for a study of the loss of ants as a defense in the island populations of the ant–plant *Cecropia* (Janzen, 1973; McKey and Janzen, 1977). On the other hand, a major pragmatic barrier to a study of this sort is the widespread tragedy of goat, pig, and rat introduction to even the smallest islands that were naturally free of large herbivores.

A second way of studying the interactions between extinct large vertebrates, at least in the New World, is through careful examination of the interactions between introduced large herbivores and the New World flora. Range horses, cows, burros, sheep, and goats have gone a long way toward replacing the fauna of large browsing and grazing vertebrates that become extinct between ten and twenty thousand years ago. For example, many of the chemical and physical traits of uneaten ripe large fruits in the Costa Rican lowland forests can be understood if we assume that they were coevolved for seed dispersal with this very recently extinguished fauna (Janzen and Martin, 1980). For example, in Guanacaste province, the large, round, hard, ripe fruits of the native tree *Crescentia alata* are essentially ignored by all native potential dispersal agents but very eagerly eaten by range horses. Range cattle likewise ignore the fruits of *C. alata* but avidly eat the fruits of *Pithecellobium saman,* which are in turn studiously ignored by range horses. Both animals are very effective dispersers of the seeds of the fruits they eat of these two species. The differences in secondary-compound and nutrient fruit chemistry of these two fruits could hardly be expected to be explicable in terms of contemporary native faunas but in the context of horses and cows may be very clear. It is ironic that I cannot locate a single study on browsing or wild fruit eating of free ranging populations of these two commonest large neotropical herbivores today. In the same context, the 3-in.-long, ovoid, juicy fruits of *Simaba cedron* (Simaroubaceae) fall to the rain forest floor in Corcovado National Park, southwestern Costa Rica. There the fruit pulp rots off, leaving a large, fibrous nut to germinate on the ground surface. There appears to be no dispersal agent at present. Although the fruit pulp has a fragrant smell and a sweet taste, it is conspicuously ignored by local animals and has the reputation of being lethal if eaten in large quantities by human beings. The ground under a *Simaba cedron* reminds me exactly of that under a *Balanites wilsoniana* (Balanitaceae) tree in the Ugandan forests where all the elephants have been shot. African elephants are extremely fond of the large fruits of *B. wilsoniana* and defecate the nuts unharmed, but they are eaten by none of the other large forest understory mammals. The fruit pulp of four species of African *Balanites* is 4–8% dry weight diosgenin, the precursor molecule for the manufacture of cortisone, sex hormones, and antibiotics (Hardmann, 1969, Janzen, 1978a). It is very difficult to avoid the conclusion that *S. cedron* was dispersed by mastodons and predict that the secondary compounds in the ripe fruit pulp will be atoxic to contemporary elephants.

VI. HOW DOES ONE MEASURE THE IMPACT OF HERBIVORY?

What happens to a plant when its defenses are broached or removed? Such a simple-sounding question involves the most difficult question in evolutionary biology, population biology, and ecology. The only real answer to this question depends on what happens to a plant's fitness when an herbivore takes a bite or is potentially present to take a bite. I gave one example of the complexity of the problem when discussing the possibility that selection for facultatively produced protease inhibitors was generated not through a reduction in herbivory by these compounds but rather through causing the herbivore to distribute its damage in such a manner as to minimize its impact. We can already see that the impact of herbivory cannot be blithely measured by the caloric content of what is eaten, by the grams dry or wet weight of what is removed, by the area of leaf eaten, etc. The removal of 10 cal or grams of leaf tissue means quite a different thing than the removal of the same quantity of shoot tips, since in the former case, the lost tissue can be replaced with a new leaf; but in the latter case there may well be irreparable loss in competitive status through a lowered rate of shoot elongation (e.g., Janzen, 1966). The loss of 10 cm^2 of leaf area early in the life of a leaf means an entirely different thing to the plant than the same loss well after the leaf has been amortized (e.g., Chester, 1950). Likewise, if a 40% leaf blade loss results in a leaf that still functions well enough to pay for itself, a 50% leaf blade loss may not and turn into a 100% loss through leaf abortion. In short, herbivory through the eyes of the plant cannot be measured in units of harvestable productivity. It must be measured in units of reduced fitness of the plant, and this is very difficult, to put it mildly.

How do herbivores affect the fitness of a plant? As in automobile ownership, there are two big costs related to hostiles. There is the cost of insurance and the cost of repairing the actual damage. The cost of insurance for a plant is conspicuously the standing chemical defenses, the resources and programming for facultative defenses, and the tactical losses in competition and other activities that come about through having resources tied up in defenses that cannot be mobilized to deal with other contingencies. Needless to say, herbivores can have a very large effect on a plant's ability to compete and deal with the physical environment through draining its resources for insurance and repair of damage.

There is only one way to know how this drain affects the fitness of plants, at least given our current state of knowledge: Lower the amount spend on insurance by favoring mutants or broach the defenses through artificial herbivory, and observe what happens to standard measures of fitness (seed production on maternal plants or plant parts, pollen donation by paternal plants or plant parts). Effects of herbivory on wild seed plant production have been investigated to some extent (e.g., Rockwood, 1973; Janzen, 1976; Cavers, 1973), but increased seed yields by less well-protected mutants have been examined only very indirectly in the sense that crop plants are generally more poorly protected than are their wild relatives and generally have greater seed yield. However, they have also had other portions of their resource budget diverted into seed production.

This area is wide open for study and guaranteed to yield results of interest to both agriculturalists and more esoteric biologists. Will the new lectin-free strains of soybeans (Pull *et al.*, 1978) have increased yield per plant with respect to some other trait besides seed yield? The resource that was being used in lectin production is now presumably available to the soybean plant for other functions. The increased protein content in high-protein strains of potatoes is in great part made up of the very protease inhibitors that are part of the defense of the potato (Ryan and Pearce, 1978). How does the morphine-free strain of opium poppy differ from the normal poppy in competitive ability and seed production per plant lifetime? These kinds of questions must be asked of wild plants, where the ultimate equalizer is the contribution to future generations through seeds or pollen. Would a phytolith-free strain of grass be fed on selectively by grazers (see Walker *et al.*, 1978) but have higher leaf production than normal grass? A toxin-free mutant may use the newly available resource in competitive growth and turn out to have a much higher fitness than the wild type, or it may have to use this resource to repair the increased damage incurred by the less well protected plant and thereby end up with the same fitness as the wild type. This kind of field biology is so poorly understood that I can cite no examples.

However, measuring the impact of herbivory has led us into one very confusing area of the biology of the interactions of animals and plants. One class of herbivory, involving ripe fruit, pollen grain, and nectar consumption, clearly raises the fitness of the plant. I caution immediately, however, that not all fruit consumers raise the fitness of the plant (Janzen, 1975b, 1977d, 1979; Howe, 1977; Howe and Estabrook, 1977), and certainly many pollen and nectar consumers can be regarded as nothing but thieves. A second class of herbivory appears to raise the fitness of the plant, but much more field work is needed to solve the puzzle. These are the cases in which the plant's physiological response to certain kinds of herbivory actually appears to produce a plant that has a higher fitness than its uneaten conspecifics. This effect is to be expected in a small number of cases when any complex system is perturbed many times (e.g., the natural history of mutation). The question is, under what circumstances will the situation persist whereby the plant depends on an external agent to perform a physiological task that is usually handled by the plant itself and therefore lacks control over its own actions? This differs from the systems involved in outcrossing pollination and seed dispersal since most plants cannot do that themselves.

Some examples will probably make this clearer. In at least two complex and well-developed ant–plant mutualisms, African *Barteria* trees (Janzen, 1972) and neotropical *Cecropia* trees (Janzen, 1973), the ants maintain a standing crop of scale insects or other homopterans inside the hollow stems. These animals feed on the plant and provide a major food source for the ants with their bodies or their honeydew exudates. The ants are obligate occupiers of the trees and protect the trees from herbivores and vines. The homopterans are zoological devices used by the plants to maintain an ant colony, the ants being directly analogous to the chemical defenses maintained (and paid for) by more ordinary plants. I would not expect there to be selection for traits that reduce the "damage" done by the Homoptera to the level that would debilitate the ant colony and its protection of the tree. The ant–plant has lost some, but not all, control over its use by an herbivore.

Simberloff *et al.* (1978) presented a reasonable case that the normal multiple root-branching pattern of the mangrove *Rhizophora mangle* is the outcome of repeated and frequent invertebrate attacks of the aerial root tips. They argued that this damage may actually raise the fitness of the plant, since a large number of aerial roots aid the plant in support, nutrient gathering, and oxygen uptake. Is it possible that the mangrove would produce its own root bifurcations to generate the number needed if the invertebrates were excluded (and therefore if the damage associated with invertebrate-induced bifurcations were a drain with no return on the resource budget of this plant)? Also, even if the presence of the invertebrates can be viewed as a "bifurcation-inducing mutation," it would be very exceptional if such damage generated the exact growth response needed to raise the fitness as would a physiological mutation producing the same effect without relinquishing control to an external agent. However, the point of Simberloff *et al.*, that a plant's postdamage responses must be considered in the equation when one is tallying the effect of an herbivore on a plant, is well taken.

The sentiments expressed in the following quotation (from Mattson and Addy, 1975) occur periodically in the ecological literature and can almost always be attributed to those who focus on ecosystems rather than their parts (e.g., Harris, 1973; Owen and Wiegert, 1976; Hendry *et al.*, 1976): "Normal insect grazing (from 5 to 30 percent of annual foliage crops) usually does not impair plant (primary) production. In fact, it may accelerate growth. . . . after an outbreak has subsided, there is evidence that the residual vegetation is more productive than the vegetation that was growing immediately before the outbreak." In evaluating such comments with respect to understanding whether herbivores select for chemical defense mechanisms in plants, there are two considerations. Fitness is not measured in units of "production"; only a very small portion of a plant's total resource budget goes into what is commonly measured in "production" (leaf weight, wood increment, etc.), and a plant is quite likely to maintain certain activities at normal levels very much at the expense of others in responding to damage by herbivores [e.g., see reviews of responses to defoliation by Chester (1950), Kulman (1971), and Rockwood (1973)]. For example, a plant that has a medium-aged static crown stripped of its leaves by caterpillars may well turn on a growth phase to replace them and be therefore much more "productive" than the undamaged crown, but consideration of the entire resource budget of the plant makes the error in this reasoning obvious. Such a statement also ignores the insurance expenditures made by plants on defenses, expenditures that otherwise could have gone into growth or reproduction.

Crop plants are particularly deceptive when one is assessing responses to herbivory. For example, light trimming, browsing, or chemical defoliating can increase seed yield in cotton, soybeans, and beans. This damage probably breaks apical dominance of the main shoot, leading to a more bushlike plant. Such a plant catches sunlight better only in nicely spaced crop systems. In wild vegetation with tight intercrown competition, such a loss of height could lead to an irreparable loss of competitive status.

We badly need field experiments with wild plants designed to show how the fitness of these plants is affected by herbivory of different types applied in a variety of competitive and edaphic circumstances. Since plants are well enough defended so that much of this herbivory is very unlikely to occur in a replicated and controlled manner, the arena is very open for the artificial herbivore, especially one that acts in a manner that mimics an herbivore immune to a particular portion of the plant's defenses.

VII. PITFALLS

The biology of plant defenses is beset with more than the usual number of methodological and conceptual pitfalls for an area in science. Much of the confusion could be derived from the melding of two areas in which the practitioners know little of each other's assumptions and weak points. The confusion is not helped by the persistence of researchers lacking strong familiarity with plant or herbivore natural history or with the natural history of secondary compounds (e.g., Jermy, 1976).

Natural history fine points are not trivial. They can be the turning points in many systems. A leaf-cutter ant colony may strip a tree crown of its leaves in 24 hours, whereas a cohort of caterpillars may require 10 days to do the same thing. A facultative defense may be strongly selected for by the latter and strongly selected against by the former. Which kinds of chemical defenses one finds interesting or important, then, may depend on which of these two herbivores one happens to be working with for quite capricious reasons. A person noting that very new foliage of woody plants is commonly red or otherwise strongly pigmented might decide to do feeding tests with new versus old foliage on generalized herbivores as a way of arguing that these pigments are defensive against herbivores. Such a study would undoubtedly show that the very new foliage is avoided by many generalist herbivores. However, the conclusion that the pigments are responsible is preposterous. It ignores the considerable body of information suggesting that new foliage can be very different from mature foliage in chemistry through the possession of a wide variety of known compounds well known to be toxic to animals (e.g., McKey, 1974) and the fact that there is no hint in the literature that the anthocyanins, commonly responsible for the coloring of new leaves, have any effect on herbivores when ingested. When Fowden (1972) isolated several new uncommon amino acids at extremely low concentrations from extremely large quantities of sugar beet waste, he appeared to have made a case for the existence of very small quantities of some secondary compounds in many plants. This indicates that perhaps these compounds had a physiological role in the plant and then later in evolutionary time were produced at higher concentrations through natural selection by herbivores. It is striking how tenaciously natural products chemists cling to the traditional belief that the compounds found in a plant must be of importance within that plant, a belief that no person entering the field of biology today would ever dream of holding if he or she had access to what we now know about plant biology. The uncommon amino acids found at extremely low concentrations in sugar beets could just as easily have been from living or dead environmental contaminants of the initial refuse, from degradation or accidental synthesis in the incredible soup of enzymes and substrates represented by the 10^9 kg of mashed sugar beets, from ordinary biochemical mistakes being made in the cells at the instant of mashing, or from protein amino acids modified after polypeptide synthesis as part of the final tailoring process for the molecule (e.g., Uy and Wold, 1977). Certainly the presence of these compounds is very weak evidence for the existence of genetic programming in the sugar beet genome for the production of very small amounts of these uncommon amino acids.

VIII. ONE LINERS

No plant is an island. It is quite clear that the arrival and departure of herbivores from a given plant depends not only on the defenses intrinsic to that plant, but also on the alternate foods available to the herbivore and the obfuscating nature of the chemical environment generated by the neighboring plants (e.g., McNaughton, 1977; Atstatt and O'Dowd, 1976).

Parasitic plants are herbivores; virtually everything discussed here—host specificity, consumer secondary-compound detoxification, problems in estimating damage, and facultative defenses—applies directly to the attack of plants by parasitic plants (e.g., Atstatt, 1977).

An herbivore is a walking compost heap; virtually all herbivores [with the exception of a wood-boring isopod (Boyle and Mitchell, 1978)] are very dependent on their gut flora for aid in the degradation of secondary compounds in plant parts (Freeland and Janzen, 1974; Langham and Smith, 1970; Oh *et al.*, 1967; Westermarck, 1959). The rumen of wild herbivores should probably be seen more in this light than as a device for gathering calories and proteins (even cellulose can be viewed as a secondary defensive compound, since its extreme indigestibility to both higher plants and animals is probably not a piece of seredipitous natural history).

An herbivore species may consist of many populations, each adapted to broach the defenses of a particular host population. Scale insects (extremely sedentary as adults, aerial plankton as immatures) may even have populations adapted to the defenses of an individual tree (Edmunds and Alstad, 1978), whereas one widespread "species" of bruchid beetle or sphinx moth may have as many as 20 allopatric populations, each with larvae that are locally adapted to the defenses of one or two different species of plant.

Why do rain forest understory shrubs contain a large amount and diversity of so-called trace elements such as boron and cobalt (Golley *et al.*, 1978)? I hypothesized that the heavily shaded rain forest understory is a resource-poor habitat where chemical defenses are of utmost importance (Janzen, 1974). Large quantities and many kinds of secondary compounds may require large quantities and many kinds of coenzymes for proper synthesis, and coenzymes normally contain a molecule of a so-called trace element (Janzen, 1977b). Different habitats are likely to have different overall levels of chemical defenses in their floras. On the worst soils or otherwise stressful sites, foliage should be the richest in chemical defenses because its loss to the plant should cause the most severe reduction in fitness (Janzen, 1974; McKey *et al.*, 1979) whereas in habitats at the opposite extreme, high interspecific variability in the intensity of defense is expected owing to mixes of life forms (for an example of ecogeographic patterns in alkaloids, see Levin and York, 1978).

I conclude with the comment that understanding the biology of secondary compounds requires the cooperative efforts (or at least the thoughts) of a greater variety of biologists and chemists than does any other area of biology. The new horizons lie in taking what we already know and applying it in straightforward experimental manipulations of living plants and herbivores in the field, in the laboratory, and on paper.

ACKNOWLEDGMENTS

This study was supported by NSF Grant DEB77-04889. D. E. Gladstone offered constructive comments.

REFERENCES

Atstatt, P. R. (1977). *Amer. Nat.* **111,** 579–586.
Atstatt, P. R., and O'Dowd, D. J. (1976). *Science* **193,** 24–29.
Bakker, R. T. (1978). *Nature (London)* **274,** 661–663.
Berenbaum, M. (1978). *Science* **201,** 532–534.
Blau, P. A., Feeny, P., Contardo, L., and Robson, D. S. (1978). *Science* **200,** 1296–1298.
Boyle, P. J., and Mitchell, R. (1978). *Science* **200,** 1157–1159.

Cavers, P. B. (1973). *Proc. Int. Symp. Biol. Control Weeds,* Misc. Publ. No. 6, pp. 140–144.
Chester, K. S. (1950). *Plant Dis. Rep., Suppl.* **193,** 190–362.
Clark, D. A. (1979). *Ecology* (in press).
Cruickshank, I. A. M. (1963). *Ann. Rev. Phytopathol.* **1,** 351–374.
del Moral, R. (1972). *Oecologia* **9,** 289–300.
Drew, S. W., and Demain, A. L. (1977). *Ann. Rev. Microbiol.* **31,** 343–356.
Edmunds, G. F., and Alstad, D. N. (1978). *Science* **199,** 941–945.
Fowden, L. (1972). *Phytochemistry* **11,** 2271–2276.
Freeland, W. J., and Janzen, D. H. (1974). *Am. Nat.* **108,** 269–289.
Glander, K. E. (1975). Habitat and resource utilization: an ecological view of social organization. Ph.D. thesis, University of Chicago, Chicago, Illinois.
Glander, K. E. (1978). *Folia Primatologia* **29,** 206–217.
Glander, K. E. (1979). "The ecology of arboreal folivores, (G. Montgomery, ed.). Smithsonian Press, Washington, D.C. pp. 561–574.
Golley, F. B., Richardson, T., and Clements, R. G. (1978). *Biotropica* **10,** 144–151.
Harborne, J. B. (1978). "Introduction to Ecological Biochemistry." Academic Press, New York.
Hardmann, R. (1969). *Trop. Sci.* **11,** 196–228.
Harris, P. (1973). *Symp. R. Entomol. Soc. London* **6,** 201–209.
Hendry, L. B., Kostell, J. G., Hindenlang, D. M., Wichmann, J. K., Fix, C. J., and Korzeniowski, S. H. (1976). *Recent Adv. Phytochem.* **10,** 351–384.
Howe, H. F. (1977). *Ecology* **58,** 539–550.
Howe, H. F., and Estabrook, G. F. (1977). *Am. Nat.* **111,** 817–832.
Janzen, D. H. (1966). *Evolution* **20,** 249–275.
Janzen, D. H. (1972). *Ecology* **53,** 885–892.
Janzen, D. H. (1973). *Biotropica* **5,** 15–28.
Janzen, D. H. (1974). *Biotropica* **6,** 69–103.
Janzen, D. H. (1975a). *Ecology* **56,** 1009–1013.
Janzen, D. H. (1975b). "Ecology of Plants in the Tropics." Arnold, London.
Janzen, D. H. (1976). *Am. Midl. Nat.* **95,** 474–478.
Janzen, D. H. (1977a). *Tropical Ecology* **18,** 162–176.
Janzen, D. H. (1977b). *Ann. Mo. Bot. Gard.* **64,** 706–736.
Janzen, D. H. (1977c). *Ecology* **58,** 921–927.
Janzen, D. H. (1977d). *Am. Nat.* **111,** 586–589.
Janzen, D. H. (1978a). *In* "The ecology of Arboreal Folivores" (G. Montgomery, ed.). Smithsonian Press, Washington, D.C. pp. 73–84.
Janzen, D. H. (1978b). *Brenesia* **14/15,** 325–335.
Janzen, D. H. (1979). *Nature* (submitted for publication).
Janzen, D. H., and Martin, P. (1980). Neotropical anachronisms: fruits the mastodonts left behind (manuscript).
Janzen, D. H., Juster, H. B., and Liener, I. E. (1976). *Science* **192,** 795–796.
Janzen, D. H., Juster, H. B., and Bell, E. A. (1977). *Phytochemistry* **16,** 223–227.
Jermy, T. (1976). *Symp. Biol. Hung.* **16,** 109–113.
Kulman, H. M. (1971). *Annu. Rev. Entomol.* **16,** 289–324.
Langham, G. W., and Smith, L. W. (1970). *Aust. J. Agric. Res.* **21,** 493–500.
Laycock, W. A. (1975). *J. Range Manage.* **28,** 257–259.
Levin, D. A., and York, B. M. (1978). *Biochem. System. Ecol.* **6,** 61–76.
Liener, I. E. (1976). *Annu. Rev. Plant Physiol.* **27,** 291–319.
McKey, D. (1974). *Amer. Nat.* **108,** 305–320.

McKey, D. (1978). *In* "Arboreal Folivores" (G. Montgomery, ed.). Smithsonian Press, Washington, D.C. p. 423–437.
McKey, D., and Janzen, D. H. (1977). *Biotropica* **9,** 57.
McKey, D., Waterman, P. G., Mbi, C. N., Gartlan, J. S., and Struhsaker, T. T. (1979). *Science* **202,** 61–64.
McNaughton, S. J. (1977). *Science* **199,** 806–807.
Mattson, W. J., and Addy, N. D. (1975). *Science* **190,** 515–522.
Milne, A. A. (1928). The House at Pooh Corner. Dutton, New York.
Moore, L. R. (1978a). *Oecologia* **34,** 185–202.
Moore, L. R. (1978b). *Oecologia* **34,** 203–223.
Ogutuga, D. B. A. (1975). *Ghana J. Agric. Sci.* **8,** 121–125.
Oh, H. K., Sakai, T., Jones, M. B., and Longhurst, W. M. (1967). *Appl. Microbiol.* **15,** 777–784.
Owen, D. F., and Wiegert, R. G. (1976). *Oikos* **27,** 488–492.
Pickard, J. (1976). *Aust. J. Ecol.* **1,** 103–114.
Pull, S. P., Pueppke, S. G., Hymowitz, T., and Orf, J. H. (1978). *Science* **200,** 1277–1279.
Regal, P. J. (1977). *Science* **196,** 622–629.
Rockwood, L. L. (1973). *Ecology* **54,** 1363–1369.
Rosenthal, G. A. (1977). *Q. Rev. Biol.* **52,** 155–178.
Rosenthal, G. A. Dahlman, D. L., and Janzen, D. H. (1978). *Science* **202,** 528–529.
Ryan, C. A. (1978). *TIBS Trends Biochem. Res.* July:148–150.
Ryan, C. A., and Pearce, G. (1978). *Am. Potato J.* **55,** 351–358.
Sigleo, A. C. (1978). *Science* **200,** 1054–1056.
Simberloff, D., Brown, B. J., and Lowrie, S. (1978). *Science* **201,** 630–632.
Smiley, J. (1978). *Science* **201,** 745–747.
Stoessl, A. (1970). *Recent Adv. Phytochem.* **3,** 143–180.
Strong, D. R. (1977). *Science* **197,** 1071.
Struhsaker, T. T. (1975). "The Red Colobus Monkey." University of Chicago Press, Chicago, Illinois.
Sussman, R. W., and Raven, R.-H. (1978). *Science* **200,** 731–736.
Uy, R., and Wold, F. (1977). *Science* **198,** 890–896.
Walker, A., Hoeck, H. N., and Perez, L. (1978). *Science* **201,** 908–910.
Wender, S. H. (1970). *Recent Adv. Phytochem.* **3,** 1–29.
Westermark, H. (1959). *Scta Vet. Scand.* **1,** 67–73.

Part II

The Chemical Participants

Chapter **9**

Naturally Occurring, Toxic Nonprotein Amino Acids

GERALD A. ROSENTHAL *And* E. A. BELL

I. INTRODUCTION

The amino acids, with which we include the imino acids, comprise a group of natural products that are extremely important both for their

ISBN 0-12-597180-X

intrinsic properties and as precursors of other plant metabolites. In addition to the 20 amino acids that are distributed universally as protein constituents, over 400 others have been isolated from natural sources, including more than 260 from higher plants. Nonprotein amino acids usually occur in the free state or as γ-glutamyl derivatives. A few, such as homoserine and ornithine, are widely distributed as primary metabolites, but the majority are more limited in their occurrence, often being restricted to a particular family, genus, or even a single species.

The majority of the naturally occurring nonprotein amino acids are α-amino acids of the L configuration, although some D-amino acids have been reported (Robinson, 1976). Like the protein amino acids, the nonprotein amino acids can be divided arbitrarily into groups on the basis of their structure (e.g., aliphatic, aromatic, or heterocyclic), the number and nature of their ionizable groups (e.g., acidic, neutral, or basic), or their physiological properties and effects in various organisms.

Much of the available information on their toxicity has been obtained in studies of human beings and domestic animals. Although this information provides unequivocal evidence that these compounds can be quite toxic to certain herbivores, it does not, of course, establish their role in plant chemical defense. Keeping this restriction in mind, we present evidence that certain of these nonprotein amino acids are toxic to a wide range of animal species, possibly including some herbivores that might otherwise prey on the plants that synthesize these secondary plant metabolites. We also present evidence that a successful predator may be effective because it has evolved a biochemical capacity to reply to the chemical challenge presented by a toxic nonprotein amino acid. We also discuss other observable aspects of plant–animal coevolution that are consistent with the belief that some nonprotein amino acids function in plant chemical defense.

Properties and Isolation

Amino acids are generally quite soluble in aqueous solvents, but many, including hydrophobic, aromatic, and heterocyclic members, are much less soluble. Both in the crystalline state and in solution they can exist as dipolar ions or zwitterions with ionizable carboxyl and amino groups associated with the α-carbon atom; the various R groups can also have ionizable groups. At low pH, amino acids are cationic but gradually deprotonate as the pH increases to become progressively more anionic. At a particular pH, termed the isoelectric point, the amino acid is neutral. The nature of the R group and its effect on the ionizable groups of the α-carbon determines the degree of protonation of these amphoteric substances. Thus, they may have dissimilar charge at a given pH, migrate differentially in an electrical field, and exhibit distinctive exchange properties with various ionic-exchange resins.

$$R-\underset{\displaystyle NH_3^+}{\underset{|}{CH}}-COO^-$$

Toxic, nonprotein amino acids can be found in all parts of the plant, but the seed is generally the preferred source for isolating these compounds. As a group, they are obtained readily by simple extraction with aqueous alcohol, which is generally acidified so that less protein is dissolved and, occasionally, to stabilize the isolated amino acid. Once extracted, these compounds can be purified by such techniques as ion-exchange chromatography, paper or thin-layer partition chromatography, and/or electrophoresis.

Very helpful discussions of amino acid methodology in general are provided by Robinson (1975) and Harborne (1973). The fundamental chemistry of some nonprotein amino acids is considered in the three-volume treatise of Greenstein and Winitz (1961). Among the several excellent sources for basic information on these secondary compounds are the texts of Meister (1965) and the review articles of Fowden *et al.* (1967), Thompson *et al.* (1969), Fowden (1970), Bell (1971), Hegarty and Peterson (1973), and Lea and Norris (1976).

II. LATHYROGENS AND NEUROTOXINS

Consumption of the seeds of certain species of *Lathyrus* (Leguminosae) can produce a malady that afflicts various animals. This disease, known as lathyrism, has two recognized pathological forms. The first, designated neurolathyrism by Selye (1957), refers to a nervous disorder that occurs in a variety of animals, including human beings, and is caused by consumption of *Lathyrus sativus, L. cicera,* and *L. clymenum. Lathyrus latifolius, L. sylvestris,* and *L. sphaericus* also produce neurological effects (Barrow *et al.,* 1974). The primary symptom of neurolathyrism is muscular weakness, especially in the extremities; irreversible paralysis and death are the extreme and ultimate manifestations of this disease. The second pathological form, which Selye (1957) termed osteolathyrism, is characterized by anomalies of bone and mesenchymal tissue. In contrast to human neurolathyrism, osteolathyrism can be produced experimentally in laboratory animals by ingestion of *L. odoratus* seeds. In this regard, the term "angiolathyrism" is employed to denote vascular tissue disorders that occur in conjunction with osteolathyrism. Attempts at dissociating clinically the bone deformities of osteolathyrism from the vascular ramifications of angiolathyrism have been unsuccessful (Barrow *et al.*, 1974).

The toxic principle of *L. odoratus* seeds is β-aminopropionitrile, but many synthetic nitriles can elicit osteolathyritic and angiolathyritic symptoms. The γ-glutamyl derivative of β-aminopropionitrile was isolated from *L. odoratus* seeds (Schilling and Strong, 1954, 1955) and shown to produce skeletal changes in animals under laboratory conditions. The presence of the γ-glutamyl moiety is not necessary for the induction of osteolathyrism. Lysyl oxidase mediates the oxidation of the lysyl residues of collagen and elastin to produce a δ-semialdehyde known as allysine. Covalent cross-linkages which stabilize collagen and elastin are initiated by allysine group interaction (O'Dell *et al.*, 1966; Siegel and Martin, 1970). β-Aminopropionitrile inhibits this enzyme so that the resulting collagen and elastin are allysine-deficient (Siegel *et al.*, 1970). This produces connective tissues having decreased tensile strength and excessively soluble collagen (Ressler, 1975).

A. β-Cyano-L-alanine

In support of synthetic peptide studies, Ressler and Ratzkin (1961) synthesized the novel nonprotein amino acid β-cyanoalanine, which they viewed initially as a likely intermediate in the conversion of asparagine to L-α,γ-diaminobutyric acid. Since decarboxylation of β-cyanoalanine produces β-aminopropionitrile, an active lathyritic substance, β-cyanoalanine was evaluated for its biological activity. Consumption of a laboratory diet containing 1% β-cyanoalanine by male rats produced hyperirritability, tremors, convulsions, and even death within 3 to 5 days (Ressler *et al.*, 1961). This discovery of the neurotoxic properties of β-cyanoalanine, coupled with the observation of Lewis *et al.* (1948) and Lewis and Schulert (1949) that *L. latifolius* and *L. sylvestris Wagneri* are neurotoxic to rats and mice, led Ressler to examine these species for β-cyanoalanine. Ultimately, the neurotoxic constituent was disclosed to be α,γ-diaminobutyric acid, a product of β-cyanoalanine reduction (Fig. 1).

β-Cyanoalanine was isolated from *Vicia sativa* and shown to be identical to the chemically synthesized material (Ressler, 1962). Bell and Tirimanna (1965) subsequently surveyed 48 species of *Vicia* and established the presence of this toxic nitrile in 16 species. Administration of β-cyanoalanine (15 mg per 100 gm fresh weight) to a weanling male rat

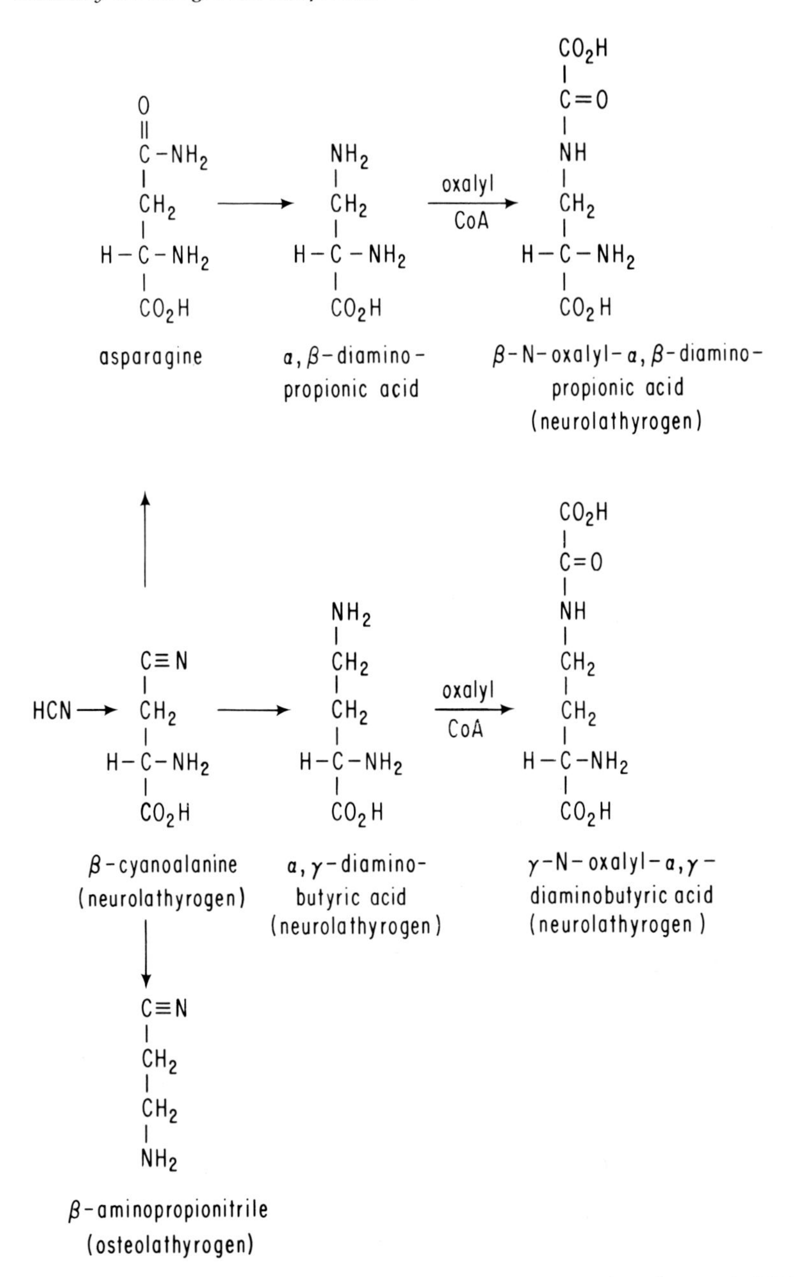

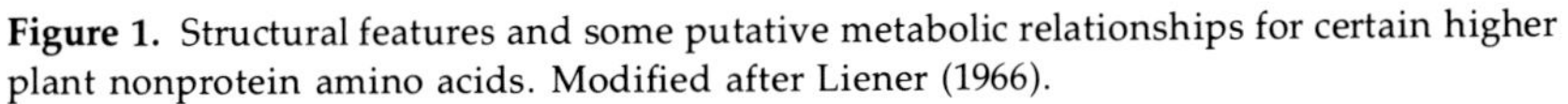

Figure 1. Structural features and some putative metabolic relationships for certain higher plant nonprotein amino acids. Modified after Liener (1966).

by stomach tube produces reversible hyperactivity, tremors, convulsions, and rigidity. When the dose is provided subcutaneously and increased to 20 mg per 100 gm, it evokes convulsions, rigidity, prostration, and death (Ressler, 1962). The predominant form of β-cyanoalanine in *V. sativa* is a dipeptide, namely, *N*-(γ-L-glutamyl)-β-cyano-L-alanine, which constitutes 0.6% of the seed dry weight but 1.7–2.6% of the young seedling dry weight (Ressler *et al.*, 1963). On a molar basis, the dipeptide is as potent as the free amino acid to male Sherman rats, but it is only about one-half as toxic to White Leghorn chicks. A mixture of the dipeptide and free amino acid, at one-half their level in *V. sativa* seeds (0.075%), was incorporated into a basal ration and provided to young chicks. Within 1 week, such treated White Leghorn chicks enter into a terminal convulsive state and exhibit tetanic spasms of the back muscles that force the head and lower limbs to bend backward while the trunk arches forward (opisthotonus) (Ressler *et al.*, 1963).

Additional studies of β-cyanoalanine by Blumenthal-Goldschmidt *et al.*, (1963), Tschiersch (1964), and Fowden and Bell (1965) revealed that cyanide serves as a carbon and nitrogen source for asparagine synthesis via β-cyanoalanine (Fig. 1). An interesting application of this type of biochemical reaction has been suggested for *Sitophilus granarius* since this insect detoxifies cyanide by diverting it into amino acid pools (Bond, 1961).

Insects conserve water by reabsorbing it from the hindgut before the elimination of fecal matter. Administration of β-cyanoalanine to the adult locust *Locusta migratoria migratorioides* impairs its capacity to reabsorb water into the hemolymph. This decreases the hemolymph fluid volume within 1 day and produces acute dehydration after 2 to 3 days (Schlesinger *et al.*, 1976). This neurotoxic nitrile is accumulated in locust hemolymph and dispatched efficiently into the nervous system (Schlesinger *et al.*, 1977). The free amino acid, at a concentration of 0.1% (w/w), proved lethal to larvae of the seed beetle *Callosobruchus maculatus* when included in a larval diet of ground *Vigna unguiculata* (Janzen *et al.*, 1977). The reader is directed to this study of the interaction of numerous nonprotein amino acids with this seed predator.

The toxicity of β-cyanoalanine reflects not only its adverse neurological properties but also its inhibitory action on various enzymes. It is a highly effective inhibitor of rat liver cystathionase; this probably accounts for the cystathioninuria that occurs in rat after its ingestion (Pfeffer and Ressler, 1967). In addition, this nonprotein amino acid inhibits aspartic decarboxylase (Tate and Meister, 1969) as well as asparaginase and glutaminase of certain prokaryotes (Ressler, 1975).

B. β-*N*-Oxalyl-L-α,β-diaminopropionic Acid

The seeds of three species of *Lathyrus*, namely, *L. sativus*, *L. cicera*, and *L. clymenum*, that are known to cause classic neurolathyrism also contain oxalyldiaminopropionic acid (Fig. 1) (Bell, 1964a). This compound was isolated and characterized independently by two groups of workers in India from seeds of *L. sativus* (Rao *et al.*, 1964; Murti *et al.*, 1964).

The concentration of oxalyldiaminopropionic acid in seeds of *L. sativus* obtained from different sources is very variable; Roy and Narasinga Rao (1968) reported values ranging from 0.1 to 2.5% of the dry seed weight. This variability may explain the conflicting reports on the toxicity of *L. sativus* seeds. In a recent survey of free amino acids in the seeds of species from more than 250 genera of the Leguminosae, Qureshi *et al.* (1977) found oxalyldiaminopropionic acid in the seeds of 17 species of *Acacia* and 13 members of *Crotalaria* and confirmed its occurrence in 21 species of *Lathyrus* as reported by Bell (1964a).

Purified oxalyldiaminopropionic acid has proved to be toxic to rodents and birds as well as primates in laboratory trials, but the response of treated animals varies with their age, their nutritional status, and the method of compound administration (Johnston, 1974). Although most research on this compound has been concerned with its toxicity to mammals and particularly human beings, preliminary experiments indicate that it is also toxic to the grasshopper *Anacridium* (C. S. Evans, personal communication).

C. L-α,γ-Diaminobutyric Acid

Another neurotoxic amino acid found in the seeds of certain species of *Lathyrus* (but not *L. sativus*, *L. cicera*, or *L. clymenum*) is α,γ-diaminobutyric acid (Fig. 1). This lower homolog of ornithine was first identified in *Polygonatum* (Fowden and Bryant, 1958) and subsequently isolated from *L. latifolius* seeds (Ressler *et al.*, 1961); it has since been identified in the seeds of 11 other *Lathyrus* species (Bell, 1962) and in the seeds of other leguminous species (E. A. Bell, unpublished observations). In 10 of 13 *Lathyrus* species, α,γ-diaminobutyric acid is accompanied by oxalyldiaminopropionic acid and the γ-oxalyl derivative of the former compound; trace amounts of the α-oxalyl derivative of diaminobutyric acid also occurs in these same plants (Bell and O'Donovan, 1966).

Ressler *et al.* (1961) reported that α,γ-diaminobutyric acid produces weakness, tremors, and convulsions when administered orally to rats, and O'Neal *et al.* (1968) obtained similar effects by intraperitoneal injection. The latter workers also demonstrated that α,γ-diaminobutyric acid can act as an inhibitor of ornithine carbamoyltransferase in mammalian liver, disrupting ornithine–urea cycle reactions and inducing ammonia toxicity. In addition, this compound may exercise a neurotoxic effect by interfering with the function of γ-aminobutyric acid, an important component of the mammalian central nervous system (Johnston, 1974).

D. γ-*N*-Oxalyl-L-α,γ-diaminobutyric Acid

As mentioned in Section II,C, oxalyldiaminobutyric acid (Fig. 1) accumulates in the seeds of at least 10 species of *Lathyrus.* Like oxalyldiaminopropionic acid, oxalyldiaminobutyric acid is neurotoxic in animals as are the oxalyl derivatives prepared synthetically from glycine, alanine, or lysine (Rao and Sarma, 1966). The α-*N*-oxalyl derivatives of both α,β-diaminopropionic acid and α,γ-diaminobutyric acid can be detected in extracts of species that synthesize the β- and γ-oxalyl derivatives. These α derivatives occur in relatively low concentration, however, and are probably formed by nonenzymatic isomerization of the β and γ derivatives in the plant itself or during the extraction process.

E. β-*N*-Methyl-L-α,β-diaminopropionic Acid

Methyldiaminopropionic acid **(1)**, an isomer of diaminobutyric acid, was first isolated from seeds of *Cycas circinalis* (Vega and Bell, 1967) and has since been detected in other species of the genus (Dossaji and Bell, 1973). Administration of this compound intraperitoneally to young chicks produced convulsions within 6 to 8 hours (Vega *et al.*, 1968). The relatively low concentration of methyldiaminopropionic acid in cycad seeds suggests that it is not responsible for the neurological changes seen in cattle that have fed on cycad plants or for the development of amyotrophic lateral sclerosis among native peoples of Guam, who use seeds of *C. circinalis* as a foodstuff (Whiting, 1963; Masson and Whiting, 1966).

$$CH_3—NH—CH_2—CH(NH_2)COOH$$

1

In the conclusions to his review on neurotoxic amino acids, Johnston (1974) points out that "most of the neurotoxic amino acids of higher plant and fungal origin are structurally related to the amino acids that are likely to function as important synaptic transmitters in the mammalian central nervous system." For example, oxalyldiaminopropionic and ibotenic acids are related structurally to glutamic acid; diaminobutyric acid is similar to γ-aminobutyric acid; and β-cyanoalanine is related to aspartic acid. Whether similar factors are valid when the toxicity of these compounds is considered in invertebrates and nonmammalian animals is not fully known. Indeed, the toxicity of these compounds does not result solely from their capacity to function as antimetabolites of transmitter amino acids.

III. BASIC AMINO ACIDS

A. L-Canavanine

Canavanine **(2)** is the guanidinooxy structural analog of arginine characterized by the replacement of the terminal methylene group with oxygen. Canavanine's distribution is limited to certain species of the Lotoideae (Papilionaceae, Fabaceae), a subfamily of the Leguminosae (for the distributional pattern, see Rosenthal, 1977a). This arginine analog has been reported to exist in nonleguminous plants, for example, onion, *Allium cepa,* and the edible fungus *Agaricus campestris,* but these reports either have not been corroborated or have been challenged (Rosenthal and Davis, 1975; Rosenthal, 1977a). In canavanine-storing legumes, this compound can be the principal seed nitrogen-storing metabolite (Tschiersch, 1961; Bell, 1972; Rosenthal, 1974); the seed usually contains 0.5% canavanine by dry weight, but much higher levels occur (Bell, 1972; Rosenthal, 1977b).

$$H_2N—C(=NH)—NH—O—CH_2—CH_2—CH(NH_2)CO_2H$$

2

The potent antimetabolic properties of canavanine have been documented for an array of experimental systems ranging from viruses to human placental tissues and including many mammals (Rosenthal, 1977a). Except for invertebrates, detailed whole-animal studies of this natural product are lacking. The marked insecticidal properties of canavanine have been demonstrated in silkworm (Isogai *et al.,* 1973), boll weevil (Vanderzant and Chremos, 1971), and fruit fly (Harrison and Holliday, 1967). Consumption by the tobacco hornworm *Manduca sexta* of an artificial diet containing only 0.02% (w/v) or 1.0 m*M* canavanine causes significant developmental aberrations and decreases survival rates for all developmental stages (Rosenthal and Dahlman, 1975). Exposure of *M. sexta* or the southern armyworm *Prodenia eridania* to the level of canavanine occurring in many seeds produces larval mortality (Rehr *et al.*, 1973a; Dahlman and Rosenthal, 1975).

An important and perhaps principal basis for the antimetabolic properties of canavanine results from the ability of certain arginyl-tRNA synthetases to aminoacylate canavanine. In species unable to discriminate against canavanine, it is subsequently incorporated into the nascent polypeptide chain in place of arginine; this substitution can cause structural and functional protein alterations (Attias *et al.*, 1969; Prouty *et al.*, 1975).

The bruchid beetle *Caryedes brasiliensis* is the only known insect predator of the seed of the neotropical legume *Dioclea megacarpa* (Janzen, 1971); this plant stores as much as 13% of its seed dry weight as canavanine (Fig. 2). This concentration means that canavanine contains 55% of the total seed nitrogen and 94% of the free amino acid nitrogen (Rosenthal, 1977b). The formidable chemical barrier represented by so much toxic material is breached by this herbivore since it possess an arginyl-tRNA synthetase that discriminates against canavanine. Canavanine is neither charged nor are canavanyl proteins produced, and the bruchid beetle circumvents a major cause of canavanine toxicity (Rosenthal *et al.*, 1976). Moreover, this specialized seed predator not only detoxifies canavanine but does so in a manner that generates dietary nitrogen by converting this arginine analog to canaline and urea and then degrading urea to ammonia (Rosenthal *et al.*, 1977).

In many canavanine-sensitive organisms, canavanine initially affects macromolecular metabolism; RNA metabolism is affected and canavanyl proteins are formed. Both DNA metabolism and overall protein production are affected secondarily. Canavanyl protein formation may influence mRNA synthesis; once existing mRNA molecules are degraded, translation of proteins required for DNA replication and general protein synthesis are inhibited. Canavanyl protein production can also diminish DNA metabolism directly, possibly by disrupting the formation of proteins functioning in DNA replication.

Canavanine can also potentiate its disruptive biological effects by mechanisms other than aberrant protein production. The substitution of canavanine for arginine in histones decreases histone basicity, which can affect their interaction with nucleic acids and ultimately genome expression. In this regard, canavanine has been shown to disrupt histone synthesis in HeLa cells (Ackermann *et al.*, 1965) and cultured hamster and mouse cells (Hare, 1969). Canavanine is also an effective antagonist of virtually all aspects of arginine metabolism. In addition, Tschiersch (1966) demonstrated that canavanine noncompetitively inhibits enzymes, such as alcohol dehydrogenase, β-glucosidase, and

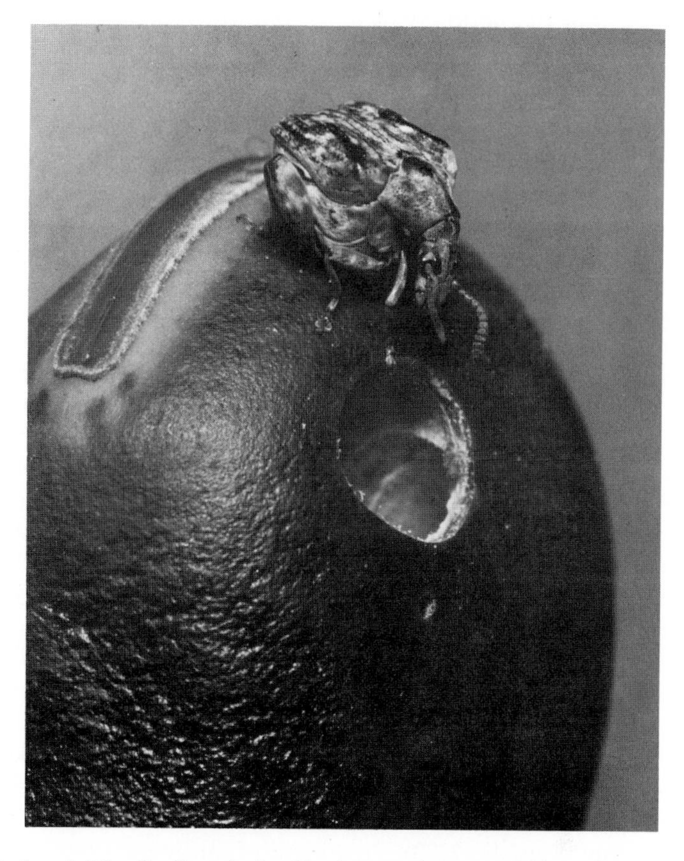

Figure 2. An adult beetle, *Caryedes brasiliensis*, recently emerged from a *Dioclea* seed. The adult's exit hole and the smaller entrance hole created by the first-stadium larva are shown. The developing larvae destroy the toxic canavanine-laden cotyledonous tissues of the seed. After Rosenthal *et al.* (1976); photo by D. Janzen.

oxynitrilase, whose substrates bear little apparent structural relation to this compound. Thus, the highly toxic nature of canavanine reflects its disruptive effects on many fundamentally important biological processes (for pertinent references, see Rosenthal, 1977a; Lea and Norris, 1976).

B. L-Canaline

Arginase-mediated hydrolytic cleavage of canavanine produces urea and canaline **(3)**. This compound is the only naturally occurring amino acid containing the terminal aminooxy group. Canaline has been isolated from jack bean, *Canavalia ensiformis* (Miersch, 1967), and the unripened seed of *Astragalus sinicus* (Inatomi *et al.*, 1968). This nonprotein amino acid could possibly occur in all canavanine-producing legumes having arginase activity; however, information on the nature and extent of stored canaline in plant materials is not available.

$$H_2N—O—CH_2—CH_2—CH(NH_2)CO_2H$$

3

Incorporation of canaline into the artificial diet of the tobacco hornworm causes depauperate larvae, reduces successful larval–pupal ecdysis, increases the severity of pupal and adult malformation, decreases survival of all developmental stages, and significantly attenuates ovarial mass production (Rosenthal and Dahlman, 1975). Canaline injected into the moth causes almost continuous motor activity. The moth flys normally at first but soon becomes disoriented; muscle activity is less patterned, even though wing movement continues. Postsynaptic potential of flight muscle fibers is prolonged, but after 20 to 40 min the electrical activity of muscle fibers is normal. This neuropharmacological agent clearly affects the activity of the central nervous system, but its exact mode of action remains to be resolved (Kammer *et al.*, 1978). Canaline, a very potent inhibitor of pyridoxal phosphate-containing enzymes, severely inhibits rat glutamate–oxaloacetate, glutamate–pyruvate, and ornithine–keto acid transaminase activities (Kekomäki *et al.*, 1969; Katunuma *et al.*, 1965). It is believed to undergo oxime formation with pyridoxal phosphate by reacting with the aldehydic carbon of the coenzyme (Rahiala, 1973). Treatment of rat liver cystathionase with canaline disrupts the Schiff base linkage essential for maintaining pyridoxal phosphate–holoenzyme integrity (Beeler and Churchich, 1976).

Several independent studies have demonstrated that canaline is not an effective inhibitor of ornithine metabolism. For example, canaline strongly impedes the activity of all seven pyridoxal phosphate-containing enzymes evaluated by Rahiala *et al.* (1971) but does not affect three ornithine-utilizing enzymes lacking a B_6 cofactor. The *inability* of canaline to adversely affect ornithine metabolism also has been shown in human liver (Natelson *et al.*, 1977) and vaccinia virus (Archard and Williamson, 1974). Certain interrelationships between canaline and canavanine metabolism have emerged, and their possible function in mammalian intermediary metabolism has been investigated (Rosenthal, 1978a,b).

C. L-Indospicine

Indospicine **(4)** is the only known higher plant natural product having the unusual ϵ-amidino group, although this amidine system is found in certain microbial antibiotics (Hegarty and Pound, 1968). Indospicine was isolated originally from the legume *Indigofera spicata,* where it constitutes 0.04–0.15% of the mature leaves and 0.5–2.0% of the seeds (Hegarty and Pound, 1970). A subsequent survey of 17 species of *Indigofera* failed to disclose an additional source of this compound (Miller and Smith, 1973), but has been identified recently in other *Indogofera* species (Charlwood and Bell, 1977).

$$H_2N—C(=NH)—CH_2—CH_2—CH_2—CH_2—CH(NH_2)CO_2H$$

4

Indospicine, unusual in being one of two naturally occurring hepatotoxic nonprotein amino acids, produces liver damage in sheep, cows, and rabbits (Nordfeldt *et al.,* 1952). An early report of indospicine-induced liver lesions in rabbits and mice was provided by Hutton *et al.* (1958a,b). Enlarged livers are obtained from rats treated for only 16 hours with 2 gm indospicine per kilogram body weight. Liver DNA, RNA, and protein contents increase, but nearly 75% of the enhancement in organ biomass results primarily from water retention (Christie *et al.,* 1969). Detailed histological studies of rats maintained on an artificial diet containing 85–600 mg indospicine per kilogram body weight show that the enlarged, fatty livers develop characteristic lesions and cellular necrosis before the onset of cirrhosis (Christie *et al.,* 1975). Administration of 1.6 gm canavanine per kilogram body weight to rats does not affect the liver of treated animals (Hegarty and Pound, 1970).

Rabbits offered 200 gm of fresh *I. spicata* leaves, as a supplement to their daily pellet rations, experience diminished food intake, loss in body weight, and an average survival period of only 3 weeks. Autopsy of such treated animals reveals gross degenerative swelling and some liver cell necrosis. Cirrhosis is common to all treated rabbits (Hutton *et al.,* 1958a). Comparable liver damage has been reported with indospicine-treated mice; in addition, indospicine consumption by sheep produces corneal cells that become opaque and, in severe cases, progress to corneal ulceration (Hutton *et al.,* 1958b).

The teratogenic nature of this basic compound was noted first by Pearn (1967a), who observed that it causes cleaving of the secondary palate and somatic dwarfism. Indospicine-treated pregnant rats ultimately produce live neonates that expire several hours after birth. Except for the cleft palate, all other body parameters are unaltered; thus, this teratogen appears to be highly site specific. In addition, cleft palate formation was related directly to the level of indospicine consumed (Pearn, 1967b).

Several properties of indospicine are shared by canavanine. For example, indospicine is charged by arginyl-tRNA synthetase and subsequently incorporated into protein (Leisinger *et al.*, 1972). Arginine production in *Pseudomonas aeruginosa* is subject to negative feedback inhibition by indospicine (Leisinger *et al.*, 1972). Certain enzymes of the biosynthetic pathway culminating in arginine production by *Escherichia coli* are regulated allosterically by canavanine (Maas, 1961). Indospicine-treated human lymphocytes exhibit decreased [^{3}H]-thymidine incorporation rates relative to untreated cells; it was suggested that indospicine inhibited arginine metabolism, which ultimately disrupted protein synthesis and DNA metabolism (Christie *et al.*, 1971). The possibility that functionally anomalous indospicyl proteins were responsible for diminished DNA metabolism was raised. A similar situation exists in several canavanine-sensitive organisms in which canavanyl protein formation is responsible for curtailed macromolecular metabolism. Finally, indospicine is a very effective competitive inhibitor of arginine for rat liver arginase. The indospicine K_i value of 1.88 m*M* is close to the arginine apparent K_m value of 1.14 m*M* (Madsen and Hegarty, 1970). Similarly, arginase obtained from *Tribolium* beetle is inhibited completely by only 10 m*M* canavanine in spite of an arginine apparent K_m value of about 200 m*M* (Harry *et al.*, 1976).

Unfortunately, information is not available on the catabolic or anabolic reactions involved in the metabolism of this interesting nonprotein amino acid.

D. Other Arginine Derivatives

L-Homoarginine **(5)** was isolated and characterized first from *Lathyrus cicera* seeds by Bell (1962) and from *L. sativus* by Rao *et al.* (1963). Drawing on Bell's discovery of lathyrine **(7)** both Rao *et al.* (1963) and Bell (1963) proposed the following metabolic relationships, in which γ-hydroxyhomoarginine **(6)** represents a postulated reaction intermediate:

$$\underset{\mathbf{5}}{H_2N{-}C({=}NH){-}NH{-}(CH_2)_4{-}CH(NH_2)CO_2H} \rightarrow$$

γ-Hydroxyhomoarginine was isolated subsequently from *L. tingitanus* (Bell, 1963). Introduction of L-[*guanidino*-^{14}C]homoarginine into *L. tin-*

$H_2N—C(=NH)—NH—CH_2—CH_2—CH(OH)—CH_2—CH(NH_2)CO_2H \rightarrow$

6

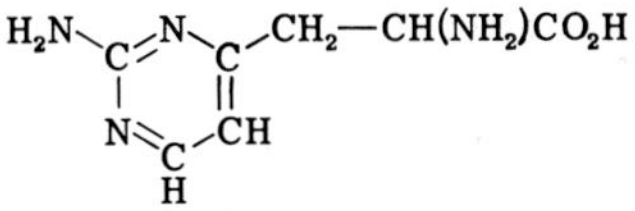

$H_2N—C{=}N—C(—CH_2—CH(NH_2)CO_2H){=}CH—CH{=}N—$ (ring closure to C)

7

gitanus fruit resulted in rapid and appreciable labeling of the isolated hydroxylated homoarginine but no significant labeling of lathyrine. Believing that the final reaction step might occur in the roots, Bell and Przybylska (1965) fed L-[*guanidino*-^{14}C]hydroxyhomoarginine and found two compounds of high specific activity. One was unreacted substrate while the other was lathyrine. Consistent with the above-proposed scheme, all species known to contain significant amounts of lathyrine also have detectable γ-hydroxyhomoarginine (Bell, 1964b). The incorporation of labeled orotic acid and uracil into lathyrine (Brown and Al-Baldawi, 1977) suggests that an alternate pathway may exist.

γ-Hydroxyarginine has been isolated from the marine cucumber *Polycheira rufescens* (Fujita, 1959), the sea anemone *Anthopleura japonica* (Makisumi, 1961), and from many species of the leguminous genus *Vicia* (Bell and Tirimanna, 1965). The last-named authors made the interesting generalization that various *Vicia* species store relatively high concentrations (about 1% by dry weight) of both canavanine and the six-carbon amino acids arginine and hydroxyarginine. On the other hand, *Lathyrus* seeds contain the seven-carbon compounds homoarginine, γ-hydroxyhomoarginine, and lathyrine (Bell and Tirimanna, 1965).

The existing information on homoarginine toxicity is sparse and somewhat conflicting. In studies of the bruchid beetle *Callosobruchus maculatus*, Janzen *et al.* (1977) demonstrated that incorporation of 5% homoarginine into the larval diet prevented adult emergence. However, when this compound is added to the diet at a level in accord with its typical concentration in nature (1% or less), it does not affect adult emergence; moreover, it exhibits a strong sparing effect on the action of canavanine. Homoarginine consumption by larval boll weevil, *Anthonomus grandis*, only slightly affects adult emergence, and this effect is readily reversed by arginine (Vanderzant and Chremos, 1971). In addition, homoarginine replaces arginine in counteracting the adverse effects of canavanine consumption.

Addition of 25 m*M* homoarginine to the diet of fifth stadium *Manduca sexta* larvae produces severely malformed pupae. When the dietary concentration of this arginine analog is reduced to 5 m*M*, pupal malformation no longer occurs (D. L. Dahlman, personal communication). At a 5 mM dietary concentration, canavanine is very poisonous to this insect. Fifth instars generally expire prior to larval–pupal ecdysis and any surviving pupae are massively deformed (Rosenthal and Dahlman, 1975).

Administration of 5×10^{-3} mole of homoarginine per kilogram body weight to male albino rats fails to produce any discernible symptoms. When the dose is increased to 1×10^{-2} mole, the rats die after 15–20 hours (O'Neal *et al.*, 1968). Under conditions in which α,γ-diaminobutyrate significantly reduces ammonia utilization and urea production, homoarginine has no apparent effect on liver tissue slices.

In microorganisms, homoarginine ranges from an effective growth inhibitor to being innocuous (Walker, 1955a,b). Studies in Rosenthal's laboratory reveal that 10μM homoarginine does not affect *Lemna minor* growth, although exposure of this aquatic microphyte to equivalent canavanine is lethal (unpublished observation). Thus, there appears to be considerable variation in the susceptibility of different organisms to homoarginine. Unfortunately, no biological studies of hydroxylated basic amino acids considered in this section have been reported. Such information would be a valuable contribution to our understanding of these secondary metabolites.

IV. HETEROCYCLIC AMINO AND IMINO ACIDS

A. L-Mimosine

Mimosine (8) possesses the structure β-[N-(3-hydroxy-4-oxopyridyl)]-α-aminopropionic acid and has been isolated from *Mimosa pudica* and *Leucaena leucocephala* (previously *L. glauca* Benth.).

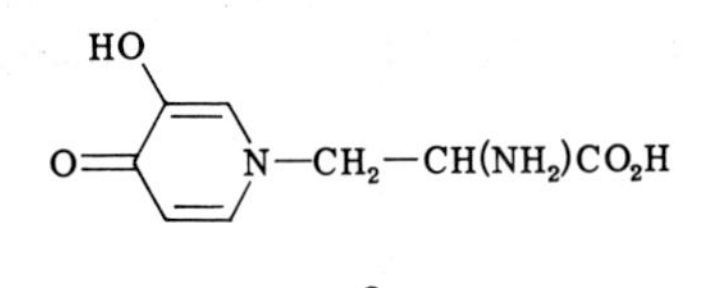

8

Approximately 2–5% of the leaf dry weight of *L. leucocephala* is contributed by mimosine (Gonzalez *et al.*, 1967), and the seed may contain as much as 9% of this amino acid (Takahashi and Ripperton, 1949). This pandemic weed of tropical and subtropical areas was cultivated as an animal food source and is planted frequently for shade or as a hedgerow in coffee and tea plantations (Hylin, 1969). Chemical evidence that mimosine from *M. pudica* is identical with leucaenol or leucaenine isolated from *L. leucocephala* was provided by Kleipool and Wibaut (1950). Thus, a single toxic agent exists that is presently found in two genera of the Mimosaceae.

A striking biological consequence of minosine consumption is the sudden loss of hair. Depilation of the mane and tail of horses and the complete loss of body hair in swine consuming *Leucaena* foliage have been documented (Owen, 1958). This author also reports that mimosine is not toxic to goats or beef and dairy cattle and provides considerable information on its action in a number of ruminant and nonruminant animals. Crounse *et al.* (1962), in their investigation of the effect of purified mimosine and ground *L. leucocephala* seeds on mice, removed hair from a large area of the body; thick coats appeared in the epilated areas of control animals within 8 to 10 days. Mice fed 5% ground seeds or 0.5% mimosine also regrew hair normally. No new pubescence developed, however, when the mimosine content of the diet was increased to 1.0% or the ground seed content to 10%. Resting hairs are not affected by mimosine, for hair damage occurs only during the period of active hair growth. Since the active phase of hair growth is short (Hegarty *et al.*, 1964), this factor may account for the contradictory reports of mimosine's depilatory capacity (Owen, 1958). In this regard, Hylin (1969) noted that mimosine is a strong inhibitor of rat liver cystathionine synthetase and cystathionase activities; these enzyme function in the conversion of methionine to cysteine via cystathionine. Animal hair follicles have a high cysteine content, and this compound is supplied, in part, from methionine. Thus, the observed failure of epilated areas to become hirsute may be related to diminished cysteine availability for hair protein production (Hylin, 1969).

Maintaining female Sprague–Dawley rats on a prolonged regime of a diet containing only 0.5% mimosine produces irregular estrous cycling. When the mimosine dietary content is increased to 1%, the estrous cycle terminates, and the rats are infertile (Hylin and Lichton, 1965). Mimosine's capacity to disrupt reproduction is also demonstrated by reduced litter size of rabbits and swine that consume *Leucaena* leaf meal (cited in Dewreede and Wayman, 1970) and diminished egg production by poultry (Thanjan, 1967). Rats fed a diet supplemented with 0.5 or 0.7% mimosine produce fetuses with deformed cranium, thorax, and pelvis. These deformities are associated with perforated uterine walls and fetal parts, enclosed in their fetal membranes, that protrude through the uterine wall. The sacral vertebrae of fetuses, developing from females fed 0.7% mimosine, fail to develop (Dewreede and Wayman, 1970). Mimosine also causes ocular lesions and cataracts (von Sallmann *et al.*, 1959).

As part of collagen biosynthesis, protocollagen, a polypeptide progenitor of collagen, undergoes proline hydroxylation; collagen is not formed from free hydroxyproline (Peterkofsky and Udenfriend, 1963). Mimosine markedly inhibits the synthesis of hydroxyproline from proline in embryonic cartilage tissues. This causes a pronounced intracellular accumulation of protocollagen, thereby denying adequate collagen to the developing tissues (Tang and Ling, 1975). In addition, hydroxyproline-deficient collagen is more rapidly degraded by collagenase (Hurych *et al.*, 1967). The resulting diminution in the collagen content of mimosine-treated animal organs may account for the observed uterine perforations of treated rats (Tang and Ling, 1975).

Some insight into the biochemical bases for mimosins's toxicity has been gained. As in the case of canaline, mimosine reacts with the aldehydic carbon of pyridoxal phosphate (Lin *et al.*, 1962a). This reaction decreases pyridoxal phosphate availability and disrupts the catalytic action of B_6-containing enzymes (Lin *et al.*, 1965). Mimosine-mediated inhibition of pyridoxal phosphate-dependent glutamic–aspartic transaminase of swine heart (Lin *et al.*, 1962b) and L-3,4-dihydroxyphenylalanine (L-dopa) decarboxylase of swine kidney (Lin *et al.*, 1963) has been demonstrated.

Either ferrous sulfate or dietary intake of mimosine as the ferrous salt reduces its intrinsic toxicity (Matsumoto *et al.*, 1951; Lin and Ling, 1961). The sparing effect of Fe^{2+} may result from interference in the formation of a complex between mimosine and pyridoxal phosphate (Lin *et al.*, 1962a). This heterocyclic amino acid may also chelate metallic ions required for enzyme activation (Tang and Ling, 1975). In this regard, Chang (1960) has reported a sparing effect for magnesium in mimosine-mediated inhibition of alkaline phosphatase. It may also function as a tyrosine antimetabolite since it suppresses tyrosine decarboxylase and tyrosinase activities (Crounse *et al.*, 1962) and other tyrosine utilizing enzymes (Prabhakaran *et al.*, 1973). However, both tyrosine and phenylalanine counteract its adverse effect on animals (Lin *et al.*, 1964). Finally, the phenylalanyl- and tyrosyl-tRNA synthetases of mung bean have been partially purified (Smith and Fowden, 1968). Mimosine, aminoacylated only by the phenylalanyl-tRNA synthetase, does not curtail esterification of phenylalanine to its cognate tRNA even when the mimosine level is 60 times greater than that of phenylalanine. Thus, at least in this higher plant, its toxicity does not appear to be related to impeded phenylalanine or tyrosine incorporation into protein.

B. L-Azetidine-2-carboxylic Acid

In 1955, Fowden reported the isolation of a novel imino acid, azetidine-2-carboxylic acid **(9)** which is a lower homolog of proline. Chemical synthesis of this compound confirmed its postulated struc-

ture. This imino acid is an important nitrogen metabolite of the monocot *Convallaria majalis,* where it constitutes over 75% of the total imino plus amino nitrogen (Fowden, 1956). In *Polygonatum multiflorum,* it accounts for at least 75% of the nonprotein nitrogen of the rhizome, a storage organ (Fowden and Bryant, 1958). Overall, these plants accumulate 3–6% of their dry weight as azetidine-2-carboxylic acid.

Azetidine-2-carboxylic acid occurs in about one-third of the examined liliaceous genera (Fowden and Steward, 1957), in the legume *Delonix regia* (Sung and Fowden, 1969), and in low levels in members of the Agavaceae—a family phylogenetically close to the Liliaceae. Fowden (1972) recounted his experience with a commercial preparation of sugar beet that produced an extract derived from approximately 10^9 kg of plant material. About 30 kg of azetidine-2-carboxylic acid were obtained from this preparation of a plant not believed previously to store this compound. Fowden made the germane point that, although a plant may produce a particular substance, the paucity of material could preclude an awareness of its occurrence. In the above instance, the level of this proline analog was only .03 ppm. This factor must be considered when one is delineating the distribution of secondary compounds.

Azetidine-2-carboxylic acid is a potent growth inhibitor of lower and higher plants (Fowden, 1963; Fowden and Richmond, 1963). It is also an effective antimetabolite of proline in animals, where it inhibits proline incorporation into protocollagen possibly by competing with proline for uptake by the cell (Takeuchi and Prockop, 1969). This imino acid not only affects hydroxylation of protocollagenous proline in embryonic cartilage, but is incorporated into the protocollagen of growing chick embryos. The resulting collagen, which contains azetidine-2-carboxylic acid, cannot be extruded at the required rate, and this reduces the amount of cellular collagen (Lane *et al.*, 1971).

The liliaceous plant *Urginea maritima* is very resistant to the lepidopteran predator *Spodoptera littoralis;* it contains about 1.7% azetidine-2-carboxylic acid on a fresh weight basis (Hassid *et al.*, 1976). After the foliage of this plant was extracted with 80% aqueous methanol, the solubilized components were fractionated extensively to segregate various groups of pharmacologically active compounds. These procedures ultimately disclosed that it was the isolated foliar azetidine-2-carboxylic acid, which incidentally accounted for 85% of the total free amino acids, that caused the observed severe insect mortality.

Peterson and Fowden (1965) partially purified the prolyl-tRNA synthetase of mung bean seed, an azetidine-2-carboxylic acid. This enzyme activated azetidine-2-carboxylic acid at about one-third the rate observed with proline. *Convallaria* and *Polygonatum,* which store this imino acid, do not esterify azetidine-2-carboxylic acid to the cognate tRNA of proline. In a similar vein, a higher homolog of proline, pipecolic acid **(10),** and 4,5-dehydropipecolate lack appreciable growth-inhibiting action and are not believed to be incorporated into proteins. In contrast, 3,4-dehydroproline, an effective proline antagonist in several microorganisms, replaces proline in the protein synthesized by several lower and higher plants (Fowden *et al.,* 1963). These findings support an important fundamental basis for the antimetabolic properties not only of azetidine-2-carboxylic acid but also other nonprotein amino acids that bear structural analogy to the components of common proteins. Both herbivores resistant to nonprotein amino acids and producer species avoid this adverse consequence of analog consumption or production if their activating and charging system can discriminate between the analog and its protein-containing counterpart (see Chapter 3, Section I by Fowden and Lea).

V. SELENIUM-CONTAINING AMINO ACIDS

The chemical similarity between selenium and sulfur has given rise to analogous groups of higher plant metabolites containing one or more atoms of selenium or sulfur (Shrift, 1972). Certain seleniferous soil-inhabiting plants, particularly members of the leguminous genus *Astragalus,* efficiently sequester selenium, and some plants can contain as much as 15,000 ppm of this trace element (Peterson, 1970). Analysis of the soluble fraction of these plants revealed that *Se*-methyl-L-selenocysteine **(11)** and selenocystathionine **(12)** are the principal causal agents for this element's toxicity. The former compound occurs in such selenium-sequestering plants as *A. bisulcatus,* and *Oonopsis condensata* (Compositae), whereas the latter predominates in *Neptunia amplexicaulis* (Mimosaceae) and *Morinda reticulata* (Rubiaceae). Both of these amino acids are found in the legume *Stanleya pinnata* (Cruciferae) (Peterson, 1970).

On the other hand, although some higher plants are capable of synthesizing selenium-containing compounds, they do not house appreciable cellular selenium and are consequently designated as nonaccumulators (Shrift, 1969). These nonaccumulating plants consistently form such compounds as L-selenomethionine **(13);** its methylated derivative, *Se*-methyl-L-selenomethionine; and L-selenocystine **(14)** (Shrift, 1972; Hegarty and Peterson, 1973).

$$CH_3—Se—CH_2—CH(NH_2)COOH$$

11

$$HOOC—CH(NH_2)—CH_2—Se—CH_2—CH_2—CH(NH_2)CO_2H$$

12

Selenium accumulators

$$CH_3—Se—CH_2—CH_2—CH(NH_2)COOH$$

13

$$HOOC—CH(NH_2)—CH_2—Se—Se—CH_2—CH(NH_2)COOH$$

14

Nonaccumulators

Selenocystathionine is synthesized by *Lecythis ollaria,* a deciduous tree of Central and South America. Human consumption of the nutlike seed causes abdominal pain, nausea, vomiting, and diarrhea. Some time thereafter, a reversible loss of scalp and body hair occurs (Aronow and Kerdel-Vegas, 1965). The pharmacologically active principle was purified from the seed by utilizing the cytotoxic properties of this compound as a biological assay. The isolated material possesses the biological and chemical properties of chemically prepared selenocystathionine.

Selenocystathionine and *Se*-methyl-L-selenocysteine are believed to cause "blind staggers," a chronic selenosis of range herbivores that eat such plants as members of the *Astragalus* and *Machaeranthera* (Hylin, 1969) (Fig. 3). In contrast, L-selenomethionine and L-selenocysteine cause a livestock malady similar to "alkali disease" that results from consumption of forage grasses and cereals growing on seleniferous soils. The symptoms associated with this disease include depilation as well as deformity and sloughing of the hooves (Hylin, 1969) (Fig. 3). Chronic poisoning of human beings has been shown to occur after ingestion of grain grown on selenium-containing soils in Columbia (Rosenfeld and Beath, 1964).

Trace levels of selenium (about 0.01 ppm) are beneficial and stimulatory to animals (Schultze, 1960); however, when forage containing more than 1 ppm is eaten, it becomes highly toxic (Peterson and Butler, 1962). For example, corn, wheat, barley, oats, grasses, and hay having only 10 to 30 ppm can cause "selenium alkali disease" (Hylin, 1969). Recently, a

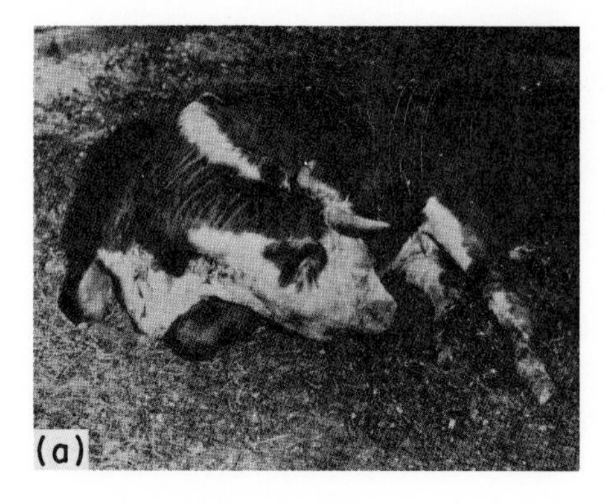

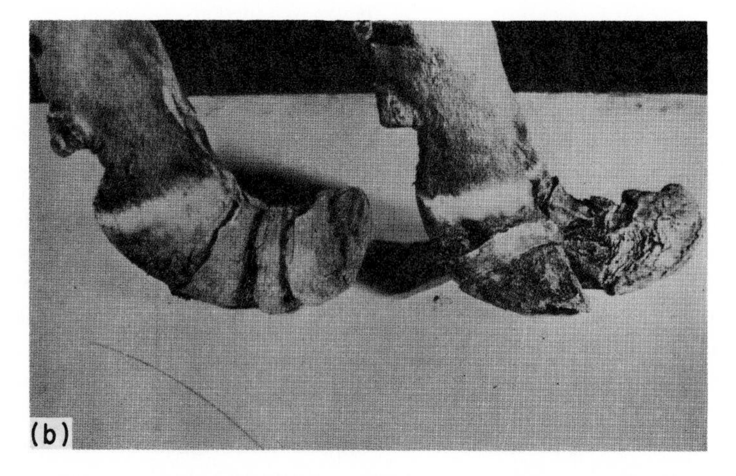

Figure 3. Consumption of range weeds containing certain selenoamino acids is reputed to cause the symptoms of "blind staggers" represented in the upper photograph. Other selenoamino acids of forage grasses and cereals inhabiting seleniferous soils cause "alkaline disease" in which a deformity and sloughing of the hooves results (bottom). After Rosenfeld and Beath (1964).

cysteinyl-tRNA synthetase from mung bean was partially purified and shown to activate selenocysteine (Shrift *et al.*, 1976). Cowie and Cohen (1957) and Butler and Peterson (1967) had previously demonstrated L-selenomethionine placement into protein. The formation of proteins containing selenocysteine and selenomethionine has led to the belief that the production of proteins containing selenoamino acids is a most important basis for this element's toxicity (Peterson and Butler, 1967;

Shrift, 1972). Thus, the exclusion of selenoamino acids from the protein of selenium-containing plants is critical to their tolerance of this harmful substance.

VI. SUBSTITUTED AROMATIC AMINO ACIDS

A. L-3,4-Dihydroxyphenylalanine

L-Dopa occurs in higher plants as both the free amino acid and as a β-glucoside (Pridham and Saltmarsh, 1963). As long ago as 1913, Torquati reported that the pods of *Vicia faba* contained 0.25% dopa, and Miller (1920) observed high concentrations in the velvet bean *Stizolobium (Mucuna) deeringianum.* Extremely high concentrations of the amino acid (5–10% dry seed weight) have since been noted in the seeds of all *Mucuna* species analyzed (Bell and Janzen, 1971). The same authors drew attention to the very uniform concentration of dopa found not only in different seeds of the same species, but in seeds from different species within the genus. This high and constant concentration was consistent, these authors suggested, with a defensive role for this compound, possibly that of protecting the seeds from insect predation. Subsequent experiments showed that dopa is toxic to the larvae of *Prodenia eridania* (Rehr *et al.,* 1973b) and to the larvae of the seed-eating beetle *Callosobruchus maculatus* (Janzen *et al.,* 1977). Addition of 5% dopa to the diet of *C. maculatus* was lethal to all test larvae; clearly, the greater concentration of dopa present in all tested *Mucuna* seeds would afford complete protection against this predator.

Large, daily doses of L-dopa (2–8 gm) have proved to be of value in assuaging the akinesis, rigidity, and tremors of Parkinson's disease (Calne and Sandler, 1970). A symptom of this malady is markedly attenuated levels of brain dopamine (an important catecholamine) (Hornykiewicz, 1966). Unlike dopamine, dopa is able to cross the mammalian blood–brain barrier, where it is decarboxylated to produce dopamine. This metabolic formation of dopamine from dopa is believed to be the clinical basis for the therapeutic activity of dopa (Calne and Sandler, 1970).

B. 5-Hydroxy-L-tryptophan

This nonprotein amino acid accumulates in the seeds of at least three species of the West African legume genus *Griffonia* (*Bandeiraea*), values as high as 14% of the dry seed weight being reported for *G. simplicifolia*

(Bell *et al.*, 1976). This amino acid, a precursor of 5-hydroxytryptamine (serotonin) in the mammalian central nervous system, is able to cross the mammalian blood–brain barrier. When injected into dogs or rats, it results in as much as a 10-fold increase in the level of brain 5-hydroxytryptamine since 5-hydroxytryptophan is decarboxylated readily in this organ. This increase in serotonin level elicits "tremors, pupillary dilation, loss of light reflex, apparent blindness, salivation, marked hyperpnea and tachycardia" (Udenfriend *et al.*, 1956, 1957).

When supplied to the larvae of *Callosobruchus maculatus* at a concentration of 1.0% in the insect diet, 5-hydroxytryptophan proved lethal (Janzen *et al.*, 1977). It seems likely, then, that the very high concentrations of both dopa and 5-hydroxytryptophan may protect the seeds of *Mucuna* and *Griffonia*, respectively, from predation by insects and other herbivores.

VII. MISCELLANEOUS

A. L-Hypoglycin

Hypoglycin A **(15)** is stored in the fruit and arilli of the sapindaceous tree *Blighia sapida* at a concentration of about 0.008% (Hassal and Reyle,

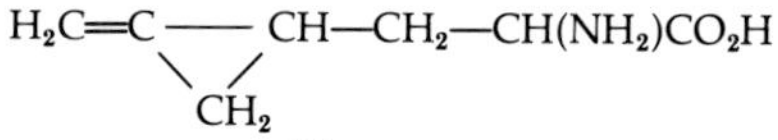

$$H_2C{=}C\text{——}CH\text{—}CH_2\text{—}CH(NH_2)CO_2H$$
$$\diagdown\ \diagup$$
$$CH_2$$

15

1955). This plant is indigenous to coastal West Africa where it is known as "ishin." In 1778, it was introduced into Jamaica and eventually the "ackee tree" became cultivated throughout the New World tropics (Hill, 1952). The edible arilli are ubiquitous in Jamaica, where their consumption has caused a serious disease known as "vomiting sickness." Incidences of this malady have a seasonal aspect; it reaches an epidemic level during the dormant portion of the growing season, when food is scarce and the unripened ackee fruit becomes an important food source. Unfortunately, the unripened arilli contain considerably more hypoglycin A (about 0.11%) than does the ripened fruit (Hassal and Reyle, 1955).

The symptoms of vomiting sickness include violent retching and vomiting, which are usually followed by convulsions, coma, and death. The most striking clinical consequence of ingesting this compound is a precipitous decline in blood glucose level from a normal range of 80 to 100 mg % to 10 mg % or less. Glucose, stored in the liver as glycogen, is depleted drastically and fatty metamorphosis of liver, kidney, and other organs follows (Hill, 1952). In the past, mortality levels reached 80%, with thousands of recorded fatalities, but modern treatment with intravenous glucose is completely successful as long as the hypoglycemia does not produce irreversible shock (Bressler *et al.*, 1969).

A γ-glutamyl derivative of hypoglycin A, designated hypoglycin B, occurs in the *Blighia* fruit (Hassal and Reyle, 1955). Hassal and John (1960) suggested that hypoglycin A be referred to simply as hypoglycin, whereas the dipeptide be designated γ-L-glutamyl hypoglycin. Hypoglycin also occurs in *Billia hippocastanum* (Hippocastanaceae), where it is a major constituent of the free amino acid pool; smaller amounts of γ-glutamylhypoglycin as well as a lower homolog of hypoglycin are also produced (Eloff and Fowden, 1970).

Cat, rabbit, monkey, and guinea pig readily exhibit hypoglycaemia. The LD_{50} for rat is 90–100 mg per kilogram body weight; in contrast, mice survive a dose of 160 mg/kg, whereas rabbits die from an intravenous injection of only 10–20 mg per kilogram body weight (Bressler *et al.*, 1969). In rat, γ-glutamylhypoglycin elicits *in vivo* effects comparable to those of hypoglycin, provided that the concentration is doubled (Hassal and Reyle, 1955). Many hypoglycemic compounds are related structurally to hypoglycin, whereas others lack its biological activity, in spite of certain common structural features (Bressler *et al.*, 1969).

In an incisive review of hypoglycin, Bressler *et al.* (1969) stated that the hypoglycemic principle may actually be a hypoglycin degradation product, methylenecyclopropane acetate, and that hypoglycemia appears to result from impaired gluconeogenesis. Hypoglycin reacts with acetyl-CoA; the resulting acetylated compound competes for carnitine acetyltransferase, which depresses the level of cofactors required for long-chain fatty acid oxidation. Consistent with this assertion, Entman and Bressler (1967) demonstrated that carnitine, which stimulates mitochondrial-dependent fatty acid oxidation, spares the hypoglycemic property of this toxic compound. In essence, curtailment of oxidative reductions denies acetyl-CoA, $NADH_2$, and ATP required for gluconeogenesis. In addition, fatty acid oxidation products regulate glycolytic reactions; thus, their loss further exacerbates glucose depletion via enhanced glycolytic activity. Persaud (1970) determined that hypoglycin is teratogenic in pregnant rats and proposed that hypoglycin uncoupling of oxidative phosphorylation causes a diminution in cellular energy charge that may account for the teratogenicity of hypoglycin.

B. L-Djenkolic Acid

The djenkol bean (*Pithecolobium lobatum*), indigenous to Java, and a related species, *P. bubalinum*, of Sumatra store djenkolic acid **(16).** These plants served as the initial source of this toxic nonprotein amino acid. The seed of the former plant stores about 1–2% of its dry weight as djenkolic acid, whereas in the latter plant it occurs at a concentration of up to 4%. Dried young leaves and twigs of the djenkol tree appear to lack this compound, as determined by paper chromatographic techniques (Van Veen and Latuasan, 1949). Djenkolic acid also occurs in certain species of the Mimosaceae, and its N-acetylated derivative has been isolated from *Acacia farnesiana* (Gmelin *et al.*, 1962).

$$HOOC{-}CH(NH_2){-}CH_2{-}S{-}CH_2{-}S{-}CH_2{-}CH(NH_2)COOH$$

16

Human consumption of this toxic seed, at least in Indonesia, is not an unavoidable consequence of an impovished diet, since well-fed Javanese actively seek out the ripened seed (Van Veen, 1973). This natural product is sparsely soluble under acidic conditions and crystallizes readily from urine. In young children, such crystals can obstruct the ureter and cause painful swelling of the external genitalia (Suharjono and Sadatun, 1968). Its consumption also causes acute kidney malfunction that results in anuria, i.e., impaired or blocked urine flow (Van Veen, 1973).

VIII. CONCLUSIONS

This chapter has assembled information that is concerned primarily with the documented toxicity and antimetabolic properties of the nonprotein amino acids. The accumulation of natural products toxic to a potential predator does not in itself establish that these compounds are produced as an evolutionary response to predation. Obviously, we cannot venture back in time to study directly the ecological and evolutionary forces that shaped the natural selection of plants storing toxic secondary compounds. If we are to suggest, as we do, that certain nonprotein amino acids are part of the chemical defense of higher plants that protects them against the feeding ravages of phytophagous insects and other herbivores, then we must support this contention with all of the available evidence.

Our belief in the protective role of nonprotein amino acids rests on the following seven lines of evidence. First, many nonprotein amino acids are highly toxic to a large number of herbivores. The potent antimetabolic properties of these compounds are wholly consistent with their function as deterrents to predator feeding activity.

Second, the very high concentration of these secondary compounds frequently accumulated in plants represents a significant expenditure of metabolic resources. It is difficult to rationalize the synthesis and storage of massive quantities of nitrogen-rich compounds without some commensurate benefit in increased fitness. In terms of the plant's overall biological fitness, it may be more beneficial to use some of the metabolic resources for synthesizing an insecticidal nonprotein amino acid in order to minimize loss of photosynthetic foliage or a significant amount of the vital seed crop to predation.

In this regard, it has become widely accepted that some nonprotein amino acids function as nitrogen-storing metabolites; we suggest, however, that this cannot be their unique function. The toxic nonprotein amino acids are often related structurally to protein amino acids; a genome for the biosynthesis and turnover of the latter compounds already exists. Thus, it is difficult to interpret the storage of compounds such as canavanine or azetidine-2-carboxylic acid solely in terms of nitrogen storage when their protein amino acid counterpart can perform the same function. In addition, perpetuation of a genome for the biosynthesis and catabolism of both sets of related amino acids places in additional metabolic burden on the plant, and it is reasonable to propose some overall biological benefit from this expenditure.

Toxic nonprotein amino acids such as canavanine, azetidine-2-carboxylic acid, and 5-hydroxy-L-tryptophan are manufactured and stored in very ample amounts even within the vegetative tissues. Conservation of plant nitrogen resources would be achieved more efficiently if these compounds were produced at the time of flower anthesis and development. These factors, constituting the third line of evidence, are explicable in terms of nonprotein amino acid capacity to provide an effective chemical barrier to predation.

Fourth, the fixed concentration of certain nonprotein amino acids within the seed population of a given species suggests that their concentration may be the result of two opposing forces. The upper concentration limit may be imposed by plant–plant competition, which culminates in the selection of those individuals with a nitrogen resource allocation just sufficient to achieve the necessary protective effects. The lower level may be maintained by predator pressure that eliminates individuals from the population with less than the necessary nonprotein amino acid content required to effectively deter predation.

Fifth, if our contention is correct, certain herbivores should exist that react to the chemical barrier represented by toxic nonprotein amino acids by developing effective mechanisms of biochemical detoxification. The interaction of *Caryedes brasiliensis* with *Dioclea megacarpa* (see Section III,A) thoroughly documents the development of an insect biochemical response to a formidable chemical challenge by the plant.

A sixth line of evidence is provided by the studies of Janzen (1969) demonstrating that the weight of a seed crop per unit area of canopy is inversely proportional to the seed size. Chemical examination of the appropriate seeds disclosed that the larger seeds contain nonprotein amino acids whereas the smaller seeds do not. The smaller seeds are subject generally to bruchid beetle attack, whereas the larger seeds enjoy much greater freedom from predation. This finding suggests that two strategies are operative in curtailing beetle feeding activity. The first is the production of a copious amount of small seeds, thereby ensuring that a portion of the total seed crop will escape predator attention. The second relies on production and accumulation of insecticidal seed amino acids. In these instances, the metabolic cost to the plant for synthesizing these compounds is probably less than the expense of producing a larger seed crop to compensate for predator attrition.

Finally, Janzen (1969) also observed that seed-eating beetles which forage for seeds that store a particular nonprotein amino acid do not survive when reared on seeds containing other members of this chemical group. This finding suggests that insect adaptation to a particular nonprotein amino acid is highly specific. A general resistance to these secondary plant metabolites is not imparted, and these compounds continue to afford protection against the great majority of potential herbivores. The general question of seed toxins has been considered recently by Bell (1978).

Thus, the evidence strongly supports our contention that some nonprotein amino acids provide an adaptive advantage to the plant in the continued struggle for survival that has characterized plant–herbivore coevolution. We are mindful, however, that pertinent field data demonstrating the capacity of these secondary natural products to reduce herbivore feeding activity, minimize microbial activity, or enhance overall plant fitness have not, as yet, been secured.

Many lacunae exist in our present knowledge of the nonprotein amino acids. There is relatively little known of the metabolic reactions culminating in the production or degradation of these compounds by the plant and the circumstances favoring the production and perpetuation of the genome of a particular nonprotein amino acid. The biochemical bases for the toxicity, degradation, and detoxification of nonprotein amino acids in herbivores are incompletely or inadequately understood.

The sporadic distributional pattern of these compounds even within a single genus, the fact that a particular plant may store 0.5% of its seed dry weight as a nonprotein amino acid whereas another plant contains 25 times as much, and the ecological and physiological rationale for the allocation of as much as 90% of the plant nitrogen resource into a single nonprotein amino acid are not fully understood. We know too little of the role of nonprotein amino acids in plant–plant interactions and in microbial infection and the overall disease process. In spite of these limitations, the employment of L-dopa in controlling clinical manifestations of Parkinson's disease, the use of canavanine for producing aberrant proteins for studies of protein turnover and the biochemistry of aging, and the use of α,γ-diaminobutyric acid and β-cyanoalanine in neurological studies are but a few examples of the utilization of these compounds. The marked experimental potential of the nonprotein amino acids augurs well for their continued and increased application in a host of future biological and biochemical problems.

ACKNOWLEDGMENTS

The senior author acknowledges the support of National Science Foundation Grants GB-40198 and BMS-75-19770, National Institutes of Health Grants AM-13830 and AM-17322, the Research Committee of the University of Kentucky, and NIH Biomedical Support Grant 5-S05-RR07114-08 in the undertaking of certain studies presented in this chapter. The critical reading of the manuscript by Dr. T. Robinson, Dr. P. Lea, and Dr. S. F. Conti is also gratefully acknowledged.

REFERENCES

Ackermann, W. W., Cox, D. C., and Dinka, S. (1965). *Biochem. Biophys. Res. Commun.* **19,** 745–750.

Archard, L. C., and Williamson, J. D. (1974). *J. Gen. Virol.* **24,** 493–501.

Aronow, L., and Kerdel-Vegas, F. (1965). *Nature (London)* **205,** 1185–1186.

Attias, J., Schlesinger, M. J., and Schlesinger, S. (1969). *J. Biol. Chem.* **244,** 3810–3817.

Barrow, M. V., Simpson, C. F., and Miller, E. J. (1974). *Q. Rev. Biol.* **49,** 101–128.

Beeler, T., and Churchich, J. E. (1976). *J. Biol. Chem.* **251,** 5267–5271.

Bell, E. A. (1962). *Nature (London)* **193,** 1078.

Bell, E. A. (1963). *Nature (London)* **199,** 70–71.

Bell, E. A. (1964a). *Nature (London)* **203,** 378–380.

Bell, E. A. (1964b). *Biochem. J.* **91,** 358–361.

Bell, E. A. (1971). *In* "Chemotaxonomy of the Leguminosae" (J. B. Harborne, D. Boulter, and B. L. Turner, eds.), pp. 179–206. Academic Press, New York.

Bell, E. A. (1972). *In* "Phytochemical Ecology" (J. B. Harborne, ed.), pp. 163–177. Academic Press, New York.

Bell, E. A. (1978). *In* "Biochemical Aspects of Plant and Animal Co-evolution" (J. B. Harborne, ed.). Academic Press, New York 143–161.
Bell, E. A., and Janzen, D. H. (1971). *Nature (London)* **229,** 136–137.
Bell, E. A., and O'Donovan, J. P. (1966). *Phytochemistry* **5,** 1211–1219.
Bell, E. A., and Przybylska, J. (1965). *Biochem. J.* **94,** 35P.
Bell, E. A., and Tirimanna, A. S. L. (1965). *Biochem. J.* **97,** 104–111.
Bell, E. A., and Fellows, L. E., and Qureshi, Y. M. (1976). *Phytochemistry* **15,** 823.
Blumenthal-Goldschmidt, S., Butler, G. W., and Conn, E. E. (1963). *Nature (London)* **197,** 718.
Bond, E. J. (1961). *Can. J. Biochem. Physiol.* **39,** 1793–1802.
Brown, E. G., and Al-Baldawi, N. F. (1977). *Biochem. J.* **164,** 589–594.
Bressler, R., Corredor, C., and Brendel, K. (1969). *Pharmacol. Rev.* **21,** 105–130.
Butler, G. W., and Peterson, P. J. (1967). *Aust. J. Biol. Sci.* **20,** 77–86.
Calne, D. B., and Sandler, M. (1970). *Nature (London)* **226,** 21–24.
Chang, L. T. (1960). *J. Formosan. Med. Assoc.* **59,** 882.
Charlwood, B. V., and Bell, E. A. (1977). *J. Chromatog.* **135,** 377–384.
Christie, G. S., Madsen, N. P., and Hegarty, M. P. (1969). *Biochem. Pharmacol.* **18,** 693–700.
Christie, G. S., deMunk, F. G., Madsen, N. P., and Hegarty, M. P. (1971). *Pathology* **3,** 139–144.
Christie, G. S., Wilson, M., and Hegarty, M. P. (1975). *J. Pathol.* **117,** 195–205.
Cowie, D. B., and Cohen, G. N. (1957). *Biochim. Biophys. Acta* **26,** 252–261.
Crounse, R. G., Maxwell, J. D., and Blank, H. (1962). *Nature (London)* **194,** 694–695.
Dahlman, D. L., and Rosenthal, G. A. (1975). *Comp. Biochem. Physiol. A* **51,** 33–36.
Dewreede, S., and Wayman, D. (1970). *Teratology* **3,** 21–28.
Dossaji, S. F., and Bell, E. A. (1973). *Phytochemistry* **12,** 143–144.
Eloff, J. N., and Fowden, L. (1970). *Phytochemistry* **9,** 2423–2424.
Entman, M., and Bressler, R. (1967). *Mol. Pharmacol.* **3,** 333.
Fowden, L. (1955). *Biochem. J.* **64,** 323–332.
Fowden, L. (1963). *J. Exp. Bot.* **14,** 387–398.
Fowden, L. (1970). *Prog. Phytochem.* **2,** 203–265.
Fowden, L. (1972). *Phytochemistry* **11,** 2271–2276.
Fowden, L., and Bell, E. A. (1965). *Nature (London)* **206,** 110–112.
Fowden, L., and Bryant, M. (1958). *Biochem. J.* **70,** 626–629.
Fowden, L., and Richmond, M. H. (1963). *Biochim. Biophys. Acta* **71,** 459–461.
Fowden, L., and Steward, F. C. (1957). *Ann. Bot. (London)* [N.S.] **21,** 60–67.
Fowden, L., Neale, S., and Tristram, H. (1963). *Nature (London)* **199,** 35–38.
Fowden, L., Lewis, D., and Tristram, H. (1967). *Adv. Enzymol.* **29,** 89–163.
Fujita, Y. (1959). *Bull. Chem. Soc. Jpn.* **32,** 439–442.
Gmelin, R., Kjaer, A., and Larsen, P. O. (1962). *Phytochemistry* **1,** 233–236.
Gonzalez, V., Brewbaker, J. L., and Hamill, D. E. (1967). *Crop Sci.* **7,** 140–143.
Greenstein, J. P., and Winitz, M. (1961). "Chemistry of the Amino Acids." Wiley, New York.
Harborne, J. (1973). "Phytochemical Methods. A Guide to Modern Techniques of Plant Analysis." Chapman & Hall, London.
Hare, J. D. (1969). *Exp. Cell Res.* **58,** 170–174.
Harrison, B. J., and Holliday, R. (1967). *Nature (London)* **213,** 990–992.
Harry, P., Dror, Y., and Applebaum, S. W. (1976). *Insect Biochem.* **6,** 273–279.
Hassal, C. H., and John, D. I. (1960). *J. Chem. Soc.* p. 4112.
Hassal, C. H., and Reyle, K. (1955). *Biochem. J.* **60,** 334–338.

Hassid, E., Applebaum, S. W., and Birk, Y. (1976). *Phytoparasitica* **4,** 173–183.
Hegarty, M. P., and Peterson, P. J. (1973). *In* "Chemistry and Biochemistry of Herbage" (G. W. Butler and R. W. Bailey, eds.), Vol. I, pp. 1–62. Academic Press, New York.
Hegarty, M. P., and Pound, A. W. (1968). *Nature (London)* **217,** 354–355.
Hegarty, M. P., and Pound, A. W. (1970). *Aust. J. Biol. Sci.* **23,** 831–842.
Hegarty, M. P., Schinckel, P. G., and Court, R. D. (1964). *Aust. J. Agric. Res.* **15,** 153–167.
Hill, K. R. (1952). *West Indian Med. J.* **1,** 243.
Hornykiewicz, O. (1966). *Pharmacol. Rev.* **28,** 925–964.
Hurych, J., Chvapil, M., Tichy, O. M., and Beniač, F. (1967). *Eur. J. Biochem.* **3,** 242–247.
Hutton, E. M., Windrum, G. M., and Kratzing, C. C. (1958a). *J. Nutr.* **64,** 321–333.
Hutton, E. M., Windrum, G. M., and Kratzing, C. C. (1958b). *J. Nutr.* **65,** 429–440.
Hylin, J. W. (1969). *J. Agric. Food. Chem.* **17,** 492–496.
Hylin, J. W., and Lichton, I. J. (1965). *Biochem. Pharmacol.* **14,** 1167–1169.
Inatomi, H., Inugai, F., and Murakami, T. (1968). *Chem. Pharm. Bull.* **16,** 2521.
Isogai, A., Murakoshi, S., Suzuki, A., and Tamura, S. (1973). *J. Agric. Chem. Soc. Jpn.* **47,** 449–453.
Janzen, D. H. (1969). *Evolution* **23,** 1–27.
Janzen, D. H. (1971). *Am. Nat.* **105,** 97–102.
Janzen, D. H., Juster, H. B., and Bell, E. A. (1977). *Phytochemistry* **16,** 223–227.
Johnston, G. A. R. (1974). *In* "Neuropoisons" (L. L. Simpson and D. R. Curtis, eds.), pp. 179–205. Plenum, New York.
Kammer, A., Dahlman, D. L., and Rosenthal, G. A. (1978). *J. Exp. Biol.* **75,** 123.
Katunuma, N., Okada, M., Matsuzawa, T., and Otsuka, Y. (1965). *J. Biochem. (Tokyo)* **57,** 445–449.
Kekomäki, M., Rahiala, E.-L., and Räihä, N. C. R. (1969). *Ann. Med. Exp. Fenn.* **47,** 33–38.
Kleipool, R. J. C., and Wibaut, J. P. (1950). *Recl. Trav. Chim. Pays-Bas* **69,** 459–462.
Lane, J. M., Delivi, P., and Prockop, D. J. (1971). *Biochim. Biophys. Acta* **236,** 517–527.
Lea, P. J., and Norris, R. D. (1976). *Phytochemistry* **15,** 585–595.
Leisinger, T., Haas, D., and Hegarty, M. P. (1972). *Biochim. Biophys. Acta* **262,** 214–219.
Lewis, H. B., and Schulert, A. R. (1949). *Proc. Soc. Exp. Biol. Med.* **71,** 440–441.
Lewis, H. B., Fajans, R. S., Esterer, M. B., Shen C.-W., and Oliphant, M. (1948). *J. Nutr.* **36,** 537–559.
Liener, I. E. (1966). *In* "Toxicants Occurring Naturally in Foods," Vol. I, p. 42, Natl. Acad. Sci., Washington, D.C.
Lin, J.-Y., and Ling, K.-H. (1961). *J. Formosan Med. Assoc.* **60,** 657–665.
Lin, J.-Y., Shih, Y.-M., and Ling, K.-H. (1962a). *J. Formosan Med. Assoc.* **61,** 997–1003.
Lin, J.-Y., Shih, Y.-M., and Ling, K.-H., (1962b). *J. Formosan Med. Assoc.* **61,** 1004–1010.
Lin, J.-Y., Lin, K.-T., and Ling, K.-H. (1963). *J. Formosan Med. Assoc.* **62,** 587–590.
Lin, J.-K., Ling, T.-A., and Tung, T. C. (1965). *J. Formosan Med. Assoc.* **64,** 265–272.
Lin, K.-T., Lin, J.-K., and Tung, T.-C. (1964). *J. Formosan Med. Assoc.* **63,** 10–25.
Maas, W. K. (1961). *Cold Spring Harbor Symp. Quant. Biol.* **26,** 183–191.
Madsen, N. P., and Hegarty, M. P. (1970). *Biochem. Pharmacol.* **19,** 2391–2393.
Makisumi, S. (1961). *J. Biochem. (Tokyo)* **49,** 284–291.
Masson, M. M., and Whiting, M. G. (1966). *Fed. Am. Soc. Exp. Biol.* **25,** 533–540.
Matsumoto, H., Smith, E. G., and Sherman, G. D. (1951). *Arch. Biochem. Biophys.* **33,** 201–211.
Meister, A. (1965). "Biochemistry of the Amino Acids," Academic Press, New York.
Miersch, J. (1967). *Naturwissenschaften* **54,** 169–170.
Miller, E. R. (1920). *J. Biol. Chem.* **44,** 481–486.
Miller, R. W., and Smith, C. R. (1973). *Agric. Food Chem.* **21,** 909–912.

Murti, V. V. S., Seshadri, T. R., and Venkitasubramanian, T. A. (1964). *Phytochemistry*, **3**, 73–78.
Natelson, S., Koller, A., Tseng H.-Y., and Dods, R. F. (1977). *Clin. Chem.* **23**, 960–966.
Nordfeldt, S., Henke, L. A., Morita, K., Matsumoto, H., Takahashi, M., Younge, O. R., Willers, E. H., and Cross, R. F. (1952). *Hawaii Agric. Exp. Stat., Tech. Bull.* **15.**
O'Dell, B. L., Elsden, D. F., Thomas, J., Partridge, S. M., Smith, R. H., and Palmer, R. (1966). *Nature (London)* **209**, 401–402.
O'Neal, R. M., Chen, C.-H., Reynolds, C. S., Meghal, S. K., and Koeppe, R. E. (1968). *Biochem. J.* **106**, 699–706.
Owen, L. N. (1958). *Vet. Rec.* **70**, 454–456.
Pearn, J. H. (1967a). *Nature (London)* **215**, 980–981.
Pearn, J. H. (1967b). *Br. J. Exp. Pathol.* **48**, 620–626.
Persaud, T. V. N. (1970). *Experientia* **27**, 414.
Peterkofsky, B., and Udenfriend, S. (1963). *J. Biol. Chem.* **238**, 3966–3977.
Peterson, P. J. (1970). *Phytochemistry* **9**, 916–917.
Peterson, P. J., and Butler, G. W. (1962). *Aust. J. Biol. Sci.* **15**, 126–146.
Peterson, P. J., and Butler, G. W. (1967). *Nature (London)* **213**, 599–600.
Peterson, P. J., and Fowden, L. (1965). *Biochem. J.* **97**, 112–124.
Pfeffer, M., and Ressler, C. (1967). *Biochem. Pharmacol.* **16**, 2299–2308.
Prabhakaran, K., Harris, E. B., and Kirchheimer, W. F. (1973). *Cytobios.* **7**, 245.
Pridham, J. B., and Saltmarsh, M. J. (1963). *Biochem. J.* **87**, 218–224.
Prouty, W. F., Karnovsky, M. J., and Goldberg, A. L. (1975). *J. Biol. Chem.* **250**, 1112–1122.
Qureshi, Y. M., Pilbeam, D. J., Evans, C. S., and Bell, E. A. (1977). *Phytochemistry* **16**, 477–479.
Rahiala, E.-L. (1973). *Acta Chem. Scand.* **27**, 3861–3867.
Rahiala, E.-L., Kekomäki, M., Jänne, J., Raina, A., and Räihä, N. C. R. (1971). *Biochim. Biophys. Acta* **227**, 337–343.
Rao, S. L. N., and Sarma, P. S. (1966). *Indian J. Biochem.* **3**, 57–58.
Rao, S. L. N., Ramachandran, L. K., and Adiga, P. R. (1963). *Biochemistry* **2**, 298–300.
Rao, S. L. N., Adiga, P. R., and Sarma, P. S. (1964). *Biochemistry* **3**, 432–436.
Rehr, S. S., Bell, E. A., Janzen, D. H., and Feeny, P. P. (1973a). *Biochem. Syst. Ecol.* **1**, 63–67.
Rehr, S. S., Janzen, D. H., and Feeny, P. P. (1973b). *Science* **181**, 81–82.
Ressler, C. (1962). *J. Biol. Chem.* **237**, 733–735.
Ressler, C. (1975). *Recent Adv. Phytochem.* **9**, 151–166.
Ressler, C., and Ratzkin, H. (1961). *J. Org. Chem.* **26**, 3356–3360.
Ressler, C., Redstone, P. A., and Erenberg, R. H. (1961). *Science* **134**, 188–190.
Ressler, C., Nigam, S. N., Giza, Y.-H., and Nelson, H. (1963). *J. Am. Chem. Soc.* **85**, 3311–3312.
Robinson, T. (1975). "The Organic Constituents of Higher Plants, Their Chemistry and Interrelationships," 3rd ed. Cordus Press, North Amherst, Massachusetts.
Robinson, T. (1976). *Life Sci.* **19**, 1097–1102.
Rosenfeld, I., and Beath, O. A. (1964). "Selenium," 2nd ed. Academic Press, New York.
Rosenthal, G. A. (1974). *J. Exp. Bot.* **25**, 609–613.
Rosenthal, G. A. (1977a). *Q. Rev. Biol.* **52**, 155–178.
Rosenthal, G. A. (1977b). *Biochem. Syst. Ecol.* **5**, 219–220.
Rosenthal, G. A. (1978a). *Life Sci.* **23**, 93.
Rosenthal, G. A. (1978b). *J. Theor. Biol.* **71**, 265.
Rosenthal, G. A., and Dahlman, D. L. (1975). *Comp. Biochem. Physiol. A* **52**, 105–108.
Rosenthal, G. A., and Davis, D. L. (1975). *Phytochemistry* **14**, 1117–1118.

Rosenthal, G. A., Dahlman, D. L., and Janzen, D. H. (1976). *Science* **192,** 256–258.
Rosenthal, G. A., Janzen, D. H., and Dahlman, D. L. (1977). *Science* **196,** 658–660.
Roy, D. B., and Narasinga Rao, B. S. (1968). *Curr. Sci.* **37,** 395–399.
Schilling, E. D., and Strong, F. M. (1954). *J. Am. Chem. Soc.* **76,** 2848.
Schilling, E. D., and Strong, F. M. (1955). *J. Am. Chem. Soc.* **77,** 2843–2845.
Schlesinger, H. M., Applebaum, S. W., and Birk, Y. (1976). *J. Insect Physiol.* **22,** 1421–1425.
Schlesinger, H. M., Applebaum, S. W., and Birk, Y. (1977). *J. Insect Physiol.* **23,** 1311–1313.
Schultze, M. O. (1960). *Annu. Rev. Biochem.* **29,** 391–412.
Selye, H. (1957). *Rev. Can. de Biol.* **16,** 1–82.
Shrift, A. (1969). *Annu. Rev. Plant Physiol.* **20,** 475–494.
Shrift, A. (1972). *In* "Phytochemical Ecology" (J. B. Harborne, ed.), pp. 145–161. Academic Press, New York.
Shrift, A., Bechard, D., Harcup, C., and Fowden, L. (1976). *Plant Physiol.* **58,** 248–252.
Siegel, R. C., Pinnell, S. R., and Martin, G. R. (1970). *Biochemistry* **9,** 4486.
Siegel, R. C., and Martin, G. R. (1970). *J. Biol. Chem.* **245,** 1653.
Smith, I. K., and Fowden, L. (1968). *Phytochemistry* **7,** 1065–1075.
Suharjono and Sadatun (1968). *Paediatr. Indones.* **8,** 20–29.
Sung, M.-L., and Fowden, L. (1969). *Phytochemistry* **8,** 2095–2096.
Takahashi, M., and Ripperton, J. C. (1949). *Hawaii, Agric. Exp. Stn., Bull.* **100.**
Takeuchi, T., and Prockop, D. J. (1969). *Biochim. Biophys. Acta* **175,** 142–155.
Tang, S.-Y., and Ling, K.-H. (1975). *Toxicon* **13,** 339–342.
Tate, S. S., and Meister, A. (1969). *Biochemistry* **8,** 1660–1668.
Thanjan, D. K. (1967). M.S. Thesis, University of Hawaii, Honolulu.
Thompson, J. F., Morris, C. J., and Smith, I. K. (1969). *Annu. Rev. Biochem.* **38,** 137–158.
Tschiersch, B. (1961). *Flora Jena* **150,** 87–94.
Tschiersch, B. (1964). *Flora Jena* **154,** 445–471.
Tschiersch, B. (1966). *Tetrahedron Lett.* **28,** 3237–3241.
Udenfriend, S., Titus, E., Weissbach, H., and Paterson, R. E. (1956). *J. Biol. Chem.* **219,** 335–344.
Udenfriend, S., Weissbach, H., and Bogdanski, D. F. (1957). *J. Biol. Chem.* **224,** 803–810.
Vanderzant, E. S., and Chremos, J. H. (1971). *Ann. Entomol. Soc. Am.* **64,** 480–485.
Van Veen, A. G. (1973). *In* "Toxicants Occurring Naturally in Foods," 2nd ed., Nat. Acad. Sci., Washington, D.C.
Van Veen, A. G., and Latuasan, H. E. (1949). *Indones. J. Nat. Sci.* **105,** 288–289.
Vega, A., and Bell, E. A. (1967). *Phytochemistry* **6,** 759–762.
Vega, A., Bell, E. A., and Nunn, P. B. (1968). *Phytochemistry* **7,** 1885–1887.
von Sallmann, L., Grimes, P., and Collins, E. (1959). *Am. J. Ophthalmol.* **47,** 107–117.
Walker, J. B. (1955a). *J. Biol. Chem.* **212,** 207–215.
Walker, J. B. (1955b). *J. Biol. Chem.* **212,** 617–622.
Whiting, M. G. (1963). *Econ. Bot.* **17,** 271–302.

Chapter 10

Cyanide and Cyanogenic Glycosides

ERIC E. CONN

I. INTRODUCTION

Knowledge of the metabolism of secondary plant compounds, both in the plants that produce these substances and in the animals that may consume them, can contribute to an understanding of herbivore–secondary plant metabolite relationships. Considerable information on the metabolism, both in plants and in animals, of the cyanide-producing compounds of plants has been acquired in recent years. Consequently, this group of secondary metabolites may serve as a particu-

ISBN 0-12-597180-X

larly useful model for the study of herbivore–plant relationships. In this chapter, the chemical nature of cyanogenic glycosides and their distribution will be described. Then, the cyanogenic process will be discussed, followed by a review of some of the factors that influence the toxicity of cyanide-producing plants to animals.

II. CHEMICAL NATURE OF CYANOGENIC GLYCOSIDES

Hydrogen cyanide (HCN), being a volatile gas, does not exist free to any extent within the cyanophoric plant. Instead, it is released by the action of enzymes acting on one or more compounds that serve as precursors in the plant. These compounds are usually carbohydrate derivatives known as cyanogenic glycosides; specifically, they are *O*-β-glycosides of α-hydroxynitriles (cyanohydrins). Hydrogen cyanide can also be produced by the action of enzyme(s) on cyanogenic lipids (see Chapter 12), a class of cyanogenic substances that have been reported only in the Sapindaceae. Another small group of natural products known as pseudocyanogenic glucosides release HCN when treated with alkali. These substances are restricted to one family of plants, the Cycadaceae, and are not further treated in this chapter.

Twenty-three cyanogenic glycosides have been isolated from higher plants. A detailed discussion of the chemical properties of these compounds would not serve the purpose of this chapter; for the interested reader, recent articles by Eyjolfsson (1970), Seigler (1977), and Conn (1969, 1978) should be consulted. In this section, we shall discuss a few of the glycosides for which there is a body of information concerning their metabolism in plants and/or animals.

The structures of two chemically related cyanogenic glycosides are shown in Fig. 1. Both compounds contain the sugar D-glucopyranose bound in a β-glycosidic linkage to the aglycone, which is an α-hydroxynitrile (cyanohydrin). Thus, the two compounds are β-glucosides. Since the carbon atom in the α-hydroxynitrile bearing the hydroxyl group is chiral (asymmetric), this α-hydroxynitrile can occur in two forms that are mirror-image isomers termed the *R* and *S* epimers. In dhurrin, it is the *S* epimer that is glucosylated; in taxiphyllin it is the *R* epimer, and the two glucosides are stereoisomers. That is, they contain the identical number of carbon, hydrogen, and oxygen atoms arranged into the same functional groups; they differ only in location of those groups in space (see Fig. 1). Dhurrin is a cyanogenic glucoside that occurs mainly in grasses; taxiphyllin was first isolated from members of the yew family (Taxaceae) but also occurs in grasses.

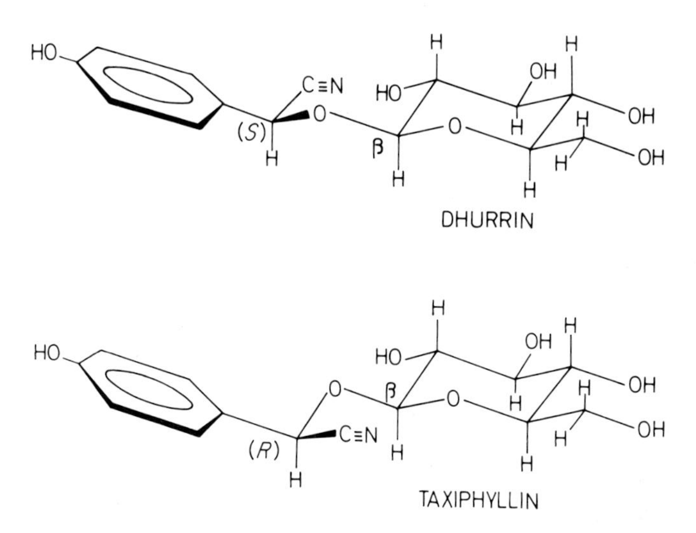

Figure 1. Structures of dhurrin and taxiphyllin.

Two related glycosides, known as prunasin and amygdalin, are found in the Rosaceae (Fig. 2). These substances differ from each other in an important way. Prunasin contains only a single carbohydrate moiety, D-glucopyranose; it therefore is a glucoside. Amygdalin contains two glucose units linked by a β-1,6-glucosidic bond; the free disaccharide is

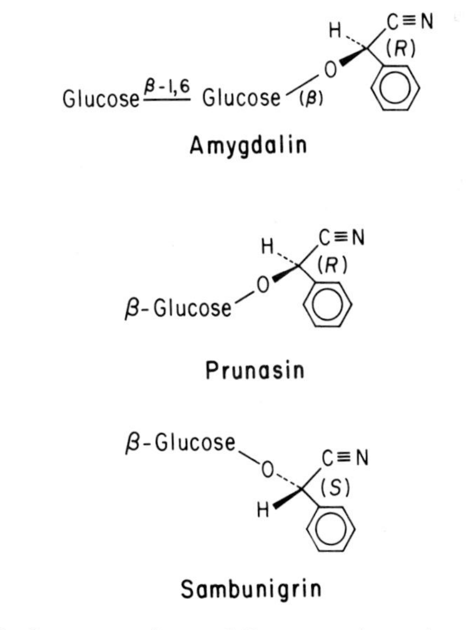

Figure 2. Structures of amygdalin, prunasin, and sambunigrin.

known as gentiobiose, and amygdalin therefore can be termed a *cyanogenic* gentiobioside. The α-hydroxynitrile in both compounds is (R)-mandelonitrile. Prunasin and amygdalin frequently occur in the same species, with prunasin being more common in vegetative tissue whereas both may be found in fruits and seeds. The S epimer corresponding to prunasin is known as sambunigrin (Fig. 2). This cyanogenic glucoside was first isolated from *Sambucus nigra* and subsequently was found in *Ximenia americana* and *Acacia cunninghamii*. The S epimer of amygdalin, known as neoamygdalin, does not occur in nature. All three compounds on enzymatic hydrolysis yield benzaldehyde, a compound that has the characteristic aromatic odor associated with bitter almonds and oil of almonds.

Two cyanogenic glycosides that differ significantly from the ones already described are linamarin and lotaustralin (Fig. 3). These β-glucosides contain the aliphatic cyanohydrins of acetone and 2-butanone as their respective aglycones. They occur frequently in the Leguminosae and are more widely distributed than any of the other cyanogenic glycosides. The compounds always occur in the same species, although not necessarily in the same ratio. The co-occurrence in a single species is attributed to the existence of a single set of biosynthetic enzymes that can act on either valine or isoleucine as precursors, converting them to linamarin and lotaustralin, respectively (Conn, 1973).

The cyanogenic glycosides as a group react in a predictable manner to acid and alkali. The β-glycosidic bond is cleaved in dilute acid at elevated temperatures to yield the sugar and a cyanohydrin; the latter dissociates nonenzymatically to produce HCN and an aldehyde or ketone. Dilute alkali typically hydrolyzes the nitrile group in many cyanogenic glycosides to form the carboxylic acid and NH_3 without

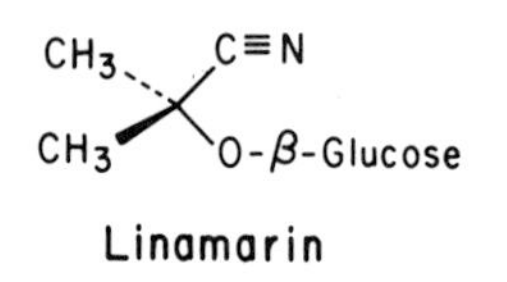

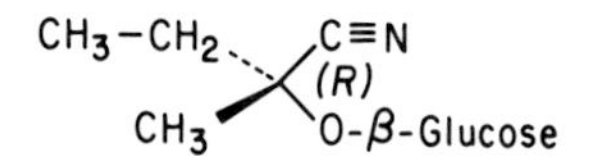

Figure 3. Structures of linamarin and lotaustralin.

cleaving any other bonds. However, dhurrin (Mao and Anderson, 1965) and taxiphyllin (Schwarzmaier, 1976) are unstable in dilute alkali and decompose to the same products obtained on acid hydrolysis. Those glucosides such as amygdalin, prunasin, and sambunigrin having electron-withdrawing groups adjacent to the carbinol carbon atoms are also epimerized by dilute alkali (Nahrstedt, 1975).

The cyanogenic glycosides show the expected physical properties of glycosides in that they are readily soluble in water and difficult to crystallize. They can be acetylated to form derivatives that are soluble in less polar solvents, and these show sharp melting points. Tables describing these and other physical data are available in Seigler's review (1977).

III. DISTRIBUTION, DETECTION, AND QUANTITATIVE DETERMINATION OF CYANOGENIC GLYCOSIDES

A. Natural Distribution

Approximately 1000 species of plants have been reported to be capable of producing HCN when their tissues are crushed or their cellular structure is otherwise disrupted. A tabulation by Gibbs (1974) distributes these cyanophoric species among 500 genera comprising 100 families, thus demonstrating that cyanogenesis is a widespread phenomenon in the plant kingdom. Nevertheless, in a vast majority of these species, only the potential to produce HCN is recorded, with little or no information as to the chemical nature of the cyanogenic precursor. Families noted for the phenomenon of cyanogenesis are the Rosaceae (150 species), Leguminosae (125), Gramineae (100), Araceae (50), Compositae (50), Euphorbiaceae (50), and Passifloraceae (30).

Although the number of cyanophoric species is large, the number of known cyanogenic compounds is not, numbering only 32 including the known cyanolipids and pseudocyanogenic glycosides (Seigler, 1977). Since the latter two classes of compounds are restricted to the Sapindaceae and Cycadaceae, respectively, the cyanogenic potential of most cyanophoric plants may obviously be ascribed either to one of the 23 known cyanogenic glycosides or to other unknown cyanogenic substances.

Seigler (1977) prepared a list of the known cyanogenic glycosides and the plants in which they have been detected (Table I). It discloses that several glycosides (amygdalin, cardiospermin, deidaclin, gynocardin, lucumin, and tetraphyllins A and B) have been found in only a single

Table I. Distribution of naturally occurring cyanogenic glycosides[a]

Compound	Species	Family
Acacipetalin	*Acacia constricta* Benth.	Leguminosae
	Acacia giraffae Burch.	
	Acacia hebeclada DC. (≡ *A. stolonifera* Burch.)	
	Acacia sieberiana DC. var. *woodii* (Burtt Davy) Keay and Brenan (≡ *A. lasiopetala* in sensu Steyn and Rimington, not Oliv.)	
Amygdalin	*Cotoneaster praecox* Vilm., *C. bullata* Bois, *C. dielsiana* Pritz, *C. intergerrima* Medic.	Rosaceae
	Cydonia vulgaris Pers.	
	Eriobotrya japonica Lindl.	
	Malus sylvestrius Mill., *M. pumila* Mill.	
	Photinia serrulata Lindl.	
	Prunus amygdalus Stokes	
	Prunus armeniaca L.	
	Prunus laurocerasus Lindl.	
	Prunus padus L.	
	Prunus persica (L.) Batsch.	
	Prunus sphaerocarpa Sw.	
	Sorbus aucuparia L.	
Cariospermin	*Cardiospermum hirsutum* Willd.	Sapindaceae
	Heterodendron oleaefolium Desf.	
Deidaclin	*Deidamia clematoides*	Passifloraceae
Dhurrin	*Sorghum almum*	Gramineae
	Sorghum bicolor Pers.	
	Sorghum halepense (L.) Pers.	
	Macadamia ternifolia F. Muell.	Proteaceae
	Stenocarpus sinuatus Endl.	
Dihydroacacipetalin	*Acacia sieberiana* DC. var. *woodii* (Burtt Davy) Keay et Brenan	Leguminosae
	Heterodendron olaefolium Desf.	Sapindaceae
Gynocardin	*Gynocardia odorata* R.Br.	Flacourtiaceae
	Pangium edule Reinw.	
Holocalin	*Sambucus nigra* L.	Caprifoliaceae
	Holocalyx balansae Mich.	Leguminosae
p-Hydroxymandelonitrile	*Nandina domestica* Thunb.	Berberidaceae
	Goodia latifolia Salisb.	Leguminosae
	Thalictrum aquilegifolium	Ranunculaceae
	Thalictrum polycarpum	

(*continued*)

Table I (*Continued*)

Compound	Species	Family
Isotriglochinin	*Alocasia macrorrhiza* Schott.	Araceae
Linamarin	*Dimorphotheca ecklonis* DC., *D. barberiae* Harv.	Compositae
	Dimorphotheca spectabilis Schl't., *D. zeyheri* Sond., *D. cuneata* Less., *D. fruticosa* DC.	
	Osteospermum jucundum Norlindh.	
	Cnidoscolus texanus (Muell. Arg.) Small	Euphorbiaceae
	Hevea brasiliensis Muell. Arg.	
	Manihot carthaginensis Muell. Arg.	
	Manihot esculenta Crantz, *M. palmata* Muell. Arg.	
	Lotus australis Andr., *L. arenarius* Brot., *L. corniculatus* L., *L. creticus* L., *L. maroccanus* Ball. *L. arabicus* L., *L. parviflorus* Desf., *L. tenuis* Waldst. et Kit. ex. Willd.	Leguminosae
	Phaseolus lunatus L.	
	Trifolium repens L.	
	Linum usitatissimum L., *L. grandiflorum* Desf., *L. perenne* L., *L. narbonense* L.	Linaceae
	Papaver nudicaule L.	Papaveraceae
Lotaustralin	*Trifolium repens* L.	
	Dimorphotheca ecklonis DC., *D. barberieae* Harv.	Compositae
	Osteospermum jucundrum Norlindh.	
	Manihot carthaginensis Muell. Arg.	Euphorbiaceae
	Lotus australis Andr., *L. arabicus* L., *L. arenarius* Brot., *L. corniculatus* L., *L. creticus* L., *L. maroccanus* Ball. *L. parviflorus* Desf., *L. tenuis* Waldst. et. Kit. ex. Willd.	Leguminosae
	Phaseolus lunatus L.	
	Linum usitatissimum L., *L. grandiflorum* Desf., *L. perenne* L., *L. narbonense* L.	Linaceae
	Papaver nudicaule L.	
Lucumin	*Lucuma mammosa* Gaertn. [≡ *Calocarpum sapota* (Jacq.) Merr]	Sapotaceae

(continued)

Table I (*Continued*)

Compound	Species	Family
Prunasin	*Sambucus nigra* L.	Caprifoliaceae
	Acacia paucijuga F. Muell. ex. Wakefield, *A. parramattensis* Tindale, *A. pulchella* R., Br.	Leguminosae
	Eremophila maculata (Ker.) F. Muell.	Myoporaceae
	Eucalyptus cladocalyx F. Muell. (≡ *E. corynocalyx* F. Muell.)	Myrtaceae
	Cystopteris fragilis Bernh.	Polypodiaceae
	Pteridium aquilinum (L.) Kühn.	
	Cotoneaster praecox Vilm., *C. bullata* Bois., *C. intergerrima* Medic., *C. congesta* Baker, *C. dammeri* Schneid., *C. dielsiana* Pritz., *C. hydrida*	Rosaceae
	Cotoneaster multiflora Bunge	
	Cotoneaster microphylla Wall.	
	Cydonia oblonga Brill.	
	Prunus avium L.	
	Prunus laurocerasus Lindl.	
	Prunus padus L.	
	Prunus persica (L.) Batsch.	
	Prunus macrophylla Sieb. & Zucc.	
	Prunus serotina Ehrh.	
	Jamesia americana T & G	Saxifragaceae
	Linaria minor Desf. (≡ *Chaenorrhinum minus* Lange). *L. striata* D. C.	Scrophulariaceae
Sambunigrin	*Sambucus nigra* L.	Caprifoliaceae
	Acacia glaucesens Willd., *A. cheeli* Blakely	Leguminosae
	Acacia cunninghamii Hook	
	Ximenia americana L.	Olacaceae
Taxiphyllin	*Juniperus oxycedrus* L.	Cupressaceae
	Phyllanthus gastroemii Muell.	Euphoriaceae
	Taxus baccata L.	Taxaceae
	Taxus cuspidata S. et Z., *T. canadensis* Marsh and *T.* × *media* Rehder (= *T. baccata* × *T. cuspidata* L.)	
	Metasequoia glyptostroboides Hu et Cheng	Taxodiaceae

(*continued*)

Table I (*Continued*)

Compound	Species	Family
Tetraphyllin A	*Tetrapathaea tetrandra*	Passifloraceae
Tetraphyllin B (≡ barterin?)	*Barteria fistulosa* Mast.	Passifloraceae
	Tetrapathaea tetranda	
Triglochinin	*Alocasia macrorrhiza*	Araceae
	Scheuzeria palustris L.	Juncaginaceae (Scheuzeriaceae)
	Triglochin maritima L., *T. palustris* L.	Juncaginaceae
	Lilaea scilloides (Poir.) Haum.	Lilaeaceae
Triglochinin methyl ester	*Thalictrum aquilegifolium* L.	Ranunculaceae
Vicianin	*Vicia angustifolia* Roth.	Leguminosae
	Vicia macrocarpa Bertol.	
	Davallia fijeensis Hook., *D. bullata* Wall., *D. denticulata* (Burm.) Nett.	Polypodiaceae
Zierin	*Sambucus nigra* L.	Caprifoliaceae
	Zieria laevigata Sm.	Rutaceae

[a] This table is taken, with the permission of the author and publisher, from Table 16 of "The Distribution of Naturally Occurring Cyanogenic Glycosides," by D. S. Seigler, *in* "Progress in Phytochemistry" (L. Reinhold, J. B. Harborne, and T. Swain, eds.), Vol. 4. Pergamon Press, Oxford, 1977. For additional information, the interested reader can consult the references and footnotes cited in the original table.

family and that one (acacipetalin) occurs only within the genus *Acacia*. Others (linamarin, lotaustralin, and prunasin) occur in numerous and unrelated families.

In contrast to the information in Table I, there is no single tabulation or list of all plant species that have been reported to be cyanophoric. To obtain such information on any given species, two approaches are advised. The chemotaxonomic treatises by Gibbs (1974) and Hegnauer (1962–1973) state whether cyanogenesis has been reported for a given species. Hegnauer has thoroughly covered the earlier literature on cyanogenesis in plants (with the exception of the Leguminosae), and the continuing studies from his laboratory represent major contributions in the use of cyanogenesis and cyanogenic compounds in chemotaxonomy. Gibbs' book, in addition to citing many literature reports, contains many of the author's personal records of cyanogenic species.

The other source of information regarding which plants are cyanophoric includes the several books and numerous bulletins on poisonous plants. Such publications list species that have either been proved to be toxic to animals because of their cyanophoric potential or that are suspected of poisoning animals by HCN. Three excellent books on poisonous plants are those of Everist (1974), Kingsbury (1964), and Watt and Breyer-Brandwijk (1962).

B. Detection

The presence of a cyanogenic glycoside in a plant specimen may be indirectly detected by performing qualitative tests for HCN released on hydrolysis (Eyjolfsson, 1970). The picrate test is one of several simple tests that can be performed. A piece of plant tissue (10–25 mg) is placed in a small vial and crushed with a glass stirring rod or a small stick. A strip of filter paper, which has been soaked in saturated picric acid and dried, is inserted into the tube so as not to touch the macerated tissue. The vial is then stoppered. The cellular disruption of the tissue presumably brings the glycoside into contact with a β-glycosidase usually also present in the plant (see Sections IV,A and IV,B), and HCN is released. The HCN then turns the picrate paper to an orange-brown or red-brown, usually within an hour or less if a glycoside is present in significant concentrations in the plant.

The picrate test as described above has its limitations. It is not specific; certain volatile aldehydes and ketones as well as H_2S and SO_2 have been reported to discolor the picrate paper (Farnsworth, 1966). We have also noted in our laboratory that volatile substances, apparently not HCN, released on hydrolysis of the glucosinolate compound(s) (mustard oil glycosides) in *Thalaspi arvense* and *Lepidium sativum* give a positive picrate test (M. P. Stone, unpublished work, 1977).

In addition, the picrate test as described does not work if the plant tissue lacks the β-glucosidase that releases the HCN. As noted in Section IV,C, several species are polymorphic for cyanogenesis and may lack the enzyme but still possess the cyanogenic glycoside. To cover this possibility a solution (0.1 ml) containing a small amount of almond emulsin and linamarase (Secor *et al.*, 1976) can be added to the macerated plant tissue before the picrate paper is inserted. Both enzymes are required to ensure hydrolysis of any aromatic (by emulsin) or aliphatic (by linamarase) glycosides that might be present.

The picrate test can be conveniently applied in the field to fresh plant material. It can also be applied to specimens dried at another place and time (e.g., herbarium specimens) and brought to the laboratory for examination. In the case of dried material it is imperative that fresh enzyme be incorporated into the test since the drying procedure may have inactivated any enzyme present originally. It should also be appreciated that a negative test on dried specimens is not conclusive regarding the cyanogenic nature of the original plant since any cyanogenic glycoside present in the fresh plant material may have been destroyed during the drying procedure.

Procedures exist for the identification of the cyanogenic glycosides found in plants. These involve paper, thin-layer, and gas–liquid chromatography techniques usually applied to partially purified material. These procedures have been reviewed by Seigler (1977) and Conn (1978).

C. Quantitative Determination

A continuing interest in the cyanogenic potential of plant foodstuffs has led to official methods for analysis of HCN by the Association of Official Agricultural Chemists (Horwitz, 1965). These procedures are based on the enzymatic hydrolysis of a cyanogenic glycoside by endogenous enzymes in a closed system, steam distillation of the HCN released into appropriate traps, and determination of the cyanide produced. Another method, which utilizes the cyanide electrode, is based on measuring the HCN released when plant tissue is macerated and the cyanogenic glycoside is hydrolyzed enzymatically (Blaedel *et al.*, 1971).

In connection with our studies on the biosynthesis of cyanogenic glycosides (Section V,A), we have developed a micro method for the determination of cyanogenic glycosides (Hahlbrock and Conn, 1970; Reay and Conn, 1970). The procedure, which is also based on measurement of the HCN released by enzymatic hydrolysis, involves the diffusion of HCN into a trap of NaOH in a closed vessel. The NaCN is then determined colorimetrically by Epstein's procedure (1947) or that of Lambert *et al.* (1975). The procedure is designed to determine 0.02–2.0 μmoles of cyanogenic glucoside; 1 μmole is the amount that would be contained in 250 mg of fresh plant material assuming a concentration of 0.1% fresh weight and a molecular weight of 247 (i.e., the molecular weight of linamarin).

The intact cyanogenic glycosides can also be quantitated by means of gas–liquid chromatography. The procedure involves conversion to the trimethyl silyl ethers, and the separation of epimeric pairs such as prunasin/sambunigrin and taxiphyllin/dhurrin is possible. Seigler (1977) has tabulated the information for this technique.

IV. PROCESS OF CYANOGENESIS

A. Enzymatic Hydrolysis

The release of HCN by a cyanogenic plant is the process known as cyanogenesis. The process is nearly always enzymatic and thus dependent on enzymes present in the cyanogenic species. The enzymatic process for prunasin is shown in Fig. 4. The initial step involves the hydrolysis of the β-glucosidic bond by a β-glucosidase to yield D-glucose and the α-hydroxynitrile (cyanohydrin) (*R*)-mandelonitrile. Cyanohydrins, being unstable, dissociate nonenzymatically to produce HCN and, in the case of prunasin, benzaldehyde. However, cyanogenic plants contain an enzyme, hydroxynitrile lyase, that catalyzes the cyanohydrin equilibrium [step (b) in Fig. 4]. The progress of the combined reaction can be followed by measuring either HCN or the benzaldehyde produced.

The action of β-glycosidase on a cyanogenic glycoside [step (a) in Fig. 4] can also be followed by measuring the sugar as it is released. This can be done chemically by determination of the sugar with specific chemical reagents or, in the case of D-glucose, by analysis with glucose oxidase (Mao and Anderson, 1967). At least one cyanogenic glycoside, found in *Nandina domestica, Goodia latifolia,* and *Thalictrum* sp. (Table I), does not require the action of a β-glucosidase as shown in Fig. 4 to release HCN because of its unusual structure (Seigler, 1977).

The release of HCN, represented in Fig. 4, occurs at a significant rate only after maceration of the cyanogenic plant tissue has occurred and the

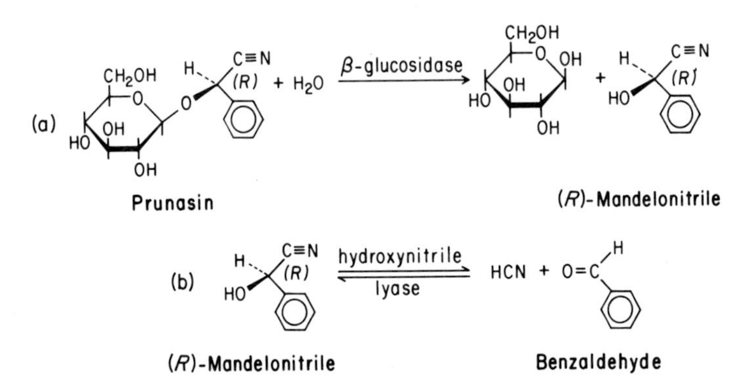

Figure 4. Enzyme-mediated reactions of cyanogenesis. The initial reaction [step (a)] involves the hydrolysis of the glucoside by a β-glucosidase. The next reaction [step (b)] is the dissociation, both enzymatic and nonenzymatic, of the α-hydroxynitrile (cyanohydrin), (*R*)-mandelonitrile.

cyanogenic substrate is brought into contact with its catabolic enzymes. This statement presumes a spatial separation of enzymes and glucoside in the intact plant (Section IV,B). The cyanogenic phenomenon represented in Fig. 4 is also obviously dependent on the co-occurrence of both substrate and enzymes in the same tissue; this can be influenced by genetic factors (Section IV,C).

The occurrence of β-glucosidases in plants is well established (Nisizawa and Hashimoto, 1970). However, information in the literature regarding whether these enzymes can utilize cyanogenic glycosides as substrates is regrettably inadequate. Recently, there have been several thorough studies of the β-glycosidases of cyanogenic plants including the emulsin complex of almonds. The significance of these studies as they relate to the metabolism of cyanogenic glycosides in plants has been reviewed (Conn, 1978) and can be summarized as follows.

With the exception of those species for which cyanogenesis is polymorphic (Section IV,C), a cyanogenic plant presumably contains the necessary enzymes for release of HCN from a cyanogenic precursor. In *fresh* plant material the enzymes should be sufficiently active to permit the detection of HCN qualitatively; if the enzymes are present in excess the quantitative determination of the cyanogenic glycoside should be possible. To ensure the hydrolysis of any glycosides present in fresh or, more importantly, in dried specimens, enzymes can be added to the plant material. Commercial preparations of almond emulsin contain enzymes that hydrolyze amygdalin, prunasin, sambunigrin, dhurrin, and taxiphyllin at different rates (Ng, 1975) but are essentially inactive on linamarin, a glycoside having an aliphatic aglycone. To ensure the hydrolysis of this widely distributed cyanogen, a partially purified preparation of linamarase from flaxseed has been employed (Butler and Conn, 1964). Acacipetalin, structurally related to linamarin and lotaustralin, is not hydrolyzed by flax linamarase but is hydrolyzed by almond emulsin (Secor *et al.*, 1976).

B. Cellular Localization of Cyanogenic Glycosides

The co-occurrence of a cyanogenic glycoside and the enzymes required for its catabolism in the same organ (e.g., stem, leaf, cotyledon, root, flower, coleoptile) is circumstantial evidence for the spatial separation of substrate from enzymes in that tissue. A reasonable explanation would be that one cell type (e.g., the parenchyma cells of a leaf) might contain the glucoside, whereas a different cell type (e.g., mesophyll) could contain the catabolic enzymes (glucosidase and hydroxynitrile lyase). It is also possible that the cyanogenic glycoside might be located within one compartment (e.g., the vacuole) of a cell and therefore separated from the catabolic enzymes located elsewhere in the same cell (e.g., cytoplasm, plasmalemma, or cell wall). It is well established that natural products can occur and may be concentrated in specialized cells/structures within a specific tissue of a plant. Little is known however regarding the localization of the enzymes involved in biosynthesis or catabolism of natural products.

Only recently has information on the localization of cyanogenic glycosides become available. Saunders *et al.* (1977a) described autoradiographic studies which indicate that 95% or more of the cyanogenic glucoside (dhurrin) in dark-grown sorghum shoots is located in the vacuoles of the cells in that tissue. Subsequently Saunders and Conn (1978) isolated vacuoles from protoplasts obtained by treating green sorghum leaves with cellulytic enzymes and confirmed that they contained dhurrin. Extending this work Kojima *et al.* (1979) were able to isolate preparations of epidermal and mesophyll protoplasts and bundle sheath strands having only minor cross contamination. Analysis of these three fractions disclosed that dhurrin was located entirely in the epidermal layers of the leaf blade whereas the two enzymes responsible for its catabolism resided almost exclusively in the mesophyll tissue. The bundle sheath strands did not contain significant amounts of dhurrin or of these enzymes. This compartmentation of dhurrin and its catabolic enzymes in different tissues clearly prevents its large-scale hydrolysis under normal physiological conditions. When the tissues are crushed and the contents of the ruptured epidermal and mesophyll cells are mixed, cyanogenesis will occur.

C. Genetic Factors Influencing Cyanogenesis

The genetic polymorphism of cyanogenesis, which has been well studied for *Trifolium repens* and *Lotus corniculatus* (Nass, 1972), must be appreciated by those interested in the relationships between herbivores and cyanogenic plants. The polymorphism is indicated by the existence of both cyanogenic and noncyanogenic individuals in these plant species. The genetic relationship for *T. repens* is indicated in Fig. 5 and can be summarized as follows. The synthesis of the cyanogenic

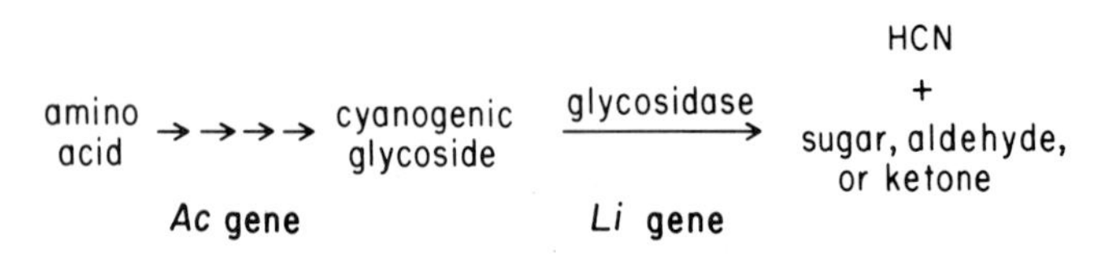

Figure 5. Genetics of cyanogenesis.

glucosides from their amino acid precursors (Section V,A) is determined by alleles of the gene *Ac*. The hydrolysis of the cyanogenic glycosides is determined by alleles of another, independently inherited gene *Li*. Only plants that possess at least one dominant allele of both genes are cyanogenic. By selective breeding, it is possible to obtain the four homozygous genotypes *AcAcLiLi, AcAclili, acacLiLi, acaclili,* only the first of which is cyanogenic, i.e., it releases HCN when crushed. The second genotype contains the glucosides but lacks the hydrolytic enzyme system, the third possesses the enzymes but not their substrates, and the fourth lacks both substrates and enzymes. Individuals heterozygous for *Ac* or *Li* have intermediate levels of glucosides or enzymes, respectively, when compared with homozygous individuals.

Although cyanogenic polymorphism may appear to complicate studies on herbivore–cyanogenic plant relationships, Jones (1971) has stressed the advantages that polymorphic species offer for studies on chemical coevolution. As Jones points out, comparison of two types of plants, presumably differing *only* in their content of a cyanogenic glycoside, in an interbreeding population offers a nearly ideal situation for determining whether such compounds may act as defensive substances. Such studies have been carried out on *Lotus corniculatus* (Jones, 1973, Jones *et al.*, 1978).

Cyanogenic polymorphism clearly occurs in other species since one can observe individual plants within a species that are strongly cyanogenic, are only weakly so, or may produce no HCN. These include the well-known example of almond trees (*Prunus amygdalus*) which produce either bitter or sweet almonds. Similarly, some macadamia trees (*Macadamia ternifolia*) produce bitter nuts. Varieties of cassava (*Manihot esculenta*) having bitter and sweet tubers are also known, as are "high-cyanide" and "low-cyanide" forms of *Sorghum bicolor*. Recently, Fikenscher and Hegnauer (1977) reported cyanogenic polymorphism in *Achillea macrophylla, A. millefolium, Centaurea scabiosa, Taxus baccata, Sedum* spp., *Sorbus aucuparia, Campanula cochlearifolia,* and *C. Rotundifolia*. Clearly, the phenomenon is sufficiently common that it must be expected in any cyanogenic population.

V. BIOSYNTHESIS OF CYANOGENIC GLYCOSIDES AND HCN

A. Cyanogenic Glycosides

The biosynthesis of cyanogenic glycosides in plants has been studied most extensively in *Sorghum bicolor, Linum usitatissimum,* and *Prunus* spp. These studies, reviewed by Conn (1978), support the biosynthetic pathway shown in Fig. 6, in which the amino acids tyrosine (in *Sorghum*), valine and isoleucine (in flax), and phenylalanine (in *Prunus*) are converted to the corresponding aglycones of dhurrin, linamarin and lotaustralin, and prunasin or amygdalin, respectively, and then glucosylated. As indicated, the pathway requires the loss of the carboxyl carbon of the amino acid and the oxidation of the α-carbon atom to the level of a nitrile with retention of the nitrogen atom. In addition, the β-carbon atom of the amino acid is oxidized to yield a hydroxyl group, which then can bear the sugar characteristic of the glycoside. In sorghum, the biosynthetic sequence, except for the final glycosylation step, is catalyzed by a membrane-bound enzyme system closely associated with the endoplasmic reticulum of the cell (Saunders *et al.*, 1977b).

Considerable information is available regarding the enzymes involved as well as the stereochemistry of cyanogenic glycoside biosynthesis (Conn, 1973, 1978). Perhaps of greatest interest to the readers of this volume is information, only now being acquired, suggesting that the biosynthetic process is a highly channeled system in the sorghum membranes. The term "channeled" is used to imply that the enzymes indicated in Fig. 6 are organized physically or arranged in the membrane in a manner that provides for the highly efficient production and subsequent utilization of the postulated intermediates. Data are available showing that the aldoxime and nitrile produced from the amino

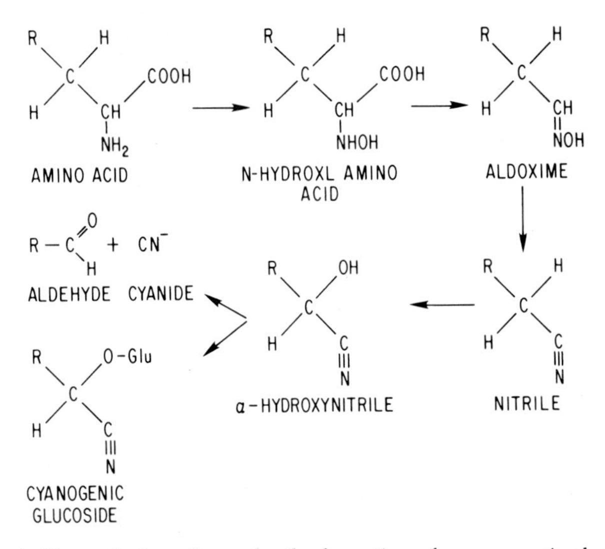

Figure 6. Biosynthetic pathway for the formation of a cyanogenic glucoside.

acid by the membrane-bound enzymes are used preferentially over aldoxime or nitrile added externally. The aldoxime and nitrile produced in the membrane do not accumulate to a significant extent, so that one does not detect metabolic pools of aldoxime or nitrile in the plant. Instead, the sorghum membrane system appears to be appropriately designed for channeling the flow of carbon atoms from tyrosine into the cyanogenic glucoside. This is perhaps not surprising since dry sorghum seed contains no detectable cyanogenic glucoside, whereas shoots of 2-day-old dark-grown seedlings may contain 25% by dry weight (Saunders *et al.*, 1977a). Clearly, there is a rapid synthesis of cyanogenic glycoside in such seedlings; moreover, the channeled nature of the process suggests that the plant is attempting to keep the cost of such synthesis at a minimum. A minimal cost for dhurrin biosynthesis is essential because the plant is utilizing a primary amino acid (tyrosine), which is also needed for the synthesis of protein, lignin, and other phenolic compounds (Stafford, 1969) during germination.

B. Production of HCN by Animals

Certain moths and polydesmoid millipedes produce HCN by processes strikingly similar to those encountered in higher plants. In millipedes, the HCN is derived either from a glucoside of mandelonitrile (Blum and Woodring, 1962) or from mandelonitrile itself (H. E. Eisner *et al.*, 1963). The mandelonitrile, which is derived from L-phenylalanine by intermediates previously demonstrated in higher plants (Duffey *et al.*, 1974), is dissociated rapidly into HCN and benzaldehyde as a defense mechanism by an elaborate apparatus in this species.

The cyanogenic apparatus consists of paired, serially arranged glands, each having two compartments (T. Eisner *et al.*, 1963). Mandelonitrile is stored in one of these compartments (reservoir) as an aqueous emulsion. A valve, operated by a muscle, separates the reservoir from the smaller compartment (vestibule), which contains the enzyme hydroxynitrile lyase (see Fig. 4). At the moment of discharge, the muscle acts to open the valve and mandelonitrile flows into the vestibule, whereupon its dissociation is initiated and HCN is produced.

In addition to mandelonitrile, millipedes produce the chemically related compounds benzoylcyanide, mandelonitrile benzoate, ethyl benzoate, and benzoic acid (Conner *et al.*, 1977; Duffey *et al.*, 1977). Although the first two have been described as "cyanogenetic" (cyanide-producing) compounds, this cannot be so unless there are esterases present that catalyze the hydrolysis of these esters to produce HCN. It is reasonable to infer that the benzoyl moiety is derived from mandelonitrile as a precursor. More importantly, the benzoyl compounds may be as effective defense compounds as the more notorious HCN.

Jones *et al.* (1962) identified HCN in the crushed tissues of all stages in the life cycle of the burnet moths *Zygaeninae* spp. These authors also refer to the crushed tissues as having the odor of bitter almonds. Since the more characteristic odor of bitter almonds is the *aromatic* odor of benzaldehyde rather than the *pungent* odor of HCN, it may well be that these moths produce mandelonitrile or a glycoside of mandelonitrile and that such compound(s) are degraded on crushing of the animal and yield HCN.

VI. TOXICITY OF CYANOGENIC PLANTS

It is generally believed that the toxicity of cyanogenic plants is due to the HCN released from the cyanogenic glycosides contained in such plants. Hydrogen cyanide is a moderately toxic substance because of its capacity to form complexes reversibly with heme proteins, notably cytochrome oxidase, the enzyme that catalyzes the terminal step in aerobic respiration. Animals, including human beings, when acutely poisoned by ingestion of cyanogenic plants, show the classic symptoms of cyanide poisoning described below. If treated sufficiently rapidly, they respond to accepted procedures for the treatment of cyanide poisoning. Nevertheless, reports in the literature are sufficiently varied to indicate that several factors must influence possible HCN poisoning by cyanogenic plants. Although some of these are discussed below, one is dealt with here.

The hydrolysis of a cyanogenic glycoside usually produces an equimolar mixture of HCN *and* an aldehyde or ketone, together with 1 or more moles of a sugar. Because of the notoriety of HCN as a poison, there is a tendency to consider it responsible for any toxic symptoms observed. However, equal attention should be given to the acetone, 2-butanone, benzaldehyde, *p*-hydroxybenzaldehyde, etc., which are simultaneously released. Although the aliphatic ketones may be of less toxicological concern, certainly the aromatic aldehydes and their acid oxidation products deserve greater attention. At the very least, such aromatic substances are not readily metabolized in animal tissues, and energy presumably is expended in their detoxification and excretion. Little attention has been paid to these other products of cyanogenic glycoside catabolisms, except in the case of cyanide-producing insects, in which the subject has been discussed in general terms (Duffey *et al.*, 1977).

Although the potential of a plant to produce HCN (and other products) from its cyanogenic glycoside is important in determining its potential toxicity, other factors must be considered. These include the size and kind of animal that may eat the plant, the speed of ingestion, the type of food simultaneously ingested with the toxic plant, the possibility of the plant's degradative enzymes remaining active in the digestive tract of the animal, the possibility of intracellular enzymes in the animal hydrolyzing any glycoside that is absorbed into the boodstream, and the ability of the animal to detoxify any HCN that it encounters. Unfortunately, there are few detailed studies on any of these topics.

A. Toxicity of HCN

The toxicity of HCN is due to its ability to combine reversibly with enzymes associated with cellular respiration, notably cytochrome oxidase. The lethal dose of HCN for human beings has been given as 0.5–3.5 mg per kilogram body weight taken orally; this corresponds to 1.0–7.0 mg/kg of KCN and refers to a single dose taken at one time (Montgomery, 1969). The heart, brain, and nervous system are rapidly affected, and death ensues within minutes to a few hours. The clinical signs of acute poisoning in human beings are well described (Thienes and Haley, 1972; Gosselin *et al.*, 1976). They include headache, dizziness, mental confusion and stupor, cyanosis with twitching and convulsions, followed by terminal coma. The respiratory rate and depth are increased initially, and then breathing becomes slow, gasping, and eventually ceases.

A registry of toxic substances (Christensen, 1976) cites the following values for oral lethal doses of HCN: mouse, 3.7 mg/kg; dog, 4.0 mg/kg; cat, 2.0 mg/kg. The value for KCN for the rat is given as 10 mg/kg. Moran (1954) cited 2.0 mg per kilogram body weight as the lethal dose of HCN for cattle and sheep. These figures give some indication of the amount of cyanogenic plant that must be consumed to be hazardous. Since HCN comprises only 5–10% of the molecular weight of the known cyanogenic glycosides, the "lethal dose" for the glycoside would be approximately 10–20 times the values just cited. Moreover, plant material containing this amount of glycoside would have to be eaten in the interval of a few minutes, and enzymatic hydrolysis by the plant enzymes would have to be sufficiently rapid to hydrolyze all of the glycoside contained in the plant. That such conditions often obtain is borne out by the numerous accounts of livestock poisoning by sorghum (Kingsbury, 1964; Gibb *et al.*, 1974), members of the rose family (wild cherries, chokecherries, mountain mahogany) (Kingsbury, 1964), arrow grass (Marsh *et al.*, 1929), and acacias (Kingsbury, 1964; Hurst, 1942; Steyn and Rimington, 1935).

If the ingestion of a cyanogenic plant produces acute poisoning, this suggests that the plant is providing both the cyanogenic glycoside and the enzymes necessary for HCN production. Moreover, HCN production has to occur within the short time interval between mastication of the plant material and its arrival in the stomach, since, for at least in human beings and other monogastric animals, the normally acidic contents of the stomach presumably would significantly inhibit the action of plant enzymes having a pH optimum of 5 to 6. Later, when the plant material moves into the duodenum and is mixed with the alkaline bile juice, the plant enzymes may regain activity.

This situation does not occur with ruminants when the cyanogenic plant material remains in the rumen at neutral pH. Such conditions strongly favor continued hydrolysis by plant catabolic enzymes that are present and may account for the reported greater susceptibility of ruminants to poisoning by cyanogenic plants (Moran, 1954). It is also possible that rumen bacteria, with their capacity to hydrolyze the β-glycosidic linkages of cellulose, may also act on β-glycosidic linkages of the cyanogenic glycosides. The rumen floras of sheep are capable of hydrolyzing lotaustralin (Coop and Blakeley, 1949).

Acute poisoning by plants containing cyanogenic glycosides is less likely if the catabolic enzymes are lacking in the plant or if they have been inactivated by heating or other treatment. Such plants in theory could be eaten with impunity by a nonruminant, provided that other plant material containing β-glucosidases were not consumed at the same time. Reference was previously made to species that contain a cyanogenic glycoside but not the catabolic enzymes. However, some edible plants (lettuce, celery, mushrooms) commonly taken in salads contain glycosidases but not the glycoside. Clearly, the two should not be consumed concurrently.

Cooking to inactivate the degradative enzymes is not a feasible option in the case of domestic animals or wild populations. However, cooking by frying, boiling, stewing, baking, or roasting is a common culinary procedure employed by man. The cyanogenic cassava tuber is also rendered edible by drying in the sun, by smoking and curing in a kiln, or by scraping or grating and then soaking in water to leach out the cyanogen and its hydrolysis products (Coursey, 1973). In instances in which the tuber is inadequately processed before cooking, the cassava preparations can retain significant quantities of unhydrolyzed cyanogenic glycoside. The ingestion of large amounts of such material is apparently responsible for a syndrome known as tropical ataxic neuropathy (TAN) encountered in West Africa (Osuntokun, 1973). Individuals suffering this condition exhibit lesions of the skin, mucous membranes, optic and auditory nerves, spinal cord, and peripheral nerves. The disease can be correlated with the frequency and quantity of cassava meals and with increased levels of plasma thiocyanate. The latter presumably results from the detoxification of cyanide in the body by the enzyme rhodanese (see next section). Patients showing the TAN syndrome also exhibit an increased frequency of goiter.

A simplistic explanation of TAN is that it is the result of chronic cyanide poisoning due to the release of HCN from certain cassava food products over an extended period of time. However, there is only circumstantial evidence relating human disease with extended exposure to cyanide (Montgomery, 1969; Wilson, 1973). Since TAN invariably occurs among poorly nourished people, the conditions may well result from unidentified nutritional deficiencies or excesses as well as possible chronic cyanide toxicity.

B. HCN Detoxification

A major factor influencing the toxicity of cyanogenic plants to animals is the ability of animals to detoxify the HCN released on ingestion. Although HCN is rapidly absorbed into the body, it is detoxified by one or possibly more routes. A major mechanism for detoxification is conversion to thiocyanate, catalyzed by the enzyme rhodanese (Fig. 7). Rhodanese (thiosulfate sulfur transferase, EC 2.8.1.1) is widespread in animal tissues, with highest concentrations in liver and kidney. It is responsible for the readily observed conversion of cyanide anion to thiocyanate anion. This process also provides the basis for one of the two treatments administered to animals exhibiting acute cyanide poisoning; they are administered sodium thiosulfate intravenously. This ensures an adequate supply of thiosulfate ion as a substrate to react with and sequester HCN.

Animals suffering acute poisoning by cyanide must be treated within minutes if they are to survive, and for this purpose detoxification by enzymatic conversion to thiocyanate is not sufficiently rapid. Instead,

$$\underset{\text{thiosulfate}}{S_2O_3^{2-}} + \underset{\text{cyanide}}{CN^-} \xrightarrow{\text{rhodanese}} \underset{\text{sulfite}}{SO_3^{2-}} + \underset{\text{thiocyanate}}{SCN^-}$$

Figure 7. Reaction catalyzed by rhodanese.

sodium nitrite is injected in amounts calculated to oxidize some of the animal's hemoglobin to methemoglobin. Methemoglobin forms an even stronger complex with cyanide anion than does cytochrome oxidase and therefore causes the cytochrome oxidase–cyanide complex to dissociate. This permits the oxidase to regain its enzymatic activity. Care must be taken not to oxidize all of the blood hemoglobin to methemoglobin; otherwise, the animal will remain anoxic due to a lack of the oxygen-carrying form of the protein (Chen and Rose, 1952).

Sulfur, in the form of sulfur-containing amino acids, has been implicated as being beneficial in the conversion of cyanide to thiocyanate by intact animals. Studies examining this possibility in the case of chronic exposure to HCN have been carried out using the rat as an experimental animal and KCN or linamarin as a source of HCN (Barrett *et al.*, 1977). However, the production of thiocyanate as an end product of HCN detoxifications is also of concern, since this anion is an established goitrogen. Goiter is endemic in several parts of Africa, and studies exist that relate its occurrence to the consumption of cassava in those regions (Ekpechi, 1973; Delange *et al.*, 1973; Ermans *et al.*, 1973).

Another route for the metabolism and presumed detoxification of HCN in an animal was demonstrated in the polydesmoid millipede, *Harpaphe haydeniana*. This involves the conversion of HCN to β-cyanoalanine and, presumably, asparagine catalyzed by the enzymes β-cyanoalanine synthase and β-cyanoalanine hydrase (Fig. 8). This reaction sequence, first elucidated in higher plants (Conn and Butler, 1969), is of interest from the standpoint of comparative biochemistry, for rhodanese also occurs in insects and, in at least one case, is responsible for the detoxification of HCN by parasites that attack cyanogenic moths (Jones *et al.*, 1962). As the role of cyanogenesis in coevolution is examined in nature, the ability of animals to survive on cyanogenic plants should be correlated with their content of rhodanese and/or β-cyanoalanine synthase.

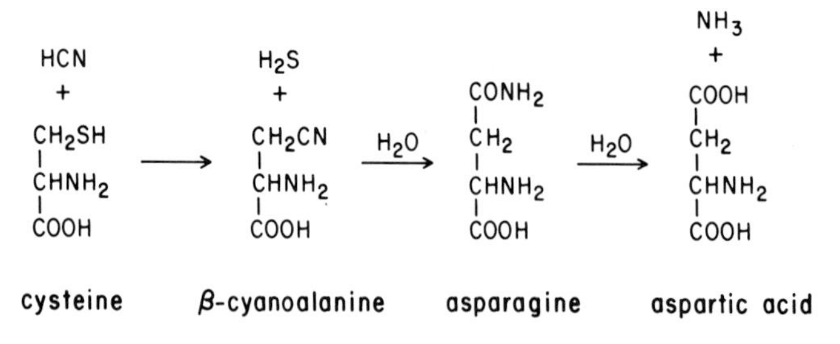

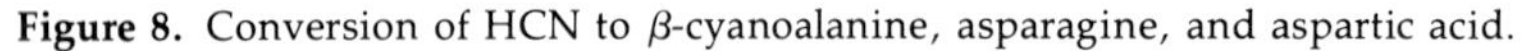

Figure 8. Conversion of HCN to β-cyanoalanine, asparagine, and aspartic acid.

C. Hydrolysis by Animal Enzymes

An unanswered question in this field is the capacity of animal tissues to hydrolyze cyanogenic glycosides. This question becomes important if an animal has a dietary source of unhydrolyzed cyanogenic glycoside. Such a situation will occur when an animal eats a plant that contains a cyanogenic glycoside but lacks the enzymes catalyzing its hydrolysis (see Section IV,C). This situation resembles that of human populations that eat cassava preparations prepared in such a way that they still contain the cyanogenic glycoside of cassava but not the catabolic enzymes (Nestel and MacIntyre, 1973).

It is clear that an intact cyanogenic glycoside can be absorbed into the circulatory system of an animal. Barrett *et al.* (1977), administering single doses of linamarin to rats by stomach tube, showed that 19% of the linamarin administered appeared as the intact glucoside in the urine in the first 24 hours. Another 24% was excreted as urinary thiocyanate, but no linamarin was found in the feces. Qualitatively similar results were obtained by G. W. Newton (unpublished observations, 1977), who administered amygdalin by stomach tube to rats and found intact amygdalin in the urine and high levels of urinary thiocyanate after dosing.

Although it is obvious that some of the linamarin and amygdalin administered in the studies cited was hydrolyzed, it is not at all clear where that hydrolysis occurred. It could have been carried out by bacteria in the rat's lower digestive tract, at which point HCN would be absorbed and subsequently detoxified. It could also have occurred following absorption of the intact glycoside through the intestinal wall and action by β-glycosidases in cells of the intestine, liver, and other organs. β-Glycosidases, including β-glucosidases, are well known in animal tissues. There is, however, only one record of amygdalin, linamarin, or any other cyanogenic glycoside having been examined as a substrate for animal glycosidases *in vitro*. In that study (Ng, 1975), a rabbit liver β-glucosidase that had as its preferred substrate a steroid glucoside was shown to be capable of hydrolyzing prunasin and linamarin. The rates of the reaction, however, were low; moreover, the affinity of the animal enzyme for the cyanogenic glucosides as substrates was not large.

Only indirect evidence suggests that intracellular hydrolysis by animal enzymes occurs. When rats were dosed with lethal levels of linamarin (Barrett *et al.*, 1977), the animals died within 4 hours. This is a short period of time, possibly inadequate to permit movement of the glycoside into the large intestine, where bacteria could hydrolyze the glycoside. There are obviously direct approaches to determine whether animal tissues can catalyze the hydrolysis of cyanogenic glycosides. Homogenates of various organs and tissues, organelle preparations of lysosomes or endoplasmic reticulum, or soluble protein fractions should be prepared and their ability to hydrolyze amygdalin and linamarin examined. Continuing interest in the possible harmful role of cyanogenic glycosides in animal nutrition clearly calls for such experiments.

VII. CONCLUSION

The cyanogenic glycosides represent an especially interesting example of secondary plant compounds influencing relationships between plants and their herbivores. They obviously are degraded by endogenous plant enzymes to produce a toxic compound that must be primarily involved in the plant–herbivore interaction. In this regard they differ from other secondary compounds such as alkaloids that occur in a nonglycosidic form in the plant and must be directly reckoned with. However, the cyanogenic glycosides should serve as models for related studies on cyanolipids and the mustard oil glycosides (glucosinolates), which presumably are degraded to HCN and thiocyanates, respectively, and strongly influence plant–herbivore interactions.

The polymorphism encountered in the cyanogenic process can complicate studies of the plant–animal interaction, but it also offers an opportunity to determine whether the cyanogenic glucoside per se instead of its breakdown products is more important in determining that interaction. To study this question, the experimentalist must be certain of the genetic nature of his plant material. He must also determine whether the herbivore being studied has the capacity to carry out the catabolic processes that the plant was unable to do. Since the occurrence of other secondary compounds in plants is also frequently polymorphic, greater attention should be paid to this factor in studies on such species and their herbivores.

ACKNOWLEDGMENT

Our work cited in this article was supported by the National Institute of General Medical Science, U.S. Public Health Service (GM-05301), and the National Science Foundation (BMS-7411997).

REFERENCES

Barrett, M. D., Hill, D. C., Alexander, J. C., and Zitnak, A. (1977). *Can. J. Physiol. Pharmacol.* **55,** 134–136.

Blaedel, W. J., Easty, D. B., Anderson, L., and Farrell, T. R. (1971). *Anal. Chem.* **43,** 890–894.

Blum, M. S., and Woodring, J. P. (1962). *Science* **138,** 512–513.

Butler, G. W., and Conn, E. E. (1964). *J. Biol. Chem.* **239,** 1674–1679.

Chen, K. K. and Rose, C. L. (1952). *J. Am. Med. Assn.* **149,** 113–119.

Christensen, H. E. (1976). "Registry of Toxic Effects of Chemical Substances." Nat. Inst. Occup. Safety Health, US Public Health Serv., Rockville, Maryland. HEW Publ. No. (NIOSH)76–191.

Conn, E. E. (1969). *Agric. Food Chem.* **17,** 519–526.

Conn, E. E. (1973). *Biochem. Soc. Symp.* **38,** 277–302.

Conn, E. E. (1978). *Encycl. Plant Physiol., New Ser.* **8,** (in press).

Conn, E. E., and Butler, G. W. (1969). *In* "Perspectives in Phytochemistry" (J. B. Harborne and T. Swaine, eds.), pp. 47–74. Academic Press, New York.

Connor, W. E., Jones, T. H., Eisner, T., and Meinwald, J. (1977). *Experientia* **33,** 206–207.

Coop, I. E., and Blakeley, R. L. (1949). *N.Z. J. Sci. Technol., Sect. A* **30,** 277–291.

Coursey, D. G. (1973). *In* "Chronic Cassava Toxicity" (B. Nestel and R. McIntyre, eds.), Monogr. IDRC-010e, pp. 27–36. Int. Dev. Res. Cent., Ottawa.

Delange, F., van der Velden, M., and Ermans, A. M. (1973). *In* "Chronic Cassava Toxicity" (B. Nestel and R. McIntyre, eds.), Monogr. IDRC-010e, pp. 147–151. Int. Dev. Res. Cent., Ottawa.

Duffey, S. S., Underhill, E. W., and Towers, G. H. N. (1974). *Comp. Biochem. Physiol. B* **47,** 753–766.

Duffey, S. S., Blum, M. S., Fales, H. M., Evans, S. L., Roncadori, R. W., Tiemann, D. L., and Nakagawa, Y. (1977). *J. Chem. Ecol.* **3,** 101–113.

Eisner, H. E., Eisner, T., and Hurst, J. J. (1963). *Chem. Ind. (London)* pp. 124–125.

Eisner, T., Eisner, H. E., Hurst, J. J., Kafatos, F. C., and Meinwald, J. (1963). *Science* **139,** 1218–1220.

Ekpechi, O. L. (1973). *In* "Chronic Cassava Toxicity" (B. Nestel and R. McIntyre, eds.), Monogr. IDRC-010e, pp. 27–36. Int. Dev. Res. Cent., Ottawa.

Epstein, J. (1947). *Anal. Chem.* **19,** 272–274.

Ermans, A. M., van der Velden, M., Kinthaert, J., and Delange, F. (1973). *In* "Chronic Cassava Toxicity" (B. Nestel and R. McIntyre, eds.), Monogr. IDRC-010e, pp. 153–157. Int. Dev. Res. Cent., Ottawa.

Everist, S. L. (1974). "Poisonous Plants of Australia." Angus & Robertson, Sydney, Australia.

Eyjolfsson, R. (1970). *Fortschritte der Chemie. Org. Naturst.* **28,** 74–108.

Farnsworth, N. R. (1966). *J. Pharm. Sci.* **55,** 225–276.

Fikenscher, L. H., and Hegnauer, R. (1977). *Pharm. Weekbl.* **112,** 11–20.

Gibb, M. C., Carbery, J. T., Carter, R. G., and Catalinac, S. (1974). *N.Z. Vet. J.* **22,** 127.

Gibbs, D. (1974). "Chemotaxonomy of Flowering Plants." McGill-Queen's Univ. Press, Montreal.

Gosselin, R. E., Gleason, M. N., and Hodge, H. C. (1976). "Clinical Toxicology of Commercial Products," 4th ed. Williams & Wilkins, Baltimore, Maryland.

Hahlbrock, K., and Conn, E. E. (1970). *J. Biol. Chem.* **245,** 917–922.

Hegnauer, R. (1962–1973). "Chemotaxonomie den Pflanzen," Vols. I–VI. Birkhaeuser, Basel.

Horwitz, W., ed. (1965). "Official Methods of Analysis of the Association of Official Agricultural Chemists," 10th ed., p. 341. AOAC, Washington DC.

Hurst, E. (1942). "The Poison Plants of New South Wales." Snelling Printings Work, Sydney, Australia.

Jones, D. A. (1971). *Science* **173,** 945.

Jones, D. A. (1973). *In* "Taxonomy and Ecology" (V. H. Heywood, ed.), Chapter 11, pp. 213–242. Academic Press.

Jones, D. A., Keymer, R. J., and Ellis, W. M. (1978). *In* "Biochemical Aspects of Plant and Animal Coevolution," J. B. Harborne, Ed. pp. 21–34. Academic.

Jones, D. A., Parsons, J., and Rothschild, M. (1962). *Nature (London)* **193,** 52–53.

Kingsbury, J. M. (1964). "Poisonous Plants of the U.S. and Canada." Prentice-Hall, Englewood Cliffs, New Jersey.

Kojima, M., Poulton, J. E., Thayer, S. S., and Conn, E. E. (1979). *Plant Physiology* (in press).

Lambert, J. L., Ramasamy, J., and Paukstelis, J. V. (1975). *Anal. Chem.* **47,** 916–918.

Mao, C.-H., and Anderson, L. (1965). *J. Org. Chem.* **30,** 603–607.

Mao, C.-H., and Anderson, L. (1967). *Phytochemistry* **6,** 473–483.

Marsh, C. H., Clawson, A. B., and Roe, G. C. (1929). *U.S., Dep. Agric., Tech. Bull.* **113.**

Montgomery, R. D. (1969). *In* "Toxic Constituents of Plant Foodstuffs" (I. E. Liener, ed.), pp. 143–157. Academic Press, New York.

Moran, E. A. (1954). *Am. J. Vet. Res.* **15,** 171–176.

Nahrstedt, A. (1975). *Arch. Pharm. (Weinheim, Ger.)* **308,** 903–910.

Nass, H. G. (1972). *Crop Sci.* **12,** 503–506.

Nestel, B., and MacIntyre, R., eds. (1973). "Chronic Cassava Toxicity." Monogr. IDRC-010e. Int. Dev. Res. Cent., Ottawa.

Ng, S.-C. J. (1975). M. S. Thesis, University of California, Davis.

Nisizawa, K., and Hashimoto, Y. (1970). *In* "The Carbohydrates" (W. Pigman and D. Horton, eds.), 2nd ed., Chap. Vol. 2A, 33. Academic Press, New York.

Osuntokun, B. O. (1973). *In* "Chronic Cassava Toxicity" (B. Nestel and R. McIntyre, eds.), Monogr. IDRC-010e, pp. 127–138. Int. Dev. Res. Cent., Ottawa, Canada.

Reay, P. F., and Conn, E. E. (1970). *Phytochemistry* **9,** 1825–1827.

Saunders, J. A., and Conn, E. E. (1978). *Plant Physiol.* **61,** 154–157.

Saunders, J. A., Conn, E. E., Lin, C. H., and Stocking, C. R. (1977a). *Plant Physiol.* **59,** 647–652.

Saunders, J. A., Conn, E. E., Lin, C. H., and Shimada, M. (1977b). *Plant Physiol.* **60,** 629–634.

Schwarzmaier, U. (1976). *Chem. Ber.* **109,** 3379–3389.

Secor, J. B., Conn, E. E., Dunn, J. E., and Seigler, D. S. (1976). *Phytochemistry* **15,** 1703–1706.

Seigler, D. S. (1977). *Prog. Phytochem.* **4,** 83–120.

Stafford, H. A. (1969). *Phytochemistry* **8,** 743–752.

Steyn, D. R., and Rimington, C. (1935). *Onderstepoort J. Vet. Sci. Anim. Ind.* **4,** 151.

Thienes, C. H., and Haley, T. J. (1972). "Clinical Toxicology," 5th ed. Lea & Febiger, Philadelphia, Pennsylvania.

Watt, J. M., and Breyer-Brandwijk, M. G. (1962). "The Medicinal and Poisonous Plants of Southern and East Africa," 2nd ed. Livingstone, Edinburgh.

Wilson, J. (1973). *In* "Chronic Cassava Toxicity" (B. Nestel and R. McIntyre, eds.), Monogr. IDRC-010e, pp. 121–125. Int. Dev. Res. Cent., Ottawa, Canada.

Chapter **11**

The Evolutionary Ecology of Alkaloids

TREVOR ROBINSON

I. INTRODUCTION: WHAT IS AN ALKALOID?

Structurally, the alkaloids are a very diverse group of compounds, and definitions beyond the most general must include many exceptions. All alkaloids contain nitrogen as part of a heterocyclic ring. In order to assert that the nitrogen atom must be part of a heterocyclic ring, it is necessary to place certain compounds classically called "alkaloids" into another

ISBN 0-12-597180-X

group known as "protoalkaloids." The latter group includes such compounds as mescaline, hordenine, and colchicine.

The presence of a nitrogen atom makes most of the alkaloids basic, as their name indicates. That is, they tend to become protonated in aqueous solution. However, the name is misleading for a few alkaloids that have electron-withdrawing groups adjacent to the nitrogen and hence are not readily protonated. This group of "neutral" alkaloids includes ricinine and colchicine. Most alkaloids are protonated at normal biological pH values and therefore exist as water-soluble cations that can be obtained as salts or, from alkaline solution, as free bases. A few alkaloids have quaternary nitrogen atoms and are therefore positively charged at all pH values; they are highly water soluble and are obtained only as salts.

Once alkaloids are defined as being heterocyclic nitrogen compounds, their further classification follows from the type of ring system that is present. In Table I only an outline of this classification with a few examples is presented. Comprehensive works are available that consider in greater detail the chemistry of all the classes of alkaloids and individual representatives of each class (Manske and Holmes, 1951–1968; Pelletier, 1970; Glasby, 1975).

II. CHEMICAL METHODS

This section is not intended to offer an exhaustive review of alkaloid chemistry but rather to provide the biological scientist with some useful methodology for determining whether alkaloids are present in certain biological material and to a limited extent for separating and characterizing any alkaloids that are found. In almost every case it is important to utilize information already published for related biological materials. Finding out whether alkaloids are known from the same family or genus, and what alkaloids they are, can certainly provide shortcuts in laboratory work. For example, among the Solanaceae one might test immediately for steroid alkaloids, tropane alkaloids, and pyridine alkaloids while assuming the absence of isoquinoline or indole alkaloids. In the genus *Senecio* one can almost count on any alkaloid being a pyrrolizidine. A second piece of biological information that one must keep in mind while proceeding with chemical methods is that, within a given species, the content and pattern of alkaloids frequently vary from tissue to tissue and from time to time. Any worthwhile survey, then, should examine all parts of the plant at different times during its development. Further discussion of this biological background can be found in Section III.

Table I. Classification of alkaloid structures with some examples

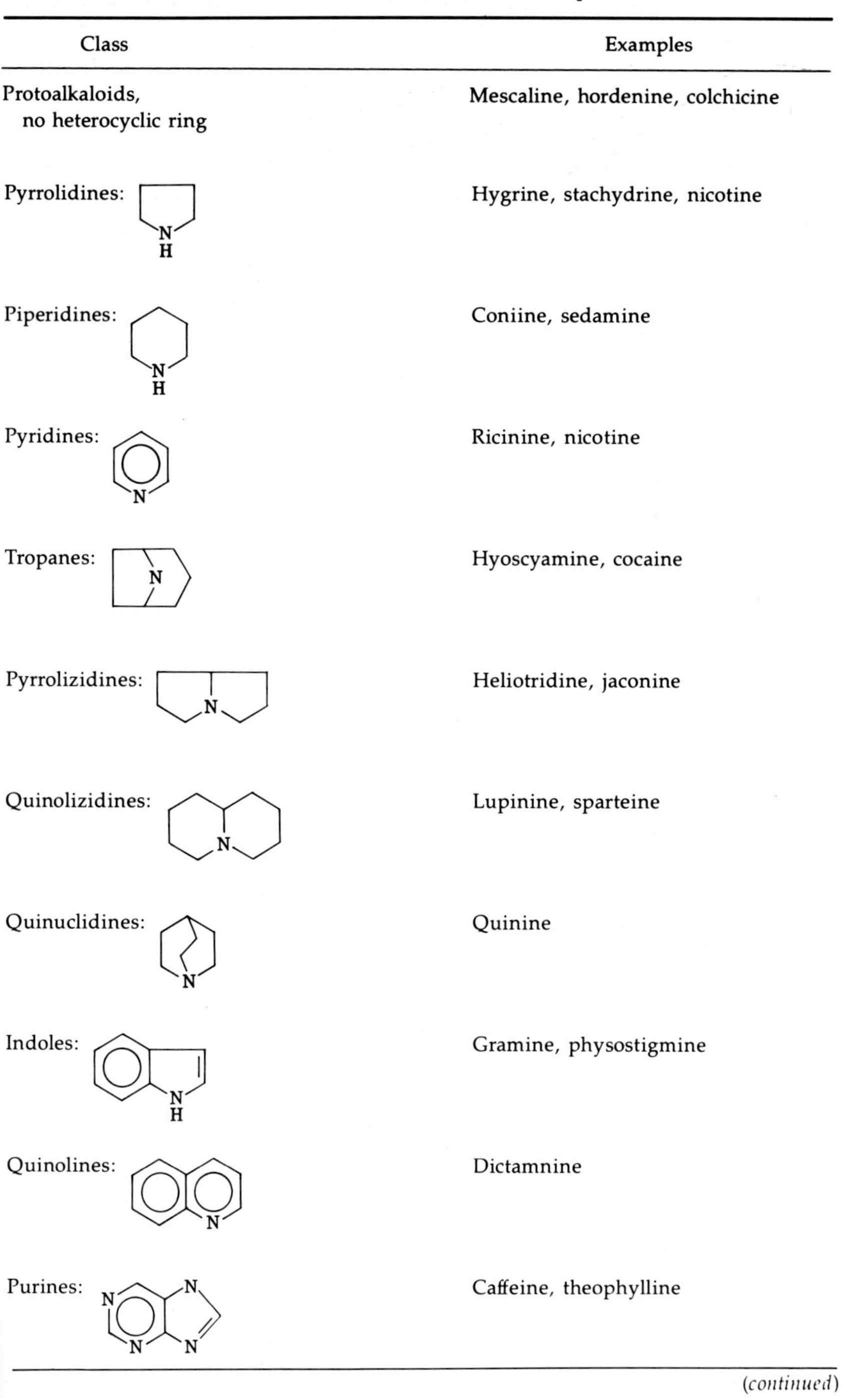

Class	Examples
Protoalkaloids, no heterocyclic ring	Mescaline, hordenine, colchicine
Pyrrolidines:	Hygrine, stachydrine, nicotine
Piperidines:	Coniine, sedamine
Pyridines:	Ricinine, nicotine
Tropanes:	Hyoscyamine, cocaine
Pyrrolizidines:	Heliotridine, jaconine
Quinolizidines:	Lupinine, sparteine
Quinuclidines:	Quinine
Indoles:	Gramine, physostigmine
Quinolines:	Dictamnine
Purines:	Caffeine, theophylline

(*continued*)

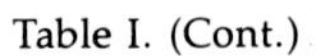
Table I. (Cont.)

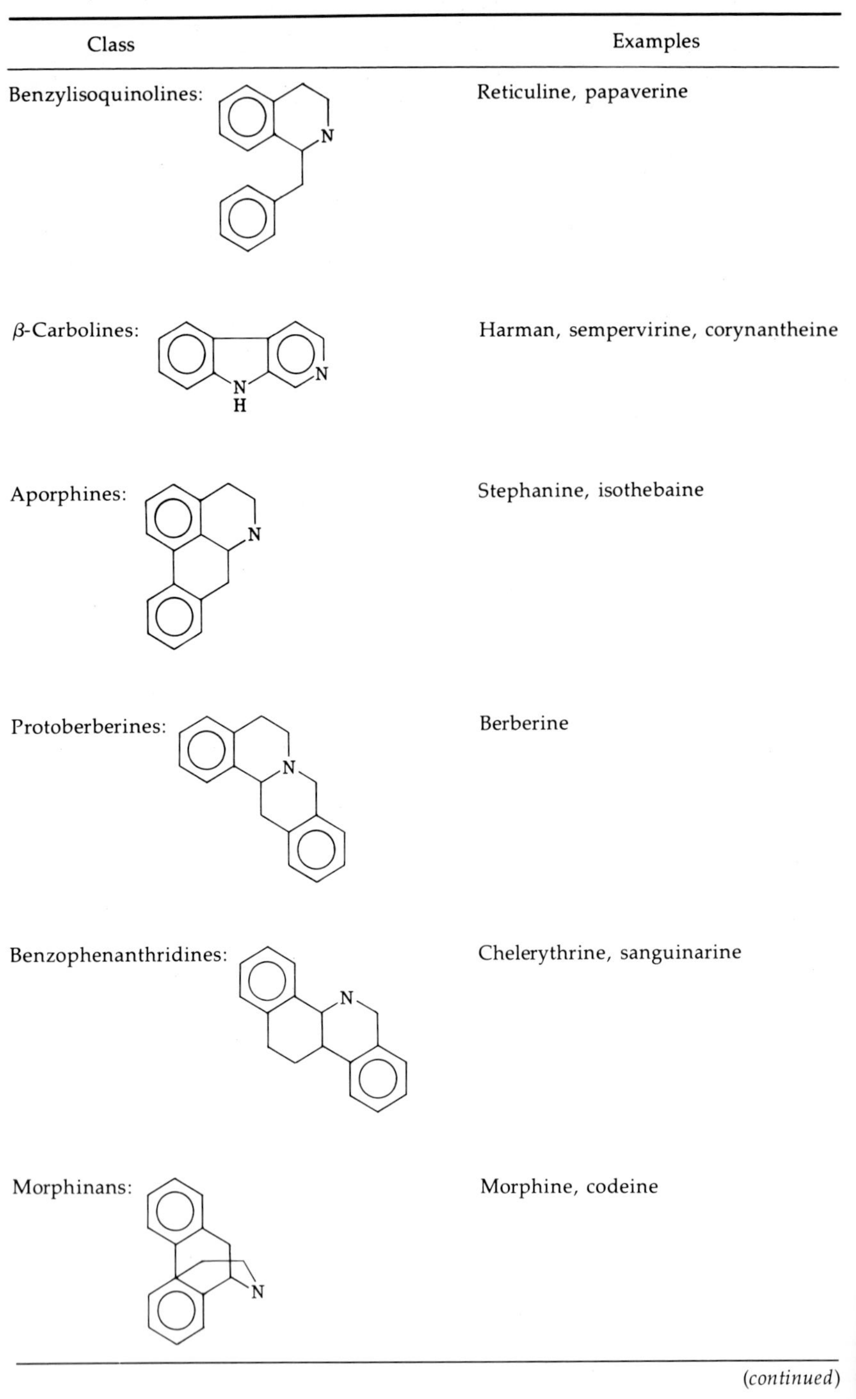

Class	Examples
Benzylisoquinolines:	Reticuline, papaverine
β-Carbolines:	Harman, sempervirine, corynantheine
Aporphines:	Stephanine, isothebaine
Protoberberines:	Berberine
Benzophenanthridines:	Chelerythrine, sanguinarine
Morphinans:	Morphine, codeine

(*continued*)

Table I. (Cont.)

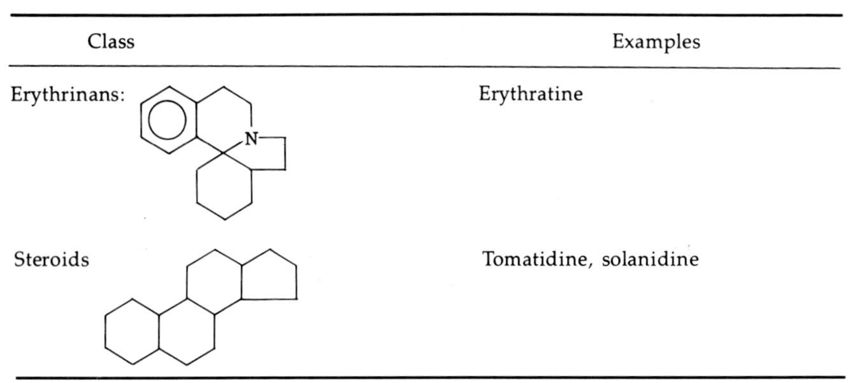

Class	Examples
Erythrinans:	Erythratine
Steroids	Tomatidine, solanidine

A. Preliminary Tests

The first step is to ascertain whether alkaloids are present. For this purpose many qualitative tests have been developed for application to plant extracts, plant saps, or even crushed tissue. A number of special "alkaloid reagents" have been devised, and, as with all simple tests, they can be very useful as long as their limitations are recognized. Dragendorff's reagent is a solution of potassium tetraiodobismuthate ($KBiI_4$) that reacts to give a red-orange complex with many basic compounds, including some nonalkaloids such as choline (Bregoff *et al.*, 1953). It also reacts with a few nonbasic substances, particularly conjugated carbonyl compounds such as pyrones, cinnamaldehyde, and ninhydrin (Farnsworth *et al.*, 1962). It reacts with neutral alkaloids such as ricinine (Robinson and Fowell, 1959) but does not react with any of the common amino acids (Winek and Fitzgerald, 1961). Some plant pigments, if present, mask the color. A quick spot test for alkaloid can be done by applying plant juice (or an extract from dry plant material) to filter paper along with a drop of Dragendorff's reagent (Raffauf, 1962) or by crushing plant material on filter paper already impregnated with reagent (Kraft, 1953).

The second most common reagent for alkaloid testing is Mayer's reagent, a solution of potassium tetraiodomercurate (Cromwell, 1955; Szasz and Buda, 1971). It gives an off-white precipitate with nearly all alkaloids but may react with other types of plant constituents, so some preliminary purification is advisable. The test works most effectively in slightly acidic solution, and, for some alkaloids that are decomposed by strong acids, citric acid is recommended for acidification (Tinker and Lauter, 1956). Several other tests are described by Cromwell (1955), and a method for determining the location of alkaloid in fresh plant tissue is described by White and Spencer (1964).

B. Screening Procedures

Usually the screening programs that have been employed to detect alkaloids in plants rely on a somewhat more elaborate procedure than simple application of the tests described in Section II,A. Two widely employed methods are those of Smolenski *et al.* (1972) and Hultin and Torssell (1965). A special procedure is necessary when pyrrolizidine alkaloids are suspected, because of their lability (Bull *et al.*, 1968). A micro method may be useful when only a small amount of plant material is available, and one such procedure is described by Weber and Ma (1976).

Many screening procedures employ chromatography applied to such extracts as described in the previous paragraphs. In this way an indication can be obtained as to the number of alkaloids present rather than a simple positive or negative result (Hultin, 1966; Farnsworth and Euler, 1962).

C. Chromatographic Analysis

All the standard chromatographic techniques have been applied to alkaloids. On paper chromatograms neutral solvent mixtures are suitable only for very strongly or very weakly basic alkaloids, i.e., quarternary bases that always have a positive charge and "neutral" alkaloids that show no tendency to become protonated. For most alkaloids, therefore, acidic solvents are used to ensure protonation, or basic solvents are used to ensure deprotonation (Bräuniger, 1954).

In most thin-layer separations, layers of silica gel have been used (Farnsworth and Euler, 1962; Hultin, 1966). Solvent systems are similar to those used in paper chromatography, with the possibility of incorporating acid, base, or buffer mixtures into the thin-layer medium.

General techniques for detecting compounds on chromatograms, such as observation in ultraviolet light and exposure to iodine vapor, may be useful but are not specific for alkaloids. The most widely used reagent for alkaloid detection is Dragendorff's reagent, and various modified recipes have advantages when used as a chromatogram spray (Robles, 1959; Thies and Reuther, 1954; Vágújafalvi, 1960). The limit of detection varies with the alkaloid, but in some cases 0.5 μg can be detected.

A number of other detection reagents have been applied to alkaloid chromatography—for example, Mayer's reagent (Csoban and Hegedüs, 1954; Pan and Wagman, 1959), potassium iodoplatinate (Paul and Conine, 1973), phosphomolybdic acid followed by reduction (List, 1954), and 7,7,8,8-tetracyanoquinodimethane (Rücker and Taha, 1977).

More specialized spray reagents can be utilized to indicate certain structural features or types of alkaloid—for example, pyridine derivatives (Lang and Lang, 1974), indoles (Toneby, 1974; Knowlton *et al.*, 1960), the methylene dioxy group (Gunner and Hand, 1968), *N*-oxides (Dann, 1960), steroidal alkaloids (Schilling and Zobel, 1966), tropines (Bräuniger, 1954), 3-pyrrolines (Mattocks, 1967), and alkaloids containing the grouping —$NHCH_2CH_2$— (McCaldin, 1960). Attention is also called to the section "Chromatographic Data" in the *Journal of Chromatography*. Alkaloids are included frequently in this section.

D. Other Methods of Characterization

The systematic application of specific reagents has been a common method of alkaloid characterization for many years and may still be useful in field testing when it is inconvenient to process chromagrams. Two general schemes of this sort are described by Gornyi (1955) and Fulton (1939).

If sufficient, pure alkaloid can be obtained, for instance, by extraction of a zone on a thin-layer plate; physical methods of characterization can be used for identification. Many alkaloids show ultraviolet absorption spectra, and very little material is needed for this analysis. Several reviews and tabulations of alkaloid spectra are available (Sangster, 1960; Holubek and Štrouf, 1965–1973; Sangster and Stuart, 1965). Certain specific features can also be indicated by ultraviolet spectra—for example, substitution patterns and the difference between methoxyl and methylenedioxy groups (Hruban and Šantavý, 1967; Hruban *et al.*, 1970).

E. Isolation of Alkaloids

The chromatographic methods described so far are useful more for characterization than for isolation of alkaloids, although for some purposes it may be possible to isolate enough material from a paper sheet or thin-layer plate.

Extraction procedures rely on the solubility of most alkaloids in a polar organic solvent, then their extraction into a less polar organic solvent from an alkaline aqueous solution, and their return to an acidic aqueous phase from an immiscible organic solvent. Quaternary alkaloids and *N*-oxides can be lost because they may not be extracted efficiently from an alkaline aqueous solution, and neutral alkaloids may be missed because they remain in the organic solvent rather than go into an acidic aqueous solution. A general method for the isolation of quaternary alkaloids involves precipitation with Mayer's reagent, redissolving the precipitate in acetone/methanol/water (6 : 2 : 1), and regenerating the alkaloids as their chlorides by passing through an anion-exchange column (Jordan and Scheuer, 1965). Some alkaloids occur as complexes with tannins, which makes them difficult to extract (Schneider and Kleinert, 1972). Other "bound forms" exist and make extraction difficult, but their exact chemical nature is unknown (Fairbairn and Ali, 1968; Weeks and Bush, 1974; Wold, 1978).

At present it appears that the best routine method for obtaining alkaloids from aqueous plant extracts is ion-exchange chromatography on cation-exchange resins with a low degree of cross-linking or on cellulose–phosphonic acid (Berggren *et al.*, 1958). By adjustment of pH, quaternary alkaloids can be separated from all others, and stronger bases can be separated from weaker bases (Stark, 1962; Christianson *et al.*, 1960). Only neutral alkaloids are not amenable to ion-exchange separation. Many other column chromatography methods have also been used for purifying alkaloids.

III. TAXONOMIC, ANATOMICAL, AND GEOGRAPHICAL DISTRIBUTION

The grouping together under a single heading of three very different types of distribution is intended to emphasize the fact that it is only a first approximation to say that a certain plant species contains a particular alkaloid. Beyond this approximation, it is important to realize that the distribution of alkaloids frequently varies from tissue to tissue within the same plant, that geographical races characterized by different alkaloid content are known within a single species, and that there is often ontogenetic variation of alkaloids (Waller and Nowacki, 1978).

A. Taxonomic Distribution

It is well known that certain plant families are characterized by the presence of alkaloids, whereas certain other families are free of alkaloids. One estimate (Cromwell, 1955) is that the presence of alkaloids has been established in about one-seventh of the total number of flowering plant families. The majority of these are dicotyledons. Another estimate (Hegnauer, 1963) is that 15–20% of all vascular plants contain alkaloids. Still another estimate (Willaman and Schubert, 1955) is that nearly one-third of the 300 or so angiosperm families have alkaloids. In any case, the majority of alkaloidal plants are dicotyledons; a few are found among the monocotyledons and gymnosperms. Among non-seed-bearing plants alkaloids occur in club mosses (Lycopodiaceae) and horsetails (Equisetaceae), but they have never been reported from bryophytes and their existence is questionable in the ferns (Panvisavas *et al.*, 1968). Li and Willaman (1972) suggest some likely taxa in which to find alkaloids on the basis of past experience.

A taxonomic correlation in a very broad sense is that alkaloids occur in herbaceous plants more frequently than in trees and that the average molecular weight of alkaloids from trees is lower than that of alkaloids from herbs (McNair, 1935b). Annual plants have a higher incidence of alkaloids than perennial plants (Levin, 1976). Alkaloids are absent in water plants except for the family Nymphaeaceae (McNair, 1943).

The decision as to whether a given plant contains alkaloids is not as clear-cut as one might wish because some plants may contain very small amounts that are detectable only when prodigious quantities of plant material are processed (e.g., 40 mg of narcotine from 100 kg of cabbage) (Hegnauer, 1963). Thus, it may be that many more plants have the ability to synthesize alkaloids than have the ability to accumulate alkaloids, and it is the latter property that is of prime interest in the interaction of plants and herbivores. For practical purposes, an alkaloidal plant can be defined as one that has more than 0.05% alkaloid by dry weight.

Generalizations about the taxonomic correlations of alkaloid occurrence are by no means absolute. However, Hegnauer (1966) has pointed out that the main groups of alkaloidal plants are found in the Centrospermae, Magnoliales, Ranunculales, Papaveraceae, Leguminosae, Rutaceae, Buxaceae, Gentianales, Campanlulaceae, Compositae, Tubiflorae, and Liliiflorae. It will be noted that certain of these taxa are orders and some are families. In contrast, plants very unlikely to contain alkaloids are found among the amentiferous group, Proteales, Cucurbitales, and monocotyledons (except for Liliaceae). In the present chapter no attempt is made to consider fine points of taxonomy in relation to alkaloid content since several publications deal with these questions (Hegnauer, 1962–1973, 1963, 1966; Mabry and Mears, 1970; Tétényi, 1970). For complete information as to which alkaloids are reported from a particular plant or, conversely, which plants contain a particular alkaloid several valuable publications are available (Willaman and Schubert, 1961; Willaman and Li, 1970; Raffauf, 1970).

Attempts have been made to correlate patterns of alkaloid occurrence with patterns of other plant constituents along taxonomic lines. Although an unequivocal statement is not possible, one of the broadest correlations is a negative one between the presence of alkaloids and the excretion of volatile oils (Treibs, 1955). Such a correlation would suggest a common function of the two types of compound. In the context of this chapter, it is tempting to think that such a function might be the repulsion of herbivores. A more limited correlation with a similar implication was observed by Tallent *et al.* (1955), who found that species of alkaloid-containing pine lacked the usual volatile terpenes but did contain saturated, aliphatic hydrocarbons. In the genus *Erythrina* a negative correlation between free amino acid content and alkaloid content was observed (Romeo and Bell, 1974). Because of the precursor–product relationship between amino acids and many alkaloids, it is interesting that amino acids in general or specific amino acids have been observed to fluctuate in concentration parallel with alkaloids (Michels-Nyomarkay, 1970; Paris and Girre, 1969).

At the subspecies level, different so-called chemical races are well recognized among alkaloid-containing plants; natural or artificial hybridization can give rise to different alkaloidal patterns, with certain alkaloids appearing as a dominant character (Mabry and Mears, 1970). It seems, however, that infraspecific variation never results in the appearance of a completely alkaloid-free plant within a characteristically alkaloid-containing species (Tétényi, 1970). This fact, incidentally, would argue for some essential function of alkaloids in their plants. Theoretical discussion of infraspecific taxonomy and the various proposals for naming chemical races are outside the scope of this chapter. Tétényi (1970) gives both a general discussion as well as extensive tables of data showing both quantitative and qualitative chemical differences observed within a species. Examination of these data shows that 10-fold differences in total alkaloid concentration are not uncommon when one compares one race to another and that the complete absence of one alkaloid of a mixture is also observed frequently. However, the major alkaloid of a species is almost always present in all races of that species. In the case of the opium poppy (*Papaver somniferum*) intensive breeding programs over many centuries have created oil seed varieties and varieties high in alkaloids, yet the oil varieties still contain a minimum of 1 to 2% morphine in the dried capsule. Moreover, morphine is their major alkaloid (Nyman and Hall, 1974). When two varieties of *Solanum dulcamara* are crossed, the hybrids contain alkaloids of both parents plus tomatidine, which is not present in either parent (Rönsch *et al.*, 1968). Differences in ploidy within a given species may also produce differences in alkaloid content. For instance, in both *Atropa* and *Datura* spp. tetraploid plants have a higher alkaloid content than diploid plants (Rowson, 1954; Mechler and Kohlenbach, 1978).

Interspecific hybridization is possible in a few species. In the example of *Papaver bracteatum* × *P. somniferum*, the F_1 hybrids have alkaloids of both parents, but the morphine characteristic of the latter parent dominates over the thebaine characteristic of the former (Böhm, 1965). With *Nicotiana tabacum* × *N. sylvestris*, it appears that the alkaloid pattern of the F_1 generation is determined more by the "mother" plant, but in later generations *N. tabacum* characteristics tend to dominate (Lovkova *et al.*, 1973).

The problem of sorting out geographical races from chemical races is considered more fully in Section III,C. Seemingly decisive evidence that infraspecific alkaloid patterns are determined strictly genetically is the observation that *in vitro* root cultures from different clones of *Atropa belladonna* produce atropine in amounts varying with the clone (Mitra, 1972).

The biochemical basis for differing patterns of alkaloids within a related taxonomic grouping is presumably to be found in small differences in enzyme activity, specificity, or compartmentation. Precise data on this point are scarce, but methylation processes are likely to be significant variables. A specific methylation reaction occurring early in the pathway can determine the following direction of the pathway. Different strains of barley (*Hordeum vulgare*) can be grouped as to whether they contain *N*-methyltyramine or *N*-dimethyltyramine (Kirkwood and Marion, 1950). A serrate-leaved variety of *Croton flavens* contains flavinine and norsinoacutine, whereas the entire-leaved variety has the corresponding *N*-methyl derivatives (Chambers and Stuart, 1968). The related species *Lophophora williamsii* and *Trichocereus pachanoi* differ in their metabolism of hydroxyphenethylamines because the former has a system for ortho methylation, whereas the latter has a system for meta methylation (Agurell and Lundström, 1968).

B. Anatomical Distribution

The general rules of alkaloid distribution within a plant stress the accumulation of alkaloids in (1) very active tissues, (2) epidermal and hypodermal tissues, (3) vascular sheaths, and (4) latex vessels (James, 1950, 1953).

For thoroughness, considerable amplification and criticism of each of these rules would be necessary. One important fact is that the sites of accumulation may not be the sites of synthesis. Nicotine, for example, is synthesized in tobacco roots but is then translocated to and accumulated in the leaves. Tropane alkaloids, made in the roots of *Datura* spp., are extensively modified in the leaves (Mothes, 1960). The alkaloids present in the seeds of *Ipomoea violacea* are synthesized in the leaves (Mockaitis *et al.*, 1973). Other cases also show the cooperation of several different plant tissues in completing the synthesis of an alkaloid.

Although alkaloids are prominent in actively growing tissues, they are not usually found at highest concentration in the youngest cells of these tissues but rather in somewhat older cells that are becoming vacuolated. Accumulation of alkaloids in vacuoles is frequently seen (Roddick, 1976, 1977; Matile *et al.*, 1970), but certain species also show specialized sites of accumulation such as special cell types (Koblitz *et al.*, 1975; Neumann and Müller, 1972, 1974), latex vesicles (Fairbairn *et al.*, 1974; Dickenson and Fairbairn, 1975), and xylem cell walls (Müller *et al.*, 1971). Quite different alkaloids may also be found in various parts of the same plant (Kennedy, 1971; Schneider and Wightman, 1974). Many more examples of alkaloid localization can be found in James (1950) and Mothes (1960).

Changes in alkaloid concentration or pattern during the development of a plant are commonly observed, and some of these changes are extreme. The most generally applicable principle is that alkaloid content increases rapidly at the time of cell enlargement and vacuolization and then declines slowly during senescence (James, 1950; Sinden *et al.*, 1973; Lee and Waller, 1972; Ziyaev *et al.*, 1975). This pattern most often applies to leaves. If concentration is measured, the maximum comes earlier than if total amount is measured. Tobacco roots synthesize nicotine only in mature cells, not during the elongation phase (Lovkova and Obroucheva, 1975). Important morphological events such as floral initiation and fruiting can affect alkaloid formation (Mothes, 1960; Verzár-Petri, 1966; Moursi and Ahmed, 1973). In developing seeds and germinating seedlings there is no universal pattern, although unfertilized ovules of alkaloidal plants usually contain alkaloids. In *Nicotiana, Papaver, Hordeum, Datura,* and *Erythroxylon* spp. the mature seeds contain little or no free alkaloid, but seeds of *Lupinus, Physostigma, Coffea,* and *Delphinium* spp. are noted for their high content of alkaloids (James, 1950). During germination the appearance of alkaloids may begin within a few days (*Hordeum, Lycopersicon, Nicotiana*) or only after several weeks (*Datura*). The alkaloid content of alkaloid-rich seeds may decline in total amount and concentration during the early stages of germination, indicating some metabolic transformation (see Section IV). In well-developed plants there are many examples of cyclic variations in alkaloid content sometimes over periods of weeks, but even diurnal cycles have been observed (Robinson, 1974). All this ontogenetic variation should be a warning to anyone who searches for a role of alkaloids in ecological interactions, for it is the alkaloid content at the exact time and place of any presumed interaction that must be determined. An extreme case, to emphasize this warning, is that of the well-known alkaloidal plant *Catharanthus roseus,* which lacks alkaloids in the seeds and in the growing plants when they are about 6 weeks old (Mothes *et al.,* 1965). Another difficulty is raised by the presence of "bound" alkaloids. As noted above, although seeds of tobacco and poppy do not show the presence of alkaloid by the usual tests, hydrolysis releases alkaloid from unknown precursors (Weeks and Bush, 1974; Fairbairn and Ali, 1968; Fairbairn and El-Masry, 1968). Thus, it is conceivable that bound alkaloids could be released when herbivores digest certain plant materials.

C. Geographical Distribution

As observed in Section III,A, it is sometimes difficult to sort out geographic variations from truly genetic variations, since many so-called geographic races may be, in fact, genetically different. Natural populations of a species that extend over a wide geographical area can be genetically heterogeneous while having particular ecotypes favored in certain parts of the overall range. This geographically graded variation, termed a "cline," is widespread among flowering plants (Heslop-Harrison, 1963), and one might expect to find examples of it in alkaloidal variation. Such examples are not readily forthcoming, however. In the case of *Baptisia leucophaea,* the results of analysis of two geographic populations were similar on the average, although individual plants showed considerable variation from the average (Mabry and Mears, 1970).

The classic studies of McNair (1931, 1935a) on the occurrence of alkaloids in relation to the climate of habitat and the more recent review of Levin (1976) on the biogeography of alkaloid-bearing plants provide much interesting statistical analysis of correlations between climate, latitude, and altitude, on the one hand, and alkaloid content, on the other. As a bare statement of correlation, it can be said that a greater proportion of alkaloid-producing plants can be found in the tropics and at low altitudes. In other words, the most productive habitats have the highest percentage of alkaloidal plants. A more limited correlation is that island populations have a lower incidence of alkaloids than would be predicted from latitude alone (Levin, 1976). Interesting as these correlations are, they may be quite irrelevant to the question of geographic races or ecotypes because they involved comparisons at all levels of taxonomic categories. In one view the correlations may simply show that plants growing under the best conditions can be permitted the extravagance of useless alkaloid accumulation. Levin, however, presents a cogent argument based on these data to the effect that alkaloid content is a response to pest pressure and that, since the most favorable habitat for plant growth is also the most favorable habitat for their pests, alkaloid accumulation and more toxic alkaloids confer a selective advantage (Levin and York, 1978). Such a broad generalization calls for corroboration from many narrow instances and, if true, should be evidenced by observations on geographical races and a correlation between clinal variation and predator population. An example of such observations is provided by a study of Colorado populations of *Lupinus* spp. described in Section VII.

In addition to apparent geographical races, which have at base a genetic explanation, it is evident that climate, soil, and other environmental factors can modify alkaloid content within a genetically homogeneous group. The most general statement that can be made in this context is that those conditions that are most favorable to overall metabolism or growth are also most favorable to alkaloid biosynthesis (James, 1950, 1953; Mothes, 1960). Many such factors, for example, mineral metabolism and light intensity, have been tested under controlled conditions, and results tend to support the general statement in spite of many exceptions. Photoperiodic influences on alkaloid content have been studied in *Datura* (Cosson, 1969, 1975), *Lycopersicon* (Sander, 1958), and *Nicotiana* spp. (Tso *et al.*, 1970). In some of these cases the beneficial effect of long-day photoperiod on alkaloid synthesis is certainly a phytochrome-mediated effect, but in others it may be that longer day lengths simply allow for more photosynthesis and thus more substrates for alkaloid biosyntheses. Many examples of effects of mineral nutrition on alkaloid content are cited by James (1950) and Mothes (1960). From these results it can be concluded that an adequate, but not excessive, nitrogen supply is important and that ammonia is better than nitrate for stimulating alkaloid formation.

Several examples cited by Tétényi (1970) suggest the occurrence of actual geographical races in alkaloidal plants, i.e., chemical differences within a genetically homogeneous group growing in different habitats. However, in none of these cases have the controlling environmental factors been determined, and, until they are, it seems best to reserve judgment regarding the relative importance of heredity and environment.

For the field ecologist it is important to recognize that one population of a species may have a different alkaloid content than another population growing at a different location. It is hoped that the foregoing discussion demonstrates that the explanation for the difference is unlikely to be immediately obvious.

IV. METABOLISM OF ALKALOID IN THEIR PLANTS

A. Biosynthesis and Catabolism of Alkaloids

There are several comprehensive books and reviews that describe the metabolic pathways used by plants to synthesize alkaloids (Mothes and Schütte, 1969; Robinson, 1968; Spenser, 1970). In this chapter the intent is to deal with general ideas of alkaloid biosynthesis rather than to consider detailed pathways leading to specific alkaloids.

It is now generally accepted that the vast majority of alkaloids are derived from amino acids, in some cases by transformations that are almost self-evident but in other cases by long series of reactions that leave the original amino acid structure scarcely recognizable. The relation of amino acid structures to alkaloid structures was recognized in a few cases early in the twentieth century, and comprehensive proposals relating dozens of alkaloids to common amino acids were first put forward 60 years ago by Sir Robert Robinson (1917, 1955). According to these proposals a few simple condensation reactions as well as oxidations, dehydrations, decarboxylations, etc., would suffice to generate alkaloid structures from amino acid precursors. A great number of experiments, beginning in about 1954, have for the most part confirmed that such reactions do account for the conversion of amino acids to alkaloids. A few new types of reaction have been recognized, and a few details of enzymology have been studied during the last 20 years, but there have not been many important surprises to anyone who understood Robinson's 1917 proposals. The most important pathways from amino acid to alkaloid are summarized in Fig. 1. Figure 2 illustrates in more detail the relation of a few alkaloids to their amino acid precursors.

It seems that important early steps in the conversion of an amino acid to an alkaloid are decarboxylation to an amine and subsequent oxidation of the amine to an aldehyde by an amine oxidase (Mann and Smithies, 1955; Smith, 1975). The condensation of an amino group with aldehyde produces the heterocyclic rings characteristic of most alkaloids. For the formation of such protoalkaloids as mescaline and ephedrine no ring closures are needed (Paul, 1973; Yamasaki *et al.*, 1973). Intramolecular condensations, in which aldehyde and amino groups are on opposite ends of the same molecule, account for the formation of pyrrolidine and piperidine rings (Leistner and Spenser,

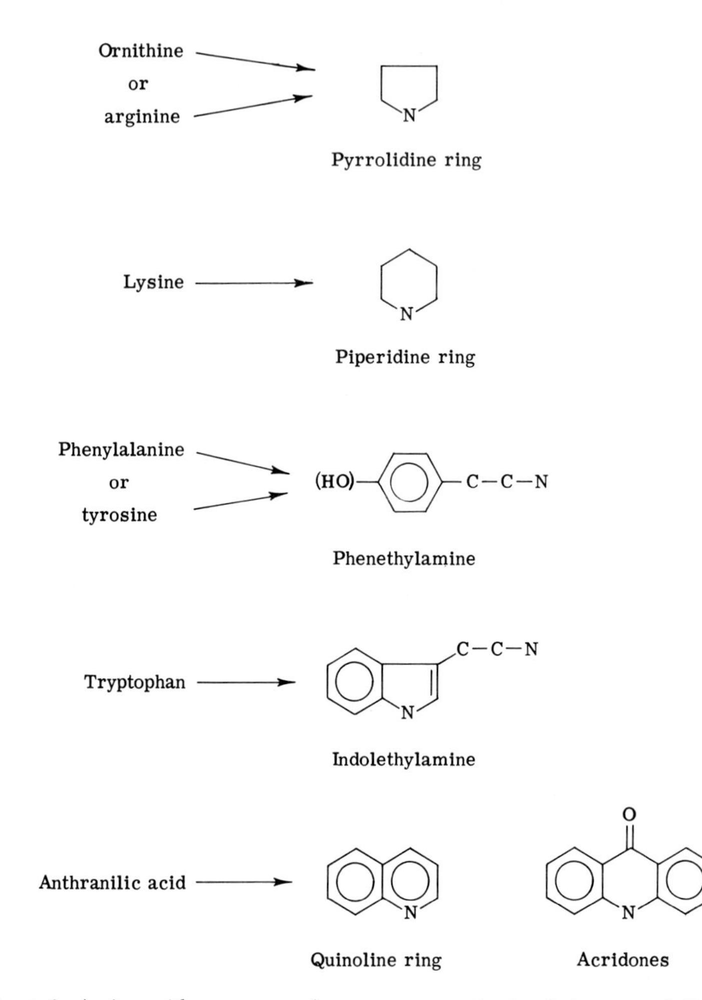

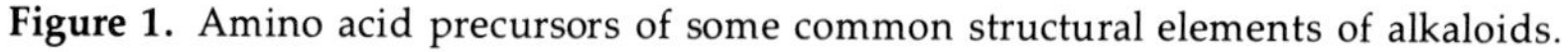

Figure 1. Amino acid precursors of some common structural elements of alkaloids.

1973; Mizusaki *et al.*, 1972). Intermolecular condensations of this type account for the building of such dimeric structures as the benzylisoquinolines from one phenethylamine-type unit and one phenylpyruvic-type unit (Wilson and Coscia, 1975; Battersby *et al.*, 1975).

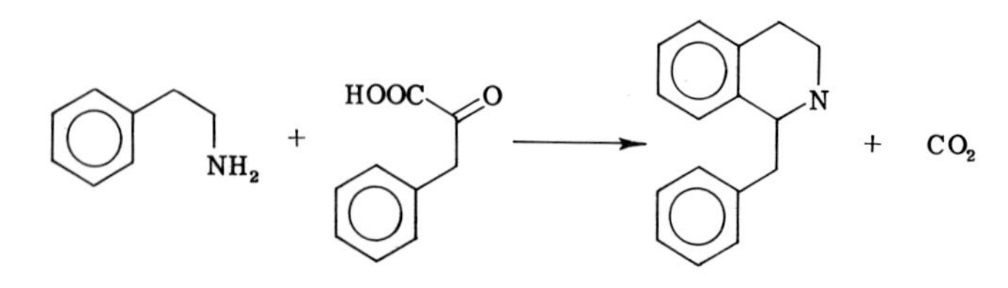

Many aromatic alkaloids show phenolic hydroxyl or methoxyl substitution. In some of these there is a single oxygen substituent occupying the position corresponding to the hydroxyl group of tyrosine and simply carried over from there into the alkaloid structure. In other cases additional oxygen substitution has occurred, presumably from the action of phenol oxidases, microsomal oxygenases, or peroxidases, although the active enzymes have not been absolutely identified (Roberts, 1973; Blaschke-Cobet and Luckner, 1973). The *O*-methyl and *N*-methyl groups that are present in many alkaloids have been shown by tracer and enzyme experiments to be derived from the methyl group of methionine (Antoun and Roberts, 1975; Basmadjian and Paul, 1971). In most cases the stage at which methylation occurs is not known. In other cases it can be inferred that site-specific methylation determines the later direction of the pathway, and therefore it must occur at a specific step (Robinson, 1968; Khanna *et al.*, 1970; Lundström and Agurell, 1971).

A feature of many alkaloid structures is the attachment of two aromatic rings to each other. The general explanation for all such structures is the occurrence of oxidative, free-radical phenol coupling reactions (Barton and Cohen, 1957). It is not known which oxidizing enzymes are responsible for phenol coupling. Both laccase and peroxidase function by a free-radical mechanism. Although there is some evidence correlating the presence of phenol oxidase activity with the content of morphinan alkaloids in *Paper somniferum* (Jindra *et al.*, 1966; Roberts, 1974), there has been little success in showing that a specific coupling reaction in a normal alkaloid pathway is catalyzed by a specific enzyme (Schenck *et al.*, 1965; Brossi *et al.*, 1973).

Although the majority of alkaloids are derived from amino acids, there are several types of alkaloids that come from other kinds of precursors. More information can be found in the books and articles cited at the beginning of this section.

Much less is known about the catabolism of alkaloids in plants than is known about their pathways of biosynthesis. Indeed, only a few years ago it was often presumed that, once formed, alkaloids merely accumulated as inert end products. Experimental questioning of this presumption has revealed that there are a few alkaloids that may be described as metabolically inactive, whereas many others undergo an active turnover (Robinson, 1974; Waller and Nowacki, 1978). In some cases the total alkaloid content stays constant or may increase. However, by the use of isotopically labeled alkaloid a definite turnover can be seen. Rates of turnover vary widely from species to species and from alkaloid to alkaloid, and half-lives ranging from 4 hours to 6 days have been calculated (Robinson, 1974). From alkaloid turnover rates it is possible theo-

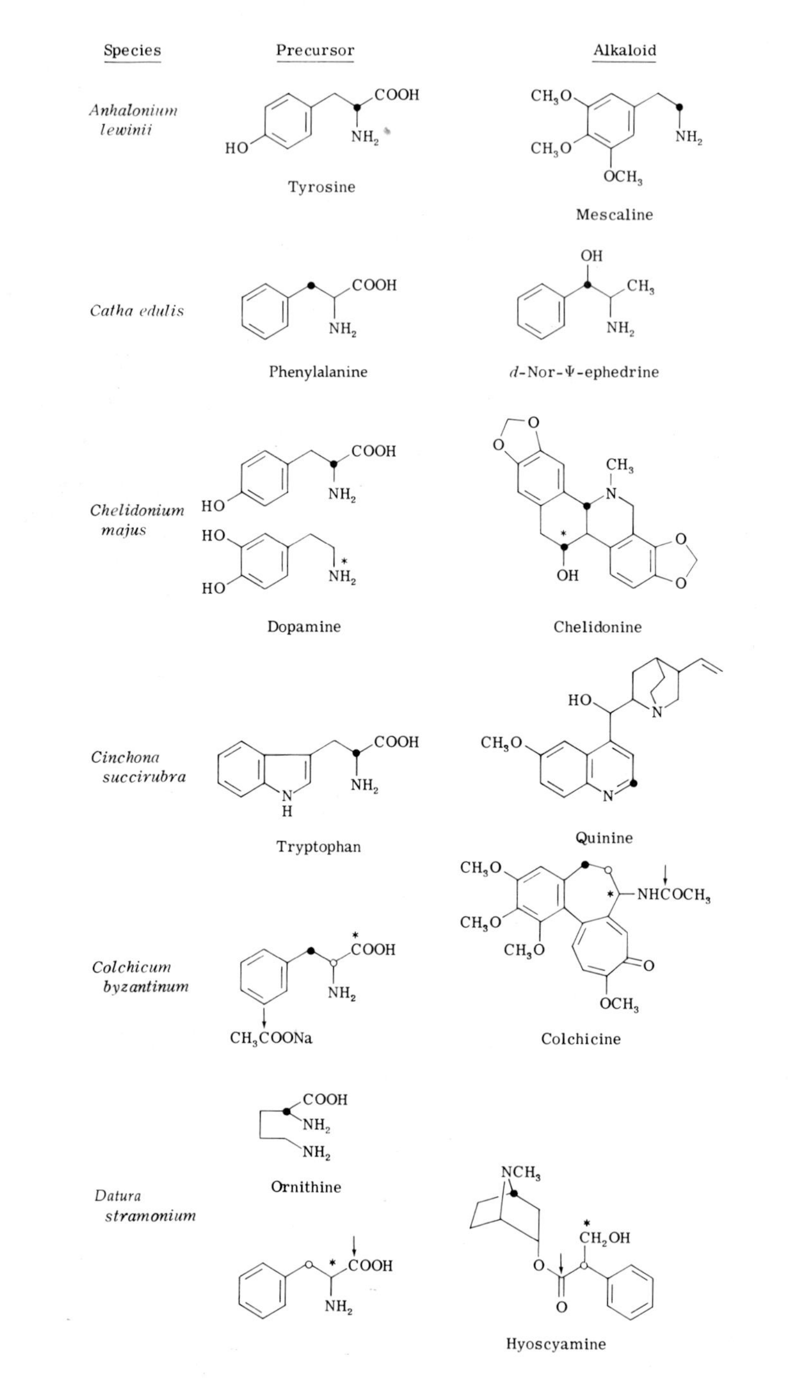

Figure 2. Some alkaloids and their precursors. Location of isotopically labeled atoms indicated by ● ⊙ * ↓. After Leete (1965). Reproduced with permission of the publisher, Copyright 1965 by the American Association for the Advancement of Science.

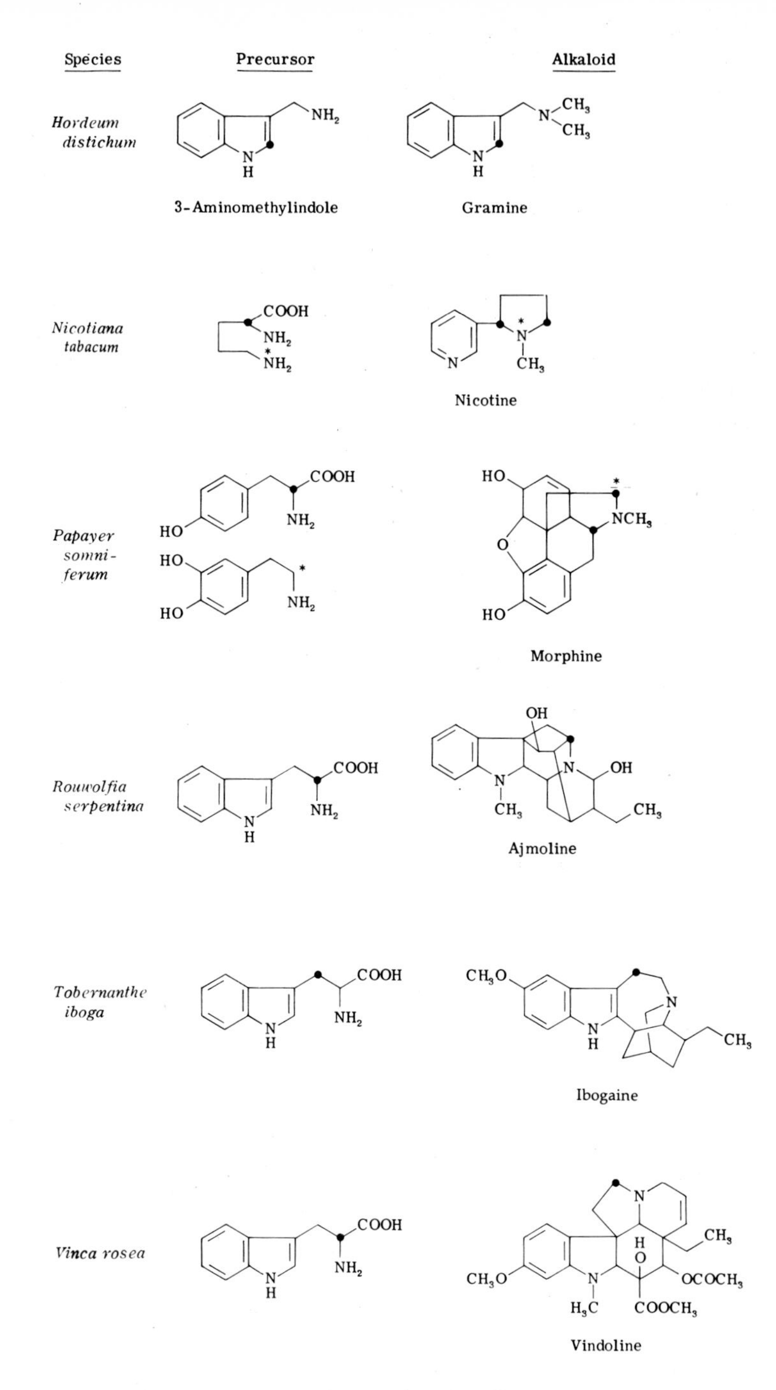
Species
Precursor
Alkaloid
Hordeum distichum
3-Aminomethylindole
Gramine
Nicotiana tabacum
Nicotine
Papayer somniferum
Morphine
Rouwolfia serpentina
Ajmoline
Tobernanthe iboga
Ibogaine
Vinca rosea
Vindoline

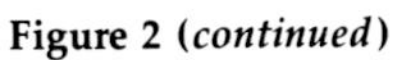

Figure 2 (*continued*)

retically to estimate what fraction of a plant's metabolism is concerned with alkaloid metabolism. The only plant for which there seem to be complete data on photosynthetic, respiratory, and alkaloid turnover rates is tobacco. By the use of available tobacco data it can be estimated that for every 35 mmoles of net CO_2 fixation 6 mmoles pass through the nicotine pathway (Robinson, 1974). This is a startlingly high proportion! Results for other plants are eagerly awaited, but it is clear that alkaloid metabolism can be a major part of a plant's metabolic expenditure. It is difficult to dismiss such activity as functionally trivial, for if it were one would expect that there would have been a selective disadvantage in maintaining it.

The fate of degraded alkaloid molecules in the plant is very poorly known. There is some conversion to primary metabolites and to carbon dioxide (Il'in and Lovkova, 1966; Fairbairn and El-Masry, 1967; Nowacki and Waller, 1975; Dyar, 1975; Waller *et al.*, 1965; Wanner and Kalberer, 1966). In addition, interconversions between alkaloids occur. Sometimes these interconversions can be seen as complementary variations in concentration of related alkaloids over a short period of time. Speculation about the possible significance of such variation in alkaloid content is given in Section VII.

B. Evolution of Alkaloid Metabolism

This section is highly speculative and must be placed into the broader context of speculation about the evolution of secondary pathways of metabolism generally (see Part I of this volume). With regard to alkaloids specifically, there is certainly a wide range of views concerning their place in evolution. Dawson (1960) suggested that alkaloid synthesis is a process gradually lost during evolution by most plants. On the other hand, McNair (1935b) believed that the synthesis of complex alkaloids was an expression of advanced status in the plant kingdom. Most current taxonomists avoid sweeping generalizations and seem content to use alkaloids as a useful character in classification at lower taxonomic levels.

It is important to note that there are very few completely unprecedented types of reaction in the pathways leading to alkaloids. The types of condensation, oxidation, and reduction that are important in the transformation of amino acids to alkaloids have many analogies in primary metabolism. Many of these reactions have also been brought about nonenzymatically under mild conditions of temperature, pH, and concentration (Robinson, 1955). It has also been shown that certain transformations within a particular alkaloid pathway can be catalyzed by enzymes from plants that never make that alkaloid (Nowacki, 1958a, 1964; Groeger, 1963; Szklarek and Mazan, 1960; Mothes, 1966; Veliky and Barber, 1975). In addition, some enzymes of an alkaloid pathway may be present at a stage of development when the complete biosynthetic pathway is not yet functional (Böhm, 1971). On the other hand, some plants provided with unnatural amino acid analogs convert them to unnatural analogs of their normal alkaloids (Kirby *et al.*, 1972; Leete *et al.*, 1971; Rosenberg and Stohs, 1976; Rueppel and Rapaport, 1971). These various observations lend credence to the viewpoint that all higher plant life is poised at the edge of a capability for alkaloid accumulation. The precursors are ubiquitous, most of the enzymes needed are widespread and nonspecific, and some reactions may occur nonenzymatically. The appearance of a complete pathway for alkaloid biosynthesis, in this view, would require a moderate genetic alteration. This change might result in a shift in metabolite compartmentation, a slight decrease in enzyme specificity (e.g., of a methylase), or loss of regulation resulting in high concentration of some amino acid (Mothes, 1969; Rosenberg and Stohs, 1974). Alternately, there could be heritable loss of a degradative enzyme in a plant that normally synthesizes an alkaloid but degrades it just as rapidly. The problem is more in understanding the environmental pressures that would lead to selection of such a mutation after it occurred, but this is a problem that concerns all secondary plant constituents, not just alkaloids.

Further evidence for the delicacy of control in alkaloid biosynthesis is the considerable difference frequently observed in alkaloid content when an intact plant is compared with tissue cultures derived from it (Barz *et al.*, 1977). In some cases cells cultured from an alkaloid plant contain little or no alkaloid (Staba and Laursen, 1966; Garnier and Morel, 1972; Tabata and Hiraoka, 1978; Furuya *et al.*, 1978; Neumann and Müller, 1971), but sometimes alkaloid content can be increased by treatment with specific inhibitors, growth regulators, or precursors (Krueger and Carew, 1978). In other cases alkaloids are produced in the tissue cultures, but they are not those characteristic of the whole plant (Furuya *et al.*, 1972; Ikuta *et al.*, 1974; Akasu *et al.*, 1976).

V. METABOLISM OF ALKALOIDS IN HERBIVORES

Alkaloids present in plants that are consumed by animals frequently undergo some metabolic transformation before being excreted. As a general rule it appears that the metabolism of alkaloids in animals does not proceed to a complete breakdown yielding carbon dioxide. Rather, a few small modifications of the structure are produced by reactions such as hydroxylation, reduction, hydrolysis, or condensation with other molecules. Most of the interest, and consequently most of the publications in this area, have been directed to pharmacologically useful alkaloids rather than alkaloids that may be encountered randomly by herbivores. Still, it appears to be possible to draw some general principles from the cases at hand (Hathway, 1977).

A. Oxidation

Oxidation is one of the commonest processes acting on compounds ingested by animals. Under this general heading removal of hydrogen atoms can be included as well as reactions that actually add oxygen atoms to the molecule. However, oxygenation is the more usual pathway for alkaloid molecules, and the most important enzymatic system is present in hepatic endoplasmic reticulum. This system depends on a special cytochrome, called P-450, because in its reduced form it produces a carbon monoxide complex absorbing light at 450 nm. The P-450 system catalyzes the oxidation of many compounds, including some alkaloids (Beckett and Sheikh, 1973; Demisch and Seiler, 1975; Murphy, 1973).

In some cases in which only whole-animal experiments have been done, it seems likely that microsomal oxidation is responsible for the products observed—for example, in the formation of *N*-oxides. It is also of interest that pretreatment of rats with certain alkaloids results in altered activity of the P-450 system toward other substrates (Johnson *et al.*, 1971; Weber *et al.*, 1974; Mitoma *et al.*, 1968). Although very little has been learned about the metabolism of alkaloids by insects, some insects that feed on tobacco plants convert nicotine to cotinine and nornicotine, and these conversions could result from activity of a P-450 oxygenase. Since the products are nontoxic to insects, this transformation appears to be advantageous (Brattsten *et al.*, 1977; Self *et al.*, 1964a).

Not all oxidations of alkaloids can be ascribed to microsomal oxygenase systems, and the best-known alternative system is the one formerly called rabbit liver "quinine oxidase". This oxidase system is widespread in mammalian liver and is a flavoprotein active against a wide variety of substrates (Lang and Kever, 1957). The enzyme responsible is in fact identical to liver aldehyde oxidase, and its mechanism of action on heterocyclic nitrogen compounds probably involves addition of hydroxide ion to a suitable ring position followed by enzymatic dehydrogenation (Rajagopalan and Handler, 1964; Stanulović and Chaykin, 1971). The oxidase specifically oxidizes pyridine derivatives at the α and γ positions of the ring, but in the case of quinine it can act only at the α position since the γ position is blocked. In human beings additional oxidation of the quinuclidine ring of quinine occurs, but this may be the result of microsomal oxygenase (Watabe and Kiyonaga, 1972; Carroll *et al.*, 1974). Whole-animal experiments have shown the formation of oxidized products from many other alkaloids, with no evidence as to the enzymes specifically involved. For instance, methylated xanthines are converted partially to methylated uric acid (Rao *et al.*, 1973).

Loss of *N*- and *O*-methyl groups is frequently observed, but it is by no means certain that microsomal oxygenase is always involved (Hawks *et al.*, 1974; Jenne *et al.*, 1975; Fishman *et al.*, 1976). Similarly, *N*-oxides are metabolic products of several alkaloids (e.g., nicotine, arecoline), but it is only a presumption that the P-450 system is responsible (Nery, 1971; Booth and Boyland, 1970). In the metabolism of nicotine several oxidative steps follow the formation of cotinine, but, although the products are known, the mechanism of their formation is not (Dagne and Castagnoli, 1972; Dagne *et al.*, 1974; Harke *et al.*, 1974; Langone *et al.*, 1974; Testa *et al.*, 1976). Mescaline (a primary amine rather than a true alkaloid) is oxidized by rats and rabbits, but not by human beings, first to trimethoxyphenylacetic acid and then to trimethoxybenzoic acid (Musacchio and Goldstein, 1967; Seiler and Demisch, 1974; Demisch and Seiler, 1975; Riceberg, 1975).

B. Hydrolysis

Several alkaloids are esters, and one of the early steps in their metabolism is hydrolysis of the ester bond. Following hydrolysis additional reactions may occur to complete the catabolic sequence.

The tropane alkaloids cocaine and hyoscyamine, on hydrolysis, yield the bases ecgonine and tropine along with the appropriate acids. The bases may be excreted as such or converted to derivatives (Hawks *et al.*, 1974; Bastos *et al.*, 1974). Esterases relatively specific for esters of tropic acid have been characterized in rabbit serum, and it has been suggested that these esterases permit rabbits to subsist on a diet of belladonna (*Atropa belladonna*) leaves (Werner and Brehmer, 1967; Stormont and Suzuki, 1970).

The major alkaloid of betel nut (*Areca catechu*) is arecoline, the methyl ester of arecaidine. The human practice of chewing betel nut with lime catalyzes some hydrolysis of the ester, but this hydrolysis is also catalyzed by liver esterases. Arecaidine has more physiological activity than arecoline and is also metabolized to additional active products

(Boyland and Nery, 1969a; Nery, 1971; Nieschulz and Schmersahl, 1967). The ability of arecaidine to alkylate thiol groups may account for carcinogenic properties of betel nut (Boyland and Nery, 1969b; Nery, 1971).

Most of the pyrrolizidine alkaloids are esters of amino alcohols ("necines") and necic acids. The toxicity of these alkaloids to mammalian liver seems to depend on their initial metabolism. The most toxic members have ester groups in unstable conjugation with pyrrole nitrogen and have this system conserved during their metabolism. Thus, if the ester group is hydrolyzed, the alkaloid is detoxified (Mattocks and White, 1971).

C. Conjugation

Conjugation describes a process in which some foreign molecule becomes joined to a common metabolite before being excreted. The addition of acetyl, sulfate, or glucuronic acid groups to alkaloids are examples of conjugation. Such a conjugation process may follow the oxidative introduction of hydroxyl groups as described in Section V,A or the appearance of hydroxyl groups by ester hydrolysis as described in Section V,B. Some examples of conjugation are given by Belpaire and Bogaert (1975), Ho *et al.* (1973) and Yeh *et al.* (1971).

In a few cases methyl groups are added to alkaloid molecules by animals. Methylation is not usually classed with other examples of conjugation, but it does fit the general definition given above. Examples are methylation of tetrahydroisoquinoline alkaloids in rat liver (Collins *et al.*, 1973) and the excretion of nicotine by human beings and dogs partly as derivatives in which the pyridine ring has been N-methylated (McKennis *et al.*, 1963).

VI. MECHANISMS OF PHYSIOLOGICAL ACTION OF ALKALOIDS

An enormous body of literature is available on the pharmacological effects of alkaloids on animals. Until recently most of these effects were described in terms of overall behavioral or anatomical changes. Understanding of effects at the molecular or biochemical level is advancing rapidly, and even though there have been few studies of alkaloid effects on either insects or higher herbivores it appears that fundamental mechanisms of action should be applicable to virtually all organisms. For example, a complete explanation for the action, on a molecular level, of an anticholinergic alkaloid at the membrane receptor of *Torpedo* electric organ should be useful in understanding how this alkaloid functions in an herbivorous insect possessing cholinergic receptors. Following this argument, the present discussion is focused as much as possible on biochemical processes. For broader effects of alkaloids (at least those of medical interest) the reader is referred to such comprehensive texts of pharmacology as that of Goodman and Gilman (1975).

The usual explanations for the physiological action of small molecules at low concentration are concerned with the following:

1. Mechanisms of DNA replication, RNA transcription, and protein synthesis
2. Membrane transport processes, both active and passive
3. Enzyme inhibition and activation
4. Blocking of receptor sites for endogenous chemical transmitters
5. Effects on conformation of macromolecules not included in 1–4

An important complication in our understanding of how a foreign chemical acts on an organism is the fact that in multicellular organisms many barriers exist before the chemical can reach the site where it exerts its effect. These may be passive barriers to its migration, or the chemical may be actively metabolized or excreted. Another consideration that cannot be ignored is the influence of concentration on activity. Many chemicals have multiple pharmacological effects, with one effect predominating at one concentration but quite a different effect predominating at another.

Because of the various complications outlined above, the transfer of *in vitro* experimental results to a complex living system is fraught with difficulty. An alkaloid at 1 m*M* might inhibit an enzyme *in vitro,* but that inhibition might not have anything to do with the pharmacological action *in vivo,* where 1 μM at the right receptor could be disasterous. Still, it may be useful to understand the exact mechanism of one kind of interaction in order to try to explain the possible mechanism in a different system. A grouping found in many different proteins may be specific for binding an alkaloid, but the rest of the protein and its response to this binding may be required to complete the action (Gero and Capetola, 1975).

The area of physiology that provides the greatest number of examples of specific alkaloid effects is neurotransmission in the autonomic system. In this system there exist cholingergic fibers (whose chemical transmitter is acetylcholine) and adrenergic fibers (whose chemical transmitter is noradrenaline). The transmitter substances act in the transmission of impulses across synapses and in the junctions between

Table II. Selected biochemical effects of alkaloids

Alkaloid	Effect	Reference
Bicuculline	Antagonizes at γ-aminobutyric acid receptor	Möhler and Okada (1978)
Caffeine	Inhibits nucleic acid synthesis; releases Ca from membranes	Lehmann (1972), Zuk and Swietlinska (1973), Ogawa (1970)
Colchicine	Binds to microtubule protein; inhibits mitosis	Hains *et al.* (1978), Twomey *et al.* (1974), Fitzgerald (1976)
Emetine	Inhibits RNA and protein synthesis	Gilead and Becker (1971), Perlman and Penman (1970)
Harmans	Inhibit membrane transport; act on serotonin receptor	Sepulveda and Robinson (1974), Symthies *et al.* (1970)
Lycorine	Inhibits ascorbic acid synthesis	Arrigoni *et al.* (1975, 1976)
Mescaline	Inhibits protein synthesis; stimulates ribosome breakdown; stimulates serotonin synthesis	Datta *et al.* (1971, 1974), Shein *et al.* (1971)
Morphine	Acts on enkephalin receptor, inhibits adenyl cyclase	Goldstein (1976), Simantov and Snyder (1976), Bradbury *et al.* (1976), Sharma *et al.* (1977)
Narciclasine	Inhibits DNA polymerase, protein synthesis	Papas *et al.* (1973), Jimenez (1975)
Pyrrolizidines	Alkylate macromolecules; are carcinogens	Mattocks and White (1971)
Reserpine	Inhibits oxidative phosphorylation, uptake of Ca, serotonin, and dopamine	Maina (1974), Bridges and Baldini (1966), Roth and Stone (1969)
Sanguinarine	Inhibits Na–K-ATPase	Straub and Carver (1975)
Strychnine	Binds to glycine receptor	Snyder *et al.* (1973)
Tomatine	Breaks down membrane compartmentation	Roddick (1975)
Vinca alkaloids	Bind to microtubular protein; inhibit mitosis	McClure and Paulson (1977), Bhattacharyya and Wolff (1977)

the nerves and the tissues that they control. Drugs can interfere with nerve transmission by a variety of mechanisms. Cocaine and related synthetic compounds inhibit electrical conductance by competing with Ca^{2+} for sites on the membrane (Goldman and Blaustein, 1966). *Veratrum* alkaloids also affect electrical transmission perhaps by increasing Na^{+} conductivity and delaying repolarization. Alkaloids from *Rauwolfia*, especially reserpine, cause a slow release of noradrenaline, so that it is depleted and adrenergic responses are diminished. Several well-known alkaloids such as nicotine, pilocarpine, and arecoline mimic acetylcholine but show slightly different effects at different acetylcholine receptors. Ephedrine is both a releaser and a mimic of noradrenaline at receptor sites. Curare, *Erythrina* alkaloids, and (±)-hyoscyamine block acetylcholine receptors, whereas yohimbine blocks the α adrenergic receptor. Inhibitors of acetylcholinesterase are very well known as synthetic "nerve gases," but several alkaloids also have activity of this kind, the most active being physostigmine. Noradrenaline is removed from action more by reuptake than by enzymatic destruction, but the enzyme monoamine oxidase may be important for destroying other amines that can mimic noradrenaline action. Harmaline and releated alkaloids inhibit monoamine oxidase.

This brief outline will serve to point out the possible impingement of many alkaloids on neural processes. A more complete discussion of this area can be found in Robinson (1968), Goodman and Gilman (1975), and Triggle and Triggle (1976). A study especially relevant to plant–herbivore relationships is that of Orgell (1963), who showed inhibition of human serum cholinesterase by several plant extracts and many pure alkaloids. The range of activities covered four orders of magnitude, and the alkaloid next to physostigmine in activity (sempervirine) had only 1% the activity of physostigmine. Several steroidal alkaloids were still a little less active.

From the many other kinds of biochemical effects of alkaloids, several have been selected for inclusion in Table II. These listings suggest the range of possibilities but should never be used as certain explanations for gross effects of any particular alkaloid.

VII. ALKALOIDS IN ECOSYSTEMS

The title of this section promises more than it can deliver because there are very few well-documented cases of a convincing role for alkaloids in ecosystems. There is no denying that in a broad sense alkaloids are physiologically active toward many organisms, and one would expect them to account in part for toxic or repellent properties of plants containing these compounds. The actual literature of the subject, however, contains much more conjecture than experimentally verified statements. What does come through to this author is that sweeping generalizations about the protective function of alkaloids are groundless. However, in certain very specific cases the presence of an alkaloid can make a difference in the relationship between a plant and a particular enemy at a particular time. The possible protective role of alkaloids with regard to attack by microorganisms and a possible allelopathic role for alkaloids are excluded from consideration in this presentation.

It is important to distinguish between repellent and toxic plant constituents. As shown by feeding preference studies with a polyphagous grasshopper (*Melanoplus bivittatus*), some lethal diets may be accepted (Harley and Thorsteinson, 1967), and compounds responsible for this toxicity could have a protective role only by reducing the population of predators, not by deterring individual predators. Among the alkaloids tested in this work solanine and tomatine were most toxic, but they were acceptable to the grasshopper. Lupinine was less toxic but was also deterrent. Hyoscyamine, hordenine, lobeline, veratrine, and gramine had low toxicity but deterred feeding. Nornicotine dipicrate was highly toxic and repellent, but it is impossible to ascribe either effect to the alkaloid because picric acid cannot be assumed to be inert. A useful appendix to this publication lists plants containing the tested compounds and growing in the range of the insect used for testing.

Another screening type of study on the toxicity of plant constituents (Janzen *et al.*, 1977) used larvae of the beetle *Callosbruchus maculatus* and showed that, of all the compounds tested, alkaloids constituted the most consistently toxic class. Correlations were drawn between alkaloid content and host specificity, thus implying a further (negative) correlation between toxicity and choice.

The steroid alkaloids have apparently received more attention than any other group with regard to having a role in insect–plant interactions. Thirty years ago it was reported that the presence of demissin prevents attack of potato beetle larvae on *Solanum demissum* (Kuhn and Loew, 1947). Fourteen years later the same workers showed that the strongest repellents in this group of compounds were the leptines, glycosides of acetylsolanidine (Kuhn and Loew, 1961). Later work on this relationship has been reviewed by Roddick (1974), and it also makes up a major part of more general reviews on the role of alkaloids in insect–plant interrelations (Levin, 1976; Levinson, 1976). It is widely believed that glycosidic steroid alkaloids are both toxic and repellent to many insects, but in fact the effect appears to be strongly specific. α-Tomatine is highly repellent toward the Colorado beetle (*Leptinotarsa decemlineata*) and the potato leafhopper (*Empoasca fabae*), but it is ineffective with beetles of the genus *Epilachna* and, as noted above, with a polyphagous grasshopper. Tomatine is, of course, the characteristic alkaloid not of potato plants but of tomato plants, and neurophysiological measurements have shown that the receptor in potato beetles is 10 times less sensitive to solanine than to tomatine—an apparent adaptation to their host plant. Thus, rather than having a general repellency, steroid alkaloids may by their special distribution determine the selection of host plant by a particular insect (Levinson, 1976). Steroidal alkaloids are also toxic to higher animals (Roddick, 1974; Jadhav and Salunkhe, 1976), but their contribution to natural relationships seems not to have been investigated. The biochemical explanation for their toxicity is not certain since *in vitro* they show multiple effects (Roddick, 1974).

The danger of making blanket generalizations about the protective value of alkaloids can be illustrated by some examples involving lupine alkaloids. Twenty years ago it was stated by a leading specialist in this area that these compounds have no protective value against microorganisms, insects, or mollusks (Nowacki, 1958b). A few years later it was demonstrated that the presence of sparteine was actually a feeding stimulus to aphids on broom (*Sarothamnus scoparius*) and that these insects could be induced to feed on abnormal hosts that had been painted with sparteine (Smith, 1966). Nevertheless, the more recent work of Dolinger *et al.* (1973) showed that when a specific situation is examined carefully these alkaloids can be found to provide a chemical defense mechanism against an insect in a very subtle way. Observations by these workers on several populations of lupines suggested that the most effective chemical defense against the larvae of a flower-feeding butterfly (*Glaucopsyche lygdanus*) was achieved in a population that had a small number of different alkaloids in relatively high concentration and, most important, had high individual variability. The variability is thought to reduce the possibility of insects developing an efficient protective mechanism, which they could do more easily against a high but uniform level of alkaloid. Thus, plants that have been bred artificially for high alkaloid content are likely to have lost this variability and are therefore susceptible to insect attack. The importance of individual variation stressed in this chapter can perhaps be extended to include temporal variation within an individual. That is, it may be more difficult for a predator to adapt to a host if the host contains toxic substances that change drastically in concentration or qualitative composition either through its life cycle or diurnally. The existence of such temporal variation has been documented in Section III,B. The lupine alkaloids also offer a case of protection against higher animals, for it has been observed (Arnold and Hill, 1972) that sheep avoid grazing varieties of lupine that have a high alkaloid content, but they graze readily on the "sweet" varieties that contain little alkaloid.

There are several studies showing that the palatability of reed canarygrass (*Phalaris arundinacea*) to sheep and meadow voles is negatively correlated with total alkaloid content, rather than with any particular alkaloid (Simons and Marten, 1971; Marten *et al.*, 1973; Kendall and Sherwood, 1975).

The case of insects that feed on tobacco plants is interesting because of the well-known toxicity of nicotine to insects. Three distinct mechanisms for avoiding nicotine toxicity have been distinguished (Self *et al.*, 1964a,b; Yang and Guthrie, 1969):

1. Selective feeding on phloem, which does not contain nicotine
2. A barrier to absorption from the digestive tract or to penetration into the nervous system
3. Enzymatic detoxification (conversion to cotinine)

Two other cases of plants that contain alkaloids toxic to most insects but that are inhibited or attacked by single species are the senita cactus (*Lophocereus schotti*) and *Cocculus trilobus*. A species of fruit fly (*Drosophila pachea*) breeds in rotting stems of the cactus and obtains a nutritionally essential steroid from the plant (Kircher *et al.*, 1967). The latter plant is attacked by two species of moth (*Oraesia excavata* and *O. emarginata*) in spite of the erythrinan-type alkaloid that it contains (Wada and Munakata, 1967). It will be interesting to learn in these two cases what special mechanisms these insects have developed to enable them to deal with their toxic situations.

Finally, one of the strangest examples of a plant–insect relationship involving alkaloids is provided by insects that utilize alkaloids of their host plants as pheromones, or defensive substances. Plants containing dehydropyrrolizidine alkaloids are utilized by certain danaid butterflies as a source of raw materials for the synthesis of their courtship pheromones (Meinwald *et al.*, 1974; Edgar and Culvenor, 1975). The structure of one of these pheromones is

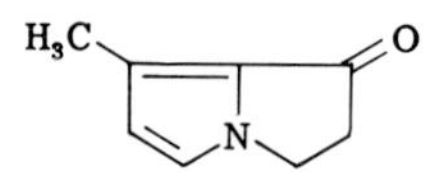

and its relationship to the pyrrolizidine alkaloids is apparent. The cinnabar moth (*Callimorpha jacobaeae*) also feeds on plants that contain pyrrolizidine alkaloids, and, although the moths contain a repellent factor, it is not certain that it is related to the alkaloids in their diet (Aplin *et al.*, 1968).

REFERENCES

Agurell, S., and Lundström J. (1968). *J. Chem. Soc. D* pp. 1638–1639.
Akasu, M., Itokawa, H., and Fujita, M. (1976). *Phytochemistry* **15,** 471–473.
Antoun, M. D., and Roberts, M. F. (1975). *Planta Med.* **28,** 6–11.
Aplin, R. T., Benn, M. H., and Rothschild, M. (1968). *Nature (London)* **219,** 747–748.
Arnold, G., and Hill, J. (1972). *In* "Phytochemical Ecology" (J. Harborne, ed.), pp. 17–101. Academic Press, New York.

Arrigoni, O., Liso, R. A., and Calabrese, G. (1975). *Nature (London)* **256,** 513–514.
Arrigoni, O., Arrigoni-Liso, R., and Calabrese, G. (1976). *Science* **194,** 332–333.
Barton, D. H. R., and Cohen, T. (1957). *Festschr. Prof. Dr. Arthur Stoll Siebzigsten Geburtstag* pp. 117–143.
Barz, W., Reinhard, E., and Zenk, M. H., eds., (1977). "Plant Tissue Culture and its Bio-Technological Application," Springer-Verlag, Berlin.
Basmadjian, G. P., and Paul, A. G. (1971). *Lloydia* **34,** 91–93.
Bastos, M. L., Jukofsky, D., and Mulé, S. J. (1974). *J. Chromatogr* **89,** 335–345.
Battersby, A. R., Jones, R. C. F., and Kazlauskas, R. (1975). *Tetrahedron Lett.* pp. 1873–1876.
Beckett, A. H., and Sheikh, A. H. (1973). *J. Pharm. Pharmacol.* **25,** 171P.
Belpaire, F. M., and Bogaert, M. G. (1973). *Biochem. Pharmacol.* **22,** 59–66.
Berggren, A., Björling, C. O., and Willmann-Johnson, B. (1958). *Acta Chem. Scand.* **12,** 1521–1527.
Bhattacharyya, B. and Wolff, J. (1977). *FEBS Letters* **75,** 159–162.
Blaschke-Cobet, M., and Luckner, M. (1973). *Phytochemistry* **12,** 2393–2398.
Böhm, H. (1965). *Planta Med.* **13,** 234–240.
Böhm, H. (1971). *Biochem. Physiol. Pflanz.* **162,** 474–477.
Booth, J., and Boyland, E. (1970). *Biochem. Pharmacol.* **19,** 733–742.
Boyland, E., and Nery, R. (1969a). *Biochem. J.* **112,** 33p.
Boyland, E., and Nery, R. (1969b). *Biochem. J.* **113,** 123–130.
Bradbury, A. F., Smyth, D. G., and Snell, C. R. (1976). *Nature (London)* **260,** 165–166.
Brattsten, L. B., Wilkinson, C. F., and Eisner, T. (1977). *Science* **196,** 1349–1352.
Bräuniger, H. (1954). *Pharmazie* **9,** 643–654.
Bregoff, H. M., Roberts, E., and Delwiche, C. C. (1953). *J. Biol. Chem.* **205,** 565–574.
Bridges, J. M., and Baldini, M. (1966). *Nature (London)* **210,** 1364–1365.
Brossi, A., Ramel, A., O'Brien, J., and Teitel, S. (1973). *Chem. Pharm. Bull.* **21,** 1839–1840.
Bull, L. B., Culvenor, C. C. J., and Dick, A. T. (1968). "The Pyrrolizidine Alkaloids." North-Holland Publ., Amsterdam.
Carroll, F. I., Smith, D., Wall, M. E., and Moreland, C. G. (1974). *J. Med. Chem.* **17,** 985–987.
Chambers, C., and Stuart, K. L. (1968). *Chem. Commun.* pp. 328–329.
Christianson, D. D., Wall, J. S., Dimler, R. J., and Senti, F. R. (1960). *Anal. Chem.* **32,** 874.
Collins, A. C., Cashaw, J. L., and Davis, V. E. (1973). *Biochem. Pharmacol.* **22,** 2337–2348.
Cosson, L. (1969). *Phytochemistry* **8,** 2227–2233.
Cosson, L. (1975). *Proc. Int. Bot. Congr., 12th, 1975 Abstract* 353.
Cromwell, B. T. (1955). *In* "Modern Methods of Plant Analysis" (K. Paech and M. V. Tracey, eds.), Vol. IV, pp. 367–516. Springer-Verlag, Berlin and New York.
Csoban, G., and Hegedüs, I. (1954). *Magy. Kem. Foly.* **60,** 121–122. *Chem. Abstr.* **52,** 6718 (1958).
Dagne, E., and Castagnoli, N., Jr. (1972). *J. Med. Chem.* **15,** 356–360.
Dagne, E., Gruenke, L., and Castagnoli, N., Jr. (1974). *J. Med. Chem.* **17,** 1330–1333.
Dann, A. T. (1960). *Nature (London)* **186,** 1051.
Datta, R. K., Antopol, W., and Ghosh, J. J. (1971). *Biochem. J.* **125,** 213–219.
Datta, R. K., Ghosh, J. J., and Antopol, W. (1974). *Biochem. Pharmacol.* **23,** 1687–1692.
Dawson, R. F. (1960). *Am. Sci.* **48,** 321–340.
Demisch, L., and Seiler, N. (1975). *Biochem. Pharmacol.* **24,** 575–580.
Dickenson, P. B., and Fairbairn, J. W. (1975). *Ann. Bot. (London)* [N.S.] **39,** 707–712.
Dolinger, P. M., Ehrlich, P. R., Fitch, W. L., and Breedlove, D. E. (1973). *Oecologia* **13,** 191–204.

Dyar, J. (1975). *Plant Physiol.* **56,** Suppl., 6.
Edgar, J. A., and Culvenor, C. C. J. (1975). *Experientia* **31,** 393–394.
Fairbairn, J. W., and Ali, A. A. E. R. (1968). *Phytochemistry* **7,** 1593–1597.
Fairbairn, J. W., and El-Masry, S. (1967). *Phytochemistry* **6,** 499–504.
Fairbairn, J. W., and El-Masry, S. (1968). *Phytochemistry* **7,** 181–187.
Fairbairn, J. W., Hakim, F., and ElKheir, Y. (1974). *Phytochemistry* **13,** 1133–1139.
Farnsworth, N. R., and Euler, K. L. (1962). *Lloydia* **25,** 186–195.
Farnsworth, N. R., Pilewski, N. A., and Draus, F. J. (1962). *Lloydia* **25,** 312–319.
Fishman, J., Hahn, E. F., and Norton, B. I. (1976). *Nature (London)* **261,** 64–65.
Fitzgerald, T. J. (1976). *Biochem. Pharmacol.* **25,** 1383–1387.
Fulton, C. C. (1939). *Am. J. Pharm.* **111,** 184–192; *Chem. Abstr.* **33,** 7486 (1939).
Furuya, T., Nakano, M., and Yoshikawa, T. (1978). *Phytochemistry* **17,** 891–893.
Furuya, T., Ikuta, A., and Syono, K. (1972). *Phytochemistry* **11,** 3041–3044.
Garnier, J., and Morel, G. (1972). *Physiol. Veg.* **10,** 617–625.
Gero, A., and Capetola, R. J. (1975). *Life Sci.* **16,** 1821–1822.
Gilead, Z., and Becker, Y. (1971). *Eur. J. Biochem.* **23,** 143–149.
Glasby, J. S. (1975). "Encyclopedia of the Alkaloids," 2 vols. Plenum, New York.
Goldman, D. E., and Blaustein, M. P. (1966). *Ann. N.Y. Acad. Sci.* **137,** 967–981.
Goldstein, A. (1976). *Science* **193,** 1081–1086.
Goodman, L. S., and Gilman, A., eds. (1975). "The Pharmacological Basis of Therapeutics," 5th ed. Macmillan, New York.
Gornyi, G. Ya. (1955). *Ukr. Khim. Zh.* **21,** 646–654; *Chem. Abstr.* **50,** 7396) (1956).
Groeger, D. (1963). *Planta Med.* **11,** 444–449; *Chem. Abstr.* **61,** 9784 (1964).
Gunner, S. W., and Hand, T. B. (1968). *J. Chromatogr.* **37,** 357–358.
Hains, F. O., Dickerson, R. M., Wilson, L., and Owellen, R. J. (1978). *Biochem. Pharmacol.* **27,** 71–76.
Harke, H.-P., Chevalier, H.-J., and Frahm, B. (1974). *Experientia* **30,** 883–884.
Harley, K. L. S., and Thorsteinson, A. J. (1967). *Can. J. Zool.* **45,** 305–319.
Hathway, D. E., ed. (1977) "Foreign Compound Metabolism in Mammals," **4,** Chem. Soc., London.
Hawks, R. L., Kopin, I. J., Colburn, R. W., and Thoa, N. B. (1974). *Life Sci.* **15,** 2189–2195.
Hegnauer, R. (1962–1973). "Chemotaxonomie der Pflanzen," 6 vols. Birkhaeuser, Basel.
Hegnauer, R. (1963). *In* "Chemical Plant Taxonomy" (T. Swain, ed.), pp. 389–427. Academic Press, New York.
Hegnauer, R. (1966). *In* "Comparative Phytochemistry" (T. Swain, ed.), pp. 211–230. Academic Press, New York.
Heslop-Harrison, J. (1963). *In* "Chemical Plant Taxonomy" (T. Swain, ed.), pp. 17–40. Academic Press, New York.
Ho, B. T., Pong, S. F., Browne, R. G., and Walker, K. E. (1973). *Experientia* **29,** 275–277.
Holubek, J., and Štrouf, O. (1965–1973). "Spectral Data and Physical Constants of Alkaloids," 8 vols. Heydon, London.
Hruban, L., and Šantavý, F. (1967). *Collect. Czech. Chem. Commun.* **32,** 3414–3426.
Hruban, L., Šantavý, F., and Hegerová, S. (1970). *Collec. Czech. Chem. Commun.* **35,** 3420–3444.
Hultin, E. (1966). *Acta Chem. Scand.* **20,** 1588–1592.
Hultin, E., and Torssell, K. (1965). *Phytochemistry* **4,** 425–433.
Ikuta, A., Syono, K., and Furuya, T. (1974). *Phytochemistry* **13,** 2175–2179.
Il'in, G. S., and Lovkova, M. Y. (1966). *Biokhimiya* **31,** 174–181.
Jadhav, S. J., and Salunkhe, D. K. (1976). *Adv. Food Res.* **21,** 308–354.
James, W. O. (1950). Alkaloids (*N.Y.*) **1,** 15–90.
James, W. O. (1953). *J. Pharm. Pharmacol.* **5,** 809–822.
Janzen, D. H., Juster, H. B., and Bell, E. A. (1977). *Phytochemistry* **16,** 223–227.

Jenne, J., Nagasawa, H., McHugh, R., MacDonald, F., and Wyse, E. (1975). *Life Sci.* **17,** 195–198.
Jimenez, J., Sanchez, L., and Vazquez, D. (1975). *FEBS Lett.* **55,** 53–56.
Jindra, A., Kovács, P., Pittnerová, Z., and Psenák, M. (1966). *Phytochemistry* **5,** 1303–1315.
Johnson, R. L., Mazel, P., Donohue, J. D., and Jondorf, W. R. (1971). *Biochem. Pharmacol.* **20,** 955–966.
Jordan, W., and Scheuer, P. J. (1965). *J. Chromatogr.* **19,** 175–176.
Kendall, W. A. and Sherwood, R. T. (1975). *Agron. J.* **67,** 667–671.
Kennedy, G. S. (1971). *Phytochemistry* **10,** 1335–1337.
Khanna, K. L., Takido, M., Rosenberg, H., and Paul, A. G. (1970). *Phytochemistry* **9,** 1811–1815.
Kirby, G. W., Massey, S. R., and Steinreich, P. (1972). *J. Chem. Soc., Perkin Trans. 1* pp. 1642–1647.
Kircher, H. W., Heed, W. B., Russell, J. S., and Grove, J. (1967). *J. Insect Physiol.* **13,** 1869–1874.
Kirkwood, S., and Marion, L. (1950). *J. Am. Chem. Soc.* **72,** 2522–2524.
Knowlton, M., Dohan, F. C., and Sprince, H. (1960). *Anal. Chem.* **32,** 666–668.
Koblitz, H., Schumann, U., Böhm, H., and Franke, J. (1975). *Experientia* **31,** 768–769.
Kraft, D. (1953). *Pharmazie* **8,** 170.
Krueger, R. J. and Carew, D. P. (1978). *Lloydia* **41,** 327–331.
Kuhn, R., and Loew, I. (1947). *Ber.* **80,** 406–410.
Kuhn, R., and Löw, I. (1961). *Chem. Ber.* **94,** 1088–1095.
Lang, K., and Keuer, H. (1957). *Biochem. Z.* **329,** 277–282.
Lang, E., and Lang, H. (1974). *Naturwissenschaften* **61,** 504–505.
Langone, J. J., Franke, J., and Van Vanakis, H. (1974). *Arch. Biochem. Biophys.* **164,** 536–543.
Lee, H. J., and Waller, G. R. (1972). *Phytochemistry* **11,** 965–973.
Leete, E. (1965). *Science* **147,** 1000–1006.
Leete, E., Bodem, G. B., and Manuel, M. F. (1971). *Phytochemistry* **10,** 2687–2692.
Lehmann, A. R. (1972). *Biophys. J.* **12,** 1316–1325.
Leistner, E., and Spenser, I. D. (1973). *J. Am. Chem. Soc.* **95,** 4715–4725.
Levin, D. A. (1976). *Am. Nat.* **110,** 261–284.
Levin, D. and York, B. M., Jr. (1978). *Biochem. Systematics Ecol.* **6,** 61–76.
Levinson, H. Z. (1976). *Experientia* **32,** 408–413.
Li, H.-L., and Willaman, J. J. (1972). *Econ. Bot.* **26,** 61–67.
List, P. H. (1954). *Naturwissenschaften* **41,** 454.
Lovkova, M., and Obroucheva, N. V. (1975). *Proc. Int. Botan. Congr., 12th, 1975* p. 364.
Lovkova, M. Y., Ilyin, G. S., Györffy, B., and Minozhedinova, N. S. (1973). *Izv. Akad. Nauk SSSR, Ser. Biol.* pp. 576–581.
Lundström, J., and Agurell, S. (1971). *Acta Pharm. Suec.* **8,** 261–274.
Mabry, T. J., and Mears, J. A. (1970). *In* "Chemistry of the Alkaloids" (S. W. Pelletier, ed.), pp. 719–746. Van Nostrand-Reinhold, New York.
McCaldin, D. J. (1960). *Can. J. Chem.* **38,** 1229–1231.
McClure, W. O. and Paulson, J. C. (1977). *Molec. Pharmacol.* **13,** 560–575.
McKennis, H., Turnbull, L. B., and Bowman, E. R. (1963). *J. Biol. Chem.* **238,** 719–723.
McNair, J. B. (1931). *Am. J. Bot.* **18,** 416–423.
McNair, J. B. (1935a). *Bull. Torrey Bot. Club* **62,** 219–226.
McNair, J. B. (1935b). *Bull. Torrey Bot. Club* **62,** 515–532.
McNair, J. B. (1943). *Lloydia* **6,** 1–17.
Maina, G. (1974). *Biochim. Biophys. Acta* **333,** 481–486.
Mann, P. J. G., and Smithies, W. R. (1955). *Biochem. J.* **61,** 89–100.
Manske, R. H. F., and Holmes, H. L., eds. (1951–1968). "The Alkaloids," 10 vols. Academic Press, New York.

Marten, G. C., Barnes, R. F., Simons, A. B., and Wooding, F. J. (1973). *Agron. J.* **65,** 199–201.
Matile, P., Jans, B., and Rickenbacher, R. (1970). *Biochem. Physiol. Pflanz.* **161,** 447–458.
Mattocks, A. R. (1967). *J. Chromatogr.* **27,** 505–508.
Mattocks, A. R., and White, I. N. H. (1971). *Nature (London), New Biol.* **231,** 114–115.
Mechler, E., and Kohlenbach, H. W. (1978). *Planta Medica* **33,** 294.
Meinwald, J., Boriack, C. J., Schneider, D., Boppré, M., Wood, W. F., and Eisner, T. (1974). *Experientia* **30,** 721–723.
Michels-Nyomarkay, K. (1970). *Herba Hung.* **9,** 43–49, *Chem. Abstr.* **75,** 31375d (1971).
Mitoma, C., Sorich, T. J., II, and Neubauer, S. E. (1968). *Life Sci.* **7,** 145–151.
Mitra, G. C. (1972). *Indian J. Exp. Biol.* **10,** 217–218.
Mizusaki, S., Tanabe, Y., Noguchi, M., and Tamaki, E. (1972). *Phytochemistry* **11,** 2757–2762.
Mockaitis, J. M., Kivilaan, A., and Schulze, A. (1973). *Biochem. Physiol. Pflanz.* **164,** 248–257.
Möhler, H. and Okada, T. (1978). *Molec. Pharmacol.* **14,** 256–265.
Mothes, K. (1960). *Alkaloids (N.Y.)* **6,** 1–29.
Mothes, K. (1965). *Naturwissenschaften* **52,** 571–585.
Mothes, K. (1966). *Naturwissenschaften* **53,** 317–323.
Mothes, K. (1969). *Experientia* **25,** 225–239.
Mothes, K., and Schütte, H. R. (1969). "Biosynthese der Alkaloide." VEB Dtsch. Verlag Wiss., Berlin.
Mothes, K., Richter, I., Stolle, K., and Gröger, D. (1965). *Naturwissenschaften* **52,** 431.
Moursi, M. A., and Ahmed, S. S. (1973). *Pharmazie* **28,** 58–61 and 62–64.
Müller, E., Nelles, A., and Neumann, D. (1971). *Biochem. Physiol. Pflanz.* **162,** 272–294.
Murphy, P. J. (1973). *J. Biol. Chem.* **248,** 2796–2800.
Musacchio, J. M., and Goldstein, M. (1967). *Biochem. Pharmacol.* **16,** 963–970.
Nery, R. (1971). *Biochem. J.* **122,** 503–508.
Neumann, D., and Müller, E. (1971). *Biochem. Physiol. Pflanz.* **162,** 503–513.
Neumann, D., and Müller, E. (1972). *Biochem. Physiol. Pflanz.* **163,** 375–391.
Neumann, D., and Müller, E. (1974). *Biochem. Physiol. Pflanz.* **165,** 271–282.
Nieschulz, O., and Schmersahl, P. (1967). *Naturwissenschaften* **54,** 21.
Nowacki, E. (1958a). *Bull. Acad. Pol. Sci., Ser. Sci. Biol.* **6,** 11; *Chem. Abstr.* **53,** 22260 (1959).
Nowacki, E. (1958b). *Rocz. Nauk Roln., Ser. A* **79,** 33–42.
Nowacki, E. (1964). *Genetica Polon.* **4,** 161–202.
Nowacki, E. K., and Waller, G. R. (1975). *Phytochemistry* **14,** 155–159.
Nyman, U., and Hall, O. (1974). *Hereditas* **76,** 49–54.
Ogawa, Y. (1970). *J. Biochem. (Tokyo)* **67,** 667–683.
Orgell, W. H. (1963). *Lloydia* **26,** 36–43 and 59–66.
Pan, S. C., and Wagman, G. H. (1959). *J. Chromatogr.* **2,** 428.
Panvisavas, R., Worthen, L. R., and Bohm, B. A. (1968). *Lloydia* **31,** 63–69.
Papas, T. S., Sandhaus, L., Chirigos, M. A., and Furusawa, E. (1973). *Biochem. Biophys. Res. Commun.* **52,** 88–92.
Paris, R. R., and Girre, R.-L. (1969). *C. R. Hebd. Seances Acad. Sci.* **268,** 62–64.
Paul, A. G. (1973). *Lloydia* **36,** 36–45.
Paul, J., and Conine, F. (1973). *Microchem. J.* **18,** 142–145.
Pelletier, S. W. (1970). "Chemistry of the Alkaloids." Van Nostrand-Reinhold, New York.
Perlman, S., and Penman, S. (1970). *Biochem. Biophys. Res. Commun.* **40,** 941–948.
Raffauf, R. F. (1962). *Econ. Bot.* **16,** 171–172.
Raffauf, R. F. (1970). "A Handbook of Alkaloids and Alkaloid-Containing Plants." Wiley
Ragazzi, E. and Veronese, G. (1965). *Mikrochim. Ichnoanal. Acta,* 966–975.
Rajagopalan, K. V. and Handler, P. (1964). *J. Biol. Chem.* **239,** 2022–2026.

Rao, G. S., Khanna, K. L., and Cornish, H. H. (1973). *Experientia*, **29,** 953–955.
Riceberg, L. J., Simon, M., Van Vunakis, H., and Abeles, R. H. (1975). *Biochem. Pharmacol.* **24,** 119–125.
Roberts, M. F. (1973). *J. Pharm. Pharmacol.* **25,** 115P.
Roberts, M. F. (1974). *Phytochemistry* **13,** 119–123.
Robinson, R. (1917). *J. Chem. Soc.* **11,** 876–899.
Robinson, R. (1955). "The Structural Relations of Natural Products," Oxford University Press, Oxford.
Robinson, T. (1968). "The Biochemistry of Alkaloids," Springer-Verlag, Berlin.
Robinson, T. and Fowell, E. (1959). *Nature* **183,** 833–834.
Robles, M. A. (1954). *Pharm. Weekblad* **94,** 178–179. (C. A. **53** 12588)
Roddick, J. G. (1974). *Phytochemistry* **13,** 9–25.
Roddick, J. G. (1975). *J. Exper. Botany* **26,** 221–227.
Roddick, J. G. (1976). *Phytochemistry* **15,** 475–477.
Roddick, J. G. (1977). *Phytochemistry* **16,** 805–807.
Romeo, J. T. and Bell, E. A. *Lloydia* **37,** 543–580.
Rönsch, H., Schreiber, K., and Stubbe, H. (1968). *Naturwiss.* **55,** 182.
Rosenberg, H., and Stohs, S. J. (1974). *Phytochemistry* **13,** 1861–1863.
Rosenberg, H., and Stohs, S. J. (1976). *Phytochemistry* **15,** 501–503.
Roth, R. H., and Stone, E. A. (1969). *Biochem. Pharmacol.* **17,** 1581–1590.
Rowson, J. M. (1954). *J. Pharm. Belg.* **5–6,** 195–221.
Rücker, G., and Taha, A. (1977). *J. Chromatogr.* **132,** 165–167.
Rueppel, M. L., and Rapoport, H. (1971). *J. Am. Chem. Soc.* **93,** 7021–7028.
Sander, H. (1958). *Planta* **52,** 447–466.
Sangster, A. W. (1960). *J. Chem. Educ.* **37,** 454–459.
Sangster, A. W., and Stuart, K. L. (1965). *Chem. Rev.* **65,** 69–130.
Schenck, G., Froemming, K. H., and Schneller, H. G. (1965). *Arch. Pharm. (Weinheim, Ger.)* **298,** 855–860.
Schilling, J., and Zobel, M. (1966). *Pharmazie* **21,** 103–105.
Schneider, E. A., and Wightman, F. (1974). *Can. J. Biochem.* **52,** 698–705.
Schneider, G., and Kleinert, W. (1972). *Planta Med.* **22,** 109–116.
Seiler, N., and Demisch, L. (1974). *Biochem. Pharmacol.* **23,** 259–271.
Self, L. S., Guthrie, F. E., and Hodgson, E. (1964a). *Nature (London)* **204,** 300–301.
Self, L. S., Guthrie, F. E., and Hodgson, E. (1964b). *J. Insect Physiol.* **10,** 907–914.
Sepulveda, F. V., and Robinson, J. W. L. (1974). *Biochim. Biophys. Acta* **373,** 527–531.
Sharma, S. K., Klee, W. A., and Nirenberg, M. W. (1977). *Proc. Nat. Acad. Sci. U.S.A.* **74,** 3365–3369.
Shein, H. M., Wilson, S., Larin, F., and Wurtman, R. J. (1971). *Life Sci.* **10,** Part 2, 273–282.
Simantov, R., and Snyder, S. H. (1976). *Life Sci.* **18,** 781–788.
Simons, A. B., and Marten, G. C. (1971). *Agron. J.* **63,** 915–919.
Sinden, S. L., Goth, R. W., and O'Brien, M. J. (1973). *Phytopathology* **63,** 303–307.
Smith, B. D. (1966). *Nature (London)* **212,** 213–214.
Smith, T. A. (1975). *Phytochemistry* **14,** 865–890.
Smolenski, S. J., Silinis, H., and Farnsworth, N. R. (1972). *Lloydia* **35,** 1–34.
Smythies, J. R., Benington, F., and Morin, R. D. (1970). *Int. Rev. Neurobiol.* **12,** 207–233.
Snyder, S. H., Young, A. B., Bennett, J. P., and Mulder, A. H. (1973). *Fed. Proc., Fed. Am. Soc. Exp. Biol.* **32,** 2039–2047.
Spenser, I. D. (1970). *In* "Chemistry of the Alkaloids" (S. W. Pelletier, ed.), pp. 669–718. Van Nostrand-Reinhold, New York.
Staba, E. J., and Laursen, P. (1966). *J. Pharm. Sci.* **55,** 1099–1101.
Stanulović, M., and Chaykin, S. (1971). *Arch. Biochem. Biophys.* **145,** 27–34.
Stark, J. B. (1962). *Anal. Biochem.* **4,** 103–109.

Stormont, C., and Suzuki, Y. (1970). *Science* **167,** 200–202.

Straub, K. D., and Carver, P. (1975). *Biochem. Biophys. Res. Commun.* **62,** 913–922.

Szász, G., and Buda, L. (1971). *Z. Anal. Chem.* **253,** 361–363.

Szklarek, B. D., and Mazan, A. (1960). *Bull. Acad. Pol. Sci., Ser. Sci. Biol.* **8,** 167–173; *Chem. Abstr.* **54,** 19866 (1960).

Tabata, M. and Hiraoka, N. (1976). *Physiol. Plantarum* **38,** 19–23.

Tallent, W. H., Stromberg, V. L., and Horning, E. C. (1955). *J. Am. Chem. Soc.* **77,** 6361–6364.

Testa, B., Jenner, P., Beckett, A. H., and Gorrod, J. W. (1976). *Xenobiotica* **6,** 553–556.

Tétényi, P. (1970). "Infraspecific Chemical Taxa of Medicinal Plants." Chem. Publ. Co., New York.

Thies, H., and Reuther, F. W. (1954). *Naturwissenschaften* **41,** 230–231.

Tinker, R. B., and Lauter, W. M. (1956). *Econ. Bot.* **10,** 254–257.

Toneby, M. I. (1974). *J. Chromatogr.* **97,** 47–55.

Treibs, W. (1955). *Perfum. Essent. Oil Rec.* **46,** 222–225.

Triggle, D. J., and Triggle, C. R. (1976). "Chemical Pharmacology of the Synapse." Academic Press, New York.

Tso, T. C., Kasperbauer, M. J., and Sorokin, T. P. (1970). *Plant Physiol.* **45,** 330–333.

Twomey, S. L., Raeburn, S., and Baxter, C. F. (1974). *Brain Res.* **66,** 509–518.

Vágújfalvi, D. (1960). *Planta Med.* **8,** 34–43; *Chem. Abstr.* **54,** 15836 (1960).

Veliky, I. A., and Barber, K. M. (1975). *Lloydia* **38,** 125–130.

Verzár-Petri, G. (1966). *Pharmazie* **21,** 48–54.

Wada, K., and Munakata, K. (1967). *Agric. Biol. Chem.* **31,** 336–339.

Waldi, D., Schnackerz, K., and Munter, F. (1961). *J. Chromatogr.* **6,** 61–73.

Waller, G. R., and Nowacki, E. (1978). "Alkaloid Biology and Metabolism in Plants," Plenum, New York.

Waller, G. R., Tang, M. S., Scott, M. R., Goldberg, F. J., Mayes, J. S., and Auda, H. (1965). *Plant Physiol.* **40,** 803–807.

Wanner, H., and Kalberer, P. (1966). *Abh. Dtsch. Akad. Wiss. Berlin, Kl. Chem., Geol. Biol.* pp. 607–608.

Watabe, T., and Kiyonaga, K. (1972). *J. Pharm. Pharmacol.* **24,** 625–630.

Weber, J. M., and Ma, T. S. (1976). *Mikrochim. Acta* **1976**(I) 217–225.

Weber, R. P., Coon, J. M., and Triolo, A. J. (1974). *Science* **184,** 1081–1083.

Weeks, W. W., and Bush, L. P. (1974). *Plant Physiol.* **53,** 73–75.

Werner, G., and Brehmer, G. (1967). *Hoppe-Seyler's Z. Physiol. Chem.* **348,** 1640–1642.

White, H. A., and Spencer, M. (1964). *Can. J. Bot.* **42,** 1481–1483.

Willaman, J. J., and Li, H.-L. (1970). *Lloydia* **33,** No. 3A.

Willaman, J. J., and Schubert, B. G. (1955). *Econ. Bot.* **9,** 141–150.

Willaman, J. J., and Schubert, B. G. (1961). "Alkaloid-Bearing Plants and Their Contained Alkaloids," US Dep. Agric., Washington, D.C.

Wilson, M. L., and Coscia, C. J. (1975). *J. Am. Chem. Soc.* **97,** 431–432.

Winek, C. L., and Fitzgerald, T. J. (1961). *J. Pharm. Sci.* **50,** 976.

Wold, J. K. (1978). *Phytochemistry* **17,** 832–833.

Yamasaki, K., Tamaki, T., Uzawa, S., Sankawa, U., and Shibata, S. (1973). *Phytochemistry* **12,** 2877–2882.

Yang, R. S. H., and Guthrie, F. E. (1969). *Ann. Entomol. Soc. Am.* **62,** 141–146.

Yeh, S. Y., Chernov, H. I., and Woods, L. A. (1971). *Proc. Soc. Exp. Biol. Med.* **136,** 782–784.

Ziyaev, R., Abdusamatov, A., and Yunusov, S. Y. (1975). *Chem. Nat. Compd. (Engl. Transl.)* **11,** 478–481.

Zuk, J., and Swietlinska, Z. (1973). *Mutat. Res.* **17,** 207–212.

Chapter **12**

Toxic Seed Lipids

DAVID S. SEIGLER

I. INTRODUCTION

Seeds perform several functions; most important among these are energy storage, dispersal, and perennation of the plant taxon. Energy

ISBN 0-12-597180-X

storage usually involves lipids (primarily triglycerides), carbohydrates, and proteins. Upon germination of the seed, lipids or carbohydrates are normally broken down to smaller compounds (ATP, acetate, glucose, etc.) and utilized by the embryo, whereas proteins are catabolized to amino acids (Levin, 1974). Upon germination of the seed, secondary compounds may be catabolized or translocated to other parts of the plant (D. O. Clegg, E. E. Conn, and D. Janzen, unpublished data). On a percent weight basis, glycerides are the most efficient energy storage compounds. Most plants seem to make either glycerides or carbohydrate storage materials, but few produce large quantities of both. Thus, the percentage of lipids in seeds varies greatly. Several mechanisms for seed dispersal have been reviewed, especially those related to the presence of sizable quantities of seed lipids (Levin, 1974). It is obvious that most seed lipids provide an excellent source of food not only for plant embryos but also for a variety of animals, fungi, and bacteria. There has been evolutionary selection for mechanisms to protect seeds from predation by several mechanisms (Levin, 1974). The seed oils and fats from a large number of plants are edible and are important (if not required) in the diet of all animals. Of primary interest in this chapter is the production of lipid secondary metabolites that deter general predation on seeds bearing these nutrient reserves.

The effect of many of these toxic secondary metabolites is probably to prevent or limit feeding rather than to kill the predator outright, although several could certainly accomplish the latter if sufficient amounts were consumed. In the former effect, the seeds may be made "distasteful" or "nonnutritious" by a certain mechanism. In this chapter, I shall review those substances that deter or limit predation or possess the capability of killing a particular predator. Coevolved predators or herbivores often have the capability of detoxifying defense compounds that are generally toxic. Insecticides, antibiotics, herbicides, and antifungal agents are useful to human beings only because of these selective effects.

I will not review those types of toxicity caused by the absence of a nutritional or other important metabolic factor, although some of these may play significant roles in plant-host selection. For example, many insects must have specific steroidal or triterpenoid precursors available to them, since they generally cannot synthesize steroidal nuclei from acetate or other simple precursors.

Many types of lipid materials are found in seeds. Among these are fatty acids, phenols, terpenes, saponins and steroidal glycosides, acetylenic compounds, hydrocarbons, glycerides, phospholipids, glycolipids, phenylpropanoid-derived compounds, wax esters, and breakdown products of glucosinolates such as isothiocyanates, thiocyanates, nitriles, and alkaloids. Several types of fatty acids may be significant as toxic substances; among these are long-chain fatty acids, hydroxy fatty acids, keto fatty acids, epoxy fatty acids, cyclopropene and cyclopropane fatty acids, cyclopentene fatty acids, fatty acids with phenolic substituents, and those with acetylenic or unusual double-bond types.

It is difficult to determine whether lipid substances such as alkaloids, terpenes, and the breakdown products of glucosinolates are to be found admixed in intact seed materials. Upon expression of the oil, improper storage of seeds, disruption of the cellular integrity of the seeds, or eating the seeds, these components are often found together with the glycerides, waxes, and other lipids in which they are soluble.

In the following sections, I shall discuss the distribution and biological and toxicological properties of some of the major groups of toxic seed lipids.

II. GLYCERIDES AND COMPONENT FATTY ACIDS

The major lipid components of almost all seeds are triglycerides. These usually contain palmitic, stearic, palmitooleic, oleic, linoleic, and linolenic acids. These compounds may be consumed by any animal without eliciting evidence of toxicity. In human beings, most of these oils have mild laxative properties, especially when consumed in large amounts (Claus, 1956). The effect is probably due to incomplete absorption or nonabsorption coupled with lubricant properties of the oils.

In a manner perhaps comparable to that of carbohydrates as feeding stimulants (Dethier, 1953; Thorsteinson, 1960), glycerides have been shown, in at least two cases, to serve as compounds that elicit aggregation in insects (Tamaki *et al.*, 1971; Muto and Sugawara, 1965; Muto *et al.*, 1968). 1-Palmito-2,3-diolein, 2-linoleo-1,3-dipalmitin, and 1-palmito-2-linoleo-3-olein were identified in three active fractions from wheat germ that elicit aggregation of the confused flour beetle (Tamaki *et al.*, 1971). Similarly, 1,3-diolein was shown to "attract" houseflies (*Musca domestica*) to the fruiting bodies of *Amanita muscaria* (Muto *et al.*, 1968), and similar fractions were shown to "attract" houseflies to the fruiting bodies of other mushrooms of the Tricholomataceae and Amanitaceae (Muto and Sugawara, 1965).

III. LIPIDS CONTAINING ERUCIC ACID AND HIGHER HOMOLOGS

Erucic (*cis*-13-docosenoic) acid is the most characteristic component of seed oils of the Cruciferae. Three-fourths of seed oils derived from the Cruciferae contain this compound (Smith, 1970; Mikolajczak *et al.*, 1961; Mattson, 1973; Appelqvist, 1976). The presence of this acid can be detected by standard gas–liquid chromatographic methods for the identification of the methyl esters of fatty acids.

Other cruciferous seed oils contain sizable quantities of even higher homologs (Appelqvist, 1976); e.g., *Lunaria biennis* seed oil contains 21% of *cis*-15-tetracosenoic acid. Nasturtium seed oil (*Tropaeolum* sp., Tropaeolaceae) is a rich source of erucic acid (Jamieson, 1943). More than 90% of the acyl groups in *Limananthes douglassi* (Limanthaceae) seed oil are comprised of *cis*-5-eicosenoate, *cis*-13-docosenoate, and *cis*-5,*cis*-13-docosadienoate (Phillips *et al.*, 1971).

Fats that are not liquid at intestine temperature in mammals are poorly absorbed and usually excreted (Hilditch and Williams, 1964). Higher molecular weight fatty acids are not utilized efficiently. Evidence of toxicity produced by C_{22} and C_{24} fatty acids has been reported.

Myocarditis was observed in rats that were fed 50–70% of their calories as erucic acid (Roine *et al.*, 1960), and this finding was later confirmed in rats that obtained as little as 15% of their calories in this manner. The accumulation of lipids in the heart increased, and eventually myocardial fibrosis and other abnormalities were observed. Feeding of oil seeds with low concentrations of erucic acid largely alleviates these problems (Mattson, 1973).

A breeding program has been initiated to produce new varieties of rapeseed (which is used for margarine production) in which most or all of the erucic acid is replaced by oleic acid (Vles *et al.*, 1976).

IV. UNSATURATED FATTY ACIDS

Many unusual nonconjugated olefinic acids have been isolated from plants (Smith, 1970; Hilditch and Williams, 1964). Most of these compounds are uncommon, but others, such as petroselenic acid (*cis*-6-octadecenoic acid), are more widely distributed. Both cis and trans isomers occur; several conjugated olefinic systems are also known (Smith, 1970). These "unusual" substitutions may make the storage lipids of these plants less profitable as a food source for predators.

Although few olefinic or acetylenic fatty acids are highly toxic, there are large variations in the ease with which they can be utilized by different organisms. Gunstone (1976) has reviewed some of these effects.

The seed oil of *Aleurites fordii,* which contains large amounts of eleostearic acid, has considerable laxative action and is considered nonedible. *Ricinodendron africanum* seed oil also contains eleostearic acid, but it is consumed (Jamieson, 1943). The seed fat of the Mexican pinonchillo tree (*Garcia nutans,* Euphorbiaceae) may contain up to 90% eleostearic acid (Hilditch and Williams, 1964).

V. ACETYLENIC AND ALLENIC LIPIDS

A. Distribution

Acetylenic compounds in plants are distributed among at least 19 families (all dicotyledonous with one exception) (Bohlmann *et al.*, 1973); the Umbelliferae and Compositae are by far the richest in these compounds and have been most thoroughly investigated. More than 1100 species from 267 genera of the Compositae and 195 species from 89 genera of the Umbelliferae have been examined, largely by Bohlmann and co-workers. Acetylenes have also been found in many species of Basidiomycetes.

The types of structures present have been reviewed by Bohlmann *et al.* (1973). Acetylenic fatty acids are mostly from four families: Santalaceae, Olacaceae, Loranthanceae, and Opiliaceae, which many phylogenists have considered to be more or less closely related. These fatty acids are most commonly encountered in seed lipids, where they are usually components of triglycerides. Acetylenic compounds are derived from acetate–malonate pathways, and most if not all compounds involve the intermediacy of oleic acid (Bohlmann *et al.*, 1973; Bohlmann and Schulz, 1968).

As a general trend, acetylenic acids of seed fats have less unsaturation than vegetative lipids (Hatt *et al.,* 1967). In developing *Santalum acuminatum* seeds more highly unsaturated fatty acids have been shown to develop as the seedlings mature (Bu'Lock and Smith, 1963).

Allenic fatty acids are rare (Hageman *et al.*, 1967). They are known to occur in the seed oils of three mints (*Leonotis nepetaefolia, Lamium purpureum* L., and *L. amplexicaule* L.) as well as the roots and leaves of *Dicoma zeyheri* (Compositae) and seed wax of *Sapium sebiferum* (Euphorbiaceae) (Sprecher *et al.*, 1965).

B. Biological Properties

Several naturally occurring acetylenic compounds are known to possess potent biological properties (Bohlmann *et al.*, 1973). In South America, *Ichthyothere terminalis* (Compositae) has been used as a piscicide (Cascon *et al.*, 1965). The compound responsible for this action has been isolated from other members of the Compositae (Bohlmann *et al.*, 1973).

The highly toxic compounds cicutoxin (from the roots of *Cicuta virosa*) and oenanthetoxin (from *Oenanthe crocata*) are well known (Bohlmann *et al.*, 1973; Crosby, 1969; Kingsbury, 1964). The lethal dose of cicutoxin for a laboratory mouse is about 4 μg. These plants have been responsible for the deaths of numerous human beings and livestock (Crosby, 1969). Less well known is carotatoxin from the common carrot (*Daucus carota*) (Crosby and Aharonson, 1967) and celery (*Apios graveolens*) (Crosby, 1969). This compound produces symptoms similar to those produced by cicutoxin, although it is much less toxic and occurs at low concentrations.

An antifungal acetylenic furanoid keto ester, wyerone **(1)**, was isolated from the shoots of *Vicia faba* L. (Leguminosae), the broad bean (Bohlmann *et al.*, 1973). The role of wyerone epoxide as a phytoalexin in *Vicia faba* and its metabolism have been reviewed (Hargreaves *et al.*, 1976).

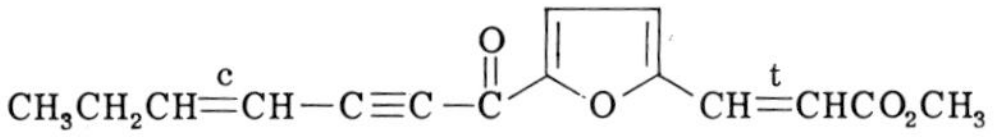

$$CH_3CH_2CH\overset{c}{=}CH-C\equiv C-\overset{O}{\overset{\|}{C}}-\text{(furan-2,5-diyl)}-CH\overset{t}{=}CHCO_2CH_3$$

1

Lipid amides, especially isobutylamides, tyramides, and phenylethylamides, with characteristic olefinic and acetylenic unsaturation patterns have been isolated from members of the Compositae, Piperaceae, and Rutaceae. They frequently possess insecticidal activity. Two other insecticidal isobutylamides, neoherculin (from *Zanthoxylum clavaherculis*, Rutaceae) and affinin (from *Heliopsis longipes*, Compositae), do not possess triple bonds but are structurally related (Bohlmann *et al.*, 1973).

Most of the acetylenic compounds of plants occur in the leaves, stems, and roots, although others are found in fruits and seeds. Diurnal and seasonal variations have been observed (Bohlmann *et al.*, 1973) and a half-life of 6 to 48 hours appears to be common.

Acetylenes present in different plant parts often differ in structure; e.g., in *Pastinaca sativa* L. (Umbelliferae) a C_{18} aldehyde occurs in the seeds, whereas only falcarinone-type compounds are present in the roots. The seeds of *Crepis foetida* L. contain large amounts of crepenynic acid in their triglycerides, whereas no acetylenic compounds could be detected in the vegetative portions of the plant.

In summary, many acetylenic compounds are known to be toxic or have pronounced biological activities. Small variations in structure often lead to nearly nontoxic compounds. Several acetylenes occur in common plants used both for foods and for flavoring.

VI. HYDROXY FATTY ACIDS

A. Distribution

Mono- and dihydroxy acids are found in a variety of plant families (Smith, 1970; Hilditch and Williams, 1964; Pohl and Wagner, 1972a & b Wolff, 1966). They are quite diverse in botanical origin and have few unifying features. Several are known from seed oils.

Among the naturally occurring hydroxy acids perhaps the best known is ricinoleic acid (*d*-12-hydroxy-*cis*-9-octadecenoic acid), which comprises up to 90% of the fatty acids in the seed oil of *Ricinus communis* (Euphorbiaceae) (Smith, 1970). In *Ricinus communis,* ricinoleic acid occurs as a component of the seed triglycerides. It also occurs in several other unrelated plant seed oils (Badami and Kudari, 1970; Kleiman and Spencer, 1971; Mikolajczak *et al.,* 1962).

Seeds of the genus *Lesquerella* (Cruciferae) are an additional source of hydroxyolefinic acids. In the genus *Lesquerella,* all of 22 species studied contained substantial amounts of monohydroxy fatty acids (Appelqvist, 1976).

The biosynthesis of hydroxy fatty acids has been reviewed (Hitchcock and Nichols, 1971).

B. Biological Properties

Intravenous feeding of labeled 9-mono- and 12-monohydroxystearic acids to rats indicated that they were readily incorporated into palmitic, stearic, and octadecenoic acids. Distribution of label suggested that breakdown to acetate and resynthesis was responsible for this action (Risser, 1965).

Castor oil, from *Ricinus communis* (Euphorbiaceae), contains up to 90% ricinoleic acid. A study relating cathartic activity to structural modifications of ricinoleic acid has been reported (Risser, 1965; Masri *et al.,* 1962).

Considerable amounts may be absorbed and utilized when fed as a part of the diet of many animals. Paul and McCay (1942) reported 92% utilization by rabbits, 99% by sheep, and 98% by rats. Rats fed on ricinoleic acid grew well, and about 7% of the fatty acids of the depot fats were shown to be ricinoleic acid (Stewart and Sinclair, 1945).

Several euphorbiaceous seed oils are used for laxatives and are suspected to contain hydroxy fatty acids. *Joannesia princeps* seed oil has a laxative effect four times as great as that of castor oil but produces no irritation or nausea (Jamieson, 1943). *Omphalea megacarpa* seed oil is also a nonirritating cathartic. *Jatropha curcas* oil has been used for both medicinal and soap-making purposes. It is a more potent laxative than castor oil.

VII. EPOXY LIPIDS

Epoxy fatty acids are found in the seed oils of several taxonomically unrelated plant families and are sporadically distributed within these families. Their distribution has been reviewed (Earle, 1970; Earle *et al.*, 1966).

cis-12,13-Epoxyoleic (vernolic) acid is biosynthesized from linoleic acid in the seeds of *Xeranthemum annuum* and *Euphorbia lagascae* (Hitchcock and Nichols, 1971). *Cephalocroton cordofanus* seeds are eaten by the natives of the western Sudan. The seed oil contains epoxy fatty acids and ricinoleic acid (Eckey, 1954). As much as 5% in the diet of methyl hydroxystearate, methyl dihydroxystearate, methyl *cis*-epoxystearate, or methyl oleate hydroperoxide has no effect on the growth or appearance of the animals (Kaunitz, 1962). Kritchevsky *et al.* (1961) found that prefeeding rats with methyl epoxystearate produces an increased inhibition of mitochondrial oxidation of cholesterol over that observed in rats prefed methyl stearate.

VIII. CYCLOPROPENE AND CYCLOPROPANE-CONTAINING LIPIDS

A. Distribution

Cyclopropene fatty acids are found in both seed and vegetative lipids of many plants of the order Malvales (Smith, 1970; Mattson, 1973; Pohl and Wagner, 1972a,b; Hilditch and Williams, 1964; Boekenoogen, 1967). Their distribution has been reviewed (Carter and Frampton, 1964). The two most common cyclopropene fatty acids are malvalic acid **(2)** and sterculic acid. Small amounts of cyclopropanoid fatty acids (1–2%) sometimes accompany the corresponding unsaturated acids. In *Euphorbia longana* seed oil, however, up to 17% of the total fatty acids is comprised of dihydrosterculic acid (Kleiman *et al.*, 1969).

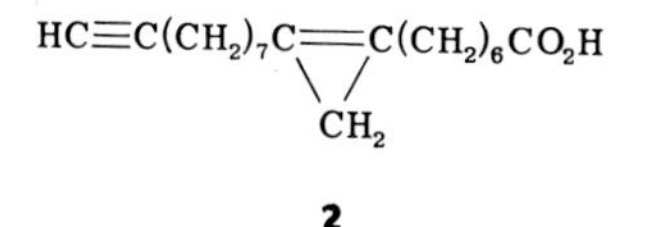

2

The biosynthesis of cyclopropane and cyclopropene fatty acids in plant tissues has been investigated (Hitchcock and Nichols, 1971; Yano *et al.*, 1972).

B. Biological Properties

Cyclopropenoid fatty acids inhibit fatty acid desaturation in several species including chicken (Johnson *et al.*, 1969), rat (Raju and Reiser, 1967), pig, cow (Phelps *et al.*, 1965), and trout (Roehm *et al.*, 1970), causing a rise in the stearate/oleate ratio. This may have an effect on the composition and function of membrane structures since variations in lipid composition are known to alter permeability of membrane systems. A high incidence of prenatal and postnatal mortality in the progeny of rats fed cyclopropenoid fatty acids was also observed (Miller *et al.*, 1969). Miller *et al.* suggested that altered membrane permeability and increased capillary fragility were responsible for these detrimental effects. Other biological effects, such as retarded growth and retarded sexual development, have been reviewed by Phelps (Phelps *et al.*, 1965). Laying hens fed 250 mg/day of cyclopropenoid fatty acids ceased to lay, although normal laying was resumed in 12 to 32 days when these acids were removed from the diet (Shenstone and Vickery, 1959). Kapok seed oil, at 10% levels in the diet, caused a suppression of growth and at 40% level, all of the animals died in a few weeks (Thomasson, 1969). *Sterculia foetida* oil at the 10% level resulted in death in weanling rats (Schneider *et al.*, 1968). The consumption of diets containing 3% of *S. foetida* oil completely inhibited reproductive performance in female rats, but the males were not affected. Cyclopropene fatty acids can also be carried in food chains since the animals that eat these acids often store them in their body products or in products such as eggs (Shenstone and Vicery, 1959; Kratzer *et al.*, 1955). Pigs fed rations containing cottonseed meal, which often contains residual lipids, produce fat with a high melting point (Deuel, 1955). When cyclopropenoid fatty acids are fed to these animals, there is a decrease in the tissue levels of monounsaturated fatty acids and a corresponding increase in saturated fatty acids (Johnson *et al.*, 1967). The increase in tissue saturated fatty acids could not be overcome by supplementing the diet with oleic or linoleic acid (Evans *et al.*, 1963).

The most important edible oil that contains cyclopropenoid fatty acids is cottonseed oil from *Gossypium hirsutum*. Others such as kapok seed oil (*Ceiba pentandra*) or oil from *Bombax malavaricum* (Cornelius *et al.*, 1970) are prepared in several countries, and at least small amounts are consumed. Much of the cyclopropenoid content of cottonseed oil is removed in processing, especially in the process of deodorization (Bailey *et al.*, 1966). The range in commercial cottonseed salad oil ranges from 0.1 to 0.5% (Rocquelin and Cluzan, 1968).

A cyclopropane fatty acid from *Sterculia foetida* oil was totally lethal to larvae of *Callosobruchus maculatus* when incorporated into an artificial diet at the level of 0.1% (Janzen *et al.*, 1977). Sinnhuber and co-workers (1968; Lee *et al.*, 1968), demonstrated that cyclopropene fatty acids possess cocarcinogenic activity when fed with aflatoxins to rainbow trout.

IX. CYCLOPENTENE FATTY ACIDS

A. Distribution

Fatty acids containing cyclopentene moieties such as hydnocarpic **(3)**, chaulmoogric, and gorlic acids are found only in seed oils of the tropical family Flacourtiaceae (Smith, 1970; Jamieson, 1943; Hilditch and Williams, 1964; Eckey, 1954). The oils of these seeds have long been used by natives of Africa, South America, and Asia for the treatment of leprosy, skin diseases, and wounds (Jamieson, 1943).

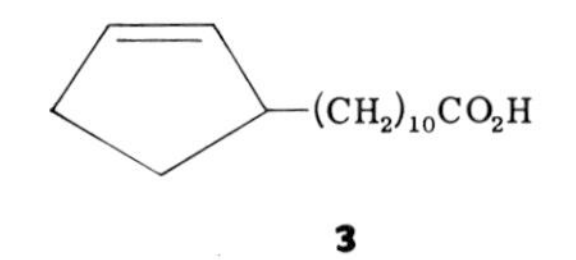

3

In an investigation of chaulmoogra oil from *Hydnocarpus kurzii* and *H. wightiana*, the principal compounds found were hydnocarpic [11-(2-cyclopenten-1-yl)undecanoic], chaulomoogric [D-13-(2-cyclopenten-1-yl)tridecanoic], and gorlic [13-[2-cyclopenten-1-yl) 6-trudecenoic acid] acids with a smaller amount of the lower homologs: aleprestic (10 : 1), aleprylic (12 : 1), and alepric (14 : 1) acids (Sengupta *et al.*, 1973). In contrast to some earlier reports, these workers also found small percentages of co-occurring oleic, stearic, and palmitic acids. *Hydnocarpus odorata* (≡ *Gynocardia odorata*) was devoid of cyclopentenoid fatty acids.

The biosynthesis of cyclopentene fatty acids has only recently been investigated. These fatty acids are synthesized by extension of a small cyclic compound, aleprolic acid [2-cyclopentene-1-carboxylic acid], which serves as a precursor (Cramer and Spener, 1976, Spener and Mangold, 1975.)

B. Biological Properties

Although the use of oils containing cyclopentenoid fatty acids and their derivatives has continued to the present time, more effective materials have now largely replaced chaulmoogra oil for the treatment of leprosy (Eckey, 1954). Cyclopentenoid fatty acids are highly toxic to both *Mycobacterium leprae* and *M. tuberculosis*. Their *in vitro* activity is 10 times more powerful than that of phenol (Claus, 1956). The oils or the ethyl esters are administered by mouth or by injection, but the quantity that can be administered orally is limited by the fact that chaulmoogra acid and its derivatives are somewhat toxic (Bernard and Müller, 1938). The oils are hardly edible, and their production is principally for medicinal use (Eckey, 1954).

X. FLUOROLIPIDS

Dichapetalum toxicarium (Dichapetalaceae) is highly toxic to livestock in West Africa. Peters and co-workers (1959, 1960; Dear and Pattison, 1963) discovered that the seed oil of this plant contains ω-fluoroctadec-*cis*-9-enoic (ω-fluorooleic) acid. These ω-fluoro acids are the only known naturally occurring fatty acids in which a halogen is incorporated (Smith, 1970).

Several other species of *Dichapetalum* are known to be poisonous in Africa (Watt and Breyer-Brandwijk, 1962): *D. cureonitens* Engl., *D. braunii* Engl. and Krause, *D. cymosum* Engl., *D. mossambicense* Engl., *D. rublandii* Engl., *D. stublmanii,* and *D. venenatum* Engl. and Gilg.

XI. PHENOLIC SEED OIL LIPIDS

A. Distribution

Many plants of the Anacardiaceae possess resinous exudates that are intensely irritating to human beings and other mammals (Kingsbury,

1964; Boekenoogen, 1967). For example, *Anacardium occidentale* L. (cashew) fruits are very irritating when unshelled and unroasted (Kingsbury, 1964). The toxic compounds have been isolated and characterized (Symes and Dawson, 1953). They are cardol, anacardic acid, and anacardol. *Semecarpus anacardium* is also known to contain anacardic acid and cardol (Steinmetz, 1957).

Most members of the genus *Toxicodendron* are covered with a similar resinous material, which produces a pronounced irritating effect and darkens upon exposure to air. This toxic material is also found on *Metopium toxiferum* (L.) Krug and Urban, poison wood, which grows in Florida and the Caribbean area. Several other species of this genus are found in Central America and the Caribbean Islands (Kingsbury, 1964). One component of this mixture is urushiol **(4)**.

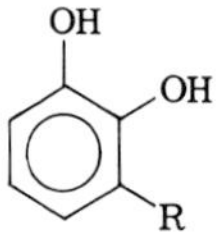

OH OH R

$R = (CH_2)_{14}CH_3,\ (CH_2)_7CH{=}CH(CH_2)_5CH_3,$
$(CH_2)_7CH{=}CHCH_2CH{=}CH(CH_2)_2CH_3,$
$(CH_2)_7CH{=}CHCH_2CH{=}CHCH_2CH{=}CH_2$

4

Alkyl- and alkenylresorcinols have been discovered in several unrelated families (Madrigal *et al.*, 1977). The biosynthesis of anacardic acids in *Ginkgo biloba* seeds has been investigated (Gellerman *et al.*, 1974, 1976).

B. Biological Properties

The milky white sap of virtually all of the members of the Anacardiaceae has been found to have some vesicant or blistering effect (Kingsbury, 1964). Although the seed kernel of *Anacardium occidentale* is edible, it is poisonous unless roasted and all the pericarp oil has exuded (Watt and Breyer-Brandwijk, 1962). Material from the bark of the tree is also vesicant (Chopra *et al.*, 1955). In India the shell oil is used medicinally and for preserving floors, timbers, and books from the white ant (Rimington, 1934). The shell oil is toxic to mosquito larvae, certain grain weevils, and moth larvae. A mucilage made from the tree gum has good adhesive properties and has insect repellent properties (Irvine, 1955; Greenway, 1941a,b).

Members of the genus *Toxicodendron* are often known as poison ivy, poison oak, or poison sumac. The toxin is a colorless or milky substance contained in resin canals throughout the entire plant except in the pollen. The chemistry of the toxic materials has been examined, and the major compound, urushiol **(4)**, is similar to those previously described (Kurtz and Dawson, 1971). The level of catecholic lipids in poison ivy is much higher in the young tender leaves of the plant than in the more mature leaves later in the year (Gellin *et al.*, 1971).

Anacardic acids occur in both the leaves and "nuts" of the ginkgo tree (Gellerman and Schlenk, 1968). Antimold, antibacterial, and antiviral activity has been ascribed to constituents of *Ginkgo biloba*, and the phenolic components are responsible for at least some of these effects (Eichbaum, 1946). These compounds are usually removed from the seed of ginkgo by fermentation or roasting before human consumption.

XII. TERPENES

A. Monoterpenes, Sesquiterpenes, and Essential Oils

Many seed oils from numerous species of plants contain terpenoid materials when expressed. The seed oils of many species of the family Umbelliferae contain essential oils. These essential oils are often monoterpenoid or sesquiterpenoid in nature (Jamieson, 1943), although several contain phenylpropenoid compounds. Most have a strong aromatic flavor.

Several mono- and sesquiterpenes (as components of essential oils) have been used for anthelmintics including sabinol, cadinol, pinene, ascaridole, *p*-cymene, citronellol, caryophyllene, limonene, phellandrene, cadinene, camphor, triacontane, pyrethrin, carvone, linalool, linallyl alcohol, tagetone, thymol, and eucalyptol (Bandoni *et al.*, 1972; Wren, 1956; Steinmetz, 1957; Watt and Breyer-Brandwijk, 1962; Stahr, 1933).

Various citrus oils and one of their components, D-limonene, appear to have weak tumor-promoting activity for mouse skin (Homburger and Boger, 1968; Roe and Pierce, 1960).

The leaves of the California bay laurel tree (*Umbellularia californica* Nutt., Lauraceae) in the past have had considerable use as food (Hall, 1973). They contain 0.5–4% of an irritating oil, of which 40–60% is umbellulone. Its vapors produce severe headache, skin irritation, and in some cases unconsciousness (Drake and Stuhr, 1935).

Thujone, a major component of wormwood, *Artemisia absinthum* L., was once a major flavoring ingredient of the liquor absinthe (Hall, 1973). At doses of 30 mg per kilogram body weight, thujone produces convulsions associated with lesions of the cerebral cortex (Hall, 1973). Thujone is a major constituent of cedar leaf oil (*Thuja occidentalis* L.) (Wallach,

1893) and an important component of sage (*Salvia officinalis* L.) (Vernazza, 1957; Brieskorn and Wenger, 1960). β-(+)-Thujone occurs in tansy, *Tanacetum vulgare* L. (Braun, 1949; Maizite *et al.*, 1950), and in yarrow (*Achillea millefolium* L.) (Kremers, 1921, 1925).

It does not appear that specific studies of the toxicity of seed oils containing mono- and sesquiterpenes have been conducted.

B. Diterpenes

Several diterpene constituents of seed oils have been thoroughly examined. These are found in two families: the Euphorbiaceae and the Thymeliaceae.

Croton seed oil and latexes from plants of the Euphorbiaceae contain potent tumor promoters for the skin of mice following initiation with single small doses of carcinogenic hydrocarbons (Miller, 1973). Hecker and co-workers studied the structures of the active components of croton oil and showed them to be 12,13-diesters of the complex tetracyclic alcohol, phorbol (**5**) (Hecker, 1968, 1970, 1971a,b). Phorbol itself is inactive as a promoter, but it has been demonstrated that it has leukemogenic activity in mice (Berenblum and Lonnai, 1970).

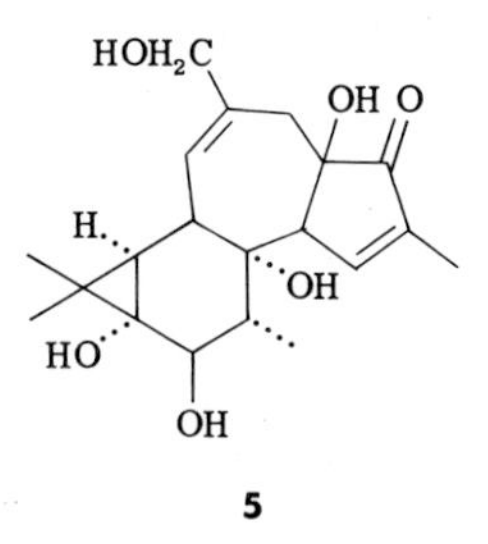

5

The oil obtained by pressing the seeds of *Croton tiglium* L. (Euphorbiaceae) has a number of unpleasant effects. It is a violent cathartic and causes severe vomiting, diarrhea, gastroenteritis, and delirium. On the skin it causes severe irritation and blistering. It has been established as a potent cocarcinogen (Berenblum, 1941a,b). It has also been demonstrated that it can act as a cocarcinogen long after exposure to the carcinogen (Orris *et al.*, 1964) in mice. When 7,12-dimethylbenzanthracene was applied to mice, no tumors developed in a year. Application of croton oil at that time initiated squamous cell epidermis carcinomas. A hydrophilic fraction of croton oil that made up only 5% of the starting material comprised most of the inflammatory and cocarcinogenic activity.

The largest genus of the Euphorbiaceae is *Euphorbia* with about 2000 species (Airy-Shaw, 1973). These plants occur throughout the world and most contain an irritating milky sap. Several have been used medicinally (Hecker, 1970, 1971a,b).

Derivatives of 12-deoxyphorbol were isolated from *Euphorbia triangularis,* a South African treelike species (Gschwendt and Hecker, 1969). Although several of these compounds possess irritant properties, they lack cocarcinogenic capacities.

The ingenol skeleton **(6)** is found in the irritants and cocarcinogens from *Euphorbia ingens* and *E. lathyris* (Hecker, 1968; Adolf *et al.*, 1968). Irritant diterpenes have also been reported from other *Euphorbia* species (Cordell and Farnsworth, 1977).

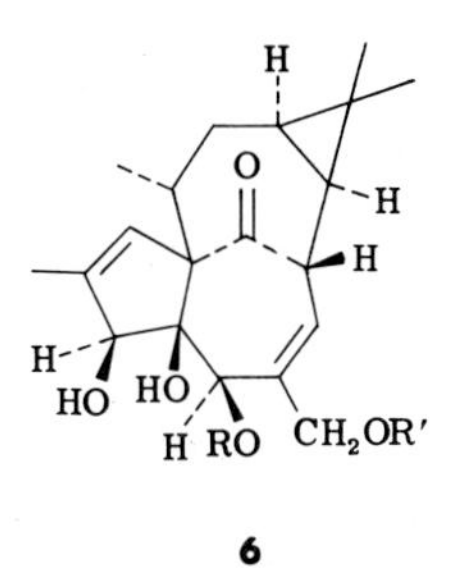

6

Other genera of euphorbiaceous plants are known to contain related compounds. Difficulties in the detoxification of *Aleurites fordii* seed meals are at least partly related to the presence of phorbol esters (Okuda *et al.*, 1975). Another irritant diterpene was isolated from *Hippomane mancinella* (Adolf and Hecker, 1975), a highly toxic plant of Florida and the Caribbean area.

The New Zealand plant *Pimelia prostrata* (Thymeliaceae) has long been recognized as poisonous. An acetate of 12-deoxyphorbol, the parent diterpene of esters from *Euphorbia triangularis,* was isolated from this plant (Gschwendt and Hecker, 1969). Daphnetoxin **(7)** was isolated from several *Daphne* species (Stout *et al.*, 1970).

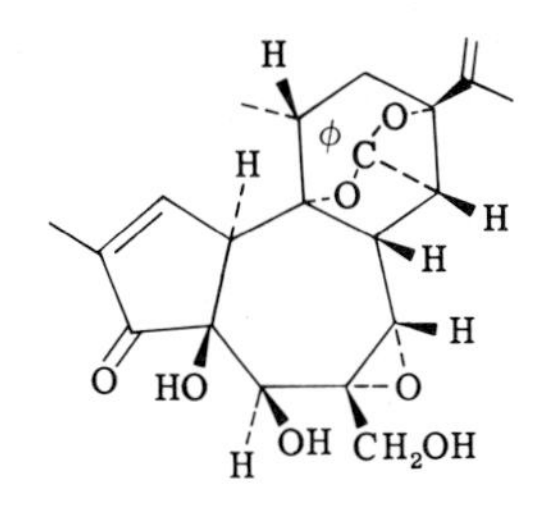

7

XIII. PHENYLPROPANOID COMPOUNDS

Phenylpropanoid-derived compounds are frequently found in essential oils from seeds. The presence of phenylpropanoid constitutents of essential oils has been reviewed (Friedrich, 1976).

Safrole is one of the most widespread of all essential oil constituents (Hall, 1973). It is the major constituent of oil of sassafras (85%), the oil of the root bark of *Sassafras albidum* (Dodsworth, 1945). *Cinnamomum microanthum* oil is 95% safrole, and it occurs in several other related species (Guenther, 1950).

Calamus oil derived from the roots of the sweet flag (*Acorus calmus*) contains about 5–90% of asarone.

Myristicin **(8)**, a component of dill, celery, parsely, parsnip, and mint, is best known as a constituent of nutmeg oil (*Myristica fragrans,* Myristicaceae). Myristicin is also found in black pepper (Richard and Jennings, 1971). Taken in large quantities, nutmeg and mace exhibit pronounced narcotic and psychomimetic properties, somewhat comparable to alcohol intoxication (Hall, 1973). Sesamol, sesamolin, and sesamin are minor components of edible sesame oil.

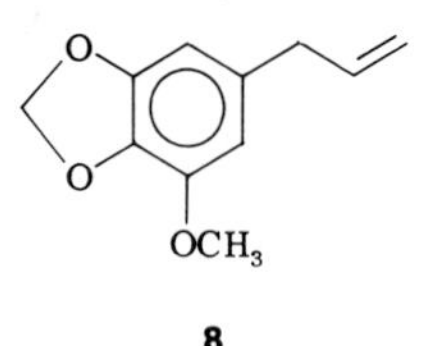

8

Safrole and myristicin are common essential oil ingredients. They also have pronounced physiological effects on animals. Both safrole and isosafrole are active synergists for pyrethrum and 1-naphthyl methylcarbamate (Sevin) as well as inhibitors of the hydroxylation of naphthalene in houseflies (Fishbein *et al.*, 1967).

Only a few compounds related to safrole have been tested for carcinogenic activity. Isosafrole (1,2-(methylenedioxy)-4-propenylbenzene) was only weakly active in inducing liver tumors when fed to male mice (Innes *et al.*, 1969). Lehman (1961) and Long *et al.* (1963) reported that continuous administration of safrole at 5000 ppm in the total diet of rats caused liver tumors and that lower levels produced lesser, noncancerous damage. Studies on dogs showed extensive liver damage at 40 and 80 mg/kg, but no tumors (Hagan *et al.*, 1967). Root beer formerly contained about 20 ppm safrole, but because of its carcinogenic capabilities safrole and essential oils containing sizable proportions of safrole cannot be used for food purposes (Hall, 1973). Myristicin is hallucinogenic, and as little as 500 mg of raw nutmeg may produce a detectable psychic response (Weil, 1965; Crosby, 1969). Myristicin also acts as both an insecticide and an insecticide synergist (Lichtenstein and Casida, 1963). A puzzling aspect of the action of nutmeg and mace is that the effects of the spice are greater than the effects of an equivalent amount of myristicin and another component, elemicin (5′-allyl-1,2,3-trimethoxybenzene), even though these are the only constituents that separately produce significant physiological response (Hall, 1973). Myristicin is apparently a monoamine oxidase inhibitor (Truitt and Ebersberger, 1962; Truitt, 1967). For many years calamus oil has had occasional use as a minor ingredient in certain bitter flavors such as in vermouth (Hall, 1973). Hagan *et al.* found that 500–10,000 ppm of the Jammu type produced growth depression, increased mortality, accumulation of fluid in the abdominal cavity, heart and liver changes, and, in a few cases, malignant tumors in rats (Hagan *et al.*, 1967; Taylor *et al.*, 1967).

Sesamol (1-hydroxy-3,4-methylenedioxybenzene) appeared to increase the incidence of benign "proliferative lesions" in rats administered 1% of this compound in the diet for several months (Ambrose *et al.*, 1958).

XIV. CYANOLIPIDS

A. Distribution

Cyanolipids are found, often in copious amounts, only in the seed oils of the Sapindaceae. Their chemical properties, biosynthesis, and distribution have been reviewed (Mikolajczak, 1977). Four basic structures are known; the fatty acid composition may vary somewhat from plant to plant. Compounds such as **9** contain mostly C_{20} fatty acids (Mikolajczak, 1977). Yields of cyanolipids range up to 58% of the seed oil.

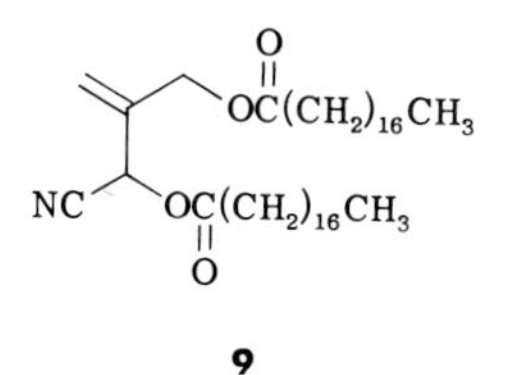

9

On the basis of pathways involved in the biosynthesis of cyanogenic glycosides (Conn, 1974), it was suggested that the structures of the non-fatty acid portions such as compound **9** might be derived from leucine (Mikolajczak *et al.*, 1969). It was subsequently demonstrated that L-[U-^{14}C]leucine is a precursor for cyanolipids in *Koelreuteria paniculata* (Seigler and Butterfield, 1976), in *Cardiospermum grandiflorum* (Seigler and Kennard, 1977), and in *Ungnadia speciosa* seed oils (D. S. Seigler, S. Saupe, and L. Blehm, unpublished data).

B. Biological Properties

Extensive studies of the toxic properties of cyanolipids have not been conducted. In one set of experiments, Janzen *et al.* (1977) fed cyanolipids to larvae of *Callosobruchus maculatus* in artificial diets. These compounds proved to be lethal at 1 and 5% levels in all cases and, although most compounds were lethal at the 0.1% level, a mixture of compounds from *Koelreuteria paniculata* enhanced larval growth at that concentration (Janzen *et al.*, 1977).

Several sapindaceous plants are recognized as poisonous. In Australia, *Atalaya hemiglauca* fruits were fed to horses and shown to be toxic, whereas dried leaf material produced no ill effects. As little as 20 gm per day of the fruits was toxic over a period of time (Everist, 1974). The South African tree *Pappea capensis* produces fruits which yield an oil that has been considered edible but purgative and used for medicinal purposes (Watt and Breyer-Brandwijk, 1962).

Numerous publications deal with efforts to detoxify kusum oil (*Schleicheria trijuga*) (Mikolajczak, 1977). These publications suggest that *Schleicheria* oil would be quite toxic if untreated. Pulasan fat (*Nephelirim mutabile*) is an edible oil (Eckey, 1954). *Ungnadia Speciosa* seed oil reportedly keeps well and has desirable physical properties for use as a salad oil as well as for soap making (Cheel and Penfold, 1919; Bachtez, 1947), but the use of *Ungnadia speciosa* seed oil as a salad oil has been rejected because of its toxicity.

REFERENCES

Adolf, W., and Hecker, E. (1975). *Tetrahedron Lett.* p. 1587.

Adolf, W., Opferkuch, H. J., and Hecker, E. (1968). *Fette, Seifen, Anstrich.* **70,** 850.

Airy-Shaw, H. K. (1973). "Dictionary of Flowering Plants and Ferns," 8th ed. Cambridge Univ. Press, London and New York.

Ambrose, A. M., Cox, A. J., and De Eds, F. (1958). *J. Agric. Food Chem.* **6,** 600.

Appelqvist, L. (1976). *In* "The Biology and Chemistry of the Cruciferae" (J. G. Vaughan, A. J. Macleod, and B. M. G. Jones, eds.), p. 221. Academic Press, New York.

Bachtez, M. (1947). *Ciencia (Mexico City)* **8,** 57.

Badami, R. C., and Kudari, S. M. (1970). *J. Sci. Food Agric.* **21,** 248.
Bailey, A. V., Harris, J. A., Skau, E. L., and Kerr, T. (1966). *J. Am. Oil Chem. Soc.* **43,** 107.
Bandoni, A. L., Mendiondo, M. E., Rondina, R. V. D., and Coussio, J. D. (1972). *Lloydia* **35,** 69.
Berenblum, I. (1941a). *Cancer Res.* **1,** 44.
Berenblum, I. (1941b). *Cancer Res.* **1,** 807.
Berenblum, I., and Lonnai, V. (1970). *Cancer Res.* **30,** 2744.
Bernhard, K., and Müller, L. (1938). *Hoppe-Seyler's Z. Physiol. Chem.* **256,** 85.
Boekenoogen, H. A. (1967). *Chem. Ind. (London)* p. 387.
Bohlmann, F., and Schulz, H. (1968). *Tetrahedron Lett.* p. 4795.
Bohlmann, F., Burkhardt, T., and Zdero, C. (1973). "Naturally Occurring Acetylenes." Academic Press, New York.
Braun, H. (1949). *Med. Monatsschr.* **3,** 528.
Brieskorn, C. H., and Wenger, E. (1960). *Arch. Pharm. (Weinheim, Ger.)* **293,** 21.
Bu'Lock, J. D., and Smith, G. N. (1963). *Phytochemistry* **2,** 289.
Carter, F. L., and Frampton, V. L. (1964). *Chem. Rev.* **64,** 497.
Cascon, S. C., Mors, W. B., Tursch, B. M., Aplin, R. T., and Durham, L. J. (1965). *J. Am. Chem. Soc.* **87,** 5237.
Cheel, E., and Penfold, A. R. (1919). *J. Soc. Chem. Ind., London* **38,** 74T.
Chopra, I. C. *et al.* (1955). "A Review of Work on Indian Medicinal Plants." New Delhi.
Claus, E. P. (1956) "Pharmacognosy", 3rd. Ed., Lea and Febiger, Philadelphia.
Conn, E. E. (1974). *Biochem. Soc. Symp.* **38,** 277.
Cordell, G. A., and Farnsworth, N. R. (1977). *Lloydia* **40,** 1.
Cornelius, J. A., Hammonds, T. W., Leichster, J. B., Ndalbahweji, J. K., Rosie, D. A., and Shone, G. G. (1970). *J. Sci. Food Agric.* **21,** 49.
Cramer, U., and Spener, F. (1976). *Biochim. Biophys. Acta* **450,** 261.
Crosby, D. G. (1969). *J. Agric. Food Chem.* **17,** 532.
Crosby, D. G., and Aharonson, N. (1967). *Tetrahedron* **23,** 465.
Dear, R. E. A., and Pattison, F. L. M. (1963). *J. Am. Chem. Soc.* **85,** 622.
Dethier, V. G. (1953). *In* "Insect Physiology" (K. Roeder, ed.), pp. 568–571. Wiley, New York.
Deuel, H. J., Jr. (1955). "Lipids," Vol. 2. Wiley (Interscience), New York.
Dodsworth, M. R. (1945). *Biol. Divulgação Inst. Oleos* **3,** 21.
Drake, M. E., and Stuhr, E. T. (1935). *J. Am. Pharm. Assoc., Sci. Ed.* **24,** 196.
Earle, F. R. (1970). *J. Am. Oil Chem. Soc.* **47,** 510.
Earle, F. R., Barclay, A. S., and Wolff, I. A. (1966). *Lipids* **1,** 325.
Eckey, E. W. (1954). *A C S Monog.* **123.**
Eichbaum, F. W. (1946). *Mem. Inst. Butantan, Sao Paulo* **19,** 71.
Evans, R. J., Davidson, J. A., LaRue, J. N., and Bademer, S. L. (1963). *Poult. Sci.* **42,** 875.
Everist, S. L. (1974). "Poisonous Plants of Australia." Angus & Robertson, Sydney, Australia.
Fishbein, L., Fawkes, J., Falk, H. L., and Thompson, S. (1967). *J. Chromatogr.* **27,** 153.
Friedrich, H. (1976). *Lloydia* **39,** 1.
Gellerman, J. L., and Schlenk, H. (1968). *Anal. Chem.* **40,** 739.
Gellerman, J. L., Anderson, W., and Schlenk, H. (1974). *Lipids* **9,** 722.
Gellerman, J. L., Anderson, W. H., and Schlenk, H. (1976). *Biochim. Biophys. Acta* **431,** 16.
Gellin, G., Wolf, C. R., and Milby, T. (1971). *Arch. Environ. Health* **22,** 280.
Greenway, P. J. (1941a). *East Afr. Agric. J.* **6,** 127, 132, 199, and 241.
Greenway, P. J. (1941b). *East Afr. Agric. J.* **7,** 96.
Gschwendt, M., and Hecker, E. (1969). *Tetrahedron Lett.* p. 3509.

Guenther, E. (1950). "The Essential Oils," Vol. 4. Van Nostrand-Reinhold, New York.
Gunstone, F. D. (1976). *Chem. Ind. (London)* p. 243.
Hagan, E. C., Hansen, W. H., Fitzhugh, O. G., Jenner, P. M., Jones, W. I., Taylor, J. M., Long, E. L., Nelson, A. A., and Brouwer, J. B. (1967). *Food Cosmet. Toxicol.* **5,** 141.
Hageman, J. M., Earle, F. R., and Wolff, I. A. (1967). *Lipids* **2,** 371.
Hall, R. L. (1973). *In* "Toxicants Occurring Naturally in Foods," 2nd ed., p. 448 Natl. Acad. Sci., Washington, D.C.
Hargreaves, J. A., Mansfield, J. W., Coxon, D. T., and Price, K. R. (1976). *Phytochemistry* **15,** 1119.
Hatt, H. H., Meisters, A., Triffett, A. C. K., and Wailes, P. C. (1967). *Aust. J. Chem.* **2,** 2285.
Hecker, E. (1968). *Cancer Res.* **28,** 2338.
Hecker, E. (1970). *Methods Cancer Res.* **6,** 439.
Hecker, E. (1971a). *In* "Pharmacognosy and Phytochemistry" (H. Wagner and L. Hörhammer, eds.), p. 147. Springer-Verlag, Berlin and New York.
Hecker. E. (1971b). *Proc. Int. Cancer Congr., 10th, 1970* Vol. V.
Hilditch, T. P., and Williams, P. N. (1964). "The Chemical Constitution of Natural Fats," 4th ed. Spottiswoode Ballantyne, London.
Hitchcock, C., and Nichols, B. W. (1971). "Plant Lipid Biochemistry." Academic Press, New York.
Homburger, F., and Boger, E. (1968). *Cancer Res.* **28,** 2372.
Innes, J. R. M., Ulland, B. M., Valerio, M. G., Petrucelli, L., Fishbein, L., Hart, E. R., Pallotta, A. J., Bates, R. R., Falk, H. L., Gart, J. J., Klein, M., Mitchell, I., and Peters, J. (1969). *J. Natl. Cancer Inst.* **42,** 1101.
Irvine, F. R. (1955). *Colon. Plant Anim. Prod.* **5,** 34.
Jamieson, G. S. (1943). *ACS Monogr.* **58.**
Janzen, D. H., Juster, H. B., and Bell, E. A. (1977). *Phytochemistry* **16,** 223.
Johnson, A. R., Pearson, J. A., Shenstone, F. S., and Fogerty, A. C. (1967). *Nature (London)* **214,** 1244.
Johnson, A. R., Fogerty, A. C., Pearson, J. A., Shenstone, F. S., and Bernsten, A. M. (1969). *Lipids* **4,** 265.
Kaunitz, H. (1962). *In* "Symposium on Foods, Lipids and their Oxidation" (H. W. Schultz, E. A. Day, and R. O. Sinnhuber, eds.), p. 269 Avi Publ. Co., Westport, Connecticut.
Kingsbury, J. M. (1964). "Poisonous Plants of the United States and Canada." Prentice-Hall, Englewood Cliffs, New Jersey.
Kleiman, R., and Spencer, G. F. (1971). *Lipids* **6,** 962.
Kleiman, R., Earle, F. R., and Wolff, I. A. (1969). *Lipids* **4,** 317.
Kratzer, F. H., Davis, P. N., and Marshall, B. J. (1955). *Poult. Sci.* **34,** 462.
Kremers, R. E. (1921). *J. Am. Pharm. Assoc.* **10,** 252.
Kremers, R. E. (1925). *J. Am. Pharm. Assoc.* **14,** 399.
Kritchevsky, D., Cottrell, M. C., Whitehouse, M. W., and Staple, E. (1961). *Fed. Proc., Fed. Am. Soc. Exp. Biol.* **20,** 283.
Kurtz, A. P., and Dawson, C. R. (1971). *J. Med. Chem.* **14,** 729.
Lee, D. J., Wales, J. H., and Sinnhuber, R. O. (1968). *Cancer Res.* **28,** 2312.
Lehman, A. J. (1961). *Assoc. Food Drug Off. U.S., Q. Bull.* **25,** 194.
Levin, D. A. (1974). *Am. Nat.* **108,** 193.
Lichtenstein, E. P., and Casida, J. E. (1963). *J. Agric. Food Chem.* **11,** 410.
Long, E. L., Nelson, A. A., Fitzhugh, O. G., and Hansen, W. H. (1963). *Arch. Pathol.* **75,** 595.
Madrigal, R. V., Spencer, G. F., Plattner, R. D., and Smith, C. R., Jr. (1977). *Lipids* **12,** 402.

Maizite, Ya., Klyava, A., and Kluga, L. (1950). *Latv. PSR Zinat. Akad., Kim. Inst. Zinat. Raksti* **1,** 101.
Masri, M. S., Goldblatt, L. A., De Eds, F., and Kohler, G. O. (1962). *J. Pharm. Sci.* **51,** 999.
Mattson, F. H. (1973). *In* "Toxicants Occurring Naturally in Foods," 2nd ed., p. 189. Natl. Acad. Sci., Washington, D.C.
Mikolajczak, K. L. (1977). *Prog. Chem. Fats Other Lipids* **15,** 97.
Mikolajczak, K. L., Miwa, T. K., Earle, F. R., Wolff, I. A., and Jones, Q. (1961). *J. Am. Oil Chem. Soc.* **38,** 678.
Mikolajczak, K. L., Earle, F. R., and Wolff, I. A. (1962). *J. Am. Oil Chem. Soc.* **39,** 78.
Mikolajczak, K. L., Seigler, D. S., Smith, C. R., Jr., Wolff, I. A., and Bates, R. B. (1969). *Lipids* **4,** 617.
Miller, A. M., Sheehan, E. T., and Vavich, M. G. (1969). *Proc. Soc. Exp. Biol. Med.* **131,** 61.
Miller, J. A. (1973). *In* "Toxicants Occurring Naturally in Foods," 2nd ed., p. 508. Natl. Acad. Sci., Washington, D.C.
Muto, T., and Sugawara, R. (1965). *Agric. Biol. Chem.* **29,** 949.
Muto, T., Sugawara, R., and Mizoguchi, K. (1968). *Agric. Biol. Chem.* **32,** 624.
Okuda, T., Yoshida, T., Koike, S., and Toh, N. (1975). *Phytochemistry* **14,** 509.
Orris, L., van Duuren, B. L., Arroyo, E., and Langseth, L. (1964). *Proc. Am. Assoc. Cancer Res.* **5,** 49.
Paul, H., and McCay, C. M. (1942). *Arch. Biochem.* **1,** 247.
Peters, R. A., Wakelin, R. W., Martin, A. J. P., Webb, J., and Birks, F. T. (1959). *Biochem. J.* **71,** 245.
Peters, R. A., Hall, R. J., Ward, P. F. V., and Sheppard, N. (1960). *Biochem. J.* **77,** 17.
Phelps, R. A., Shenstone, F. S., Kemmerer, A. R., and Evans, R. (1965). *Poult. Sci.* **44,** 358.
Phillips, B. E., Smith, C. R., Jr., and Tallent, W. H. (1971). *Lipids* **6,** 93.
Pohl, P., and Wagner, H. (1972a). *Fette, Seifen, Anstrichm.* **74,** 424.
Pohl, P., and Wagner, H. (1972b). *Fette, Seifen, Anstrichm.* **74,** 541.
Raju, P. K., and Reiser, R. (1967). *J. Biol. Chem.* **242,** 379.
Richard, H. M., and Jennings, W. G. (1971). *J. Food Sci.* **36,** 584.
Rimington, C. (1934). *J. S. Afr. Vet. Med. Assoc.* **5,** 227.
Risser, N. M. (1965). Ph.D. Dissertation, Dept. of Food Science, University of Illinois, Urbana.
Rocquelin, G., and Cluzan, R. (1968). *Ann. Biol. Anim., Biochim., Biophys.* **8,** 395.
Roe, F. J. C., and Pierce, W. E. H. (1960). *J. Natl. Cancer Inst.* **24,** 1389.
Roehm, J. N., Lee, D. J., Wales, J. H., Polityka, S. D., and Sinnhuber, R. O. (1970). *Lipids* **5,** 80.
Roine, P., Uksila, E., Teir, H., and Rapola, J. (1960). *Z. Ernaehrungswiss.* **1,** 118.
Schneider, D. L., Sheehan, E. T., Vavich, M. G., and Kemmerer, A. R. (1968). *J. Agric. Food Chem.* **16,** 1022.
Seigler, D. S., and Butterfield, C. S. (1976). *Phytochemistry* **15,** 842.
Seigler, D. S., and Kennard, D. (1977). *Phytochemistry* **16,** 1826.
Sengupta, A., Gupta, J. K., Dutta, J., and Ghosh, A. (1973). *J. Sci. Food Agric.* **24,** 669.
Shenstone, F. S., and Vickery, J. R. (1959). *Poult. Sci.* **38,** 1055.
Sinnhuber, R. O., Lee, D. J., Wales, J. H., and Ayres, J. L. (1968). *J. Natl. Cancer Inst.* **41,** 1293.
Smith, C. R., Jr. (1970). *Prog. Chem. Fats Other Lipids* **11,** 139.
Spener, F., and Mangold, H. K. (1975). *Phytochemistry* **14,** 1369.
Sprecher, H. W., Maier, R., Barber, M., and Holman, R. T. (1965). *Biochemistry* **4,** 1856.

Stahr, E. T. (1933). "Manual of Pacific Coast Drug Plants." Science Press, Lancaster, Pennsylvania.

Steinmetz, E. F. (1957). "Codex Vegetabilis." E. F. Steinmetz Publ., Amsterdam.

Stewart, W. C., and Sinclair, R. G. (1945). *Arch. Biochem.* **8,** 7.

Stout, G. H., Balkenhol, W. G., Poling, M., and Hickernell, G. L. (1970). *J. Am. Chem. Soc.* **92,** 1070.

Symes, W. F., and Dawson, C. R. (1953). *J. Am. Chem. Soc.* **75,** 4952.

Tamaki, Y., Loschiavo, S. R., and McGinnis, A. J. (1971). *J. Agric. Food Chem.* **19,** 285.

Taylor, J. M., Jones, W. I., Hagan, E. C., Gross, M. A., Davis, D. A., and Cook, E. L. (1967). *Toxicol. Appl. Pharmacol.* **10,** 405.

Thomasson, H. J. (1955). *J. Nutr.* **56,** 455.

Thorsteinson, A. J. (1960). *Annu. Rev. Entomol.* **5,** 193.

Truitt, E. B., Jr. (1967). *In* "Ethnopharmacologic Search for Psychoactive Drugs" (D. H. Efron, ed.), Publ. 1645. USPHS, Washington, D.C.

Truitt, E. B., Jr., and Ebersberger, E. M. (1962). *Fed. Proc., Fed. Am. Soc. Exp. Biol.* **21,** 418.

Vernazza, N. (1957). *Acta Pharm. Jugosl.* **7,** 163.

Vles, R. O., Bijister, G. M., Klinekoort, J. S. W., Timmer, W. G., and Zaalberg, J. (1976). *Fette, Seifen, Anstrichm.* **78,** 128.

Wallach, O. (1893). *Justus Liebigs Ann. Chem.* **272,** 99.

Watt, J. M., and Breyer-Bandwijk, C. M. G. (1962). "The Medicinal and Poisonous Plants of Southern and Eastern Africa," 2nd ed. Livingstone, Edinburgh.

Weil, A. T. (1965). *Econ. Bot.* **19,** 194.

Wolff, I. A. (1966). *Science* **154,** 1140.

Wren, R. C. (1956). "Potter's New Cyclopaedia of Botanical Drugs and Preparations." Pitman, London.

Yano, I., Morris, L. J., Nichols, B. W., and James, A. T. (1972). *Lipids* **7,** 35.

Chapter **13**

Chemistry and Biological Effects of Glucosinolates

C. H. VAN ETTEN *And* **H. L. TOOKEY**

I. INTRODUCTION

Man has long been interested in cruciferous plants such as mustard and horseradish because of their pungency. Early investigators recognized that water must be present in order for pungency to appear. In 1840 Bussy isolated sinigrin, a thioglucoside, from black mustard seed. He also found a component, myrosinase, necessary for the hydrolysis of sinigrin to the pungent mustard oil allyl isothiocyanate (Kjaer, 1960). The hydrolytic enzyme myrosinase [thioglucosidase, EC 3.2.3.1 (Commission on Biochemistry Nomenclature, 1973)] is always present in plants that contain thioglucosides but is separated from them so that

HERBIVORES

ISBN 0-12-597180-X

crushing of the plant cells in the presence of water is required to bring about hydrolysis of the thioglucosides (Scheme 1). Before the end of World War II, only two other thioglucosides had been isolated as crystalline products: sinalbin from white mustard and glucocheirolin from

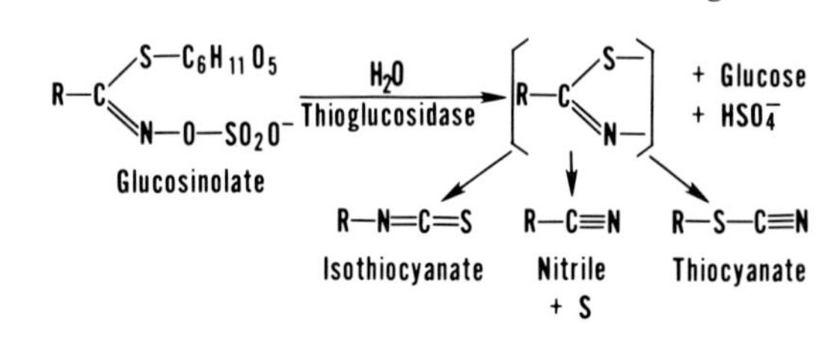

wallflower seed. The latter compound illustrates a system of naming in which "gluco" is prefixed to the name of the species from which the thioglucoside was first isolated (Kjaer, 1960). Not long after the establishment of the correct structure of these thioglucosides by Ettlinger and Lundeen (1956), the name "glucosinolate" was proposed (Ettlinger and Dateo, 1961). The suffix "ate" identifies the compound as an anion, and an appropriate chemical prefix describes the side chain "R" (see Scheme 1). Close to 80 natural glucosinolates are now known (Kjaer, 1977), 74 of which are listed by Ettlinger and Kjaer (1968) and Underhill *et al.* (1973). They occur in all species of Cruciferae and in some species from other plant families. The taxonomic relationship of plants containing glucosinolates, their biosynthesis, and the chemistry of their hydrolytic aglucon products are all much better understood than they were 20 years ago. The biological effects of the glucosinolates and their breakdown products are of practical importance, since they are consumed by man and by domestic animals. To provide orientation and perspective, each of these areas of knowledge is briefly reviewed and excellent reviews by other investigators cited. Because methods of analysis have been neglected in previous review articles, analytical methods are explored here in considerable detail. General information on the biology and chemistry of the Cruciferae is available in a recent book (Vaughan *et al.*, 1976).

II. DISTRIBUTION OF GLUCOSINOLATES IN PLANTS

All species of the Cruciferae family that have been investigated contain one or more glucosinolates. Included in this family are forages (*Brassica* spp.), rapeseed (*B. napus, B. campestris*), cole crops (*B. oleracea*), mustards (*B. nigra, B. hirta*), and horseradish (*Armoracia* spp.).

Glucosinolates are widespread throughout the plant families of the order Capparales: Capparaceae, Cruciferae, Resedaceae, and Moringaceae (Kjaer, 1973). Other plant families such as Caricaceae, Tropaeolaceae, and Limnanthaceae also contain glucosinolate-bearing species. A detailed discussion of the taxonomic distribution of glucosinolates is given by Ettlinger and Kjaer (1968).

The major horticultural crop containing glucosinolates is cabbage (*B. oleracea*). The 1974 commercial cabbage crop in the United States alone was 1 million tons for the fresh market and 0.3 million tons for sauerkraut (U.S. Department of Agriculture, 1975).

Brassica species are commonly grown as forage in certain parts of the world (Morrison, 1959; Tapper and Reay, 1973), e.g., Dwarf Essex kale (*B. napus*) and marrow-stem kale (*B. oleracea*). Crucifers also occur as weeds in pastures and so serve as inadvertent forage (Bachelard and Trikojus, 1960; Walker and Gray, 1970). *Brassica* oilseeds are important commercial crops. World production of rapeseed (*B. napus* and *B. campestris*) was 7.2 million metric tons in 1974 (Food and Agriculture Organization of the United Nations, 1974). Much of the seed meal is used as livestock feed. Condiment mustard seed (*B. hirta, B. juncea, B. nigra*) production was 0.3 million metric tons (Food and Agriculture Organization, 1974).

Glucosinolates that have been identified in common plants are listed in Table I. They are grouped according to chemical structure. Because their trivial names often appear in the literature, they are also given here. Glucosinolates 1, 2, 3, 6, 7, 14, 15, 18, 19, and 21 may hydrolyze to form pungent, steam-volatile isothiocyanates. Isothiocyanates from glucosinolates 20, 22, 23, and 24 may further degrade to release thiocyanate ion (SCN^-). The remaining glucosinolates may be hydrolyzed to form nonvolatile isothiocyanates or oxazolidinethiones. However, depending on conditions (see Section III), glucosinolates may form nitriles instead of isothiocyanates.

Table II lists the content of the principal glucosinolates in the vegetative parts of selected plants. Most of the plants are crucifers. Qualitative data have been collected over the years, but only recently have many quantitative data appeared. Data on some crops, e.g., cabbage, are drawn from enough varieties to make a reasonable assessment of glucosinolate content, but other data may be derived from only a few samples. Glucosinolate content of domestic cabbage varies over a considerable range. "Cabbage" includes both *Brassica oleracea* cultivars and the *B. campestris* cultivar called Chinese cabbage. White cabbage contains large amounts of glucosinolates with three-carbon chains, allyl glucosinolate (allyl-GS), and 3-methylsulfinylpropyl-GS. Red cabbage,

Table I. Glucosinolates in common plants

No.	Trivial name	Chemical name[a]	Structure of R group[b]
1.	Glucocapparin	Methyl-GS	CH_3
2.	Sinigrin	Allyl-GS	$CH_2{=}CH{-}CH_2$
3.	Glucoibervirin	3-Methylthiopropyl-GS	$CH_3{-}S{-}(CH_2)_3$
4.	Glucoiberin	3-Methylsulfinylpropyl-GS	$CH_3{-}SO{-}(CH_2)_3$
5.	Glucoconringiin	2-Methyl-2-hydroxypropyl-GS	$(CH_3)_2COH{-}CH_2$[c]
6.	Gluconapin	3-Butenyl-GS	$CH_2{=}CH{-}(CH_2)_2$
7.	Glucoerucin	4-Methylthiobutyl-GS	$CH_3{-}S{-}(CH_2)_4$
8.	Glucoraphanin	4-Methylsulfinylbutyl-GS	$CH_3SO(CH_2)_4$
9.	Glucoerysolin	4-Methylsulfonylbutyl-GS	$CH_3SO_2(CH_2)_4$
10.	Progoitrin	2(*R*)-Hydroxy-3-butenyl-GS	$CH_2{=}CH{-}\overset{*}{C}HOHCH_2$[c,d]
11.	*epi*Progoitrin	2(*S*)-Hydroxy-3-butenyl-GS	$CH_2{=}CH{-}\overset{*}{C}HOHCH_2$[c,d]
12.	—	4-Methylthio-3-butenyl-GS	$CH_3S{-}CH{=}CH{-}(CH_2)_2$
13.	Glucocleomin	2-Methyl-2-hydroxybutyl-GS	$CH_3{-}CH_2{-}C(CH_3)OH{-}CH_2$[c]
14.	Glucobrassicanapin	4-Pentenyl-GS	$CH_2{=}CH{-}(CH_2)_3$
15.	Glucoberteroin	5-Methylthiopentyl-GS	$CH_3{-}S{-}(CH_2)_5$
16.	Glucoalyssin	5-Methylsulfinylpentyl-GS	$CH_3{-}SO{-}(CH_2)_5$
17.	—	2-Hydroxy-4-pentenyl-GS	$CH_2{=}CH{-}CH_2{-}CHOH{-}CH$
18.	Gluconasturtiin	2-Phenylethyl-GS	$C_6H_5CH_2CH_2$
19.	Glucotropaeolin	Benzyl-GS	$C_6H_5CH_2$
20.	Sinalbin	*p*-Hydroxybenzyl-GS	*p*-$HOC_6H_4CH_2$[e]
21.	Glucolimnanthin	*m*-Methoxybenzyl-GS	*m*-$CH_3OC_6H_4{-}CH_2$
22.	Glucobrassicin	3-Indolylmethyl-GS	CH_2[e]; N—H
23.	Neoglucobrassicin	3-(*N*-Methoxy)indolylmethyl-GS	CH_2[e]; N—OCH_3
24.	Sulfoglucobrassicin	3-(*N*-Sulfo)indolylmethyl-GS	CH_2[e]; N—SO_{3-}

[a] GS, glucosinolate.

[b] See Scheme 1.

[c] Cyclic oxazolidinethione forms if GS is hydrolyzed under conditions expected to isothiocyanate.

[d] Asterisk denotes center of asymmetry.

[e] SCN^- forms as a decomposition product from unstable isothiocyanate.

Table II. Glucosinolate content of vegetative part of plants

Plant name	Plant part	Glucosinolate[a]	No. varieties tested	Concentration[b] (μg/gm) Mean	Range	Reference
Brassica oleracea L. var. *gongylodes,* kohlrabi	Stem	3-Indolylmethyl-GS[c]	—	133	—	Michajlovskij *et al.* (1969)
		3-Methylsulfinylpropyl-GS		89	—	
		2-Butyl-GS		105	—	
		3-Butenyl-GS		45	—	
var. *capitata* white cabbage	Head (leaves)	Allyl-GS	19	128	14–530	VanEtten *et al.* (1976)
		3-Methylsulfinylpropyl-GS		148	52–302	
		3-Butenyl-GS		10	1–40	
		4-Methylsulfinylbutyl-GS		30	1–125	
		2(*R*)-Hydroxy-3-butenyl-GS		17	4–78	
		3-Indolylmethyl-GS[c]		215	123–463	
		Total[d]		595	275–975	
red cabbage	Head (leaves)	Allyl-GS	7	33	7–62	VanEtten *et al.* (1976), C. H. VanEtten (unpublished)
		3-Methylsulfinylpropyl-GS		49	21–86	
		3-Butenyl-GS		36	16–56	
		4-Methylsulfinylbutyl-GS		215	138–290	
		2(*R*)-Hydroxy-3-butenyl-GS		33	18–61	
		3-Indolylmethyl-GS[c]		275	—	
		Total[d]		675	378–940	
broccoli, brussels sprouts, cauliflower, collards	stem, bud, leaf	Similar to cabbage	—	—	—	Mullin and Sahasrabudhe (1978), C. H. VanEtten (unpublished)
var. *accephala* kale, marrow stem	Forage	2(*R*)-Hydroxy-3-butenyl-GS	1	3	—	Josefsson *et al.* (1972)
		3-Indolylmethyl-GS[c]		230	—	
		Other		115	—	

(*continued*)

Table II (*continued*)

Plant name	Plant part	Glucosinolate[a]	No. varieties tested	Concentration[b] (μg/gm)		Reference
				Mean	Range	
Brassica campestris L. ssp. *pekinensis*, Chinese cabbage	Head	3-Butenyl-GS	14	48	3–228	Daxenbichler *et al.* (1979)
		2(*R*)-Hydroxy-3-butenyl-GS		34	3–178	
		4-Pentenyl-GS		80	11–250	
		5-Methylsulfinylpentyl-GS		59	5–140	
		2-Phenylethyl-GS		54	15–248	
		3-Indolylmethyl-GS[c]		185	87–490	
		Total[d]		495	159–1242	
ssp. *rapifera*, white turnip	Root	2(*R*)-Hydroxy-3-butenyl-GS	—	300	—	Mullin and Sahasrabudhe (1978)
		3-Indolylmethyl-GS[c]		123	—	
		Others, as butyl-GS		625		
	Root	2-Phenylethyl-GS	—	S[e]		Astwood *et al.* (1949), Lichtenstein *et al.* (1962)

	Leaf	2(*R*)-Hydroxy-3-butenyl-GS	L[e]	—	Tapper and MacGibbon (1967)
		2(*R*)-Hydroxy-4-pentenyl-GS	S[e]		
B. napus L., rutabaga	Root	2(*R*)-Hydroxy-3-butenyl-GS	330	—	Mullin and Sahasrabudhe (1978)
		3-Indolylmethyl-GS[c]	60		
		Others, as butyl-GS	490		
Carica papaya L., papaya	Latex (dry)	Benzyl-GS		67,000–106,000	Tang (1973)
Raphanus sativus L., radish					
	Root	4-Methylthio-3-butenyl-GS	L[e]		Friis and Kjaer (1966), Gmelin and Virtanen (1961)
		3-Indolylmethyl-GS[c]	S[e]		
Armoracia lapathifolia and *rusticana*, horseradish					
	Root	Allyl-GS	4500		Stoll and Seebeck (1948), Stahmann *et al.* (1943)
		2-Phenylethyl-GS	S		

[a] GS, glucosinolate.
[b] Calculated as glucosinolate ion on fresh weight basis for vegetables.
[c] Includes both 3-indolylmethyl-GS and 3-(*N*-methoxy)indolylmethyl-GS.
[d] Calculated from enzymatically released glucose by using an average molecular weight of 418 for GS ion.
[e] When concentration is not reported, S means small; L, large.

Table III. Glucosinolate content of seed meals

Plant name	No. cultivars tested	Glucosinolate[a]	Concentration (%)[b]		Reference
			Mean	Range	
Brassica campestris L., rape					
Winter type	4	3-Butenyl-GS	2.92	2.64–3.45	Josefsson and Appelqvist (1968)
		2(*R*)-Hydroxy-3-butenyl-GS	0.39	0.18–0.51	
Summer type	2	3-Butenyl-GS	1.83	1.60–1.93	Josefsson and Appelqvist (1968)
		2(*R*)-Hydroxy-3-butenyl-GS	1.03	0.97–1.06	
Brassica napus L., rape					
Winter type	17	3-Butenyl-GS	1.54	0.98–2.05	Josefsson and Appelqvist (1968)
		2(*R*)-Hydroxy-3-butenyl-GS	3.59	2.26–4.68	
Summer type	5	3-Butenyl-GS	1.34	1.15–1.60	Josefsson and Appelqvist (1968)
		2(*R*)-Hydroxy-3-butenyl-GS	3.02	2.60–3.40	
Bronowski	5	3-Butenyl-GS	0.40	0.65–1.15	Josefsson and Appelqvist (1968)
		2(*R*)-Hydroxy-3-butenyl-GS	0.48	0.09–3.40	
Brassica hirta Moench (*Sinapis alba* L.), white mustard	5	*p*-Hydroxybenzyl-GS	8.9	7.9–10.1	Josefsson (1970)
B. nigra (L.) Koch, black mustard	2	Allyl-GS	—	4.2–5.0	Ettlinger and Thompson (1962)

[a] GS, glucosinolate

[b] Calculated as glucosinolate ion on air-dry basis of defatted seed meal.

however, contains more of the four-carbon glucosinolates (e.g., 4-methylsulfinylbutyl-GS) than does white cabbage. Chinese cabbage has no three-carbon glucosinolates but does have many of the five-carbon glucosinolates, 4-pentenyl-GS, and 5-methylsulfinylpentyl-GS. All three cabbages have 3-indolylmethyl-GS's as the most abundant glucosinolates.

Glucosinolates occur in all parts of the plant but are usually more concentrated in the seed (Table III). Indolylmethyl-GS's usually occur only in small amounts in seeds.

The *B. napus* varieties of rapeseed (excluding Bronowski) are much higher than the *B. campestris* rapeseeds in 2-hydroxy-3-butenyl-GS, the source of goitrin (5-vinyloxazolidine-2-thione). The response of animals fed rapeseed is difficult to interpret if the kind (*B. napus* or *B. campestris*) is not known. The Bronowski variety recently developed for its low total glucosinolate content gives a much better animal feeding response than the standard varieties from either *B. campestris* or *B. napus*. More detailed information on the amounts of glucosinolates in the more common plants that are used for food or feed is reported elsewhere (Josefsson, 1967a; VanEtten *et al.*, 1976; Daxenbichler *et al.*, 1979; Tookey *et al.*, 1979).

III. CHEMISTRY AND BIOCHEMISTRY OF GLUCOSINOLATES

A. Structure

The glucosinolate structure shown in Scheme 1 was confirmed by the synthesis of benzyl-GS (glucotropaeolin) by Ettlinger and Lundeen (1957). Glucosinolates are anions and thus occur in nature as salts. They are usually regarded as potassium salts, although a complex organic cation, sinapine, occurs in the crucifers (Schultz and Gmelin, 1952, 1953). All glucosinolates contain β-D-thioglucose (Ettlinger and Kjaer, 1968) and sulfate moieties. The configuration of the sulfate moiety relative to the R group is anti (Waser and Watson, 1963; Marsh and Waser, 1970).

B. Plant Thioglucosidase

Glucosinolates may be destroyed by heat, as during processing of oilseeds, or they may be enzymatically hydrolyzed. Since all glucosinolates in a plant appear to be accompanied by thioglucosidase, enzymatic hydrolysis is the rule. The enzyme and its substrate glucosinolate appear to be localized in different parts of the cellular structure so that glucosinolates are apparently stable in the intact plant (Kjaer, 1960; Pihakashi and Iversen, 1976). For example, benzyl-GS of *Carica papaya* seed is concentrated in the endosperm, whereas thioglucosidase is concentrated in the integument (Tang, 1973). Plant materials cannot be crushed or ground without glucosinolate hydrolysis unless the material is dry enough to preclude hydrolysis. Air-dried leaves or seed meals are usually dry enough to be stable.

In normal plant metabolism, little is known of the dynamic state of glucosinolates, i.e., whether there is constant biosynthesis and catabolism. There is, however, some evidence that glucosinolates are utilized in the metabolism of germinating seeds. In *Crambe abyssinica* nearly all the glucosinolates are hydrolyzed during the first 3 days of seed germination (Tookey and Wolff, 1970).

Before 1959, it was thought that two enzymes were necessary for glucosinolate hydrolysis: a thioglucosidase to release glucose and a separate sulfatase to release HSO_4^- (Sandberg and Holly, 1932; Neuberg and Schoenbeck, 1933). A sulfatase enzyme from snails or marine animals does indeed form desulfoglucosinolates (Takahashi, 1960), but it is not a requirement for glucosinolate hydrolysis. Evidence showing that only one enzyme, a thioglucosidase, is necessary was provided by Nagashima and Uchiyama (1959). A theory consistent with this single enzyme was proposed by Ettlinger *et al.* (1961) and reiterated by Miller (1965). Thioglucosidases from crucifers are dependent on sulfhydryl groups in the enzyme protein for their activity (Nagashima and Uchiyama, 1959; Tsuruo and Hata, 1967). Their activity is enhanced by the presence of ascorbate (Ettlinger *et al.*, 1961; Tsuruo and Hata, 1967; Tookey, 1973a). Several isozymes (proteins that vary in size but have the same enzyme activity) have been isolated (Lonnerdal and Janson, 1973; Henderson and McEwan, 1972; Vose, 1972).

C. Products from Enzymatic Hydrolysis

D-Glucose and HSO_4^- are always formed during thioglucosidase hydrolysis of the glucosinolates (Scheme 1), but factors other than thioglucosidase determine which of the aglucon products will predominate—isothiocyanate, nitrile, or thiocyanate (VanEtten and Tookey, 1978; Tookey *et al.*, 1979). Isothiocyanates may be formed by a Lössen type of rearrangement following the removal of glucose (Ettlinger *et al.*, 1961). This product, R–NCS, was long regarded as the "normal" hydrolytic product. Semipurified thioglucosidase separated from a crucifer can be expected to hydrolyze glucosinolates to isothiocyanates at nearly neutral pH values. Such an enzyme from white mustard, *Brassica hirta,* has been extensively used to hydrolyze glucosinolates for analytical purposes.

Most of the known glucosinolates form stable organic isothiocyanates or nitriles when hydrolyzed by the accompanying enzyme (Kjaer, 1960). Certain isothiocyanates are, however, unstable and react further to form either ring compounds or SCN^-, as will be discussed separately. Organic thiocyanate formation has been reported from glucosinolates in several species, e.g., *Thlaspi arvense* L. and species of *Lepidium* (Gmelin and Virtanen, 1959).

Early investigators isolated organic nitriles instead of the expected isothiocyanates. Schmid and Karrer (1948) isolated 1-cyano-4-methylsulfinyl-3-butene from radish seed. They stated that the isothiocyanate was formed from finely ground seed meal, but plant material that was handled very little contained a nitrile instead of isothiocyanate. Prior treatment of the plant material and the conditions under which the enzyme hydrolysis is carried out influence the course of hydrolysis (Table IV). Autolysis of fresh plant material in the presence of small amounts of water is the most favorable condition for nitrile formation. The studies cited in Table IV did not include a measure of products from the indolylmethyl-GS's of cabbage leaves.

Table IV. Factors influencing the products of glucosinolate hydrolysis[a]

	Products	
Plant part	Nitriles	Isothiocyanates
	Favored by prior plant treatment	
Seed	Fresh	Long storage
	Low-temperature storage	High-temperature storage
Leaves	Fresh	Air-dried
Seed	Air-dried	Dried with heat
Seed meal	Air-dried	Dried with heat
	Favored by conditions of hydrolysis	
Seed meal	Autolysis of fresh material	Hydrolysis with prepared thioglucosidase
	Low pH	Neutral pH
	Low temperature (0°–25°C)	High temperature (up to 75°C)
	Low moisture	High dilution with water

[a] Conditions observed by study of crambe seed and rapeseed, defatted seed, leaves, and cabbage leaves (VanEtten *et al.*, 1966; VanEtten and Daxenbichler, 1971; Daxenbichler *et al.*, 1977).

If nitriles are formed, a terminal double bond in the nitrile may be converted to an epithio group (Scheme 2). Such epithionitriles were first identified from progoitrin and *epi*-progoitrin (Daxenbichler *et al.*, 1967, 1968). A protein favoring the formation of epithio compounds but con-

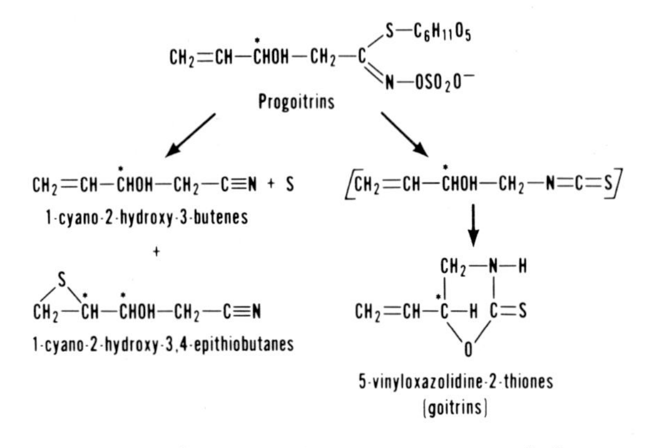

taining no thioglucosidase activity was separated from crambe seed meal (Tookey, 1973b). When this fraction is added to crude thioglucosidase and the substrate *epi*-progoitrin (no. 11 of Table I), epithionitriles are formed instead of the unsaturated nitrile. Formation of epithionitriles is apparently widespread, having been shown also in *Brassica, Nasturtium, Allysum, Cakile,* and others (Tookey *et al.*, 1979, and references therein).

Under certain conditions of hydrolysis, some 3-hydroxypent-4-enethionamides may also be formed from the progoitrin if iron salts are present (Austin *et al.*, 1968).

The commonly found progoitrins (Greer, 1962b; Daxenbichler *et al.*, 1965) do not yield stable isothiocyanates. If these glucosinolates (nos. 10 and 11, Table I) are hydrolyzed under conditions expected to form isothiocyanates, a further reaction with the hydroxyl group forms the ring compound 5-vinyloxazolidine-2-thione (5-vinyl-OZT) (Scheme 2). Other glucosinolates with a hydroxyl in the side chain also form oxazolidinethiones.

Glucobrassicin and neoglucobrassicin (nos. 22 and 23, Table I) may hydrolyze to give unstable 3-indolylmethyl isothiocyanates or nitriles (Scheme 3) (Gmelin and Virtanen, 1961, 1962). Elliott and Stowe (1970) identified a third member of this group (no. 24, Table I). The SCN^- formed from these unstable isothiocyanates is used as a measure of the three glucosinolates. The most common, 3-indolylmethyl-GS, is found at levels up to 0.23% of the fresh weight in seeds of woad (*Isatis tinctoria* L.); 1-week-old seedlings also contain substantial amounts of the re-

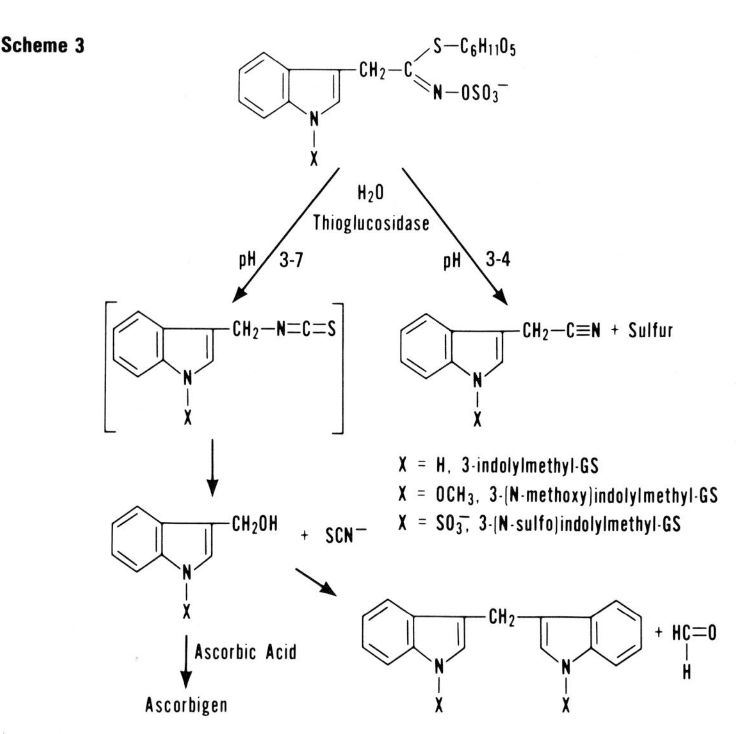

maining two indolylmethyl-GS's (Elliott and Stowe, 1971). These glucosinolates are found in low concentrations in other seeds but appear to be high in rapidly growing tissue (Josefsson, 1967b; Chang and Bible, 1974). According to recent work (Mahadevan and Stowe, 1972) woad leaves also contain "desthioglucobrassicin," the nonsulfated 3-indolylmethyl-GS.

p-Hydroxybenzyl-GS (found in seed of white mustard) forms the unstable *p*-hydroxybenzyl isothiocyanate that decomposes to give SCN^- and *p*-hydroxybenzyl alcohol (Kawakishi *et al.*, 1967). Estimation of SCN^- is an indirect method for quantitative analysis.

D. Biosynthesis

It was suggested (Kjaer, 1954; Ettlinger and Lundeen, 1956) that natural amino acids may be precursors of the aglucon part of glucosinolates because they have similar structures. During the past 15 years, biosynthesis of 15 of the glucosinolates has been examined experimentally. The studies involved feeding to plants amino acids labeled with ^3H, ^{14}C, ^{15}N, or ^{35}S and determining the extent of incorporation of the isotopes in the glucosinolates as they were synthesized by the plants.

The results of this experimental approach indicate that all glucosinolates are derived from amino acids. Thus, tryptophan is a precursor for 3-indolylmethyl-GS, and phenylalanine for benzyl-GS and 2-phenylethyl-GS. In many cases, the amino acid undergoes carbon chain elongation before incorporation into the glucosinolate. This is illustrated in Scheme 4. Methionine (A) is lengthened by two carbon atoms to form (B), which is then oxidized stepwise to an aldoxime. Sulfur is added to the aldoxime, converting it to a thiohydroximic acid (C), where X represents an unknown group. Glucose is added to form the desulfo-GS, which is finally converted to a glucosinolate (D), in this case 4-methylthiobutyl-GS. The thio group may be oxidized to 4-methylsulfinylbutyl-GS or to 4-methylsulfonylbutyl-GS. The related 3-butenyl-GS is also derived from 4-methylthiobutyl-GS by loss of the methylthio group. More detail on biosynthesis is available in reviews by Kjaer and Larsen (1973) and Underhill *et al.* (1973).

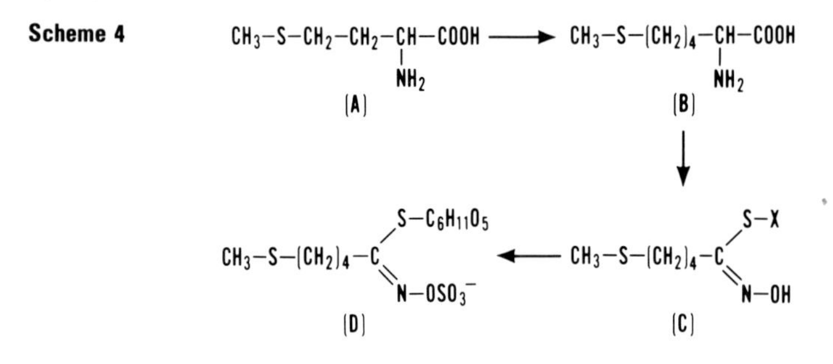

IV. METHODS OF ANALYSIS

A. Early History

The first analytical methods consisted of isolation and identification of organic isothiocyanates (mustard oils) by classic procedures. After the seed meal or plant part was crushed and allowed to stand in water, the volatile isothiocyanates were steam-distilled from the mixture and collected in aqueous alcohol, from which they could be separated as a neat oil (Stoll and Jucker, 1955). Further characterization involved reaction of the oil with ammonia to give a monosubstituted crystalline thiourea or reaction with phenylhydrazine to give a phenylalkyl thiosemicarbazide. Isothiocyanates that were not steam volatile were separated from the aqueous plant slurry by extraction with a water-insoluble organic solvent.

Quantitative estimation of the isothiocyanates as a group consisted primarily of reaction of their thiourea derivatives with an excess of standard silver nitrate followed by titration of the excess silver ion with standard ammonium thiocyanate (Stoll and Jucker, 1955; Wetter, 1955). Methods based on ultraviolet absorption of 5-vinyl-OZT (Astwood *et al.*, 1949) and of thiourea derivatives for the isothiocyanates (Kjaer *et al.*, 1953) were more sensitive than the titration methods. The ultraviolet methods were adapted to rapeseed by Wetter (1957) and Appleqvist and Josefsson (1967). The latter authors describe the separation of oxazolidinethione from isothiocyanates by extraction into isooctane, followed by quantitative conversion to thiourea with ammonia. The thioureas as a group were determined from their characteristic ultraviolet absorption.

Early measurements of sinigrin were made by estimation of the glucose or sulfate released by thioglucosidase hydrolysis (Sandberg and Holly, 1932). Anthrone reagent was later used in an estimation of the glucose released from all glucosinolates (Schultz and Gmelin, 1954).

B. Influence of Newer Methods

The development of separation techniques such as paper, thin-layer, column, and gas–liquid chromatography (glc) enabled workers to isolate and characterize many new glucosinolates as well as various aglucon hydrolytic products. Not only did the methods of separation assist in the isolation work, but they were also useful in the analyses for specific glucosinolates and in chemotaxonomic studies (Rodman, 1978).

Paper chromatographic methods for the separation of glucosinolates include those described by Schultz and Wagner (1956); for thioureas of isothiocyanates by Kjaer and Rubinstein (1953), Kjaer *et al.* (1953), and Ettlinger and Thompson (1962); and for 5-vinyl-OZT (goitrin) by Kreula and Kiesvaara (1959). Thin-layer chromatography was used for the separation of indolyl compounds in the identification of 3-(*N*-methoxy)indolylmethyl-GS (Gmelin and Virtanen, 1962).

Separation of isothiocyanates by glc was reported by Jart (1961). Similar glc methods were applied to allyl isothiocyanate (allyl-NCS) (Kirk *et al.*, 1964; Miller, 1965). A proposed glc method for allyl-NCS in mustard seed was recommended by Anderson (1970) for adoption as the official method of the Association of Official Analytical Chemists. Problems to overcome in methods of analysis should be appreciated from that study. Although recovery of added allyl-NCS by the glc method averaged only 90.8% (standard deviation 5.9), the allyl-NCS from mustard seed by the method averaged as much as 22% higher than by three other methods. The proposed method was more specific and gave slightly better standard deviation and consistent and significantly higher results. A method for the determination of butenyl- and pentenyl-NCS in rapeseed was reported by Youngs and Wetter (1967) and was later modified by Wetter and Youngs (1976). A gas chromatographic method was used by Daxenbichler *et al.* (1970) to determine the organic nitriles and 5-vinyl-OZT from 2-hydroxy-3-butenyl-GS's (progoitrins).

Direct glc separation and determination of trimethylsilyl glucosinolate derivatives or the corresponding desulfoglucosinolates were reported for use on seed meals by Underhill and Kirkland (1971) and Persson (1974). Thies (1977) described modifications in which the glucosinolates were separated from accompanying sugars by adsorption and elution on an anion-exchange resin. This modification permitted the use of the method on sugar-containing extracts such as those from leaves or stems.

Methods in vogue involving high-pressure column chromatography should find use in the analysis for individual glucosinolates as well as in their isolation. It would be improper in discussing new techniques not to mention physical methods of identification such as the use of nuclear magnetic resonance and glc combined with mass spectrophotometry.

C. Current Methods

1. Total Glucosinolates

Quantitative measurement of either glucose or HSO_4^- may be used to measure total glucosinolates (see Section III,C). For such a measurement, one must be certain that none of the glucosinolates in the sample have been hydrolyzed during preparation. Usually this is accomplished by wet heat inactivation of the thioglucosidase before any hydrolysis of the glucosinolates occurs.

Analysis for HSO_4^- (McGhee *et al.*, 1965; VanEtten *et al.*, 1965; Josefsson and Appelqvist, 1968) did not provide the speed or permit the small sample size required for use in plant breeding. The method of Lein and Schön (1969) and Lein (1970, 1972) utilized a specific and sensitive enzyme system for glucose: hexokinase plus ATP followed by glucose-6-phosphate dehydrogenase and NAD^+. The end product, NADH, is readily measured at 340 nm in a spectrophotometer. The method gives a precision of better than $\pm 1\%$ and may be applied to one cotyledon of a single rapeseed.

Another enzyme system specific for glucose is also used to estimate total glucosinolates. Glucose released by thioglucosidase hydrolysis reacts with glucose oxidase to form hydrogen peroxide, which, in the presence of peroxidase, reacts with a chromogen to form a colored product (Bjorkman, 1972). The reagent is subject to interferences found in plants, such as phenolic compounds and ascorbic acid. Some of these interferences may be removed with charcoal before the glucose enzyme system is utilized (VanEtten *et al.*, 1974). Test paper designed for urinary glucose measurement has been used successfully to detect rapeseed of low glucosinolate content (Lein, 1970; VanEtten *et al.*, 1974; McGregor and Downey, 1975). A recent method specific for glucosinolate glucose in plant extracts involves adsorption of glucosinolate ions on an anion-exchange resin. The glucosinolates are treated with thioglucosidase while on the resin, and the released glucose is then measured by a glucose-specific enzyme system (VanEtten and Daxenbichler, 1977).

2. *Measurement of Thiocyanate Ions*

A review of the many analytical methods for determining SCN^- has been published (Ashworth, 1975). Of these methods, the preferred procedure for SCN^- in cruciferous plants is to treat the plant extract (after hydrolysis by thioglucosidase) with acidic ferric nitrate and measure the resulting colored ferric thiocyanate complex. Substances that interfere are either colored or react with ferric iron to form a color. These interfering colors remain when a drop of mercuric chloride is added to decompose the ferric thiocyanate complex. This residual absorbance in the sample is subtracted from the absorbance present after the first addition of ferric nitrate to give the net absorbance caused by the SCN^-. Suitable methods specifically designed for the estimation of SCN^- from *p*-hydroxybenzyl-GS of seed are described by Ettlinger and Thompson (1962) and Josefsson (1968). Methods used for the estimation of SCN^- from the 3-indolymethyl-GS's of vegetative tissue are described by Michajlovskij and Langer (1958), Johnston and Jones (1966), and Josefsson (1967b).

It has been assumed that SCN^- is formed from the indolylmethyl- or *p*-hydroxybenzyl-GS's. However, as reported by Srivastava and Hill (1975), the small amount of SCN^- found in a low- or a high-glucosinolate rapeseed was about the same from the autolyzed seed in which the remaining glucosinolates had hydrolyzed to give nitriles as from the seed hydrolyzed at pH 7 with added myrosinase to form isothiocyanates and goitrin. The source of SCN^- in rapeseed has not been positively identified.

3. *Measurement of Individual Glucosinolates*

Determination of individual glucosinolates may be either by analysis of aglucon hydrolytic products or by assay of derivatized glucosinolates. Estimation of glucosinolates by analysis of aglucon products may be accomplished by first adsorbing glucosinolates on an anion-exchange resin to separate them from interfering materials. The glucosinolates are then enzymatically hydrolyzed while on the resin under conditions in which they are converted entirely to isothiocyanates (or further react to oxazolidinethione or SCN^- as described in Section III). The mixture of isothiocyanates and oxazolidinethiones are extracted into a suitable solvent such as dichloromethane and then analyzed by glc (Daxenbichler *et al.*, 1970; Daxenbichler and VanEtten, 1977). Analysis of isothiocyanates and oxazolidinethiones from the ion-exchange resin and SCN^- by a separate procedure usually accounts for more than 80% of total glucosinolates as determined by free glucose formed during the hydrolysis (VanEtten and Daxenbichler, 1977).

Analysis of trimethylsilyl derivatives of glucosinolates by glc has been adapted to plant breeding work (Thies, 1977). Thies (1978) recently reported a modification of the method in which desulfonated glucosinolates are analyzed. The method consists of sulfate removal from glucosinolates adsorbed on an ion-exchange resin by the use of an enzyme from the snail *Helix pomatia*. The desulfonated glucosinolates are neutral and easily washed from the resin for silylation and subsequent glc analysis. Advantages over Thies' 1977 method are claimed to be better separation and better reproducibility. The separation of the glucosinolates by adsorption on an ion-exchange resin followed by (1) hydrolysis to give aglucon products (Daxenbichler and VanEtten, 1977) or (2) hydrolysis to give desulfonated glucosinolates (Thies, 1978) provides procedures by which it is hoped all the glucosinolates can be determined accurately.

4. *Analysis of Food and Feed*

In the preparation of food or feed, processing of the material has a marked effect on glucosinolates and thioglucosidase activity. The organic aglucon portion of the molecule (see Section III) may form isothiocyanates or nitriles. Depending on their structure, labile isothiocyanates may form SCN^- and other products or, if they have a hydroxyl in the proper position, they may form oxazolidinethiones. The nitriles may form an epithio group if the aglucon has terminal unsaturation. Since these compounds vary in their biological effect, it is important to know of the presence of specific nitriles, isothiocyanates, oxazolidinethiones, and SCN^-, as well as any thioglucosidase activity and unhydrolyzed glucosinolates. Isothiocyanates, oxazolidinethiones, and nitriles in the food material may be directly extracted into dichloromethane and determined by glc as reported for autolyzed cabbage (Daxenbichler *et al.*, 1977). Unhydrolyzed glucosinolates may be separated on an anion-exchange column and then determined by glc (VanEtten *et al.*, 1976). Analyses of SCN^- from intact glucosinolates and SCN^- existing free would require testing aqueous extracts treated with thioglucosidase and corresponding extracts not enzyme treated. Extraneous glc peaks become a problem if the glc method is further modified to detect nitriles and oxazolidinethiones at a few parts per million in body tissues of cattle (VanEtten *et al.*, 1977).

The information on the interaction of herbivores and plants containing glucosinolates in the native environment is limited. In order to show the relationship at the molecular level between herbivores and plants containing glucosinolates, the current analytical methods such as those we have reviewed will have to be adapted to determine which specific glucosinolate hydrolytic products are consumed by the animal. For example, it is very likely that the glucosinolate hydrolytic products will be different if live green foliage is consumed than if dry hay from the foliage is fed (VanEtten and Daxenbichler, 1971).

V. BIOLOGICAL EFFECTS

A. General

Thyroid enlargement or hypothyroid goiter has long been associated with the consumption of certain foods or feeds from crucifers (Roche and Lissitzky, 1960). In fact, enlarged thyroids in human beings in areas where the iodine content of the diet is marginal were reported to be enhanced by the consumption of cabbage and related *Brassica* (Kelley and Snedden, 1960; Michajlovskij *et al.*, 1969).

Evidence suggested that enlarged thyroids in children in Tasmania might be caused by the consumption of milk from cows fed high levels of marrow-stem kale (Clements and Wishart, 1956). Enlarged thyroids were observed as increased amounts of kale were fed to the dairy cattle, in spite of an increase in the iodine content of the diet of the children. Calves from the cows also showed hyperplasia of the thyroid. Although assignment of the cause of the hypothyroidism to specific compounds in the milk is still controversial, the observation remains that some malfunction of the thyroid developed that could be explained by high consumption of the marrow-stem kale forage by dairy cows in the area. It should be noted that human goiter resulting from cruciferous foods or feeds has not generally been regarded as a public health problem. Greer (1962a) estimated that 96% of all human goiter (hypothyroid) was related to iodine deficiency.

Physiological effects of glucosinolates are not confined to the thyroid; damage to liver or kidneys has been attributed to nitriles from glucosinolate hydrolysis. General reviews on the glucosinolates as a source of natural toxicants include those by VanEtten and Wolff (1973), VanEtten and Tookey (1978), VanEtten (1979), and Tookey *et al.* (1979).

B. Effect of Specific Organic Compounds

1. *(R)- and (S)-5-Vinyl-OZT (Goitrin)*

Inhibition of thyroid uptake of radioiodine by goitrin from *Brassica* was shown by Astwood *et al.* (1949). Following synthesis of the *R* and *S* forms by Ettlinger (1950), Greer (1962a) found both forms to be equally potent as antithyroid agents. Measurement of the reduction of radioiodine uptake by the thyroid is a rapid method of estimating the goitrogenic effect of a substance in the animal's ration (VanderLaan and Bissell, 1946; Astwood *et al.*, 1945; Stanley and Astwood, 1947). Peltola (1965) reported that levels of 5-vinyl-OZT in the ration that are too low to affect the radioiodine test may cause minor enlargement of the thyroid of rats.

Rats fed (*R*)-5-vinyl-OZT as 0.23% of the ration for 90 days developed enlarged thyroids (twice as large as controls) and had body weight gains of 85% of controls (VanEtten *et al.*, 1969a). Similar results were found by feeding synthetic and isolated 5-vinyl-OZT to poultry (Clandinin *et al.*, 1966; Matsumoto *et al.*, 1968).

These goitrins act by inhibiting the incorporation of iodine into thyroxine precursors and interfering with secretion of thyroxine (Matsumoto *et al.*, 1968; Akiba and Matsumoto, 1973). The goitrogenic effect is not readily overcome by increasing iodine content of the diet (Greer *et al.*, 1964). Goitrins are reported to be nonteratogenic in the rat (Khera, 1977).

2. *Thiocyanate Ion*

By feeding human beings 0.5 gm KSCN per day, the SCN^- content of the blood was increased 5–10 mg per 100 ml (Aswood, 1943), at which level the iodine uptake of the thyroid was inhibited (Stanley and Astwood, 1947). Thyroid enlargement in animals caused by SCN^- is prevented by adding additional iodine to the diet unless the SCN^- is fed at very high levels (Greer *et al.*, 1966).

The feeding of 1.5 to 3.0 gm/day of SCN^- to milk cows, which is the amount expected in the daily feeding of 15 to 30 kg of *Brassica* fodder, gives an increase in SCN^- content of the milk from 1.7–3.5 mg/liter to 6.2–9.3 mg/liter (Piironen and Virtanen, 1963). This amount was not considered harmful to those consuming the milk. However, as the SCN^- increased in the milk, the iodine content of the milk decreased. Iodine increased to normal levels again after the feeding of SCN^- was stopped. Similar observations have been reported by others (Garner *et al.*, 1960; Miller *et al.*, 1965). Thus, SCN^- in the diet of the mother could cause goiter in the young that consumed iodine-deficient milk. Goiter in young chicks also may be caused by low iodine content of eggs from hens on a ration containing rapeseed meal (Goh and Clandinin, 1977).

Thiocyanate ion is a common component of foods and feeds from crucifers; it is derived from the spontaneous decomposition of 3-indolylmethyl-NCS's and *p*-hydroxybenzyl-NCS. In addition to SCN^- from these glucosinolates, it is also formed by the mammalian liver as a product of detoxification of organic nitriles and isothiocyanates (Wood, 1975) (see also Section V,C).

3. *Isothiocyanates*

Isothiocyanates are irritating to skin and mucous membranes. They inhibit thyroid uptake of radioiodine (Langer, 1964; Bachelard and Trikojus, 1960). Since they can be metabolically converted to SCN^-, the antithyroid effects ascribed to isothiocyanates may be partially caused by SCN^-. One isothiocyanate, 3-methylsulfonylpropyl-NCS, is converted by rumen fluid to a goitrogenic disubstituted thiourea (Bachelard and Trikojus, 1963).

4. *Organic Nitriles*

Little is known of the physiological effects of most of the nitrile hydrolytic products. Only the nitriles derived from 2-hydroxy-3-butenyl-GS (progoitrin and *epi*-progoitrin) have been studied to any extent.

Poor growth accompanied by pathological changes in the liver and kidney was observed in rats fed crambe seed meal containing *epi*-progoitrin and thioglucosidase that, under the conditions of autolysis, formed nitriles. A mixture of these nitriles (1-cyano-2-hydroxy-3-butene and 1-cyano-2-hydroxy-3,4-epithiobutanes) fed to rats as 0.1% of their ration caused severe growth repression and major lesions of liver and kidney (VanEtten *et al.*, 1969a). The livers showed necrosis and bile duct hyperplasia; the kidneys showed damage to the tubular epithelium. Similar results were obtained when rats were fed rapeseed meals containing nitriles (Lo and Hill, 1972). Acute toxicity tests in mice (LD_{50}) are 170 mg/kg for 1-cyano-2-hydroxy-3-butene and 178 and 240 mg/kg for *erythro-* and *threo*-1-cyano-2-hydroxy-3,4-epithiobutane, respectively (Van Etten *et al.*, 1969b). By comparison, 5-vinyl-OZT has an LD_{50} of 1260 to 1415 mg/kg.

5. Glucosinolates

It was first thought that, if foods or feeds containing glucosinolates were cooked without hydrolysis by the accompanying thioglucosidase, the goitrogenic properties of the food would be lost (Greer, 1957). However, Greer (1962a) found that radioiodine uptake by the thyroid was also suppressed by feeding progoitrin (2-hydroxy-3-butenyl-GS). The goitrogenic activity developed more slowly than if 5-vinyl-OZT were fed (Greer and Deeney, 1959). The oxazolidinethione itself appeared in the blood and in the urine. Several intestinal bacteria including *Paracolobactrum* (Oginsky *et al.*, 1965) and *Enterobacter* (Tani *et al.*, 1974) were found that are capable of hydrolyzing glucosinolates. These bacteria are important from the viewpoint of animal nutrition or possible toxicity to animals. Thus, intact glucosinolates fed to nonruminants are probably hydrolyzed in the lower gut to release aglucon products. Tests with rats in which isolated *epi*-progoitrin was fed gave a different pathological picture of the liver and kidney than did those with animals fed the pure goitrin (Van Etten *et al.*, 1969a). Thioglucosidase activity of the gut content of poultry increases when heat-treated rapeseed meal is fed (Marangos and Hill, 1974).

No published work has been found on the formation of aglucon hydrolytic products from the progoitrins in the digestive tract of ruminants except that reporting the formation of polymerized products and small amounts of 5-vinyl-1,3-thiazolidin-2-one (instead of goitrin) when progoitrin was treated with rumen fluid from sheep (Lanzani and Piana, 1974).

C. Effect of Specific Plants

1. Horticultural Crops

Experimental goiter was first produced by feeding cabbage to rabbits (Chesney *et al.*, 1928). The experiment could not always be repeated by others (Greer, 1950). Analyses of the edible part of fresh cabbage gave SCN^- concentrations ranging from 0.7 to 10.2 mg per 100 gm. Concentration was highest in the spring and varied little as to where the cabbage was grown, but varied from plant to plant within a field (Michajlovskij and Langer, 1958, 1959). A single feeding of raw cabbage to guinea pigs or rats on a low-iodine ration caused a reduced iodine uptake by the thyroid greater than that caused by the SCN^- known to be present in the cabbage (Langer, 1961; Langer and Stolc, 1964). A mixture of SCN^-, allyl-NCS, and 5-vinyl-OZT fed to rats on a low-iodine ration (at levels of the compounds equal to those they would receive from an *ad libitum* diet of cabbage) caused similar increases in thyroid weight, depression of iodine uptake by the thyroid, and lowered synthesis of iodoamino acids (Langer, 1966). Allyl-NCS administered daily to a rat by stomach tube at a level of 2 to 4 mg for 60 days caused a marked increase of SCN^- in the blood and decreased radioiodine uptake by the thyroid (Langer, 1964).

During autolysis of cabbage (which may occur in finely shredded raw cabbage) many of the glucosinolates hydrolyze to give organic nitriles instead of isothiocyanates (Daxenbichler *et al.*, 1977). Certain nitriles are toxicants to liver and kidney (Tookey *et al.*, 1965; VanEtten *et al.*, 1969a), but they may also be partially metabolized to SCN^- as are the isothiocyanates (Wood, 1975). If the initial hydrolysis formed either nitrile or isothiocyanate, some conversion to SCN^- by the body tissues would occur and the SCN^- would act as a goitrogen, as reported by Langer (1966).

Rutabaga and turnip roots both suppress radioiodine uptake by the thyroid of the rat (Greer, 1962a). These vegetables contain relatively large amounts of progoitrin and other glucosinolates (Table II).

2. *Condiments*

Included in the common plants from the Cruciferae that are used as flavoring agents is seed from white mustard (*Brassica hirta* or *Sinapis alba*). It contains sinalbin (*p*-hydroxybenzyl-GS), which on hydrolysis would be expected to form SCN^- and *p*-hydroxybenzyl alcohol (Kawakishi and Muramatsu, 1966). Acute toxicity in dairy cows has been attributed to the consumption of white mustard seed (Holmes, 1965).

Other mustard seeds and horseradish roots contain sinigrin (allyl-GS), which forms allyl-NCS. Allyl-NCS is such a lachrymator and vesicant that the compound is unlikely to be consumed in such amounts to be harmful in other ways. However, if the unhydrolyzed glucosinolate were eaten and then hydrolyzed in the intestinal tract (see Section V,B) to form the isothiocyanate, large amounts could be absorbed from the intestinal tract.

3. *Crucifer Oilseeds*

In man's continued efforts to become more efficient in growing and processing food and feed, he has been confronted with the unpalatability and toxicity of oilseed meal from crucifers. Attempts to explain and overcome this problem have led to extensive research into the composition and the biological effect of rapeseed and related meals (Bowland *et al.*, 1965; Rutkowski, 1971; Appelqvist and Ohlson, 1972; Tallent, 1972).

Major attention in early work was given to the goitrogenic properties of seed meals of rape. Some improvement in the quality of both rape and crambe meals for animal feeding was achieved by extraction of the glucosinolates themselves or their hydrolytic products or by chemical treatment of the meal to inactivate toxicants (Bowland *et al.*, 1965; Tookey *et al.*, 1979). A critical examination of the feeding results suggests that toxic substances other than goitrogenic isothiocyanates and oxazolidinethiones are involved (Kennedy and Purves, 1941; Nordfeldt *et al.*, 1954; Holmes and Roberts, 1963; Tookey *et al.*, 1965). The apparent part that nitriles play in the toxicity of the glucosinolates led to the preparation and testing of seed meals, fractions of the meals of known composition, and some pure compounds. Initial work along these lines has been reported (Van Etten *et al.*, 1969b; Srivastava *et al.*, 1975).

Feeding of rapeseed meal to poultry has been associated with growth depression, perosis, thyroid enlargement, lowered egg production, and liver damage (Bowland *et al.*, 1965). Thyroid malfunction is clearly linked to glucosinolate content, but the causes of the other problems are not well understood. Perosis in chickens fed rapeseed meal is prevented in part by removal of alcohol-soluble material from the meal, but the toxicants have not been further identified (Holmes and Roberts, 1963). Liver hemorrhage in poultry is associated with the presence of rapeseed in the ration (Olomu *et al.*, 1975). The cause of liver damage is not known (Hall, 1974), but organic nitriles from glucosinolates are suspect (Smith and Campbell, 1976).

Swine have been fed rapeseed meals, but the amounts have been limited in order to prevent growth depression and enlargement of thyroid, liver, and kidney (Nordfeldt *et al.*, 1954; Bowland *et al.*, 1963). Ruminants are more tolerant of glucosinolates than are monogastric animals (Josefsson, 1972); meals from the standard varieties of rapeseed can be fed to cattle at 10% of the ration with no ill effect.

The practical solution to these problems has been met, in part, by the development of new rapeseed varieties that are low in total glucosinolate content (Srivastava and Hill, 1975). The standard varieties of rapeseed could be fed only in limited amounts to ruminants and in lesser amounts to monogastric animals. New low-glucosinolate varieties may be fed to broiler chicks as 20% of the ration or to cattle as a source of supplemental protein instead of soybean meal (Rapeseed Association of Canada, 1978).

4. Forages, Pasture, and Wild Plants

Dwarf Essex rape forage (*B. napus*) is the major *Brassica* forage grown in the United States, although some marrow-stem kale (*B. oleracea*) is grown in the northwestern part (Morisson, 1959). Marrow-stem kale is grown in Great Britain and Sweden. Analyses showed that kale with thick stems (Table II) was lower in total glucosinolate than was kale with thin stems (Josefsson *et al.*, 1972). After the green forage was ensiled for 5 months, the 3-indolylmethyl-GS's decreased 38–73%. There are reports of a goitrogenic effect from feeding kale to ewes (Wright, 1958) and cattle (Clements and Wishart, 1956). On the basis of the classification of Dwarf Essex rape forage as a *B. napus* and analysis for progoitrin hydrolysis products (VanEtten and Daxenbichler, 1971), this variety would be expected to be goitrogenic. However, we find no reports in the literature that it is.

There are reports that kale anemia in ruminants may be caused by glucosinolate hydrolytic products. However, recent work (Smith, 1974) demonstrates that the problem is caused by the presence of *S*-methylcysteine sulfoxide or its derived products.

In a review, Tapper and Reay (1973) discuss a number of pasture weeds that contain glucosinolates. Plants containing glucosinolates that form oxazolidinethiones, other than (*R*)- and (*S*)-5-vinyl-OZT from *Crambe abyssinica* and *Brassica napus*, include (−)-5-allyl-OZT from *B. campestris* and *B. napus*, 5,5-dimethyl-OZT from *Cochlearia officinalis* L. and *Conringia orientalis* L., (*R*)-5-phenyl-OZT from *Barbarea vulgaris* R. Bn and *Reseda luteola* L., and (*R*)-5-ethyl-5-methyl-OZT from *Cleome spinosa* Jacq. (Ettlinger and Kjaer, 1968). More recently found oxazolidinethiones include (*S*)-5-phenyl-OZT from *Sibara virginica* (L.) Rollins (Gmelin *et al.*, 1970) and (*S*)-5-(*p*-methoxyphenyl)-OZT from *Arabis hirsuta* (L.) Scop. (Kjaer and Schuster, 1972). Cheirolin, 3-methylsulfonylpropyl-NCS, found in a pasture weed, *Rapistrum rugosum*, when injected into rats also inhibits radioiodine uptake by the thyroid (Bachelard and Trikojus, 1960).

VI. EPILOGUE

Much of the literature on the glucosinolates is concerned with practical problems encountered in the use of glucosinolate-containing plants for food or feed. Such an approach leads to rather fragmented knowledge. Fortunately, there has also been enough basic chemical knowledge discovered to suggest some general principles of the interrelationship between these kinds of plants and the herbivores that consume them. The existence of 80 glucosinolates, each of which may give rise to at least two aglucon products, makes it impossible to draw a precise picture of glucosinolate–herbivore interrelationship at present. The responses of mammals to specific aglucon products are unknown except for a few, such as goitrin and allyl-NCS; others are still under active investigation.

There is considerable interest in insects as herbivores, especially those insects in which the host range is limited to certain plants, e.g., *Hylemya brassicae* (cabbage maggot) and *Pieris rapae* (cabbage butterfly). These insects may be attracted to the plant by the glucosinolates or their hydrolytic products. Such chemical attraction may serve in the detection of the plant, ovipositioning of the female, and feeding by the larvae. These secondary chemicals from the plant are a small part of the environment to which the insect is exposed and may be only a part of the attractiveness of a particular plant. For this reason, extreme care is required to relate cause and effect in studies linking glucosinolates and insects. The relative importance of the secondary plant metabolites and other environmental factors is still a subject of debate (see Nair and McEwen, 1976; Nair *et al.*, 1976; Slansky and Feeny, 1977).

REFERENCES

Akiba, Y., and Matsumoto, T. (1973). *Poult. Sci.* **52,** 562–567.

Anderson, D. L. (1970). *J. Assoc. Off. Anal. Chem.* **53,** 1–3.

Appelqvist, L. A., and Josefsson, E. (1967). *J. Sci. Food Agric.* **18,** 510–519.

Appelqvist, L. A., and Ohlson, R., eds. (1972). "Rapeseed. Cultivation, Composition, Processing and Utilization." Elsevier Amsterdam.

Ashworth, M. R. F. (1975). *In* "Chemistry and Biochemistry of Thiocyanic Acid and its Derivatives" (A. A. Newmann, ed.), pp. 258–337. Academic Press, New York.

Astwood, E. B. (1943). *J. Pharmacol. Exp. Ther.* **78,** 78–89.

Astwood, E. B., Bissell, A., and Hughes, A. M. (1945). *Endocrinology* **37,** 456–481.

Astwood, E. B., Greer, M. A., and Ettlinger, M. G. (1949). *J. Biol. Chem.* **181,** 121–130.

Austin, F. L., Gent, C. A., and Wolff, I. A. (1968). *Can. J. Chem.* **46,** 1507–1512.

Bachelard, H. S., and Trikojus, V. M. (1960). *Nature (London)* **185,** 80–82.

Bachelard, H. S., and Trikojus, V. M. (1963). *Aust. J. Biol. Sci.* **16,** 166–176.

Bjorkman, R. (1972). *Acta Chem. Scand.* **26,** 1111–1116.

Bowland, J. P., Zivkovíc, S., and Manns, J. G. (1963). *Can. J. Anim. Sci.* **43,** 279–284.

Bowland, J. P., Clandindin, D. R., and Wetter, L. R., eds. (1965). "Rapeseed Meal for Livestock and Poultry—A Review," Publ. 1257. Can. Dep. Agric., Ottawa.

Chang, C., and Bible, B. (1974). *J. Am. Soc. Hortic. Sci.* **99,** 159–162.

Chesney, A. M., Clawson, T. A., and Webster, B. (1928). *Bull. Johns Hopkins Hosp.* **43,** 261–277.

Clandinin, D. R., Bayly, L., and Caballero, A. (1966). *Poult. Sci.* **45,** 833–838.

Clements, F. W., and Wishart, J. W. (1956). *Metab., Clin. Exp.* **5,** 623–639.

Commission on Biochemical Nomenclature, I.U.P.A.C. and I.U.B. (1973). "Enzyme Nomenclature, 1972." Am. Elsevier, New York.
Daxenbichler, M. E., and VanEtten, C. H. (1977). *J. Assoc. Off. Anal. Chem.* **60,** 950–953.
Daxenbichler, M. E., VanEtten, C. H., and Wolff, I. A. (1965). *Biochemistry* **4,** 318–323.
Daxenbichler, M. E., VanEtten, C. H., Tallent, W. H., and Wolff, I. A. (1967). *Can. J. Chem.* **45,** 1971–1974.
Daxenbichler, M. E., VanEtten, C. H., and Wolff, I. A. (1968). *Phytochemistry* **7,** 989–996.
Daxenbichler, M. E., Spencer, G. F., Kleiman, R., VanEtten, C. H., and Wolff, I. A. (1970). *Anal. Biochem.* **38,** 374–382.
Daxenbichler, M. E., VanEtten, C. H., and Spencer, G. F. (1977). *J. Agric. Food Chem.* **25,** 121–124.
Daxenbichler, M. E., VanEtten, C. H., and Williams, P. H. (1979). *J. Agric. Food Chem.* **27,** 34–37.
Elliott, M. C., and Stowe, B. B. (1970). *Phytochemistry* **9,** 1629–1631.
Elliott, M. C., and Stowe, B. B. (1971). *Plant Physiol.* **48,** 498–503.
Ettlinger, M. G. (1950). *J. Am. Chem. Soc.* **72,** 4792–4796.
Ettlinger, M. G., and Dateo, G. P., Jr. (1961). "Studies of Mustard Oil Glucosides," Contract No. DA 19-129 QM-1059, Proj. No. 7-84-06-032, Simplified Food Logistics, Final Rep. Dep. Chem., Rice Inst., Houston, Texas.
Ettlinger, M. G., and Kjaer, A. (1968). *Recent Adv. Phytochem.* **1,** 59–144.
Ettlinger, M. G., and Lundeen, A. J. (1956). *J. Am. Chem. Soc.* **78,** 4172–4173.
Ettlinger, M. G., and Lundeen, A. J. (1957). *J. Am. Chem. Soc.* **79,** 1764–1765.
Ettlinger, M. G., and Thompson, C. B. (1962). "Studies of Mustard Oil Glucosides II," Final Rep. Contract DA 19-129-QM-1689, Off. Tech. Serv., US Dep. Commerce, Washington, D.C.
Ettlinger, M. G., Dateo, G. P., Jr., Harrison, B. W., Mabry, T. J., and Thompson, C. P. (1961). *Proc. Natl. Acad. Sci. U.S.A.* **47,** 1875–1880.
Food and Agriculture Organization of the United Nations (1974). "Production Yearbook," Vol. 28-1, pp. 101–103 and 133. FAO, Rome.
Friis, P., and Kjaer, A. (1966). *Acta Chem. Scand.* **20,** 698–705.
Garner, R. J., Sanson, B. F., and Jones, H. G. (1960). *J. Agric. Sci.* **55,** 283–286.
Gmelin, R., and Virtanen, A. I. (1959). *Acta Chem. Scand.* **13,** 1474–1475.
Gmelin, R., and Virtanen, A. I. (1961). *Ann. Acad. Sci. Fenn., Ser. A2* **107,** 1–25.
Gmelin, R., and Virtanen, A. I. (1962). *Acta Chem. Scand.* **16,** 1378–1384.
Gmelin, R., Kjaer, A., and Schuster, A. (1970). *Acta Chem. Scand.* **24,** 3031–3037.
Goh, Y. K., and Clandinin, D. R. (1977). *Br. Poult. Sci.* **18,** 705–710.
Greer, M. A. (1950). *Physiol. Rev.* **30,** 513–548.
Greer, M. A. (1957). *Am. J. Clin. Nutr.* **5,** 440–444.
Greer, M. A. (1962a). *Recent Prog. Horm. Res.* **18,** 187–219.
Greer, M. A. (1962b). *Arch. Biochem. Biophys.* **99,** 369–371.
Greer, M. A., and Deeney, J. M. (1959). *J. Clin. Invest.* **38,** 1465–1474.
Greer, M. A., Kendell, J. W., and Smith, M. (1964). *In* "Thyroid Gland" (R. Pitt-Rivers and W. R. Trotter, eds.), pp. 357–389. Butterworth, London.
Greer, M. A., Stott, A. K., and Milne, K. A. (1966). *Endocrinology* **79,** 237–247.
Hall, S. A. (1974). *Vet. Rec.* **94,** 42–44.
Henderson, H. M., and McEwan, T. J. (1972). *Phytochemistry* **11,** 3127–3133.
Holmes, R. G. (1965). *Vet. Rec.* **77,** 480–481.
Holmes, W. B., and Roberts, R. (1963). *Poult. Sci.* **42,** 803–809.
Jart, A. (1961). *Acta Chem. Scand.* **15,** 1223–1230.
Johnston, T. D., and Jones, D. I. H. (1966). *J. Sci. Food Agric.* **17,** 70–72.

Josefsson, E. (1967a). *Phytochemistry* **6,** 1617–1627.
Josefsson, E. (1967b). *J. Sci. Food Agric.* **18,** 492–495.
Josefsson, E. (1968). *J. Sci. Food Agric.* **19,** 192–194.
Josefsson, E. (1970). *J. Sci. Food. Agric.* **21,** 94–97.
Josefsson, E. (1972). *In* "Rapeseed: Cultivation, Composition, Processing, and Utilization" (L. A. Appelqvist and R. Ohlson, eds.), pp. 354–378. Elsevier, Amsterdam.
Josefsson, E., and Appelqvist. L. A. (1968). *J. Sci. Food Agric.* **19,** 567–570.
Josefsson, E., Ellerström, S., and Sjödin, J. (1972). *Z. Pflanzenzuecht.* **67,** 353–359.
Kawakishi, S., and Muramatsu, K. (1966). *Agric. Biol. Chem.* **30,** 688–692.
Kawakishi, S., Namiki, M., and Watanabe, H. (1967). *Agric. Biol. Chem.* **31,** 823–830.
Kelley, F. C., and Snedden, W. W. (1960). *W.H.O., Monogr. Ser.* **44,** 227–235.
Kennedy, J. H., and Purves, H. D. (1941). *Br. J. Exp. Pathol.* **22,** 241–244.
Khera, K. S. (1977). *Food Cosmet. Toxicol.* **15,** 61–62.
Kirk, L. P., Black, L. T., and Mustakas, G. C. (1964). *J. Am. Oil Chem. Soc.* **41,** 599–602.
Kjaer, A. (1954). *Acta Chem. Scand.* **8,** 1110.
Kjaer, A. (1960). *In* "The Chemistry of Organic Natural Products" (L. Zechmeister, ed.), pp. 122–176. Springer-Verlag, Berlin and New York.
Kjaer, A. (1973). *Nobel Symp.* **25,** 229–234.
Kjaer, A. (1977). *Pure Appl. Chem.* **49,** 137–152.
Kjaer, A., and Larsen, P. (1973). *Biosynthesis* **2,** 71–105.
Kjaer, A., and Rubenstein, K. (1953). *Acta Chem. Scand.* **7,** 528–536.
Kjaer, A., and Schuster, A. (1972). *Acta Chem. Scand.* **26,** 8–14.
Kjaer, A., Conte, J., and Larsen, I. (1953). *Acta Chem. Scand.* **7,** 1276–1283.
Kreula, M., and Kiesvaara, M. (1959). *Acta Chem. Scand.* **13,** 1375–1382.
Langer, P. (1961). *Hoppe-Seyler's Z. Physiol. Chem.* **323,** 194–198.
Langer, P. (1964). *Hoppe-Seyler's Z. Physiol. Chem.* **339,** 33–35.
Langer, P. (1966). *Endocrinology* **79,** 1117–1122.
Langer, P., and Stolc, V. (1964). *Hoppe-Seyler's Z. Physiol. Chem.* **335,** 216–220.
Lanzani, A., and Piana, G. (1974). *J. Am. Oil Chem. Soc.* **51,** 517.
Lein, K. A. (1970). *Z. Pflanzenzuecht.* **63,** 138–153.
Lein, K. A. (1972). *Angew. Bot.* **46,** 263–284.
Lein, K. A., and Schön, W. J. (1969). *Angew. Bot.* **43,** 87–92.
Lichtenstein, E. P., Strong, F. M., and Morgan, D. G. (1962). *J. Agric. Food Chem.* **10,** 30–33.
Lo, M. T., and Hill, D. C. (1972). *Can. J. Physiol. Pharmacol.* **50,** 962–966.
Lonnerdal, B., and Janson, J. C. (1973). *Biochim. Biophys. Acta* **315,** 421–429.
McGhee, J. G., Kirk, L. P., and Mustakas, G. C. (1965). *J. Am. Oil Chem. Soc.* **42,** 889–891.
McGregor, D. I., and Downey, R. K. (1975). *Can. J. Plant Sci.* **55,** 191–196.
Mahadevan, S., and Stowe, B. B. (1972). *Plant Physiol.* **50,** 43–50.
Marangos, A., and Hill, R. (1974). *Proc. Nutr. Soc.* **33,** 90A.
Marsh, R. E., and Waser, J. (1970). *Acta Crystallogr., Sect. B* **26,** 1030–1037.
Matsumoto, T., Itoh, H., and Akiba, Y. (1968). *Poult. Sci.* **47,** 1323–1330.
Michajlovskij, N., and Langer, P. (1958). *Hoppe-Seyler's Z. Physiol. Chem.* **312,** 26–30.
Michajlovskij, N., and Langer, P. (1959). *Hoppe-Seyler's Z. Physiol. Chem.* **317,** 30–33.
Michajlovskij, N., Sedlak, J., and Kostekova, O. (1969). *Rev. Czeck. Med.* **15,** 132–144.
Miller, H. E. (1965). Master's Thesis, pp. 1–68. Rice University, Houston, Texas.
Miller, J. K., Swanson, E. W., and Gagle, R. G. (1965). *J. Dairy Sci.* **48,** 1118–1121.
Morrison, F. B. (1959). *In* "Feeds and Feeding," 2nd ed., pp. 394–396. Morrison Publishing, Clinton, Iowa.
Mullin, W. J., and Sahasrabudhe, M. R. (1978). *Can. Inst. Food Sci. Technol. J.* **11,** 50–52.
Nagashima, Z., and Uchiyama, M. (1959). *Nippon Nogei Kagaku Kaishi* **33,** 478, 484, 881, 980, 1068, and 1144–1149.

Nair, K. S. S., and McEwen, F. L. (1976). *Can. Entomol.* **108,** 1021–1030.
Nair, K. S. S., McEwen, F. L., and Snieckus, V. (1976). *Can. Entomol.* **108,** 1031–1036.
Neuberg, C., and Schoenbeck, O. (1933). *Biochem. Z.* **265,** 233.
Nordfeldt, S., Gellerstedt, N., and Falkmer, S. (1954). *Acta Pathol. Microbiol. Scand.* **35,** 217–236.
Oginsky, E. L., Stein, A. E., and Greer, M. A. (1965). *Proc. Soc. Exp. Biol. Med.* **119,** 360–364.
Olomu, J. M., Robblee, A. R., Clandinin, D. R., and Hardin, R. T. (1975). *Can. J. Anim. Sci.* **55,** 219–222.
Peltola, P. (1965). *Curr. Top. Thyroid Res., Proc. Int. Thyroid Conf., 5th, 1965* pp. 872–876.
Persson, S. (1974). *Proc. Int. Rapeseed Conf., 4th, 1974,* 381–384.
Pihakashi, K., and Iversen, T.-H. (1976). *J. Exp. Bot.* **27,** 242–258.
Piironen, E., and Virtanen, A. I. (1963). *Z. Ernaehrungswiss.* **3,** 140–147.
Rapeseed Association of Canada (1978). *In* "Canadian Rapeseed Meal: Poultry and Animal Feeding," Publ. No. 51. Rapeseed Assoc. Can., Winnepeg, Manitoba, Canada.
Roche, J., and Lissitzky, S. (1960). *W.H.O., Monogr. Ser.,* **44,** 351–368.
Rodman, J. E. (1978). *Phytochem. Bull.* **11,** 6–30.
Rutkowski, A. (1971). *J. Am. Oil Chem. Soc.* **48,** 863–868.
Sandberg, M., and Holly, D. M. (1932). *J. Biol. Chem.* **96,** 443–447.
Schmid, H., and Karrer, P. (1948). *Helv. Chim. Acta* **31,** 1087–1092.
Schultz, O. E., and Gmelin, R. (1952). *Z. Naturforsch., Teil B* **7,** 500–506.
Schultz, O. E., and Gmelin, R. (1953). *Z. Naturforsch., Teil B* **8,** 151–156.
Schultz, O. E., and Gmelin, R. (1954). *Z. Naturforsch., Teil B* **9,** 27–29.
Schultz, O. E., and Wagner, W. (1956). *Z. Naturforsch., Teil B* **11,** 73–78.
Slansky, F., Jr., and Feeney, P. (1977). *Ecol. Monogr.* **47,** 209–228.
Smith, R. H. (1974). *Annu. Rep. Stud. Anim. Nutr. Allied Sci.* (*Rowett Res. Inst.* **30,** Reprint No. 751, 112–131.
Smith, T. K., and Campbell, L. D. (1976). *Poult. Sci.* **55,** 861–867.
Srivastava, V. K., and Hill, D. C. (1975). *Can. J. Biochem.* **53,** 630–633.
Srivastava, V. K., Philbrick, D. J., and Hill, D. C. (1975). *J. Anim. Sci.* **55,** 331–335.
Stahmann, M. A., Link, K. P., and Walker, J. C. (1943). *J. Agric. Res.* **67,** 49–63.
Stanley, M. H., and Astwood, E. B. (1947). *Endocrinology* **41,** 66–84.
Stoll, A., and Jucker, E. (1955). *In* "Moderne Methoden der Pflanzen analye" (K. Paech and M. V. Tracey, eds.), Vol. IV, pp. 689–718. Springer-Verlag, Berlin and New York.
Stoll, A., and Seebeck, E. (1948). *Helv. Chim. Acta* **31,** 1432–1434.
Takahashi, N. (1960). *J. Biochem.* (*Tokyo*) **47,** 230–237.
Tallent, W. H. (1972). *J. Am. Oil Chem. Soc.* **49,** 15–19.
Tang, C. S. (1973). *Phytochemistry* **12,** 769–773.
Tani, N., Ohtsuru, M., and Hata, T. (1974). *Agric. Biol. Chem.* **38,** 1623–1630.
Tapper, B. A., and MacGibbon, D. B. (1967). *Phytochemistry* **6,** 749–753.
Tapper, B. A., and Reay, P. F. (1973). *In* "Chemistry and Biochemistry of Herbage" (G. W. Butler and R. W. Bailey, eds.), Vol. 1, pp. 447–452. Academic Press, New York.
Thies, W. (1977). *Z. Pflanzenzuecht.* **79,** 331–335.
Thies, W. (in press). *Proc. Int. Rapeseed Conf., 5th, 1978.*
Tookey, H. L. (1973a). *Can. J. Biochem.* **51,** 1305–1310.
Tookey, H. L. (1973b). *Can. J. Biochem.* **51,** 1654–1660.
Tookey, H. L., and Wolff, I. A. (1970). *Can. J. Biochem.* **48,** 1024–1028.
Tookey, H. L., VanEtten, C. H., Peters, J. E., and Wolff, I. A. (1965). *Cereal Chem.* **42,** 507–514.
Tookey, H. L., VanEtten, C. H., and Daxenbichler, M. E. (1979). *In* "Toxic Constituents of Plant Foodstuffs" (I. E. Liener, ed.), 2nd ed. Academic Press, New York (*in press*).

Tsuruo, I., and Hata, T. (1967). *Agric. Biol. Chem.* **31,** 27–32.
Underhill, E. W., and Kirkland, D. F. (1971). *J. Chromatogr.* **57,** 47–54.
Underhill, E. W., Wetter, L. R., and Chisholm. M. O. (1973). *Biochem. Soc. Symp.* **38,** 303–326.
U.S. Department of Agriculture (1975). "Agricultural Statistics," p. 157. USDA, Washington, D.C.
VanderLaan, W. P., and Bissel, A. (1946). *Endocrinology* **38,** 308–314.
VanEtten, C. H. (1979). *In* "CRC Handbook of Nutrition and Food" (M. Rechcigl, Jr., ed.). Chemical Rubber Publ., West Palm Beach, Florida (in press).
VanEtten, C. H., and Daxenbichler, M. E. (1971). *J. Agric. Food Chem.* **19,** 194–195.
VanEtten, C. H., and Daxenbichler, M. E. (1977). *J. Assoc. Off. Anal. Chem.* **60,** 946–949.
VanEtten, C. H., and Tookey, H. L. (1978). *In* "Effects of Poisonous Plants on Livestock" (R. F. Keller, K. R. VanKempen, and L. F. James, eds.), pp. 507–520. Academic Press, New York.
VanEtten, C. H., and Wolff, I. A. (1973). *In* "Toxicants Occurring Naturally in Foods," 2nd ed., pp. 210–234. Natl. Res. Counc., Washington, D.C.
VanEtten, C. H., Daxenbichler, M. E., Peters, J. E., Wolff, I. A., and Booth, A. N. (1965). *J. Agric. Food Chem.* **13,** 24–27.
VanEtten, C. H., Daxenbichler, M. E., Peters, J. E., and Tookey, H. L. (1966). *J. Agric. Food Chem.* **14,** 426–430.
VanEtten, C. H., Gagne, W. E., Robbins, D. J., Booth, A. N., Daxenbichler, M. E., and Wolff, I. A. (1969a). *Cereal Chem.* **46,** 145–155.
VanEtten, C. H., Wolff, I. A., and Daxenbichler, M. E. (1969b). *J. Agric. Food Chem.* **17,** 483–491.
VanEtten, C. H., McGrew, C. E., and Daxenbichler, M. E. (1974). *J. Agric. Food Chem.* **22,** 483–487.
VanEtten, C. H., Daxenbichler, M. E., Williams, P. H., and Kwolek, W. F. (1976). *J. Agric. Food Chem.* **24,** 452–455.
VanEtten, C. H., Daxenbichler, M. E., Schroeder, W. S., Princen, L. H., and Perry, T. W. (1977). *Can. J. Anim. Sci.* **57,** 75–80.
Vaughan, J. G., MacLeod, A. J., and Jones, B. M. G. (1976). "The Biology and Chemistry of the Cruciferae." Academic Press, New York.
Vose, J. R. (1972). *Phytochemistry* **11,** 1649–1653.
Walker, N. J., and Gray, I. K. (1970). *J. Agric. Food Chem.* **18,** 346–352.
Waser, J., and Watson, W. H. (1963). *Nature (London)* **198,** 1297–1298.
Wetter, L. R. (1955). *Can. J. Biochem. Physiol.* **33,** 980–984.
Wetter, L. R. (1957). *Can. J. Biochem. Physiol.* **35,** 293–297.
Wetter, L. R., and Youngs, C. G. (1976). *J. Am. Oil Chem. Soc.* **53,** 162–164.
Wood, J. L. (1975). *In* "Chemistry and Biochemistry of Thiocyanic Acid and Its Derivatives" (A. A. Newman, ed.), pp. 156–221. Academic Press, New York.
Wright, E. (1958). *Nature (London)* **181,** 1602–1603.
Youngs, G. C., and Wetter, L. R. (1967). *J. Am. Oil Chem. Soc.* **44,** 551–554.

Chapter **14**

Sesquiterpene Lactones and Other Terpenoids

TOM J. MABRY *And* **JAMES E. GILL**

ISBN 0-12-597180-X

I. INTRODUCTION

Terpenoids represent one of the largest and biologically most important classes of natural products; thus, it is not surprising that they have been found in all living organisms so far examined for them. Terpenoids exhibit remarkable structural and functional diversity, especially in view of their common origin from the same C_5 isopentenoid units. Although most terpenoids occur free in nature, often being accumulated in specific tissues or special glands, many are conjugates of organic acids, sugars, chlorophylls, proteins, and many other primary metabolic constituents.

Many terpenoids function as attractants for insect pollinators and in various defensive roles. Others are involved in complex and vital metabolic processes, whereas some serve special roles in membrane structures and surface coatings. Since this chapter cannot deal with all these aspects of terpenoids, it focuses on their origin and function in plants, with emphasis on the role of sesquiterpene lactones in plant–herbivore interactions. In addition, the feeding-deterrent and toxic properties of other selected terpenoids are considered.

II. MEVALONIC ACID BIOGENETIC PATHWAY

The polymerization of C_5 isopentenoid pyrophosphate units gives the distinct classes of terpenoids: mono- (C_{10}), sesqui- (C_{15}), di- (C_{20}), sester- (C_{25}), tri- (C_{30}), and tetra- (C_{40}) terpenes. Perhaps the two most important features of this biogenetic scheme are, first, that once the carbon–carbon bonds between the C_5 units are formed they are remarkably

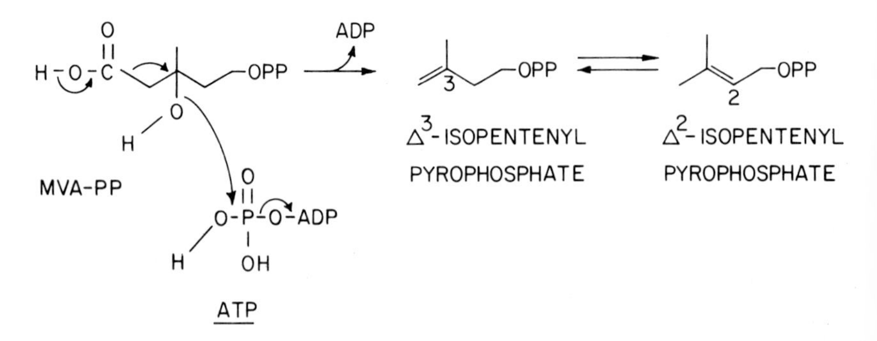

Figure 1. Conversion of mevalonic acid pyrophosphate (MVA-PP) to the pair of isopentenyl pyrophosphates involved in all terpenoid biogenesis.

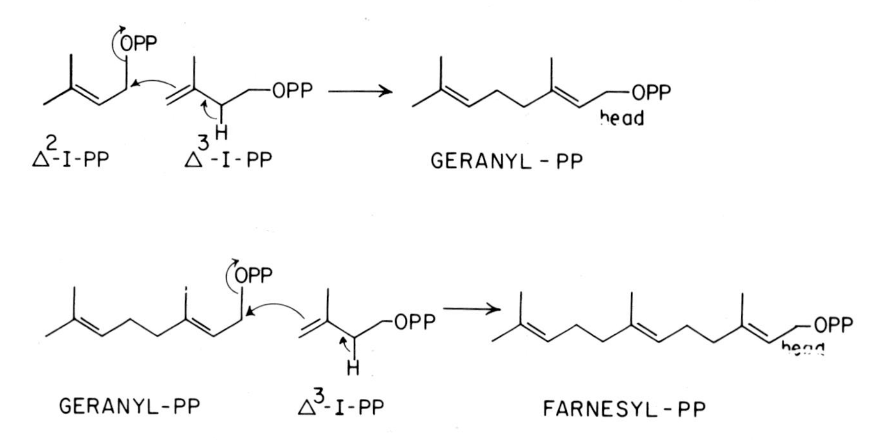

Figure 2. Formation of geranyl and farnesyl pyrophosphates, precursors for the mono- C_{10}) and sesqui- (C_{15}) terpenes, respectively, from the pair of C_5 isopentenoid units.

stable and, second, that these polymers of C_5 units undergo innumerable cyclizations, rearrangements, and oxidations to yield thousands of structurally unique terpenoids.

Although acetate (acetyl-CoA) is the basic building block of all terpenoids, mevalonic acid represents the branch point separating terpenoid biogenesis from other metabolic pathways. The discovery that mevalonic acid pyrophosphate undergoes decarboxylation with simultaneous elimination of a hydroxyl group (Fig. 1) to yield Δ^3-isopentenyl

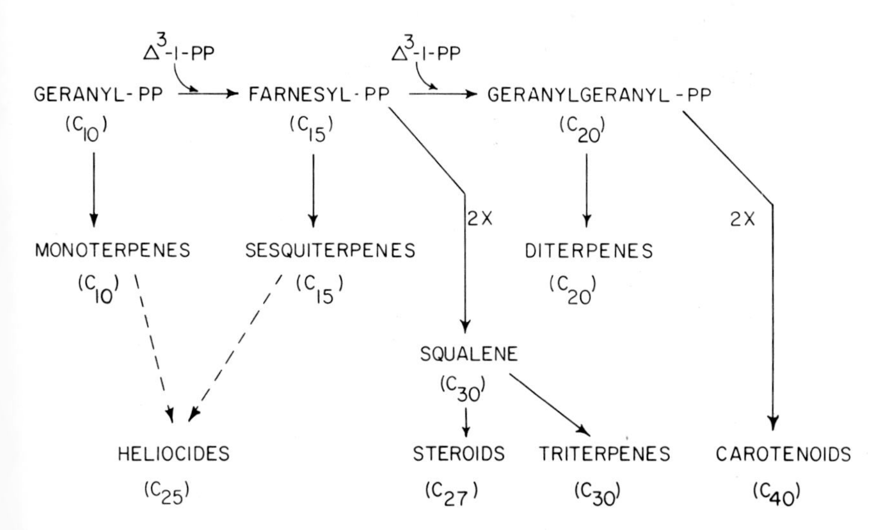

Figure 3. Biogenetic scheme for the major classes of terpenoids.

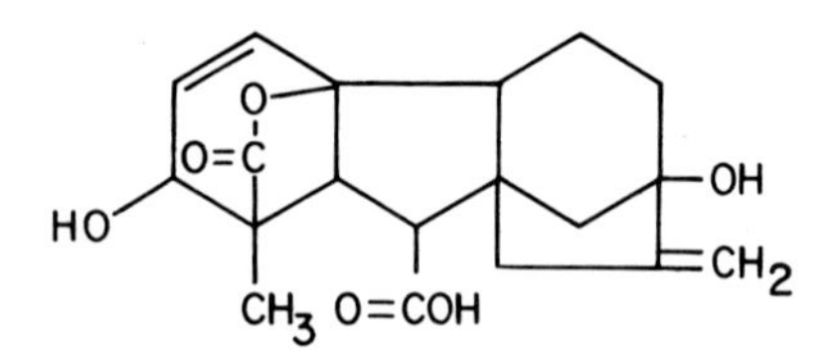

Figure 4. Gibberellic acid (C_{19}), a highly modified diterpene plant hormone.

pyrophosphate (Δ^3-I-PP) represented a major breakthrough in our understanding of how all classes of terpenoids are formed. Isomerization of the double bond in the Δ^3-I-PP gives a Δ^2 isomer, and it is the condensation of this pair of isopentenyl pyrophosphates that leads to all isopentenoid biogenesis (Fig. 2). Most classes of terpenoids result from "head-to-tail" polymerizations of the pair of Δ^2- and Δ^3-isopentenyl pyrophosphates. Thus, there is general acceptance today that all terpenoids are the end products of a metabolic pathway that can be described as follows (Figs. 2 and 3): acetate → mevalonate → Δ^2- and Δ^3-isopentenyl pyrophosphates → geranyl pyrophosphate → farnesyl pyrophosphate → $(C_5)_x$ compounds.

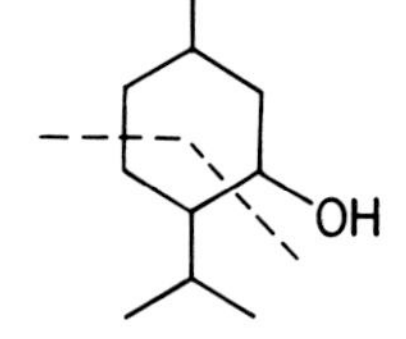

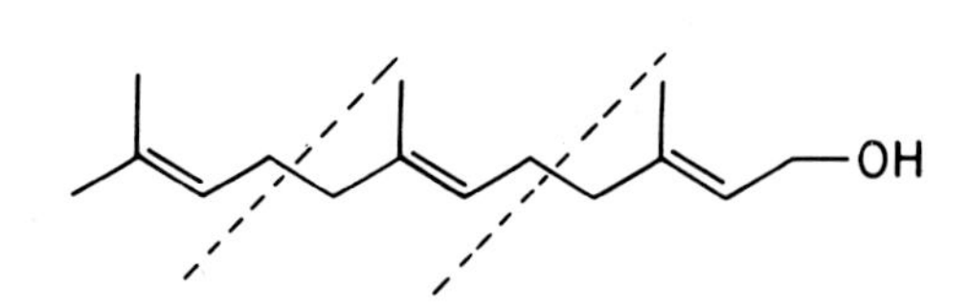

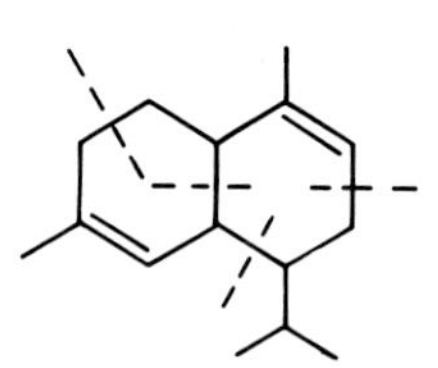

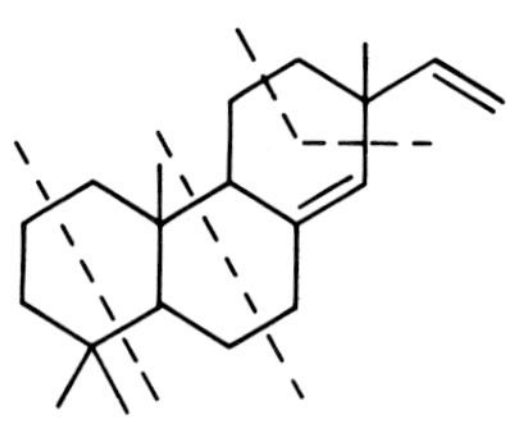

Figure 5. The biogenetic origin of a terpenoid often can be recognized by segmenting its structures into repeating C_5 subunits each having the carbon skeleton of isoprene; in addition, the C_5 subunits are usually joined in a head-to-tail fashion.

When speculating about how a particular terpenoid has been formed *in vivo*, one must consider two transformation processes: the basic isopentenoid polymerization pathway and the reactions that can occur once the skeleton is formed. The latter steps, which include ring closures, rearrangements, introduction of oxygen functionality, and loss and gain of carbon atoms, may result in a structure so modified that its origin from C_5 isopentenoid units cannot be easily recognized (for example, see Fig. 4).

Nevertheless, the origin of all terpenoids from Δ^2- and Δ^3-isopentenyl pyrophosphates gives them an inherent structural similarity based on repeating five-carbon subunits that have the skeleton of isoprene:

$$CH_2{=}\underset{\displaystyle CH_3}{\underset{|}{C}}-CH{=}CH_2$$

Indeed, terpenoids were at one time thought to arise from the polymerization of isoprene units, leading to the term "isoprenoid" to describe them. The "isoprene rule," which involves segmenting the skeleton of a terpenoid into five-carbon subunits, each exibiting the isoprene skeleton (Fig. 5), is useful for understanding the formation of many terpenoids.

III. ANALYSIS OF SESQUITERPENE LACTONES AND OTHER TERPENOIDS

In structural properties, terpenoids range from nonpolar hydrocarbons to acidic and glycosidic types. Thus, their physical and chemical properties overlap with those of many other classes of natural products, making it difficult to determine the terpenoid content of a given plant without isolating, purifying, and quantifying each individual compound present. The methods presented here for analyzing sesquiterpene lactones and a few other subgroups of terpenoids are generally applicable to most members of this group of plant constituents.

A. Sesquiterpene Lactones

1. *Extraction*

Two procedures for isolating sesquiterpene lactones from plant material are routinely employed in our laboratory. They are given below.

a. Procedure I. Twenty grams of dried leaves and stems are ground in a Waring Blendor; then 100 ml of methylene dichloride are added, and the blending is continued for about 2 minutes. The slurry is filtered, and the filtrate plus methylene dichloride washings are evaporated to dryness *in vacuo*. The residue thus obtained is dissolved in 25 ml of 95% (v/v) ethanol, and then 25 ml of 4% aqueous lead acetate are added to remove phenolics and other nonterpenoids.

Next, the solution is filtered, and the filtrate is concentrated *in vacuo* until only water and oil remain. The oil is extracted into methylene dichloride, which is subsequently dried over anhydrous magnesium sulfate; the solution is filtered and evaporated *in vacuo*.

b. Procedure II. Because sesquiterpene lactones occur in trichomes, they can be extracted by washing the unground dried plant material once or twice in ether or methylene dichloride for about 5 minutes. The material thus obtained sometimes requires lead acetate treatment as in Procedure I.

2. *Chromatographic Separation*

The oil obtained by either of the extraction procedures described above is analyzed by fluorescent thin-layer chromatography, usually with silica gel plates, developed with various combinations of such solvents as benzene, ethyl acetate, ether, methanol, and chloroform. Sesquiterpene lactones can be detected by observing the plates over ultraviolet light. Crystalline samples are usually obtained by column chromatography of the crude syrup over silica gel, eluting first with chloroform and then with mixtures of chloroform and ether or ethyl acetate.

Perhaps the most important recent development in the field of chromatographic isolation of polar terpenoids such as sesquiterpene lactones is preparative high-pressure liquid chromatography (hplc). Analytical hplc has long been used for detecting trace components in complex mixtures and now preparative hplc, especially with reverse-phase techniques (C_{18}-bonded silica gel columns), permits the isolation of gram-size quantities of polar terpenoids directly from plant extracts (Hostettmann *et al.*, 1977).

3. *Structure Determination*

Pure sesquiterpene lactones obtained by chromatography are analyzed primarily by ^1H and ^{13}C nuclear magnetic resonance (nmr). The ^1H nmr spectra of most of these compounds exhibit characteristic doublets ($J \simeq 2.5$ Hz) between 5 and 7 ppm for the protons of the exomethylene group in the α,β-unsaturated lactone function (for a compilation of 295 nmr spectra of sesquiterpene lactones, see Yoshioka *et al.*, 1973); these and other signals can be used to detect sesquiterpene lactones even in mixtures. For the final structure determinations, ^{1}H nmr spin-decoupling experiments in conjunction with infrared (ir) and mass spectroscopy (ms) often can clarify the skeleton, although single-crystal X-ray analysis is usually required to obtain the absolute stereochemistry.

B. Volatile Terpenes

1. *Extraction*

The classic procedure for the isolation of essential oils, which usually consists of many mono- and sesquiterpenes along with other types of plant constituents, is steam distillation of fresh plant material. This is a reliable and reproducible method for the rapid isolation of large quantities of volatiles. For such studies, plants are collected fresh in the field and maintained at a cold temperature until subjected to steam distillation. We employ a circulatory steam distillation apparatus that permits continuous collecting of the oil in ether (von Rudloff, 1969). The mono- and sesquiterpenes may also be extracted directly from the plant tissue with ether, petroleum ether, or acetone; for small samples, this is often the method of choice.

2. *Analysis*

Mixtures of volatile terpenoids are usually analyzed by gas chromatography (gc) or by combined gas chromatography–mass spectrometry. Initial gc analysis of an essential oil typically utilizes a 2.5-m × 0.32-cm o.d. stainless steel column packed with 3% Carbowax 20-m on Chromosorb G; the gc may be operated isothermally or temperature programmed. It is often useful to separate initially the total oil into fractions by preparative gas–liquid chromatography using, for example, a 2.5-m × 0.64-cm o.d. stainless steel column packed with 25% Apiezon L on 80/100 mesh Chromosorb W. The combination of an analytical gas chromatograph, a mass spectrometer, and a computerized data acquisition system is the principal means of structure determination for essential oils and has revolutionized the analysis of substances that can be volatized or converted to volatile derivatives.

C. Higher Terpenoids

In general, diterpenoids and higher terpenoids are separated by the same techniques described above including especially silica gel chromatography (thin- and thick-layer and column). Identities are established by ms, ir, nmr, and, if necessary, X-ray analysis.

IV. TOXIC MONOTERPENES

Many monoterpenes are volatile and account for much of the odor of a plant, often acting as attractants to pollinators and serving defensive functions through olfactory repellence and direct toxicity. Many plants accumulate complex mixtures of monoterpenes in resin ducts and glands, from which they can in some instances be exuded through trichomes onto the surfaces of leaves and flower parts.

A. Pyrethroids

Perhaps the best-known terpenoids with insecticidal properties are the pyrethroids from species of *Chrysanthemum* (Compositae). Pyrethroids have been used since ancient times in the form of a powder prepared from dried flowers, and both natural and synthetic compounds are presently employed as commercial insecticides. Six active components have been isolated from the flowers of *C. cinerariaefolium;* all are esters (Fig. 6) of two cyclopropane terpenic acids, namely, chrysanthemic and pyrethric acids (Matsui and Yamamoto, 1971; Casida, 1973). The acids are derived from the mevalonate pathway and represent rearranged monoterpenes (Epstein and Poulter, 1973); the biogenetic origin of the alcohol moieties has not yet been settled. Pyrethroids have low mammalian toxicity but show excellent activity against many insects, exhibiting a characteristic rapid paralytic "knockdown effect" on flying insects (e.g., houseflies). Pyrethrin I has a particularly

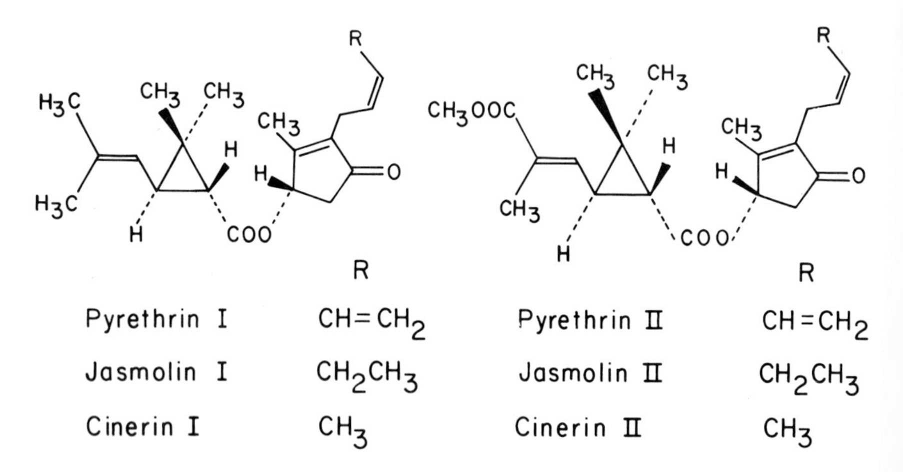

Figure 6. Six insecticidal pyrethroids isolated from *Chrysanthemum cinerariaefolium* (Compositae).

high kill activity (LD_{50} 0.6 μg per housefly) (Elliott and Janes, 1973), whereas pyrethrin II produces a faster knockdown effect. The toxicity of the pyrethroids is greatly enhanced when they are applied with synergists such as piperonyl butoxide, and some synthetic types have greater activity than the natural ones (Elliott, 1977).

B. Toxic Thujaplicins and Related Compounds in Cedar

The wood of western red cedar (*Thuja plicata*) is renowned for its ability to withstand insect attack both in timber stands and as sawmill products. This resistance has been ascribed to the presence of monoterpenes including β-thujaplicin, which was found to be highly insecticidal in tests against larvae of the old-house borer *Hylotrupes bajulus* (Cerambycidae), a common pest of structural timber in Europe (Becker, 1965). In assays with larvae of the termite *Reticulothermes flavipes*, both β- and γ-thujaplicin showed low termicidal activity in contrast to another monoterpene from red cedar, methyl thujate (Arndt, 1968). Methyl thujate is also toxic to larvae of the black carpet beetle, the furniture carpet beetle, and the case-making moth. Although individual components vary in their toxicity toward different pests, the collective mixture of monoterpenes produced by western red cedar apparently plays a primary role in defense against a variety of phytophagous insects.

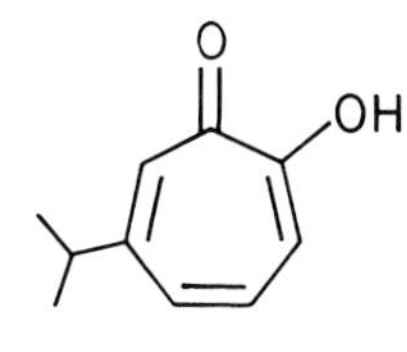

β-Thujaplicin
(*Thuja*)

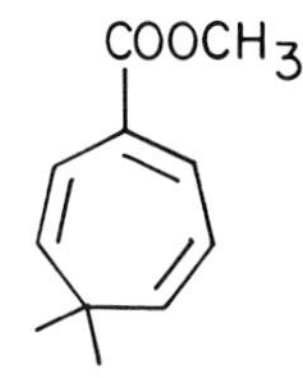

Methyl Thujate
(*Thuja*)

C. Conifer Resistance to Bark Beetles

Tree-killing bark beetles (Coleoptera; Scolytidae) are among the most destructive insects in the coniferous forests of North America, causing annual losses of millions of dollars. Concentrated beetle attacks normally result in the death of the tree, which is generally a prerequisite for the successful development of the beetle. Compounds in the resin of conifers, primarily monoterpenes, provide the major defense against these attacking beetles.

Some trees respond to attack by producing increased levels of monoterpenes that are more toxic, repellent, or inhibitory to the insects than the normal resin constituents. Such a secondary resin produced by the grand fir *Abies grandis* in response to attack by the beetle *Scolytus ventralis* and its associated pathogenic fungus *Trichosporium symbioticum* contains higher levels of myrcene, α-pinene, and Δ^3-carene than normally present in the plant. This same response can be obtained by inoculation with the fungus alone (Russell and Berryman, 1976). Δ^3-Carene, myrcene, and limonene are highly toxic to the western pine beetle *Dendroctonus brevicomis,* and, significantly, the latter two compounds comprise nearly 30% of the volatiles of resistant Ponderosa pine trees (*Pinus ponderosa*) (Smith, 1965, 1966). Smith (1975) suggested that the resistance to attack by these beetles is related to both monoterpene content and the level of resin flow.

Bioassays of 13 monoterpenes that occur in shortleaf (*Pinus echinata*) and loblolly (*P. taeda*) pines demonstrated that limonene and some of its derivatives were toxic to the southern pine beetle *Dendroctonus frontalis.* These tests demonstrated that beetles might respond quite differently to structurally similar compounds; for example, α-phellandrene was highly toxic, whereas the β isomer was only slightly toxic (Coyne and Lott, 1976).

Douglas fir (*Pseudotsuga menziesii*) populations in western North America show clinal variation in their terpene content, particularly for

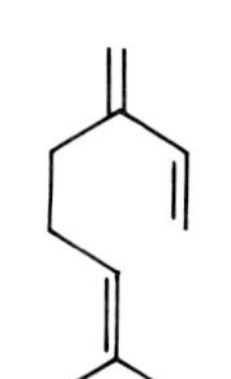
Myrcene

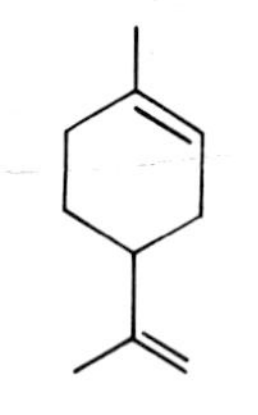
Limonene

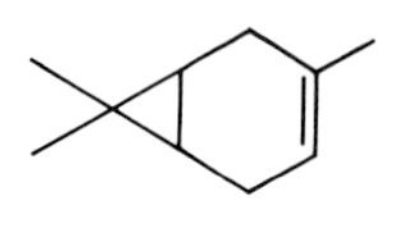
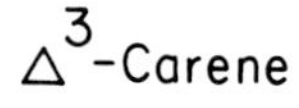
Δ^3-Carene

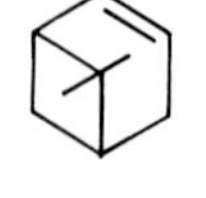
α-Pinene

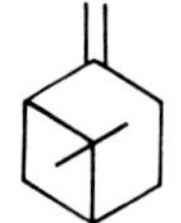
β-Pinene

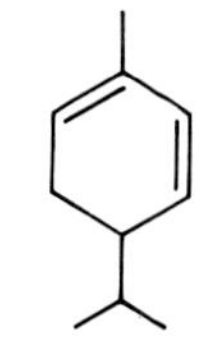
α-Phellandrene

α- and β-pinene (von Rudloff, 1973; Zavarin and Snajberk, 1973, 1975). The douglas fir beetle (*Dendroctonus pseudotsugae*) is known to be attracted by α-pinene and repelled by the β isomer (Heikkenen and Hrutfiord, 1965). Thus, it is presumably the ratio of these two isomers that determines the natural resistance of different populations of this species to this pest.

D. Other Monoterpenes with Defensive Roles

The repellent properties of nepetalactone, a cyclopentanoid monoterpene from the catnip plant *Nepeta cataria*, have been demonstrated with 17 insects (Eisner, 1964). The leaf-cutting ant *Atta cephalotes* is repelled by limonene in the peel oil of grapefruit (Cherrett, 1972). Monoterpenes and other terpenoids in the peel oil of various types of citrus fruits are toxic to the eggs and larvae of the Caribbean fruit fly *Anastrepha suspensa* (P. Greany, personal communication).

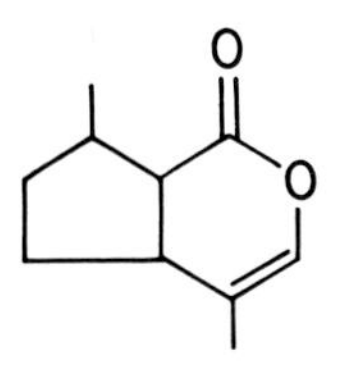

Nepetalactone
(*Nepeta*)

Xylomollin, a secoiridoid isolated from the bitter unripe fruits of *Xylocarpus molluscensis* (Meliaceae), is an effective deterrent to feeding by the African armyworm *Spodoptera exempta* at the 100 ppm level in leaf disc tests and strongly inhibits respiration in rat liver mitochondria (Kubo *et al.*, 1976c; Kubo and Nakanishi, 1977).

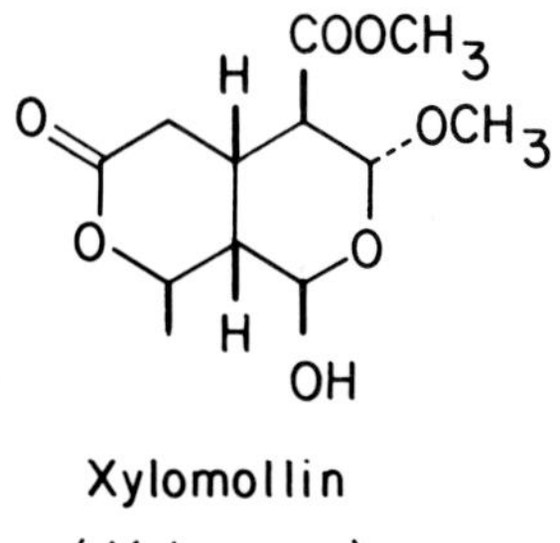

Xylomollin
(*Xylocarpus*)

Monoterpenes also apparently provide protection against browsing animals. For example, investigations of Douglas fir (*Pseudotsuga menziesii*) with respect to browsing by black-tailed deer (*Odocoileus hemionus columbianus*) demonstrated that resistant clones produce considerably higher total monoterpene concentrations (Radwan and Ellis, 1975). Volatile terpenes inhibit bacterial activity in the rumen of deer, resulting in reduced food consumption.

V. SESQUITERPENE LACTONES IN THE COMPOSITAE

Most of the approximately 900 known sesquiterpene lactones are constituents of members of the Compositae, occurring only infrequently in a few other higher and lower plants (e.g., Magnoliaceae, Lauraceae, and Umbelliferae; the basidiomycete *Lactarius vellereus* and liverworts of the genera *Frullania* and *Diplophylum*) (Yoshioka *et al.*, 1973; Burnett *et al.*, 1978a). Indeed, no other class of natural products, with the possible exception of the polyacetylenes, better characterizes this family of over 1,400 genera and 25,000 species (Mabry and Bohlmann, 1977). Sesquiterpene lactones are commonly referred to as bitter principles, which suggests that they may serve as deterrents to mammalian feeding. Michael additions, which can link proteins to the α,β-unsaturated γ-lactone moiety present in most sesquiterpene lactones, may play some role in their having feeding-deterrent properties; such reactions have already been implicated in their cytotoxic, allergenic, and antibiotic properties (Fig. 7).

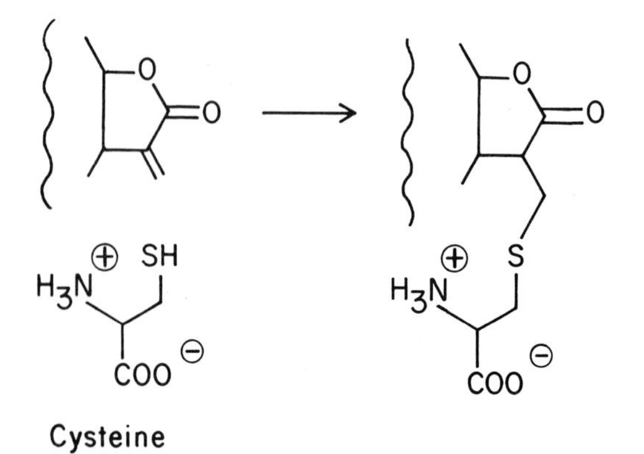

Figure 7. Sulfhydryl, hydroxyl, and amine functions are known to add *in vitro* to the α, β-unsaturated carbonyl system present in most sesquiterpene lactones (see, for example, Kupchan *et al.*, 1970).

Sesquiterpene lactones may occur throughout a plant, but they are most commonly associated with leaves and flower parts, sometimes being accumulated in glandular trichomes (Rodriguez *et al.*, 1976) (Fig. 8).

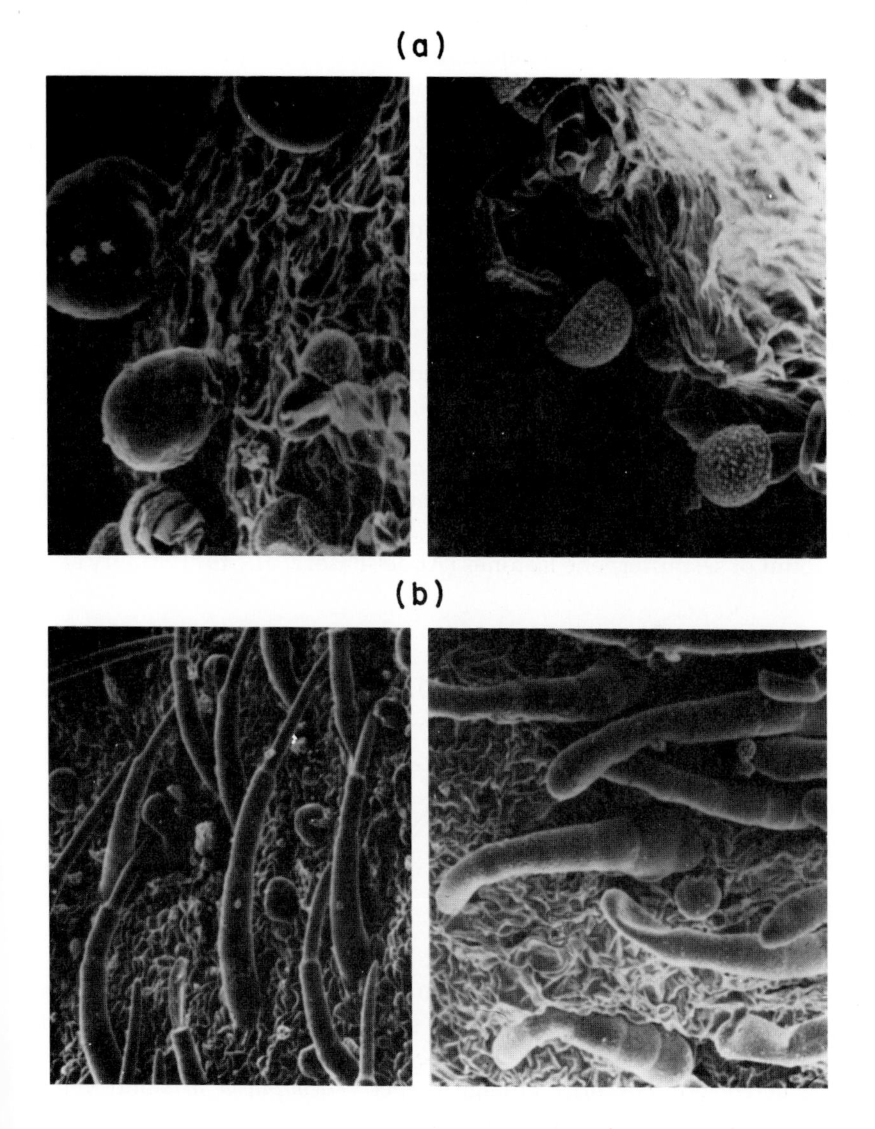

Figure 8. (a) Glandular and (b) nonglandular trichomes on leaves and flower parts of *Parthenium hysterophorus*, some of which contain sesquiterpene lactones (Rodriguez *et al.*, 1976).

A. Biogenesis

Classic experiments using labeled precursors have not been extensively applied to sesquiterpenoid biogenesis. Nevertheless, knowledge of the terpenoid pathway coupled with the structures of sesquiterpene lactones that co-occur supports the transformation of farnesyl pyrophosphate into germacranolides and the subsequent cyclization of these 10-membered carbocyclic compounds to give the 5/7 and 6/6 ring systems present in guaianolides, pseudoguaianolides (with methyl migration), and eudesmanolides (Fig. 9) (Herout, 1974; Herz, 1977).

B. Defensive Properties of Glaucolide-A

In establishing the feeding-deterrent and insect-growth inhibition properties of glaucolide-A, a germacranolide from *Vernonia,* laboratory and field investigations utilizing six species of Lepidoptera as well as two species of mammals were conducted with three closely related species of *Vernonia* (Burnett, 1974; Burnett *et al.,* 1974, 1977a,b, 1979a, b, 1979; Mabry *et al.,* 1977).

The three sympatric species of *Vernonia* have differences in their sesquiterpene lactone content: *Vernonia gigantea* and *V. glauca* contain glaucolide-A, whereas the third, *V. flaccidifolia,* produces no detectable amount of sesquiterpene lactones (Abdel-Baset *et al.,* 1971; Mabry *et al.,* 1975).

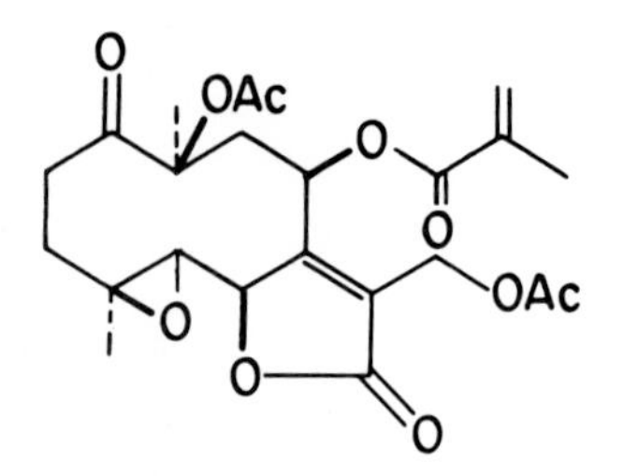

Glaucolide-A (*Vernonia*)

1. *Insect Feeding Deterrence Tests*

The feeding-deterrent properties of glaucolide-A were tested with six species of lepidopterous larvae. All six insects are sympatric with the species of *Vernonia,* and three of them, the yellow-striped armyworm (*Spodoptera ornithogalli*), cabbage looper (*Trichoplusia ni*), and yellow woolly bear (*Diacrisia virginica*), feed in nature on *Vernonia* species that contain glaucolide-A. A fourth species, the saddle-back caterpillar (*Sibine stimulea*), feeds on *V. flaccidifolia,* the plant that does not produce

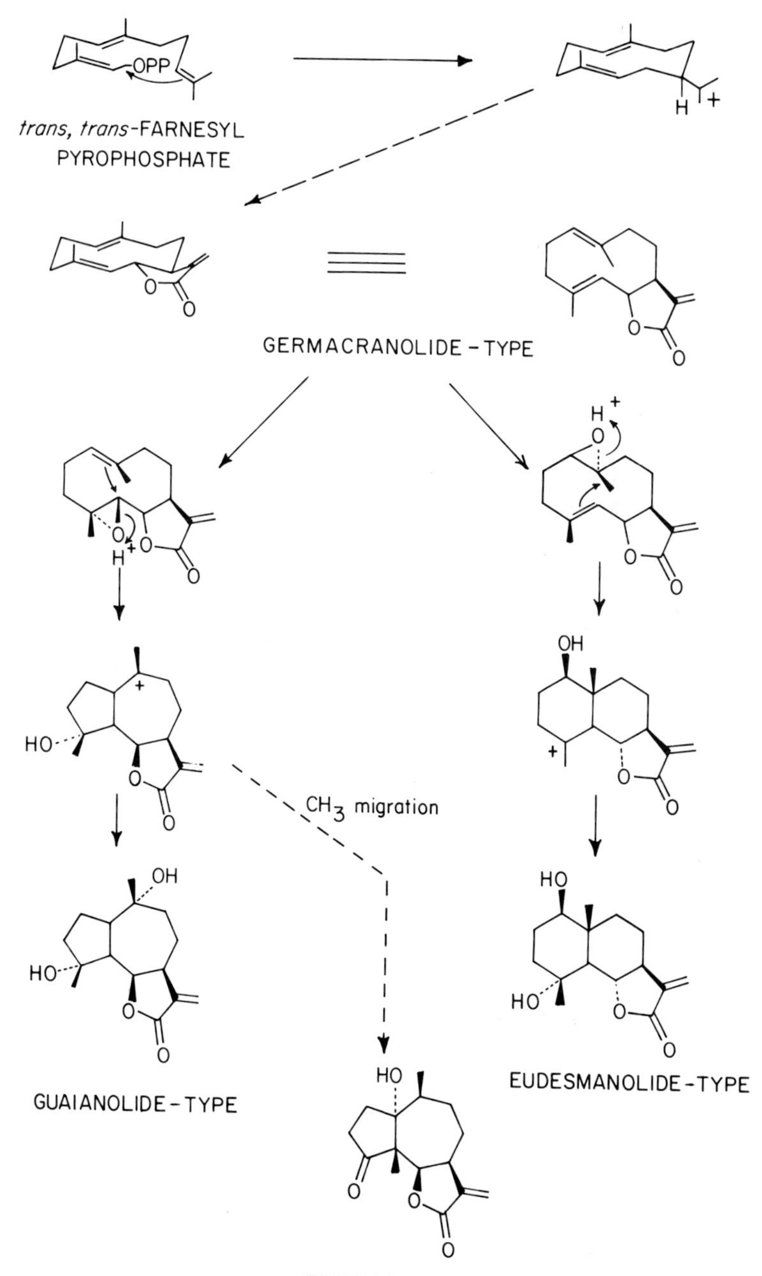

Figure 9. Possible pathways from farnesyl pyrophosphate to germacranolides and then to advanced types of sesquiterpene lactones with 5/6 and 6/6 ring systems.

sesquiterpene lactones. The southern (*Spodoptera eridania*) and fall (*S. frugiperda*) armyworms were not observed on *Vernonia* species in the field but are sympatric with them and, like most other armyworms, are polyphagous.

The fall, southern, and yellow-striped armyworms and the saddle-back caterpillar all avoided leaf sections of *Vernonia* species containing glaucolide-A (Fig. 10). In contrast, the yellow woollybear and cabbage looper both prefer the glaucolide-A species. Thus, with the exception of the yellow-striped armyworm, which sometimes feeds on sesquiterpene lactone-containing *Vernonia* species in nature, all the larvae follow the same feeding pattern in the laboratory as in the field. However, when offered agar pellets prepared from freeze-dried leaf powder of *V. flaccidifolia,* all insects chose pellets that did not contain added glaucolide-A.

Because the southern armyworm larvae exhibited a dramatic preference for diets without glaucolide-A, additional assays were conducted using various concentrations of glaucolide-A added to the *Vernonia flaccidifolia* medium. As the concentration of glaucolide-A in the pellet increased (up to approximately 1%), feeding decreased correspondingly. These feeding tests established that, when larvae are offered a choice between fresh leaves from sesquiterpene lactone-containing species and from *V. flaccidifolia,* they tend to exhibit the same feeding pattern observed in nature. However, when offered *V. flaccidifolia* leaf material in the form of a pellet, all prefer pellets without added sesquiterpene lactone.

Further tests were conducted to determine the effect of glaucolide-A on the growth and development of the larvae. Newly hatched larvae were reared on artificial diets containing added glaucolide-A in concentrations up to 0.5% dry weight (Fig. 11). The reduced growth of the fall and southern armyworms was more pronounced than could be accounted for by reduced feeding. The insect species most adversely affected by the ingestion of glaucolide-A are those that always avoid diets containing this compound. The southern and fall armyworms also showed reduced survival, especially during the early larval instars (Burnett *et al.,* 1978b). With the exception of the yellow woollybear, the time required to reach pupation was lengthened for all larvae reared on glaucolide-A diets. It was also determined that glaucolide-A on the leaves of *Vernonia* species did not deter oviposition by the Lepidoptera species investigated, except for the fall armyworm (Burnett *et al.,* 1979).

Since glaucolide-A could deter the feeding of some but not all insect larvae in laboratory tests, several garden and field plots of *Vernonia*

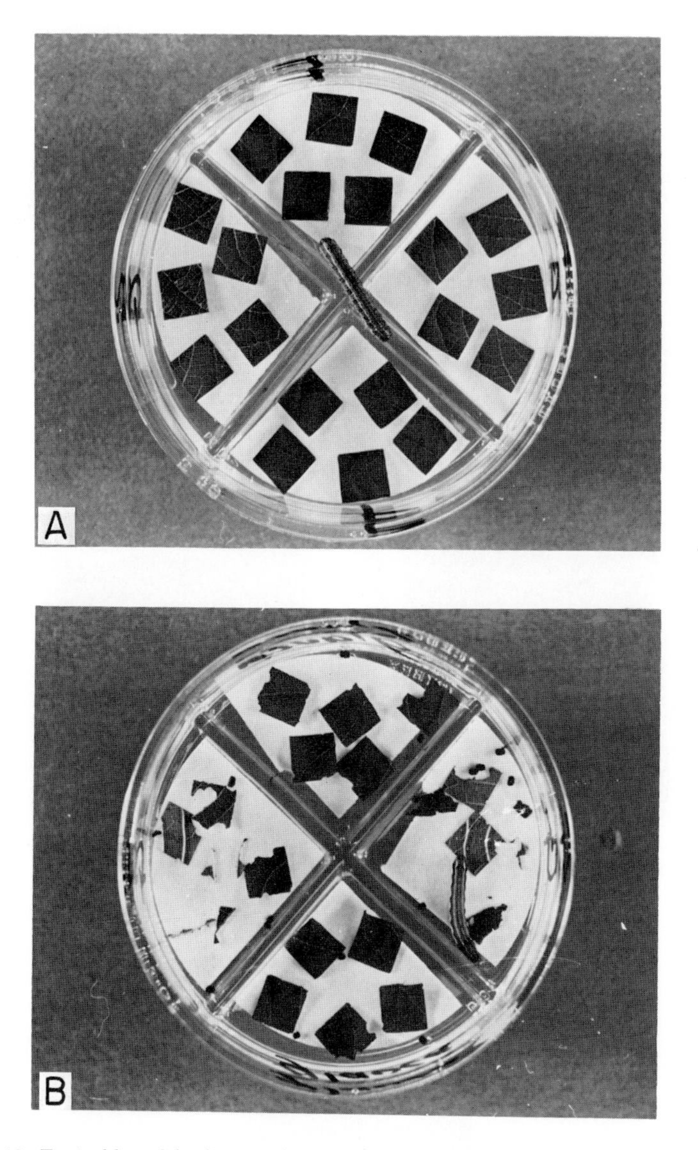

Figure 10. Typical larval feeding preference tests on fresh *Vernonia* foliage. A—initiation of test; B—completed test (from Burnett, 1974).

species were observed for relative amounts of herbivory. Surprisingly, in all cases, *Vernonia flaccidifolia* was fed on less by insects than were the glaucolide-A-containing species, a puzzling observation that may reflect the importance of glaucolide-A in deterring feeding by mammals.

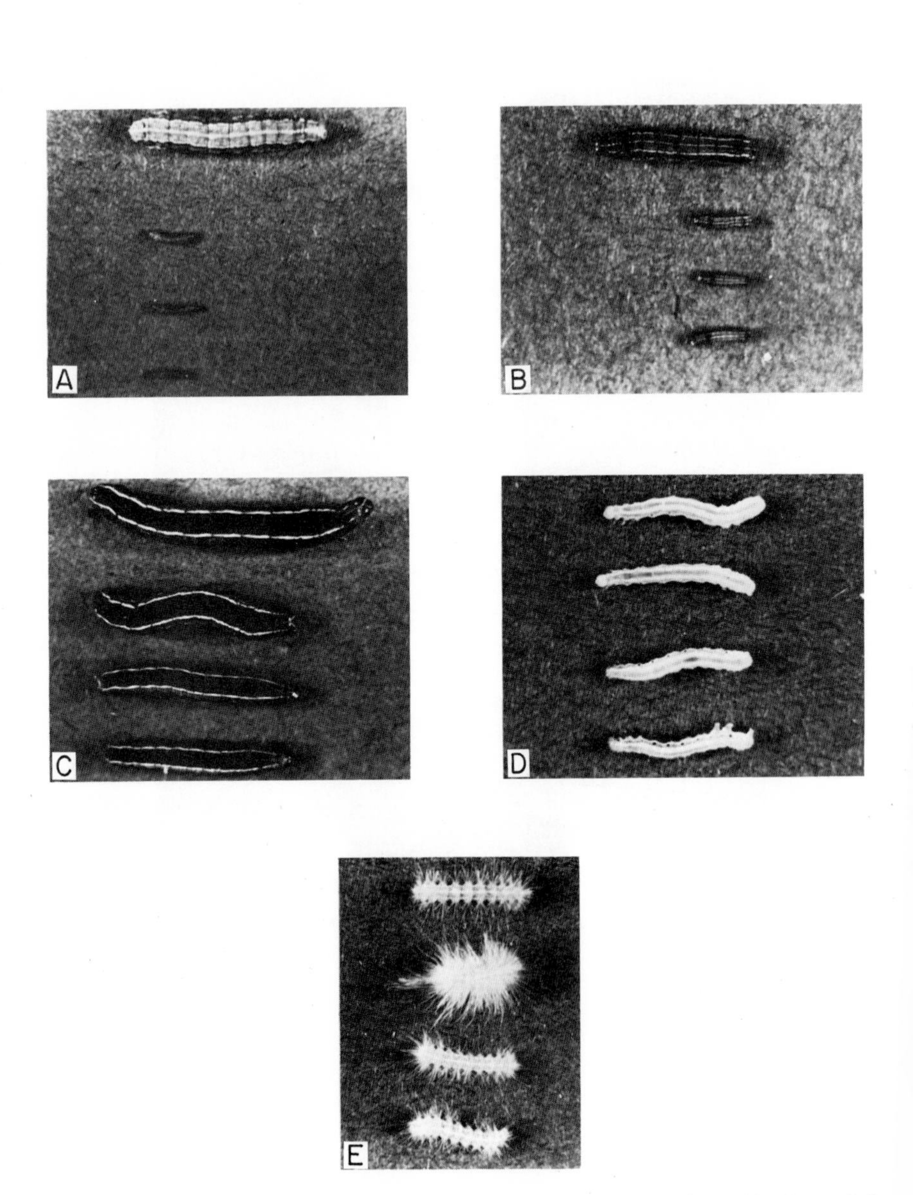

Figure 11. Relative size of larvae reared on diets containing glaucolide-A. (Top insect fed on control diet followed by increasing glaucolide-A concentrations of 1/8%, 1/4%, and 1/2% by dry weight.) A-fall armyworm; B-southern armyworm; C-yellow-striped armyworm; D-cabbage looper; E-yellow woollybear (from Burnett, 1974).

2. Mammalian Feeding Deterrence Tests

Species of *Vernonia* and other Compositae that produce high concentrations of bitter sesquiterpene lactones are largely untouched in otherwise heavily grazed pastures, suggesting a general defensive role for sesquiterpene lactones against mammalian herbivores.

Wild rabbits were observed in garden plots of *Vernonia* species to feed heavily on only *Vernonia flaccidifolia*. When wild cottontail rabbits (*Sylvilagus floridanus*) were placed in test cages, they consistently fed on *V. flaccidifolia* while only sampling and rejecting the glaucolide-A-containing *V. gigantea* (Fig. 12). An additional series of tests compared glaucolide-A-coated *V. flaccidifolia* with untreated plants. Again, the rabbits preferred the plants lacking the bitter lactone.

Analogous results were obtained with the whitetail deer (*Odocoileus virginianus*). By smelling, licking, and then sampling each plant the deer learned to discriminate between *Vernonia* species, consistently choosing those lacking glaucolide-A. The different species of *Vernonia* tested were apparently recognized by smell, although sesquiterpene lactones themselves are not volatile. Additionally, the deer were able to distinguish between *V. flaccidifolia* coated and uncoated with glaucolide-A but only by first tasting them. These results suggest that the bitterness of the lactone was a primary factor in determining the feeding preference of deer on *Vernonia* species. It was difficult to monitor mammalian feeding in the field since the glaucolide-A species are not eaten, whereas *V. flaccidifolia* would probably be totally consumed when encountered.

In summary, glaucolide-A apparently protects such species as *Vernonia gigantea* and *V. glauca* against mammals and some insects. In contrast, *V. flaccidifolia* has no chemical protection against mammals but does possess an undefined defense against insects. Since most sesquiterpene lactones are bitter, it may be this property that accounts for their being mammalian feeding deterrents.

C. Other Activities of Sesquiterpene Lactones

Sesquiterpene lactones (Fig. 13) have received considerable attention in recent years due to their cytotoxic and antitumor activity (see, for example, Kupchan, 1976; Hartwell, 1976). The conjugated exocyclic double bond in the lactone group of compounds such as helenalin and vernolepin is normally required for this activity (Kupchan *et al.*, 1971), although at least one antitumor compound (Jamieson *et al.*, 1976) lacks this functionality.

Tenulin, the major sesquiterpene lactone constituent of *Helenium amarum* (bitter sneezeweed), imparts a bitter taste to milk (Ivie *et al.*,

Figure 12. Cottontail rabbits placed in a cage with *Vernonia flaccidifolia* (left) and *Vernonia gigantea* (right) eat only the *V. flaccidifolia,* which does not contain glaucolide-A (from Burnett, 1974). (a) Plants before rabbit placed in cage; (b) Plants after rabbit placed in cage.

CYTOTOXIC AND ANTITUMOR TYPES

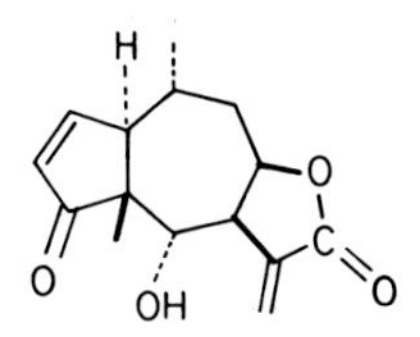

Helenalin

Vernolepin

MAMMALIAN TOXINS

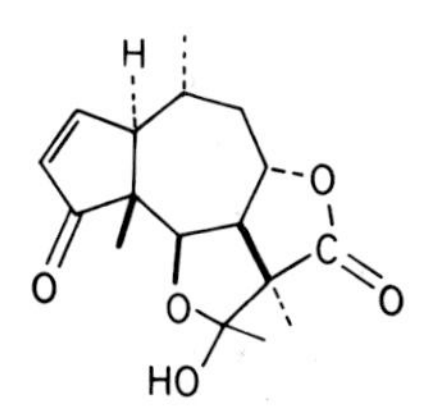

Tenulin

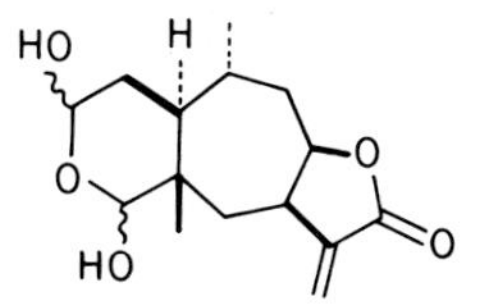

Hymenovin

ALLERGIC CONTACT DERMATITIS TYPES

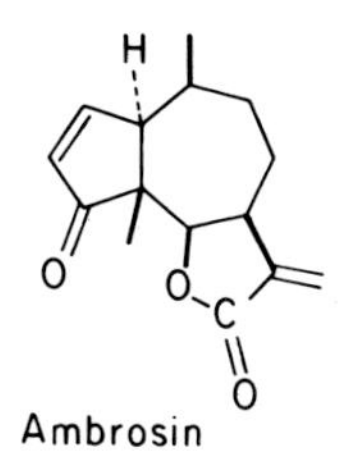

Ambrosin

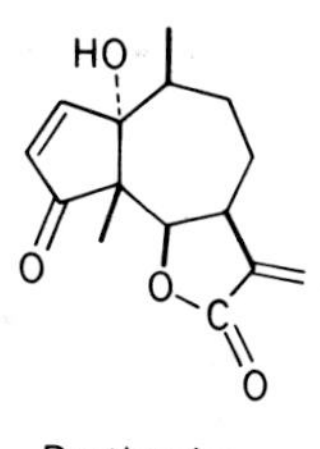

Parthenin

ALLELOPATHIC TYPES

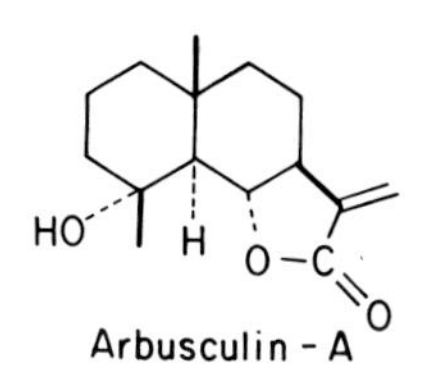

Arbusculin - A

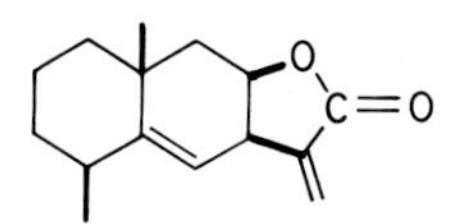

Alantolactone

Figure 13. A few of the many sesquiterpene lactones known to exhibit various types of biological activities.

1975b) and is toxic to livestock in larger doses. Helenalin from *H. microcephalum* is toxic to several mammalian species in oral doses as low as 85 mg/kg (Witzel *et al.*, 1976). *Hymenoxys odorata* (bitterweed), which occurs in the southwestern United States, contains an economically significant livestock toxin, hymenovin, which primarily affects sheep and goats. Annual losses from *Hymenoxys odorata* in Texas alone were reported to be over $3 million. Hymenovin is poisonous to sheep in orally administered doses as low as 100 mg/kg (Ivie *et al.*, 1975a; Kim *et al.*, 1975). A similar poisoning has been noted among sheep grazing on South African species of *Geigeria*, which contains the sesquiterpene lactone vermeerin (de Kock *et al.*, 1968).

Sesquiterpene lactones are one of the important causes of allergic contact dermatitis in human beings (Mitchell, 1969; Mitchell *et al.*, 1970, 1971a,b, 1972; Bleumink *et al.*, 1973). Most members of the Compositae for which allergenic activity has been reported are known to contain sesquiterpene lactones (Mitchell, 1975). Liverworts of the genus *Frullania* (Knoche *et al.*, 1969; Perold *et al.*, 1972) also contain sesquiterpene lactones that produce a severe allergic rash.

Among the most highly allergenic compounds known are parthenin and ambrosin, two pseudoguaianolides from *Parthenium hysterophorus* (Mitchell and Dupuis, 1971), a species introduced into India in the early 1950's. Today, this plant represents one of the major causes of allergic contact dermatitis in that subcontinent, affecting thousands of adults (Lonkar *et al.*, 1974). Sesquiterpene lactones are accumulated in fragile surface trichomes in *P. hysterophorus* (Rodriguez *et al.*, 1976), which accounts for the easy exposure of human beings to the compounds (see Fig. 8).

Sesquiterpene lactones such as arbusculin-A from species of sagebrush (*Artemisia*) are active allelopathic agents, inhibiting radicle, hypocotyl, and lateral root growth of *Cucumis sativus* (McCahon *et al.*, 1973). Alantolactone from *Inula* inhibits the growth of *Chlorella* cells (Kwon *et al.*, 1973) and restricts the germination and growth of seeds and seedlings from various species (Dalvi *et al.*, 1971). Vernolepin from *Vernonia hymenolepis* reduced the elongation of *Avena* coleoptiles (Sequeria *et al.*, 1968).

VI. OTHER SESQUITERPENES WITH FEEDING-DETERRENT PROPERTIES

In addition to sesquiterpene lactones, feeding-deterrent activity has been demonstrated for a number of sesquiterpenoids (Fig. 14). For ex-

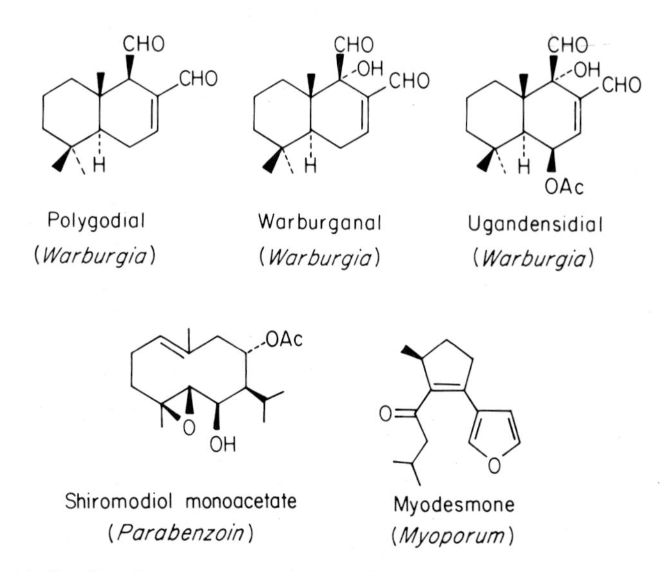

Figure 14. Feeding deterrent sesquiterpenoids from African and Australian Plants.

ample, polygodial, ugandensidial, and warburganal from the bark of the East African spice trees *Warburgia ugandensis* (Canellaceae) and *W. stuhlmannii* inhibit feeding by two African armyworms, *Spodoptera exempta* and *S. littoralis* (Kubo *et al.*, 1976b; Kubo and Nakanishi, 1977). Choice tests using leaf discs of *Zea mays* coated with various concentrations of these compounds established that warburganal, the most active feeding inhibitor of the three, was effective against *S. exempta* larvae at concentrations of 0.1 ppm but relatively ineffective against the tobacco hornworm *Manduca sexta* and the grasshopper *Schistocerca vaga*. A fourth related compound, muzigadial, was reported to be comparable with warburganal in its feeding inhibition properties (Kubo *et al.*, 1977d; Meinwald *et al.*, 1978). Electrophysiological investigations demonstrated that application of warburganal and other feeding inhibitors to the chemoreceptive sensilla of *S. exempta* resulted in the loss of the taste response. The inhibitory action of these compounds against *S. exempta* is presumably due to Michael additions to the α,β-unsaturated carbonyl function (Fig. 7) since altering this group can eliminate the activity. Whether the "hot" taste experienced by human beings with these compounds involves a similar reaction is not known.

Shiromodiol mono- (Fig. 14) and diacetate and shiramool from *Parabenzoin trilobum* deter the feeding of several insect species (Wada *et al.*, 1970; Munakata, 1977). All three compounds exhibited similar feeding inhibitory activity against *Spodoptera littoralis,* with nearly complete inhibition at the 0.5% concentration level. The mono- and diacetates also inhibit feeding by the tobacco cutworm *Spodoptera litura*, with the monoacetate being active at levels as low as 0.12% in leaf disc choice tests. When tested against the oligophagous insect *Trimeresia miranda*, the diacetate exhibited the greatest activity, with 70% feeding inhibition at 0.03% concentration and 100% inhibition at 0.25% concentration. In a separate series of experiments using a broad spectrum of compounds, the effectiveness of terpenoid feeding inhibitors was found to vary with insect age and diet (Wada and Munakata, 1971).

Resistance of tomatoes to the two-spotted spider mite (*Tetranychus urticae*) has been attributed to sesquiterpenoids, which are both repellent and toxic to this insect (Patterson *et al.*, 1975).

Myoporum deserti (Myoporaceae), a widespread Australian shrub, has been responsible for serious livestock losses in that region (Sutherland and Park, 1967). These plants contain a group of toxic furanoid sesquiterpene ketones such as myodesmone (Fig. 14), substances that are known to cause liver and kidney lesions.

VII. TOXIC DITERPENES

A. Diterpene Acids as Larval Growth Inhibitors

Varieties of sunflower (*Helianthus annuus*) that are resistant to attack by larvae of the sunflower moth *Homeosoma electellum* contain in their florets high concentrations of two diterpene acids, trachyloban-19-oic acid and (−)-16-kauren-19-oic acid. Since sunflower florets containing only small amounts of these compounds are a major portion of the diet for the first instar larvae of *H. electellum*, it is likely that these acids serve as feeding inhibitors (Elliger *et al.*, 1976; Waiss *et al.*, 1977). Indeed, larval growth inhibition tests involving their incorporation into artificial diets demonstrated that at the 1% level both kaurenoic and

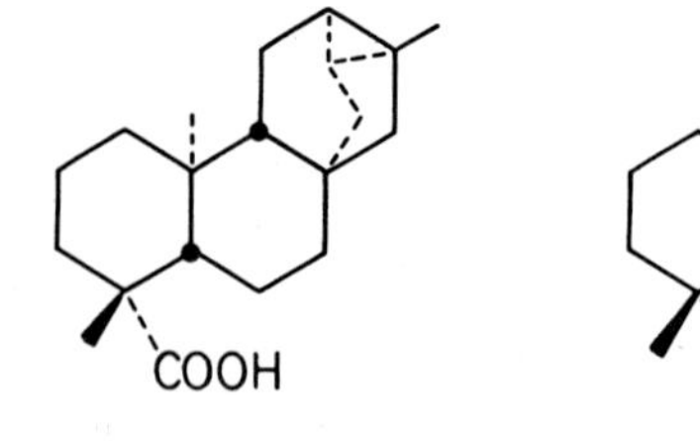

Trachyloban-19-oic Acid (*Helianthus*)

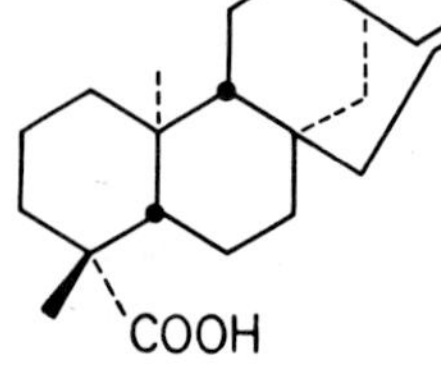

(−)-16-Kauren-19-oic Acid (*Helianthus*)

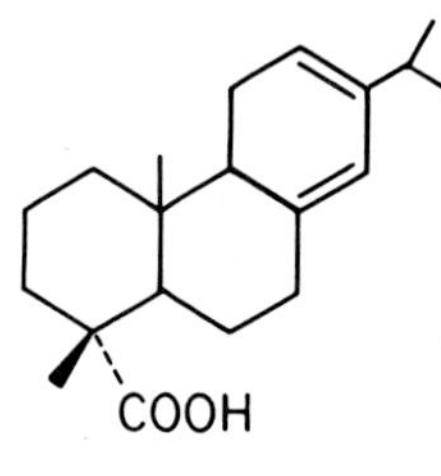

Levopimaric Acid (*Pinus*)

trachylobanoic acids decreased the growth of sunflower moth larvae and the tobacco budworm (*Heliothis virescens*) by about 50%. At the 0.5% concentration level, both reduced larval growth of the cotton bollworm (*H. zea*) and the pink bollworm (*Pectinophora gossypiella*) to less than 5% (Elliger *et al.*, 1976). Several tricyclic terpenic acids (e.g., levopimaric acid) also substantially inhibited growth of the pink bollworm larvae.

B. Kaurenoid Diterpenes from *Isodon* Species

Several diterpenes from *Isodon* (Labiatae) possess feeding-deterrent properties. Inflexin (from *I. inflexus*), isodomedin (from *I. shikokianus* var. *intermedius*), and kamebanin (from *I. kameba*) all exhibit specific toxicity toward lepidopteran larvae including the African armyworm *Spodoptera exempta* (Kubo and Nakanishi, 1977). They are also cytotoxic, having LD_{50} values (KB test) of 5.4, 4.0, and 5.1 μg/ml, respectively (Kubo *et al.*, 1977a,b,c). All three diterpenes possess an exocyclic α,β-unsaturated carbonyl group capable of undergoing a Michael addition, a functionality already implicated in their antimicrobial properties (Kubo *et al.*, 1974).

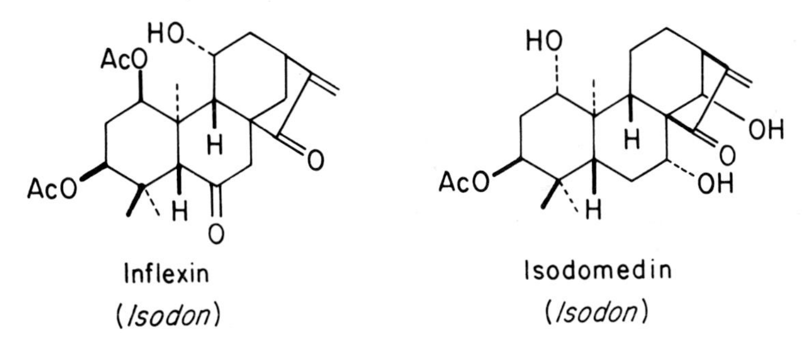

C. Diterpenes from the Verbenaceae

Kato *et al.* (1972) isolated two feeding-deterrent *ent*-clerodane diterpenoids, clerodendrins A and B, from mature leaves of *Clerodendron tricotomum*. Their activity as well as that of clerodin, a closely related diterpenoid isolated previously from *Clerodendron infortunatum* (Barton *et al.*, 1961; Paul *et al.*, 1962), was established by the leaf disc assay method. Both clerodendrins A and B gave 100% feeding inhibition in a 2-hour test against *Spodoptera litura* at about 250 ppm, whereas clerodin was considerably more active. Although many of the clerodane diterpenoids have a bitter taste, no clear correlation between feeding-deterrent activity and bitterness could be established.

In tests with other larvae, clerodendrins A and B gave 100% feeding-deterrent activity in the 2-hour test against larvae of the Oriental tussock moth *Euproctis subflava* at 1000 ppm and against the European corn borer *Ostrinia nubilalis* at 5000 ppm. The latter insect has long been controlled with the use of powdered stems from *Ryania speciosa* (Flacourtiaceae), a plant containing the insecticidal diterpene ryanodine (Weisner, 1963; Srivastava and Przybylska, 1968).

Extracts containing clerodane and *ent*-clerodane diterpenes from other members of the Verbenaceae (*Clerodendron fragrans, C. calamitosum, C. cryptophyllum, Callicarpa japonica,* and *Caryopteris divaricata*) exhibited feeding-deterrent activity at the 1% level (Hosozawa, 1974a,b; Munakata, 1977). *Caryopteris divaricata* yielded eight toxic diterpenes (clerodin, dihydroclerodin, clerodin hemiacetal, caryoptin, dihydrocaryoptin, caryoptin hemiacetal, caryoptinol, and dihydrocaryoptinol), whereas *Clerodendron fragrans* afforded only the single, weakly active diterpene alcohol phytol. Clerodendron A was identified as the active feeding-deterrent principle of *Clerodendron cryptophyllum,* whereas *Clerodendron calamitosum* gave a new feeding-deterrent compound, 3-epicaryoptin.

Against larvae of *Spodoptera litura,* the clerodin series was more active than the caryoptin types. For example, for 100% inhibition in the 2-hour test with leaf discs, about 50 ppm of the clerodin series of compounds was required, whereas for most caryoptin derivatives concentrations of 200 ppm or greater were necessary for similar activity. When these feeding-deterrence assays were allowed to continue for 24 hours, the larvae still would not bite the diterpene-coated leaves and eventually starved to death. The term "absolute feeding inhibitor" is used to describe this type of activity, as distinguished from "relative feeding inhibitor," which produces only short-term feeding inhibition.

The observation that leaves of *Ajuga remota* (Labiatae) are not eaten by African armyworms led to the isolation of several other feeding-

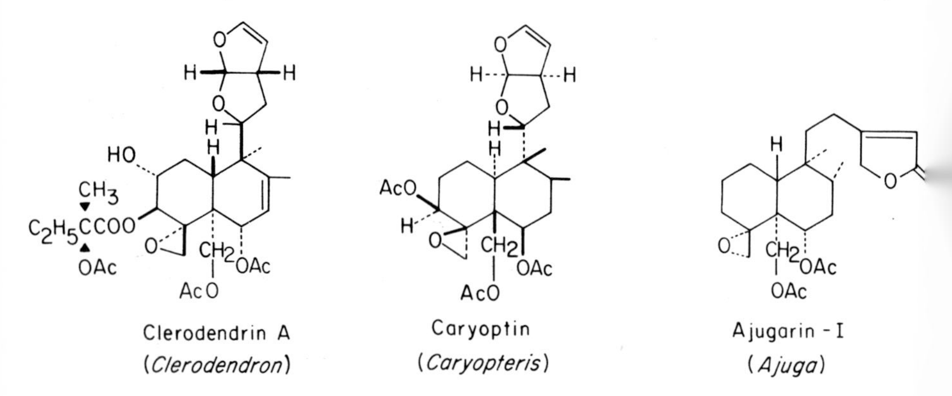

Clerodendrin A (*Clerodendron*) Caryoptin (*Caryopteris*) Ajugarin-I (*Ajuga*)

deterrent *ent*-clerodanes, the ajugarins, which in leaf disc tests exhibited activity against *Spodoptera exempta* at 100 ppm and against *S. littoralis* at 300 ppm. Initially, the ajugarins were thought to possess antiecdysone properties against the armyworms, but this was later disproved (Kubo *et al.*, 1976a; Kubo and Nakanishi, 1977).

D. Poisonous Andromedane Diterpenes from the Ericaceae

Many ericaceous trees are known to contain andromedane diterpenoids that cause poisoning in mammals, including human beings. For example, *Rhododendron japonicum* produces a series of four closely related toxic diterpenoids, rhodojaponins I–IV, in its flowers (Hikino *et al.*, 1969, 1970a,b), and the leaves of the poisonous shrub *Leucothoe grayana* contain several toxic andromedane diterpenes, grayanotoxins I–VII (Hikino *et al.*, 1970a,b). Similarly, the toxicity of *Pieris japonica* has been attributed to asebotoxins I–VII, a group of diterpenoids isolated from the flowers of this species (Hikino *et al.*, 1969, 1970a, 1971). Leaves of *Agauria salicifolia* and other East African ericaceous plants also contain diterpenes toxic to goats and sheep.

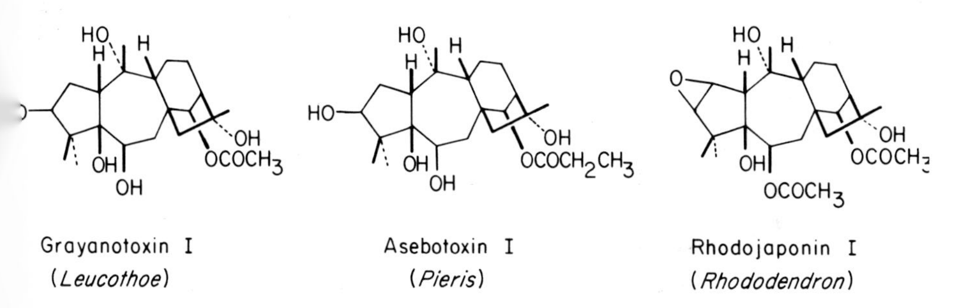

Grayanotoxin I (*Leucothoe*) Asebotoxin I (*Pieris*) Rhodojaponin I (*Rhododendron*)

E. Toxic Diterpenes in the Thymelaeaceae and the Euphorbiaceae

Pimelea prostrata (Thymelaeaceae), a New Zealand shrub responsible for livestock poisoning, contains the tetracyclic diterpene prostratin (Cashmore *et al.*, 1976). Intraperitoneal injections of 3 μg per gram body weight of this toxin were fatal to mice within 2 hours. Other *Pimelea* species from Australia have also been reported to be toxic to cattle.

The fruit and bark of *Daphne mezereum* (Thymelaeaceae) are highly toxic to mammals due in part to the diterpenoid daphnetoxin (Ronlan and Wickberg, 1970). Seeds contain another toxic diterpene, mezerein. Both mezerein and daphnetoxin can produce irritant dermatitis in human beings, and mezerein possesses cocarcinogenic activity (Hecker, 1975), a property typical of many other structurally similar diterpene esters including the phorbols from *Euphorbia*. Odoracin, a structurally similar diterpenoid from *Daphne odora*, possesses nematicidal activity in immersion tests against the rice white-tip nematode *Aphelenchoides besseyi*. Odoracin gave 100% nematicidal activity at 5 ppm and 70% after 5 days at the 1 ppm level (Kogiso *et al.*, 1976).

In a screening of 60 *Euphorbia* (Euphorbiaceae) species, 12-deoxyphorbol was found to be one of the most common toxic diterpenoids present in the latex, normally occurring as a mono- or diester of aliphatic acids and, more rarely, with aromatic acids (Schmidt and Evans, 1976; Hirata, 1975). The latex of *Euphorbia poisonii* contains a series of *o*-ester tricyclic types and *o*-acyl esters of 12-deoxyphorbol (Schmidt and Evans, 1976; Evans and Schmidt, 1976), which can produce severe skin lesions in human beings and livestock.

Tung oil and meal produced from the fruits of *Aleurites* (Euphor-

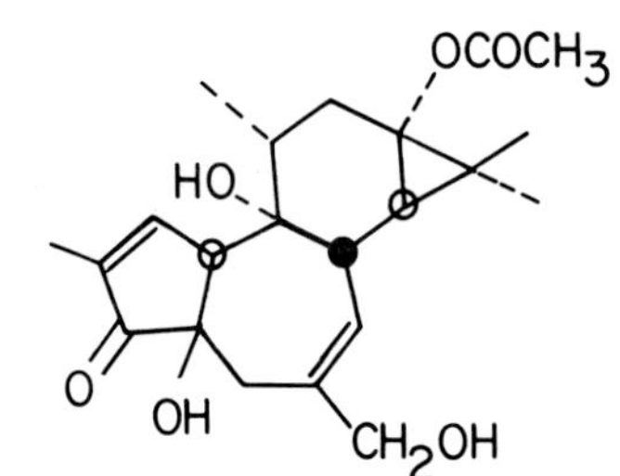

Prostratin
(*Pimelea*)

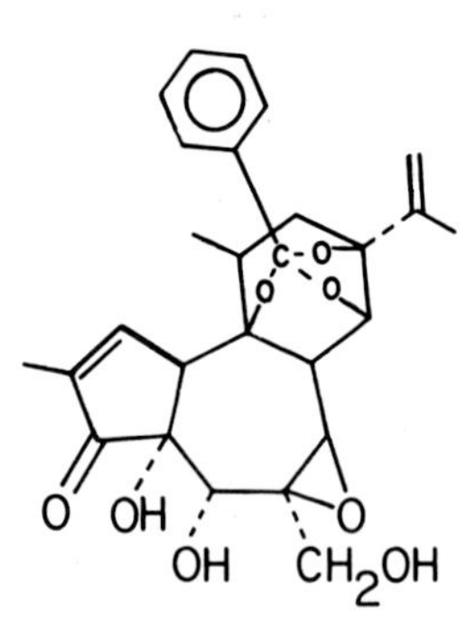

Daphnetoxin
(*Daphne*)

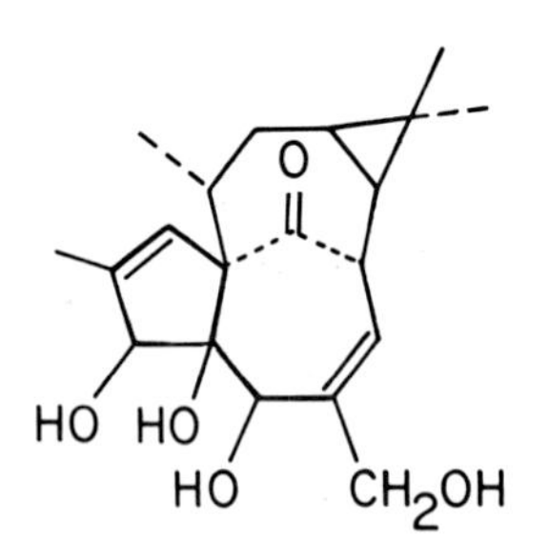

Ingenol
(*Elaeophorbia*)

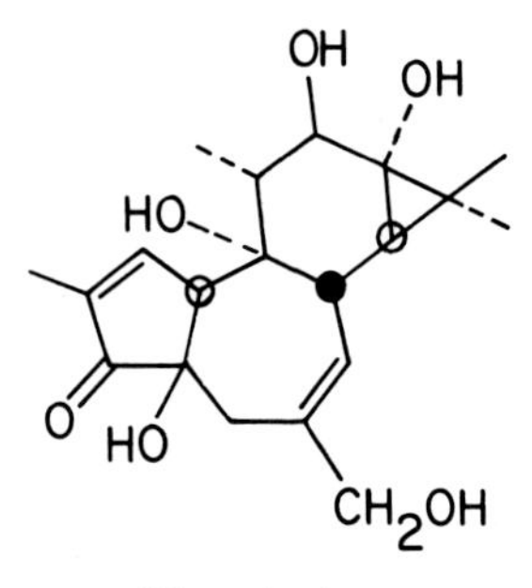

Phorbol
(*Euphorbia*)

biaceae) trees are toxic to mammals, causing irritation to the skin and internal organs. Attempts to detoxify this protein-rich meal led to the isolation of a toxic terpenoid diester, 12-*O*-palmityl-13-*O*-acetyl-16-hydroxyphorbol (Okuda *et al.*, 1975). The monoester is not toxic.

Ingenol derivatives occur throughout the Euphorbiaceae including *Euphorbia* spp. (Hirata, 1975; Evans and Kinghorn, 1974; Upadhyay *et al.*, 1976) and *Elaeophorbia* spp. (Kinghorn and Evans, 1974). They exhibit toxicological and cocarcinogenic properties similar to those of the phorbol compounds. (−)-Methyl barbascoate, a clerodane diterpenoid isolated from *Croton californicus* (Euphorbiaceae), has been shown to possess insecticidal and antitumor activity (Wilson *et al.*, 1976).

VIII. HELIOCIDES IN *Gossypium* SPECIES

It has long been known that varieties of cotton (*Gossypium*) with gossypol-containing glands are avoided by bollworm larvae, rabbits, rodents, and other herbivores. In the 1960's, low-gossypol varieties of cotton were developed in an effort to obtain cottonseeds that could be used as a food source; however, these low-gossypol varieties were highly susceptible to insect attack. Thus, from these early studies it appeared that gossypol provided the major chemical defense of cotton against herbivory since resistance in races of *G. hirsutum* increased with increasing number of glands and with elevated gossypol content. Indeed, gossypol was shown to have detrimental effects on several insect pests of cotton. However, further analysis of the pigment glands, especially in wild varieties, revealed an array of toxic C_{15}, C_{25}, and C_{30} terpenoids, all structurally related to gossypol (Fig. 15) (Stipanovic *et al.*, 1977). The C_{25} terpenoids are now referred to as heliocides because of their insecticidal activities against *Heliothis* species and are designated H_1–H_4 and B_1–B_4, depending on whether they were first detected in *Gossypium hirsutum* or *G. barbadense*. Heliocide H_1 is the most toxic member of the series so far discovered. Knowledge of the genetic control

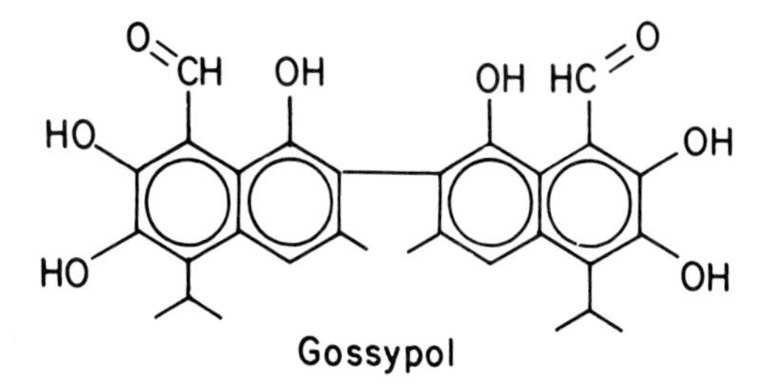

Gossypol

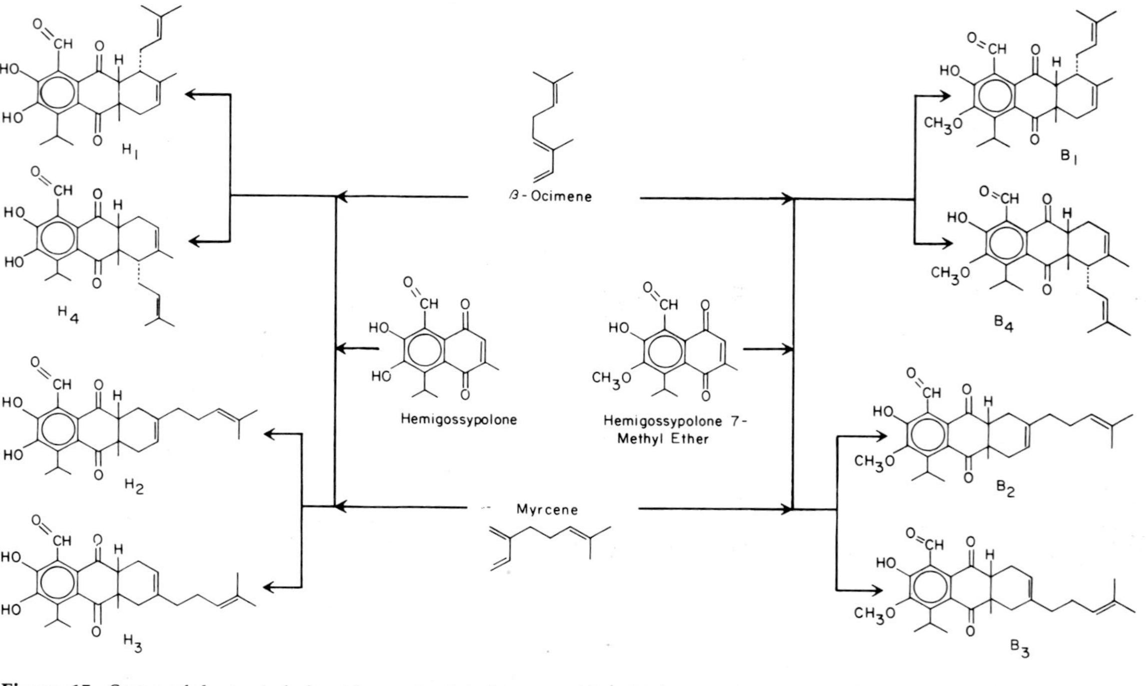

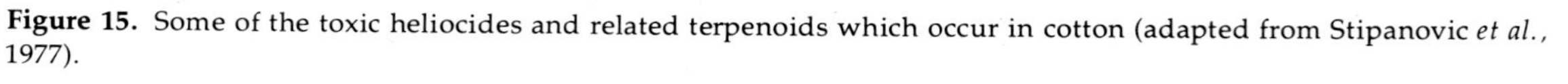

Figure 15. Some of the toxic heliocides and related terpenoids which occur in cotton (adapted from Stipanovic *et al.*, 1977).

of heliocide formation provides a sound basis for breeding resistant cultivars of cotton.

IX. TRITERPENES

A. Lantadenes A and B from *Lantana camara* (Verbenaceae)

Populations of *Lantana camara* that cause poisoning of livestock contain as major constituents toxic triterpenoids, usually lantadenes A and B. When lantadenes A and B were administered intraruminally to sheep at doses of 80 and 200 mg per kilogram body weight, respectively, the characteristic toxicity associated with ingestion of the whole *Lantana* plant was observed (Hart *et al.*, 1976a,b). Icterogenin (the C_4 hydroxy derivative of lantadene A) was isolated from a related species, *Lippia rehmanni,* and found to be toxic to rabbits.

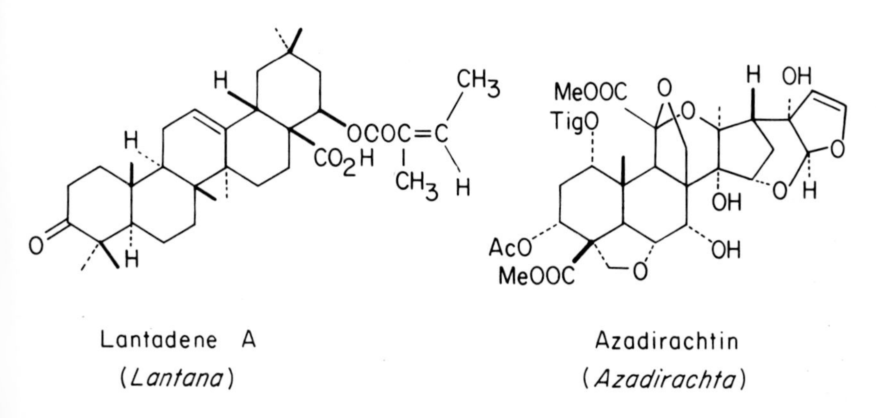

Lantadene A (*Lantana*)

Azadirachtin (*Azadirachta*)

B. Limonoids

The limonoids are a class of C_{26} bitter principles occurring in members of the Meliaceae, Rutaceae, and Simaroubaceae (Jacobson, 1977) that are thought to be derived from the oxidation of tetracyclic triterpenoids. Melianone from *Melia azedarach* (chinaberry) and limonin, nomilin, nimbin, nimbalide, azidirone, and salannin from *Azadirachta indica* (Indian neem tree) are all feeding deterrents and insecticidal limonoids.

The leaves and seeds of *Melia azedarach* have long been recognized for their feeding inhibition against locusts. Meliantriol was isolated from the fruit and shown to inhibit feeding of the desert locust *Schistocerca gregaria* (Lavie *et al.*, 1967). The potent feeding inhibitor azadirachtin, isolated from *Azadirachta indica* (Butterworth and Morgan, 1968, 1971; Zanno *et al.*, 1975) and from *M. azedarach*, causes 100% feeding inhibition against the desert locust at a concentration of 40 μg/liter (Morgan and Thornton, 1973). It is also a highly active feeding deterrent for the African armyworm *Spodoptera exempta* (Kubo and Nakanishi, 1977).

Harrisonin and the related limonoid bitter principle obacunone have been isolated from *Harrisonia abyssinica* (Simaroubaceae). Harrisonin inhibits feeding activity of *Spodoptera exempta* at the 20 ppm level in leaf disc tests (Kubo and Nakanishi, 1977) and also exhibits antibiotic and cytotoxic activity (2.2 μg/ml, KB test).

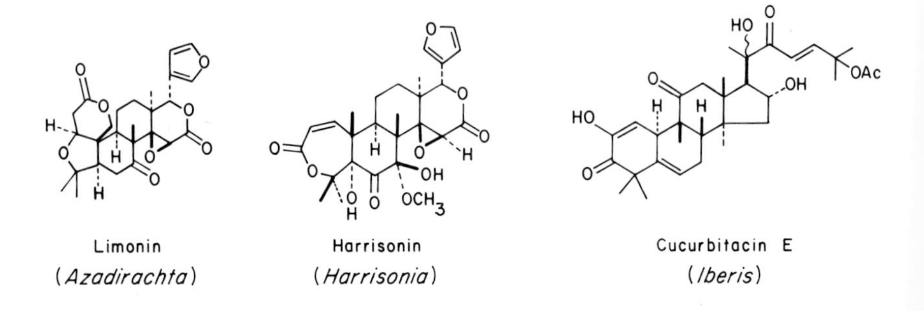

C. Cucurbitacins

Among the most bitter triterpenoids are the cucurbitacins, which are widespread throughout the Cucurbitaceae. These highly oxygenated tetracyclic compounds often contain an α,β-unsaturated ketone moiety in their side chains. All of the more than 20 known cucurbitacins exert strong physiological actions including cytotoxic and antitumor activity, purgative action, and poisoning of insects and mammals. Not surprisingly, these compounds serve as feeding deterrents to a wide variety of animals. In choice tests, the honeybee *Apis mellifera* demonstrated a strong preference for fruits lacking these bitter compounds. Cucurbitacins have been suggested to act as repellents to other bees and wasps and possibly to some lepidopterous species (Chambliss and Jones, 1966; Ehrlich and Raven, 1964). Cucurbitacins afford protection against the polyphagous mite *Tetranychus urticae*, which is often damaging to nonbitter cucurbits (Da Costa and Jones, 1971). They also inhibit feeding by the leaf beetles *Phyllotreta undulata, P. tetrastigma,* and *Phaedon cochleariae.* Most *Iberis* species produce cucurbitacins, and apparently they represent the primary chemical defense for these plants. For example, cucurbitacins E and I from the seeds and green parts of

Iberis amara (Cruciferae) are potent feeding inhibitors for the flea beetle *Phyllotreta nemorum* (Nielsen *et al.*, 1977). In addition, the presence of cucurbitacins has been correlated with red spider mite tolerance in cucumbers (Kooistra, 1971).

The fruits and roots of plants containing cucurbitacins are toxic to mammals, producing strong purgative action and poisoning; thus, these compounds also play a role in defense against mammalian herbivores.

X. SUMMARY

In addition to their many vital metabolic roles, terpenoids represent a major defense in plants against insects and mammals. However, it appears that less than 1% of the known terpenoids have been investigated for their ecologically relevant feeding-deterrent and toxic properties. Thus, as for most other classes of natural products, the role terpenoids play in plant–herbivore interactions remains a fertile field for future research.

ACKNOWLEDGMENTS

We wish to thank Dr. William C. Burnett, Jr., Dr. Nobuo Ohno, Dr. Robert Stipanovic, and Mr. Jonathan Gershenzon for helpful comments and data. Some of the research described in this chapter was supported by the Robert A. Welch Foundation, the National Science Foundation, and the National Institutes of Health.

REFERENCES

Abdel-Baset, Z. H., Southwick, L., Padolina, W. G., Yoshioka, H., and Mabry, T. J. (1971). *Phytochemistry* **10,** 2201–2204.

Arndt, U. (1968). *Holzforschung* **22,** 104.

Barton, D. H. R., Cheung, H. T., Cross, A. D., Jackman, L. M., and Martin-Smith, M. (1961). *J. Chem. Soc.* pp. 5061–5073.

Becker, G. (1965). *Holzforschung* **17,** 19.

Bleumink, E., Mitchell, J. C., and Nater, J. P. (1973). *Arch. Dermatol.* **108,** 220–222.

Burnett, W. C., Jr. (1974). Ph.D. Dissertation, University of Georgia, Athens.

Burnett, W. C., Jr., Jones, S. B., Jr., Mabry, T. J., and Padolina, W. G. (1974). *Biochem. Syst. Ecol.* **2,** 25–29.

Burnett, W. C., Jr., Jones, S. B., Jr., and Mabry, T. J. (1977a). *Taxon* **26,** 203–207.

Burnett, W. C., Jr., Jones, S. B., Jr., and Mabry, T. J. (1977b). *Plant Syst. Evol.* **128,** 277–286.

Burnett, W. C., Jr., Mabry, T. J., and Jones, S. B., Jr. (1978a). *In* "Biochemistry of Plant-Animal Coevolution" (J. B. Harborne, ed.), pp. 233–257. Academic Press, New York.

Burnett, W. C., Jr., Jones, S. B., Jr., and Mabry, T. J. (1978b). *Am. Midl. Nat.* **100,** 242–246.
Burnett, W. C., Jr., Jones, S. B., Jr., Mabry, T. J., and Betkouski, M. F. (1979). *Oecologia* **39,** 71–77.
Butterworth, J. H., and Morgan, E. D. (1968). *Chem. Commun.* pp. 23–24.
Butterworth, J. H., and Morgan, E. D. (1971). *J. Insect Physiol.* **17,** 969–977.
Cashmore, A. R., Seelye, R. N., Cain, B. F., Mack, H., Schmidt, R., and Hecker, E. (1976). *Tetrahedron Lett.* **20,** 1737–1738.
Casida, J. E., ed. (1973). "Pyrethrum: The Natural Insecticide." Academic Press, New York.
Chambliss, O. L., and Jones, C. M. (1966). *Proc. Am. Soc. Hortic. Sci.* **89,** 394–405.
Cherrett, J. M. (1972). *In* "Phytochemical Ecology" (J. B. Harborne, ed.), p. 13–24. Academic Press, New York.
Coyne, J. F., and Lott, L. H. (1976). *J. Ga. Entomol. Soc.* **11,** 301–305.
Da Costa, C. P., and Jones, C. M. (1971). *Science* **172,** 1145–1146.
Dalvi, R. R., Singh, B., and Salunkhe, D. K. (1971). *Chem.-Biol. Interact.* **3,** 13–18.
de Kock, W. T., Pachler, K. G. R., and Wessels, P. L. (1968). *Tetrahedron* **24,** 6045–6052.
Ehrlich, P. R., and Raven, P. H. (1964). *Evolution* **18,** 586–608.
Eisner, T. (1964). *Science* **146,** 1318–1320.
Elliger, C. A., Zinkel, D. F., Chan, G. B., and Waiss, A. C., Jr. (1976). *Experientia* **32,** 1364–1366.
Elliott, M., ed. (1977). "Synthetic Pyrethroids," ACS Symp. Ser. No. 42. Am. Chem. Soc., Washington, D.C.
Elliott, M., and Janes, N. F. (1973). *In* "Pyrethrum: The Natural Insecticide" (J. E. Casida, ed.), pp. 56–100. Academic Press, New York.
Epstein, W. W., and Poulter, C. D. (1973). *Phytochemistry* **12,** 737–747.
Evans, F. J., and Kinghorn, A. D. (1974). *Phytochemistry* **13,** 2324–2325.
Evans, F. J., and Schmidt, R. J. (1976). *Phytochemistry* **15,** 333–335.
Hart, N. K., Lamberton, J. A., Sionmis, A. A., Suares, H., and Seawright, A. A. (1976a). *Experientia* **32,** 412–413.
Hart, N. K., Lamberton, J. A., Sionmis, A. A., and Suares, H. (1976b). *Aust. J. Chem.* **29,** 655–671.
Hartwell, J. L. (1976). *Cancer Treat. Rep.* **60,** 1031–1067.
Hecker, E. (1975). *In* "Handbuch der allgemeinen Pathologie," Vol. 6, pp. 651–676. (H-W. Altmann, F. Büchner, H. Cottier, E. Grundmann, G. Holle, E. Letterer, W. Mashoff, H. Messen, F. Roulet, G. Seifert and G. Siebert, eds.) Springer-Verlag, Berlin and New York.
Heikkenen, H. J., and Hrutfiord, B. F. (1965). *Science* **150,** 1457–1459.
Herout, V. (1974). *In* "Chemistry in Botanical Classification" (G. Bendz and J. Santesson, eds.), pp. 55–62. Academic Press, New York.
Herz, W. (1977). *In* "The Biology and Chemistry of the Compositae" (V. H. Heywood, J. B. Harborne, and B. L. Turner, eds.), Vol. 1, pp. 337–357. Academic Press, New York.
Hikino, H., Ito, K., Ōhta, T., and Takemoto, T. (1969). *Chem. Pharm. Bull.* **17,** 1078–1079.
Hikino, H., Ogura, M., Ohta, T., and Takemoto, T. (1970a). *Chem. Pharm. Bull.* **18,** 1071–1073.
Hikino, H., Shoji, N., Koriyama, S., Ohta, T., Hikino, Y., and Takemoto, T. (1970b). *Chem. Pharm. Bull.* **18,** 2357–2359.
Hikino, H., Ogura, M., and Takemoto, T. (1971). *Chem. Pharm. Bull.* **19,** 1980–1981.
Hirata, Y. (1975). *Pure Appl. Chem.* **41,** 175–199.
Hosozawa, S., Kato, N., and Munakata, K. (1974a). *Agric. Biol. Chem.* **38,** 823–826.

Hosozawa, S., Kato, N., Munakata, K., and Chen, Y.-L. (1974b). *Agric. Biol. Chem.* **38,** 1045–1048.

Hostettmann, K., Pettei, M. J., Kubo, I., and Nakanishi, K. (1977). *Helv. Chim. Acta* **60,** 670–672.

Ivie, G. W., Witzel, D. A., Herz, W., Kannan, R., Norman, J. O., Rushing, D. D., Johnson, J. H., Rowe, L. D., and Veech, J. A. (1975a). *J. Agric. Food Chem.* **23,** 841–845.

Ivie, G. W., Witzel, D. A., and Rushing, D. D. (1975b). *J.Agric. Food Chem.* **23,** 845–849.

Jacobson, M. (1977). *ACS Symp. Ser.* **62,** 153–164.

Jamieson, G. R., Reid, E. H., Turner, B. P., and Jamieson, A. T. (1976). *Phytochemistry* **15,** 1713–1715.

Kato, N., Takahashi, M., Shibayama, M., and Munakata, K. (1972). *Agric. Biol. Chem.* **36,** 2579–2582.

Kim, H. L., Rowe, L. D., and Camp, B. J. (1975). *Res. Commun. Chem. Pathol. Pharmacol.* **11,** 647–650.

Kinghorn, A. D., and Evans, F. J. (1974). *Planta Med.* **26,** 150–154.

Knoche, H., Ourisson, G., Perold, G. W., Foussereau, J., and Maleville, J. (1969). *Science* **166,** 239–240.

Kogiso, S., Wada, K., and Munakata, K. (1976). *Agric. Biol. Chem.* **40,** 2119–2120.

Kooistra, E. (1971). *Euphytica* **20,** 47–51.

Kubo, I., and Nakanishi, K. (1977). *ACS Symp. Ser.,* **62,** 165–178.

Kubo, I., Taniguchi, M., Satomura, Y., and Kubota, T. (1974). *Agric. Biol. Chem.* **38,** 1261–1262.

Kubo, I., Lee, Y.-W., Balogh-Nair, V., Nakanishi, K., and Chapya, A. (1976a). *Chem. Commun.,* pp. 949–950.

Kubo, I., Lee, Y., Pettei, M., Pilkiewicz, F., and Nakanishi, K. (1976b). *Chem. Commun.* pp. 1013–1014.

Kubo, I., Miura, I., and Nakanishi, K. (1976c). *J. Am. Chem. Soc.* **98,** 6704–6705.

Kubo, I., Nakanishi, K., Kamikawa, T., Isobe, T., and Kubota, T. (1977a). *Chem. Lett.* **2,** 99–102.

Kubo, I., Miura, I., Nakanishi, K., Kamikawa, T., Isobe, T., and Kubota, T. (1977b). *Chem. Commun.* pp. 555–556.

Kubo, I., Miura, I., Kamikawa, T., Isobe, T., and Kubota, T. (1977c). *Chem. Lett.* pp. 1289–1292.

Kubo, I., Muira, I., Pettei, M., Lee, Y.-W., Pilkiewicz, F., and Nakanishi, K. (1977d). *Tetrahedron Lett.* pp. 4553–4556.

Kupchan, S. M. (1976). *Cancer Treat. Rep.* **60,** 1115–1126.

Kupchan, S. M., Fessler, D. C., Eakin, M. A., and Giacobbe, T. J. (1970). *Science* **168,** 376–378.

Kupchan, S. M., Eakin, M. A., and Thomas, A. M. (1971). *J. Med. Chem.* **14,** 1147–1152.

Kwon, Y. M. Wong, W. S., Woo, L. K., and Lee, M. J. (1973). *Hanguk Saenghwa Hakhoe Chi* **6,** 85–94.

Lavie, D., Jain, M. K., and Shpan-Gabrielith, S. R. (1967). *Chem. Commun.* pp. 910–911.

Lonkar, A., Mitchell, J. C., and Calnan, C. D. (1974). *Trans. St. John's Hosp. Dermatol. Soc.* **60,** 43–53.

Mabry, T. J., and Bohlmann, F. (1977). *In* "The Biology and Chemistry of the Compositae" (V. H. Heywood, J. B. Harborne, and B. L. Turner, eds.), Vol. 2, pp. 1097–1104. Academic Press, New York.

Mabry, T. J., Abdel-Baset, Z., Padolina, W. G., and Jones, S. B., Jr. (1975). *Biochem. Syst. Ecol.* **2,** 185–192.

Mabry, T. J., Burnett, W. C., Jr., Jones, S. B., Jr., and Gill, J. E. (1977). *ACS Sym. Ser.* **62,** 179–184.

McCahon, C. B., Kelsey, R. G., Sheridan, R. P., and Shafizadeh, F. (1973). *Bull. Torrey Bot. Club* **100,** 23–28.

Matsui, M., and Yamamoto, I. (1971). *In* "Naturally Occurring Insecticides" (M. Jacobson and D. G. Crosby, eds.), pp. 3–70. Marcel Dekker, New York.

Meinwald, J., Prestwich, G. D., Nakanishi, K., and Kubo, I. (1978). *Science* **199,** 1167–1173.

Mitchell, J. C. (1969). *Trans. St. John's Hosp. Dermatol. Soc.* **55,** 174–183.

Mitchell, J. C. (1975). *Recent Adv. Phytochem.* **9,** 119–138.

Mitchell, J. C., and Dupuis, G. (1971). *Br. J. Dermatol.* **84,** 139–150.

Mitchell, J. C., Fritig, B., Singh, B., and Towers, G. H. N. (1970). *J. Invest. Dermatol.* **54,** 233–239.

Mitchell, J. C., Geissman, T. A., Dupuis, G., and Towers, G. H. N. (1971a). *J. Invest. Dermatol.* **56,** 98–101.

Mitchell, J. C., Roy, A. X., and Dupuis, G. (1971b). *Arch. Dermatol.* **104,** 73–76.

Mitchell, J. C., Dupuis, G., and Geissman, T. A. (1972). *Br. J. Dermatol.* **87,** 235–240.

Morgan, E. D., and Thornton, M. D. (1973). *Phytochemistry* **12,** 391–392.

Munakata, K. (1977). *ACS Symp. Ser.* **62,** 185–196.

Nielsen, J. K., Larsen, L. M., and Sorensen, H. (1977). *Phytochemistry* **16,** 1519–1522.

Okuda, T., Yoshida, T., Koike, S., and Toh, N. (1975). *Phytochemistry* **14,** 509–515.

Patterson, C. G., Knavel, D. E., Kemp, T. R., and Rodriguez, J. G. (1975). *Environ. Entomol.* **4,** 670–674.

Paul, I. C., Sim, G. A., Hamor, T. A., and Monteath Robertson, J. (1962). *J. Chem. Soc.* pp. 4133–4145.

Perold, G. W., Muller, J. C., and Ourisson, G.)1972). *Tetrahedron* **28,** 5797–5803.

Radwan, M. A., and Ellis, W. D. (1975). *For. Sci.* **21,** 63–67.

Rodriguez, E., Dillon, M. D., Mabry, T. J., Mitchell, J. C., and Towers, G. N. H. (1976). *Experientia* **32,** 236–237.

Ronlan, A. and Wickberg, B. (1970). *Tetrahedron Lett.* **49,** 4261–4264.

Russell, C. E. and Berryman, A. A. (1976). *Can. J. Bot.* **54,** 14–18.

Schmidt, R. J., and Evans, F. J. (1976). *Phytochemistry* **15,** 1778–1779.

Sequeira, L., Hemingway, R. L., and Kupchan, S. M. (1968). *Science* **161,** 789–790.

Smith, R. H. (1965). *J. Econ. Entomol.* **58,** 509–510.

Smith, R. H. (1966). *In* "Breeding Pest-Resistant Trees" (H. D. Gerhold *et al.,* eds.), pp. 189–196. Pergamon, Oxford.

Smith, R. H. (1975). *J. Econ. Entomol.* **68,** 841–844.

Srivastava, S. N., and Przybylska, M. (1968). *Can. J. Chem.* **46,** 795–797.

Stipanovic, R. D., Bell, A. A., and Lukefahr, M. J. (1977). *ACS Symp. Ser.* **62,** 197–214.

Sutherland, M. D., and Park, R. J. (1967). *In* "Terpenoids in Plants" (J. B. Pridham, ed.), pp. 147–157. Academic Press, New York.

Upadhyay, R. R., Zorintan, M. H., and Ansarin, M. (1976). *Planta Med.* **30,** 196–197.

von Rudloff, E. (1969). *In* "Recent Advances in Phytochemistry" (M. K. Seikel and V. C. Runeckles, eds.,), Vol. 2, pp. 127–162. Appleton-Century-Crofts, New York.

von Rudloff, E. (1973). *Pure Appl. Chem.* **34,** 401–410.

Wada, K., and Munakata, K. (1971). *Agric. Biol. Chem.* **35,** 115–118.

Wada, K., Matsui, K., Enomoto, Y., Ogiso, O., and Munakata, K. (1970). *Agric. Biol. Chem.* **34,** 941–945.

Waiss, A. C., Jr., Chan, B. G., and Elliger, C. A. (1977). *ACS Symp. Ser.* **62,** 115–128.

Weisner, K. (1963). *Pure Appl. Chem.* **7,** 285–296.

Wilson, S. R., Neubert, L. A., and Huffman, J. C. (1976). *J. Am. Chem. Soc.* **98,** 3669–3674.
Witzel, D. A., Ivie, G. W., and Dollahite, J. W. (1976). *Am. J. Vet. Res.* **37,** 859–861.
Yoshioka, H., Mabry, T. J., and Timmermann, B. N. (1973). "Sesquiterpene Lactones." Univ. of Tokyo Press, Tokyo.
Zanno, P. R., Miura, I., Nakanishi, K., and Elder, D. L. (1975). *J. Am. Chem. Soc.* **97,** 1975–1977.
Zavarin, E., and Snajberk, K. (1973). *Pure Appl. Chem.* **34,** 411–434.
Zavarin, E., and Snajberk, K. (1975). *Biochem. Syst. Ecol.* **2,** 121–129.

Chapter 15

Saponins

SHALOM W. APPLEBAUM *And* YEHUDITH BIRK

I. INTRODUCTION

Saponins are secondary plant products containing a polycyclic aglycone moiety of either steroid (C_{27}) or triterpenoid (C_{30}) structure attached to a carbohydrate moiety (mono- or oligosaccharide chain) composed of pentoses, hexoses, or uronic acids. Many of the biological

ISBN 0-12-597180-X

properties for which saponins are known may be credited to this hydrophobic/hydrophilic asymmetry and consequent ability to lower surface tension. Saponins are designated thus after their characteristic "soaplike," foam-forming properties in aqueous solutions. Erythrocytes lyse in saponin solutions, and this is the immediate reason for their acute toxicity when injected intravenously. Saponins leach from plant material into water. This phenomenon has been exploited since ancient times for poisoning and catching fish. Dietary saponins are more toxic to exothermic (cold-blooded) than to endothermic animals. Many saponins have a bitter taste, which deters animals from feeding on plants in which saponins are present. Their widespread occurrence in numerous plant species has prompted the suggestion that they may serve as a chemical defense mechanism in plants, imparting resistance to pests or diseases.

II. CHEMICAL STRUCTURE OF SAPONINS

A. Sapogenins

1. General Structure

Saponins can be classified into two major groups according to the chemical nature of the aglycone moieties, which are collectively termed sapogenins. Isolated and individually characterized sapogenins of soybeans are referred to as soyasapogenols. Thus, the name soyasapogenol A denotes one of the specific aglycones that have been identified in the complex mixture of soybean sapogenins.

Steroidal sapogenins contain 27 carbon atoms. They have been investigated primarily as potential precursors for steroid hormone synthesis. Their composition and structure have been reviewed by Shoppee (1964). The triterpenoid sapogenins, most of which are pentacyclic compounds derived from oleanan (Steiner and Holzem, 1955), contain 30 carbon atoms. Triterpenoid saponins occur frequently in forage and seed crops and other plants of agricultural importance and have been investigated primarily for their potential effects on nutrition. Several surveys of composition and structure of many titerpenoid sapogenins have been published (Boiteau *et al.*, 1964; Basu and Rastogi, 1967; Agarwal and Rastogi, 1974). The basic structure of steroid and triterpenoid sapogenins is presented in Fig. 1. For reviews of saponins, see Birk (1969), Cheeke (1971), Bondi *et al.* (1973), and Birk and Peri (1978).

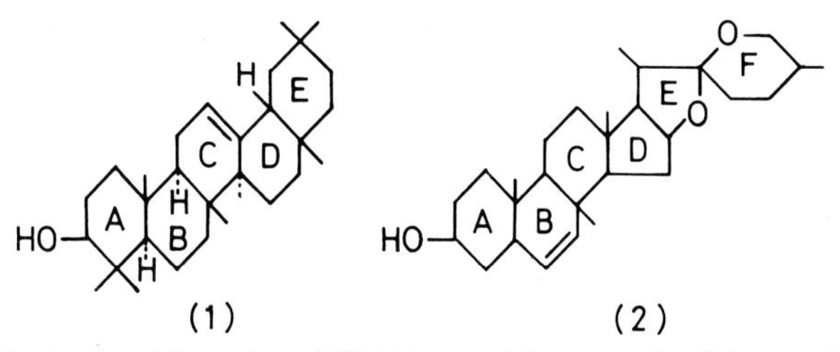

Figure 1. Basic structural formulas of (1) triterpenoid sapogenins (oleanan type), and (2) steroid sapogenins.

2. *Specific Triterpenoid Sapogenins*

The chemical structures of identified sapogenins that have been isolated from legumes are presented in Fig. 2. The first five were isolated from soybeans and are known by the common names soyasapogenols A–E (Smith *et al.*, 1958a,b; Willner *et al.*, 1964). In addition, alfalfa roots and foliage contain carboxylic acids called medicagenic acid (Djerassi *et al.*, 1957) and soyasapogenol U. The latter, identified by Shany *et al.* (1972) as hederagenin, has a CH_2OH group instead of the COOH group present at C-23 of medicagenic acid.

B. *Carbohydrate Moiety*

The carbohydrates found in the triterpenoid saponins are hexoses (mainly D-glucose and D-galactose), pentoses (mainly D- and L-arabinose and D-xylose), methyl pentoses (L-rhamnose and quinovose), and uronic acids (D-glucuronic acid and D-galacturonic acid). Different sugars, mainly glucose, galactose, rhamnose, xylose, and arabinose, have been identified as components of the carbohydrate moiety of steroid saponins (Shoppee, 1964).

C. *Integral Saponins*

The complexity of native saponins stems from the variety of combinations of the constituent sapogenins and numerous sugars. For example, the saponins of alfalfa varieties (*Medicago sativa*) differ in composition as well as in absolute amount (Pedersen *et al.*, 1966; Hanson *et al.*, 1973). DuPuits and Lahontan alfalfa varieties contain 30 different saponins, but the chemical composition of the saponins in the two cultivars is not the same (Berrang *et al.*, 1974). The aglycones of DuPuits saponins are mainly of the dicarboxylic (medicagenic acid) type, whereas Lahontan

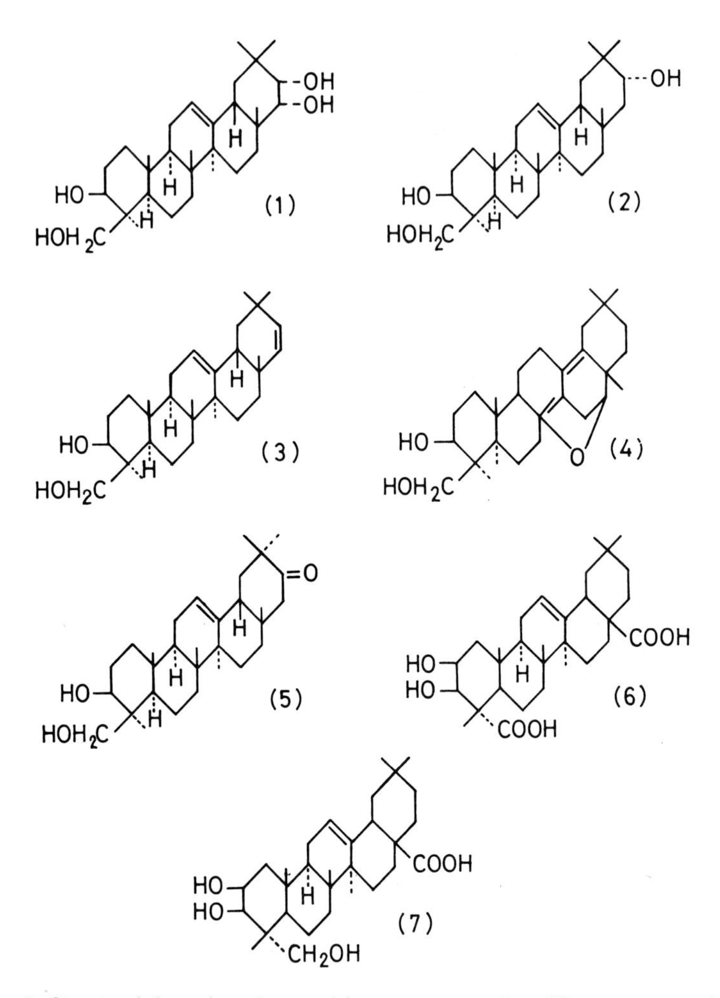

Figure 2. Structural formulas of several legume sapogenins: (1) soyasapogenol A; (2) soyasapogenol B; (3) soyasapogenol C; (4) soyasapogenol D; (5) soyasapogenol E; (6) medicagenic acid; (7) hederagenin.

saponins contain soybean sapogenins and monocarboxylic acids (hederagenin) as aglycones. In addition, DuPuits saponins lack D-galactose, which is present in Lahontan saponins. Polish cultivated varieties of alfalfa (*Medicago media*) reportedly have about the same saponin content and chemical composition (Jurzysta, 1975).

Examples of isolated triterpenoid saponins for which the chemical structure has been established are sugar beet saponin (Jeger, 1950; Wagner and Sternkopf, 1958), aescin (from horse chestnuts) (Hoppe *et al.*, 1968; R. Tschesche and G. Wulff, unpublished results, 1968), gypsoside (Kochetkov *et al.*, 1963; Kochetkov and Khorlin, 1966), and the alfalfa saponin medicagenic acid-3-β-*O*-triglucoside (Gestetner, 1971). An example of a typical steroidal saponin structure is that of dioscin (Kawasaki and Yamauchi, 1962).

III. STANDARD METHODS OF DETECTION AND DETERMINATION

This section is intended to offer general guidelines for the analysis of saponin in plant material of biological interest. For detailed procedures the reader should refer to the publications cited.

A. Determination of Total Steroidal Saponin Content

A specific spectrophotometric method for general quantitative determination of microgram amounts of steroidal sapogenins, independent of their structural characteristics, is that of Baccou *et al.* (1977). The method is based on color-producing reaction with anisaldehyde in sulfuric acid and ethyl acetate. Determination can be carried out directly on free or esterified sapogenins or integral saponins. The presence of other plant constituents such as carbohydrates, sterols, or fatty acids and oils does not affect this colorimetric assay.

B. Determination of Total Triterpenoid Saponin Content

Total saponin content of a specific plant tissue is calculated by first determining the sapogenin/sugar ratio and using this as a conversion factor after subsequent determinations of total sapogenin content alone. *Note:* This ratio must first be determined whenever plant material of unknown composition is examined.

1. Acid hydrolysis is performed on a 1% suspension of plant material (defatted and powdered) by refluxing in 1 *N* H_2SO_4 prepared in dioxane–water (1 : 3).

2. Total sapogenin content is determined after extraction into diethyl ether from the above acid hydrolyzate and single-step chromatography on Al_2O_3. A modified Liebermann–Burchard reagent (a mixture of sulfuric acid and acetic anhydride) is recommended (Gestetner *et al.*, 1966b). Chromatography on Al_2O_3 is essential in order to remove accompanying sterols and related substances, which might otherwise lead to artifactual sapogenin determination with this nonspecific reagent.

Five to ten milligrams of dried and defatted plant material suffice for a single determination.

For determination of medicagenic acid, a method based on titration of the two carboxylic groups is recommended (Tencer *et al.*, 1972). This method could be adapted, in principle, for determining hederagenin or any other sapogenin containing carboxylic groups.

3. Total carbohydrate content must be determined after acid hydrolysis of saponins that have been separated from other plant carbohydrates or glycosides [see Section III,C(1)]. A suitable method for quantifying total carbohydrate utilizes the Sumner reagent (Noelting and Bernfeld, 1948).

C. Separation Methods

1. Saponins can be separated by paper chromatography (Birk *et al.*, 1963), paper electrophoresis (Gestetner *et al.*, 1963), and thin-layer chromatography (Van Duuren, 1962). Various reagents are used for staining *in situ* after separation: the Carr–Price reagent (a saturated solution of $SbCl_3$ in chloroform) (Coulson, 1958), the Liebermann–Burchard reagent (Van Atta and Guggolz, 1958), or a 25% (w/v) aqueous solution of phosphotungstic acid (Kazerovskis, 1962).

2. Sapogenins can be isolated by circular paper chromatography (Gestetner, 1964) or thin-layer chromatography (B. Gestetner and Y. Birk, unpublished results, 1968). For laboratories with suitable equipment, separation by gas-liquid chromatography subsequent to silylation with bis(trimethylsilyl)acetamide is recommended. A suitable column packing is 1% OV-101 or OV-1 on Gaschrom Q (2 m × 3 mm) with a flame ionization detector and column temperature of 290°C. The procedure is modified from Grunwald (1970). With suitable standard sapogenins as reference compounds, a tentative identification is obtained.

3. Carbohydrates can be identified after acid hydrolysis by a variety of paper or thin-layer chromatographic methods. We have found the thin-layer chromatographic procedure of Pifferi (1965) to be very satisfactory. In this method, plates of Kieselgel G slurried in 0.02 *M* sodium acetate are developed with chloroform–methanol (6 : 4, v/v). For detection on the developed plates, we recommend the anisaldehyde spray reagent [acidified ethanolic anisaldehyde solution (Stahl, 1965)].

Carbohydrates can be quantified by gas–liquid chromatography of alditol acetates on a column of 1.5% ethylene glycol succinate plus 1.5% XE-60 on Chromosorb W (2 m × 6 mm) with a flame ionization detector and column temperature of 180°C. The procedure is modified from Sawardeker *et al.* (1965) and Shaw and Moss (1969).

D. Measurement of Foam-Forming Activity

Foam-forming activity is measured in a 25-ml graduated cylinder by shaking 5 ml of a 0.01–1% (w/v) saponin solution in *M*/15 K_2HPO_4 for 1 minute and measuring the number of milliliters of foam formed after an additional minute (O'Dell *et al.*, 1959). There is a good correlation between the amount of foam formed and the concentration of various saponin preparations. For working with unidentified saponin preparations from previously unexamined plant sources, this nonspecific, semiquantitative method can be used until other, more specific properties are recognized and a more selective method is adapted for assay.

E. Determination of *in Vitro* Hemolytic Activity

The hemolytic activity of a saponin is expressed by its hemolytic index (HI), which is the numerical ratio between the weight of reaction mixture and the smallest weight of saponin that causes complete hemolysis of washed (ram) erythrocytes after 20 hours of incubation under controlled conditions (Büchi and Dolder, 1950; Wasicky and Wasicky, 1961). As an example, the HI of 0.1 mg saponin is rated at 20,000 if this is the minimal amount, in 1 ml of isotonic buffer (0.33 *M* phosphate buffer, pH 7.4), that induces complete hemolysis of 1 ml of suspended erythrocytes after 20 hours of incubation.

For a more sensitive assay of the HI of crude saponin preparations, the method of Kofler and Adam (1972), as modified by O'Dell *et al.* (1959), is suggested. Segal *et al.* (1966) suggested that absorbancy at 540 nm of a supernatant solution from a standard reaction mixture of a suspension of erythrocytes and saponins in isotonic buffer be measured after incubation for 3 hours at 22°C. A micro method for the determination of saponins utilizes filter paper discs wetted with saponin solution and embedded in blood gelatin (Kartnig *et al.*, 1964). Hemolysis is accompanied by clearance of the blood gelatin, and within a certain range of concentration (which should be determined for each type of saponin examined) the diameter of the hemolytic zone is proportional to the concentration of saponin. Quantitative determinations of less than milligram amounts of saponin can also be performed by observing the rate of hemolysis (Schulz-Langner, 1966).

F. Bioassay

Inhibition of lettuce seed germination reflects quantitative differences in the level of alfalfa saponins (Pedersen *et al.*, 1966) and in principle could be used for quantitative determination of other seed-inhibiting saponins.

Growth inhibition of the soil-inhabiting fungus *Trichoderma viride* has been developed into a sensitive bioassay for triterpenoid alfalfa saponins containing medicagenic acid or hederagenin (Zimmer *et al.*, 1967; Leath *et al.*, 1972; Horber *et al.*, 1974; Livingston *et al.*, 1977). The basis for this bioassay is the ability of saponins containing medicagenic acid to precipitate with sterols, i.e., in the case of the mycelial membrane, to bind to membrane sterol, disrupt membrane functions, and ultimately arrest mycelial growth (Gestetner *et al.*, 1971a; Assa *et al.*, 1972). When seeking to adapt this bioassay to other, unidentified plant saponins, one should examine the interactions between these saponins and cholesterol or 7-dehydrocholesterol as a preliminary indication of the feasibility of using the *T. viride* test.

Fish are very sensitive to and are characteristically paralyzed by saponin solutions. This response serves as the basis for the bioassay of saponins with minnows (e.g., Jones and Elliott, 1969; Johnson, 1971), which can be adapted for use with any available fish species, provided that suitable basic responses, such as range of sensitivity and effect of specific saponins, are first determined.

IV. DISTRIBUTION AND DIVERSITY

Saponins are widely distributed in the plant kingdom and are probably present in most higher plants. They have been identified in 500 species belonging to more than 80 different families (Basu and Rastogi, 1967). They occur in forage legumes such as alfalfa (*Medicago sativa*), Ladino clover (*Trifolium repens*), burr clover (*Medicago hispida*), strawberry clover (*Trifolium fragiferum*), and trefoil (*Lotus corniculatus*) (Walter *et al.*, 1954, 1955; Walter, 1957, 1960, 1961) and in numerous legume seeds (Sumiki, 1929; Diekert and Morris, 1958; Applebaum *et al.*, 1969). Other saponin-containing plants include crocus, asparagus, spinach, beet root, sugar beet, yams, beech, and horse chestnut (George, 1965; Chakravarti *et al.*, 1963). They are also found in various medicinal plants (Pasich, 1961), tea (Hashizume and Sakato, 1966), and several wild or domesticated woody plants (Takahashi *et al.*, 1963).

A. Distribution in the Plant

Saponins accumulate and have been identified in all parts of the plant: leaves, stems, roots, bulbs, blossoms, and fruits. Saponins occur in all parts of the alfalfa plant (Cole *et al.*, 1945; Morris *et al.*, 1961; Morris and Hussey, 1965; Focke *et al.*, 1968), but more saponin is found in the roots than in the foliage (Morris *et al.*, 1961; Shany *et al.*, 1970a). Examination of root extracts from various tissues of two alfalfa cultivars indicated that the saponins are more concentrated in the outer portion of the alfalfa root, i.e., in the bark, than in the xylem (Quazi, 1976). Sugar beets contain up to 0.3% saponin, which is concentrated mainly in the surface layer, particularly in the root (Silin, 1964). In three consecutive harvestings, the average saponin content is lower in the first than in the second or third harvest.

The ratio of sapogenin to carbohydrate in saponins may differ in the various parts of the plant. Thus, alfalfa root saponin exhibits a higher ratio than does leaf saponin from the same plant (Gestetner *et al.*, 1970). Alfalfa seed saponins lack medicagenic acid (Jurzysta, 1973), a distinctive feature of saponin extracts from other portions of the alfalfa plant, and the composition of alfalfa seed saponins approaches that of the soybean seed.

B. Seasonal Variation

Studies on the distribution and content of saponin in plants in the course of an annual growth cycle show a dependence on temperature and season (Hein, 1959). Thus, the saponin content of underground organs of *Saponaria officinalis* and *Polemonium coeruleum* diminishes during flowering and fruit bearing and is accompanied by an accumulation of saponins in the reproductive organs (Drożdż, 1962). The saponin content of alfalfa declines as the plant ages (Pedersen and Taylor, 1962).

C. Genetic Variability

Some varietal and geographic differences were observed in the saponin content of five alfalfa varieties studied at eight different locations in the United States (Hanson and Kohler, 1961; Hanson *et al.*, 1963). The relatively high proportion of the variety variance component enhances the possibility of obtaining low-saponin plants by selection (Jones, 1969). Attempts were made to change saponin concentration by selection and breeding (Pedersen and Wang, 1971). As a result, the average saponin content of the "low" lines was reduced by about 30% and that of the "high" lines was increased by about 80%, as compared to the unselected parent variety. Progress in selection for high saponin content was more rapid than that in selection for low saponin content. The selection for high saponin concentration increased the relative concentration of carboxylic acid type of aglycones, whereas selecting for low saponin concentration increased the relative concentration of the neutral aglycones (Pedersen *et al.*, 1973). Similar results were obtained by Quazi (1975), who cultivated one variety that contains primarily saponins with the carboxylic acid type of aglycones, and a second variety that contains the neutral saponins. Seasonal influence was observed in the first cultivar with regard to the ratio of total content of saponins to saponins containing medicagenic acid.

V. BIOSYNTHESIS

The sequence of reactions and intermediates culminating in the biosynthesis of sterol and triterpenes from acetic acid via mevalonic acid and squalene is reviewed by Boiteau *et al.* (1964). This biosynthetic pathway is shown schematically in Fig. 3.

When germinating soybeans are grown on a substrate containing ^{14}C-labeled acetic acid, the specific activity is distributed equally among sapogenins, sitosterols, and squalene (Arigoni, 1958). This finding agrees with the hypothesis outlined in Fig. 3 concerning the common biosynthesis of sterols and triterpenes from squalene (Ruzicka, 1953).

Labeled saponins and respective sapogenins, obtained by administering labeled mevalonate or squalene to germinating seeds or to seedlings of soybean and alfalfa plants, can be identified by cochromatography, scanning for labeled sapogenins, and autoradiography. The formation of labeled sapogenins also supports the validity of the scheme in which mevalonic acid serves as the source of five-carbon units to form both the isoprenoid units and, via squalene, the sterol or triterpenoid structure (I. Peri, unpublished data, 1977).

Nowacki *et al.* (1976), using [^{14}C]acetate as a precursor, demonstrated that medicagenic acid is the first sapogenin synthesized in germinating seeds of *Medicago media*.

VI. BIOLOGICAL EFFECTS

In this section we describe the biological effects of saponins ingested by herbivores or otherwise encountered in their natural environment. Medical research on the pharmacological effects of saponins injected into whole organisms or added to tissue culture media are not included.

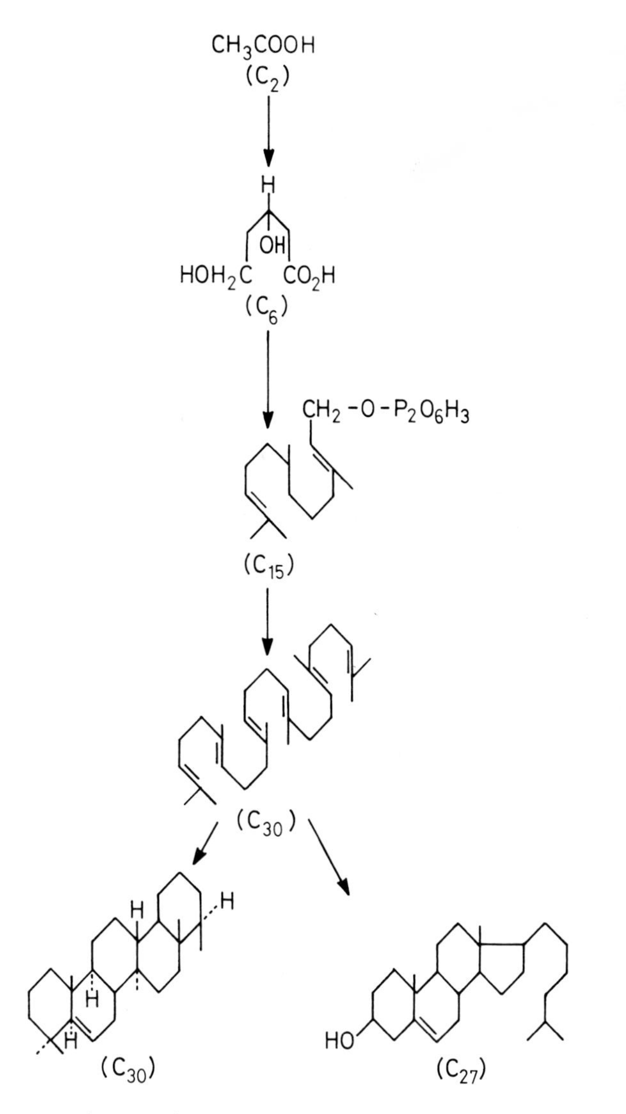

Figure 3. Sequence of intermediates in biosynthesis of sterols and triterpenes: (C_2), acetic acid; (C_6), mevalonic acid; (C_{15}), farnesyl pyrophosphate; (C_{30}), squalene; (C_{30}), triterpene; (C_{27}), sterol.

A. Nutritional Effects

1. *Ruminant Bloat*

Saponins are present in many pasture plants. In most, their level is negligible, but appreciable quantities accumulate in forage legumes. It has been suggested that they are among the factors producing ruminant bloat: By altering the surface tension of the ruminal contents, they cause gas formed in the rumen by anaerobic bacterial fermentation to be entrapped as a froth (McCandlish, 1937; Quin, 1943; Olson, 1944). Alfalfa saponins elicit typical symptoms of bloat in ruminants (Lindahl *et al.*, 1954). Experimental bloat symptoms were produced in sheep by intraruminal or intravenous administration of a mixture of alfalfa saponins (Lindahl *et al.*, 1957). These treatments also elicit an increase in the respiratory rate, which is subsequently replaced by an irregular respiratory pattern. Jackson *et al.* (1959) demonstrated that the respiratory-inhibiting activity of alfalfa forage is highly correlated with the occurrence of bloat in cattle. Taking into account the isolation procedure of the respiratory-inhibiting activity, Jackson and Shaw (1959) suggested that this factor is of a saponin nature. There are quantitative differences in saponin content of bloat-producing and nonbloat-producing alfalfa stands (Coulson and Davis, 1962). The absence of bloat after feeding on trefoil may be related to its low saponin content, about 0.06% of the dry weight (Walter, 1961).

The apparent correlation between bloat and inhibition of respiration may, however, be fortuitous; subsequent studies on the *in vitro* respiratory inhibition by alfalfa forage fractions showed no correlation to the bloat index of the forage (McNairy *et al.*, 1963). The foam strength of alfalfa saponins is pH dependent, with an optimum between pH 4.5 and 5.0. However, the pH optimum for foam strength of rumen content is 5.4–5.7 (Mangan, 1958, 1959). Obviously, other factors contribute to foam formation in the rumen content. The production of slime from alfalfa saponins by rumen bacteria is suggested to be a significant factor in bloat formation (Gutierrez *et al.*, 1959).

Other than their possible effect on bloat formation, alfalfa saponins neither exhibit growth-depressing effect, nor depress serum cholesterol levels of ruminants. This is due to saponin degradation by microbial fermentation, as mentioned above.

2. *Effect on Monogastric Animals*

There are distinct species differences in the response of monogastric animals to dietary saponins, poultry being the most sensitive. Alfalfa saponins are, to a large extent, responsible for growth depression of chicks and for the decrease in food consumption and utilization of diet (Draper, 1948; Cooney *et al.*, 1948; Lepkovsky *et al.*, 1950). Several reports deal with the effect of alfalfa meal or saponin supplement to chick diets or to diets of laying hens. Growth of chicks is depressed by 10% (w/w) alfalfa meal (Heywang, 1950), by crude (hemolytic and foam-forming) fractions of alfalfa (Peterson, 1950a,b), and by isolated alfalfa saponins in the range 0.1–0.5% (w/w) (Heywang and Bird, 1954; Ishaaya *et al.*, 1969). Egg production, but not egg weight, is depressed by similar supplements to diets of laying hens (Heywang, 1950; Heywang *et al.*, 1959). The detrimental effect disappears soon after feeding is stopped.

Alfalfa cultivars low in saponin content, and particularly in content of saponins containing medicagenic acid, make it possible to include alfalfa meal at moderate levels in poultry diets, with no detrimental effect. At the 10% (w/w) level, the growth of chickens reared for meat was even improved (Pedersen *et al.*, 1972).

The first factor to be considered in evaluating the low nutritional value of alfalfa meal for monogastric animals is its deterrent effect on feeding, which is a result of the bitter taste imparted to the diet by the constituent saponins (Cheeke, 1976). The improved growth response of chicks and rats to low-saponin alfalfa (Pedersen *et al.*, 1972) may be due simply to greater palatability. However, although meals of low-saponin lines of alfalfa were generally more palatable to the meadow vole (*Microtus pennsylvanicus*) than the corresponding high-saponin lines, only very high concentrations of saponin in the diet (4%) affected palatability to voles (Kendall and Leath, 1976).

The second factor to be considered in assessing the adverse effect of alfalfa saponins is that it can be effectively reversed by inclusion of cholesterol (or β-sitosterol) and cottonseed oil in the diet (Peterson, 1950b; Ishaaya *et al.*, 1969). One of the reasons for growth depression may therefore be due to the sequestering of cholesterol by alfalfa saponins containing medicagenic acid.

In contrast to the sensitivity of poultry, levels of 20% (w/w) alfalfa meal (dried and ground alfalfa foliage) have been incorporated into pig grower rations with no adverse effects on growth. Mice are much less tolerant of alfalfa saponins than rats; at a dietary level of 2% (w/w) alfalfa saponin, feed intake of mice is depressed markedly and results in weight loss, whereas rats are not affected (Cheeke, 1971). However, Reshef *et al.* (1976) reported that the growth of mice and quails is not affected by 2% (w/w) alfalfa foliage saponins in the diet, a concentration equivalent to the saponin concentration in a diet composed solely of alfalfa meal. Growth depression is observed only by supplementing the diet with 0.5% (w/w) alfalfa root saponin, which has a high content of saponins containing medicagenic acid. This may be one of the reasons for the discrepancy in the reported tolerance of mice toward alfalfa saponins. Soybean saponins, which lack medicagenic acid, have no effect on weight gain, growth ratio, or amount of feed ingested by chicks, mice, or rats (Ishaaya *et al.*, 1969). It seems appropriate to attribute at least some of the various biological activities of alfalfa saponins mainly to their medicagenic acid content (Shany *et al.*, 1970b; Gestetner *et al.*, 1971b).

The third factor to be considered in assessing the adverse effects of saponins is related to their surface activity and is evinced by nonspecific inhibition of many enzymes. α-Chymotrypsin, *Tribolium* larval proteases, and cholinesterase are among those inhibited to a certain extent. Different fractions of crude soybean saponin exhibit different activity toward the different enzymes; fraction c possesses a high specific anticholinesterase activity, whereas fraction e is most effective against proteases (I. Ishaaya and Y. Birk, unpublished results, 1968). These inhibitions are regarded as saponin–protein interactions, and prior incubation of soybean saponins with casein or soybean protein counteracts the inhibition of enzyme activity (Birk *et al.*, 1963; I. Ishaaya and Y. Birk, 1965, and unpublished results, 1968). Prior incubation with cholesterol known to complex with saponins containing medicagenic acid, does not abolish the subsequent inhibition of enzymes. This indicates that the active site for saponin–protein interaction differs from the binding site for saponin–cholesterol complex formation.

Little is known of the effect of dietary sugar beet saponin. An immediate influence on mineral metabolism, as a consequence of feeding limited doses of sugar beet saponin to calves, was reported (Brune and Kundlich, 1960).

Both alfalfa and soybean saponins are detrimental to tadpoles and guppies when introduced into their aquatic environment. These organisms appear to be first affected by partial paralysis, then lose their ability to swim, and shortly thereafter die. Death is probably due to the surface activity of saponins, which interferes with breathing in an aquatic medium (I. Ishaaya and Y. Birk, unpublished results, 1968; Ishaaya *et al.*, 1969).

3. *Effect of Alfalfa Saponins on Mammalian Hypercholesterolemia*

The observation that dietary alfalfa saponins form complexes with cholesterol and lower plasma cholesterol levels of chicks (for review, see Cheeke, 1971) has encouraged attempts to exploit this response to manipulate the plasma cholesterol levels of other monogastric animals, including human beings. In ruminants, neither growth depression nor reduced tissue cholesterol levels, due to saponin ingestion, have been demonstrated, possibly because of bacterial catabolism of the saponin in the rumen. Since many saponins form insoluble complexes with cholesterol, it has been assumed that dietary saponins in the guts of monogastric animals may combine with endogenous cholesterol excreted via the bile. This can prevent cholesterol reabsorption and result in reduced serum cholesterol (Cheeke, 1971). Alfalfa saponin–cholesterol complexes are formed in the gut of mice but not of quails (Reshef *et al.*, 1976). Following ingestion of alfalfa saponins by mice, a lower level of cholesterol is evident in the liver, accompanied by the appearance of cholesterol in larger amounts in the feces and followed by an increased rate of its synthesis in the liver. In quails, only an increase in the rate of cholesterol biosynthesis is observed. However, this additional amount of cholesterol is not utilized by quails for complexing dietary alfalfa saponins. However, it may serve as a precursor for the formation of bile acids, which in turn may be the cause of the slightly increased absorption of lipids.

Alfalfa saponins are not absorbed from the gastrointestinal tract in mammals (Birk, 1969) but do prevent dietary-induced hypercholesterolemia in monkeys (Malinow *et al.*, 1977). These observations suggest that a biological consequence of saponin consumption is interference with intestinal absorption of cholesterol, probably through the formation of insoluble complexes of the saponins with cholesterol. Thus, alfalfa saponins may constitute an important new therapy for human hypercholesterolemia. This suggestion is supported by the finding that alfalfa saponins are tolerated by rats and monkeys. Since saponins are present in small amounts in many plants commonly ingested by human beings, one may assume that they are tolerated to a certain extent by human beings as well.

4. Role of Saponins in Resistance of Plants to Insects

Saponins are present in a wide variety of plants, but their main function is still not clear (Birk, 1969). It has been suggested that they impart resistance against insects (Applebaum and Birk, 1972; Applebaum *et al.*, 1969) in conjunction with other secondary plant substances, which are presumed to have appeared in plants during their parallel evolution with insects (Fraenkel, 1959). On this premise, we examined the effect of saponins on the bruchid beetle *Callosobruchus chinensis*, an oligophagous insect restricted in its host range to some representatives of the legume family but unable to develop in soybeans, navy beans, peanuts, and certain varieties of garden peas. Soybean saponins, and in particular fraction c, are detrimental to developmental stages of *C. chinensis* when incorporated into the larval diet (Applebaum *et al.*, 1965). Sapogenins alone, or their sugar constituents, have no adverse effect. Saponin preparations from chickpeas, garden peas, lentils, and peanuts effectively inhibit development of *C. chinensis.* In the case of chickpeas and lentils, other seed components ameliorate this effect, and *C. chinensis* develops readily on these seeds (Applebaum *et al.*, 1969).

The growth of larvae of the rust-red flour beetle *Tribolium castaneum* is inhibited somewhat by the addition of 1% (w/w) soybean saponins to their diet (Ishaaya *et al.*, 1969). Alfalfa saponins are far more toxic under similar conditions, but this toxicity can be reversed by including appropriate amounts of cholesterol in the diet. In more detailed experiments, it was found that alfalfa root saponin, and in particular the fraction containing medicagenic acid, is more toxic to *T. castaneum* larvae than is a saponin preparation from foliage (Gestetner *et al.*, 1970). Here too, larval growth is restored to its normal rate by addition of cholesterol to the diet. The detrimental effect of these saponins is therefore apparently due to the characteristic of saponins containing medicagenic acid to complex with cholesterol. Insects themselves are unable to synthesize sterols and exhibit an absolute dietary requirement for cholesterol or one of several other closely related sterols. Shortening the oligosaccharide moiety of saponins containing medicagenic acid by partial hydrolysis increases the toxicity of the product toward *T. castaneum.* Free medicagenic acid is most toxic, whereas the soyasapogenins are harmless. The detrimental effect of alfalfa sapogenins is reversed by cholesterol, stigmasterol, β-sitosterol, or campesterol (Shany *et al.*, 1970b). Thus, by enzymatically hydrolyzing alfalfa saponins in their midgut, *T. castaneum* larvae potentiate their toxicity.

Various degrees of resistance to the white grub *Melolontha vulgaris,* a European scarabeid beetle feeding on roots of pasture plants, occur in varieties of alfalfa. The high saponin content in roots of certain strains was initially regarded as imparting resistance by antibiosis and nonpreference (Horber, 1965), although it would be difficult to directly prove either of these two possibilities. In a reappraisal of his previous experiments, Horber (1972) remarked that refusal of nonpreferred food may have reduced weight gain and molting and increased mortality, a situation akin to starvation in the event that nonpreference is adhered to strictly.

Larvae of the grass grub *Costelytra zealandica* feed on roots of a wide range of plants in pastureland of New Zealand. Resistance to this insect has been demonstrated in alfalfa both in the field and in the laboratory. Seedlings are less resistant to attack than are mature stands of alfalfa. It is assumed that larvae are unable to consume sufficient amounts of the woody tap root of the older plants and that they consequently die of starvation (Kain and Atkinson, 1970), but Farrell and Sweney (1972) suggest that the poor survival and reduced weight gain of grubs reared on alfalfa results from the phenomenon of antibiosis. Sutherland *et al.* (1975), studying the effect of saponins on the feeding response of *C. zealandica* larvae, found crude extract of alfalfa root to contain a strong feeding deterrent for the larvae, at concentrations within the range expected for saponin content in fresh alfalfa root. Finally, saponins purified from the crude extract are found to be most effective, confirming the saponin nature of the deterrent in alfalfa.

Larval mortality in populations of the leaf-cutter bee *Megachile pacifica* (*M. rotundata*) pollinating alfalfa seed fields has been examined (Thorp and Briggs, 1971). These bees line their brood cells with pieces of alfalfa leaves, and this led to the suggestion that leaf saponins are responsible for the high mortality. However, Parker and Pedersen (1974) studied the effect of leaves from the two series of strains of alfalfa, one group with a high and one with a low concentration of saponin, on the mortality of immature *M. pacifica;* they found that the average difference between strains is not significant. This seems to rule out the involvement of the saponins present in alfalfa leaf pieces and used in lining the brood cell in the high mortality of leaf-cutter bees foraging on alfalfa. Thorp and Briggs (1971) found that crude extracts and purified alfalfa saponins injected into the pollen food mass increases larval mortality. Unfortunately, no attempt was made to correlate the levels of medicagenic acid-containing saponins of the blossoms (Morris and Hussey, 1965) with larval mortality on the various strains. *Megachile pacifica* bees forage for pollen and are regarded as very efficient pollinators of alfalfa in the United States. This pollen is collected and fed to the brood, and substantial toxicity may stem from this source.

Forage legumes are unsuitable for development of several locust species (cited in Uvarov, 1966). A diet of alfalfa increases mortality and extends the duration of larval development of *Melanoplus sanguinipes* and *M. differentialis.* Adults emerging from these populations are small and have unusually short forewings. Mortality is high, particularly in the first larval instar. Similar results are observed with populations of *Schistocerca gregaria.* Some of the larvae complete their larval development after only four instars, and here too the emerging adults are small and have shortened forewings. Feeding of *Locusta* on *Trifolium pratense* also results in high mortality, particularly in the first instar, increases duration of larval development, and produces small adults. In experiments carried out in our laboratory (N. Levin, S. W. Applebaum, and Y. Birk, unpublished results, 1977) we observed that young adults of *Locusta migratoria* in the somatic growth phase lose weight when kept on a diet of alfalfa for 5 days, whereas similar adults fed a standard grass diet gain weight. The amount of foliage ingested is less on the alfalfa diet than on the standard grass diet. Addition of 0.5 to 1.0% (w/w) of alfalfa saponins containing medicagenic acid to the standard grass diet or to a synthetic diet renders it less palatable to all the larval instars and to adults of *L. migratoria*. Growth is depressed and mortality increases to a greater extent than when the insects are starved.

Fecal pellets of saponin-fed locusts are wet, in contrast to the dry pellets of locusts on a saponin-free synthetic diet. Water resorption in the hindgut is impaired and hemolymph volume decreases as a consequence of the saponin diet. With isolated rectal preparations, we demonstrated that inhibition is reversible. The mode of action of saponins appears to be more complex than that which arises from surface activity, since saponins inhibit (in *in vitro* preparations) at low molar concentrations where other detergents lack any effect. Oxygen consumption of isolated rectal preparations that were reversed inside-out and tied off at both ends was not affected by transfer from physiological saline to a 1% (w/w) saponin solution. This indicates that *in vivo* ingested saponins that reach the hindgut affect water resorption without penetrating the cuticular lining of the rectum. The surface architecture of hindgut cuticle in *L. migratoria* has been described (Klein and Applebaum, 1975). Pores transversing this cuticle are responsible for permeability to water. Their exclusion limit is below that of the saponin molecule, but these molecules can occlude the funnel-like opening of the pores on the cuticle surface and thereby hinder the passage of fluid. This may involve some type of interaction between the saponin and components of the epicuticle. The nature of such putative components is unknown. This phenomenon can lead to the choice of appropriate compounds specifically designed to control insects by disrupting their water balance.

Mortality increases in experiments with the leafhopper *Empoasca fabae* feeding on a minimal artificial diet of 1% sucrose in 0.5% agar that is supplemented with a commercial saponin (extracted from *Yucca* sp.) in the range 0.1–0.5% (w/v) (Roof *et al.*, 1972). It is not clear whether mortality is due to antibiosis or to deterrence, which would lead to starvation and dehydration. This technique was subsequently adapted for the assay of alfalfa saponins using *E. fabae* and the first two instars of the pea aphid *Acyrthosiphon pisum*. Saponins of two varieties, one rich in medicagenic acid and the other lacking this sapogenin, were separated by thin-layer chromatography into 10 fractions and then assayed. The saponins containing medicagenic acid were markedly toxic, whereas saponins containing soybean sapogenins exhibited little or no toxicity. We have conducted some comparative experiments in our laboratories with the peach aphid *Myzus persicae* and with preparations of soybean saponins, alfalfa root saponins, and a purified fraction of root saponins containing medicagenic acid-3-β-*O*-triglucoside. These experiments were of two types. "Preference" experiments offer the aphids the choice of several distinct diets, some lacking saponins (control) and some containing saponins. "No-choice" experiments offer the aphids a single diet that can be rejected completely or partially. In "no-choice" experiments with apteriform larvae (destined to lack wings in the adult stage) mortality increased significantly on the saponin diet. Surviving larvae ingested as much of a 0.1% (w/v) saponin diet as did control larvae on a saponin-free diet. Adult apterate aphids in "no-choice" experiments on diets containing alfalfa root saponin or saponin containing medicagenic acid similarly ingested as much or more than the adults on control diets. The purified saponin containing medicagenic acid is nontoxic during the initial 3 days but exhibits some toxicity to adults and progeny kept on this diet for 6 days. The crude root saponin preparation is more toxic; after 6 days, all adults were dead, and no progeny survived. In "preference" experiments, apterate adults of *M. persicae* chose not to feed on a diet containing alfalfa root saponin. All these responses are consistent with an observed incompatability of legumes for *M. persicae*, with saponins implicated in this resistance.

There is no apparent correlation between resistance of alfalfa to *Acyrthosiphon pisum* and the saponin content of foliage (Pedersen *et al.*, 1975, 1976). Resistance was shown to be correlated positively to root saponin content, but this content cannot be related to leaf saponin content or to content of other plant components. The apparent lack of correlation between leaf saponin content of different alfalfa varieties and their resistance to several alfalfa pests (the seed chalcid *Bruchophagus roddi*, the clover root curculio *Sitona hispidulus*, the spotted alfalfa aphid *Therioaphis maculata*, and the alfalfa weevil *Hypera postica*) seems to minimize the involvement of saponins in resistance to alfalfa pests. None of the 16 saponin fractions isolated from alfalfa leaves and incorporated into artificial diet adversely affected *H. postica* (Hsiao, 1969). In fact, we suggest that this lack of apparent correlation be regarded with due caution. *Bruchophagus roddi* develops in alfalfa seeds, which lack medicagenic acid as a component of their saponins. *Sitona hispidulus* larvae develop on the roots, and a correlation is not apparent between root and foliage saponin content of varieties bred for high saponin or low saponin content in foliage. Aphids feed from sieve tubes and are unaffected by components accumulating in cells. In fact, all of these insects are to various degrees specific pests of alfalfa. Consistent with the general theory of interrelations between mono- or oligophagous insects and their host plants, it is to be expected that an alfalfa-specific insect is not inhibited by alfalfa leaf saponins. It has been postulated that during coadaptive evolution, plants have developed specific compounds, such as saponins, which deter or inhibit development of most insect species while allowing or even stimulating development of a restricted number of insect species as a result of adaptive mechanisms that have evolved in those insects. Thus, one may find that alfalfa saponins even enhance development of *Hypera postica*. Finally, it should be kept in mind that there is much variation in the chemical makeup and composition of saponins in alfalfa varieties bred for differences in total saponin content (Berrang *et al.*, 1974). A much more careful examination of specific saponin composition of the plant tissue eaten and the response of alfalfa pest to changes in this composition is mandatory for resolving the role of saponins in resistance.

A class of triterpenoid saponins, gymnemic acids, has been identified in the leaves of the asclepiad vine *Gymnema sylvestre*. These saponins have been termed "taste distortion agents" (Eisner and Halpern, 1971) because of their ability to suppress the perceived sweetness of sugars and, to a lesser extent, the taste of amino acids. It is suggested that these compounds could serve to protect plants against mammalian herbivores, by camouflaging the attractiveness of sugar.

Gymnemic acid has a strong deterrent effect on larvae of the noctuid moth *Prodenia eridania* (Granich *et al.*, 1974). It is unclear whether this deterrence is also based on taste distortion, eliciting phagorepellence, or is related to a direct effect of gymnemic acid on these larvae. In choosing a host plant, polyphagous insects often rely on stimuli of primary food substances such as sugars or amino acids. Moreover, amino acids could appear as soon as digestion of protein commences. Taste distortion could then turn the insect away from what would otherwise be an attractive meal.

B. Mode of Action

Saponins, being surface-active glycosides with hydrophobic as well as hydrophilic properties, are potentially able to cause changes in the microstructure of natural cell membranes (Bangham and Horne, 1962). The chemical site of action and biochemical basis for activities of saponins is best exemplified by analysis of the mechanism whereby saponins hemolyze erythrocytes.

An attempt has been made to relate the hemolytic activity to the structure and composition of the saponin molecule. It has been shown that the higher hemolytic activity of alfalfa root extract, as compared to that of the foliage, results from the relatively higher content of medicagenic acid, as well as from the higher ratio of sapogenin to sugar in the alfalfa root saponin extract. The saponins that contain medicagenic acid and hederagenin as their aglycones can be precipitated by cholesterol and have strong hemolytic activity (Gestetner *et al.*, 1971a). The main difference between medicagenic acid and the soyasapogenols, which comprise the other aglycones of alfalfa saponins, is in the presence of two carboxylic groups. Blocking of the carboxylic groups in medicagenic acid results in complete loss of the hemolytic activity of the saponin (Gestetner *et al.*, 1971b). The alfalfa saponins, which contain soyasapogenols A, B, C, D, and E as their sapogenins and do not lyse red blood cells, differ from soybean saponins, which contain the same sapogenins but do have a considerable hemolytic activity (Gestetner *et al.*, 1971a). This discrepancy is explained by the lower sapogenin/carbohydrate ratio (1:5) in alfalfa saponins as compared to the 1:1 ratio in soybean saponins, which relates the hemolytic activity to the overall hydrophobicity of the saponin molecule.

A considerable number of saponins and aglycones of the triterpene type have been tested for hemolytic activity by the use of cattle erythrocyte preparations *in vitro* (Schlösser and Wulff, 1969). From the results, it is deduced that the optimal hemolytic activity depends on the following structural arrangements. Aglycones require a polar grouping in ring A and a moderate polar grouping in ring D or E. Compounds containing a 16α-OH or 16-keto group together with a 3β-OH group exhibit the highest hemolytic potential. The distance of 10.5 Å between 3β-OH and 16α-OH appears to be of special significance. Acylation of either hydroxyl group results in a remarkable loss of activity. With saponins that bear a sugar chain on 3β-OH, the distance between the strong and the weak polar grouping is less critical. The composition of the sugar chain influences the hemolytic potential of saponins. A polar grouping in ring D and/or E, such as a sugar chain or a number of OH groups, results in considerable inactivation of the substance.

The hemolytic activity of saponins is generally attributed to their interaction with cholesterol in the erythrocyte membrane (Dourmashkin *et al.*, 1962; Glauert *et al.*, 1962). Saponins are able to displace endogenous lipids from lipoproteins (Gurd, 1960). In the case of alfalfa saponins, this activity can be abolished by the presence of extraneous cholesterol (Shany, 1971). Alfalfa saponins and certain hemolytic proteins (e.g., streptolysin O, creolysin) share a common binding site on the erythrocyte membrane (Shany *et al.*, 1974). Treatment of membranes with any one of these agents prevents the binding of the other. The binding site of saponin is identified with membranal cholesterol (Schröeder *et al.*, 1971) and also seems to be the binding site for the hemolytic proteins. Addition of alfalfa saponins to membranes prepared by osmosis of rabbit erythrocytes and to cholesterol-containing liposomes results in the formation of pits or holes surrounded by rings, as revealed by negative staining followed by electron microscopic examination.

Assa *et al.* (1973) investigated the membranal elements that might be involved in the mechanism of hemolysis of erythrocytes by saponins. Using the unique and well-defined alfalfa saponin, medicagenic acid-3-β-*O*-triglucoside, they demonstrated that this saponin forms interaction products of different stability with membranal cholesterol, proteins, and phospholipids. No interaction was found with fatty acids of the membrane. The structural changes resulting from the interaction of membranal components with the saponin affect only slightly the activity of membranal enzymes. It was suggested that breakdown of the membrane structure results from a combination of nonspecific interactions of saponins with membrane proteins, phospholipids, and cholesterol. These interactions finally lead to hemolysis.

C. Metabolic Fate of Ingested Saponins

It was noted above that some animals increase secretion of cholesterol into the intestine as a consequence of dietary saponin. In those cases in which the cholesterol is capable of complexing with the saponin, e.g., when the saponin contains a medicagenic acid aglycone, the co-precipitation of cholesterol and saponin may be regarded as a form of detoxification. In insects, which are unable to synthesize endogenous cholesterol, this complexing mechanism is inadequate. Insects, like other organisms, are able to partly or completely hydrolyze the oligosaccharide chain of the integral saponin, depending on their digestive enzymes. In cases in which the integral saponin is specifically toxic, such as the case of the legume seed beetle *Callosobruchus chinensis,* the hydrolysis would effectively detoxify the saponin. This may be the explanation for the ability of this insect to develop on legume seed varieties containing saponins. We have demonstrated several glycosidase activities of the locust midgut that are able to degrade legume saponins (N. Levin, S. W. Applebaum, and Y. Birk, unpublished results, 1977).

Analysis of soybean saponins in the digestive tract of rats, mice, and chicks that have been kept on diets containing 20% (w/w) soybean meal reveals the presence of saponins and the absence of sapogenins in the small intestine, whereas in the cecum and colon only sapogenins are found. Neither saponins nor sapogenins can be detected in the blood of these animals, which indicates that the liberated soybean sapogenins, as well as the saponins that may have passed intact through the digestive tract, are not absorbed into the bloodstream. The inability of the upper part of the digestive tract to attack soybean saponins has also been demonstrated by incubating soybean saponins or saponin fractions with slices of the small intestine of chicks, rats, and mice; no sapogenins could be detected in the digestion mixtures. On the other hand, considerable amounts of soybean sapogenins have been found in incubates of soybean saponins and of fractions with the ceca or colons of all the animals tested. Since it is generally accepted that the walls of the cecum and colon do not secrete enzymes, it is assumed that the hydrolysis of soybean saponins is carried out by the cecal microorganisms that enter the colon as a result of the peristaltic and antiperistaltic movements of the digestive tract. To prove this assumption, soybean saponins were incubated with washed, microorganism-free cecum and colon as well as with cecal microflora, and, in fact, the cecum and colon did not decompose the saponins, whereas the microorganisms did (Gestetner *et al.*, 1968).

An attempt was also made to separate a saponin-hydrolyzing enzyme from the cecal microflora and to purify it by means of column chromatography with DEAE-cellulose and on calcium phosphate gel (hydroxyapatite). The partially purified enzymatic preparation hydrolyzed soybean as well as alfalfa saponins and liberated free sapogenins and monosaccharides.

VII. CONCLUDING REMARKS

Saponins are distributed widely in plants and have proved to be remarkably heterogenous. The variability in composition and quantity is found to depend on the phenotypic expression of genetic characteristics under the influence of environmental factors. Saponins have aroused the interest of investigators for a variety of reasons. Steroid saponins serve as precursors for steroid sex hormones in the pharmaceutical industry, and the search continues for additional parent material of increased value for this purpose. Triterpenoid saponins, and in particular the saponins containing medicagenic acid, exhibit an exciting potential for controlling the level of serum cholesterol for therapeutic purposes. The ability of saponins to interfere with normal reabsorption of water in the hindgut of insects may lead to the use of specific saponins or the development of synthetic analogs for control of certain insects. On the other hand, specific saponins seem to selectively deter feeding of some insects and are toxic to others. This property has already prompted an attempt to control stored-product weevils by dusting the grain stores with saponin preparations (Su *et al.*, 1972).

Breeding programs for the production of legume pasture plants that accumulate high levels of medicagenic acid-type saponins in the root system and have low saponin content in the foliage are feasible. These varieties could be expected to exhibit increased resistance to root-damaging insects while at the same time having an improved nutritional value for poultry. Finally, the heterogeneity of saponins and sapogenins, taken together with the subtle effects of small differences in composition on their hydrophobic–hydrophilic properties, presents us with a valuable tool for the investigation of nonspecific and specific binding phenomena on the cellular and membranal level.

ACKNOWLEDGMENT

The author's research described in this chapter was supported by Grant 161 from the United States–Israel Binational Science Foundation (B.S.F.).

REFERENCES

Agarwal, S. K., and Rastogi, R. P. (1974). *Phytochemistry* **13,** 2623–2645.
Applebaum, S. W., and Birk, Y. (1972). *In* "Insect and Mite Nutrition" (J. G. Rodriguez, ed.), pp. 629–636. North-Holland Publ., Amsterdam.
Applebaum, S. W., Gestetner, B., and Birk, Y. (1965). *J. Insect Physiol.* **11,** 611–616.
Applebaum, S. W., Marco, S., and Birk, Y. (1969). *J. Agric. Food Chem.* **17,** 618–622.
Arigoni, D. (1958). *Experientia* **14,** 153–155.
Assa, Y., Gestetner, B., Chet, I., and Henis, Y. (1972). *Life Sci.* **11,** Part II, 637–647.
Assa, Y., Shany, S., Gestetner, B., Tencer, Y., Birk, Y., and Bondi, A. (1973). *Biochim. Biophys. Acta* **307,** 83–91.
Baccou, J. C., Lambert, F., and Sauvaire, Y. (1977). *Analyst* **102,** 458–465.
Bangham, A. D., and Horne, R. W. (1962). *Nature (London)* **196,** 952–953.
Basu, N., and Rastogi, R. P. (1967). *Phytochemistry* **6,** 1249–1270.
Berrang, B., Davis, K. H., Jr., Wall, M. E., Hanson, C. H., and Pedersen, M. W. (1974). *Phytochemistry* **13,** 2253–2260.
Birk, Y. (1969). *In* "Toxic Constituents of Plant Foodstuffs" (I. E. Liener, ed.), 1st ed., pp. 169–210. Academic Press, New York.
Birk, Y., and Peri, I. (1978). *In* "Toxic Constituents of Plant Foodstuffs" (I. E. Liener, ed.), 2nd ed. Academic Press, New York.
Birk, Y., Bondi, A., Gestetner, B., and Ishaaya, I. (1963). *Nature (London)* **197,** 1089–1090.
Boiteau, P., Pasich, B., and Rakato Ratsimamanga, A. (1964). "Les triterpénöides en physiologie végétale et animale" (Gauthier-Villar, editions) 1370 pp. C.N.R.S., Paris.
Bondi, A., Birk, Y., and Gestetner, B. (1973). *In* "Chemistry and Biochemistry of Herbage" (G. W. Butler and R. W. Bailey, eds.), Vol. 1, pp. 511–528. Academic Press, New York.
Brune, H., and Kundlich, O. (1960). *Z. Tierphysiol., Tierernaehr. Futtermittelkd.* **15,** 274–284.
Büchi, J., and Dolder, R. (1950). *Pharm. Acta Helv.* **25,** 179–188.

Chakravarti, R. N., Mitra, M. N., and Chakravarti, D. (1963). *Bull. Calcutta Sch. Trop. Med.* **11,** 20.
Cheeke, P. R. (1971). *Can. J. Anim. Sci.* **51,** 621–632.
Cheeke, P. R. (1976). *Nutr. Rep. Int.* **13,** 315–324.
Cole, H. H., Huffman, C. F., Kleiber, M., Olson T. M., and Shalk, A. F. (1945). *J. Anim. Sci.* **4,** 183–236.
Cooney, W. T., Butts, J. S., and Bacon, L. E. (1948). *Poult. Sci.* **27,** 828–830.
Coulson, C. B. (1958). *J. Sci. Food Agric.* **9,** 218–287.
Coulson, C. B., and Davies, T. (1962). *J. Sci. Food Agric.* **13,** 53–56.
Diekert, J. W., and Morris, N. J. (1958). *J. Agric. Food Chem.* **6,** 930–933.
Djerassi, C., Thomas, D. B., Livingston, A. L., and Thompson, C. R. (1957). *J. Am. Chem. Soc.* **79,** 5292–5297.
Dourmashkin, R. R., Dougherty, R. M., and Harris, R. J. C. (1962). *Nature (London)* **194,** 1116–1119.
Draper, C. I. (1948). *Poult. Sci.* **27,** 659.
Drożdż, B. (1962). *Diss. Pharm.* **14,** 519.
Eisner, T., and Halpern, B. P. (1971). *Science* **172,** 1362.
Farrell, J. A. K., and Sweney, W. J. (1972). *N.Z. J. Agric. Res.* **15,** 904–908.
Focke, I., Focke, R., and Franzke, W. (1968). *Albrecht-Thaer-Arch.* **12,** 813–819.
Fraenkel, G. S. (1959). *Science* **129,** 1466–1470.
George, A. J. (1965). *Food Cosmet. Toxicol.* **3,** 85–91.
Gestetner, B. (1964). *J. Chromatogr.* **13,** 259–261.
Gestetner, B. (1971). *Phytochemistry* **10,** 2221–2223.
Gestetner, B., Ishaaya, I., Birk, Y., and Bondi, A. (1963). *Isr. J. Chem.* **1,** 460–467.
Gestetner, B., Birk, Y., and Bondi, A. (1966). *Phytochemistry* **5,** 799–806.
Gestetner, B., Birk, Y., and Tencer, Y. (1968). *J. Agric. Food Chem.* **16,** 1031–1035.
Gestetner, B., Shany, S., Tencer, Y., Birk, Y., and Bondi, A. (1970). *J. Sci. Food Agric.* **21,** 502–507.
Gestetner, B., Assa, Y., Henis, Y., Birk, Y., and Bondi, A. (1971a). *J. Sci. Food Agric.* **22,** 168–172.
Gestetner, B., Shany, S., and Assa, Y. (1971b). *Experientia* **27,** 40–41.
Glauert, A. M., Dingle, J. T., and Lucy, J. A. (1962). *Nature (London).* **196,** 953–955.
Granich, M. S., Halpern, B. P., and Eisner, T. (1974). *J. Insect Physiol.* **20,** 435–439.
Grunwald, C. (1970). *Anal. Biochem.* **34,** 16–23.
Gurd, F. R. N. (1960). *In* "Lipide Chemistry" (D. J. Hanahan, ed.), pp. 260–325. Wiley, New York.
Gutierrez, J., Davis, R. E., and Lindahl, I. L. (1959). *Appl. Microbiol.* **7,** 304–308.
Hanson, C. H., and Kohler, G. O. (1961). *Proc. Tech. Alfalfa Conf., 7th, 1961* p. 46.
Hanson, C. H., Kohler, G. O., Dudley, J. W., Sorensen, E. L., Van Atta, G. R., Taylor, K. W., Pedersen, M. W., Carnahan, H. L., Wilsie, C. P., Kehr, W. R., Lowe, C. C., Standford, E. H., and Yungen, J. A. (1963). *U.S., Dept. Agric., Tech. Bull.* 33–44.
Hanson, C. H., Pedersen, M. W., Berrang, B., Wall, M. E., and Davis, K. H., Jr. (1973). *In* "Anti-Quality Components of Forages" (A. G. Matches, ed.), Spec. Publ. No. 4, p. 33–52. Crop Sci. Soc. Am., Madison, Wisconsin.
Hashizume, A., and Sakato, Y. (1966). *Nippon Nogei Kagaku Kaishi* **40,** 8–12. *Chem. Abstr.* **64,** 13019c.
Hein, S. (1959). *Planta Med.* **7,** 185–204.
Heywang, B. W. (1950). *Poult. Sci.* **29,** 804–811.
Heywang, B. W., and Bird, H. R. (1954). *Poult. Sci.* **33,** 239–241.

Heywang, B. W., Thompson, C. R., and Kemmerer, A. R. (1959). *Poult. Sci.* **38,** 968–971.
Hoppe, W., Gieren, A., Brodherr, N., Tschesche, R., and Wulff, G. (1968). *Angew. Chem.* **80,** 563–564.
Horber, E. (1965). *Proc. Int. Congr. Entomol., 12th, 1964* pp. 540–541.
Horber, E. (1972). *In* "Insect and Mite Nutrition" (J. G. Rodriguez, ed.), pp. 611–627. North-Holland Publ., Amsterdam.
Horber, E., Leath, K. T., Berrang, B., Marcavian, V., and Hanson, C. H. (1974). *Entomol. Exp. Appl.* **17,** 410–424.
Hsiao, T. H. (1969). *Proc. Int. Symp. Insect Hostplant, 2nd, 1969,* In: *Ent. exp. & appl.* **12,** 777–788.
Ishaaya, I., and Birk, Y. (1965). *J. Food Sci.* **30,** 118–120.
Ishaaya, I., Birk, Y., Bondi, A., and Tencer, Y. (1969). *J. Sci. Food Agric.* **20,** 433–436.
Jackson, H. D., and Shaw, R. A. (1959). *Arch. Biochem. Biophys.* **84,** 411–416.
Jackson, H. D., Shaw, R. A., Pritchard, W. R., and Hatcher, B. W. (1959). *J. Anim. Sci.* **18,** 158–162.
Jeger, O. (1950). *Fortschr. Chem. Org. Naturst.* **7,** 1–86.
Johnson, I. J. (1971). *Abstr. West. Soc. Crop. Sci., 1971 Annu. Meet.* pp. 12–13.
Jones, M. (1969). Ph.D. Dissertation, Michigan State University, East Lansing.
Jones, M., and Elliott, F. C. (1969). *Crop Sci.* **9,** 688–691.
Jurzysta, M. (1973). *Acta Soc. Bot. Pol.* **42,** 201–207.
Jurzysta, M. (1975). *Pamiet. Pulawski* **62,** 99–107.
Kain, W. M., and Atkinson, D. S. (1970). *Proc. N.Z. Weed Pest Control Conf.* **23,** 180–183.
Kartnig, T., Graune, F. J., and Herbst, R. (1964). *Planta Med.* **12,** 428–439.
Kawasaki, T., and Yamauchi, T. (1962). *Chem. Pharm. Bull.* **10,** 703–708.
Kazerovskis, K. K. (1962). *J. Pharm. Sci.* **51,** 352–354.
Kendall, W. A., and Leath, K. T. (1976). *Agron. J.* **68,** 473–476.
Klein, M., and Applebaum, S. W. (1975). *J. Entomol., Ser. A* **50,** 31–36.
Kochetkov, N. K., and Khorlin, A. J. (1966). *Arznei.-Forsch.* **16,** 101–109.
Kochetkov, N. K., Khorlin, A. J., and Ovodov, Ju. S. (1963). *Tetrahedron Lett.* **8,** 477–482.
Kofler, L., and Adam, A. (1927). *Arch. Pharm. (Weinheim, Ger.)* **265,** 624–643.
Leath, K. T., Davis, K. H., Jr., Wall, M. E., and Hanson, C. H. (1972). *Crop Sci.* **12,** 851–856.
Lepkovsky, S., Shaeleff, W., Peterson, D., and Perry, R. (1950). *Poult. Sci.* **29,** 208–213.
Lindahl, I. L., Cook, A. C., Davis, R. E., and Maclay, W. D. (1954). *Science* **119,** 157–158.
Lindahl, I. L., Dougherty, R. W., and Davis, R. E. (1957). *U.S. Dep. Agric., Tech. Bull.* **1161,** 15–27.
Livingston, A. L., Whitehand, L. C., and Kohler, G. O. (1977). *J. Assoc. Off. Anal. Chem.* **60,** 957–960.
McCandlish, A. C. (1937). *Wiss. Ber. Milchwiltsch. Weltkongr., 11th, 1937* Vol. 1, pp. 410–412.
McNairy, S. A., Jr., Goetsch, G. O., Hatcher, B. W., and Jackson, H. D. (1963). *J. Anim. Sci.* **22,** 61–65.
Malinow, M. R., McLaughlin, P., Kohler, G. O., and Livingston, A. L. (1977). *Steroids* **29,** 105–110.
Mangan, J. L. (1958). *N.Z. J. Agric. Res.* **1,** 140–147.
Mangan, J. L. (1959). *N.Z. J. Agric. Res.* **2,** 47–61.
Morris, R. J., and Hussey, E. W. (1965). *J. Org. Chem.* **30,** 166–168.
Morris, R. J., Dye, W. B., and Gisler, P. S. (1961). *J. Org. Chem.* **26,** 1241–1243.
Noelting, G., and Bernfeld, P. (1948). *Helv. Chim. Acta* **31,** 286–290.

Nowacki, E., Jurzysta, M., and Dietrych-Szostak, D. (1976). *Biochem. Physiol. Pflanz.* **169,** 183–186.
O'Dell, B. L., Regam, W. O., and Beach, T. J. (1959). *Mo., Agric. Exp. Stn., Res. Bull.* **702,** 12.
Olson, T. M. (1944). *S. D., Agric. Exp. Stn., Circ.* **52,** 11.
Parker, F. D., and Pedersen, M. W. (1974). *Environ. Entomol.* **4,** 103–104.
Pasich, B. (1961). *Diss. Pharm.* **13,** 1–10.
Pedersen, M. W., and Wang, L.-C. (1971). *Crop. Sci.* **11,** 833–835.
Pedersen, M. W., Zimmer, D. E., Anderson, J. O., and McGuire, C. F. (1966). *Proc. Int. Grassl. Cong., 10th, 1966* pp. 693–698.
Pedersen, M. W., Anderson, J. O., Street, J. C., Wang, L.-C., and Baker, R. (1972). *Poult. Sci.* **51,** 458–463.
Pedersen, M. W., Berrang, B., Wall, M. E., and Davis, K. H., Jr. (1973). *Crop Sci.* **13,** 731–735.
Pedersen, M. W., Sorensen, E. L., and Anderson, M. J. (1975). *Crop Sci.* **15,** 254–256.
Pedersen, M. W., Barnes, D. K., Sorensen, E. L., Griffin, G. D., Nielson, M. W., Hill, R. R., Jr., Frosheiser, F. I., Sonoda, R. M., Hanson, C. H., Hunt, O. J., Peaden, R. N., Elgin, J. H., Jr., Devine, T. E., Anderson, M. J., Goplen, B. P., Elling, L. J., and Howarth, R. E. (1976). *Crop. Sci.* **16,** 193–199.
Peterson, D. W. (1950a). *J. Biol. Chem.* **183,** 647–653.
Peterson, D. W. (1950b). *J. Nutr.* **42,** 597–607.
Pifferi, P. G. (1965). *Anal. Chem.* **37,** 925.
Quazi, H. M. (1975). *N.Z. J. Agric. Res.* **18,** 227–232.
Quazi, H. M. (1976). *N.Z. J. Agric. Res.* **19,** 347–348.
Quin, J. I. (1943). *Onderstepoort J. Vet. Sci. Anim. Ind.* **18,** 113–117.
Reshef, G., Gestetner, B., Birk, Y., and Bondi, A. (1976). *J. Sci. Food Agric.* **27,** 63–72.
Roof, M., Horber, I., and Sorensen, E. L. (1972). *Proc. North Cent. Branch Entomol. Soc. Am.* **27,** 140–143.
Ruzicka, L. (1953). *Experientia* **9,** 357–367.
Sawardeker, J. S., Sloneker, J. H., and Jeanes, A. (1965). *Anal. Chem.* **37,** 1602–1604.
Schlösser, E., and Wulff, G. (1969). *Z. Naturforsch., Teil B* **24,** 1284–1290.
Schröeder, F., Holland, J. F., and Biber, L. I. (1971). *J. Antibiot.* **24,** 846–849.
Schulz-Langner, E. (1966). *Planta Med.* **14,** 49–56.
Segal, R., Mansour, M., and Zaitschek, D. V. (1966). *Biochem. Pharmacol.* **15,** 1411–1416.
Shany, S. (1971). Ph.D. Dissertation, Hebrew University of Jerusalem.
Shany, S., Birk, Y., Gestetner, B., and Bondi, A. (1970a). *J. Sci. Food Agric.* **21,** 131–135.
Shany, S., Gestetner, B., Birk, Y., and Bondi, A. (1970b). *J. Sci. Food Agric.* **21,** 508–510.
Shany, S., Gestetner, B., Birk, Y., Bondi, A., and Kirson, I. (1972). *Isr. J. Chem.* **10,** 881–884.
Shany, S., Bernheimer, A. W., Grushoff, P. S., and Kim, K.-S. (1974). *Mol. Cell. Biochem.* **3,** 179–186.
Shaw, D. H., and Moss, G. W. (1969). *J. Chromatogr.* **41,** 350–357.
Shoppee, C. W. (1964). "Chemistry of Steroids," 2nd ed. Butterworth, London.
Silin, P. M. (1964). "Technology of Beet-Sugar Production and Refining." Isr. Program Sci. Transl., Jerusalem.
Smith, H. M., Smith, J. M., and Spring, F. S. (1958a). *Tetrahedron* **4,** 111–131.
Smith, H. M., Smith, J. M., and Spring, F. S. (1958b). *Chem. Ind. (London)* pp. 889–890.
Stahl, E. (1965). "Thin Layer Chromatography. A Laboratory Handbook." Academic Press, New York.

Steiner, M. and Holtzem, H. (1955). *In* "Moderne Methoden der Pflanzenanalyse" (K. Paech and M. V. Tracey, eds.), Vol. III, p. 58. Springer-Verlag, Berlin and New York.

Su, H. C. F., Speirs, R. D., and Mahany, P. G. (1972). *J. Econ. Entomol.* **65,** 844–847.

Sumiki, Y. (1929). *Bull. Agric. Chem. Soc. Jpn.* **5,** 27–32.

Sutherland, O. R. W., Hood, N. D., and Hillier, J. R. (1975). *N.Z. J. Zool.* **2,** 93–100.

Takahashi, T., Miyazaki, M., Yasue, M., Imamura, H., and Honda, O. (1963). *J. Jpn. Wood Res. Soc.* **9,** 59–62.

Tencer, Y., Shany, S., Gestetner, B., Birk, Y., and Bondi, A. (1972). *J. Agric. Food Chem.* **20,** 1149–1151.

Thorp, R. W., and Briggs, D. L. (1971). *Environ. Entomol.* **1,** 399–401.

Uvarov, B. (1966). "Grasshoppers and Locusts," Vol. 1. Cambridge University Press, London and New York.

Van Atta, G. R., and Guggolz, J. (1958). *J. Agric. Food Chem.* **6,** 849–850.

Van Duuren, A. J. (1962). *J. Am. Soc. Sugar Beet Technol.* **12,** 57–63.

Wagner, J., and Sternkopf, G. (1958). *Nahrung* **2,** 338–357.

Walter, E. D. (1957). *J. Am. Pharm. Assoc.* **46,** 466–467.

Walter, E. D. (1960). *J. Am. Pharm. Assoc.* **49,** 735–736.

Walter, E. D. (1961). *J. Pharm. Sci.* **50,** 173.

Walter, E. D., Van Atta, G. R., Thompson, C. R., and Maclay, W. D. (1954). *J. Am. Chem. Soc.* **76,** 2271–2273.

Walter, E. D., Bickoff, E. M., Thompson, C. R., Robinson, C. H., and Djerassi, C. (1955). *J. Am. Chem. Soc.* **77,** 4936–4939.

Wasicky, R., and Wasicky, M. (1961). *Qual. Plant. Mater. Veg.* **8,** 65–79.

Willner, D., Gestetner, B., Lavie, D., Birk, Y., and Bondi, A. (1964). *J. Chem. Soc., Suppl.* **1,** 5885–5888.

Zimmer, D. E., Pedersen, M. W., and McGuire, C. F. (1967). *Crop Sci.* **7,** 223–224.

Chapter **16**

Phytohemagglutinins

IRVIN E. LIENER

I. INTRODUCTION

For almost a century, it has been known that many plant seed extracts clump or agglutinate animal erythrocytes, a property that led to the designation of the principle responsible for this phenomenon as "phytohemagglutinins." The first description of a phytohemagglutinin can be found in a report submitted by Stillmark in 1888 in which he describes the agglutination of red blood cells by extracts of the castor bean, which was known to be highly toxic. He named the substance responsible for this effect *ricin* after the botanical name of the castor bean, *Ricinus communis*. Shortly afterward, Hellin (1891) noted that

ISBN 0-12-597180-X

extracts of the seed of the jequirity bean *Abrus precatorius* also caused the agglutination of red blood cells, and, following the precedent of Stillmark, he named the active principle "abrin." These observations drew the attention of the eminent German bacteriologist Paul Ehrlich, who realized that he could investigate certain problems of immunology with these plant substances rather than with bacterial toxins, which were a popular study at that time. Ehrlich's studies with ricin and abrin (1891) led to the development of the most fundamental principles of immunology, namely, that animals could develop immunity to these toxic substances when they were injected into the body and that the antibodies responsible for this immunity were directed specifically against the injected toxin and no other toxin. In 1908, another pioneer of immunology, Karl Landsteiner, noted that these phytohemagglutinins were species-specific; that is, extracts from a particular plant could agglutinate cells from only certain species of animals (Landsteiner and Raubitschek, 1908).

Although as early as 1919 Sumner had crystallized concanavalin A from the jack bean *Canavalia ensiformis,* it was not until 1936 that he realized that this protein was responsible for the hemagglutingating activity of this bean (Sumner and Howell, 1936). The importance observation was also made that concanavalin A was capable of precipitating certain carbohydrates such as glycogen. The true significance of this seemingly trivial observation did not become apparent until many years later, as we shall see.

In the late forties, Boyd and Reguera (1949) discovered that the phytohemagglutinins exhibited a high degree of specificity toward human red blood cells of various blood group types. It was in fact this high degree of specificity that led Boyd and Sharpleigh (1954) to coin the word *lectin* (from the latin word *legere,* to pick or choose) to emphasize the specificity that these substances exhibit toward blood groups. Although the term lectin is frequently used interchangeably with phytohemagglutinin or phytoagglutinin, it has taken on a much broader meaning since carbohydrate-binding proteins have been shown to be universally distributed in the animal as well as the plant kingdom (Liener, 1976).

Advantage has been taken of the ability of phytoagglutinins to agglutinate red blood cells as a means of separating erythrocytes from leukocytes. While doing this Nowell (1960) made the important observation that phytoagglutinins are mitogenic; that is, they induce the transformation of lymphocytes from small, resting cells to large, actively growing cells, which ultimately undergo mitotic division. Over the ensuing years, it has become increasingly apparent that the phytoagglutinins exhibit an amazingly wide variety of other biological effects. In addition to hemagglutinating and mitogenic activities, these include the preferential agglutination of tumor cells and an inhibition of tumor growth, inhibition of the fertilization of the ovum by sperm, inhibition of fungal growth, insecticidal action, and an insulin-like effect on fat cells. All of these effects stem from a single, unique property of the phytoagglutinins, namely, their ability to bind specific sugar residues located on the surface of cells. Because of the presence of more than one binding site in each molecule of phytoagglutinin, the latter form cross-linkages between adjacent cells, giving rise to the phenonemon of agglutination. Because of their specific sugar-binding properties, the lectins have provided investigators with an extremely useful tool not only for isolating a wide variety of glycoproteins but also for probing the intimate molecular architecture of cell surfaces and the changes induced therein by carcinogenic agents. Fascinating and important as some of these facets of the problem may be, this chapter is confined largely to a description of the distribution and properties of those phytoagglutinins that have been fairly well characterized, the manner in which they may interact with higher animals, and their possible physiological function in the plant itself. More detailed reviews of the phytoagglutinins are available for the interested reader (Gold and Balding, 1975; Lis and Sharon, 1973; Sharon and Lis, 1972; Callow, 1975; Liener, 1976; Sharon, 1977).

II. GENERAL CONSIDERATIONS

A. Distribution in the Plant Kingdom

The phytoagglutinins are universally distributed throughout the plant kingdom. A recent and comprehensive list of plants in which phytohemagglutinating activity has been demonstrated can be found in a book devoted to this subject by Gold and Balding (1975). In the family Leguminosae alone, over 600 species and varieties have been shown to contain agglutinins (Toms and Western, 1971). Most of the compilations of the distribution of lectins are based on the use of crude extracts (finely ground plant material that has been extracted with physiological saline) in conjunction with a suspension of one or more different kinds of animal red blood cells. It should be emphasized that the hemagglutinins of many plants may have escaped detection by this technique simply because they were tested against blood cells for which they may not have been specific.

Table I is a compilation of those plants from which the phytoagglutinins have been isolated and characterized with respect to physicochemical properties, specificity, and biological activity. As the interest in lectins continues to grow, this list will no doubt have to be expanded in the ensuing years.

B. Purification and Assay Procedures

Until quite recently, relatively few lectins were isolated in pure form since conventional methods of protein purification often failed to yield preparations that the protein chemist would consider homogenous. With the advent of affinity chromatography and as our knowlege of the sugar specificity of lectins has increased, the purification of many lectins can now be achieved in a relatively simple and rapid fashion. In using affinity chromatography for the purification of lectins, a sugar ligand for which the lectin is specific is covalently attached to an insoluble matrix such as Sepharose or agarose. When a crude preparation of the lectin is passed through such a column, the lectin is preferentially bound to the column and can be subsequently released by washing with the appropriate sugar. Numerous examples of how this technique has been exploited for the isolation of lectins are noted in the sections that deal with the properties of individual lectins. For further details regarding the methodology for isolating lectins, the reader is referred to the review by Liener (1976).

The agglutinating activity of lectins is most commonly determined with erythrocytes from one or more of several animals such as human beings, rats, rabbits, horses, and cows. The simplest assay is the serial dilution technique in which the end point is determined by visual inspection (Kabat and Mayer, 1961). Although this technique is simple and rapid, it is, at best, only semiquantitative and will detect only those lectins that have multiple binding sites. Attempts have been made to improve the precision of such assays by spectrophotometric techniques that measure the decrease in turbidity of a suspension of cells as a consequence of agglutination (Liener, 1955). Howard and Shannon (1977) described a very sensitive assay that takes advantage of the antigenic as well as the chemical specificity of lectins. The solution containing the lectin is treated with Sepharose-bound sugar for which the lectin is specific, and all nonabsorbing impurities are removed by washing. Radioactive-labeled antibody that has been prepared against a purified preparation of the lectin is then added to the lectin–Sepharose–sugar conjugate, and the amount of radioactivity retained by

Table I. Physicochemical and biological properties of purified plant lectins

Botanical name	Common name	Molecular weight	Subunits	Sugar[a]	Human blood type[b]	Other animal blood[c]	Mitogenic activity	References
Abrus precatorius	Jequerity bean, Rosary pea, Indian licorice	65,000[d] 126,000[e]	2 4	D-Gal D-Gal	NS	ck,c		Osnes and Pihl (1978)
Amphicarpaea bracteata	hog-peanut			D-GalNAc	A_1			Blacik *et al.* (1978)
Arachis hypogeae	Peanut, ground nut	110,000	4	D-Gal-β-(1,3)-D-GalNAc	T	rb	+	Lotan *et al.*, (1975c)
Bandieraea simplicifolia		114,000	4	α-D-Gal	A_1,A_2,B		−	See Section III,G
Bauhinia purpurea alba	Camel's foot tree (seeds)	195,000		α-D-GalNAc	N	rb		Irimura and Osawa (1972), Irimura *et al.* (1975)
Canavalia ensiformis	Jack bean	105,000	4	α-D-Man, α-D-Glc	NS	ck,g,rb,sh	+	See Section III,A
Caragama aborescens	Pea tree	60,000	2	D-GalNAc	NS	ck,g,rb		Bloch *et al.* (1976)
Crotalaria juncea	Sunn hemp	120,000	4	D-Gal	NS			Ersson *et al.* (1973), Ersson and Porath (1974), Ersson (1977)
Cystisus sessilifolius	broom			(β-D-GlcNAc)$_3$	H			Matsumoto and Osawa (1972, 1974)
Dolichos biflorus	Horse gram, field bean	110,000	4	D-GalNAc	A_1	−	−	Etzler *et al.* (1977)
Euonymus europaea	Spindle tree (seeds)	166,000	4		A_2,B_1, H_1			Pacak and Kocourek (1975), Petryniak *et al.* (1977)
Glycine max	Soybean	120,000	4	D-Gal, D-GalNAc	A_1O	rb	−	See Section III,E

(continued)

Table I. *(continued)*

Botanical name	Common name	Molecular weight	Sub-units	Sugar[a]	Human blood type[b]	Other animal blood[c]	Mito-genic activity	References
Hordeum vulgare	Barley	40,000	2	D-GlcNAc				Foriers *et al.* (1975, 1976), Partridge *et al.* (1976)
Laburnum alpinum		150,000			H			Sachs *et al.* (1976)
Lens culinaris (or *esculenta*)	Lentil	60,000	2	D-Glc, D-Man	NS	g,rb	+	Fliegerova *et al.* (1974)
Lotus tetragonolobus	Asparagus pea, winged pea	A, 120,000	4	α-L-Fuc	H(O)		−	Kalb (1968), Pereira and Kabat (1974a,b), Yariv *et al.* (1973)
		B, 58,000	2	α-L-Fuc	H(O)		−	
		C, 117,000	4	α-L-Fuc	H(O)		−	
Maakia amurensis	Agglutinin	132,000	4				−	Kawaguchi *et al* (1974)
	Mitogen	75,000	2				+	
Maclura pomifera	Osage orange	43,000	4	D-GalN				Ullevitch *et al.* (1974), Chuba *et al.* (1973), Bausch and Poretz (1976)
Oryza sativa	Rice (seed)	10,000	1				+	Takahashi *et al.* (1973), Barber and Barber (1975)
Phaseolus coccineus	Scarlet runner bean	120,000	4	D-GalNAc	A,B,O	ck,g,rb,sh	+	Nowakova and Kocourek (1974)
Phaseolus lunatus (or *limensis*)	Lima bean	I, 124,000	2	D-GalNAc	A	g,rb	+	See Section III,C
		II, 247,000	4	D-GalNAc	A	sh	+	
Phaseolus vulgaris	Navy bean	128,000	4	D-GlcNAc	NS	rb	−	Andrews (1974), Pusztai and Watt (1974)
	Black bean	128,000			rb			Jaffé and Gaede (1959), Jaffé and Hannig (1965)

	Red kidney bean	138,000	4	D-GlcNAc	NS	ck,g,rb, sh,ho, sw	+	See Section III,C
	Wax bean	120,000	4	D-GalNAc	NS	ck,g,rb, sh,sw, ho		Takahaski *et al.* (1967, 1974, 1976), Takahashi and Liener (1968), Sela *et al.* (1973)
Phytolacea americanus	Pokeweed (roots)	Pa-1, 22,000; Pa-2 to Pa-5,[n] 19,000–31,000			None		+ (T+B), + (T)	See Section III,H
Pisum sativum	Garden pea	57,000	4	α-D-Glc, α-D-Man	NS	ck,ct,d,g, ho,m, rb,sh, sw	+	See Section III,F
Ricinus communis	Castor bean	60,000[d] 120,000[e]	2 4	D-Gal, D-GalNAc D-Gal	NS			See Section III,B
Robinia pseudoacacia	Black locust (seeds)	100,000			NS	ck,g,rb, sh	+	Font and Bourillon (1971), Bourillon and Font (1968), Leseny *et al.* (1972), Sharif *et al.* (1977)
Sophora japonica	Pagoda tree (seeds)	132,000	4	D-Gal, D-GalNAc	A,B,I	g,rb	–	Poretz (1973), Poretz and Barth (1976), Terao and Osawa (1973), Betail and Coulet (1974), Balding and Gold (1973), Chien *et al.* (1974,1975)

(continued)

Table I. (*continued*)

Botanical name	Common name	Molecular weight	Sub-units	Sugar[a]	Human blood type[b]	Other animal blood[c]	Mitogenic activity	References
Solanum tuberosum	Potato (tuber)	46,000	2	D-GlcNAc				See Section III,I
Triticum vulgare	Wheat (germ)	36,000	2	$(\text{D-GlcNAc})_{1-3}$	None		–	See Section III,D
Ulex europeus	Gorse	I, 68,000 II, 170,000	2 4	L-Fuc $(\text{D-GlcNAc})_2$	H,O			Matsumoto and Osawa (1970), Osawa and Matsumoto, (1973), Hořejší and Kocourek, (1974), Frost *et al.* (1975), Allen and Johnson (1977)
Vicia cracca	Tufted vetch Gerard vetch	125,000	4	D-GalNAc	A	g,rb		Aspberg *et al.* (1968), Sundberg *et al.* (1970) Kristiansen and Porath (1968), Kristiansen (1974), Rüdiger (1977)
Vicia ervilia		60,000	4	D-Glc, D-Man	NS	g,rb		Fornstedt and Porath (1975), Kristiansen (1975), Rougé *et al.* (1977)

Vicia faba	Broad bean, Fava bean	50,000	2	GlcN, Man	NS	c,ct,d,g, ho,n,m, rb,ck	+	Lee *et al.* (1974), Wang *et al.* (1974), Allen *et al.* (1976), Allen and Johnson, (1976), Ziska (1976); DeClerque *et al.* (1976)
Vicia graminea	Vetch	100,000	4		N		–	Prigent and Bourillon (1976), Lisowska *et al.* (1976)
Vicia unijuga		130,000			N	g,rb		Lutsik *et al.* (1977), Tajima *et al.* (1977)
Wistaria floribunda	Japanese wisteria	I, 64,000	2	D-GalNAc		ck,g,rb	+	Toyoshima *et al.* (1971), Toyoshima and Osawa (1975), Cheung *et al.* (1975), Kurokawa *et al.* (1976)
		II, 140,000	4	D-Gal	NS		–	

[a] Key to abbreviations: D-Glc, glucose; D-Gal, galactose; D-Man, mannose; D-GlcN, Glucosamine; D-GalN, galactosamine; D-GalNAc, *N*-acetylgalactosamine; D-GlcNAc, *N*-acetylglucosamine; L-Fuc, fucose.

[b] NS, nonspecific.

[c] Key to abbreviations of animals: c, cow, ck, chicken; ct, cat; d, dog; g, guinea pig; ho, horse; m, mouse; rb, rabbit; rt, rat; sh, sheep; sw, swine. Data taken from Toms and Western (1971).

[d] Toxic lectins named abrin and ricin in the case of *A. precatorius* and *R. communis* respectively.

[e] Exhibit agglutinating activity only.

this conjugate is a measure of the amount of lectin present in the original test solution.

The mitogenic activity of lectins is measured most commonly by their ability to stimulate the *in vitro* synthesis of DNA by peripheral lymphocytes, an activity that can be quantitated by measuring the uptake of labeled thymidine (Pellegrino *et al.*, 1973).

C. Structural Features

A cursory examination of the information shown in Table I suggests that, although there is a wide diversity in the physicochemical properties of the lectins, certain broad generalizations can nevertheless be made. Most lectins appear to have molecular weights ranging from 100,000 to 150,000 and are composed of four subunits, which may or may not be identical and which are usually held together by noncovalent forces. A lesser number of lectins, such as those of the lentil, lima bean, and wheat germ, are dimers having about one-half the molecular weight of the tetrameric lectins. With few exceptions (i.e., soybean agglutinin), each subunit has one sugar-binding site, and, as mentioned before, it is this feature of multivalency that accounts for the ability of lectins to agglutinate cells or to precipitate glycoproteins or large polysaccharide polymers. Not all lectins, however, are multivalent and hence do not exhibit agglutinating properties, but they do produce other biological effects. For example, ricin and abrin are extremely toxic, and the pokeweed is a potent mitogen.

Most of the lectins are glycoproteins containing 4–10% carbohydrate, which is covalently bound to the protein. Notable exceptions are concanavalin A and the lectins from wheat germ and peanut, which are devoid of carbohydrate. At the other extreme the lectins from rice and potato contain 25 and 50% carbohydrate, respectively. The exact role, if any, that the carbohydrate moiety may play in the biological activity of the lectins is not clear.

Metal ions appear to be essential for the sugar-binding property of concanavalin A (see Section III,A) and the lectins from the soybean (Jaffé *et al.*, 1977) and the pea and lentil (Paulova *et al.*, 1971a,b). Only in the case of concanavalin A has the role of metal ions been studied in great detail.

In those cases in which studies have been made, a common feature of lectin structure is the virtual absence of α-helical structure. The β-pleated sheet structure is the predominant secondary structural feature of concanavalin A (Pflumm *et al.*, 1971), the lectins from *Pisum sativum* (Bures *et al.*, 1972), *Ricinus communis* (Shimazaki *et al.*, 1975),

Dolichos biflorus (Pére *et al.*, 1975), *Robinia pseudoacacia* (Pére *et al.*, 1975), and *Bandeiraea simplicifolia* (Lönngren *et al.*, 1976). It has been suggested (Lönngren *et al.*, 1976) that the β-pleated sheet structure may be responsible for the interaction of the subunits that make up the oligomeric structure of most lectins. Although the complete amino acid sequence of concanavalin A has been known since 1972 (Edelman *et al.*, 1972), only now are concerted efforts being made to determine the amino acid sequence of other lectins. From preliminary data from several laboratories it appears that despite their diverse origin there is a high degree of homology among the various lectins (Van Driesche *et al.*, 1976; Etzler *et al.*, 1977; Foriers *et al.*, 1977a,b).

In view of the many apparent points of similarity of structure among the lectins, it is difficult to explain the differences in sugar specificity displayed by various lectins. Obviously, there must be subtle differences in structure that alter the nature of the sugar-binding sites, and the elucidation of these structural differences is a challenging problem that will no doubt receive increasingly greater attention in the near future.

D. Specificity

The high degree of specificity that lectins display toward red blood cells of various animal species or different blood groups of the same species* is a reflection of the difference in the sugar composition of glycoproteins located on cell surfaces. In the case of lectins preferentially agglutinating malignant cells, this property appears to depend primarily on the manner in which these receptor sites are distributed over the surface of the cell (Lis and Sharon, 1977) rather than on the number of binding sites.

Identification of the specific sugar residues with which the lectins interact is achieved most frequently by testing the ability of simple sugars to inhibit cell agglutination, the assumption being that the best inhibitor is similar or identical to the cell surface lectin receptor and that the lectin binds exclusively to terminal carbohydrate residues. Both of these assumptions now appear to be oversimplifications (Lis and Sharon, 1973). The most inhibitory sugar is not necessarily identical with the receptor recognized by the lectin. Such a sugar may mimic another

* Most lectins agglutinate the erythrocytes of all human blood groups and are referred to as "nonspecific agglutinins." The "specific agglutinins." on the other hand, show a specificity for erythrocytes of a given blood type. Thus, there are certain lectins that agglutinate one or more cells bearing blood groups A, B, O, M, N, H, etc.

carbohydrate that occurs in the receptor site, which in turn may be influenced by neighboring sugars, the nature of the sugar linkage, and the charge properties of the protein backbone to which the oligosaccharide is attached. Furthermore, the specific sugars do not necessarily have to be in terminal nonreducing positions on the receptor site (Goldstein *et al.*, 1973).

Beyond what has already been said, no general statement can be made regarding the sugar specificity of the lectins; each must be considered on an individual basis. In Table I are recorded the specificities of various purified lectins with respect to blood groups or simple sugars, if such information is available. Except in the case of concanavalin A, little is known about the structural features of the lectin molecule that govern its specificity toward sugars.

E. Distribution within the Plant

In most species of plants, the highest concentrations of lectins are found in the seeds, although they may be present to a lesser extent in the leaves, roots, and stems (Toms and Western, 1971; Rougé, 1975, 1976). There is some indication that lectin-like proteins may be present in the stem and leaves of some plants, which, although not detectable by the usual agglutination test, may cross-react with antibodies to lectin purified from the seed (Talbot and Etzler, 1977). Failure to detect lectins may also be due to the fact that they may be bound to sugar residues of the plant cells; in this instance, extraction with sugar may be necessary (Bowles *et al.*, 1976).

As far as the subcellular distribution of lectins is concerned, they appear to be localized in the cytoplasm of the cotyledon and embryonic cells (Mialonier *et al.*, 1973; Clarke *et al.*, 1975), where, at least in the case of *Ricinus communis,* they are concentrated in the protein bodies (Tully and Beevers, 1976; Youle and Huang, 1976). Lectins are also present in the spherical bodies of intracellular spaces, the cell wall, and sieve-tube sap and as membrane components of dictyosomes, endoplasmic reticulum, and mitochondria (Kauss and Ziegler, 1974; Bowles and Kauss, 1975, 1976; Bowles *et al.*, 1976).

Since the concentration of lectin in mature leaves decreases rapidly as its concentration in the cotyledon increases (Mialonier *et al.*, 1973), lectins are probably synthesized in the leaves and immediately translocated to the developing seed, where they increase in concentration during the course of seed maturation (Howard *et al.*, 1972; Rougé, 1975, 1976). As the seed germinates, another rapid decrease in lectin content occurs (Martin *et al.*, 1964; Subbulakshmi *et al.*, 1976; Chen *et al.*, 1977) at a rate that parallels the loss of reserve protein (Rougé, 1974a,b).

III. SPECIFIC LECTINS

This section is devoted to a consideration of those individual lectins for which a considerable body of information concerning their physicochemical and biological properties exists. In addition to the lectins that have been selected for special treatment, the reader is referred to Table I for a listing of other lectins that have been isolated and partially characterized but for which information is not as complete.

A. Concanavalin A

Because of the ease with which it can be purified by affinity chromatography on Sephadex (Olson and Liener, 1967; Agrawal and Goldstein, 1967) and the remarkable effects that it displays in biological systems, concanavalin A has been the most studied of all lectins. In fact, our knowledge regarding concanavalin A has become so extensive that at least two books devoted exclusively to this subject have been published (Chowdhury and Weiss, 1975; Bittiger and Schnebli, 1976).

Concanavalin A displays the greatest specificity toward polysaccharides and glycoproteins in which α-D-mannopyranosyl and α-D-glucopyranosyl residues are generally located at the terminal, nonreducing end. However, it also binds to mannose residues located in the interior of the polysaccharide molecule (Goldstein *et al.*, 1973). This probably explains why so many glycoproteins are strongly bound to concanavalin A despite the fact that their terminal position is occupied by sialic acid residues (Chase and Miller, 1973; Neri *et al.*, 1974; Pospíšilova *et al.*, 1974). Sepharose-bound concanavalin A has in fact proved to be an extremely useful tool for isolating a wide variety of biologically important glycoproteins (Liener, 1976).

The structure of concanavalin A is now known in great detail due largely to the studies by Edelman's group (Edelman *et al.*, 1972; Wang *et al.*, 1975; Reeke *et al.*, 1975; Cunningham *et al.*, 1975; Becker *et al.*, 1975) and Hardman and Ainsworth (1972, 1976). At neutral pH, concanavalin A exists as a tetramer composed of four identical subunits, each of which has a molecular weight of 26,000. Each subunit is a single polypeptide chain comprised of 237 amino acids of known sequence. Each subunit contains one sugar-binding site and one Mn^{2+}- and one

Ca^{2+}-binding site. These four subunits, which can best be described as flat-based domes, associate to form a dimer, as shown in Fig. 1. This figure also shows the Ca^{2+}- and Mn^{2+}-binding sites, which are located in a surface cavity near the apex of the dome. The exact location of the sugar-binding site has proved to be a highly controversial issue, but recent evidence (Becker *et al.*, 1976) suggests that the saccharide-binding site(s) is quite near the metal-binding sites, as depicted in Fig. 1. There also appears to be a hydrophobic-binding site (I), which binds the hydrophobic moiety of saccharides containing such residues.

The question of the precise role that metal ions play in the activity of concanavalin A has also been the object of considerable study. The fact that the binding of concanavalin A to Sephadex depends on the presence of metal ions (Karlstam, 1973) is certainly an indication of the importance of metal ions for the interaction of this lectin with saccharides. Occupation of the Mn^{2+}-binding site appears to be a prerequisite of Ca^{2+} interaction with the second metal-binding site, and the occupation of both sites is necessary before sugar binding can occur (Kalb and Levitzki, 1968). A variety of other metals can be substituted for Mn^{2+} including cadmium, cobalt, nickel, zinc, lead, and lanthanides (Shoham *et al.*, 1973; Sherry *et al.*, 1975). On the other hand, the metal

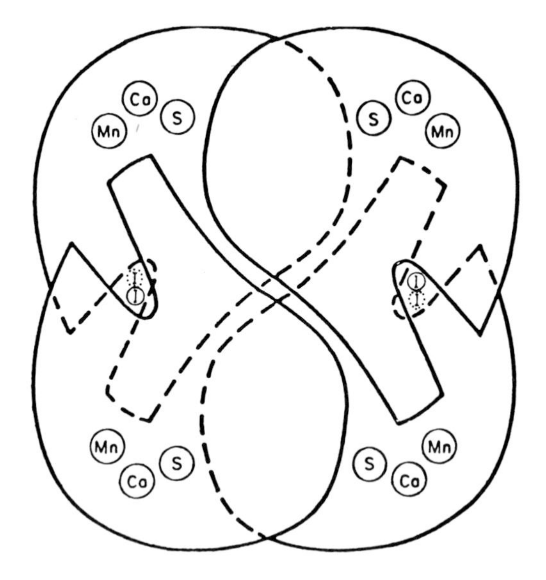

Figure 1. Schematic representation of the tetramer of concanavalin A. Each subunit is approximately 42 × 40 × 39 Å. The manganese and calcium sites are indicated by Mn and Ca, respectively. The saccharide-binding site near the metals is indicated by S and the hydrophobic-binding site in the cavity by I. Taken from Becker *et al.* (1976).

requirement for Ca^{2+} is quite specific. The exact function of the metal ions is not clear, but physical studies (Brewer *et al.*, 1974; Doyle *et al.*, 1975; Grimaldi and Sykes, 1975) have clearly shown that the binding of metal ions by concanavalin A causes profound conformational changes, which in turn influence the sugar-binding properties of the molecule.

The activity of concanavalin A is markedly influenced by pH (So and Goldstein, 1967), temperature (Gordon and Marquardt, 1974; Huet, 1975), and chemical modification (Gunther *et al.*, 1973; Young, 1974). It is significant that, under conditions in which agglutinating activity toward cells or precipitability toward oligosaccharides or glycoproteins (i.e., low pH, low temperature, and blocking of amino groups) is not manifested, concanavalin A exists in the form of a dimer rather than a tetramer. The tetrameric form must therefore be necessary for reactions involving agglutination or precipitation. Although the dimer, because of its reduced valence, can no longer provide the degree of cross-linkage necessary for agglutination, it is still capable of binding low molecular weight sugars (Hassing and Goldstein, 1970) and of binding to cell surfaces (Huet *et al.*, 1974). Agglutination may therefore represent a two-step reaction, the first step involving a monovalent or divalent binding process and a second step involving a multivalent process in which the cross-linking of cell receptor sites results in cellular agglutination.

B. Castor Bean

For many years, it was believed that ricin, the toxic protein from *Ricinus communis,* was responsible for the hemagglutinating activity exhibited by preparations of this protein. It is now clear that ricin and the agglutinin are two distinct proteins, the former being responsible for toxicity and the latter for agglutination. Although ricin and the agglutinin both interact with sugars with terminal galactose units, ricin is inhibited by *N*-acetylgalactosamine, whereas the agglutinin is not inhibited by this sugar (Nicolson *et al.*, 1974). Advantage can be taken of this subtle difference in specificity to elute ricin with *N*-acetylgalactose from agarose, which binds both ricin and the agglutinin, while the agglutinin is eluted with galactose.

Ricin has a molecular weight of about 60,000 and is composed of two polypeptide chains having molecular weights of 32,000 and 34,000, which are held together by disulfide bonds (Olsnes and Pihl, 1972a,b; 1973; Olsnes *et al.*, 1974a,b; Refsnes *et al.*, 1974; Funatsu *et al.*, 1977a,b). The smaller of these two chains, the isoleucine chain, or A chain, inhibits protein synthesis and has been referred to as the "effectomer." The alanine chain, or B chain, on the other hand, has no effect on protein synthesis but serves to bind the toxin to the cell surface via galactose-containing receptor sites. Once bound to the surface of the cell, the A chain (or possibly both chains) is transported to the cytoplasm, where it exerts its toxic effect. The toxic effect of ricin is due to an inhibition of protein synthesis (Lin *et al.*, 1971; Olsnes, 1972), an effect that accounts for its ability to inhibit the growth of tumors (Lin *et al.*, 1970) and virus-transformed cells (Nicolson *et al.*, 1975). The specific target of ricin appears to be the 60 S subunit of the ribosome, where it interferes with peptide chain elongation (Olsnes and Pihl, 1972a; Montanaro *et al.*, 1973, 1975; Greco *et al.*, 1974; Sperti *et al.*, 1973, 1975; Carrasco *et al.*, 1975). Ricin is one of the most toxic substances known. The lethal dose of ricin is about 1 μg per kilogram body weight in the mouse, rat, and dog, whereas the rabbit is about 10 times more sensitive (Olsnes and Pihl, 1978).

The nontoxic castor bean agglutinin has a molecular weight of 120,000 and is a tetramer comprised of two different subunits, one having a molecular weight of 33,000 to 37,000 and the other, a molecular weight of 27,500 to 31,000 (Gürtler and Horstman, 1973; Nicolson *et al.*, 1974; Olsnes *et al.*, 1974a). Chemical and immunochemical evidence (Pappenheimer *et al.*, 1974; Olsnes *et al.*, 1974c; Olsnes and Saltvedt, 1975; Funatsu *et al.*, 1977a) indicates that the heavier of these two chains is probably identical or very homologous to the B chain of ricin, and the smaller chain to the toxic A chain of ricin. Whatever structural difference exists between the A chain of ricin and the lighter chain of the agglutinin must be responsible for the toxicity of the former. More detailed structural information is obviously needed to answer this question, and notable progress is being made in this direction (Funatsu *et al.*, 1977a,b; Kimura *et al.*, 1977).

C. *Phaseolus* Lectins

Tom and Western (1971) present a list of over 500 varieties of *Phaseolus* that have been reported to exhibit hemagglutinating activity. These lectins can be broadly distinguished on the basis of their specificity toward different blood groups. One large group, represented by *P. vulgaris,* reacts nonspecifically with human erythrocytes irrespective of blood groups and is very poorly inhibited by simple sugars. Another group, represented by *P. lunatus,* is specific for type A blood group cells and can be inhibited by simple sugars. There is also a third group, which displays no hemagglutinating activity at all (Brücher *et al.*, 1969; Jaffé *et al.*, 1972), but this could simply be a consequence of a failure to select appropriate cells for testing. In addition to hemagglutinating activity, extracts of some of the legumes possess lectins capable of specifically agglutinating leukocytes and stimulating the mitosis of lymphocytes. Although most of the lectins isolated from *Phaseolus* are similar in size and subunit structure, differences in carbohydrate content, other physicochemical properties, and biological activities have been reported. The extent to which these differences may be due to genetic factors, different methods of isolation, improper identification of species or varieties, or imprecise criteria of homogeneity is difficult to assess.

1. *Red Kidney Bean (Phaseolus vulgaris)*

Early studies on lectins from this bean were characterized by laborious attempts to separate the various components responsible for erythrocyte and leukocyte agglutination and mitogenic activity, and the results were often contradictory and confusing. This problem has been considerably clarified in recent years, however, largely due to studies in the laboratories of Yachnin (Yachnin and Svenson, 1972; Miller *et al.*, 1973, 1975) and Felstedt (Felstedt *et al.*, 1975, 1977; Leavitt *et al.*, 1977). All of the diverse biological activities of kidney bean extracts can be satisfactorily accounted for by a family of five heterogeneous proteins ("isolectins") composed of four subunits designated as L and R, as shown in Fig. 2. Each of these subunits has approximately the same molecular weight, about 34,000, but differs in isoelectric point and to some extent in amino acid sequence. More importantly, however, the L subunit has strong mitogenic activity and a high affinity for receptors of lymphocyte membranes but not for erythrocyte membranes. The R subunit exhibits a converse specificity for lymphocytes and erythrocytes. As a consequence of this difference in specificity, the tetramer with four L subunits (L-PHAP) is a potent leukoagglutinin with weak hemagglutinating activity. The hybrid tetramers with two or more R subunits (2R/2L, 3R/1L, and 4R) exhibit potent hemagglutinating activity and modest leukoagglutinating activity and have therefore been des-

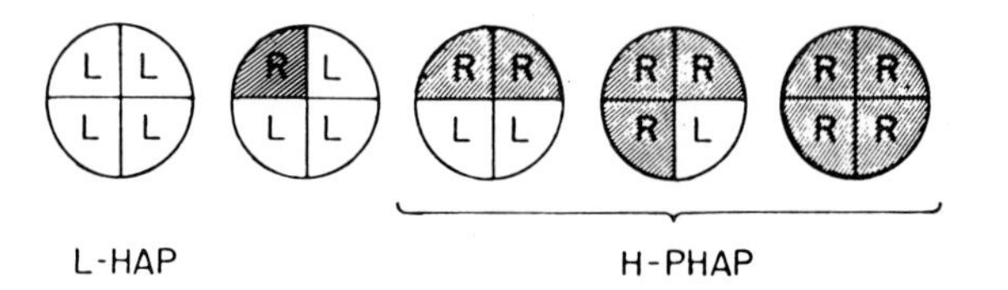

Figure 2. Schematic representation of the tetrameric structure of the five lectins in the red kidney bean, where L and R are the subunits responsible for leukoagglutinating and hemagglutinating activities, respectively. Taken from Miller *et al.* (1975).

ignated as H-PHAP. The hybrid molecules (3R/1L, 2R/1L, and 1R/3L) are mitogenic, indicating that both L and R subunits are required for this kind of activity. The hybrid 1R/3L is devoid of hemagglutinating activity, presumably because it is monovalent and hence does not provide the cross-linkages necessary for agglutination.

Harms-Ringdahl and co-workers (1973, Harms-Ringdahl and Jönvall, 1974) isolated a mitogenic factor from red kidney beans. This factor appears to be quite different from any of the isolectins predicted by Yachnin's model since it has a molecular weight of 10,000 and, unlike the lectins previously described, contains a very high percentage of cystine residues.

2. *Lima Bean* (*Phaseolus lunatus*)

The lima bean agglutinin was the first lectin shown to exhibit blood group anti-A specificity (Boyd and Reguera, 1949). Since its interaction with blood group substance A from human ovarian cyst or hog mucin is inhibited by α- and β-glycosides of *N*-acetyl-D-galactosamine, its specificity is most likely directed to this sugar residue of blood group A substance (Galbraith and Goldstein, 1972). The lima bean lectin may in fact be purified by absorption to polyleucyl blood group A substance followed by elution with *N*-acetyl-D-galactosamine.

Two lima bean lectins have been described with molecular weights of 124,000 and 247,000 (Galbraith and Goldstein, 1972; Bessler and Goldstein, 1974). Both are composed of identical subunits with a molecular weight of 31,000 linked by interchain disulfide bonds to form a larger subunit of 62,000. The two lectins may therefore be regarded as a dimer and tetramer of the 62,000 subunit. The tetramer has four times the agglutinating activity of the dimer and is severalfold more potent as a mitogenic agent (Ruddon *et al.*, 1974). This enhanced activity of the tetramer is no doubt a reflection of the greater valence, which affords the opportunity for more extensive cross-linking of receptor sites.

D. Wheat Germ Agglutinin

Following the serendipitous discovery of Aub *et al.* (1963) that wheat germ lipase contains an agglutinin specific for malignant cells, the lectin responsible for this effect was purified by Burger and Goldberger (1967) and Allen *et al.* (1973). Several laboratories have now reported the purification of wheat germ agglutinin by methods that primarily take advantage of the specificity of this lectin toward *N*-acetyl-D-glucosamine or glycoproteins containing this sugar moiety. For example, the following absorbents have been employed for this purpose: immobilized ovomucoid (Marchesi, 1973), chitin (Bloch and Burger, 1974), and Sepharose-bound derivatives of *N*-acetyl-D-glucosamine (Lotan *et al.*, 1973a; Shaper *et al.*, 1973; Bouchard *et al.*, 1976). Once purified by any of these techniques, the lectin can be readily crystallized, thereby being rendered suitable for X-ray crystallographic studies (Wright, 1974, 1977; Wright *et al.*, 1974).

The amino acid composition of wheat germ agglutinin is characterized by a remarkably high content of glycine residues (22%) and half-cystine residues (18–20%). All of the latter are involved in disulfide bonds rather than free cysteine, thus imparting a high degree of heat stability as well as resistance to denaturing agents. Unlike many other lectins, the wheat germ agglutinin does not contain any covalently bound carbohydrate (Privat *et al.*, 1974a).

At neutral pH, the wheat germ agglutinin is a dimer composed of two identical subunits, each of which has a molecular weight of 18,000 (Nagata and Burger, 1974; Rice and Etzler, 1974). Rather curiously, each subunit has two sugar-binding sites rather than one, and these sites show a higher degree of specificity toward oligomers of β-(1,4)-*N*-acetylglucosamine (Privat *et al.*, 1974b,c) than they do toward the simple sugars, which is indicative of an extended binding site. There is evidence implicating tyrosine, tryptophan, and carboxyl side chains in subunit interactions and agglutinating activity (Rice and Etzler, 1975; Privat and Monsigny, 1975; Jordan *et al.*, 1977). There are indications that the wheat germ lectin may also be capable of binding to sialic acid residues (Janson and Burger, 1973; Greenaway and LeVine, 1973; Adair and Kornfeld, 1974; Winquist *et al.*, 1976).

Until recently, mitogenic activity had not been observed with any preparations of wheat germ agglutinin. Brown *et al.* (1976), however, reported the isolation from wheat germ of a lectin that not only had agglutinating activity but was also mitogenic toward human peripheral lymphocytes. The exact nature of the difference between this mitogen and the more common agglutinin remains to be elucidated.

E. Soybean Agglutinin

The protein responsible for the agglutination of erythrocytes by soybean extracts was first isolated and characterized by Liener and coworkers (Liener and Pallansch, 1952; Pallansch and Liener, 1953, Wada *et al.*, 1958) and more extensively studied in later years by Sharon's group (Lis *et al.*, 1964, 1966a,b, 1969; Gordon *et al.*, 1972; Lotan *et al.*, 1974). The isolation of soybean agglutinin can now be achieved in a relatively simple fashion by affinity chromatography on immobilized derivatives of galactose (Gordon *et al.*, 1972, 1973) or galactosamine (Allen and Neuberger, 1975).

The soybean lectin has a molecular weight of 120,000 and is composed of four subunits. Although each has a molecular weight of 30,000 (Lotan *et al.*, 1974), there appears to be some difference in these subunits as reflected by the ratio of charge to frictional coefficient (Lotan *et al.*, 1975a). Equilibrium dialysis experiments indicate two binding sites per tetramer for *N*-acetylgalactosamine. Thus, the soybean lectin appears to be an example of one of the few lectins in which the number of binding sites is not equal to the number of subunits. Sequence studies have revealed a high degree of homology with the peanut lectin and the L subunit of the kidney bean lectin (Foriers *et al.*, 1977).

Because of its reduced valence in comparison with other lectins, it is perhaps not surprising that the agglutinating and mitogenic activities of the soybean lectin are enhanced by introducing additional cross-linkages by treatment with glutaraldehyde (Lotan *et al.*, 1973b). An increase in agglutinating and mitogenic activities may also occur spontaneously as a result of aggregation when the soybean lectin is stored in a lyophilized condition (Lotan *et al.*, 1975a; Schechter *et al.*, 1976).

The soybean lectin was one of the first plant proteins shown to be a glycoprotein (Wada *et al.*, 1958) whose carbohydrate moiety was subsequently characterized (Lis *et al.*, 1964, 1969). The fact that this carbohydrate is rich in mannose accounts for the fact that it is bound by concanavalin A (Bessler and Goldstein, 1973). Turner and Liener (1975) took advantage of this property to prepare a lectin-free soybean extract for nutritional studies (see Section IV). Periodate oxidation of the carbohydrate moiety of the soybean lectin, followed by reduction with [^{3}H]borohydride, yields a labeled product that retains full activity (Lotan *et al.*, 1975b), demonstrating the nonessentiality of the carbohydrate for biological activity.

On the basis of a detailed examination of its interaction with blood group substances, the soybean lectin was found to be specific for *terminal* α-linked *N*-acetylgalactosamine or to α- or β-D-galactopyranosyl residues (Pereira *et al.*, 1974). This behavior is quite different from that of concanavalin A, which, as pointed out earlier (Section III,A), interacts quite readily with internal sugar residues. That the soybean lectin prefers terminal galactosyl residues is further indicated by the fact that the enzymatic removal of terminal sialic acid residues from the surface of mouse lymphocytes, so as to expose terminal galactosyl residues, renders such cells more responsive to stimulation by the soybean lectin (Novogrodsky and Katchalski, 1973).

F. Pea Lectin

The garden pea (*Pisum sativum*) contains two closely related lectins, which have been purified and characterized in great detail (Entlicher *et al.*, 1970; Paulova *et al.*, 1971a; Marik *et al.*, 1974). These lectins are composed of two heavy and two light chains, which have molecular weights of 17,000 and 7000, respectively (Trowbridge, 1974; Van Driesche *et al.*, 1976). In addition to being a strong agglutinin, the pea lectin is also mitogenic (Trowbridge, 1973).

There are several points of similarity and dissimilarity between the pea lectin and other lectins. It resembles concanavalin A in that it is inhibited by α-glycosides of mannose and glucose, although there are minor differences in detailed specificity (Van Wauwe *et al.*, 1973, 1975; Kaifu *et al.*, 1975). It further resembles concanavalin A in its requirement for Ca^{2+} and Mn^{2+} for activity and in being devoid of carbohydrate. Unlike concanavalin A, however, and despite the fact that it is probably a tetramer, it only has two sugar-binding sites per molecule, thus resembling in this respect the soybean lectin. Also in contrast to concanavalin A, succinylation does not reduce the tetrameric structure to dimers (Trowbridge, 1973). A tryptophan residue that is essential for activity can be found in the heavy chain of the pea lectin in a sequence identical to amino acids 83–90 in concanavalin A (Burés *et al.*, 1972; Čermáková *et al.*, 1976).

G. *Bandeiraea simplicifolia*

Special mention should be made of the lectin from *Bandeiraea simplicifolia* since it represents one of the few lectins that exhibit specificity toward type B human erythrocytes (Hayes and Goldstein, 1974). Its sugar specificity is directed toward terminal nonreducing α-D-galactopyranosyl residues (Ross *et al.*, 1976), and it can be purified by affinity chromatography on an immobilized matrix containing melibiose (Hayes and Goldstein, 1974) or guaran (Horisberger, 1977).

This lectin is actually a family of five tetrameric isolectins with molecular weights of about 114,000 and is comprised of two subunits, A and B, of similar size but of differing sugar specificity (Murphy and Goldstein, 1977). Thus, the model of this lectin is quite analogous to that of the kidney bean (see Fig. 2). The specificity of the A subunit is directed to α-D-*N*-acetylgalactose as well as α-D-galactose, whereas that of the B subunit is directed exclusively toward α-D-galactose. Each subunit has one galactose-binding site, which requires Ca^{2+} for binding activity (Hayes and Goldstein, 1975).

H. Pokeweed Mitogen

One of the most potent plant mitogens is that which is found in the roots of the pokeweed (*Phytolacca americana*). In contrast to most of the other plant mitogens, which are active only with T lymphocytes, the pokeweed mitogen is active toward both T and B cells (Janossy and Graves, 1971, 1972; Shortman *et al.*, 1975). Actually, the pokeweed mitogen can be resolved into five different mitogen proteins, each of which has distinctive physicochemical and biological properties (Waxdal, 1974; Waxdal and Basham, 1974; Yokoyama *et al.*, 1976). The most potent mitogen, Pa-1, has hemagglutinating activity as well and is a polymer of subunits with a molecular weight of 22,000. It is mitogenic for both murine B and T cells. The other four mitogenic proteins (Pa-2 to Pa-5) are monomers with molecular weights ranging from 19,000 to 31,000. Unlike Pa-1, they are nonhemagglutinating and are specific for T cells. Mitogens Pa-2 to Pa-5, unlike Pa-1, are very rich in cystine (18% of the total amino acid residues). Since polymeric antigenic determinants are believed to be important for the stimulation of B cells (Basham and Waxdal, 1975; Janossy *et al.*, 1976), the polymeric nature of Pa-1 probably accounts for its activity toward B cells as well as its hemagglutinating activity. At present, neither the sugar specificity nor the number of binding sites of any of the pokeweed mitogens are known.

Two mitogenic proteins have also been isolated from a Japanese species of pokeweed, *Phytolacca esculenta* (Tokuyama, 1973). Their molecular weights, 18,000 and 32,000, and their high content of cystine suggest that they are similar to the mitogens Pa-2 to Pa-5 described by others from a different species of pokeweed.

I. Potato Lectin

Aside from the wheat germ agglutinins, relatively few lectins have been purified from plants other than legumes. The agglutinin from the potato is described here as an example of a lectin from a nonleguminous plant.

The agglutinin isolated from potatoes by Allen and Neuberger (1973) has some rather unique properties. It has a molecular weight of 46,000 and is presumably a dimer of two identical subunits. Its most unusual feature, however, is the fact that it contains almost 50% carbohydrate and 16% of its amino acids are hydroxyproline and 11.5% half-cystine. In the latter respect, it resembles the wheat germ agglutinin and the pokeweed mitogen, which also have a very high cystine content. The potato lectin displays a specificity toward oligosaccharides containing terminal *N*-acetylglucosamine residues but not toward the monosaccharide unless it is attached to an aromatic aglycone. On the basis of this specificity, the potato lectin can be purified by affinity chromatography on chitin (Van Driesche and Kanarek, 1975) or Sepharose to which *p*-aminobenzyl-1-thio-*N*-acetyl-β-glucosamine has been attached (Delmotte *et al.*, 1975).

IV. NUTRITIONAL SIGNIFICANCE

The fact that lectins are found in those legumes such as beans and peas that constitute an important source of dietary protein for many segments of the world's population raises the question as to their possible nutritional significance. That the toxicity and growth-inhibiting effect produced by feeding animals raw beans, particulary *Phaseolus vulgaris,* is due, at least in part, to the effects of lectins has been well documented elsewhere (Liener, 1974). More recently this conclusion was corroborated by Pusztai and Palmer (1977), who found that kidney bean protein from which the lectin had been removed by affinity chromatography was nontoxic. It should be emphasized, however, that not all lectins are necessarily toxic or even growth inhibitory when fed to animals. For example, Turner and Liener (1975) fed rats a crude soybean extract from which the lectins had been removed by adsorption to concanavalin A (see Section III,E) and were unable to detect any significant improvement in growth performance compared to an extract from which the lectin had not been removed. It is conceivable, of course, that the failure to observe any improvement in nutritive value of the protein in an experiment such as this may be due to a masking effect of other antinutritional factors such as protease inhibitors known to be present in soybeans (Liener and Kakade, 1969).

Jaffé (1960) proposed many years ago that the toxic effect of lectins when ingested orally may be due to their ability to bind receptor sites on the surface of the intestinal epithelial cells. Support for this hypothesis comes from the studies of Etzler (Etzler and Branstrator, 1974; Etzler, 1977), who found that lectins do bind to the crypts and villi of the intestine. Since surface-bound lectins are known to produce profound physiological effects in the cells with which they interact, one of these effects could be a serious impairment in the ability of these cells to absorb nutrients from the gastrointestinal tract, thus causing inhibition of growth and, in extreme cases, even death.

An alternative effect on intestinal cells is suggested by the studies of Jayne-Williams and Burgess (1974), who observed that germ-free Japanese quail were much better able to tolerate the toxic effects of raw navy beans than conventional birds. It was theorized that the binding of lectins to the cells lining the intestine may interfere with the normal defense mechanism of these cells whereby normally innocuous intestinal bacteria are prevented from passing from the lumen of the gut into the lymph, blood, and other tissues of the animal body.

V. PHYSIOLOGICAL FUNCTION IN THE PLANT

Finally, and almost inevitably, we are forced to address ourselves to the all-important question as to the possible role that lectins might play either in the metabolic processes of the plant itself or in its interaction with its environment. The mere fact that lectins make up from 2 to 10% of the protein in most plants does not necessarily mean that these lectins must play an essential physiological role in the plant. Nevertheless, there is now sufficient evidence to suggest that the lectins are not merely adventitious components of plant tissue but may in fact have some very important functions necessary for their survival in a hostile environment and, perhaps, may even be necessary for certain essential growth processes. Proposals advanced to explain the occurrence of lectins in plants have been many and varied and include their involvement in (1) the symbiotic relationship between legumes and nitrogen-fixing bacteria (Bohool and Schmidt, 1974), (2) defense mechanisms against pathogenic bacteria (Sequeira and Graham, 1976), (3) stimulation of pollen germination (Southworth, 1975), and (4) mediation of plant cell extension growth (Kauss and Glaser, 1974).

The possibility that lectins may afford protection against insect predators has received scant attention. There has been one report, however, that points to such a role for lectins. Janzen *et al.* (1976) showed that a lectin purified from black beans (*Phaseolus vulgaris*) was toxic to the larvae of the bruchid beetle (*Callosobruchus maculatus*). The significance of this observation lies in the fact that it may explain why bruchids do not prey on seeds from this particular species of beans. Until more species of legumes and their lectins are tested against their natural insect predators, it is difficult to say how general this phenomenon may be.

There is some evidence to indicate that lectins may serve as mediators of recognition between compatible pollen and the female stigma in plants. The surface of the stigma appears to possess a glycoprotein receptor site whose interaction with the pollen of the same species is blocked by concanavalin A (Watson *et al.*, 1974; Knox *et al.*, 1976). This would suggest that perhaps pollen possess a lectin whose specificity is similar to that of concanavalin A and that it is the specificity of this lectin which is responsible for compatible pollen–stigma interaction.

REFERENCES

Adair, W. L., and Kornfeld, S. (1974). *J. Biol. Chem.* **249,** 4696–4704.

Agrawal, B. B. L., and Goldstein, I. J. (1967). *Biochim. Biophys. Acta* **147,** 262–271.

Allen, A. K., and Neuberger, A. (1973). *Biochem. J.* **135,** 307–314.

Allen, A. K., and Neuberger, A. (1975). *FEBS Lett.* **50,** 362–364.

Allen, A. K., Neuberger, A., and Sharon, N. (1973). *Biochem. J.* **131,** 155–162.

Allen, A. K., Desai, N. N., and Neuberger, A. (1976). *Biochem. J.* **155,** 127–135.

Allen, H. J., and Johnson, E. A. Z. (1976). *Biochim. Biophys. Acta* **444,** 374–385.

Allen, H. J., and Johnson, E. A. Z. (1977). *Carbohydr. Res.* **58,** 253–265.

Andrews, A. T. (1974). *Biochem. J.* **139,** 421–429.

Aspberg, K., Holman, H., and Porath, J. (1968). *Biochim. Biophys. Acta* **160,** 116–117.

Aub, J. C., Tieslau, C., and Lankester, A. (1963). *Proc. Natl. Acad. Sci. U.S.A.* **50,** 613–616.

Balding, P., and Gold, E. R. (1973). *Z. Immunitaetsforsch., Expl. Klin. Immunol.* **145,** 156–165.

Barber, C. B. and Barber, S. (1975). 60th Ann. Meet. Am. Assoc. Cereal Chem., p. 104. Kansas City, Mo.

Basham, T. Y., and Waxdal, M. J. (1975). *J. Immunol.* **114,** 715–716.

Bausch, J. N., and Poretz, R. D. (1976). *Fed. Proc., Fed. Am. Soc. Exp. Biol.* **35,** 888.

Becker, J. W., Reeke, G. N., Wang, J. L., Cunningham, B. A., and Edelman, G. M. (1975). *J. Biol. Chem.* **250,** 1513–1524.

Becker, J. W., Reeke, G. N., Cunningham, B. A., and Edelman, G. M. (1976). *Nature (London)* **259,** 406–409.

Bessler, W., and Goldstein, I. J. (1973). *FEBS Lett.* **34,** 58–62.

Bessler, W., and Goldstein, I. J. (1974). *Arch. Biochem. Biophys.* **165,** 444–445.

Betail, G., and Coulet, M. (1974). *C. R. Seances Soc. Biol. Ses Fils.* **168,** 295–299.

Bittiger, H., and Schnebli, H. P., eds. (1976). "Concanavalin A." Wiley, New York.

Blacik, L. J., Breen, M., Weinstein, H. G., Sittig, R. A., and Cole, M. (1978). *Biochim. Biophys. Acta* **538,** 225–231.

Bloch, R., and Burger, M. M. (1974). *Biochem. Biophys. Res. Commun.* **58,** 13–19.

Bloch, R., Jenkins, J., Roth, J., and Burger, M. M. (1976). *J. Biol. Chem.* **251,** 5929–5935.

Bohlool, B. B., and Schmidt, E. L. (1974). *Science* **185,** 296–271.

Bouchard, P., Moroux, Y., Tixier, R., Privat, J.-P., and Monsigny, M. (1976). *Biochimie* **58,** 1247–1253.

Bourillon, R., and Font, J. (1968). *Biochim. Biophys. Acta* **154,** 28–39.

Bowles, D. J., and Kauss, H. (1975). *Plant Sci. Lett.* **4,** 411–418.

Bowles, D. J., and Kauss, H. (1976). *Biochim. Biophys. Acta* **443,** 360–374.

Bowles, D. J., Schnaurenberger, C., and Kauss, H. (1976). *Biochem. J.* **160,** 375–382.

Boyd, W. C., and Reguera, R. M. (1949). *J. Immunol.* **62,** 333–337.

Boyd, W. C., and Sharpleigh, E. (1954). *Science* **119,** 419.

Brewer, C. F., Marcus, D. M., and Grollman, A. P. (1974). *J. Biol. Chem.* **249,** 4614–4619.

Brown, J. M., Leon, M. A., and Lightbody, J. J. (1976). *J. Immunol.* **117,** 1976–1980.

Brücher, O., Wecksler, M., Levy, A., Palozzo, A., and Jaffé, W. G. (1969). *Phytochemistry* **8,** 1739–1743.

Burés, L., Entlicher, G., and Kocourek, J. (1972). *Biochim. Biophys. Acta* **285,** 235–242.

Burger, M. M., and Goldberg, A. R. (1967). *Proc. Natl. Acad. Sci. U.S.A.* **57,** 359–362.
Callow, J. A. (1975). *Curr. Adv. Plant Sci.* **18,** 181–193.
Carrasco, L., Fernandez-Puentes, C., and Vazquez, D. (1975). *Eur. J. Biochem.* **54,** 499–503.
Cermáková, M., Entlicher, G., and Kocourek, J. (1976). *Biochim. Biophys. Acta* **420,** 236–245.
Chase, P. S., and Miller, F. (1973). *Cell. Immunol.* **6,** 132–139.
Chen, L. H., Thacker, R. R., and Pan, S. H. (1977). *J. Food Sci.* **42,** 1666–1667.
Cheung, G., Haratz, A., Katar, M., and Poretz, R. D. (1975). *Am. Chem. Soc., Chem. Congr. North Am. Cont., 1st,* Abstract No. 19.
Chien, S.-M., Leminski, T., and Poretz, R. D. (1974). *Immunochemistry* **11,** 501–506.
Chien, S.-M., Singla, S., and Poretz, R. D. (1975). *J. Immunol. Methods* **8,** 169–174.
Chowdhury, T. K., and Weiss, K. (1975). "Concanavalin A." Plenum, New York.
Chuba, J. V., Kuhns, W. J., Nigrelli, R. F., Vandenheede, J. R., Osuga, D. T., and Feeney, R. E. (1973). *Nature (London)* **242,** 342–343.
Clarke, A. E., Knox, R. B., and Jermyn, M. A. (1975). *J. Cell Sci.* **19,** 157–167.
Cunningham, B. A., Wang, J. L., Waxdal, M. J., and Edelman, G. M. (1975). *J. Biol. Chem.* **250,** 1503–1512.
DeClerque, A., Van Wauwe, J. P., Dhaese, P., and Loontiens, F. G. (1976). *Arch. Int. Physiol. Biochim.* **84,** 150–151.
Delmotte, F., Kieda, C., and Monsigny, M. (1975). *FEBS Lett.* **53,** 324–330.
Doyle, R. J., Thomasson, D. L., Gray, R. D., and Glew, R. H. (1975). *FEBS Lett.* **52,** 185–187.
Edelman, G. M., Cunningham, B. A., Reeke, G. N., Becker, J. W., Waxdal, M. J. and Wang, J. L. (1972). *Proc. Natl. Acad. Sci. U.S.A.* **69,** 2580–2584.
Ehrlich, P. (1891). *Dtsch. Med. Wochenschr.* **17,** 1218–1222.
Entlicher, G., Kŏsteř, J. V., and Kocourek, J. (1970). *Biochim. Biophys. Acta* **221,** 272–281.
Ersson, B. (1977). *Biochim. Biophys. Acta* **494,** 51–60.
Ersson, B., and Porath, J. (1974). *FEBS Lett.* **48,** 126–129.
Ersson, B., Aspberg, K., and Porath, J. (1973). *Biochim. Biophys. Acta* **310,** 446–452.
Etzler, M. (1977). *FEBS Lett.* **75,** 231–236.
Etzler, M., and Branstrator, M. L. (1974). *J. Cell Biol.* **62,** 329–343.
Etzler, M., Talbot, D. F., and Ziaya, P. R. (1977). *FEBS Lett.* **82,** 39–41.
Felsted, R. L., Leavitt, R. D., and Bachur, N. R. (1975). *Biochim. Biophys. Acta* **405,** 72–81.
Felsted, R. L., Egorin, M. J., Leavitt, R. D., and Bachur, N. R. (1977). *J. Biol. Chem.* **252,** 2967–2971.
Fliegerová, O., Salvetová, A., Tichá, M., and Kocourek, J. (1974). *Biochim. Biophys. Acta* **351,** 416–426.
Font, J., and Bourillon, R. (1971). *Biochim. Biophys. Acta* **243,** 111–116.
Foriers, A., DeNeve, R., and Kanarek, L. (1975). *Arch. Int. Physiol. Biochim.* **83,** 362.
Foriers, A., DeNeve, R., and Kanarek, L. (1976). *Arch. Int. Physiol. Biochim.* **84,** 617–618.
Foriers, A., Wuilmart, C., Sharon, N., and Strosberg, A. D. (1977a). *Biochem. Biophys. Res. Commun.* **75,** 980–986.
Foriers, A., Van Driessche, E., DeNeve, R., Kanarek, L., and Strosberg, A. D. (1977b). *FEBS Lett.* **75,** 237–240.
Fornstedt, N., and Porath, J. (1975). *FEBS Lett.* **57,** 187–191.
Frost, R. G., Reitherman, R. W., Miller, A. L., and O'Brien, J. S. (1975). *Anal. Biochem.* **69** 170–179.
Funatsu, G., Yoshitake, S., and Funatsu, M. (1977a). *Agric. Biol. Chem.* **41,** 1225–1231.
Funatsu, G., Ueno, S., and Funatsu, M. (1977b). *Agric. Biol. Chem.* **41,** 1737–1743.
Galbraith, W., and Goldstein, I. J. (1972). *Biochemistry* **11,** 3976–3984.

Gold, E. R., and Balding, P. (1975). "Receptor-Specific Proteins. Plant and Animal Lectins." Am, Elsevier, New York.
Goldstein, I. J., Reichert, C. M., Misaki, A., and Gorin, P. A. J. (1973). *Biochim. Biophys. Acta* **317,** 500–504.
Gordon, J. A., and Marquardt, M. D. (1974). *Biochim. Biophys. Acta* **332,** 136–144.
Gordon, J. A., Blumberg, S., Lis, H., and Sharon, N. (1972). *FEBS Lett.* **24,** 193–196.
Gordon, J. A., Blumberg, S., Lis, H., and Sharon, N. (1973). *In* "Methods in Enzymology" (V. Ginsburg, ed.), Vol. 28, Part B, pp. 365–368. Academic Press, New York.
Greco, M., Montanaro, L., Novello, F., Saccone, C., Sperti, S., and Stripe, F. (1974). *Biochem. J.* **142,** 695–697.
Greenaway, P. J., and LeVine, D. (1973). *Nature (London) New Biol.* **241,** 191–192.
Grimaldi, J. J., and Sykes, B. D. (1975) . *J. Biol. Chem.* **250,** 1618–1624.
Gunther, G. R., Wang, J. L., Yahara, I., Cunningham, B. A., and Edelman, G. M. (1973). *Proc. Natl. Acad. Sci. U.S.A.* **70,** 1012–1016.
Gürtler, L. G., and Horstmann, H. J. (1973). *Biochim. Biophys. Acta* **295,** 582–594.
Hardman, K. D., and Ainsworth, C. F. (1972). *Biochemistry* **11,** 4910–4919.
Hardman, K. D., and Ainsworth, C. F. (1976). *Biochemistry* **15,** 1120–1128.
Harms-Ringdahl, M., and Jörnvall, H. (1974). *Eur. J. Biochem.* **48,** 541–547.
Harms-Ringdahl, M., Fedorcsák, I., and Ehrenberg, L. (1973). *Proc. Natl. Acad. Sci. U.S.A.* **70,** 569–573.
Hassing, G. S., and Goldstein, I. J. (1970). *Eur. J. Biochem.* **16,** 549–556.
Hayes, C. E., and Goldstein, I. J. (1974). *J. Biol. Chem.* **249,** 1904–1914.
Hayes, C. E., and Goldstein, I. J. (1975). *J. Biol. Chem.* **250,** 6837–6840.
Hellin, H. (1891). Dissertation, Dorpat.
Hořejsí, V., and Kocourek, J. (1974). *Biochim. Biophys. Acta* **336,** 329–337.
Horisberger, M. (1977). *Carbohydr. Res.* **53,** 231–237.
Howard, I. K., Sage, H. J., and Horton, C. B. (1972). *Arch. Biochem. Biophys.* **149,** 323–326.
Howard, J., and Shannon, L. (1977). *Anal. Biochem.* **79,** 234–239.
Huet, C., Lonchampt, M., Huet, M., and Bernadac, A. (1974). *Biochim. Biophys. Acta* **365,** 28–39.
Huet, M. (1975). *Eur. J. Biochem.* **59,** 627–632.
Irimura, T., and Osawa, T. (1972). *Arch. Biochem. Biophys.* **151,** 475–482.
Irimura, T., Kawaguchi, T., Terao, T., and Osawa, T. (1975). *Carbohydr. Res.* **39,** 317–327.
Jaffe, C. L., Ehrlich-Rogozinski, S., Lis, H., and Sharon, N. (1977). *FEBS Lett.* **82,** 191–196.
Jaffé, W. G. (1960). *Arzneim.-Forsch.* **12,** 1012–1016.
Jaffé, W. G., and Gaede, K. (1959). *Nature (London)* **183,** 1329–1330.
Jaffé, W. G., and Hannig, K. (1965). *Arch. Biochem. Biophys.* **109,** 80–91.
Jaffé, W. G., Brücher, O., and Palozzo, A. (1972). *Z. Immunitaetsforsch., Exp. Klin. Immunol.* **142,** 439–447.
Janossy, G., and Greaves, M. F. (1971). *Clin. Exp. Immunol.* **9,** 483–498.
Janossy, G., and Greaves, M. F. (1972). *Clin. Exp. Immunol.* **10,** 525–536.
Janossy, G., De La Coneha, E. G., and Waxdal, M. J. (1976). *Clin. Exp. Immunol.* **26,** 108–117.
Janson, V. K., and Burger, M. M. (1973). *Biochim. Biophys. Acta* **291,** 127–135.
Janzen, D. H., Juster, H. B., and Liener, I. E. (1976). *Science* **192,** 795–796.
Jayne-Williams, D. J., and Burgess, C. D. (1974). *J. Appl. Bacteriol.* **37,** 149–169.
Jordan, F., Bassett, E., and Redwood, W. R. (1977). *Biochem. Biophys. Res. Commun.* **75,** 1015–1021.
Kabat, E. A., and Mayer, M. M. (1961). "Experimental Immunology," pp. 114–115. Thomas, Springfield, Illinois.

Kaifu, R., Osawa, T., and Jeanloz, R. W. (1975). *Carbohydr. Res.* **40,** 111–117.
Kalb, A. J. (1968). *Biochim. Biophys. Acta* **168,** 532–536.
Kalb, A. J., and Levitzki, A. (1968). *Biochem. J.* **109,** 669–672.
Karlstam, B. (1973). *Biochim. Biophys. Acta* **329,** 295–304.
Kauss, H., and Glaser, C. (1974). *FEBS Lett.* **45,** 304–307.
Kauss, H., and Ziegler, H. (1974). *Planta* **121,** 197–200.
Kawaguchi, T., Matsumoto, I., and Osawa, T. (1974). *J. Biol. Chem.* **249,** 2786–2792.
Kimura, M., Funatsu, G., and Funatsu, M. (1977). *Agric. Biol. Chem.* **41,** 1733–1736.
Knox, R. B., Clarke, A., Harrison, S., Smith, P., and Marchalonis, J. J. (1976). *Proc. Natl. Acad. Sci. U.S.A.* **73,** 2788–2792.
Kristiansen, T. (1974). *Biochim. Biophys. Acta* **338,** 246–253.
Kristiansen, T. (1975). *Protides Biol. Fluids, Proc. Colloq.* **23,** 663–665.
Kristiansen, T., and Porath, J. (1968). *Biochim. Biophys. Acta* **158,** 351–357.
Kurokawa, T., Tsuda, M., and Sugino, Y. (1976). *J. Biol. Chem.* **251,** 5686–5693.
Landsteiner, K., and Raubitschek, H. (1908). *Zentralbl. Bakteriol., Parastitenkd., Infektionskr. Hyg., Abt. 1: Orig.,* **45,** 660–664.
Leavitt, R. D., Felsted, R. L., and Bacher, N. R. (1977). *J. Biol. Chem.* **252,** 2961–2966.
Lee, J. K. N., Pachtman, E. A., and Frumin, A. M. (1974). *Ann. N.Y. Acad. Sci.* **234,** 162–168.
Leseny, A. M., Bourillon, R., and Kornfeld, S. (1972). *Arch. Biochem. Biophys.* **153,** 831–836.
Liener, I. E. (1955). *Arch. Biochem. Biophys.* **54,** 223–231.
Liener, I. E. (1974). *J. Agric. Food Chem.* **22,** 17–22.
Liener, I. E. (1976). *Annu. Rev. Plant Physiol.* **27,** 291–319.
Liener, I. E., and Kakade, M. L. (1969). *In* "Toxic Constituents of Plant Foodstuffs" (I. E. Liener, ed.), pp. 8–68. Academic Press, New York.
Liener, I. E., and Pallansch, M. J. (1952). *J. Biol. Chem.* **197,** 29–36.
Lin, J.-Y., Tserng, K.-Y., Chen, C. C., Lin, L.-T., and Tung, T.-C (1970). *Nature* **227,** 292–293.
Lin, J.-Y., Shaw, Y.-S., and Tung, T.-C. (1971). *Toxicon* **9,** 97–101.
Lis, H., and Sharon, N. (1973). *Annu. Rev. Biochem.* **42,** 541–574.
Lis, H. and Sharon, N. (1977). *In* "The Antigens" (M. Sela, ed.), Vol. 4, pp. 429–529. Academic Press, New York.
Lis, H., Sharon, N., and Katchalski, E. (1964). *Biochim. Biophys. Acta* **83,** 376–378.
Lis, H., Sharon, N., and Katchalski, E. (1966a). *J. Biol. Chem.* **241,** 684–689.
Lis, H., Fridman, C., Sharon, N., and Katchalski, E. (1966b). *Arch. Biochem. Biophys.* **117,** 301–309.
Lis, H., Sharon, N., and Katchalski, E. (1969). *Biochim. Biophys. Acta* **192,** 364–366.
Lisowska, E., Szeliga, W., and Duk, M. (1976). *FEBS Lett.* **72,** 327–330.
Löngren, J., Goldstein, I. J., and Zand, R. (1976). *Biochemistry* **15,** 436–440.
Lotan, R., Gussin, A. E. S., Lis, H., and Sharon, N. (1973a). *Biochem. Biophys. Res. Commun.* **52,** 656–662.
Lotan, R., Lis, H., Rosenwasser, A., Novogrodsky, A., and Sharon, N. (1973b). *Biochem. Biophys. Res. Commun.* **55,** 1347–1355.
Lotan, R., Siegelman, H. W., Lis, H., and Sharon, N. (1974). *J. Biol. Chem.* **249,** 1219–1224.
Lotan, R., Cacon, R., Cacon, M., Debray, H., Carter, W. G., and Sharon, N. (1975a). *FEBS*
Lotan, R., Debray, H., Cacan, M., Cacan, R., and Sharon, N. (1975b). *J. Biol. Chem.* **250,** 1955–1957.
Lotan, R., Skutelsky, R., Damon, D., and Sharon, N. (1975c). *J. Biol. Chem.* **250,** 8518–8523.

Lutsik, M. D., Potapov, M. I., and Kirichenko, N. V. (1977). *Probl. Gematol. Pereliv. Krovi* **22,** 48–52.
Marchesi, V. T. (1973). *In* "Methods in Enzymology" (V. Ginsburg, ed.), Vol. 28, Part B, pp. 354–356. Academic Press, New York.
Marik, T., Entlicher, G., and Kocourek, J. (1974). *Biochim. Biophys. Acta* **336,** 53–61.
Martin, F. W., Waszczenko-Zacharczenko, E., Boyd, W. C., and Schertz, K. F. (1964). *Ann. Bot. (London)* **28,** 219–324.
Matsumoto, I., and Osawa, T. (1970). *Arch. Biochem. Biophys.* **140,** 484–491.
Matsumoto, I., and Osawa, T. (1972). *Biochem. Biophys. Res. Commun.* **46,** 1810–1815.
Matsumoto, I., and Osawa, T. (1974). *Biochemistry* **13,** 582–588.
Mialonier, G., Privat, J.-P., Monsigny, M., Kahlen, G., and Durrand, R. (1973). *Physiol. Veg.* **11,** 519–537.
Miller, J. B., Noyes, C., Heinrikson, R., Kingdon, H. S., and Yachnin, S. (1973). *J. Exp. Med.* **138,** 939–951.
Miller, J. B., Hsu, R., Heinrikson, R., and Yachnin, S. (1975). *Proc. Natl. Acad. Sci. U.S.A.* **72,** 1388–1391.
Montanaro, L., Sperti, S., and Stirpe, F. (1973). *Biochem. J.* **136,** 677–683.
Montanaro, L., Sperti, S., Mattioli, A., Testoni, G., and Stirpe, F. (1975). *Biochem. J.* **146,** 127–131.
Murphy, L. A., and Goldstein, I. J. (1977). *J. Biol. Chem.* **252,** 4739–4742.
Nagata, Y., and Burger, M. M. (1974). *J. Biol. Chem.* **249,** 3116–3122.
Neri, G., Smith, D. F., Gilliam, E. B., and Walborg, E. F., Jr. (1974). *Arch. Biochem. Biophys.* **165,** 323–330.
Nicolson, G. L., Blaustein, J., and Etzler, M. E. (1974). *Biochemistry* **13,** 196–203.
Nicolson, G. L., Lacorbiere, M., and Hunter, T. R. (1975). *Cancer Res.* **35,** 144–155.
Novogrodsky, A., and Katchalski, E. (1973). *Proc. Natl. Acad. Sci. U.S.A.* **70,** 2515–2518.
Nowakova, N., and Kocourek, J. (1974). *Biochim. Biophys. Acta* **359,** 320–333.
Nowell, P. C. (1960). *Cancer Res.* **20,** 462–468.
Olsnes, S. (1972). *Naturwissenshaften* **59,** 497–502.
Olsnes, S., and Pihl, A. (1972a). *FEBS Lett.* **20,** 327–329.
Olsnes, S., and Pihl, A. (1972b). *Nature (London)* **238,** 459–461.
Olsnes, S., and Pihl, A. (1973). *Biochemistry* **12,** 3121–3126.
Olsnes, S., and Pihl, A. (1978). *Trends Biochem. Sci.* **3,** 7–10.
Olsnes, S., and Saltvedt, E. (1975). *J. Immunol.* **14,** 1743–1748.
Olsnes, S., Saltvedt, E., and Pihl, A. (1974a). *J. Biol. Chem.* **249,** 803–810.
Olsnes, S., Refsnes, K., and Pihl, A. (1974b). *Nature (London)* **249,** 627–631.
Olsnes, S., Pappenheimer, A. M., Jr., and Meren, R. (1974c). *J. Immunol.* **113,** 842–847.
Olson, M. O. J., and Liener, I. E. (1967). *Biochemistry* **6,** 105–111.
Osawa, T., and Matsumoto, I. (1973). *In* "Methods In Enzymology" (V. Ginsburg, ed.), Vol. 28, Part B, pp. 323–327. Academic Press, New York.
Pacak, F., and Kocourek, J. (1975). *Biochim. Biophys. Acta* **400,** 371–386.
Pallansch, M. J., and Liener, I. E. (1953). *Arch. Biochem. Biophys.* **45,** 366–374.
Pappenheimer, A. M., Jr., Olsnes, S., and Harper, A. A. (1974). *J. Immunol.* **113,** 835–841.
Partridge, J., Shannon, L., and Gumpf, D. (1976). *Biochim. Biophys. Acta* **451,** 470–483.
Paulova, M., Entlicher, G., Tichá, M., Koštíř, J. V., and Kocourek, J. (1971a). *Biochim. Biophys. Acta* **237,** 513–518.
Paulova, M., Tichá, M., Entlicher, G., Koštíř, J. V., and Kocourek, J. (1971b). *Biochim. Biophys. Acta* **252,** 388–395.
Pellegrino, M. A., Furone, S., Pellegrino, A., and Reisfeld, R. A. (1973). *Clin. Immunol. Immunopathol.* **2,** 67–73.

Pére, M., Bourillon, R., and Jirgensons, B. (1975). *Biochim. Biophys. Acta* **393,** 31–36.
Pereira, M. E. A., and Kabat, E. A. (1974a). *Biochemistry* **13,** 3184–3192.
Pereira, M. E. A., and Kabat, E. A. (1974b). *Ann. N.Y. Acad. Sci.* **234,** 301–309.
Pereira, M. E. A., Kabat, E. A., and Sharon, N. (1974). *Carbohydr. Res.* **37,** 89–102.
Petryniak, J., Pereira, M. E. A., and Kabat, E. A. (1977). *Arch. Biochem. Biophys.* **178,** 118–134.
Pflumm, M. N., Wang, J. L., and Edelman, G. M. (1971). *J. Biol. Chem.* **246,** 4369–4375.
Pflumm, M. N., Wang, J. L., and Edelman, G. M. (1971). *J. Biol. Chem.* **246,** 4369–4375.
Poretz, R. D. (1973). In "Methods in Enzymology" (V. Ginsburg, ed.), Vol. 28, Part B, pp. 349–354. Academic Press, New York.
Poretz, R. D., and Barth, F. (1976). *Immunology* **31,** 187–194.
Poretz, R. D., Riss, H., Timberlake, J. W., and Chien, S.-M. (1974). *Biochemistry* **13,** 250–256.
Pospíšilova, J., Haškovic, C., Entlicher, G., and Kocourek, J. (1974). *Biochim. Biophys. Acta* **373,** 444–452.
Prigent, M. J., and Bourillon, R. (1976). *Biochim. Biophys. Acta* **420,** 112–121.
Privat, J.-P., and Monsigny, M. (1975). *Eur. J. Biochem.* **60,** 555–567.
Privat, J.-P., Delmotte, F., Mialonier, G., Bouchard, P., and Monsigny, M. (1974a). *Eur. J. Biochem.* **47,** 5–14.
Privat, J.-P., Delmotte, F., and Monsigny, M. (1974b). *FEBS Lett.* **46,** 224–227.
Privat, J.-P., Delmotte, F., and Monsigny, M. (1974c). *FEBS Lett.* **46,** 229–232.
Pusztai, A., and Palmer, R. (1977). *J. Sci. Food Agric.* **28,** 620–623.
Pusztai, A., and Watt, W. B. (1974). *Biochim. Biophys. Acta* **365,** 57–71.
Reeke, G. N., Becker, J. W., and Edelman, G. M. (1975). *J. Biol. Chem.* **250,** 1525–1547.
Refsnes, K., Olsnes, S., and Pihl, A. (1974). *J. Biol. Chem.* **249,** 3557–3562.
Rice, R. H., and Etzler, M. E. (1974). *Biochem. Biophys. Res. Commun.* **59,** 414–419.
Rice, R. H., and Etzler, M. E. (1975). *Biochemistry* **14,** 4093–4099.
Ross, T. T., Hayes, C. E., and Goldstein, I. J. (1976). *Carbohydr. Res.* **47,** 91–97.
Rougé, P. (1974a). *C. R. Hebd. Seances Acad. Sci., Ser. D* **278,** 449–452.
Rougé, P. (1974b). *C. R. Hebd. Seances Acad. Sci., Ser. D* **278,** 3083–3087.
Rougé, P. (1975). *C. R. Hebd. Seances Acad. Sci., Ser. D* **280,** 2105–2108.
Rougé, P. (1976). *C. R. Hebd. Seances Acad. Sci., Ser. D* **283,** 1823–1825.
Rougé, P., Chatelain, C., and Pére, D. (1977). *Planta Med.* **31,** 141–145.
Rudden, R. W., Weisenthal, L. M., Lundeen, D. E., Bressler, W., and Goldstein, I. J. (1974). *Proc. Natl. Acad. Sci. U.S.A.* **71,** 1848–1851.
Rüdiger, H. (1977). *Eur. J. Biochem.* **72,** 317–322.
Sachs, V., Carstens, U., and Szirmai, E. (1976). *Agressologie* **17,** 357–360.
Schechter, B., Lis, H., Lotan, R., Novogrodsky, A., and Sharon, N. (1976). *Eur. J. Immunol.* **6,** 145–149.
Sela, B.-M., Lis, H., Sharon, N., and Sachs, L. (1973). *Biochim. Biophys. Acta* **310,** 273–277.
Sequeira, L., and Graham, T. L. (1976). *Physiol. Plant Pathol.* **3,** 233–242.
Shaper, J. H., Barker, R., and Hill, R. L. (1973). *Anal. Biochem.* **53,** 564–570.
Sharif, A., Brochier, J., and Bourillon, R. (1977). *Cell. Immunol.* **31,** 302–310.
Sharon, N. (1977). *Sci. Am.* **236,** 108–119.
Sharon, N., and Lis, H. (1972). *Science* **177,** 949–959.
Sherry, A. D., Newman, A. D., and Gutz, C. G. (1975). *Biochemistry* **14,** 2191–2196.
Shimazaki, K., Walborg, E. F., Jr., Neri, G., and Jirgensons, B. (1975). *Arch. Biochem. Biophys.* **169,** 731–736.
Shoham, M., Kalb, A. J., and Pecht, I. (1973). *Biochemistry* **12,** 1914–1917.

Shortman, K., von Bochmer, A., Lipp, J., and Hopper, K. (1975). *Transplant. Rev.* **25,** 163–167.
So, L. L., and Goldstein, I. J. (1967). *J. Biol. Chem.* **242,** 1617–1622.
Southworth, D. (1975). *Nature (London)* **258,** 600–602.
Sperti, S., Montanaro, L., Mattioli, A., and Stirpe, F. (1973). *Biochem. J.* **136,** 813–815.
Sperti, S., Montanaro, L., Mattioli, A., and Testoni, G. (1975). *Biochem. J.* **148,** 447–451.
Stillmark, H. (1888). Dissertation, Dorpat.
Subbulakshmi, G., Kumar, K. G., and Venkatamaran, L. V. (1976). *Nutr. Rep. Int.* **13,** 19–31.
Sumner, J. B. (1919). *J. Biol. Chem.* **37,** 137–144.
Sumner, J. B., and Howell, I. F. (1936). *J. Bacteriol.* **32,** 227–237.
Sundberg, L., Porath, J., and Aspberg, K. (1970). *Biochim. Biophys. Acta* **221,** 394–395.
Tajima, T., Tokita, T., Sakurai, Y., Gotoh, A., and Ikemoto, S. (1977). *Chikusan No Kenkyu* **31,** 423–424.
Takahashi, T., and Liener, I. E. (1968). *Biochim. Biophys. Acta* **154,** 160–164.
Takahashi, T., Ramachandramurthy, P., and Liener, I. E. (1967). *Biochim. Biophys. Acta* **133,** 123–133.
Takahashi, T., Yamada, N., Iwamoto, K., Shimabayashi, Y., and Izutsu, K. (1973). *Agric. Biol. Chem.* **37,** 29–36.
Takahashi, T., Yagi, T., Oda, T., and Liener, I. E. (1974). *Agric. Biol. Chem.* **38,** 865–867.
Takahashi, T., Shimabayashi, Y., and Oda, T. (1976). *Agric. Biol. Chem.* **40,** 2027–2031.
Talbot, C. F., and Etzler, M. E. (1977). *Fed. Proc., Fed. Am. Soc. Exp. Biol.* **36,** 2755.
Terao, T., and Osawa, T. (1973). *J. Biochem. (Tokyo)* **74,** 199–201.
Tokuyama, H. (1973). *Biochim. Biophys. Acta* **317,** 338–350.
Toms, G. C., and Western, A. (1971). *In* "Chemotaxonomy of Leguminosae" (J. B. Harborne, D. Boulter, and B. L. Turner, eds.), pp. 367–462. Academic Press, New York.
Toyoshima, S., and Osawa, T. (1975). *J. Biol. Chem.* **250,** 1655–1660.
Toyoshima, S., Akiyama, Y., Nakano, K., Tonomura, A., and Osawa, T. (1971). *Biochemistry* **10,** 4457–4463.
Trowbridge, I. S. (1973). *Proc. Natl. Acad. Sci. U.S.A.* **70,** 3650–3654.
Trowbridge, I. S. (1974). *J. Biol. Chem.* **249,** 6004–6012.
Tully, R. E., and Beevers, H. (1976). *Plant Physiol.* **58,** 710–715.
Turner, R. H., and Liener, I. E. (1975). *J. Agric. Food Chem.* **23,** 484–487.
Ulevitch, R. J., Jones, J. M., and Feldman, J. D. (1974). *Prep. Biochem.* **4,** 273–281.
Van Driesche, E., and Kanarek, L. (1975). *Arch. Int. Physiol. Biochim.* **83,** 414–415.
Van Driesche, E., Foriers, A., Shosberg, A. D., and Kanarek, L. (1976). *FEBS Lett.* **71,** 220–222.
Van Wauwe, J.-P., Loontiens, F. G., Carchon, H. A., and De Bruyne, C. K. (1973). *Carbohydr. Res.* **30,** 249–256.
Van Wauwe, J.-P., Loontiens, F. G., and De Bruyne, C. K. (1975). *Biochim. Biophys. Acta* **379,** 456–461.
Wada, S., Pallansch, M. J., and Liener, I. E. (1958). *J. Biol. Chem.* **233,** 395–400.
Wang, J. L., Becker, J. W., Reeke, G. N., and Edelman, G. M. (1974). *J. Mol. Biol.* **88,** 259–262.
Wang, J. L., Cunningham, B. A., Waxdal, M. J., and Edelman, G. M. (1975). *J. Biol. Chem.* **250,** 1490–1502.
Watson, L., Knox, R. B., and Creaser, E. H. (1974). *Nature (London)* **249,** 574–576.
Waxdal, M. (1974). *Biochemistry* **13,** 3671–3676.
Waxdal, M., and Basham, T. Y. (1974). *Nature (London)* **251,** 163–164.

Winquist, L., Eriksson, L., Dallner, G., and Ersson, B. (1976). *Biochem. Biophys. Res. Commun.* **68,** 1020–1026.

Wright, C. S. (1974). *J. Mol. Biol.* **87,** 835–841.

Wright, C. S. (1977). *J. Mol. Biol.* **111,** 439–441.

Wright, C. S., Keith, C., Nagata, Y., Burger, M. M., and Longridge, R. (1974). *J. Mol. Biol.* **87,** 843–846.

Yachnin, S., and Svenson, R. H. (1972). *Immunology* **22,** 871–883.

Yariv, J., Kalb, A. J., and Blumberg, S. (1973). *In* "Methods in Enzymology" (V. Ginsburg, ed.), Vol. 28, Part B, pp. 356–360. Academic Press, New York.

Yokoyama, K., Yano, O., Terao, T., and Osawa, T. (1976). *Biochim. Biophys. Acta* **427,** 443–452.

Youle, R. J., and Huang, A. H. C. (1976). *Plant Physiol.* **58,** 703–709.

Young, N. M. (1974). *Biochim. Biophys. Acta* **336,** 46–52.

Ziska, P. (1976). *Acta Biol. Med. Ger.* **35,** 1575–1576.

Chapter 17

Proteinase Inhibitors

C. A. RYAN

I. INTRODUCTION

Natural proteinase inhibitors are proteins or polypeptides that bind very specifically and tightly to enzymes that split peptide bonds of proteins, resulting in the inhibition of the proteolytic activities of these enzymes. Natural proteinase inhibitors are found throughout all life forms, where they are significant components of cytoplasm, secretions, and intercellular fluids of many organs and tissues. A number of reviews and books concerning the major advances in the field have been published within the last few years (Birk, 1976; Dechary, 1970; Fritz and Tschesche, 1971; Fritz *et al.*, 1974; Kassell, 1970; Laskowski and Sealock, 1971; Liener and Kakade, 1969; Means *et al.*, 1974; Peeters, 1975; Puztai, 1967; Reich *et al.*, 1975; Richardson, 1977; Royer, 1975; Ryan, 1973; Tschesche, 1974; Umezawa, 1976; Vogel *et al.*, 1968; Werle and Zickgraf-Rüdel, 1972; Whitaker and Feeney, 1973). The majority of proteinase inhibitor research is relatively new, and an increasing number of reports dealing with the occurrence, structure, and functional roles of these inhibitors in living systems have appeared.

The major functions of proteinase inhibitors appear to be either regulation of proteolytic enzymes or protection of tissues or fluids from proteolytic attack, although it is often difficult to clearly separate roles of regulation from protection. Many interesting biochemical and physiological processes that involve proteinase inhibitors have been studied

ISBN 0-12-597180-X

(Fritz *et al.*, 1974; Tschesche, 1974; Peeters, 1975), such as their role in protecting the pancreas from autodigestion, in regulating and protecting blood proteins, and in fertilization. Proteinase inhibitors have been found in the venom of certain snakes and insects (Takahashi *et al.*, 1974; Strydom, 1973; Shkenderov, 1973), where they apparently protect proteinaceous toxins from degradation by the victim's proteinases.

In addition to the large proteinaceous inhibitors is a class of low molecular weight inhibitors isolated from culture filtrates of actinomycetes (Umezawa, 1976). These inhibitors usually contain three or more amino acids, peptide in nature and having an aldehyde at the C-terminal end, that effectively inhibit various proteinases, depending on the R group of the C-terminal aldehyde. Since these inhibitors are secreted by microbes and differ from the large proteinaceous or polypeptide inhibitors in their mode of action, they will not be further treated in this chapter. However, they have been interesting and valuable in both medical and physiological studies dealing with proteolytic enzymes of animal and bacterial origins.

II. METHODS OF ASSAY

The most common methods for determining the concentration of proteinase inhibitors are based on the ability of the inhibitors to complex with proteolytic enzymes. Loss of proteolytic enzyme activity is the most frequently employed assay method. It is conducted by following either the hydrolysis of artificial ester or peptide substrates spectrophotometrically or the digestion of protein substrates by various techniques (Tan and Stevens, 1971; Bergmeyer, 1970). Ordinarily any method for quantitating proteolytic activity can be employed to determine inhibitor activity. Synthetic substrates are very convenient since they can often be used with very small quantities of enzymes and inhibitors; most assays can be performed with just a few microliters of either crude extracts or purified inhibitors.

For quantitative estimates of inhibitor concentrations, it is necessary to know the exact enzyme concentrations. This requires the titration of the active site of the enzyme under study. Some titrants are known for enzymes with trypsin and chymotrypsin specificities (Kezdy and Kaiser, 1970; Chase and Shaw, 1970), but with some enzymes such titrants are not available and the exact stoichiometry of enzyme–inhibitor interactions cannot always be determined easily. In titrating an enzyme with natural proteinase inhibitors to determine stoichiometry, the titration is usually carried out with inhibitor concentrations that inhibit the enzyme less than 50% and then extrapolated to 100% inhibition since, at higher inhibitor concentrations, the relationship often becomes nonlinear.

Methods for estimating inhibitor concentrations using measurements other than enzyme activity can be employed, such as physicochemical methods involving, for example, ultracentrifugation or gel filtration chromatography. These methods ordinarily require several milligrams of pure inhibitor protein. Immunological methods have been used successfully to study proteinase–inhibitor interactions, and very small amounts of inhibitor protein can usually be assayed (Ryan, 1977). However, several milligrams of the highly pure inhibitors are necessary to inject into animals to produce the antibodies.

Assays of proteinase inhibitors are ordinarily done indirectly by measuring loss of enzyme activity. Since several inhibitor species may be present in crude, partially purified, or even pure preparations, the study of individual inhibitors using enzyme assays can be a problem. For example, in the potato, at least seven different inhibitor species have been identified, and several of these have the same specificities (Richardson, 1977). Immunological methods have proved to be very helpful in specifically identifying and quantitating the various inhibitors or "isoinhibitors" that are present in both legumes and potatoes (Ryan, 1977).

III. ISOLATION OF NATURAL PROTEINASE INHIBITORS

The isolation of proteinase inhibitors from various tissues and fluids is performed routinely by modern techniques of protein separation. Since the inhibitors are mostly typical proteins, glycoproteins, or polypeptides and many are unusually stable to extremes of pH and temperature, they do not ordinarily present problems in isolation that are not encountered with other proteins. However, isolation of plant proteinase inhibitors can be complicated by their high degree of heterogeneity in certain tissues that contain "isoinhibitor" variants, which usually differ by only a few amino acids. Examples of inhibitors with isoinhibitor forms are the lima bean inhibitor, which contains at least six isoinhibitor variants (Tan and Stevens, 1971), and potato inhibitors I and II, both of which are composed of isoinhibitor subunits (Melville and Ryan, 1972; Richardson and Cossins, 1974; Bryant *et al.*, 1976; Iwazaki *et al.*, 1972). Both ion-exchange and isoelectric focusing have been used to successfully separate the isoinhibitors.

Another problem encountered with plant tissues is the formation of polyphenols when the tissues are crushed or homogenized. Inclusion of reducing agents can be helpful, or the colored materials can sometimes be removed with DEAE-Sephadex or polyvinylpyrrolidone. Extraction of tissues with buffers of low pH can also retard formation of the colored substances.

Several techniques are used commonly for preliminary separations of inhibitor-rich protein fractions, such as gel filtration and ion-exchange chromatography. Affinity chromatography with immobilized enzymes can be very useful. With some inhibitors, however, the binding is so strong that unusual measures must be taken to remove them from the columns, for example, by elution with a high concentration of urea or guanidine. Care must be taken not to denature the inhibitors or the immobilized enzymes. In addition, some success in purifying inhibitors has been achieved with preparative electrophoresis and preparative isoelectric focusing procedures.

Many of the plant proteinase inhibitors are stable to heat. With these inhibitors, a heat treatment early in the purification has often facilitated their recovery with good yields. It is only rarely that a proteinase inhibitor cannot be purified by some means. Even inhibitors present in tissues in small amounts have yielded to purification by modern techniques.

IV. SPECIFICITY AND MECHANISM OF ACTION

All natural proteinase inhibitors contain active sites that combine with the catalytic site region of enzymes. All of the naturally occurring inhibitors, despite their differences in specificities, are thought to employ the same types of protein–protein interactions to form complexes with proteolytic enzymes. However, in order to understand the specificity and mechanism of protein proteinase inhibitors it is necessary to understand something of the nature of the specificity and mechanism of the proteolytic enzymes that are inhibited.

Proteolytic enzymes can be generally classified into four categories based on their mechanism of action (Hartley, 1960; Walsh, 1975). Although probably not inclusive of all enzymes, the classification does include most known proteinases and is very useful when one is dealing with these enzymes. The four major classes of proteinases are serine, metal, sulfhydryl, and acid. These terms indicate the chemical nature of a functional group at their catalytic site. Some further generalizations can be made concerning the classes of proteolytic enzymes found in higher plants, animals, and microorganisms. In general, proteolytic enzymes from plant tissues (Ryan, 1973) are commonly comprised of sulfhydryl endopeptidases, serine carboxypeptidases, and metalloaminopeptidases. On the other hand, among the proteinases from animals and microorganisms, serine endopeptidases are found most commonly as are metal carboxypeptidases and metalloleucine aminopeptidases. Several exceptions to the above generalizations are known.

Thus, although the catalytic specificity of plant and animal proteinases may be very similar, the mechanisms by which they cleave peptide bonds can be quite different. This difference is reflected in the specificities of the proteinase inhibitors. A natural proteinase inhibitor that inhibits, for example, the animal serine endopeptidase trypsin will not ordinarily inhibit plant sulfhydryl endopeptidases, even those that have trypsin-like specificities.

Inhibitors of the serine proteinases that have trypsin-like or chymotrypsin-like specificities are the most common types found in both animals and plants, and their mechanisms of inhibition are best understood. Much less is known of the mode of action of the natural inhibitors that inhibit metallo-, acid-, or sulfhydryl-type proteinases.

The interaction of proteinases with proteinase inhibitors is usually very strong, resulting in inhibitor constants (K_i's) of 10^{-7} to 10^{-12} *M* (Laskowski and Sealock, 1971), but the strength of the inhibition is usually pH dependent (Laskowski and Sealock, 1971). For example, serine proteinase inhibitor complexes with K_i values of 10^{-10} *M* at pH 7 usually have constants of about 10^{-3} *M* at pH 3. The mode of inhibition has been found to be competitive in virtually every case (Laskowski and Sealock, 1971). Values of K_i are difficult to measure by traditional Lineweaver–Birk kinetic analysis because the interactions are stoichiometric and the concentration of enzyme changes significantly with little change in inhibitor concentration. A comprehensive treatment of the kinetics of the interactions of inhibitors with proteinases can be found in the review of Laskowski and Sealock (1971).

An identification system for the amino acids at the reactive site of the inhibitors is shown in Fig. 1. The amino acid residue at -P_1- for any -P_1-P_1'-peptide bond determines the specificity of the inhibitors (Kowalski *et al.*, 1974). The inhibitors appear to be rather rigid molecules in which the R group of the amino acid at P_1 protrudes into the solvent and is recognized by the appropriate proteinase as a potential substrate. The R group fits as a substrate into a cleft in the enzyme that determines specificity, and this binding, together with secondary forces between the enzyme and inhibitor, results in the stabilization of the complex that

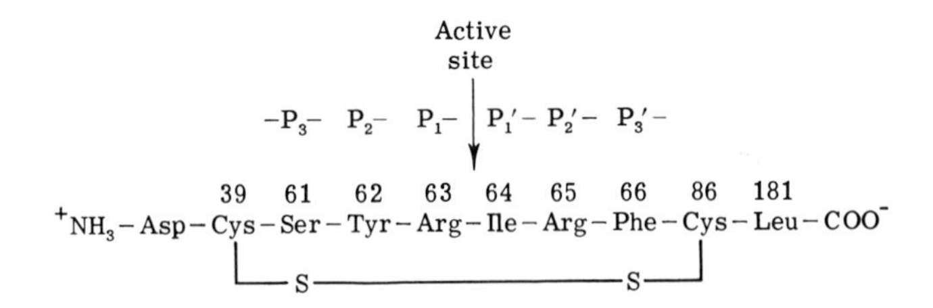

Figure 1. Amino acid sequence of soybean trypsin inhibitor illustrating nomenclature of proteinase inhibitor active sites.

inhibits the catalytic activity of the enzyme. Thus, when lysine or arginine is at the P_1 position of the inhibitor, trypsin-like proteinases are usually inhibited. When leucine, phenylalanine, or tryptophan is the amino acid at P_1, chymotrypsin-like enzymes are inhibited. If alanine is at P_1, elastase-like enzymes are inhibited.

Some examples of amino acids at the -P_1-P_1'-loci in proteinase inhibitor active sites are presented in Table I. The identification of these residues at the active sites was a result of the extraordinary observations of M. Laskowski, Jr., and his associates at Purdue University. When natural inhibitors of trypsin or chymotrypsin are incubated for several hours with about 1% (w/w) of the appropriate enzyme at about pH 3–4 (where the inhibitors bind weakly with the enzyme), the inhibitors act

Table I. Reactive inhibitory sites of several plant proteinase inhibitors[a]

Enzyme inhibited	Reactive site residues -P_1-P_1'-	Species (inhibitor)
Trypsin	Arg-Ala	Wheat, rye
	Arg-Ile	Soybean (Kunitz)
	Arg-Leu	Maize
	Arg-Ser	Soybean (Bowman–Birk)
	Lys-Ser	Chick pea, lima bean, potato (IIa), *Phaseolus vulgaris*
Chymotrypsin	Arg-Ile	Soybean (Kunitz)
	Leu-Ser	Lima bean (variant 1), soybean (Bowman–Birk)
	Phe-Ser	Lima bean (variant IV)
	Lys-Ser	Potato IIa/IIb
	Leu-Asp	Potato (inhibitor I)
	Met-Asp	Potato (inhibitor I)
Elastase	Ala-Ser	Garden bean
Nagarse	Lys-Ser	Potato IIa/IIb
Subtilisin	Met-Val	*Streptomyces albogriseolus*

[a] Richardson (1977).

as substrates and are cleaved specifically at the -P_1-P_1' site (Laskowski and Sealock, 1971). In Fig. 2, this specific cleavage at the active center is demonstrated with soybean trypsin inhibitor. Ordinarily, only a single bond at the active site position will cleave under these conditions. When the solution is readjusted to pH 7 the cleaved inhibitors still retain inhibitory activities, like the uncleaved native inhibitor.

Both kinetic and crystallographic evidence has led to an explanation of the mechanism of action during the events of proteinase inhibitor–proteinase interactions. David Blow and associates of Cambridge and Robert Huber and associates of the Max Planck Institute in Munich resolved the crystal structures of the complex between pancreatic trypsin inhibitor (PTI) and trypsin (Huber *et al.*, 1974) and the complex between soybean trypsin inhibitor (STI) and trypsin (Sweet *et al.*, 1974), respectively. These structures revealed that a tetrahedral intermediate existed in each complex between the P_1 carboxyl group of the inhibitors and the serine hydroxyl group at the active center of the enzyme. An extremely tight fit was found between four or five amino acid side chains of the active site regions of the enzymes and inhibitors, and strong binding energy resulted from van der Waals and hydrophobic interactions, which excluded water from the active site region of the complex. The tetrahedral bond was actually a destabilizing factor that was unnecessary for binding to occur. Huber and associates (1975) studied the crystal structure of a complex between PTI and anhydrotrypsin (Ako *et al.*, 1974), a catalytically inactive derivative of trypsin in which only the serine at the active site has been specifically converted to a dehydroalanine. They found that even in the absence of the reactive

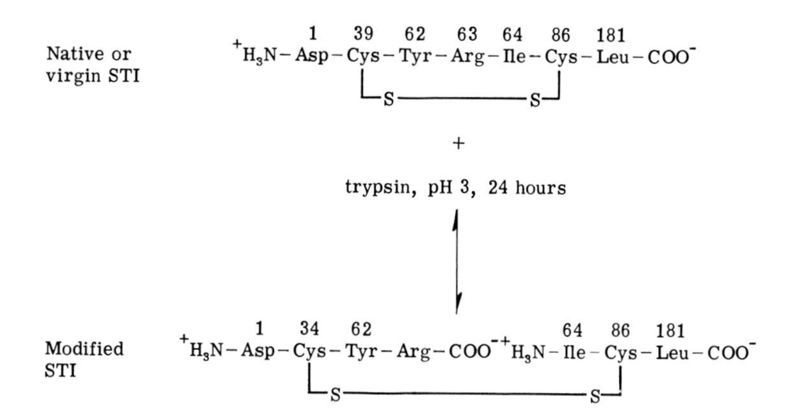

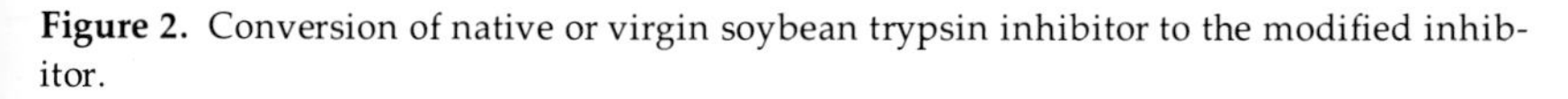

Figure 2. Conversion of native or virgin soybean trypsin inhibitor to the modified inhibitor.

serine hydroxyl of the enzyme, which now cannot form a tetrahedral intermediate, the enzyme and inhibitor formed a complex that was virtually identical to the complex of PTI and unmodified trypsin.

The determination of the crystal structures of PTI and STI and their complexes with trypsin was facilitated by knowledge of the amino acid sequence of the inhibitors. The sequences of many inhibitors are now known, and such information is useful in both evolutionary and physiological studies; it also adds to our understanding of the mechanism of action. It is now recognized from sequence analyses that families of inhibitors occur in nature (Laskowski and Sealock, 1971). Within each family are groups of related, "homologous" inhibitors that have been identified from tissues or fluids within or among various organisms. They apparently all share a common ancestral gene. For example, one family of inhibitors is comprised of those that exhibit homology with the basic pancreatic inhibitor called the Kunitz inhibitor. These homologous inhibitors have a wide distribution, such as in turtle eggs, snake venoms, and snails (Creighton, 1975). Another homologous family includes inhibitors similar to another pancreatic trypsin inhibitor called Kazal's inhibitor. These inhibitors have been isolated from avian eggs, porcine seminal plasma, and a bacterium (Kato *et al.*, 1975; Ikenaka *et al.*, 1974). In plants, a group of inhibitors from various beans are homologous with an inhibitor called the Bowman–Birk inhibitor from soybeans. This family will be discussed in greater detail in section V. Many other inhibitors are known that do not appear to be related to any others. The nonhomologous inhibitors are described as "analogous," indicating that they probably do not share the same ancestral gene.

An interesting case of homology involving gene duplication–elongation events exists in ovomucoids from avian eggs. Quail ovomucoid, a glycoprotein of molecular weight about 26,000, consists of three structurally homologous "domains" (Kato *et al.*, 1975). Each "domain" has an inhibitory site, and each is homologous to the other two domains and to the Kunitz basic pancreatic trypsin inhibitor. Two of the domains have a carbohydrate moiety attached to them. The existence of the three domains, each having an active site, explains data obtained almost a decade ago (Feeney *et al.*, 1969) which demonstrated that certain egg white ovomucoids have three distinct inhibitory sites.

None of the plant proteinase inhibitors have been shown convincingly to possess any homology with inhibitors from animals or microorganisms, although it has been suggested that the bromelain inhibitor may be homologous with a family of animal inhibitors (Reddy *et al.*, 1975). However, this hypothesis has been challenged on the basis of protein conformation (Creighton, 1975).

V. PLANT PROTEINASE INHIBITORS

The presence of proteinase inhibitors in plants was first reported by Read and Haas (1938). Since then, inhibitors present in storage organs of Leguminosae, Gramineae, and Solanaceae have attracted most attention (Ryan, 1973). This attention is probably a reflection of agricultural and nutritional interest in the proteins from these families. In addition, the availability of large quantities of source material from seeds and tubers of plants from these families is an advantage for purifying inhibitor proteins for use as biochemical tools to study the mechanism of action of both proteolytic enzymes and their inhibitors. In addition to their presence in storage organs, proteinase inhibitors have been identified in leaf extracts of a number of plant species (Walker-Simmons and Ryan, 1976; Chein and Mitchell, 1970). As more research on proteinase inhibitors continues, it is anticipated that many other species of plants will be found to contain proteinase inhibitors in their storage organs and/or vegetative tissues.

Plant proteinase inhibitors are all relatively small proteins, and a few can be classed as polypeptides. Their molecular masses range from about 4000 to 80,000 daltons. Several of the inhibitors are composed of subunits, or protomers, whereas many are single polypeptide chains. Some animal proteinase inhibitors have covalently attached carbohydrate moieties, but the plant proteinase inhibitors do not appear to have carbohydrates covalently associated with them. All plant, and most animal, inhibitors contain disulfide linkages, which may contribute to their high degree of stability. The only inhibitor proteins reported to date to be lacking disulfide bridges are those isolated from the leeche *Hirudo medicinalis* (Seemüller *et al.*, 1977). Inhibitor I from potatoes has a single disulfide bridge (Melville and Ryan, 1972), but most plant inhibitors contain several disulfide cross-links.

Although many plant inhibitors have been studied, only a few families of homologous inhibitors have been clearly identified. The most thoroughly studied family consists of a group of five inhibitors from legumes. These inhibitors, which have a molecular weight of about 9000 and whose amino acid sequences are known, display a high degree of homology (Richardson, 1977). Although homologous, they have considerable variations in their active site specificities toward three different pancreatic endopeptidases: trypsin, chymotrypsin, and elastase. These inhibitors are even more interesting in that they appear to have been gene duplicated and elongated; that is, one-half of their sequence is homologous with the other half (Tan and Stevens, 1971), indicating that their evolutionary precursors probably were much smaller molecules. The amino acid sequences of the five inhibitors are presented in Fig. 3.

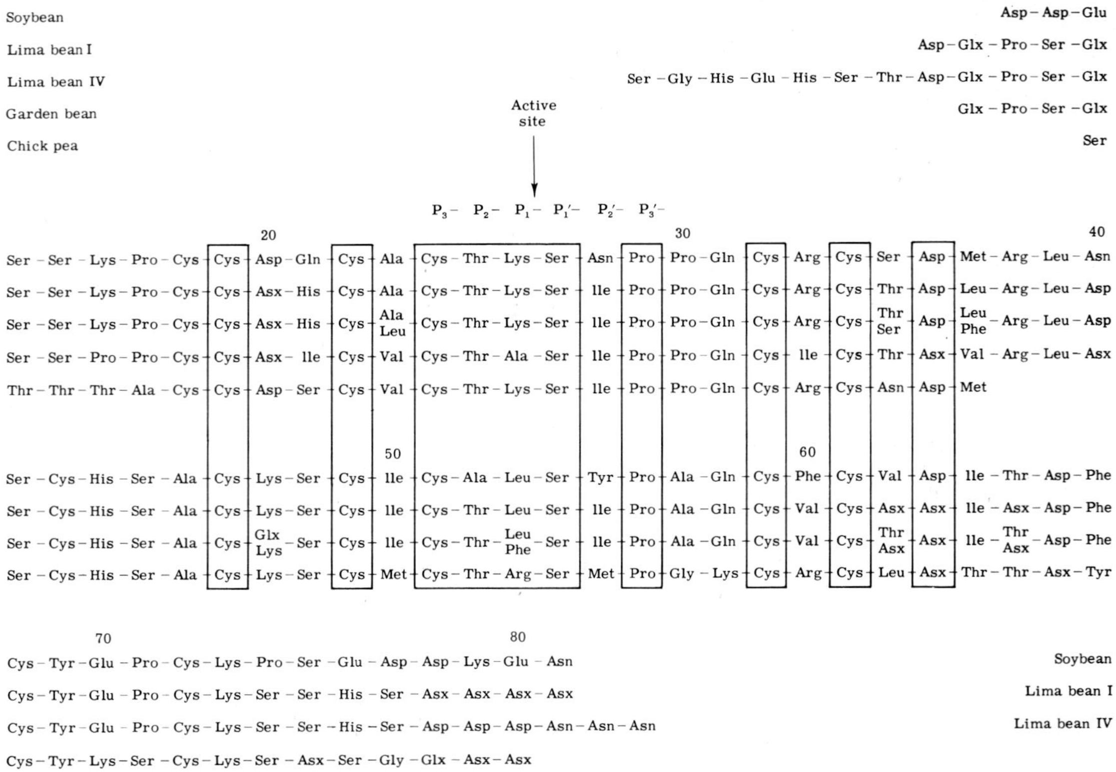

Soybean Asp–Asp–Glu

Lima bean I Asp–Glx–Pro–Ser–Glx

Lima bean IV Ser–Gly–His–Glu–His–Ser–Thr–Asp–Glx–Pro–Ser–Glx

Garden bean Glx–Pro–Ser–Glx

Chick pea Ser

Active site

P_3– P_2– P_1– P_1'– P_2'– P_3'–

20 30 40

Ser–Ser–Lys–Pro–Cys–Cys–Asp–Gln–Cys–Ala–Cys–Thr–Lys–Ser–Asn–Pro–Pro–Gln–Cys–Arg–Cys–Ser–Asp–Met–Arg–Leu–Asn

Ser–Ser–Lys–Pro–Cys–Cys–Asx–His–Cys–Ala–Cys–Thr–Lys–Ser–Ile–Pro–Pro–Gln–Cys–Arg–Cys–Thr–Asp–Leu–Arg–Leu–Asp

Ser–Ser–Lys–Pro–Cys–Cys–Asx–His–Cys–Ala/Leu–Cys–Thr–Lys–Ser–Ile–Pro–Pro–Gln–Cys–Arg–Cys–Thr/Ser–Asp–Leu/Phe–Arg–Leu–Asp

Ser–Ser–Pro–Pro–Cys–Cys–Asx–Ile–Cys–Val–Cys–Thr–Ala–Ser–Ile–Pro–Pro–Gln–Cys–Ile–Cys–Thr–Asx–Val–Arg–Leu–Asx

Thr–Thr–Thr–Ala–Cys–Cys–Asp–Ser–Cys–Val–Cys–Thr–Lys–Ser–Ile–Pro–Pro–Gln–Cys–Arg–Cys–Asn–Asp–Met

50 60

Ser–Cys–His–Ser–Ala–Cys–Lys–Ser–Cys–Ile–Cys–Ala–Leu–Ser–Tyr–Pro–Ala–Gln–Cys–Phe–Cys–Val–Asp–Ile–Thr–Asp–Phe

Ser–Cys–His–Ser–Ala–Cys–Lys–Ser–Cys–Ile–Cys–Thr–Leu–Ser–Ile–Pro–Ala–Gln–Cys–Val–Cys–Asx–Asx–Ile–Asx–Asp–Phe

Ser–Cys–His–Ser–Ala–Cys–Glx/Lys–Ser–Cys–Ile–Cys–Thr–Leu/Phe–Ser–Ile–Pro–Ala–Gln–Cys–Val–Cys–Thr/Asx–Asx–Ile–Thr/Asx–Asp–Phe

Ser–Cys–His–Ser–Ala–Cys–Lys–Ser–Cys–Met–Cys–Thr–Arg–Ser–Met–Pro–Gly–Lys–Cys–Arg–Cys–Leu–Asx–Thr–Thr–Asx–Tyr

70 80

Cys–Tyr–Glu–Pro–Cys–Lys–Pro–Ser–Glu–Asp–Asp–Lys–Glu–Asn Soybean

Cys–Tyr–Glu–Pro–Cys–Lys–Ser–Ser–His–Ser–Asx–Asx–Asx–Asx Lima bean I

Cys–Tyr–Glu–Pro–Cys–Lys–Ser–Ser–His–Ser–Asp–Asp–Asp–Asn–Asn–Asn Lima bean IV

Cys–Tyr–Lys–Ser–Cys–Lys–Ser–Asx–Ser–Gly–Glx–Asx–Asx

Figure 3. Amino acid sequences of active site regions in the N-terminal half of homologous inhibitors from legume seeds. Identical amino acids among the inhibitors are enclosed in boxes. The region is also boxed, although some amino acid replacements are present.

It is likely that a number of other inhibitors from other legume seeds, and perhaps other families, will display gene duplication and homology with those in Fig. 3.

An interesting homology was detected among potato inhibitors. A 5000 molecular weight polypeptide inhibitor specific for chymotrypsin (Hass *et al.*, 1976) appears to be related to a carboxypeptidase inhibitor of similar size (Ryan *et al.*, 1974). The N-terminal sequence of 25 amino acid residues of the 5000 molecular weight chymotrypsin inhibitor also shows a high degree of homology with the N-terminus of a much larger inhibitor from potatoes called inhibitor II (Iwasaki *et al.*, 1977). However, the amino acid sequence of the latter two inhibitors has not been completed. As more and more sequences are established in various plant families it will be interesting to build phylogenetic relationships to establish the relationships among plant proteinase inhibitors that exists and to incorporate this knowledge into ecological studies.

The physiological function of plant proteinase inhibitors has been questioned for many years. They have been considered as both regulatory and protective proteins and even as artifacts of evolutionary processes. The inhibitory activity of most plant proteinase inhibitors is usually specific for digestive proteinases, which are produced by animals and microorganisms, but occasionally these inhibitors have been found to inhibit proteinases produced by plants (Ryan, 1973). Inhibitors of endogenous proteinases are present in various seeds (Ryan, 1973; Richardson, 1977), but they appear to be present in small quantities and disappear rapidly as the seeds germinate (Ryan, 1973; Richardson, 1977). It is probable that these inhibitors control a proteolytic activity in the seeds until the onset of germination, when an increase in proteolytic activity is produced to assist the growth of the embryo. None of the proteinase inhibitors that are present in large quantities in seeds or tubers appear to have a function in regulation. As a result of their specificities, their presence in plants has often been thought of in terms of protecting the plants by arresting the digestive proteinases of attacking pests (Ryan, 1973; Applebaum, 1964).

The possibility that proteinase inhibitors play a role in the defense of plants against insects by inhibiting the digestive proteinases of insects has been studied for several decades. Mickel and Standish (1947) reported that larvae of certain insect pests were unable to develop normally on soybean products. Later, Lipke and associates (1954) studied the toxicity of soybean inhibitors in the development of the larvae of *Tribolium confusum*, a predator of stored grains. A specific inhibitor of *Tribolium confusum* larval digestive proteolysis was isolated. A similar inhibitor was found in wheat (Birk *et al.*, 1963).

Several naturally occurring inhibitors from various sources were found to inhibit proteolytic activity of the midget proteinases of *Tenebrio molitar,* another common pest of stored grains. Although the inhibitor of *Tribolium* from soybeans was inactive, soybean trypsin inhibitor, Bowman–Birk inhibitor (from soybean), lima bean inhibitor, and ovomucoid inhibited *Tenebrio* trypsin (Applebaum *et al.*, 1964). This indicated that the active site of trypsin-like enzymes of insects was similar to that of higher animals and could be potently inhibited by natural trypsin inhibitors from plants.

In 1964, Applebaum proposed that the proteinase inhibitors in legumes evolved as a defense mechanism against insects and that protein digestion in insects should be considered as a factor in host selection. More recent work on the digestive enzymes of insects has made it quite clear that enzymes are present that have specificities resembling those of higher animals and are generally inhibited by plant proteinase inhibitors (Ryan, 1973). However, the complement of enzymes in insect guts cannot always be equated with those of higher animals. It may be that significant differences in complements of proteolytic enzymes do exist among insects, and this diversity may determine whether proteinase inhibitors in various seeds are toxic toward the insects that consume them.

The possible involvement of proteinase inhibitors in plant protection received considerable support in 1972 (Green and Ryan, 1972) from the finding that Colorado potato beetle infestation induced a rapid accumulation of proteinase inhibitors in the leaves of potato and tomato plants. Even unattacked leaves distant from the attack sites accumulated inhibitors. This direct relationship between insect attack and the presence of inhibitors in plant leaves introduced a new dimension to the study of the biochemical mechanisms of insect-induced plant protection in plant tissues. Virtually any type of crushing or tearing of the vegetative tissues of tomato or potato plants by insects or any other agent releases a substance called the proteinase inhibitor inducing factor (PIIF) into the vascular system of the damaged plant, where it is rapidly transported to other tissues and initiates accumulation of proteinase inhibitors (Green and Ryan, 1973). The PIIF is released from the tissues of the plant during wounding and is not introduced into the plant from the wounding agent or organism. In this regard, PIIF fits the definition of a wound hormone and differs from the elicitors (Albersheim and Anderson-Prouty, 1975) that are introduced into plants by fungi and cause the induction of a metabolic process leading to the antifungal phytoalexins.

In young tomato plants, within 48 hours after single leaves are severely wounded, two inhibitor proteins can account for over 2% of the soluble proteins in leaves throughout the plants. The accumulation requires light (optimum above 800/ftc) and is temperature dependent (optimum of 37°C) (Green and Ryan, 1973). The two proteinase inhibitor proteins that are known to accumulate in response to wounding have been well characterized. Inhibitor I (Melville and Ryan, 1972) has a molecular weight of 39,000 and contains four subunits of identical size. Inhibitor II (Bryant *et al.*, 1976) has a molecular weight of 21,000 and is composed of two subunits of identical size. Both inhibitors I and II are potent inhibitors of chymotrypsin and subtilisin, and they both inhibit trypsin, but less strongly. The amino acid sequence of inhibitor I has been determined (Richardson and Cossins, 1974; 1975), and approximately half of the sequence of inhibitor II has been reported (Iwasaki *et al.*, 1977).

The wound-induced accumulation of the two inhibitors has presented an opportunity for studying the biochemical basis of the production of potentially poisonous substances by both wounded and unwounded plant tissues in response to insect attack. An experimental system was developed to assay PIIF and to study its biochemical effects in the leaves. When leaves of young plants 3–5 cm in height) were excised cleanly with a razor blade and the cut end was placed in a vial with water under conditions of optimal light and temperature, little PIIF was released from the cut and little accumulation of inhibitors occurred during the next several hours (Ryan, 1974). Small plants excised at the base of the main stem could also be treated this way without accumulating inhibitors. If these leaves or plants were supplied with water extracts containing PIIF for only a few minutes through the cut petiole or stem and subsequently placed in vials of water under the same light and temperature conditions, they accumulated inhibitors at a linear rate for many hours. The two inhibitors were quantified in leaf extracts by means of a convenient, rapid, and sensitive immunological radial diffusion assay with specific antibodies prepared against pure inhibitor proteins (Ryan, 1977).

The synthesis of proteinase inhibitors in response to PIIF appears to result from the production of new mRNA, which competes with other mRNA's for cellular ribosomes (Gustafson and Ryan, 1976). About 2% of the total newly synthesized proteins in PIIF-induced leaves are proteinase inhibitors I and II. However, proteinase inhibitor levels in tomato leaves are apparently regulated independently from the other cellular proteins. Instead of being degraded, the inhibitor proteins are deposited in the central vacuole of the cells during or after synthesis (Shumway *et al.*, 1970, 1976; Walker-Simmons and Ryan, 1977b). Although they represent only 2% of the newly synthesized proteins, the inhibitors accumulate rapidly while other proteins are made and degraded (Gustafson and Ryan, 1976). Leaf vacuoles that had accumulated inhibitors in response to PIIF were isolated intact and shown to contain virtually all of the inhibitors I and II present in leaf cells (Walker-Simmons and Ryan, 1976). Apparently, the sequestering of inhibitors in the vacuoles may help to ensure their very long half-lives.

Neither the chemical nature of PIIF nor its mode of action has been fully elucidated. Its activity is associated with soluble oligosaccharides that appear to be related to, or composed in part by, neutral cell wall polysaccharides. The complete identification of the chemical structure of PIIF should provide some novel approaches to the practical application of proteinase inhibitors in plant–pest interactions.

The widespread occurrence of PIIF in leaves from a number of genera ranging across the plant kingdom was established by supplying leaf extracts of the various plants to young tomato plants and measuring their capacity to induce inhibitor I accumulation (McFarland and Ryan, 1974). Thirty-seven species representing 20 families from the four major plant subdivisions exhibiting PIIF-like activities are shown in Table II.

Table II. PIIF-like activities in extracts of various plants[a]

Plant	Common name	Activity of tomato PIIF (%)
Flowering		
Anacardiaceae		
Rhus vernix	Sumac	7
Apiaceae		
Apium graveolens	Celery stalk	0
	Celery leaves	118
Brassicaceae		
Raphanus sativa	Radish	30
Brassica oleraceae	Cabbage	0
Brassica oleraceae	Cauliflower leaves	66
	Cauliflower flowers	86
Compositae		
Lactuca sativa	Lettuce	0–51
Helianthus annuus	Sunflower	95
Tragopogon dubius	Goatsbeard	94
Taraxacum officinale	Dandelion	37
Convolvulaceae		
Ipomaea sp.	Morning glory	56
Ericaceae		
Monotropa uniflora	Indian pipe	80
Gramineae		
Zea mays	Corn	96
Hordeum vulgare	Barley	113
Triticum sativum	Wheat	102

Table II *(Continued)*

Plant	Common name	Activity of tomato PIIF (%)
Leguminosae		
Medicago sativa	Alfalfa	81
Pisum sativum	Pea	80
Vicia faba	Broad bean	41
Gleditsia triacanthos	Honey locust	30
Liliaceae		
Allium sativa	Onion	33
Rosaceae		
Fragaria virginiana	Strawberry	48
Malus pumila	Apple	93
Solanaceae		
Lycopersicon esculentum	Tomato	100
Solanum tuberosum	Potato	110
Nicotiana tabacum	Tobacco	122
Solanum dulcamara	Nightshade	83
Datura stramonium	Jimson weed	127
Petunia sp.	Petunia	27
Conifers		
Pinaceae		
Pinus monticola	Western white pine	127
Pinus ponderosa	Western yellow pine	11
Pteriodophytes		
Polypodiaceae		
Athyrium felix-femina	Fern	125
Equisetaceae		
Equisetum arvense	Horsetail	105
Bryophtes		
Marchantiaceae		
Marchantia polymorpha	Liverwort	76
Aulacomniaceae		
Aulacomnium androgynum	Moss	60
Fungi		
Agaricaceae		
Agaricus campestris		148
Boletaceae		
Boletus felleus		84
Lycoperdaceae		
Lycoperdon sp.	Puffball	214
Saccharomycetaceae		
Saccharomyces cerevisiae	Bakers' yeast	91
Lichen		
Stictaceae	Lichen	120

[a] McFarland and Ryan (1974).

Extracts of tissues from only two species, celery stalks and cabbage leaves, failed to cause the tomato plants to accumulate inhibitor I. Extracts from fruiting bodies of one fungal species, *Agaricus campestris,* contained over 20 times more activity than tomato leaf extracts.

Trypsin inhibitory activity was found to accumulate in leaves of 10 of 23 species of plants that were wounded, supplied with PIIF from their own extracts, or supplied with PIIF from tomato leaf extracts (Table III) (Walker-Simmons and Ryan, 1977). Alfalfa accumulated the highest levels of trypsin inhibitors, followed by tobacco, tomato, potato, strawberry, cucumber, squash, clover, broad bean, and grape. Thus, the wound induction of proteinase inhibitors and the signal that initiates the response in unwounded tissues are present in a variety of plants from several different genera.

The concept of proteinase inhibitors as protective agents in plant leaves against animal digestive proteinases is new and relatively untested, since the presence of inhibitors in vegetative aerial plant tissues has been known for only a relatively short time. However, the rapid (within 24 hours) accumulation of proteinase inhibitors in response to wounding could have a major adverse effect on the ability of feeding insects to digest the leaf proteins. This would be particularly true with larvae that had recently hatched on the leaves and were completely dependent on proteins of the leaf for the amino acids required for their growth. With an inhibited system for digesting proteins, a severe nitrogen deficiency could result that could arrest the normal development of the larvae. Such a situation could be lethal, particularly if coupled with other defense mechanisms, unless the larvae could escape to a new food source. It is unlikely that occasional flying insects would be similarly affected since they would not be so limited in finding new sources of food if the leaves began to affect their digestion.

The process of inhibitor accumulation in tomato leaves probably does not place a severe strain on the plant, since only 2% of the protein synthesis of the leaf is utilized for the few hours necessary to synthesize and store the inhibitors. The wound-induced inhibitors can be likened to a primitive immune-like response in which an attacking organism can stimulate the synthesis of long-lived proteins that react with certain molecules (in this case digestive proteinases) of the attacker to arrest an invasion that could be lethal to the plant if unchecked.

In some plants a variety of inhibitors with different specificities and multiple forms (isoinhibitors) are present. It has been suggested that, unlike the active sites of enzymes, the active centers (specificities) of the inhibitors mutate much faster than the rest of the protein (Kato *et al.,* 1975). Thus, a wide spectrum of inhibitors of proteolytic enzymes are

Table III. Trypsin inhibitor activity in various plant leaf extracts[a]

Plant	Common name	Trypsin inhibitor activity in extracts of leaves[b]			
			Detached and treated with		
		Attached, no treatment	Water	Endogenous PIIF	Tomato PIIF
Betulaceae					
Betula pubescens	Birch	0	0	0	0
Compositae					
Lactuca sativa	Lettuce	0	0	0	0
Cucurbitaceae					
Cucurbita maxima	Squash	12	25	18	18
Cucumis sativus	Cucumber	13	26	26	22
Gramineae					
Hordeum vulgare	Barley	0	0	0	0
Triticum sativum	Wheat	3	4	4	5
Zea mays	Corn	0	0	0	2
Leguminosae					
Lens culinaris	Lentil	0	3	4	4
Medicago sativa	Alfalfa	49	180	245	338
Phaseolus vulgaris	Kidney bean	0	3	4	4
Pisum sativum	Pea	4	9	4	9
Trifolium repens	Clover	0	2	7	9
Vicia faba	Broad bean	4	9	7	9
Liliaceae					
Allium cepa	Onion	0	0	0	0
Pinaceae					
Larix occidentalis	Larch	32	—	21	26
Rosoceae					
Fragaria virginiana	Strawberry	27	42	24	45
Malus pumila	Apple	16	20	18	14
Prunus serotina	Cherry	0	0	0	0
Prunus persica	Peach	0	0	0	0
Solanaceae					
Lycopersicum esculentum	Tomato	60	62	108	120
Nicotiana tobacum	Tobacco	133	133	284	284
Solanum tuberosum	Potato	37	49	68	68
Vitaceae					
Vitis vinifera	Grape	3	8	—	6

[a] Walker-Simmons and Ryan (1977a).
[b] μg trypsin inhibited/ml leaf juice.

usually found. Such a battery of proteinase inhibitors might be effective in inhibiting the digestive proteinases of an attacking insect. On the other hand, insects appear to have a variety of proteolytic enzymes. Single mutations at or near the active center regions of the enzymes could have considerable influence on their ability to recognize the inhibitor proteins either as inhibitors or as protein substrates. Thus, one insect's poison could be another's food protein. So it is not difficult to envision that plant–insect interactions involving inhibitor–enzyme relationships may be constantly evolving through natural selection.

It is unlikely that inhibitors act alone to protect any given plant. It is more likely that inhibitors represent one line of protection among various other protective chemicals that may also be present. For some plants, depending on their ecological history, inhibitors may have provided an important selective advantage against certain pests and may still be doing so. However, a generalization cannot be made since evidence indicates that all plants do not contain proteinase inhibitors and in fact many plants are apparently surviving quite well without them.

The research area of the function and evolution of proteinase inhibitors in plants is, like many other areas dealing with chemical ecology, just developing. It is anticipated that more scientists working on plant–pest responses will investigate the possible role of proteinase inhibitors in the system they are studying. Proteinase inhibitor research not only should provide a stimulating challenge to biochemists, biologists, and chemical ecologists in determining the scope of the importance of these substances in nature but should serve as a model for studying the biochemical basis of the ability of plants to sense pest attack and respond by producing potentially toxic proteins.

The biochemistry of intracellular and intercellular communication of information in plants is poorly understood. The full understanding of all biochemical interactions depends on our ability to understand the fundamental chemistry and biochemistry of the substances involved and the processes by which they react and relay information in tissues and cells. When these processes are better understood we will be able to make more rational decisions concerning the safe, clean, and efficient use of chemicals to enhance the natural defenses of plants for man's benefit. With the onset of an era of genetic engineering it is possible to envision the transfer of natural defensive processes from one species of plant to a species in which they can deter insects that have not evolved defenses against them. The age of chemical ecology is upon us, and it is our challenge to not simply accept it but to consider it as an opportunity to pursue new, imaginative research frontiers with significant potential in areas of both fundamental and applied science.

REFERENCES

Ako, H., Foster, R. J., and Ryan, C. A. (1974). *Biochemistry* **13,** 132–139.
Albersheim, P., and Anderson-Prouty, A. J. (1975). *Annu. Rev. Plant Physiol.* **26,** 31–52.
Applebaum, S. W. (1964). *J. Insect. Physiol.* **10,** 783–788.
Applebaum, S. W., Birk, Y., Harpaz, I., and Bondi, A. (1964). *Comp. Biochem. Physiol.* **11,** 85–103.
Bergmeyer, H.-U., ed. (1970). "Methods of Enzymatic Analysis." Academic Press, New York.
Birk, Y. (1976). *In* "Methods in Enzymology" (L. Lorand, ed.), Vol. 45, Part B, pp. 695–751. Academic Press, New York.
Birk, Y., Gertler, A., and Khalef, S. (1963). *Biochim. Biophys. Acta* **67,** 326–328.
Bryant, J., Green, T. R., Gurusaddaiah, T., and Ryan, C. A. (1976). *Biochemistry* **15,** 3418–3424.
Chase, T., Jr., and Shaw, E. (1970). *In* "Methods in Enzymology" (G. E. Perlmann and L. Laszlo, eds.), Vol. 19, pp. 20–27. Academic Press, New York.
Chein, T. F., and Mitchell, H. L. (1970). *Phytochemistry* **9,** 717–720.
Creighton, T. E. (1975). *Nature (London)* **255,** 743–745.
Dechary, J. M. (1970). *Econ. Bot.* **24,** 113–122.
Feeney, R. E., Means, G. E., and Bigler, J. C. (1969). *J. Biol. Chem.* **244,** 1957–1962.
Fritz, H., and Tschesche, H., eds. (1971). "Proceedings of the First International Research Conference on Proteinase Inhibitors." de Gruyter, Berlin.
Fritz, H., Tschesche, H., Green, L., and Truscheit, E., eds. (1974). "Proteinase Inhibitors," Bayer-Symp., 5th. Springer-Verlag, Berlin and New York.
Green, T. R., and Ryan, C. A. (1972). *Science* **175,** 776–777.
Green, T. R., and Ryan, C. A. (1973). *Plant Physiol.* **51,** 19–21.
Gustafson, G., and Ryan, C. A. (1976). *J. Biol. Chem.* **251,** 7004–7010.
Hartley, B. S. (1960). *Annu. Rev. Biochem.* **29,** 45–77.
Hass, G. M., Venkatakrishnan, R., and Ryan, C. A. (1976). *Proc. Natl. Acad. Sci. U.S.A.* **73,** 1941–1944.
Huber, R., Kukla, D., Bode, W., Schwager, P., Bartels, K., Deisenhoper, J., and Steigemann, W. (1974). *J. Mol. Biol.* **89,** 73–101.
Huber, R., Bode, W., Kukla, D., Kohl, U., and Ryan, C. A. (1975). *Biophys. Struct. Mech.* **1,** 189–201.
Ikenaka, T., Odani, S., Sakai, M., Nabeshima, Y., Sato, S., and Murao, S. (1974). *J. Biochem. (Tokyo)* **76,** 1191–1209.
Iwazaki, T., Kiyohara, T., and Yoshikawa, M. (1972). *J. Biochem. (Tokyo)* **72,** 1029–1035.
Iwazaki, T., Wada, J., and Kiyohara, T. (1977). *J. Biochem. (Tokyo)* **82,** 991–1004.
Kassell, B. (1970). *In* "Methods in Enzymology" (G. E. Perlmann and L. Lorand, eds.), Vol. 19, pp. 839–932. Academic Press, New York.
Kato, I., Schrode, J., Wilson, K. A., and Laskowski, M., Jr. (1975). *Protides Biol. Fluids, Proc. Colloq.* pp. 235–243.
Kezdy, F. J., and Kaiser, E. T. (1970). *In* "Methods in Enzymology" (G. E. Perlmann and L. Lorand, eds.), Vol. 19, pp. 3–20. Academic Press, New York.
Kowalski, D., Leary, T., McKee, R. E., Sealock, R. W., Wang, D., and Laskowski, M., Jr. (1974). *In* "Proteinase Inhibitors" (H. Fritz *et al.*, eds.), Bayer Symp., 5th, pp. 311–324. Springer-Verlag, Berlin and New York.
Laskowski, M., Jr., and Sealock, R. W. (1971). *In* "The Enzymes" (P. D. Boyer, ed.), 3rd ed., Vol. 3, pp. 375–473. Academic Press, New York.

Liener, I. E., and Kakade, M. L. (1969). *In* "Toxic Constituents in Plant Foodstuffs" (I. E. Liener, ed.), pp. 8–68. Academic Press, New York.
Lipke, H., Fraenkel, G. S., and Liener, I. E. (1954). *J. Agric. Food Chem.* **2,** 410–414.
McFarland, D., and Ryan, C. A. (1974). *Plant Physiol.* **54,** 706–708.
Means, G., Ryan, D., and Feeney, R. E. (1974). *Acc. Chem. Res.* **7,** 315–320.
Melville, J. C., and Ryan, C. A. (1972). *J. Biol. Chem.* **247,** 3445–3453.
Mickel, C. E., and Standish, J. (1947). *Mich., Agric. Exp. Stan., Tech. Bull.* **178.**
Peeters, H., ed. (1975). "Protides of the Biological Fluids," pp. 27–285. Pergamon, Oxford.
Puztai, A. (1967). *Nutr. Abstr. Rev.* **37,** 1–9.
Read, J. W., and Haas, L. W. (1938). *Cereal Chem.* **15,** 59–68.
Reddy, M. N., Keirn, P. S., Heinrikson, R. L., and Kezdy, F. J. (1975). *J. Biol. Chem.* **250,** 1741–1750.
Reich, E., Refkin, D. B., and Shaw, E., eds. (1975). "Proteases and Biological Control," pp. 367–455. Cold Spring Harbor Lab., Cold Spring Harbor, New York.
Richardson, M. (1977). *Phytochemistry* **16,** 159–169.
Richardson, M., and Cossins, L. (1974). *FEBS Lett.* **45,** 11–13.
Richardson, M. and Cossins, L. (1975). *FEBS Lett.* **52,** 365.
Royer, A. (1975). *In* "Les protéines des graines" (J. Mieze, ed.), p. 159. Geneva.
Ryan, C. A. (1973). *Annu. Rev. Plant Physiol.* **24,** 173–196.
Ryan, C. A. (1977). *In* "Immunological Aspects of Foods" (N. Catsimpoolis, ed.), pp. 182–198. Avi Publ. Co., Westport, Connecticut.
Ryan, C. A., Hass, G. M., and Kuhn, R. W. (1974). *J. Biol. Chem.* **249,** 5495–5499.
Santarius, K., and Ryan, C. A. (1977). *Anal. Biochem.* **77,** 1–9.
Seemüller, U., Meier, M., Ohlsson, K., Müller, H.-P., and Fritz, H. (1977). *Hoppe-Seyler's Z. Physiol. Chem.* **358,** 1105–1117.
Shkenderov, S. (1973). *FEBS Lett.* **33,** 343–347.
Shumway, L. K., Rancour, J. M., and Ryan, C. A. (1970). *Planta* **93,** 1–14.
Shumway, L. K., Yang, V. V., and Ryan, C. A. (1976). *Planta* **129,** 161–165.
Strudom, D. J. (1973). *Nature (London) New Biol.* **243,** 88–89.
Sweet, R. M., Wright, H. T., Janin, J., Chotina, C. H., and Blow, D. M. (1974). *Biochemistry* **13,** 4212–4228.
Takahashi, H., Iwanaga, S., Kitawaga, T., Hokama, Y., and Suzuki, T. (1974). *J. Biochem. (Tokyo)* **76,** 721–733.
Tan, C. G. L., and Stevens, F. C. (1971). *Eur. J. Biochem.* **18,** 515–523.
Tschesche, H. (1974). *Angew, Chem., Int. Ed. Engl.* **13,** 10–28.
Umezawa, H. (1976). *In* "Methods in Enzymology (L. Lorand, ed.), Vol. 45, pp. 678–695. Academic Press, New York.
Vogel, R., Trautschold, I., and Werle, E. (1968). "Natural Proteinase Inhibitors." Academic Press, New York.
Walker-Simmons, M., and Ryan, C. A. (1977a). *Plant Physiol.* **59,** 437–439.
Walker-Simmons, M., and Ryan, C. A. (1977b). *Plant Physiol.* **60,** 61–63.
Walsh, K. (1975). *In* "Proteinases and Biological Control" (E. Reich, D. B. Rifkin, and E. Shaw, eds.), pp. 1–11. Cold Spring Harbor Lab., Cold Spring Harbor, New York.
Werle, E., and Zickgraf-Rüdel, G. (1972). *Z. Klin. Chem. Klin. Biochem.* **10,** 139–150.
Whitaker, J. R., and Feeney, R. E. (1973). *In* "Toxicants Occurring Naturally in Foods," pp. 276–298. Nat. Acad. Sci., Washington, D. C.

Chapter 18

Flavonoid Pigments

JEFFREY B. HARBORNE

I. INTRODUCTION

Flavonoids are aromatic compounds, and all show strong absorption in the ultraviolet region; many absorb light in the visible region as well and hence are brightly colored. They are also heterocyclic compounds derived from flavone (**1**, Fig. 1), which has a γ-pyrone ring with ether-linked oxygen as part of its structure. Another important structural feature is the substitution in the two aromatic rings by one or more hydroxyl groups. Flavonoids are polyphenolic and hence weakly acidic and in solution undergo large bathochromic spectral shifts in the presence of added base. As polyphenols, they may interact by hydrogen bonding to

ISBN 0-12-597180-X

form loose molecular complexes (i.e., anthocyanins with flavones) or may bind to other classes of plant constituents such as proteins.

A large number of naturally occurring flavonoids have been described; according to a recent estimate (Harborne *et al.*, 1975), there are about 2000 known structures. In spite of a bewildering array of structural variation, some simple patterns can be discerned. All are based on the same C_{15} skeleton, which is formed by the same pathway from three malonate units condensing with a phenylalanine-derived C_6–C_3 precursor. Also, there are only a small number of common structures, all of which are closely related to each other.

Flavonoids are conveniently divided into 12 classes according to the oxidation level of the central pyran ring (Table I). Most of these classes are mentioned here, but emphasis is given to the three most important and widespread: the anthocyanins, flavones, and flavonols. The oligomeric and polymeric proanthocyanidins are not considered since they are dealt with in detail in Chapter 19 by Swain. It should be observed (Table I) that the different classes vary in their biological properties; many (e.g., anthocyanins) are strongly pigmented, but others (e.g., flavanones) are colorless. Flavonoids also vary considerably in solubility; although many are water soluble, some compounds are highly ether soluble and occur in leaf waxes or bud exudates.

Flavonoids are most characteristic of and occur universally throughout higher plants, both angiosperms and gymnosperms. They also occur regularly in ferns, mosses, and liverworts. They are generally absent from algae, fungi, and bacteria, although some exceptions do occur.

Table I. Major known classes of flavonoids

Class	No. of known structures	Biological properties
Anthocyanins	250	Red to blue pigments
Chalcones	60	Yellow pigments
Aurones	20	
Flavones	350	Cream pigments in flowers; feeding repellents (?) in leaves
Flavonols	350	
Flavanones	150	Some have bitter tastes
Dihydrochalcones	10	
Proanthocyanidins	50	Astringent substances, some with tanning ability
Catechins	20	
Flavan-3,4-diols	20	
Biflavonoids	65	None known
Isoflavonoids	150	Estrogenic and fungitoxic

Thus, flavones are found in certain green algae (Characeae), and the chlorinated flavonol chlorflavonin is produced as an antibiotic by the fungus *Aspergillus candidus* (see Harborne, 1972; Swain, 1974).

Flavonoids have an established ecological role in providing attractive colors to flowers in order to promote pollination and fruit and seed dispersal. The functions of the widespread, often ubiquitous "colorless" flavones and flavonols, present especially in leaves, remain more obscure. However, there is evidence to indicate that these substances may be of considerable significance as feeding deterrents. The colorless isoflavonoids are particularly active biologically. One group of isoflavonoids, the rotenoids, are well known because of their insecticidal activity. The simpler isoflavones are also biologically important because of their hormonal effects in mammals. It has been estimated (Shutt, 1976) that over a million ewes fail to lamb each year in Australia because of the deleterious effects caused by the ingestion of these substances in pasture clovers.

II. STRUCTURAL VARIATION

A. Common Flavonoids

Although several hundred flavonoid aglycones have been isolated from higher plants, only eight of these occur widely (Harborne *et al.*, 1975). Any one of these eight structures may be expected to be present in the hydrolyzed extract of any higher plant studied. These substances are shown in Fig. 1. All have similar hydroxylation in the A ring and differ mainly in the oxidation level of the central pyran nucleus and in the number of hydroxyl groups (one, two, or three) in the B ring. All these common structures occur naturally in water-soluble form as *O*-glycosides (see Section II,D).

Three are anthocyanidins: the scarlet-colored pelargonidin **(2)**, the crimson cyanidin **(3)**, and the mauve delphinidin **(4)**. These three pigments, or their simple methyl ethers (see Section II,C), occur very widely in plants, being particularly abundant in colored flowers and fruits. Within the angiosperms, they are almost ubiquitous but are replaced in one group of families in the order Centrospermae by a different class of purple pigments, betacyanins, derived biosynthetically from the aromatic amino acid L-dopa. Pelargonidin and delphinidin occur especially frequently in cyanic flowers, their synthesis being correlated with natural selection by animal pollinators for bright scarlet colors

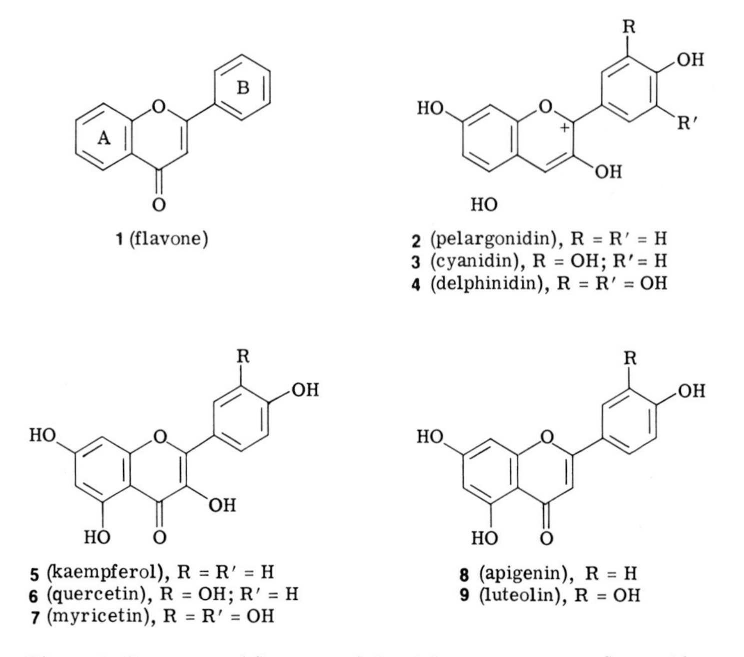

Figure 1. Structures of flavone and the eight most common flavonoids.

(hummingbirds) and deep blue colors (bees), respectively. In contrast, pigmented leaves normally contain cyanidin, the other two anthocyanidins being relatively rarely found in vegetative tissue.

The flavonols corresponding in structure to the three main anthocyanidins are kaempferol **(5)**, quercetin **(6)**, and myricetin **(7)**, in order of increasing B-ring hydroxylation. Although these flavonols usually accompany the anthocyanidins in flowers where they have an important role as copigments, they occur even more frequently, indeed almost universally, in plant leaves. A survey of the leaves of over 1000 angiosperms showed, for example, that 48% of species had kaempferol, 56% quercetin, and 10% myricetin (Swain and Bate-Smith, 1962). In all, over 60% of the sample had one or other of the three flavonols. While kaempferol and quercetin occur throughout the angiosperms, myricetin is more restricted in its occurrence, being present mainly in leaves of woody plants in association with tannins (see also Chapter 19).

Flavones lack the 3-hydroxyl group present in flavonols and anthocyanidins. Only two structures are common: apigenin **(8)** and luteolin **(9)** (see Fig. 1). These two flavones, besides occurring with flavonols, are also found on their own in many herbaceous plants, there being a tendency for them to replace flavonols in more specialized angiosperm families. Flavones, unlike the other common flavonoids, are frequently found in C-glycosidic combination (see Section II,D) and are thus often reported as their C-glycosidic derivatives, vitexin (8-C-glucosylapigenin) and orientin (8-C-glucosylluteolin). They also occur widely in the more usual O-glycosidic combinations.

Tricetin, the flavone corresponding in structure to myricetin or delphinidin, so far appears to be of rare occurrence. A methylated derivative of tricetin, namely tricin (see Section II,C), is, however, more regular in occurrence, being particularly common as a constituent of grasses.

Another peculiarity of flavones is their ability to link together by carbon–carbon bonds to form dimers. Over 60 biflavones are known, one of the most common being amentoflavone or 3′,8″-biapigenin. These substances are largely restricted to and are typical of gymnosperm tissues, but they occasionally occur elsewhere in the plant kingdom (Geiger and Quinn, 1975).

B. Variations in Hydroxylation Pattern

Many flavonoids differing differing in hydroxylation pattern from the eight common structures are known, but the great majority are of limited occurrence in nature. Some typical examples are illustrated in Fig. 2. The range extends from flavone itself and simple mono- and dihydroxyflavones (e.g., **10, 11**), which occur exclusively in farina on leaves of *Primula* species, to compounds such as digicitrin **(12)** from the foxglove, which has eight hydroxyl groups, seven of which are protected by methylation. Most of the rare flavonoids are closely related to one or other of the eight common structures, being formed from them by insertion of an extra hydroxyl function in an otherwise unoccupied site in the flavonoid nucleus. Thus, morin **(13)** is derived from kaempferol by the insertion of an extra hydroxyl in the 2′ position. This substance occurs in *Morus alba* leaves and is one of two flavonoids that contribute to the selective feeding of silkworm larvae on mulberry plants (Hamamura *et al.*, 1962).

Similar insertion of a 2′-hydroxyl into luteolin **(9)** gives the flavone isoetin **(14),** recently identified as a yellow flower pigment in a rare Canary Island endemic, *Heywoodiella oligocephala* (Compositae). However, yellow flower color of a flavonoid nature is more frequently based on the substitution of an extra hydroxyl group into the A ring. Introduction of a hydroxyl at the 6 or 8 position causes a significant shift in color, and such compounds, e.g., gossypetin **(15)** and quercetagetin **(16),** are yellow instead of pale cream, as in quercetin **(6).** These two yellow pigments contribute color to such well-known plants as the

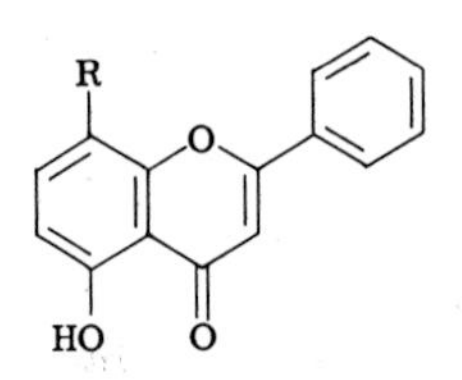

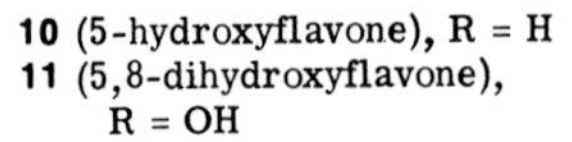
10 (5-hydroxyflavone), R = H
11 (5,8-dihydroxyflavone),
R = OH

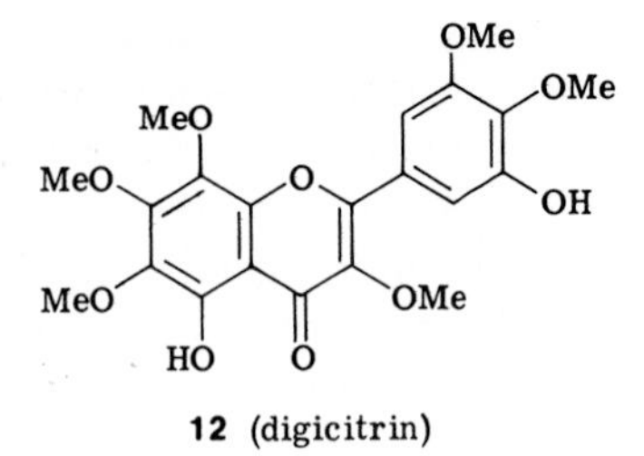

12 (digicitrin)

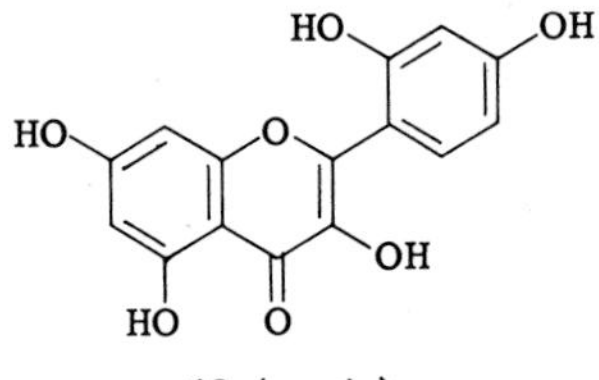

13 (morin)

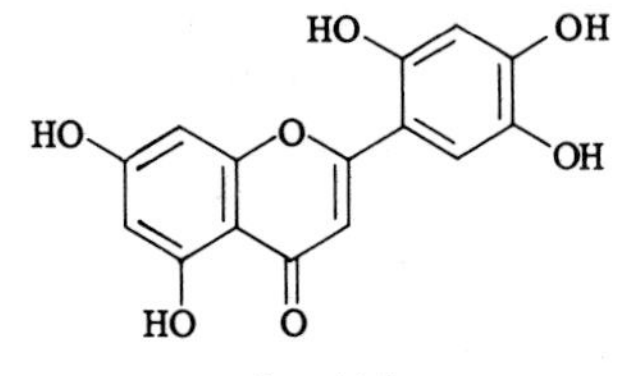

14 (isoetin)

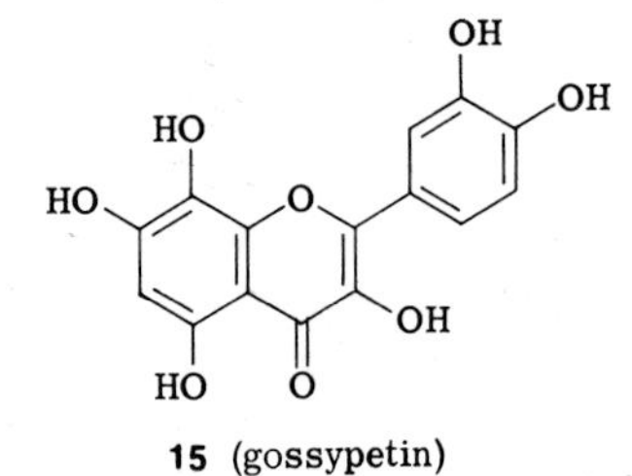

15 (gossypetin)

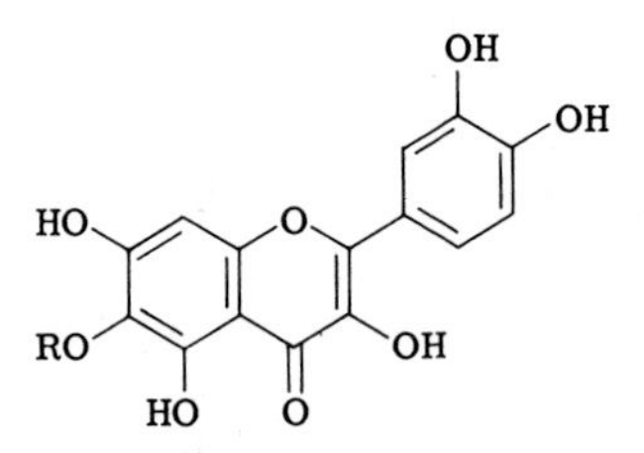

16 (quercetagetin), R = H
17 (patuletin), R = Me

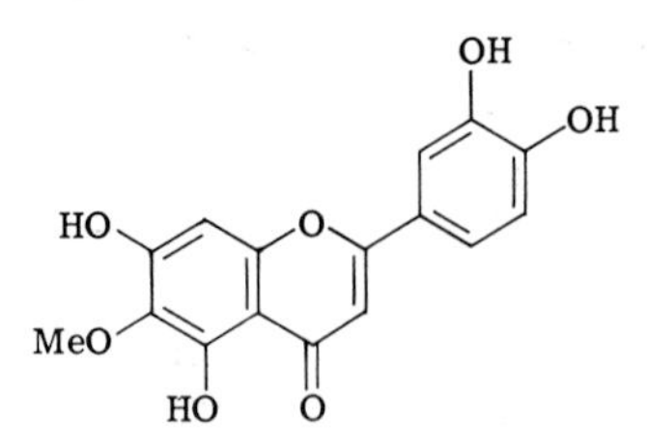

18 (6-methoxyluteolin)

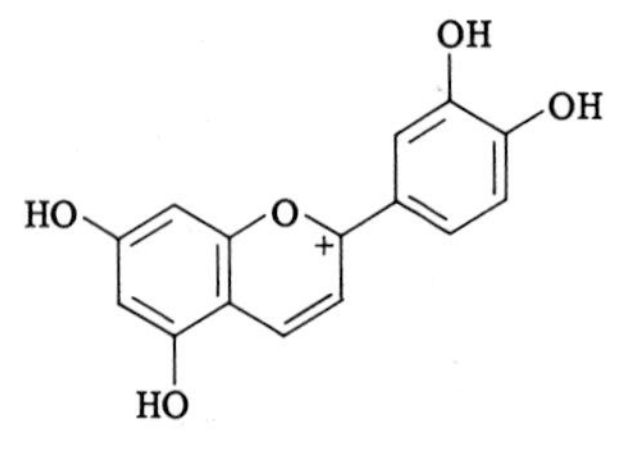

19 (luteolinidin)

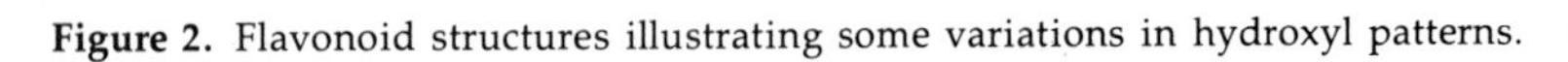
Figure 2. Flavonoid structures illustrating some variations in hydroxyl patterns.

primrose, *Primula vulgaris,* cotton flower, *Gossypium hirsutum,* and corn marigold, *Chrysanthemum segetum.* In some plants, such as the composite *Rudbeckia hirta,* these yellow pigments have a special function because of their strong absorbance in the ultraviolet (uv) region; i.e., they serve as honey guides to insect pollinators. Patuletin **(17),** a methyl ether of quercetagetin, is differentially distributed in the inner part of the ray flower of *Rudbeckia,* whereas carotenoid pigments, which reflect light in the uv region, are evenly distributed over the whole ray (Thompson *et al.,* 1972).

Flavones and flavonols with extra oxygen functions in the A ring also occur regularly in leaves of angiosperms, particularly in herbaceous plants. In such cases, they may be visually apparent to the insect, even if they are not perceived by human beings. There is evidence that 6-methoxyluteolin **(18),** which occurs as the 7-rhamnoside in alligator weed, acts as a feeding attractant to the *Agasicles* beetle (Zielske *et al.,* 1972).

Other rare flavonoids are those in which a particular hydroxyl function found in the common structures is removed rather than added to. One example is luteolinidin **(19),** which lacks the 3-hydroxyl group present in the common anthocyanidins and is referred to as a 3-deoxyanthocyanidin. Such pigments are rare in nature, but where they do occur, as in New World gesnerads, their synthesis seems to be correlated with hummingbird pollination and natural selection for scarlet or orange flower color.

C. Methylated Flavonoids

The most common flavonoid *O*-methyl ethers are peonidin **(20),** petunidin **(21),** and malvidin **(22),** which are derived from cyanidin or delphinidin by methylation of the hydroxyl groups in the B ring (see Fig. 3). Peonidin was first isolated from peony flowers, petunidin from *Petunia,* and malvidin from *Malva,* but they are by no means restricted to their original plant sources. All three are actually widely present in plant tissues with cyanic color. Petunidin and malvidin occur, for example, with delphinidin in grapes, and the color of genuine red wines is derived from these pigments. Very rarely, methylation may affect the hydroxyl groups of the A ring of anthocyanidins. One such derivative is hirsutidin **(24),** which occurs to any extent only in the Primulaceae, and there only in *Primula* and *Dionysia* species.

The three flavonol methyl ethers corresponding in structure to peonidin, petunidin, and malvidin are isorhamnetin **(25),** larycitrin **(26),** and syringetin **(27).** At one time, they were thought to be relatively rare but

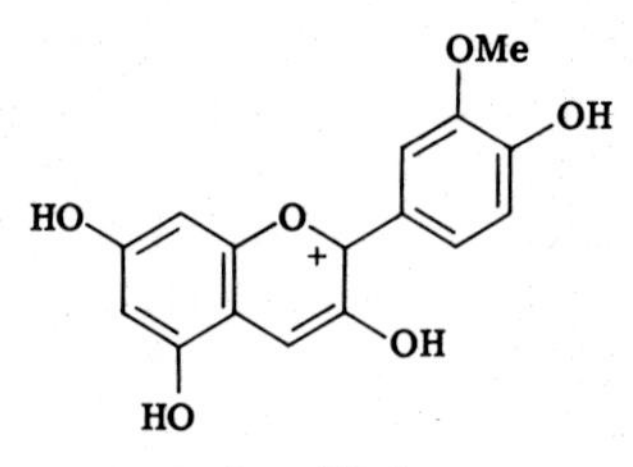

20 (peonidin)

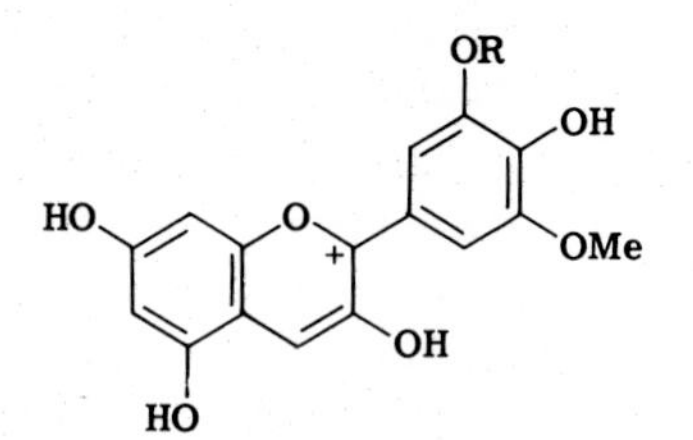

21 (petunidin), R = H
22 (malvidin), R = Me

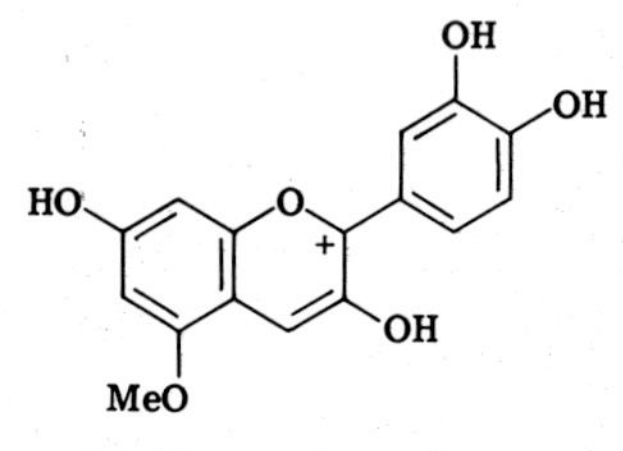

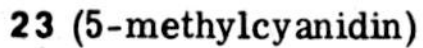

23 (5-methylcyanidin)

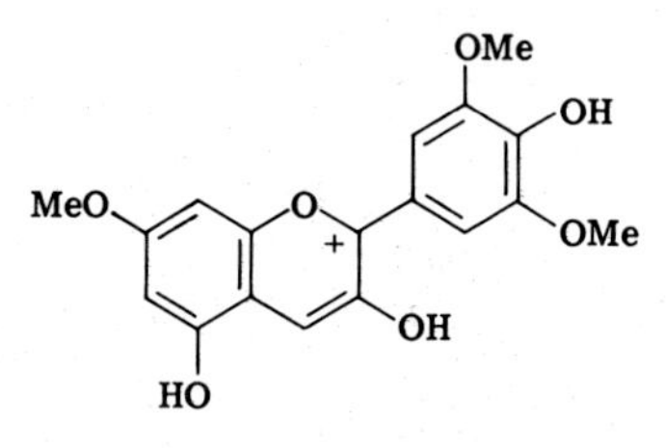

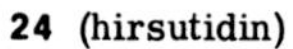

24 (hirsutidin)

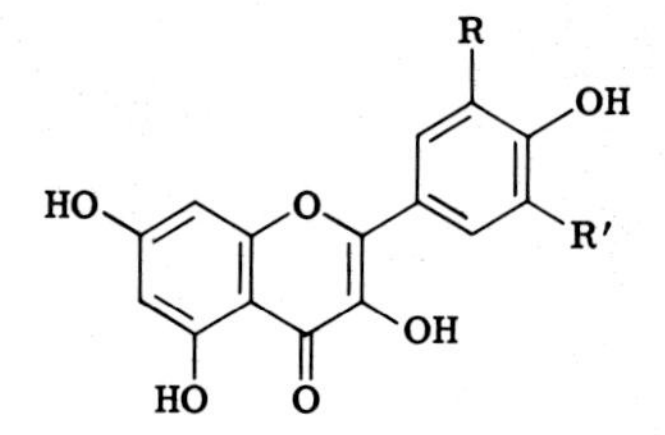

25 (isorhamnetin), R = OMe; R′ = H
26 (larycitrin), R = OMe; R′ = OH
27 (syringetin), R = R′ = OMe

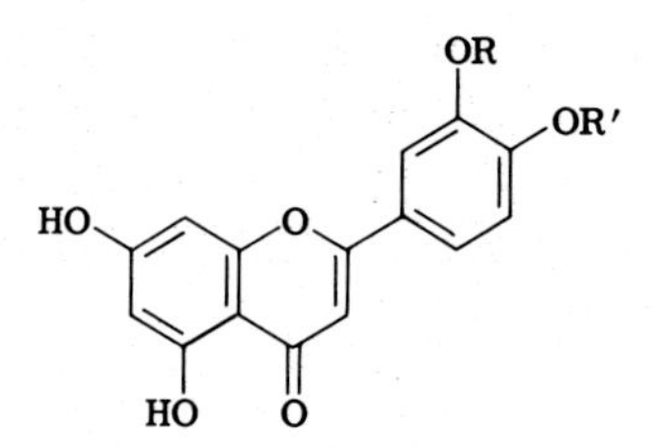

28 (chrysoeriol), R = Me; R′ = H
29 (diosmetin), R = H; R′ = Me

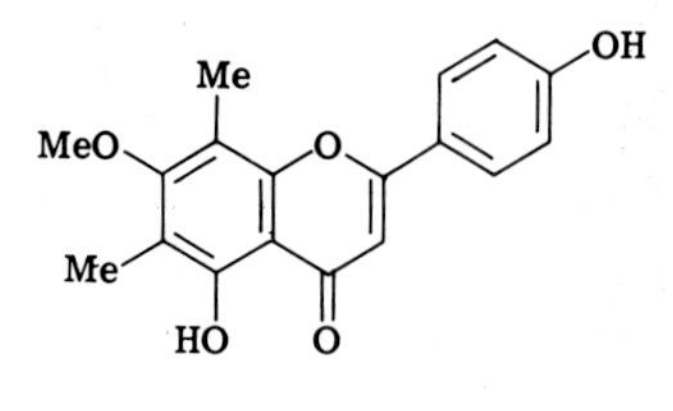

30 (sideroxylin)

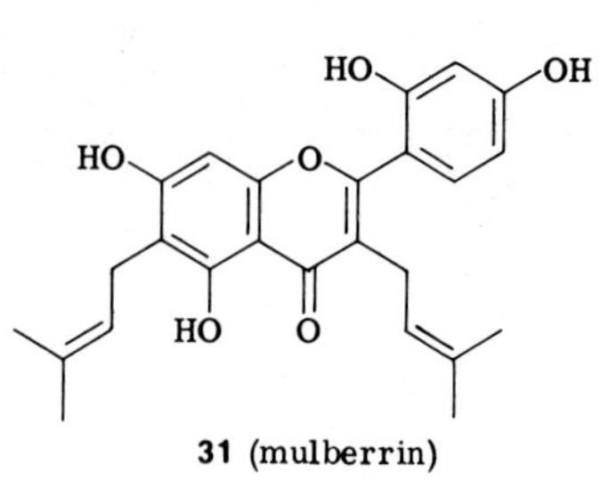

31 (mulberrin)

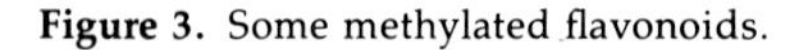

Figure 3. Some methylated flavonoids.

recent evidence suggests that they are in fact quite common. There are many other flavonol methyl ethers of more restricted natural occurrence. In the case of quercetin **(6)**, all five monomethyl ethers are known, and in addition at least seven isomeric dimethyl, two trimethyl, and three tetramethyl ethers have been described (Gottlieb, 1975). The frequency of methylation at the different hydroxyl groups of flavonols varies with their chemical reactivity and, although the 3-, 7-, and 3′-hydroxyls often carry methyl substituents, 4′-methylation is less common and 5-methylation is quite rare. Furthermore, if extra hydroxyl groups are substituted in the nucleus, e.g., in the 6 position to give quercetagetin **(16)**, this extra hydroxyl is often methylated at the same time thereby giving compounds such as patuletin **(17)**. In the case of flavones, the most common methyl ether is chrysoeriol [luteolin 3′-methyl ether **(28)**] but diosmetin [luteolin 4′-methyl ether **(29)**] and acacetin (apigenin 4′-methyl ether) are not uncommon, since both are known to exist in a range of higher plants.

Methylation in the flavonoid series masks the reactive phenolic groups and at the same time changes the physical properties significantly by introducing ether solubility into molecules that are otherwise hydrophilic. Whereas substances with one *O*-methyl group (e.g., peonidin, isorhamnetin, chrysoeriol) usually occur in glycosidic combination in the water-soluble vacuolar fraction of the leaf, those with two or more methyl groups [e.g., digicitrin **(12)**] are often present without sugar attachment in a lipid-soluble fraction. The location of such methylated flavonols and flavones within the leaf has not been fully explored, but lipid-like solubility would suggest that they are either in the cytoplasm or in the wax of the leaf surface. Indeed, highly methylated derivatives have been isolated from leaf waxes in some plants, e.g., *Eucalyptus,* and from bud exudates of *Populus* and *Alnus*. One structure from *Eucalyptus* leaf wax is sideroxylin **(30)**, which is unusual in being C- as well as O-methylated.

Finally, there is another structural feature in the flavonoid series that has the same effect as methylation on solubility: attachment of C_5 isoprenoid residues. One such compound is mulberrin **(31)**, which is present in the bark of *Morus alba*.

D. Glycosylation Patterns

Flavonoids usually occur in living cells as glycosides. This combination with sugar is important in the case of flower pigments in providing sap solubility, protection from enzymatic oxidation, and stability to light. Glycosylation may also be important in leaf flavonoids as a protective device for keeping otherwise toxic material in an inactively bound form within the plant. Certainly, there is much evidence that free flavonoids are more biologically active, e.g., as enzyme inhibitors, than the bound, glycosidic forms.

In its simplest form, glycosylation involves the attachment of a glucopyranose residue by a β linkage to the 3-hydroxyl group of, for example, quercetin to give quercetin 3-β-D-glucoside **(32)** (see Fig. 4). Sugars, however, may be occasionally attached to more than one of the many phenolic hydroxyl groups of a given flavonoid, and compounds such as quercetin 7,4′-diglucoside **(33),** present in onion scales, are known. Furthermore, other sugars besides glucose may be involved; probably the most common of all flavonoids is rutin **(34),** which has glucose and rhamnose attached as a disaccharide unit to the 3-hydroxyl of quercetin. In all, over 80 different glycosides of quercetin have been characterized so far in higher plants (Harborne and Williams, 1975), and many more undoubtedly await identification. Most of the other common flavonoids also occur in a wide variety of glycosidic combinations. Thus, there is considerable glycosidic complexity and variety in the flavonoid series.

The sugars most frequently found in association with flavonoids include all the common plant sugars in monosaccharide attachment. The mode of linkage is normally β in the case of D-sugars and α with L-sugars; an exception is arabinose, which is attached to quercetin in both α and β forms. In the case of disaccharides, some are more common than others, the two most frequent being rutinose and sophorose. A further structural complication in some glycosides is the presence of an acyl group attached to one or more of the sugar hydroxyl groups. Such acyl groups are commonly aromatic acids and are based on one of the four common hydroxycinnamic acids: *p*-coumaric, caffeic, ferulic, and sinapic acids. Anthocyanins with acyl attachment are fairly common in some plant families, a typical example being petanin **(35),** a pigment characteristically present in the potato and other Solanaceae. Derivatives of hydroxybenzoic acids are also known, as are compounds in which the acyl group is aliphatic, e.g. is acetic acid or malonic acid. One example of an acylated flavone is the acetylated acacetin tetraglycoside **(36)** from *Coptis japonica*.

One final complication in the case of flavone glycosides is that the sugar, instead of being bound in the usual way by ether linkage to a phenolic hydroxyl group, is joined directly to the flavonoid nucleus in the 6 or 8 position by a carbon–carbon bond. Such derivatives are referred to as flavone *C*-glycosides or *C*-glycosylflavones, two examples being 8-*C*-glucosylapigenin, or vitexin **(37),** and

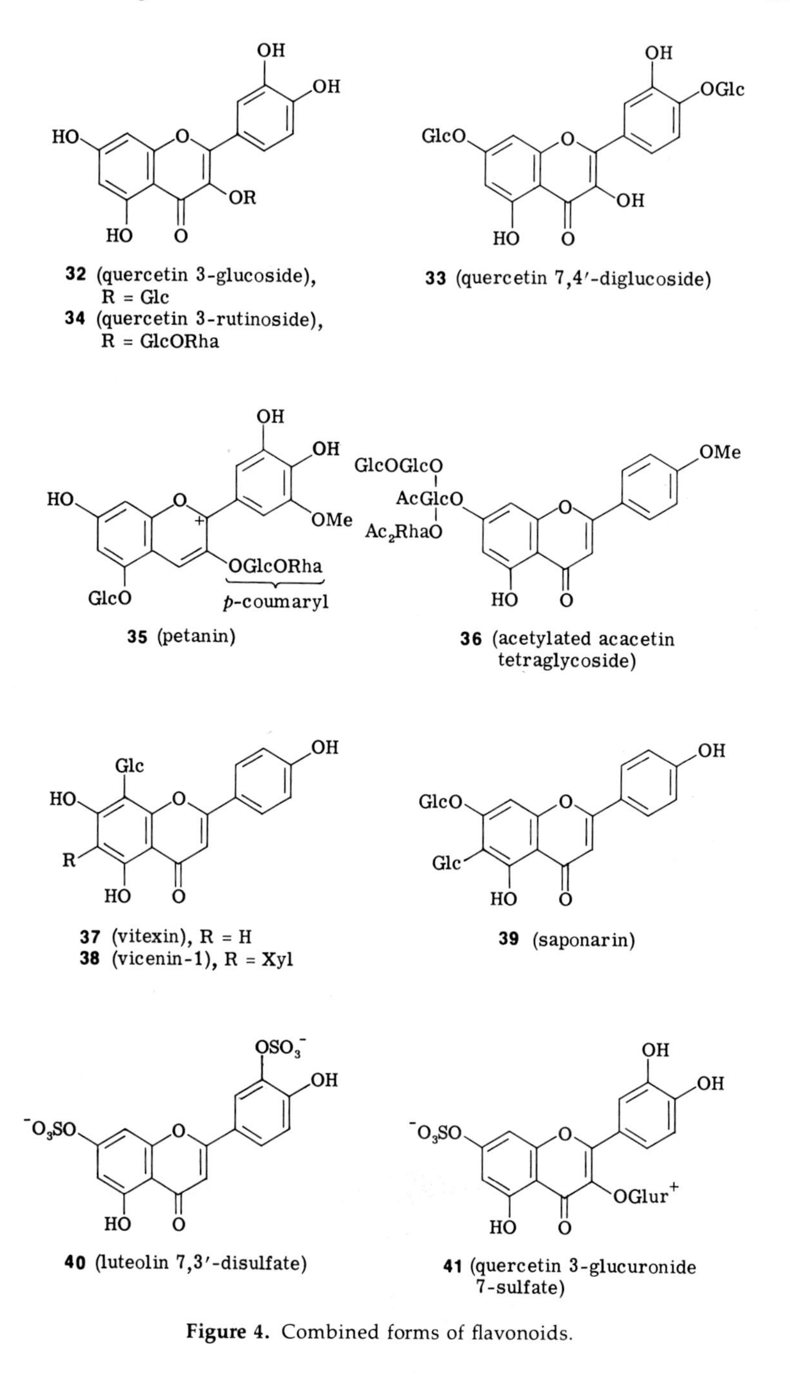

Figure 4. Combined forms of flavonoids.

6-xylosyl-8-glucosylapigenin, or vicenin-1 **(38).** *O*-Glycosides of *C*-glycosides are also possible, e.g., saponarin **(39),** the 7-*O*-glucoside of isovitexin present in barley leaves. Fortunately, C-glycosidic links are readily distinguished from O-glycosidic links by their much greater resistance to both acid and enzymatic hydrolysis (Chopin and Bouillant, 1975).

Besides being found in glycosidic combination, flavones and flavonols may also be present in a water-soluble form, covalently bound to inorganic sulfate. One such substance is luteolin 7,3′-disulfate **(40),** a major flavonoid constituent of eel grass, *Zostera marina*. Such compounds are anionic and exist with a cation, potassium, tightly bound. Because of their negative charge, these sulfates are readily distinguished from other flavonoids by electrophoresis, since they are mobile in acidic buffers of pH 2.2. Over 50 different flavone and flavonol sulfates have been characterized (Harborne, 1975b), nearly half of these having sugar attachments as well (e.g., **41**).

In spite of so much glycosidic variation, certain glycosides are much more abundant than others. In the case of the eight common aglycones (Fig. 1), these are the simple glucosides, and rutinosides are also regularly present.

E. Other Classes of Flavonoids

1. *Chalcones and Aurones*

Chalcones and aurones are of principal ecological interest because they contribute to the yellow color of the flowers of a number of plants, especially members of the Compositae. Nevertheless, they are coloring matters of other tissues, for example, heartwoods and seeds of certain legumes. They are readily detected as flower pigments by the fact that when yellow petals are fumed with the alkaline vapor of a cigar or a bottle of ammonia, there is a dramatic color change to orange or red. Because of this color response, which is not shown by the more common yellow carotenoid pigments, these flavonoids are known as anthochlor pigments.

Such anthochlors (Fig. 5) may be the only class of yellow pigment in certain flowers. For example, the chalcone isosalipurposide **(42)** is the sole yellow coloring matter of yellow carnations, whereas the aurone aureusidin, occurring as the 6-glucoside aureusin **(43),** is the major yellow pigment in the snapdragon, *Antirrhinum majus* (Scrophulariaceae). Anthochlors also are found regularly in yellow flowers in association

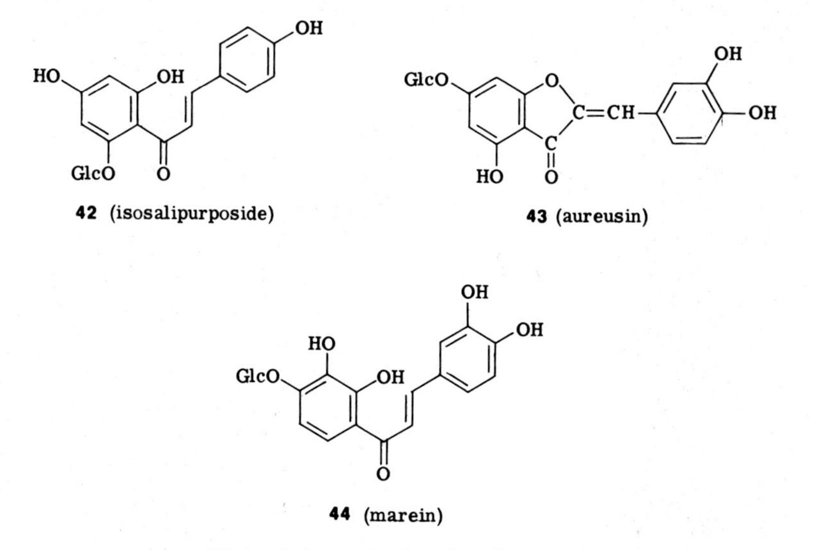

Figure 5. Some "anthochlor" flavonoids.

with lipid-soluble carotenoids. In such cases, the two types of pigment may be differentially distributed in the flower tissue in such a way that the carotenoid is spread over the whole flower and acts as a general attractant to an insect visitor, whereas the yellow chalcone is present only in the inner ray and provides a uv-absorbing honey guide to the center of the flower. This happens in *Coreopsis bigelovii,* where the chalcone involved is marein **(44)** (see Section IV,A).

2. *Flavanones and Dihydroflavonols*

Flavanones and dihydroflavonols are simple reduction products of flavones and flavonols, respectively, and in terms of biosynthesis are formed first and then undergo oxidation in the 2 and 3 positions to give rise to the more highly oxidized compounds. As biosynthetic intermediates, they are presumably universally present as trace constituents but do not usually accumulate in quantity in leaf or flower tissue. Nevertheless, the presence of the two flavanones naringenin **(45)** and eriodictyol **(46),** corresponding in structure to apigenin and luteolin, has been recorded with some frequency in plants (see Fig. 6). They are found in glycosidic form, and one series of flavanone glycosides occurring in *Citrus* fruit which have the disaccharide neohesperidose attached to the 7 position, as in naringin **(47),** are of special interest because of their bitter taste to human beings (see Section IV,B). Whether insects can

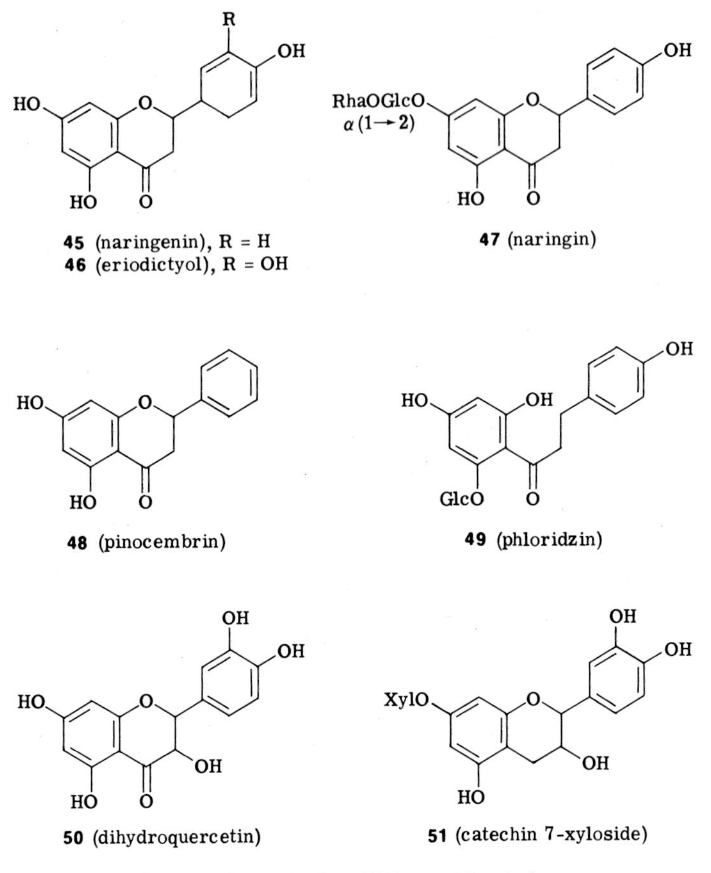

Figure 6. Some "reduced" flavonoids of plants.

perceive flavanone accumulation in tissues is not generally known, but one substance, pinocembrin **(48),** has been implicated as an attractant to a bark beetle feeding on fruit trees (Levy *et al.*, 1974).

On ring opening, flavanones give rise to the corresponding chalcones, and the two classes of flavonoids are closely interrelated. Indeed, there is an enzyme, chalcone–flavanone isomerase, present in plants that catalyzes this interconversion. One type may also be formed from the other during extraction and work-up of plant extracts unless special care is taken to use very mild (low temperatures and pH 7) conditions. Reduction of the chalcone formed by ring opening of a flavanone gives rise to yet another flavonoid class, the dihydrochalcones; such compounds are quite rare in nature. However, one member of this group, phloridzin **(49),** must be mentioned because of its great pharmacological activity in human beings in producing glucosuria due to its interference with the tubular reabsorption of glucose in the kidney. Phloridzin occurs universally in the genus *Malus* (apple) and can function in leaves and roots as an herbivore feeding deterrent because, like naringin, it has a bitter taste.

Dihydroflavonols, corresponding in structure to the three common flavonols (see Fig. 1), are all known, e.g., dihydroquercetin **(50),** and are found most frequently in the free state in the heartwood of trees. They also occur occasionally in glycosidic form in leaf tissues. More highly reduced flavonoids, often co-occurring with dihydroflavonols, are the related flavan-3,4-diols and flavan-3-ols, where the 4-carbonyl is successively reduced to OH and H. Flavan-3-ols are also known as catechins and may have significant biological activity. They are present mainly in wood tissue, where they may have an effect on arthropod feeders. Thus, catechin 7-xyloside **(51)** has been implicated as a feeding attractant to the elm bark beetle *Scolytus multistriatus* (Doskotch *et al.,* 1973).

3. *Isoflavones and Isoflavonoids*

Isoflavones such as genistein **(52)** are isomeric with the flavones (e.g., apigenin) and are derived biosynthetically from the same flavanone intermediate by aryl migration (see Fig. 7). They appear to be more restricted in their natural distribution than flavones, since they have been found regularly in only one subfamily of the Leguminosae, the Papilionoideae. There are, however, occasional records of their presence in Amaranthaceae (*Iresine*), Iridaceae (*Iris*), Myristicaceae (*Myristica*), and Rosaceae (*Prunus*). This suggests that they could be more widespread and, since there is no satisfactory method of screening plant tissues for their presence, our knowledge of their natural occurrence is certainly imperfect.

As a group, isoflavones and their derivatives are significantly more biologically active than the corresponding flavones. Although simple isoflavones may not be highly active as such, they can become active after structural modification. 6-Isopentenylgenistein, for example, together with the 2′-hydroxy derivative, luteone **(53),** is an effective antifungal agent of *Lupinus* leaves. More consistent fungitoxic activity is displayed by reduced derivatives, isoflavans, and pterocarpans such as medicarpin **(54),** which are formed as phytoalexins in many legumes. Reduction *in vivo* of isoflavones during ingestion by mammals also leads to biological activity of an estrogenic nature (see Section VII).

Finally, one other series of complex isoflavonoids, those containing

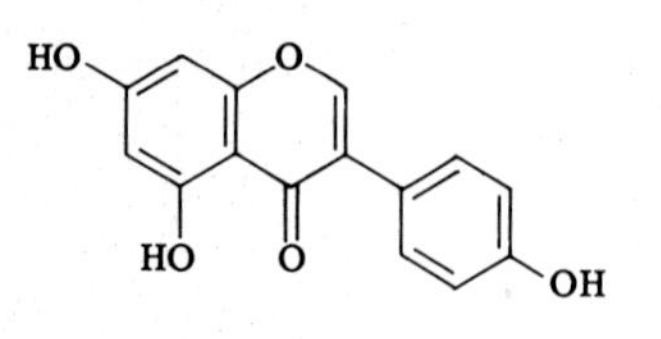

52 (genistein)

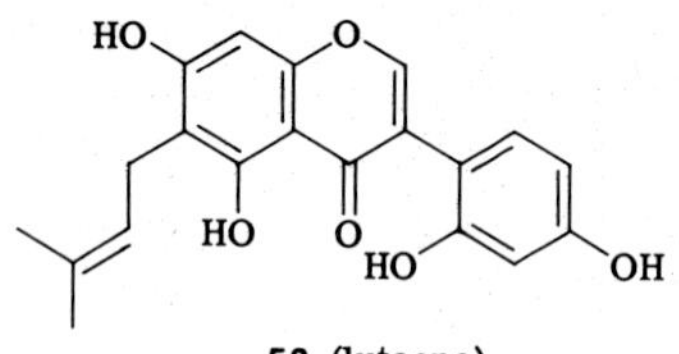

53 (luteone)

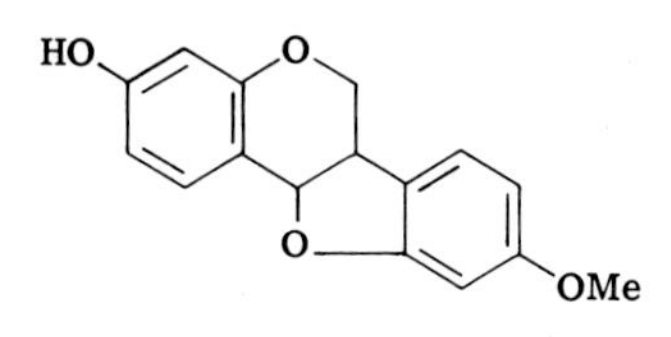

54 (medicarpin)

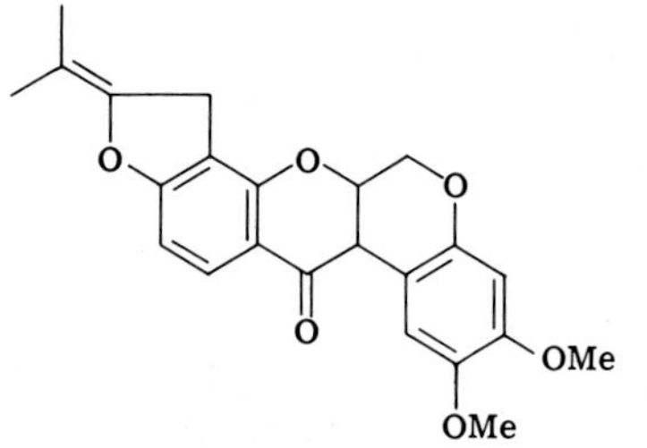

55 (rotenone)

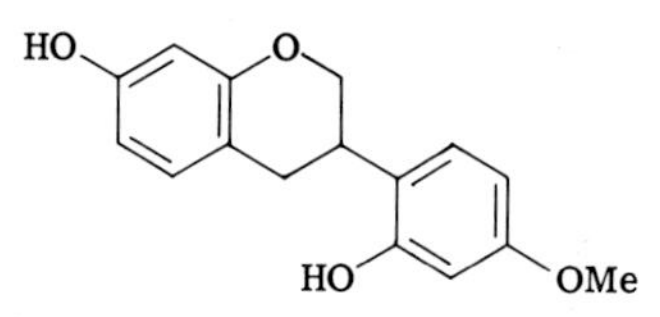

56 (vestitol)

Figure 7. Some isoflavones and related isoflavonoids.

isopentenyl substitution, must be mentioned: the rotenoids [e.g., rotenone **(55)**]. Their well-known insecticidal activity is discussed in more detail in Section IV,C.

III. ANALYTICAL ASPECTS

A. Qualitative Patterns

Because of their intense visible colors, anthocyanins are easy to identify in plant extracts. They are conveniently separated by paper or thin-layer chromatography. The color of the spots or bands so produced indicates at once whether the pigments are based on pelargonidin (orange), cyanidin, peonidin (magenta), or delphinidin and its derivatives (purple). The only important precaution to be taken in analysis, necessary because of the instability of anthocyanins at neutral or alkaline pH, is the incorporation of mineral acid in the extracting medium (e.g., 3% concentrated HCl in absolute methanol). Further details of the identification of anthocyanins are given by Harborne (1967). Other colored flavonoids (aurones, chalcones, yellow flavonols) can be extracted with neutral solvents (e.g., 95% aqueous ethanol) and are then similarly separated and identified.

Most of the flavonoids without visible color can be separated by simple chromatographic procedures, and the compounds are readily detected on chromatograms by their colors in uv light with and without ammonia vapor. Under such conditions, flavone and flavonol glycosides appear as dark-absorbing colors that become bright yellow, green, or brown when the paper is fumed with ammonia. In a few cases, the colors may be intensely fluorescent; such fluorescence is characteristic of flavonoids lacking a 5-hydroxyl or in which the 5-hydroxyl is masked.

Because of the number of different glycosides that may be present in a given plant tissue, it is usually advisable to carry out two-dimensional separation on paper or on a thin-layer plate. The patterns of spots may vary in different tissues, and it is important to separately screen extracts of the various plant organs.

It is often useful to examine, by one-dimensional chromatography, acid-hydrolyzed extracts at a fairly early stage in an investigation; this gives valuable information on the nature of the flavone and flavonol aglycones present in combined form. Once the pattern has been established by such means it is a relatively routine matter to identify the various constituents present in a given plant. Well-known substances can normally be identified by using a combination of paper chromatography, spectral techniques, and simple chemical and biochemical procedures (Harborne, 1967). For more complicated structures, increasing use has been made of nuclear magnetic resonance spectrometry (Mabry *et al.*, 1970) and mass spectral measurements (see Harborne *et al.*, 1975). Excellent reviews of analytical procedures for flavonoids have been written by Markham (1975) and Swain (1976).

B. Quantitative Patterns

The concentration of a colored flavonoid in a given plant extract can usually be determined spectrophotometrically. Anthocyanin concentrations are readily measured directly in floral tissues because there is rarely any interference from other plant pigments. Concentrations can be determined in methanol containing 1% concentrated HCl at the appropriate wavelength, depending on which of the three major anthocyanidin types are present (pelargonidin, cyanidin, and delphinidin derivatives have absorbance maxima at 520, 535, and 545 nm, respectively). Pigments can also be extracted with 1% aqueous HCl, although it takes longer and the pigment λ_{max} is at a slightly shorter wavelength (e.g., cyanidin derivatives have λ_{max} at 507 nm in aqueous solution). Anthocyanin concentrations can be similarly measured in the leaf, although the preferred extraction medium here is 1% aqueous HCl since, if methanol is used, the chlorophyll will be extracted at the same time and it interferes with the spectral measurement.

Yellow flavonoids are similarly measured in floral extracts by spectrophotometry, but some preliminary fractionation may be necessary. Thus, these flavonoids may be accompanied by carotenoids, which overlap with them in absorbance in the 400–430 nm region. In such cases, carotenoids can be removed from aqueous alcoholic petal extracts by being taken into petroleum, and measurements can then be made between 350 and 420 nm on the remaining aqueous residue. Spectra of practically all flavonoids are shifted to longer wavelengths of light by the addition of traces of aqueous alkali, alcoholic aluminum chloride, or other inorganic salts to the cell solution. Measurements of spectra in the presence of such reagents can sometimes be useful in avoiding interference by nonflavonoid substances with similar absorption maxima (Swain, 1976).

In determining the levels of colorless flavonoids in crude leaf extracts by spectrophotometry, there are considerable problems due to the presence of mixtures of closely related compounds with differing wavelength maxima and because of interference by other classes of uv-absorbing plant constitutents. Thus, fractionation is practically always necessary, except in the rare cases in which there is a very high concentration of a single flavonoid (e.g., rutin in buckwheat leaves, *Fagopyrum esculentum*). Preliminary purification can be done by paper chromatography, but fractionation on columns of, e.g., Sephadex or by electrophoresis may be useful. High-performance liquid chromatography is an analytical tool being developed for flavonoids (Van Sumere *et al.*, 1979) but, even in this procedure, some preliminary purification of the extract is necessary.

Data on quantitative aspects of flavonoid accumulation in plants are rather sparse. Such information is important when one is considering flavonoid function, and it is hoped that in the future more attention will be given to such determinations. In the case of anthocyanins in flowers, pigment concentrations vary from 0.3 to 14% dry weight. In exceptional cases of deeply pigmented flowers (e.g., black tulip) or deeply pigmented fruits (e.g., blackberry), their natural levels are probably higher and may represent up to 30% of the dry weight.

Less information is available on flavonoid concentrations in leaves. Such values are, of course, affected by variations due to stage of growth, physiological status, and environmental pressures. However, many measurements of rutin concentrations in plants have been made because of its pharmacological value, and concentrations of about 3% dry weight are commonly recorded in plants where it is a major leaf constituent. Since rutin may be only one of a variety of flavonoids present in any one species, this result suggests an average value in most angiosperms of 5% dry weight for the total flavonoid content, with lower and higher limits of about 1 and 10%. Higher values must certainly pertain on occasion; the total catechin content of tea leaves is usually about 20% dry weight or more. Since biologically active flavonoids are effective *in vitro* in solutions at concentrations of 0.1 to 1%, the above figures indicate that the natural amounts of flavonoids, as they accumulate in plant tissues, are more than sufficient to produce an effect on a grazing animal or phytophagous insect.

IV. GENERAL BIOLOGICAL PROPERTIES

A. Color

The most striking impact of flavonoids on animals is achieved through their contribution to plant color. This is true for human beings as well as for birds and insects. Man's aesthetic appreciation of ornamental plants is based largely on form, shape, and color of these plants. Likewise, in plants eaten for food, color plays a significant role in our gustatory appreciation and response.

Flower color is one of several attractants that draw a range of animals to plants for pollination purposes. The importance of anthocyanin color in flowers is to provide a contrast to the uniform green background that plants generally present to animal visitors. Certain pollinators (e.g., birds, bees, butterflies) have developed well-known color preferences, and the coevolutionary adaptation of plants to their pollinators includes modification of flower color. Animals, of course, benefit from flower visiting by the nutritional rewards of nectar and pollen (see Harborne, 1977a).

The relationship between anthocyanin chemistry and flower color is fairly straightforward. *In vitro,* increasing hydroxylation of the B ring of the anthocyanidin molecule shifts the color from orange and scarlet through red and magenta to purple and mauve. *In vivo,* it is clear that there are a number of modifying factors, especially copigmentation with flavones and the presence of metal ion, which have an effect in determining the final color of the flower (for details, see Harborne, 1976).

The attraction of ripe fruits for seed dispersal by animals is also often based on anthocyanin color. In probably about 50% of angiosperm fruits, coloration is due to flavonoids; in the remainder, carotenoids and other classes of pigments predominate. Many red colors are due to anthocyanin in medium concentrations, whereas intense purple and purple-black colors are provided by high concentrations of pigment in seed coat or skin. Other flavonoids are occasionally responsible for orange or yellow-colored fruits. Thus, the isoflavones osajin and pomiferin contribute color to the rind of the osage orange *Maclura pomifera* (Moraceae), whereas chalcones pigment the yellow fruits of *Merrillia caloxylon* (Leguminosae).

A new role for flavonoids has been discovered in the attraction to plants of insects, especially bees, that have perception in the uv range of the spectrum. It has been found in certain flowers that mainly yellow flavonoids not only contribute to visual color but also exhibit intense uv absorption. In these plants, such flavonoids are specifically located in the cells of the upper surface of the flower corolla and only in the inner petal or ray. As a result, they act as honey or nectar guides, invisible to the human eye but visible to bees. These flavonoids by their presence in surface cells mask any carotenoids present and thereby prevent their normal reflectance of uv light.

Although most of the flavonoids implicated as honey guides are yellow, some are cream or white (e.g., in the sunflower *Helianthus annuus*). Possession of yellow color per se is not, therefore, a prerequisite of this phenomenon. Indeed, there is evidence that "colorless" flavonoids act as uv guides in some white flowered plants. The fact that all flavonoids absorb strongly in the uv region means that some insects may be able to detect their presence in other tissues besides flowers, but this has yet to be demonstrated experimentally.

B. Taste

The taste properties of natural flavonoids have rarely been studied to any degree in an ecological context. Both bitterness and sweetness are, however, present in one class of flavonoids, the flavanones, and these tastes have been explored in detail with regard to the presence of flavanones in *Citrus* fruits (Horowitz, 1964). Bitterness is associated with the combination of a particular disaccharide [neohesperidose, Rha-α-(1 $\rightarrow$ 2)-Glc] with the flavanone nucleus. Any structural modification diminishes or destroys the taste property; the isomeric flavanone 7-rutinosides [Rha-α-(1 $\rightarrow$ 6)-Glc], for example, are essentially tasteless. The bitterness due to naringin (naringenin 7-neohesperidoside) is significant in comparison to the standard bitter alkaloid, quinine, and on a molar basis has one-fifth of its bitterness. Most of the soluble bitterness in *Citrus* fruits is due to naringin and related flavanones.

On the basis of the above results, it is suggested that, in the genus *Citrus*, flavanone glycosides, by imparting a bitter taste to fruits, provide a barrier to feeding by mammals with taste perceptions similar to those of human beings. No tests have been done on mammals other than human beings, and it is doubtful whether bitterness due to flavanone glycosides is produced in other plants to any appreciable extent. Flavanones with taste properties have been recorded in only a few instances in other families besides the Rutaceae. Thus, the presence pinocembrin 7-neohesperidoside, which is bitter, was recorded in fruit of *Sparattosperma vernicosum* (Bignoniaceae) (Kutney *et al.*, 1970), a plant of the Brazilian jungle, but it is not known whether it protects the fruit from being eaten by undesirable predators.

Another taste property, that of sweetness, has been recorded in a few dihydrochalcones, but intense sweetness is associated at present only with synthetic compounds derived by ring opening and reduction from the intensely bitter flavanones. The presence of both bitterness and sweetness in molecules of the flavanone series suggests that other flavonoids, more widespread in occurrence, can have taste properties and be potentially repellent to herbivores. Unfortunately, very few compounds have been tested on human beings, and it is not the normal practice of phytochemists to so examine new, natural flavonoids. Further research here, however, would seem to be important and certainly essential if a role for flavonoids as feeding deterrents is to be confirmed.

C. Toxicity

Present evidence indicates that the common flavonoids are apparently harmless to human beings. Practically all vegetables, fruits, and processed foods and drinks based on plant materials (e.g., tea) contain these compounds, so that relatively large quantities are consumed daily. No deleterious effects have been associated so far with such ingestion. Furthermore, certain bioflavonoids (hesperidin, rutin, etc.) are effective in the treatment of capillary fragility in human beings, although a vitamin-like activity, once associated with these compounds, is no longer accepted. These substances have also been proposed to medicinal use as antiinflammatory agents (Gabor, 1972).

The absence of any apparent toxicity might well be related to the presence of an efficient detoxifying system in human beings, and the fact that most flavonoids contain several phenolic groups means that they are very readily conjugated with glucuronic acid or sulfate as a first step in their detoxification. The possibility remains that those flavonoids in which most or all of the hydroxyl groups are masked by methylation can be less readily detoxified and thus be in a position to produce symptoms of toxicity. The only evidence has been obtained from studies of rats, not human beings. Stout *et al.* (1964) showed that tangeritin (5,6,7,8,4′-pentamethoxyflavone), which occurs in tangerine peel, is cytotoxic in human beings and also causes neonatal death in rats. When fed at the rate of 10 mg per kilogram body weight each day to female rats during gestation, tangeretin caused the death of 83% of the offspring, either at birth or within 3 days of birth.

A range of other methylated flavones and flavonols was shown by Kupchan and co-workers (1969) to be cytotoxic and active against malignant cells in culture. Clearly, although such compounds may function as antitumor agents, they may also be hazardous to normal cells *in vivo*, so that some degree of toxicity may well be associated with ether-soluble methylated flavonoids of one type or another. Finally, it can be observed that even unmethylated flavonoids can have long-term toxic effects on mammals, since quercetin itself has been found to have mutagenic effects in microorganisms (Bjeldanes and Chang, 1977).

With insects, toxicity of common flavonoids has been demonstrated in tobacco budworm, *Heliothis virescens*, tobacco bollworm, *H. zea*, and *Pectinophora gossypiella* (Shaver and Lukefahr, 1969). When fed at low concentrations, quercetin and three of its glycosides (the 3-rhamnoside, 3-glucoside, and 3-rutinoside) repelled feeding by these insects; when applied in the diet at concentrations of 0.2% or over, these common flavonoids killed the larvae.

Undoubtedly, other insects are better adapted to flavonoids in the diet than these three crop plant pests. For example, the silkworm *Bombyx mori* requires the flavonoids present in mulberry leaf as dietary feeding stimulants, so presumably it is well adapted to dealing with any possible toxic effects. So few data are available on this topic that a wide screening of possible toxic effects in insects is urgently needed, in order to assess the general situation and to determine whether particular structural modifications (different sugar substitutions, methoxylation, etc.) are associated with toxic symptoms.

The only group of flavonoids that are known to be especially toxic to insects are the isoflavonoid-based rotenoids, widely used commercially as insecticides. This group of about 10 structures are present, principally in the roots but also in aerial parts, in tropical legumes of the genera *Derris, Lonchocarpus, Mundulea,* and *Tephrosia.* The plant roots were originally used by native peoples as fish poisons but have been employed since about 1911 as a source of active insecticides in Western countries. The major compound is rotenone **(55)** (see Fig. 7), the active principle of derris root, *Derris elliptica.*

Rotenone has a dramatic effect on oxygen uptake because it specifically inhibits the NADH-dependent dehydrogenase step of the mitochondrial respiratory chain. This fact itself does not explain why rotenone is so potent against fish and insects but not against mammals (apart from swine). However, this apparent selectivity and the alleged nontoxicity in human beings of rotenone is often exaggerated since, in higher concentrations, rotenone can be as harmful to warm-blooded animals as to insects. Indeed, rotenone-containing roots have been employed for committing suicide by New Guinea natives and as a basis for arrow poisons in Sumatra.

As an insecticide, rotenone is potent against a wide range of pests, and its application to agricultural crops controls many leaf-chewing beetles, caterpillars, flea beetles, and aphids. Continued use of rotenoids over a number of years has led inevitably to the development of resistance in certain pests, e.g., the Mexican bean beetle. Natural resistance is also shown by a number of potential plant pests, such as the grasshopper *Melanoplus femur-rubrum.*

The toxicity of rotenone when applied to plant crops is limited by the poor water solubility of this compound, but it is nevertheless very active at low concentrations. Aphids are killed by solutions of 1 : 300,000, and the lethal dose in the silkworm is 0.003 mg per gram body weight. In the natural environment of the tropical forest, rotenoids presumably act as feeding deterrents to animals, although they apparently do not have an unpleasant taste. For example, silkworms fed on mulberry leaves treated with rotenone consume up to 30 times the lethal dose before succumbing to its harmful effects.

The relationship between structure and insecticidal activity is not entirely clear, although rotenone seems to be the most effective of the known natural rotenoids (Fukami and Nakajima, 1971). Rotenone is optically active, and its biological properties are reduced when this activity is lost. Although it is not known whether the complete rotenone skeleton is required for insecticidal activity, it is of interest that a simpler isoflavonoid molecule, the isoflavan vestitol **(56)** (Fig. 7), has been shown to be toxic to the black beetle larvae of *Heteronychus aratoo,* which feed on *Lotus* roots. Resistance of *Lotus pedunculatus* to insect attack and the corresponding susceptibility of *L. corniculatus* seem to depend on the presence of this isoflavan in the roots of the former plant. The crude root extract of this *Lotus* species is quite toxic, the LD_{50} in the larvae being 0.307 gm fresh weight (Russell *et al.*, 1978).

D. Interactions with Enzymes

The effectiveness of flavonoids in the environment as toxins may be indirect, via their effect on proteins or enzymes, for the phenolic hydroxyl group is known to have a strong affinity for proteins. The isolation of active enzyme preparations from plant cells is frequently severely hindered by the presence of flavonoids and other phenolics in the same cells; the situation can be satisfactorily ameliorated only by adding a selective adsorbent, such as polyvinylpyrrolidone, to trap the flavonoids before they are able to bind irreversibly to active enzyme. The interactions of proanthocyanidins (condensed tannins) with proteins are also well documented (see Chapter 19 by Swain), and it is known that the presence of proanthocyanidins in plant tissues can seriously reduce the nutritional value of plant proteins to herbivores.

Although it is not known how far monomeric flavonoids e.g., quercetin glycosides, can interact with protein and thus reduce plant palatability, it is clear that these compounds are potent enzyme inhibitors. Thus, they can interfere with vital processes within the animal cell, unless detoxified before any harmful effects are induced. Enzyme inhibition is a general property of many common flavonoids, particularly when they are present in unbound form without sugar attachment. Thus, quercetin at physiological concentrations (between 10^{-4} and 10^{-6} M) is a strong inhibitor of at least a dozen enzyme systems, including IAA oxidase, RNA polymerase, and ATPase (Wagner, 1979). It has also been shown to inhibit aldose reductase, but in this case its glycoside, quercitrin, is even more active than the aglycone.

Another discovery (Racker, 1975) has been that quercetin specifically inhibits calcium ion transport in Ehrlich ascites tumor cells by interfering with Ca-dependent ATPase activity. Fewtrell and Gomperts (1976) have extended these observations on calcium transport and have also shown that, in rat peritoneal mast cells, quercetin and fisetin at 100 μM completely block the secretion of antigen-induced histamine. These results indicate that common flavonols may have serious effects on cell membranes and on calcium transport in animal cells, which, although not fatal, may reduce the fitness of the individual. Such possible toxic effects via enzyme interactions must be considered when the dietary effects of flavonoids on herbivores are being assessed.

V. DETOXIFICATION IN ANIMALS

The pathway of orally fed flavonoids in mammals has been extensively investigated in human beings and in rhesus monkeys, rats, guinea pigs, and rabbits. The fate of flavonoids *in vivo* is complicated by the fact that gut microflora from the lower intestine considerably modify and degrade the flavonoids to simple phenolic acids (Fig. 8). Indeed, Griffiths and Barrow (1972) showed that flavonoid glycosides, such as rutin, are largely excreted unchanged in urine and feces when fed to germ-free rats at the rate of 100 mg per rat. Aglycones, such as catechin, similarly fed were recovered in conjugated form largely as the corresponding 3-glucuronide. Thus, all the usual metabolic products (phenolic acids) recovered in the urine after feeding flavonoids to mammals are in fact due to bacterial degradation rather than to the activities of mammalian detoxifying enzymes.

Bacteria also have the metabolic capacity to remove sugars from glycosides, such as rutin, so that some interconversion of glycosides happens, and fed rutin may be recovered in part as quercetin 3-glucuronide. However, most is catabolized by the pathway shown for myricetin in Fig. 8 and is recovered as *m*-hydroxybenzoic acid. Other classes of flavonoids follow slightly different pathways (Fig. 8), but all compounds are finally catabolized to benzoic acid or to a simple phenolic acid. Anthocyanins are similarly degraded, pelargonidin giving rise in rats to *p*-hydroxyphenyllactic acid (Griffiths and Smith, 1972). The absence of A-ring fragments from these breakdown products is due to the lability of the phloroglucinol moiety thus produced, which is so rapidly oxidized that it is never recovered as such.

The metabolic fate of isoflavonoids in mammals is also affected by the presence of microbial flora in the intestine. Although degradation to B-ring fragments occurs in the case of genistein, it is not always necessary for the C_{15} skeleton to be degraded, and C_{15} products have been identified. Thus, biochanin A and especially formononetin are converted by dehydroxylation, demethylation, and reduction to the isoflavan equol, which has long been known to be present in the urine of mares and goats. The biological significance of the incomplete breakdown of isoflavonoids in mammals is further discussed in Section VII.

In contrast to the fund of information on flavonoid metabolism in mammals, practically nothing is known of the fate of ingested flavonoids in other animals. However, something can be inferred about their fate in butterflies and moths, since Ford (1941) found that flavonoids accumulate in the body and wings of about 10% of insects surveyed. Using a simple color test, Ford found flavones in 38 of 328 butterfly genera and in 10 of 192 moth genera. Morris and Thomson

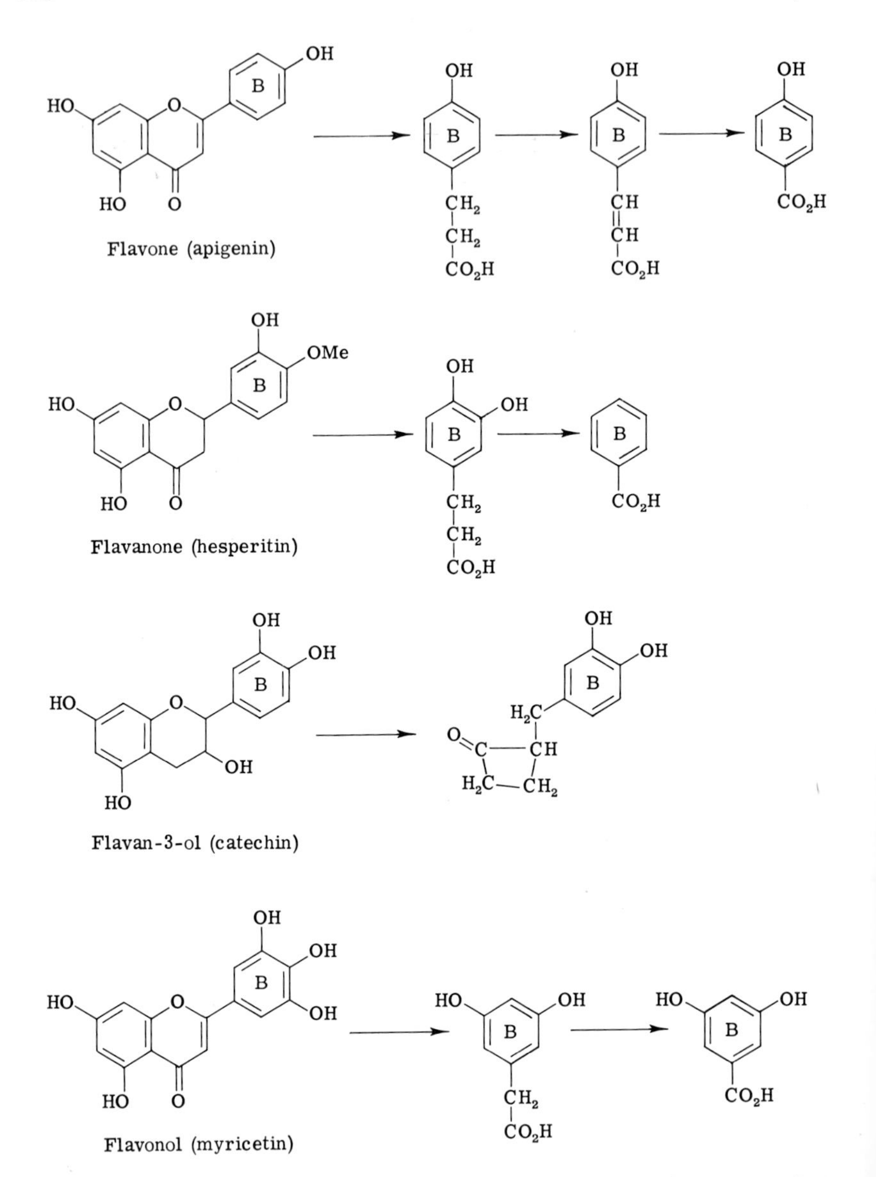

Figure 8. Degradation of flavonoids by mammalian gut flora, yielding various B-ring fragments.

(1963, 1964) investigated this situation in more detail in the case of the small heath (*Coenonymphia pamphilus*) and the marbled white (*Melanargia galathea*), and they identified the flavones tricin and orientin and their glycosides in the butterfly wings. The larvae of both these insects feed on the common grasses *Dactylis glomerata, Poa annua,* and *Festuca ovina,* the major flavonoid leaf components of which are indeed tricin and orientin derivatives. It is apparent, therefore, in these two insects that dietary ingested flavonoids are not detoxified but instead are stored in the body and wings in more or less unchanged form.

Whether the same is true for the other insects reported by Ford (1941) to be flavone-positive will be known only after further investigation. According to this author, the presence of flavones in the wings follows morphological classification, but this may not mean very much since butterflies within a particular genus or family presumably tend to feed on similar plants (Ehrlich and Raven, 1965). The most important question raised by this work is the fate of flavonoids in the 90% of Lepidoptera apparently lacking flavones in the wings. It is difficult to believe that the larvae of these insects never encounter flavonoids as dietary ingredients. It seems more likely on the basis of present evidence that they possess an efficient detoxifying system and successfully break down the flavonoid skeleton to small fragments. The mode of degradation is not known, although a potential degrading enzyme (phenolase) has been detected in the saliva of most phytophagous insects (Miles, 1968). Experimental studies along these lines are sorely needed.

VI. EFFECT OF FLAVONOIDS ON INSECT FEEDING

Although the role of flavonoids in insect feeding behavior has not been as well studied as that of other secondary constituents such as the glucosinolates, there is evidence that these substances can determine the feeding of certain phytophagous insects on particular host plants (Schoonhoven, 1972). Most interest has centered on the silkworm *Bombyx mori,* which specifically feeds on leaves of mulberry (*Morus alba*) and is attracted to them by, among other substances, morin and isoquercitrin (quercetin 3-glucoside). Other flavonoids that act similarly as feeding attractants include 6-methoxyluteolin 7-rhamnoside in alligator weed (*Alternanthera phylloxeroides*), catechin 7-xyloside in elm bark, and various flavanones in *Prunus* bark, which stimulate the feeding of different beetles (Table II).

The point can be made that nearly all these compounds are glycosides, so that the conjugation of the flavonoid with a specific sugar is a sig-

Table II. Flavonoids as feeding attractants in plants

Animal	Plant host	Flavonoid(s) implicated	Reference
Beetles			
Agasicles sp. nov.	*Alternanthera phylloxeroides* leaf (Amaranthaceae)	6-Methoxyluteolin 7-rhamnoside[a]	Zielske *et al.* (1972)
Scolytus mediterraneus	*Prunus* spp. bark (Rosaceae)	Taxifolin, pinocembrin, dihydrokaempferol	Levy *et al.* (1974)
S. multistriatus	*Ulmus europea* bark (Ulmaceae)	Catechin 7-xyloside[b]	Doskotch *et al.* (1973)
Moth			
Bombyx mori	*Morus nigra* leaf (Moraceae)	Morin, quercetin 3-glucoside[c]	Hamamura *et al.* (1962)

[a] Concentration in the plant is ca. 1 gm/5 kg fresh wt.

[b] A wax constituent, lupeyl cerotate, is needed as well to initiate feeding.

[c] A series of other constituents, including essential oils, are also involved as feeding stimulants to the silkworm.

nificant feature in the relationship between chemical structure and insect behavior. Furthermore, evidence has been obtained with silkworm larvae fed on artificially supplemented diets to show that the nature of this sugar is important in insect response. Thus, substitution of quercetin 3-glucoside by either the 3-rhamnoside or 3-rutinoside leads to repellency and a refusal to eat. A receptor site in the larval antennae that is sensitive to repellent chemicals is triggered off by the latter two compounds. It is remarkable that such a small change in chemical structure, i.e., altering the 3-sugar from glucose to rhamnose, should have such a profound effect on feeding behavior.

Such effects on insect feeding are not confined to *Bombyx mori.* The same three compounds and morin tested on other insects give a range of responses (Table III). *Anthonomus grandis* is repelled by rutin but stimulated to feed by both the 3-glucoside and the 3-rhamnoside. Larvae of a third insect, *Pectinophora gossypiella,* are actually killed if fed on diets containing sufficient amounts (0.2%) of these three flavonol glycosides. In addition to the insects listed in Table III, cabbage white butterfly (*Pieris brassicae*) larvae are also known to be highly sensitive to flavonoids when fed artificial diets (Schoonhoven, 1972).

These data indicate that it may be of selective advantage to have within the plant leaves a mixture of glycosides of a given flavonol, since these substances can be of value as general feeding deterrents. Thus, the

Table III. Effects of four dietary flavonoids on feeding behavior of different insects

Insect	Flavonoid[a]			
	Morin	Quercitrin	Rutin	Isoquercitrin
Anthonomus grandis	−	+	−	+
Bombyx mori	+	−	−	+
Heliothis virescens[b]	−	+	+	+
Pectinophora gossypiella[b]	−	−	−	−

[a] Key: −, deterrent; +, stimulant. Data from Schoonhoven (1972).
[b] In these insects, the flavonoids are toxic to the larvae when supplied in quantity.

presence of several different plant glycosides can cause a range of insect species to avoid feeding. On the other hand, the presence of a single plant flavonol glycoside can act as a barrier to fewer potential feeders. In practice, the fact that flavones and flavonols occur normally in plants as relatively complex mixtures of different glycosides (see Section II,D) supports this idea. Control of insect predation is probably only one of several possible explanations for the considerable complexity of flavonoid conjugation found in plant tissues.

Less attention has been paid to the possible deterrent value of lipid-soluble flavonoids in leaves or on leaf surfaces. However, Rhoades and Cates (1976), in an investigation of the antiherbivore chemistry of the creosote bush *Larrea tridentata,* found that the phenolic resin on the leaves seriously deterred feeding by grasshoppers, a katydid, and a geometrid moth larva both in the laboratory and in the field. Among the resin constituents are several methylated flavonoids, such as quercetin 3,7-dimethyl ether; the possible toxicity of these compounds to animals has been referred to (see Section IV,C). It was also suggested by Rhoades and Cates that, when present together with other phenolic components as in *Larrea,* lipid-soluble flavonoids form a matrix, which binds irreversibly to protein in much the same way as condensed tannins (see Chapter 19 by Swain).

One last feature of the biochemistry of flavonoids supporting the view that they contribute, together with other secondary constituents, to the antiherbivore chemistry of angiosperms is that their patterns of distribution conform to such an idea. Thus, there is a significant changeover in leaf flavonoid with evolutionary advancement, particularly with the replacement of the woody by the herbaceous form (see Harborne 1977b). In summary, it is apparent that flavonoid-based tannins, so characteristic of woody plants, disappeared from herbs to be replaced by low molecular weight flavonoids, which can accumulate in quantity. These may be flavone and flavonol glycosides in more specialized herbaceous groups (e.g., Umbelliferae) and eventually may be 6-methoxylated and more highly substituted derivatives in the most highly specialized families (e.g., Tubiflorae plants). This evidence of a coevolutionary pressure on angiosperms caused by animal feeding is supported by the response through constantly shifting alterations in flavonoid constituents. Such changes in pattern are apparent not only in the wider scope of genus, family, and order but also in the narrower confines of species populations. The considerable geographic variations in flavonoid profiles within populations of certain cosmopolitan plant species can be satisfactorily explained at present only by variations in predation (Harborne, 1975a).

VII. EFFECT OF ISOFLAVONOIDS ON MAMMALIAN REPRODUCTION

The discovery that isoflavonoids have estrogenic activity in mammals was made during the 1940's when sheep in Australia were allowed to graze for longer periods than usual on pastures containing subterranean clover, *Trifolium subterraneum.* As a result of this practice, lambing percentages were seriously reduced (to less than 30%); an active material causing this infertility was traced to the clover plant. The agent was eventually isolated and identified as a mixture of two isoflavones: genistein and formononetin (Bradbury and White, 1954). Structural comparison (Fig. 9) with the animal hormone estrone and its most active synthetic analog, diethylstilbestrol, shows why these isoflavones are estrogens—they mimic the steroidal nucleus of the natural female hormone. Although they are, in fact, rather weak estrogens on a molar basis (Biggers, 1959), they are presumably effective because of the relatively large quantity (up to 5% dry weight) in the clover fodder.

A more active substance, coumestrol, was later isolated by Bickoff (1968) from alfalfa, *Medicago sativa,* and ladino clover, *Trifolium repens.* Although coumestrol is 30 times more active than genistein or formononetin, its concentration in leguminous fodder plants is generally much lower, so that it is probably less effective *in vivo* than the isoflavones. In fact, recent research (Shutt, 1976) has shown that formononetin and not genistein is the most important estrogen in clover to sheep. This is because it is a proestrogen, being converted (by demethylation and reduction) to a more active substance, the related isoflavan equol, within the animal body (see Fig. 10). This isoflavan was actually isolated from pregnant mares' urine as long ago as 1932 by Marrian and

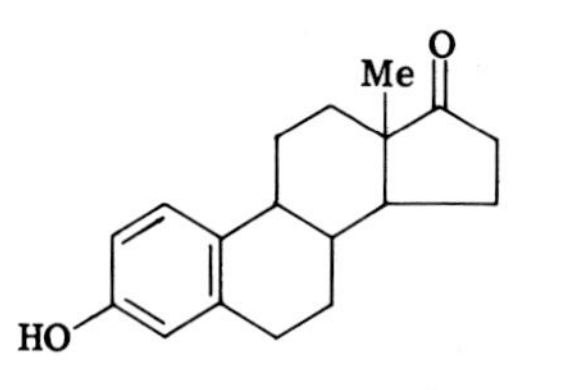

Estrone

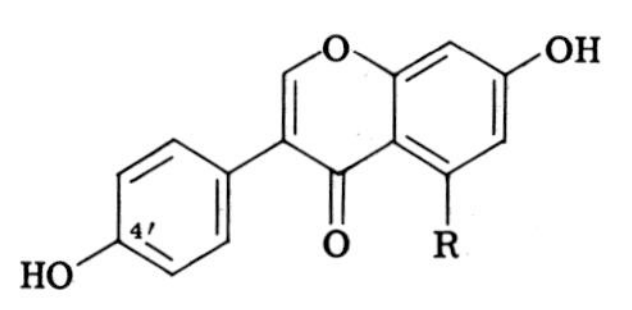

Genistein, R = OH
Daidzein, R = H
Formononetin, R = H (Me at 4′-OH)

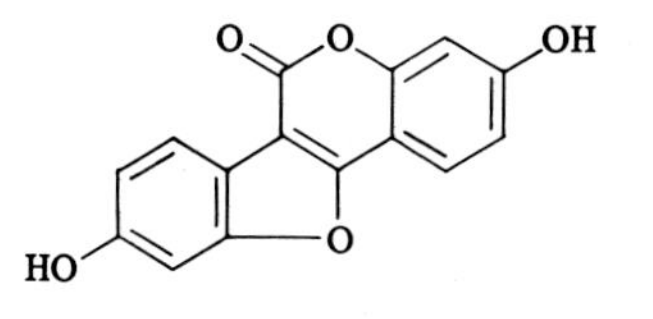

Coumestrol

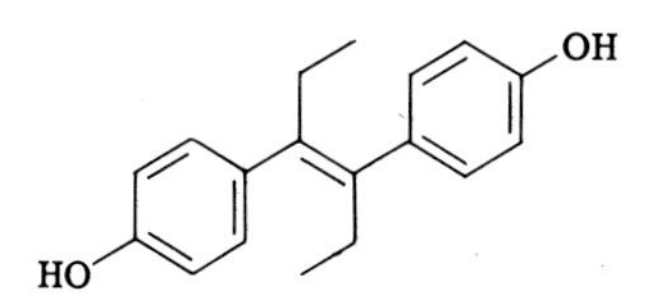

Diethylstilbestrol

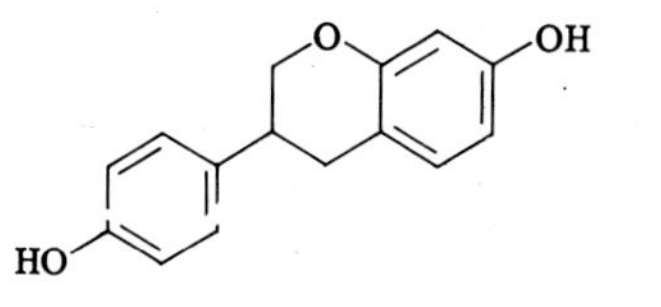

Equol

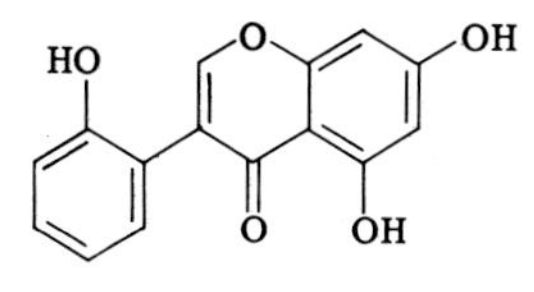

Isogenistein

Figure 9. Isoflavonoids as estrogenic mimics.

Haslewood. Comparative metabolic studies with genistein indicate that it is degraded in the rumen to inactive products, one of which is *p*-ethylphenol, so that the effective estrogenic activity of clover when eaten by sheep is largely due to formononetin content.

All isoflavones are estrogenic when given by parenteral injection to animals, and much work has been done on structure–activity relationships in the isoflavonoid series using mice (Biggers, 1959). For example, it can be shown that the two para-substituted hydroxyl groups are needed for maximal activity; transfer of the 4′-hydroxyl in genistein to the 2′ position (to give isogenistein) reduces estrogenic activity by 75% (Fig. 9). The key point to be emphasized, however, is that it is the metabolites produced *in vivo* from a given plant isoflavone that determine if estrogenic upset is likely. It is the activity of metabolites such as

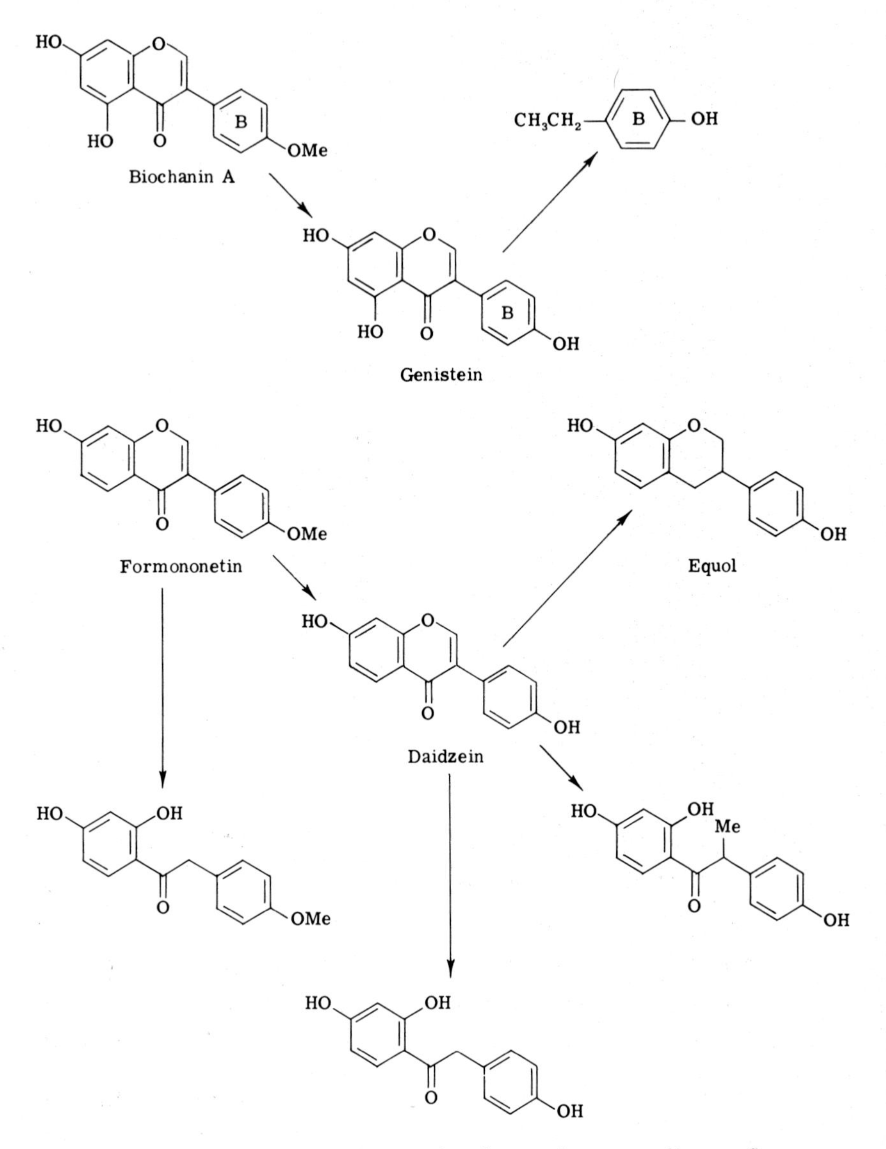

Figure 10. Degradative pathways of isoflavones by mammalian gut flora.

equol that is important in causing reproductive disturbance in farm animals and not that of the parent compounds present in the plant.

From the agricultural viewpoint, the presence of isoflavonoids in clovers and other leguminous fodder plants is a hazard to farm animals because of their effects on reproduction. Symptoms produced include difficult labor, infertility, and lactation in unbred ewes. The problems caused by estrogens of plant origin result from their high concentration in the blood plasma of the animals ingesting the plants relative to the endogenous hormone levels. Thus, a weak plant estrogen can exert a significant estrogenic effect and hormonal imbalance, even though its activity is only 10^{-3}–10^{-5} times that of estrone or estradiol. Plant estrogens can also act as antiestrogens by competing more effectively for receptor proteins in estrogen-sensitive tissues than the endogenous hormones, which may be present in lower concentration (Shutt, 1976; Labov, 1977). Although these effects are most pronounced in sheep, they also occur to a lesser extent in cows and in other farm animals. Presumably, most wild mammals would be likely to be affected by ingestion of phytoestrogens to some degree.

Efforts have been made to deal with the agricultural problem caused by the presence of estrogenic substances in clover. A survey of *Trifolium* showed that 18 species have as high an isoflavone content as *T. subterraneum,* most other species having relatively smaller amounts. Plant breeding experiments have been carried out in the pasture clovers, and strains with a safe, low isoflavone content are now available. Unfortunately, it is difficult to replace existing strains of subterranean clover in Australian pastures because they have become well adapted to their environment and have built up large reserves of seed in the soil. In spite of all attempts to reduce the feeding of breeding ewes on such pastures, it is estimated that each year 1 million Australian ewes fail to lamb because of "clover" disease. Immunization procedures are currently being developed to overcome this problem (Shutt, 1976).

Isoflavones are more or less restricted in their distribution to the Leguminosae, so that if these compounds have a deterrent function in nature, such a function can operate only in this family. There is no reason, however, why an array of quite different chemical structures should not have the same purpose in other plant families. Weak estrogenic activity has been detected, for example, in some flavones and flavonols, two classes of compound that are widely present in the angiosperms.

The following question remains: Is isoflavone synthesis purely incidental to estrogenic activity in mammals, or have these substances been deliberately produced by the plant to interfere with the reproductive capacity of grazing animals? That isoflavones do have an ecological impact is supported by a report (Leopold *et al.*, 1976) that birds (quails) are also affected by the estrogenic effects of pasture isoflavones. It appears that the birds feed on pastures rich in leguminous species, and they use the presence of isoflavones as a form of population control.

Thus, in years of good rainfall, legumes that are eaten grow luxuriously and are relatively low in isoflavone on a fresh weight basis. There are no estrogenic effects, and egg laying is normal. However, in years of poor rainfall, the leaves of the plants are less profuse, and the plants become richer in isoflavone on a fresh weight basis. An estrogenic effect is exerted on the female quails, and egg laying is curtailed. Thus, there is a self-regulating system whereby the increase in quail population is kept at a low level when the food available to the birds is limited.

Natural population limitation is a feature of many animal communities, and it is possible that phytoestrogens have a role in other species besides quails. A similar situation exists in the breeding behavior of the red kangaroo in desert areas of central Australia, since a flush of green food after rains increases the incidence of estrus and the frequency of twinning in these marsupials (Newsome, 1966); the nature of the phytoestrogens responsible here remains to be determined.

VIII. CONCLUSION

Any attempt to ascribe a particular purpose to the presence of flavonoids in plant cells is bound to be fraught with problems because, as a group, the flavonoids are so numerous and embrace such a wide range of different structures with various solubilities and biological properties. Clearly, there are many different functions; also the same substance may well serve several purposes. Furthermore, the functions of flavonoids cannot be considered in isolation from those of other secondary constituents, since they may often act in concert with them. For example, anthocyanin pigment in the purple-black fruits of the deadly nightshade *Atropa belladonna* acts as a visual warning signal to mammalian feeders that highly toxic alkaloids are present. The same signal, as has been pointed out (Rothschild, 1975), has a dual purpose and is actually an attractant to birds who can eat the berries without suffering the consequences of alkaloid poisoning. Again, flavonoid pigments and terpenoid-based carotenoids are found together fairly frequently in flowers and fruits, both classes contributing significantly to the final color of the plant tissue. Red and yellow bicolored flowers of South American gesnerads, for example, which are especially attractive to hummingbird visitors, normally contain both pigment types differentially distributed in the different flower parts.

Undoubtedly, the first encounter of most herbivores with flavonoids in the plant is a visual one. Anthocyanins and other colored flavonoids provide a feeding signal in flower (for nectar and pollen) and in fruit to a very wide range of animals. The extent to which the pigments providing the visual signal are accompanied by other "colorless" flavonoids that provide a barrier to certain groups of animal feeders is far from clear. The only definite example of bitter-tasting flavonoids are the flavanones of *Citrus* and, although the flavanones of these fruits undoubtedly provide repellency to some feeders, the actual color attractants in *Citrus* are not flavonoid based but are carotenoid in nature.

The main role of "colorless" flavonoids, which occur universally in leaf tissue, must surely lie in their adding to the plant a further line of chemical defense against overgrazing by herbivores. However, the idea that flavone and flavonol glycosides are active feeding deterrents still has not been established beyond all doubt, and this putative role is based largely on circumstantial evidence. The fact that certain flavonols are feeding attractants to particular phytophagous insects can, of course, be considered, in the broad coevolutionary context of plant–insect interactions, to indicate that the same substances are significant deterrents to many other potential feeders. The same argument is used for other secondary constituents such as alkaloids, cardiac glycosides, and glucosinolates. The roles of these toxins as feeding deterrents have been switched by opportunist feeders such as cinnabar moths, monarch butterflies, and cabbage aphids (see Rothschild, 1972), who have developed the ability to detoxify or store these substances and use them as feeding stimulants.

It seems unlikely that leaf flavonoids act as deterrents because they are highly poisonous (like alkaloids or cyanogens) or have considerable physiological effects on animals at miniscule concentrations (like cardiac glycosides). Their harmful effects on feeders would seem to lie in other directions. For example, the isoflavone estrogens have an indirect effect on lowering the fitness of mammals and birds by interfering with reproductive capacity. Such interference is not always harmful in the case of an animal species seeking some form of population control. It is possible that leaf flavonoids may well have a significant long-term deleterious effect on insect herbivores, but this has yet to be demonstrated. The most promising line for future work would seem to lie in a much closer scrutiny of the interaction between leaf flavonoids and a wide range of phytophagous insects.

REFERENCES

Bickoff, E. M. (1968). *Commonw. Bur. Pastures Field Crops (G.B.), Rev. Ser.* **1,** 1–39.

Biggers, J. D. (1959). *In* "Pharmacology of Plant Phenolics" (J. W. Fairbairn, ed.), pp. 51–69. Academic Press, New York.

Bjeldanes, L. F., and Chang, G. W. (1977). *Science* **197,** 577.

Bradbury, R. B., and White, D. E. (1954). *Vitam. Horm. (N.Y.)* **12,** 207–233.
Chopin, J., and Bouillant, M. L. (1975). *In* "The Flavonoids" (J. B. Harborne, T. J. Mabry, and H. Mabry, eds.), pp. 632–691. Chapman & Hall, London.
Doskotch, R. W., Mikhail, A. A., and Chatterji, S. K. (1973). *Phytochemistry* **12,** 1153–1156.
Ehrlich, P. R., and Raven, P. H. (1965). *Evolution* **18,** 586–608.
Fewtrell, C. M. S., and Gomperts, B. D. (1976). *Nature (London)* **265,** 635–636.
Ford, E. B. (1941). *Proc. R. Entomol. Soc. London, Ser. A* **16,** 65–90.
Fukami, H., and Nakajima, M. (1971). *In* "Naturally Occurring Insecticides" (M. Jacobson and D. G. Crosby, eds.), pp. 71–97. Dekker, New York.
Gabor, M. (1972). "The Anti-inflammatory Action of Flavonoids," p. 110. Akádemiai Kiadó, Budapest.
Geiger, H., and Quinn, C. (1975). *In* "The Flavonoids" (J. B. Harborne, T. J. Mabry, and H. Mabry, eds.), pp. 692–742. Chapman & Hall, London.
Gottlieb, O. R. (1975). *In* "The Flavonoids" (J. B. Harborne, T. J. Mabry, and H. Mabry, eds.), pp. 296–375. Chapman & Hall, London.
Griffiths, L. A., and Barrow, A. (1972). *Biochem. J.* **130,** 1161–1162.
Griffiths, L. A., and Smith, G. E. (1972). *Biochem. J.* **128,** 901–911.
Hamamura, Y., Hayashiya, K., Naito, K., Matsuura, K., and Nishida, J. (1962). *Nature (London)* **194,** 754–755.
Harborne, J. B. (1967). "Comparative Biochemistry of Flavonoids." Academic Press, New York.
Harborne, J. B. (1972). *Recent Adv. Phytochem.* **4,** 107–141.
Harborne, J. B. (1975a). *In* "The Flavonoids" (J. B. Harborne, T. J. Mabry, and H. Mabry, eds.), pp. 1056–1095. Chapman & Hall, London.
Harborne, J. B. (1975b). *Phytochemistry* **14,** 1147–1155.
Harborne, J. B. (1976). *In* "Chemistry and Biochemistry of Plant Pigments" (T. W. Goodwin, ed.), 2nd ed., Vol 1, pp. 736–779. Academic Press, New York.
Harborne, J. B. (1977a). "Introduction to Ecological Biochemistry" p. 243. Academic Press, New York.
Harborne, J. B. (1977b). *Biochem. Syst. Ecol.* **5,** 7–22.
Harborne, J. B., and Williams, C. A. (1975). *In* "The Flavonoids" (J. B. Harborne, T. J. Mabry, and H. Mabry, eds.), pp. 376–441. Chapman & Hall, London.
Harborne, J. B., Mabry, T. J., and Mabry, H., eds. (1975). "The Flavonoids." Chapman & Hall, London.
Horowitz, R. M. (1964). *In* "Biochemistry of Phenolic Compounds" (J. B. Harborne, ed.), pp. 545–572. Academic Press, New York.
Kupchan, S. M., Sigel, C. W., Knox, J. R., and Udayamurthy, M. D. (1969). *J. Org. Chem.* **34,** 1460–1463.
Kutney, J. P., Warnock, W. D. C., and Gilbert, B. (1970). *Phytochemistry* **9,** 1877–1878.
Labov, J. B. (1977). *Comp. Biochem. Physiol. A* **57,** 3–10.
Leopold, A. S., Erwin, M., Oh, J., and Browning, B. (1976). *Science* **191,** 98–100.
Levy, E. C., Ishaaya, I., Grureritz, E., Cooper, R., and Lavie, D. (1974). *J. Agric. Food Chem.* **22,** 376–383.
Mabry, T. J., Markham, K. R., and Thomas, M. B. (1970). "The Systematic Identification of Flavonoids." Springer-Verlag, Berlin and New York.
Markham, K. R. (1975). *In* "The Flavonoids" (J. B. Harborne, T. J. Mabry, and H. Mabry, eds.), pp. 1–44. Chapman & Hall, London.
Marrian, G. F., and Haslewood, G. A. D. (1932). *Biochem. J.* **26,** 1226–1232.
Miles, P. W. (1968). *Annu. Rev. Phytopathol.* **6,** 137–164.
Morris, S. J., and Thomson, R. H. (1963). *J. Insect Physiol.* **9,** 391–399.

Morris, S. J., and Thomson, R. H. (1964). *J. Insect Physiol.* **10,** 377–383.
Newsome, A. E. (1966). *CSIRO Wildl. Res.* **11,** 187–196.
Racker, E. (1975). *Biochem. Soc. Trans.* **3,** 785–802.
Rhoades, D. F., and Cates, R. G. (1976). *Recent Adv. Phytochem.* **10,** 168–213.
Rothschild, M. (1972). *In* "Phytochemical Ecology" (J. B. Harborne, ed.), pp. 1–12. Academic Press, New York.
Rothschild, M. (1975). *In* "Coevolution of Animals and Plants" (L. E. Gilbert and P. H. Raven, eds.), pp. 20–52. Univ. of Texas Press, Austin.
Russell, G. B., Sutherland, O. R. W., Hutchinson, R. F. N., and Christmas, P. E. (1978). *J. Chem. Ecol.* **4,** 571–580.
Schoonhoven, L. M. (1972). *Recent Adv. Phytochem.* **5,** 197–224.
Shaver, T. N., and Lukefahr, M. J. (1969). *J. Econ. Entomol.* **62,** 643–646.
Shutt, D. A. (1976). *Endeavour* **35,** 110–113.
Stout, M. G., Reich, H., and Huffman, M. N. (1964). *Cancer Chemother. Rep.* **36,** 23–24.
Swain, T. (1974). *Compr. Biochem. A* **29,** 125–302.
Swain, T. (1976). *In* "Chemistry and Biochemistry of Plant Pigments" (T. W. Goodwin, ed.), 2nd ed., Vol. 2, pp. 166–206. Academic Press, New York.
Swain, T., and Bate-Smith, E. C. (1962). *Comp. Biochem.* **3,** 755–809.
Thompson, W. R., Meinwald, J., Aneshansley, D., and Eisner, J. (1972). *Science* **177,** 528.
Van Sumere, C. F., Van Brussell, W., Casteele, K. V., and Van Rompey, L. (1979). *Recent Adv. Phytochem.* **12,** 1–28.
Wagner, H. (1979). *Recent Adv. Phytochem.* **12,** 589–616.
Zielske, A. G., Simons, J. N., and Silvestein, R. M. (1972). *Phytochemistry* **11,** 393–396.

Chapter **19**

Tannins and Lignins

TONY SWAIN

I. INTRODUCTION

Tannins and lignins are phenolic compounds that are found, often in high concentrations, in all classes of vascular plants. They do not occur in prokaryotes, protists, fungi, or animals. Both tannins and lignins are formed biosynthetically, either wholly or partially, from the products of the shikimic acid pathway (Fig. 1) (Haslam, 1974).

Most tannins are of moderate size (300–3000 daltons) and occur as soluble components in the sap of living cells, from which they can be extracted, at least in part, by polar solvents (Haslam, 1966). A few less polar tannins are found either in the ether-soluble resins that occur on the surfaces of certain leaves or as part of the complex mixtures of phenolic compounds in the dead heartwood cells of trees. Finally, many phenolic compounds of low molecular weight that occur in plants may undergo polymerization by either autoxidation or enzyme-catalyzed

ISBN 0-12-597180-X

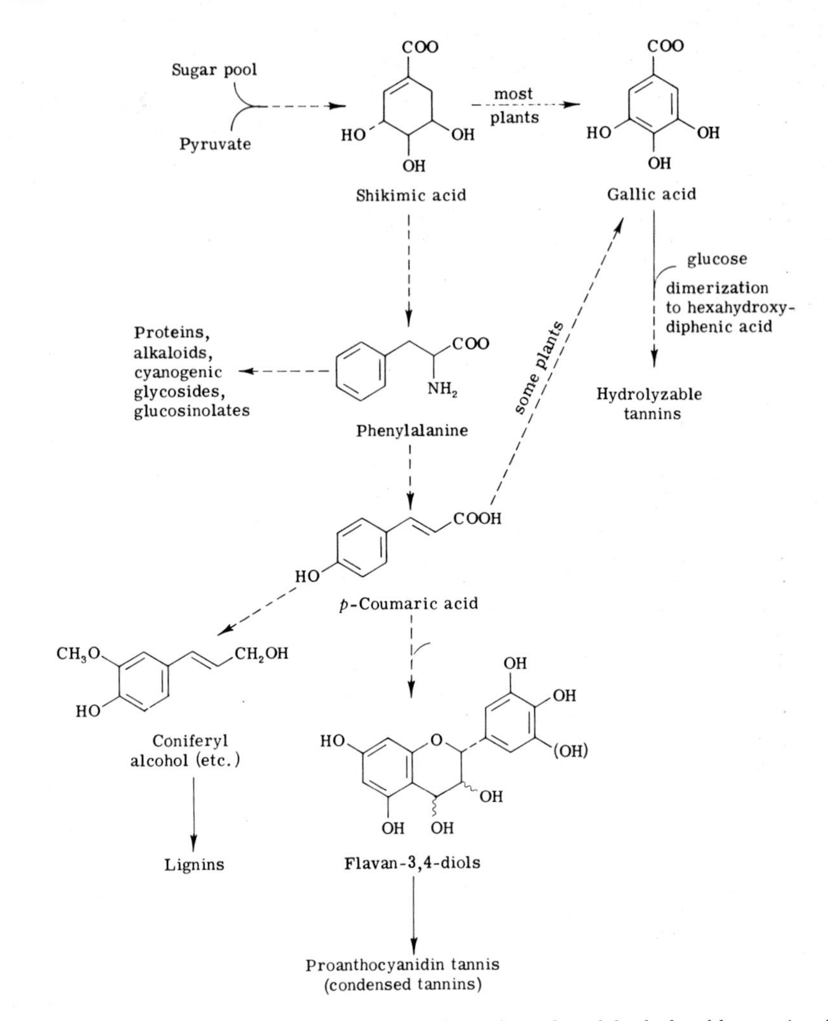

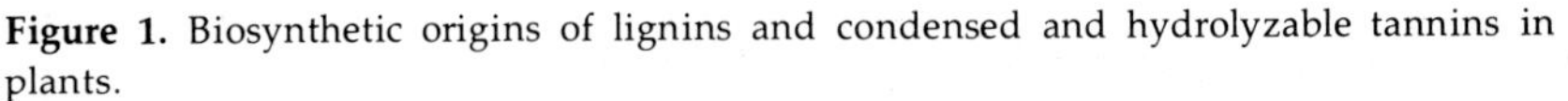
Figure 1. Biosynthetic origins of lignins and condensed and hydrolyzable tannins in plants.

oxidation to yield tannin-like compounds. As a result, tannins show a wide structural divergence and are differentially distributed both among and within individual species.

Lignins, on the other hand, are mainly polymers with molecular weight greater than 5000 that are closely associated with the polysaccharides of the cell walls of plants, from which they can be obtained only by rather drastic chemical treatments (Sarkanen and Ludwig, 1971). They show a relatively restricted variation in structural features in all members of the various subclasses of vascular plants.

Both tannins and lignins have been and are of enormous importance in determining the evolutionary success of land plants since the Devonian (430 years ago), most obviously in the development of arborescent forms (Swain, 1974b, 1978a). Here, the lignins contribute the necessary stiffening and rigidity of the cell walls and internal structures of the erect plants, whereas the tannins help to inhibit attack on lignified tissues by fungi and bacteria. Of equal value is the defense that tannins and lignins afford against herbivores. The former reduce both the nutritional availability of soluble plant proteins and polysaccharides and the activity of the digestive enzymes and symbiotic microorganisms of the herbivore's own gut; lignins increase the toughness of the plant and inhibit digestibility. By complexing, and thus preventing the breakdown of, nucleic acids, proteins, and polysaccharides, both classes also contribute greatly to the stability of plant detritus and to the formation of the important humic acid elements of the soil (Swain, 1977a).

Indeed, it seems doubtful whether plants, and hence life as we know it, could have developed on land without the acquisition of the ability to synthesize these two classes of phenolic compounds (Swain, 1974a,b, 1977a,b, 1978a).

II. TANNINS

A. Introduction

The term *tannin* was introduced at the end of the eighteenth century to define the organic substances present in water extracts of the leaves, bark, wood, fruit, or galls of certain ferns, gymnosperms, and angiosperms. These substances are responsible for transforming the raw animal skins and hides soaked in such extracts into leathers that are much more resistant to bacterial decay, heat, or abrasion than the original materials. This process of tanning has been of great importance since the dawn of civilization in affording mankind durable yet flexible polymers that can be put to many uses (Haslam, 1966).

It is still not certain how tannins act in leathering hides, but it is generally believed that sufficient hydrogen bonds are formed between the phenolic hydroxyls of the tannin and the peptide groups of the amorphous, bacterially susceptible regions of insoluble collagen fibrils to form cross-links between adjacent protein chains. Oxidation of phenolic groups in the tannins to quinones, either before their use or *in situ* in the hide, may give rise to covalent bonds with the ϵ-amino group of lysine as well as arginine in the collagen and thereby increase the longevity of the leather.

Tannins also bind to almost all soluble proteins, giving rise to insoluble copolymers at normal pH and ionic strength. Enzymes complexed in this way show a marked reduction in activity, and the bound protein moieties are also inert to hydrolysis by external proteases (Goldstein and Swain, 1965). Tannins can also form complexes with other natural polymers such as nucleic acids and polysaccharides.

The protein–tannin copolymers are stable only because sufficient bonds and cross-links are formed. It should be noted that many common phenolic compounds, such as simple flavonoids [e.g., quercetin **(1)**] or hydroxycinnamic acids [e.g., caffeic acid **(2)**], may form hydrogen bonds with NH groups of the peptide bonds of proteins but do not form stable cross-links and hence do not act as tannins. However, these compounds have many chemical and physical properties that are similar to those of tannins and are often mistakenly included as such in gross analytical determinations of plant constituents. This has led to a great deal of confusion among ecologists concerning the function of tannins in plants.

B. Types of Tannins

Compounds that act as tannins in forming complexes with proteins and other macromolecules can be divided into four groups according to chemical structure, molecular weight, water solubility, and tannin action: proanthocyanidins (condensed tannins), hydrolyzable tannins, oxytannins, and a miscellaneous group referred to here as β-tannins (Swain, 1978b) (Table I). In addition, we can distinguish a class of compounds that may be called prototannins since they are mainly precursors of the first three groups.

1. *Proanthocyanidin Tannins*

The proanthocyanidins (Haslam, 1975) are by far the most widely distributed tannins in vascular plants (Table II) (Swain, 1976). Structurally they can be regarded as being formed by the condensation of such flavan-3-ols as catechin **(5)**, epicatechin **(6)**, or the corresponding gallocatechins **(7** and **8)** to form dimers and higher oligomers, the majority of which contain the 4–8 linkage shown in **9** and **10.** Because of the variation in the stereochemistry of the 3-hydroxyl groups in the monomers (*S* in catechin and *R* in epicatechin) and of the 4–8 linkage, eight possible dimers of **9** and **10** are possible. All the isomers of **9** have been found in nature, the most common being those designated by Weinges as "B1–B4," whose stereochemistry is listed below formula **9.** Trimers and tetramers formed by further condensation can obviously

Table I. Properties of tannins

Type of tannin	Chemical structure	MW	Water solubility[a]	Effectiveness as protein precipitants[b]
Proanthocyanidin (condensed or leucoanthocyanidin)	Oligomers of flavan-3-ols and flavan-3,4-diols **(9, 10, 11)**	1000–3000	± to +++	++++
Hydrolyzable	Esters of gallic acid **(18)** or hexahydroxydiphenic acid **(20)** with glucose and other polyols **(21, 22, 23)**	1000–1500	+++	+++++
Oxytannins	Various: formed by oxidative phenolic coupling of prototannins (e.g., **5–8**) and other phenolic compounds	600–1500	++	+++
β-Tannins	Various: lignans **(25)**, stilbenes **(26)**, etc.	300–500	– to ±	+++
Prototannins	Precursors of proanthocyanidin, hydrolyzable tannins and oxytannins (e.g., **5–8, 18, 20**)	200–600	+++	±

[a] Key: –, nonsoluble; +++, very soluble;
[b] Or sequestering agents: ±, hardly effective; +++++, very effective.

Table II. Distribution of proanthocyanidins and hydrolyzable tannins in the plant kingdom

Taxon	Genera examined containing tannins[a]	
	Proanthocyanidins (%)	Hydrolyzable (%)
Psilopsida	0	0
Lycopsida	0	0
Sphenopsida	28	0
Ferns	92	0
Gymnosperms	74	0
Angiosperms	54[b]	13[c]
Dicotyledons	62	18
Monocotyledons	29	0

[a] The distribution of β-tannins has not been examined. Oxytannins probably parallel the distribution of proanthocyanidins.

[b] The percentage of genera containing procyanidin **(9)** and prodelphinidin **(10)** is 42 and 23, respectively (45 and 25 in dicots and 27 and 9 in monocots).

[c] Mainly in Hamamelididae (26%), Rosidae (28%), and Dilleniidae (18%). The Magnoliidae, Caryophyllidae, Asteridae, and Ranunculidae contain 5, 0, 1, and 3%, respectively, of genera containing hydrolyzable tannins.

exist in a multitude of stereochemical forms, but those so far examined appear to consist of either polycatechins or *epi*catechins.

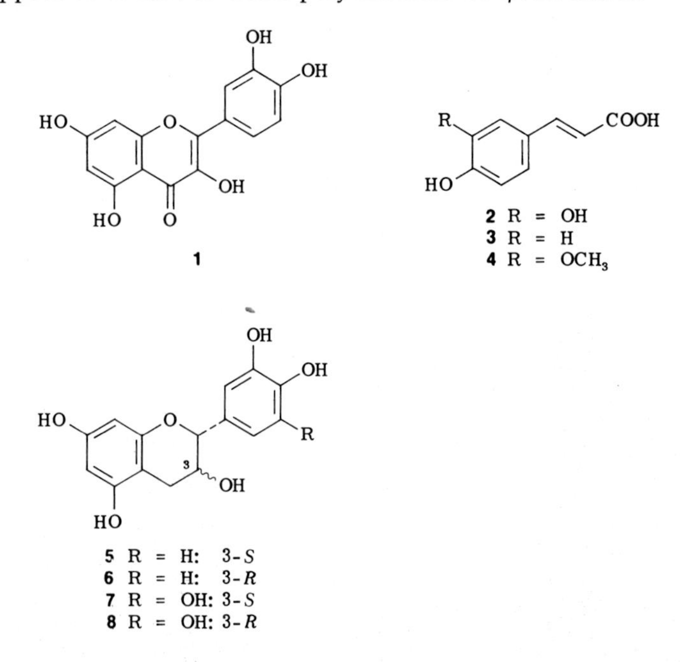

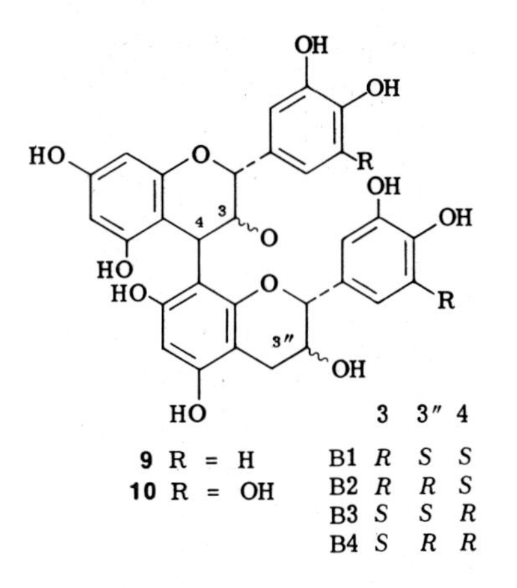

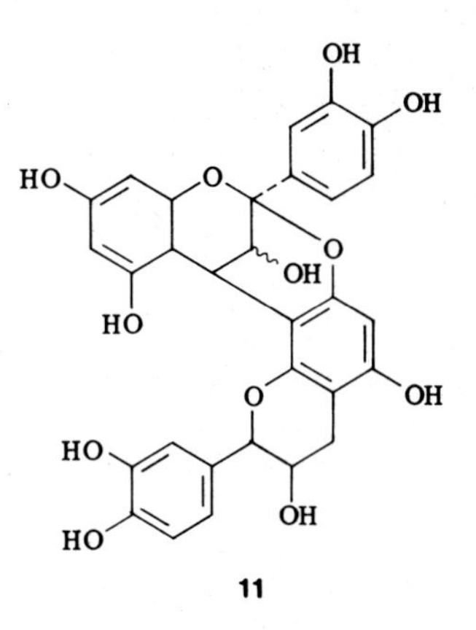

A second method of linkage found in a number of proanthocyanidins is that leading to the dimer **11.** Higher oligomers of this "A series" are formed by the addition of extra monomers by single 4–8 linkages, as for the "B series" described above.

The biosynthetic steps in the formation of dimers and higher oligomers of the proanthocyanidins are not known. It is believed that the flavan-3-ols, formed via the normal flavonoid biosynthetic pathway, form the starter unit to which corresponding flavan-3,4-diols (or suitably activated equivalents) are attached. However, no monomeric flavan-3,4-diols corresponding to the catechin or gallocatechin series **(12** or **13)** have been found in plants, although several 5-deoxyflavan-3,4-diols [e.g., leucofisetinidin **(14)** and leucorobinetinidin **(15)**] are known, along with the corresponding oligomers, mainly from legumes. These oligomers have all the properties of proanthocyanidin tannins.

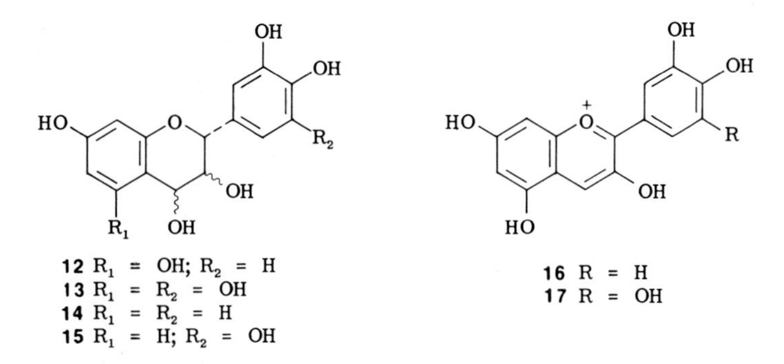

The proanthocyanidins are so called because on treatment with hot mineral acid they yield small amounts (5–15%) of the corresponding anthocyanidins [e.g., cyanidin **(16)** and delphinidin **(17)**], the remainder being further polymerized to yield insoluble phlobaphenes. It is the former side reaction that makes it possible for the proanthocyanidins to be detected and quantified and the orientation of hydroxyl groups in the monomers to be determined.

2. *Hydrolyzable Tannins*

Hydrolyzable tannins are restricted to the dicotyledons of the angiosperms, but even here only three of the seven subclasses contain a high percentage of genera containing these compounds (Table II) (Bate-Smith, 1974). This class of tannins consists of esters of glucose or, rarely, other polyols, with gallic acid **(18)**, *m*-digallic acid **(19)**, hexahydroxydiphenic acid **(20)**, or their congeners (Haslam, 1966). Typical structures include tannic acid **(21)**, corilagin (1-galloyl-3,6-hexahydroxydiphenoylglucose) **(22)**, and chebulagic acid **(23)**. All these compounds can be readily hydrolyzed by hot mineral acid or by esterases (e.g., tannase) to yield the sugar core and the constituent acids, although **20**, on release from the tannin, rapidly forms its dilactone, ellagic acid **(24)**. Depending on the acids obtained on such hydrolysis, hydrolyzable tannins are conventionally divided into gallotannins (e.g., **21**) and ellagitannins (e.g., **22** and **23**).

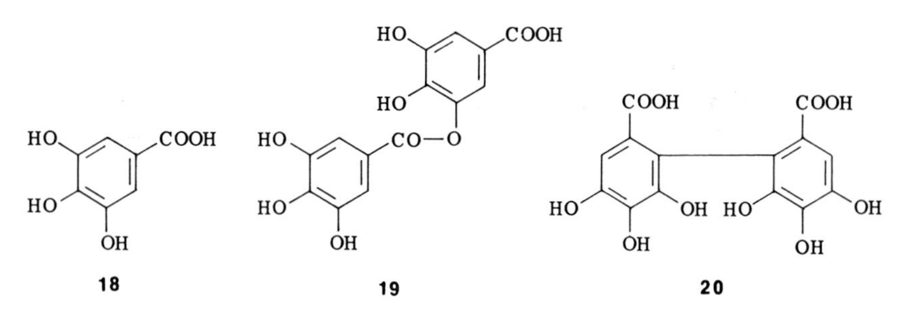

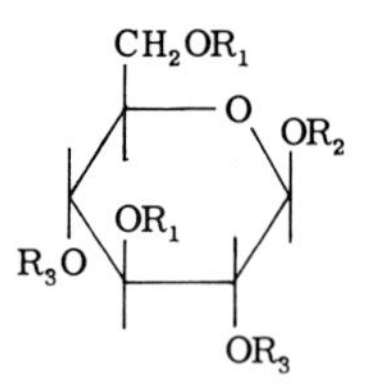

21 R_1, R_2, R_3 = **18** or **19** to give 8 or 9 galloyl groups

22 R_1 = **20**; R_2 = **18**; R_3 = H

23 R_1 = **20**; R_2 = **18**;

R_3 = [structure: COO⁻, COO⁻, O, O, HO, OH, HOOC, OH]

It is now generally agreed that gallic acid **(18)** arises in higher plants directly from shikimic acid, although pathways via the β-oxidation of cinnamic acid intermediates may also occur (Fig. 1). The mode of formation of the various glucose esters, however, is not known. Hexahydroxydiphenic acid **(20)** is believed to arise by *in situ* intramolecular coupling of two appropriately positioned galloyl groups on the glucose core. Further oxidative reactions then can lead to the formation of the moieties found in chebulagic acid **(23)** and so on.

3. *Oxytannins*

Oxytannins are not present in intact plants but may readily be formed on injury. In view of the fact that flavan-3-ols [e.g., catechin **(5)**] are more readily autoxidized than other natural phenols (e.g., **1** or **2**), it seems probable that plants containing these proanthocyanidin precursors will be more likely to produce oxytannins than other plants. The oxidation of catechins produces structures quite different from those of the proanthocyanidin polymers. Usually the central pyran ring is opened, and the polymerization involves the formation of both C—C and C—O bonds (Haslam, 1966).

4. *β-Tannins*

The β-tannins comprise a heterogenous class that include a greater number of compounds of lower molecular weight than the first three classes, although they are reportedly capable of precipitating or sequestering proteins. Among the most interesting of these are piceatannol **(26)** from the bark of spruce (Haslam, 1966) and the ether-soluble phenolic constituents of the resin which occurs on the surface of the leaves of the creosote bushes (*Larrea* spp.) (Rhoades, 1977a,b, Rhoades and Cates, 1976). The major constituent of the latter is nordihydroguaiaretic acid (NDGA) **(25),** but it is apparent from the evidence presented (Rhoades, 1977a,b) that the other compounds present (mainly *O*-methylated flavones and flavonols: e.g., 3,7-di-*O*-methylquercetin) play an equally important part in the complexing of proteins. Unlike other tannins, the resin does not give a precipitate with proteins, but the water-soluble urea-dissociable complex formed is highly resistant to proteolysis and shows greatly reduced digestibility to a variety of insects. One reason why this group of low-molecular weight components and not those of the nearly equivalent size [(+)-catechin **(5)**] have a strong tannin like action may lie in their insolubility in water. It may be presumed that once hydrogen bonds are formed between the protein and NDGA and the methylated flavonoids they are more difficult to break than those involving water-soluble compounds like the catechins. Of course oxidation, as in the case of the oxytannins, increases the tannin-like activity of the resin. Similar arguments may apply to piceatannol. Other β-tannins, whose structures are still unknown, are found in the brown algae (Glombitza *et al.*, 1976).

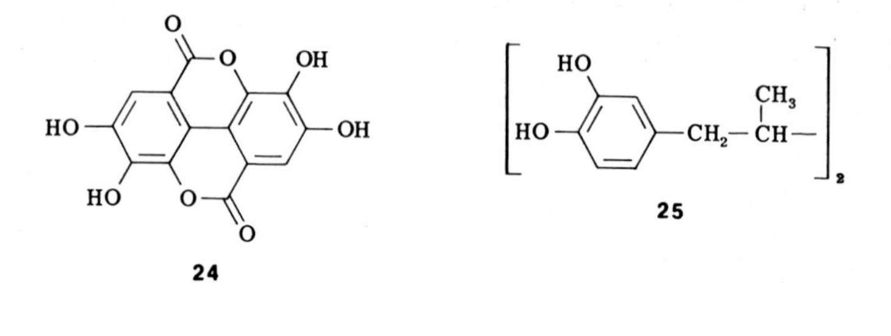

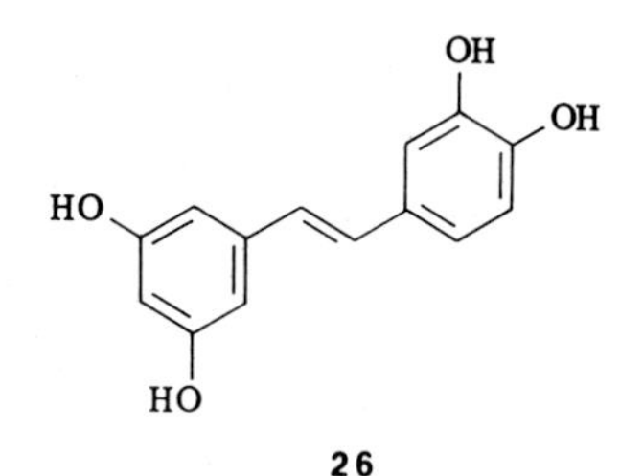

5. *Prototannins*

The prototannins, in the main, comprise the flavan-3-ols **(5–8)** and stable flavan-3,4-diols **(14** and **15),** together with the corresponding precursors of the hydrolyzable tannins, simple gallic and hexahydrodiphenic esters of glucose (Haslam, 1966). One can include a large number of commonly distributed phenolics (e.g., **2,** which is present in nearly two-thirds of all angiosperms) that are capable of undergoing autoxidation. However, in most cases, the products of such reactions lack tannin activity.

C. Analysis of Tannins

1. *Extraction*

Most tannins are polar compounds (Table I) and hence can be expected to be readily extracted from plant material by the use of such solvents as methanol, ethanol, acetone, or ethyl acetate (Haslam, 1966).

For the proanthocyanidins, hot 50–80% aqueous methanol is the solvent of choice. However, due to the strong hydrogen bonding between the tannins and the proteins, polysaccharides, and nucleic acids of the cell, only a proportion of the proanthocyanidins present are extracted [30–95% procyanidins, 0–50% prodelphinidins (Bate-Smith, 1975)]. More dilute methanol should be used for extracting dried plant material, and the latter should be prepared in the most careful way possible. In the case of leaves, for example, this means ensuring not only that the dried sample is prepared from undamaged whole fresh leaves that have not been treated with preservatives and have been dried at the lowest possible temperature, but also that surface (or internal) saprophytes, which otherwise may instigate changes in the chemistry of the material, have been killed or wholly inhibited. The reason for these precautions are plain: (1) Leaf damage usually leads to enzyme-catalyzed or other changes in cellular constituents including tannins that can affect the extractability and analytical parameters of the latter. (2) Treatment with preservatives such as formaldehyde or ethanol, which cross-link or extract tannins, respectively, may equally lead to variability in the overall analysis (Cooper-Driver and Balick, 1979). (3) High drying temperatures (50°C) can "fix" tannins irreversibly to other cell polymers. The same precautions should be taken with other tissues. For example, many ripe juicy fruits collected in the field can be preserved only by slicing selected portions directly into alcohol and then analyzing both the tissue and the alcohol surrounding them.

For hydrolyzable tannins, aqueous acetone or ethyl acetate is the best solvent, since aqueous alcohols tend to degrade these tannins by transesterification, leading to loss of gallic acid and hexahydroxydiphenic acid residues (Haslam, 1966). No reports have appeared on the relative efficiency of extraction of oxytannins and β-tannins. For the latter compounds, ether or ethyl acetate would appear to be the solvent of choice (Rhoades and Cates, 1976).

It must be remembered that in the majority of cases the extract will contain, in addition to the tannins, considerable quantities of other phenolic materials that may interfere with the qualitative and quantitative examination of the former (Harborne, 1973).

2. *Chromatographic Separation*

Tannins can be separated chromatographically from low molecular weight phenols in crude extracts by taking advantage of the affinity of tannins, through hydrogen bonding, to a number of adsorbents such as cellulose, polyamide, and Sephadex LH-20 (Haslam, 1966, 1975). For proximate qualitative analysis, however, direct two-dimensional paper chromatography [top phase of butan-2-ol–acetic acid–water (14 : 1 : 5, v/v) followed by 6% aqueous acetic acid] can be used. The tannins can be revealed after development either by examination under short-wave ultraviolet (uv) light (254 nm), where they appear as dark purple spots, or by the use of a variety of phenolic reagents (diazotized *p*-nitroaniline, potassium ferro- and ferricyanides, etc.) (Haslam, 1966).

3. Determination of Structure

The orientation of the phenolic hydroxyl groups in the monomers of proanthocyanidins and the nature of the acylating acids in hydrolyzable tannins are determined readily by first hydrolyzing the tannin extract (or separated components) in boiling 0.1 *N* mineral acid for 0.5–2 hours. As mentioned earlier, small but sufficient yields of the corresponding anthocyanidins are obtained from proanthocyanidins (e.g., **9–11**), whereas gallic acid **(18)** or ellagic acid **(24)** is produced from the bulk of hydrolyzable tannins. All these compounds can be identified easily by one-dimensional paper chromatography in Forestal solvent [acetic acid–concentrated HCl–water (30 : 3 : 10, v/v)] after extraction of the products from the hydrolysis mixture with amyl alcohol (Bate-Smith, 1977). When crude plant extracts are used, the hydrolyzed extracts usually contain one or more flavones or flavonols (e.g., **1**), produced by hydrolysis of the corresponding glycosides, and hydroxycinnamic acids (e.g., **2**, **3** or **4**), formed from esterified derivatives (Harborne, 1973). In this chromatographic system, the anthocyanidins are readily seen after development because of their visible color [cyanidin is red (R_f0.50); delphinidin is blue-red (R_f0.30)], whereas ellagic acid gives a pale violet fluorescence under long-wave uv light (R_f 0.33), and gallic acid gives a purple absorbance under short-wave uv light (R_f 0.65). Naturally, many of the other commonly occurring aglycones can be seen on the chromatogram, but they can be distinguished by their visible or fluorescent colors and R_f values (Harborne, 1973).

A distinction between procyanidins and prodelphinidins can also be made by comparing the visible spectrum of the acid hydrolyzate formed after heating the extract or tissue in *n*-butanol containing 5% concentrated hydrochloric acid: Cyanidin-yielding tannins have λ_{max} at 547 nm, whereas prodelphinidins have λ_{max} at 558 nm (Bate-Smith, 1975).

4. Quantitative Analysis

Numerous methods exist for the quantitative determination of tannins in crude extracts of plants. Most depend on estimating the amount of total phenols present in the extract before and after treatment with finely divided collagen (hide powder) or similar absorbents using a variety of colorimetric or spectrophotometric techniques. The principal drawback to such methods is that they were devised for extracts of commercial tannins and therefore require relatively large amounts of material. It is important to realize that tannins cannot be properly estimated quantitatively with any degree of accuracy using direct colorimetric measurements for total phenols on the crude extract alone. Either some estimation of the protein-binding capacity of the tannins must be made, or specific methods for individual tannin moieties must be used (Bate-Smith, 1973).

Bate-Smith (1972a,b, 1973, 1975) introduced a series of methods for analyzing the majority of both proanthocyanidins and hydrolyzable tannins in fresh or dried plant material. Briefly (see Bate-Smith, 1977), the material is dried carefully, finely ground (to pass a 100-mesh sieve, approx. 0.1 mm in diameter), and then extracted three times with hot 50% aqueous methanol (ca. 0.1 ml of solvent per milligram of material), the residue being kept. The extract is concentrated (to one-third volume), and proanthocyanidins are determined by heating an aliquot with *n*-butanol containing 5% (v/v) concentrated hydrochloric acid (conveniently, 0.5 ml extract to 4.0 ml reagent) for 2 hours at 95°C and measuring the absorbance over the range 545 nm (procyanidin) to 560 nm (prodelphinidin). Proanthocyanidins in the original or residual powder are estimated similarly, after the powder and a small amount of 50% aqueous methanol are first boiled for 4 to 5 minutes.

Ellagitannins are estimated by treating the extract with aqueous nitrous acid (0.5 ml of the extract with 2.0 ml of ca. 0.1 *M* HNO_2: from $NaNO_2$ and acetic acid) at room temperature under nitrogen; the blue color that develops (λ_{max} 600 nm) is measured when it reaches a maximum (10–15 mm at 40°C) (Bate-Smith, 1972a). Galloyl groups can be estimated by their reaction with iodate at 15°C (0.5 ml with 1.5 ml of 12% KIO_3 in 33% methanol), the red-brown color being measured at 550 nm (Bate-Smith, 1977). All these determinations are achieved more or less independently of the presence in the extract of other phenolic compounds that have no tannin activity.

The most important determination introduced by Bate-Smith (1973) is that of "relative astringency," the ability of the extract to precipitate blood proteins relative to a standard tannin [tannic acid **(17)**]. For this determination, a suitable volume of extract is made up to 1 ml and mixed with 1 ml of a fresh sample of diluted blood (1 : 50 with water). After removal of the tannin–protein precipitate by centrifugation, the residual hemoglobin is determined by its absorbance at 578 nm, and the equivalent tannic acid concentration (tannic acid equivalent, TAE) can then be calculated. The astringency of the unextracted powder can be determined similarly (10 mg powder with 3 ml blood diluted 1 : 100 with water). It should be noted that these procedures use a minimal amount of protein in relation to the amount of tannin, unlike other methods, which employ an excess of hide powder or polyvinylpyrrolindone, which can then also absorb a proportion of nontannin phenolics.

Using these methods along with determinations on model compounds, Bate-Smith (1977, and references quoted) was able to show that, whereas hydrolyzable tannins are more or less completely extractable from plants, a considerable number of proanthocyanidins exist in nonextractable forms (4–100% of the total; probably complexed with proteins, etc., in the cell) that nevertheless show tannin activity. Prodelphinidins are generally much less extractable than procyanidins, and the extractability of both decreases as tannin-containing tissues mature. Another important finding is that the yield of anthocyanidin from proanthocyanidins is dependent on structure (Bate-Smith, 1975). The yield of cyanidin from procyanidins belonging to the B series **(9)** was only one-fourth to one-half that of procyanidins belonging to the A series **(11),** although the effective tannin activity (astringency) for both is about the same. Prodelphinidins **(10),** on the other hand, give twice the yield of anthocyanidin [delphinidin **(13)**] and have double the astringency. This means that estimates of the relative amounts of proanthocyanidin tannins in a tissue made by measuring the anthocyanidins on heating extracts in butanolic HCl may be grossly inaccurate unless both relative astringency and extractability are also measured.

As an example, representative results from several *Acer* species (Bate-Smith, 1977) are given in Table III. It can be seen that the degree of extractability of proanthocyanidins varies from 26 to 100% (*A. carpinifolium* and *A. pseudoplatanus*); there is no correlation between the amount of proanthocyanidins and relative astringency (TAE) (*A. carpiniforlium* v. *A. capillipes*) even when the contribution from hydrolyzable tannins is taken into account. In addition, there is a variation in the relative astringency of the extractable and nonextractable forms of the proanthocyanidins. Since the ecological effect of tannins, in the majority of cases, is based on their ability to precipitate proteins, this observation is extremely important.

D. Ecological Significance of Tannins

The ecological significance of tannins was briefly alluded to in Section I and is detailed in numerous recent reviews (Swain, 1974a,b, 1977a,b, 1978a; Levin, 1976) and several of the chapters in Part II of this treatise. Therefore, only a few general observations are presented here.

Although tannins are effective as defenses against herbivores, it has been suggested that their major role in evolution has undoubtedly been to protect plants against fungal and bacterial attack (Swain, 1978a). This

Table III. Variation in the extractability, composition, and astringency of tannins in leaves of selected *Acer* species (dry weight basis)

Species	Total			Extractable		
	Proanthocyanidins[a] (%)	λ_{max}[b] (nm)	TAE[c]	Procyanidin[a,d] (%)	HHDG[e] (%)	Gall[f] (%)
Acer pseudoplatanus	1.3	548	0.5	1.3 (100)	0	8.5
Acer spicatum	2.7	548	18	2.0 (74)	17.0	25.0
Acer capillipes	11.6	554	11	8.0 (69)	6.0	11.0
Acer davidii	15.6	554	13	8.0 (51)	3.0	ND
Acer macrophyllum	17.3	548	13	5.0 (29)	0	0
Acer carpinifolium	40.7	548	11	10.7 (26)	0	ND

[a] Assuming $E_{1\,cm}^{1\%}$ for procyanidin is 150 (λ_{max} 548 nm) and 250 when both procyanidins and prodelphinidins are present (λ_{max} 554 nm).
[b] 548 nm means only procyanidins **(9)** are present; 554 nm, both procyanidins and prodelphinidins **(10)** (λ_{max} 558 nm) are present.
[c] TAE, tannic acid equivalent.
[d] Figures in parentheses are percent proanthocyanidins extractable.
[e] Hexahydroxydiphenic acid equivalents.
[f] Gallic acid esters; ND, not detected.

is supported by the fact that in present-day angiosperms and gymnosperms the highest concentration of these compounds is found either in nonliving cells of trees (e.g., up to 40% of the dry weight in heartwood, bark), which would otherwise readily succumb to the numerous saprophytes that exist in living systems, or in mutualistic diseased tissues (galls). The leaves of such trees usually contain much lower amounts of tannins (0.5–10% dry weight), although their concentration increases with time. Nevertheless, Bate-Smith (1972a,b, 1977) reported that on a dry weight basis leaves of *Viburnum plicatum* contain up to 33% procyanidin, whereas *Geranium sanguineum* has 20% of ellagitannins. Leaves of herbaceous species, on the whole, contain few or no tannins in their cells, although there are exceptions (e.g., *Lathyrus pratensis,* 16% soluble procyanidin).

Proanthocyanidin tannins are extremely resistant to microbial attack; only recently has an organism, *Penicillium adametzi,* been isolated that can degrade them (Grant, 1976). The stability of hydrolyzable tannins is likely to be much lower since several fungi produce enzymes (tannases) that are able to degrade these secondary compounds (Haslam, 1966). Nevertheless, their protective effect against pathogens is likely to be as high as that of proanthocyanidins because of their greater astringency (see Table III).

Thus, the presence of tannins in the leaves of plants is of undoubted importance in determining the overall susceptibility of these tissues to phytopathogens, including viruses (Swain, 1977a; Levin, 1971, 1976; Harborne, 1977). Not only are tannins capable of sequestering the nucleic acids, proteins, and polysaccharides of the plant tissues themselves, hence preventing the breakdown of these polymers by fungal and bacterial extracellular hydrolases, but many other essential enzymes of the attacking organism may be inhibited, thus affecting their overall metabolism. In addition, tannins can complex with the polymeric components of the cell walls of invading pathgens and stop cell division.

By complexing with proteins, nucleic acids, and polysaccharides in fallen leaves, tannins also affect the rate of metabolism of these polymers by soil organisms. It is likely that the release of nitrogen, phosphorus, and sugars from dead leaves and other plant tissues that fall from temperate deciduous perennials before the onset of winter and form part of the soil detritus is slowed down sufficiently to ensure a better supply of these essential nutrients for plant growth the following spring (Synge, 1975). In a similar way, the presence of high concentrations of tannins in the seed coats of many, even generally nontanniniferous species protects the overwintering seed from soil microorganisms and other predators (Bate-Smith and Ribéreau-Gayon, 1974; Swain, 1977a).

However, many more detailed investigations on the antimicrobial effect of tannins are required before the overall importance of their function can be properly elucidated.

We are on much firmer ground with regard to the value of tannins to the plant as defenses against herbivores (Swain, 1977a; Levin, 1971, 1976; Harborne, 1977). Observations and experiments with cattle and sheep have shown that the preference of these herbivores for certain food crops (e.g., sericea, *Lespedeza cuneata*) is markedly dependent on tannin content. Similarly, investigations on the food choice of such herbivorous wild animals as giant tortoises (Swain, 1976, 1978b), deer (Cooper-Driver *et al.*, 1977), *Colobus* monkeys (Oates *et al.*, 1977), insects (Rhoades, 1977a) and gorillas (Harborne, 1977) indicate that tannin concentrations higher than 2% deter feeding. Of course, tannins are not the only compounds responsible for feeding aversion to plants, but they appear to be the most important. They seem to have been overlooked by most workers because the astringency (puckeriness) they impart to the mouth is not usually recognized as a "taste." This assertion is based on the fact that no textbook dealing with sensory perception has ever mentioned them, yet tannins apparently have a similar deterrent effect on the acceptability of food in all land vertebrates, as shown by the observation that the concentration of tannins in foods that causes rejection of the foods by the Mediterranean tortoise *Testuda graeca,* is about the same as for a variety of mammals including man (Swain, 1976, 1978a). Obviously the ability to sense the presence of undesirable amounts of tannins in foods has been of significant importance in plant–animal coevolution.

Besides deterring vertebrate herbivores, tannins also affect insect predation on plants. Again, the concentration of these compounds at which the food is rejected appears to be about 2% dry weight (Feeny, 1975). No studies on the effect of tannins on the food acceptance of other herbivores have been reported, but the β-tannins of the brown algae are known to inhibit the activity of hydroids, copepods, nematodes, and other phytoplanktonic animals (Sieburth and Conover, 1965). It seems highly probable, therefore, that tannins affect the feeding of animals in most other herbivorous phyla.

Finally, a comment is necessary on the relative metabolic cost of tannins and other plant defenses in relation to the hypothesis that tannins occur mainly in "easy-to-find" (apparent) plants, whereas other defensive substances are used by unapparent plants (Feeny, 1976; Rhoades and Cates, 1976) (see Chapter 1). If one assumes that the energy cost *per carbon atom* of any secondary metabolite is approximately the same (Swain, 1979), then a typical tetrameric procyanidin (C_{60}) will require two to six times more energy than metabolites such as triterpenes (C_{30}), benzylisoquinoline alkaloids (C_{17}), and iridoids (C_{10}) (Harborne, 1977). Furthermore, the effective deterrent concentration of tannins (say 2% or more) is much higher than that of the other above-mentioned compounds (say 0.1%). On these grounds, it appears that a typical tannin might cost the plant up to 120 times more metabolic energy to produce an effective deterrent concentration than would certain other defensive products. However, this argument ignores metabolic turnover. It is well known that alkaloids, terpenoids, and simple phenolic compounds are turned over rapidly, often having half-lives of less than 1 day (in the case of volatiles, perhaps, hours) (Swain, 1979). On the other hand, polymeric tannins, once formed, are likely to be more or less stable. Even though some energy may be recovered from compounds that are turned over rapidly, it seems probable that over an average growing season (120 days) there may be little or no difference in metabolic cost between the production of tannins and the formation of other, more ephemeral and more recently evolved chemical constituents of defense (Swain, 1974b). It is more likely, therefore, that the distribution of defensive compounds in plants is dependent on the biochemical evolution of the taxa concerned (e.g., Swain, 1974a,b, 1977a,b, 1978a; Bate-Smith, 1974) than on their apparency. It is remarkable, however, that the tannins, the most ancient of successful plant chemical defensive compounds (Swain, 1974a,b, 1978a), are used by present-day apparent plants to survive. This indicates their prime importance in ecosystems and perhaps explains why we live in a world devoid of the presumably tannin-free giant lycopods and horsetails that dominated the landscape in the Mississippian ca. 330 million years ago!

III. LIGNINS

A. Introduction

Lignins (the plural term is preferable due to the multiplicity of possible structures) are phenolic heteropolymers with a molecular weight greater than 5000 that are formed by the oxidative condensation of cinnamyl (phenylpropene) alcohols **(27–29)** and occur in mature cell walls of all vascular plants (Sarkanen and Ludwig, 1971; Neish and Freudenberg, 1968).

Lignins are distributed as noncrystalline inclusions throughout the entire cell wall of woody tissue, nearly half being concentrated in the middle lamella and associated outer (primary) wall (Côte, 1977). The

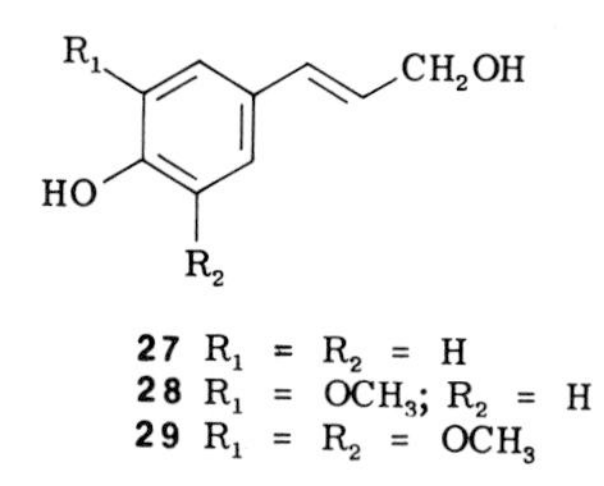

aromatic polymers are closely associated with the hemicellulose fraction of the wall and may be covalently linked to them or other wall constituents. This association gives rise to difficulties in isolating lignins in a pure state and, for a long time, led to a misunderstanding of their chemical nature.

The concentration of lignins in the cells varies from nearly zero in the submerged parts of aquatic plants to over 40% in the extractive free wood cells of the gymnosperms and angiosperms. In fact, after cellulose, lignins are the most abundant of all natural polymers (Nimz, 1974).

B. Structure of Lignins

Lignins were first recognized as being chemically defined entities, separate from the carbohydrate fraction of wood, over 140 years ago, but not until the start of this century was it suggested that they were aromatic polymers, possibly formed by the condensation of coniferyl alcohol **(28).** This proposal was based on the chemical reactivity of the latter compound and the isolation of vanillin **(31)** among the products of the alkaline oxidation of lignin (Sarkanen and Ludwig, 1971; Nimz, 1974). Even so, it took another 40 years before the idea of lignins being phenolic polymers was fully accepted; it was not until the work of Freudenberg and his collaborators in Heidelberg in the 1950's that the proximate structures of lignins could be delineated with any degree of certainty (Neish and Freudenberg, 1968; Nimz, 1974; Sakakibara, 1977). By then, it had been shown that mild alkaline oxidation of angiosperm lignins gave rise to syringaldehyde **(32)** and *p*-hydroxybenzaldehyde **(30)** in addition to vanillin **(31).** The Heidelberg school showed that lignin-like polymers could be produced by the action of a phenolase from the edible field mushroom (*Psalliota campestris*) on coniferyl and the related cinnamyl alcohols **(27–29)** having substitution patterns corresponding to the known aldehydes produced by lignin oxidation. These polymers possessed most of the attributes of the corresponding lignins, including their analytical, degradative, and other characteristics.

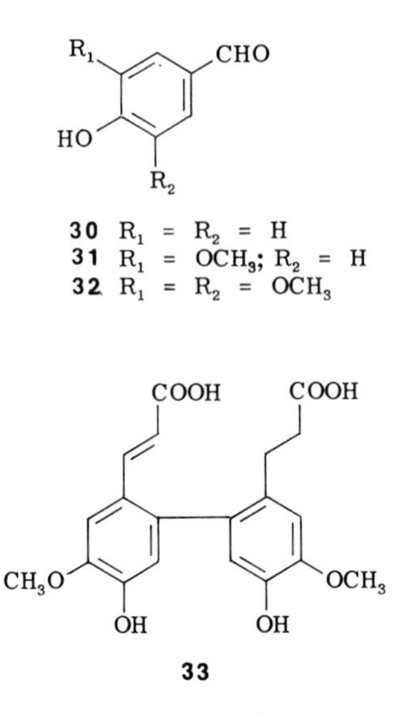

Short-term experiments produced dimers, trimers, and other low molecular weight oligomers of the alcohols whose mode of linkage could be determined readily. On the basis of these experiments, Freudenberg put forward a "type" formula for lignins indicating all the possible *linkage* types that might be involved in the formation of the polymers. However, he was careful to point out that, since the condensation involved random free-radical reactions, no single formula could represent a given lignin even within the cells of a given plant. It is now recognized that these proposals are substantially correct (Nimz, 1974; Sakakibara, 1977), and a partial "structure" of the lignin from European beech (*Fagus sylvatica*), obtained by the use of a combination of mild hydrolytic or other chemical procedures with modern spectrometric techniques (Nimz, 1974; Sakakibara, 1977), is shown in Fig. 2. As can be seen, the structure includes phenylpropane elements derived from all three cinnamyl alcohols in the approximate ratio of coniferyl **(28)** to sinapyl **(29)** to *p*-coumaryl **(27)** of 100 : 70 : 7. This is typical of the lignins from angiosperms as shown by isolation of the corresponding benzaldehydes **(27–28)** after alkaline oxidation of hardwoods. By means of such methods, gymnosperm lignins have been shown to contain much larger proportions of the guaicyl moieties arising from **28** [for example, the ratio in the lignin from Norway spruce (*Picea excelsa*) is 100 : 18 : 8], whereas lignins from monocotyledons have higher amounts of *p*-hydroxybenzene groups from **27** (in the grass *Miscanthus sacchariflorас* the ratio if 100 : 133 : 94) (Neish and Freudenberg, 1968; Higuchi *et al.*, 1967). In practice, however, the ratios of the C_9 units derived from the

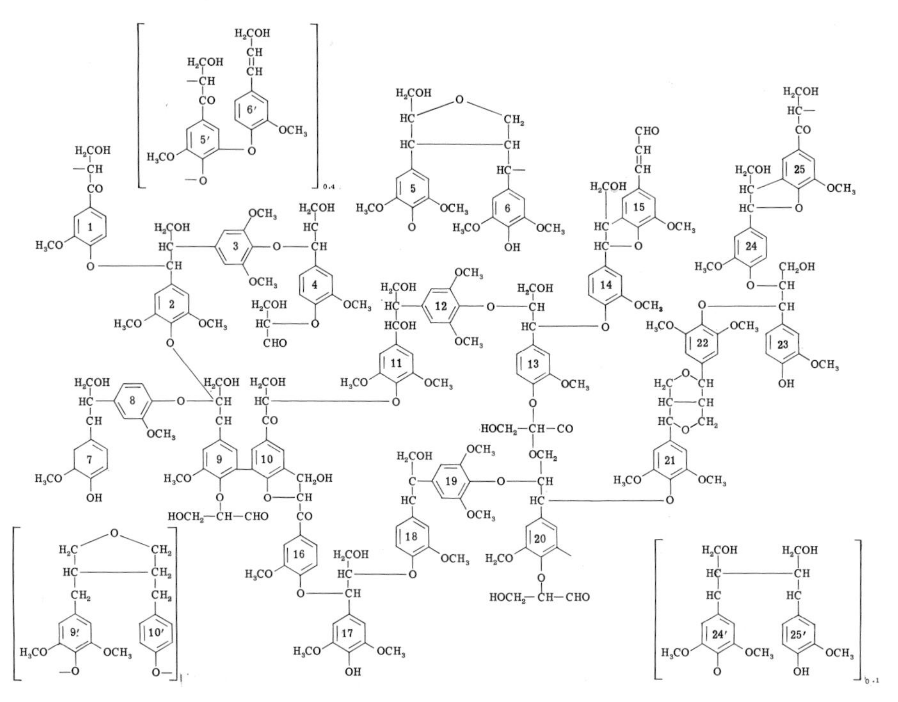

Figure 2. Proposed structure of Beech lignin (Nimz, 1974).

three precursor alcohols is almost impossible to determine with accuracy since it is known that, due to the extra 5-methoxy group, sinapyl units are less "condensed" (i.e., have fewer C—C bonds linking their aromatic rings) than are coniferyl and *p*-coumaryl units, the latter, as expected, being the most strongly bonded (Neish and Freudenberg, 1968). As a result, there are differential yields of the benzaldehydes **(30–32)**, lesser amounts being obtained as substitution decreases (e.g., from **29** > **28** > **27**). It is of interest, however, that the primitive *Selaginella* and *Isoetes* contain a high proportion of sinapyl-derived residues **(24)** like the angiosperms, whereas the lignin from *Psilotum* appears to have large amounts of *p*-hydroxyphenyl units from **25.**

Our knowledge of the structure of lignins rests not only on chemical degradations but also on biosynthetic experiments with a variety of precursors. These have confirmed the role in lignin biosynthesis of the cinnamyl alcohols **(27–29)** and their formation via the corresponding commonly occurring cinnamic acids **(2–4)** (Gross, 1977, 1978; Stafford, 1974a,b). Some questions remain, however. For example, why do all lignins, especially those from grasses, contain large amounts of *p*-coumaroyl **(3)** and feruloyl **(4)** esters (up to 10% of the lignin) (Higuchi *et al.*, 1967)? Are lignins really linked to polysaccharides in the cell wall, and if so in what way (Neish and Freudenberg, 1968; Morrison, 1974)? Does the acylation of polysaccharides of cell walls of all vascular plants examined (Friend, 1976) with hydroxycinnamic acids, especially diferulic acid **(33)** (Hartley 1972; Hartley and Jones, 1976), have any significance in the biosynthesis and evolution of lignin (Swain, 1978a)? Most of these questions have been neglected by investigators of lignin structure but are obviously of great ecological interest.

C. Analysis of Lignins

1. *Extraction*

In spite of the fact that lignin appears from its structure to be a rather intractable, insoluble polymer, part of it (up to 25%) can be directly extracted with polar solvents [usually dioxane–water (9 : 1, v/v)] from finely milled wood (Neish and Freudenberg, 1968). The secret lies mainly in the milling that is usually carried out after preextraction of low molecular weight compounds for several days in the presence of a nonswelling solvent such as toluene. The crude, milled wood lignin contains some polysaccharides from which it can be freed by their precipitation with benzene. It has a molecular weight of 8000 to 11,000 and is eminently suitable for analytical studies. Percolation of milled wood with boiling water for several days or even weeks also removes part of the lignin (Nimz, 1974). However, the only way to extract the bulk of plant lignin is to form soluble derivatives. The most important methods are variations of the commercial alkaline sulfite treatment of wood that is used to solubilize lignins in paper maufacture or the use of thioglycolic acid; these processes produce lignosulfonic and lignothioglycolic acids, respectively, which are soluble in polar solvents. Only the latter derivatives are suitable for further analysis.

2. *Quantitative Analysis*

On a large scale, lignins are usually determined by removing all other components from the tissues by first extracting proteins, nucleic acids, lipids, etc., with polar and nonpolar solvents and then treating the residue with hot 70% sulfuric acid to hydrolyze the insoluble polysaccharides. The final product is called Klason lignin and is determined gravimetrically.

For small samples, however, the most convenient method is to extract the lignins from thoroughly preextracted tissue (water, ethanol, acetone, ether, etc.) with 25% acetyl bromide in acetic acid at 70°C for 30 minutes (1 ml/10 mg of extracted material) and measure the absorbance of the resulting neutralized solution at 280 nm (Morrison, 1972). It should be noted that, although the results obtained by this method correlate quite well with the lignin content of plants determined conventionally as described above, no heed is taken of the contribution of possible acylating aromatic acids to absorbance or the degree of extractability of different lignins by the reagent. It should be further remarked that the method does give rise to absorbance at 280 nm from polysaccharides alone (presumably due to the formation of furfural), and this must be taken into account in tissues containing low amounts of lignin.

3. *Qualitative Analysis*

Ecologically, it appears to be of little importance whether a plant contains a mixed lignin (elements from the three alcohols **27–29**) or a near "pure" polymer (say over 80% guaicyl residues from **28**). But this may be due to our inability to appreciate the relative importance of highly condensed [higher amounts of *p*-coumaryl units **(27)**] and less condensed [higher amounts of sinapyl units **(29)**] lignins. If it is necessary to know something about the variation in substitution patterns, then the best method is to determine the relative amounts of benzaldehydes **30–32** obtained by alkaline nitrobenzene oxidation of exhaustively extracted and base-hydrolyzed tissue (otherwise simple cinnamic acids and even tyrosine interfere). The oxidation has to be carried out under high pressure, conveniently in a small stainless steel bomb. [Suitable quantities are 50 mg extracted tissue with 5 ml 2 *N* NaOH and 0.2 ml nitrobenzene heated at 160° for 3 hours (Bland, 1960). The products are then extracted from the acidified liquors with ether or chloroform and separated by thin-layer chromatography (Towers and Maass, 1965) on silica gel in benzene–acetic acid (9 : 1, v/v) (R_f of **30,** 0.62; **31,** 0.44; and **32,** 0.28; revealed by 2,4-dinitrophenylhydrazone)].

D. Ecological Significance of Lignins

To a large extent, ecologists take lignins for granted. They rightly assume that woody tissues are indigestible, unpalatable, and unattractive without any thought of the biochemical reasons why. This viewpoint is compounded by most modern textbooks on plant biochemistry and physiology, which either ignore the existence of lignin or mention it only in passing.

Wood, as Gertrude Stein might have said, is wood and, one might add, the quintessence of woodiness lies in the presence of lignins. Their evolutionary importance in the development of aborescent forms by imparting structural toughness to plant tissues has been mentioned. What is equally important is the protection that lignins provided to the plant against pathogenic attack and consumption by herbivores.

It seems likely, as suggested earlier, that these two ecological advantages may be ascribed both to lignins and to their possible precursors, feruloyl and *p*-coumaroyl acylated polysaccharides. Lignin and pathogenicity have been thoroughly discussed by Friend (1976). The observations he presents strongly indicate that lignin-like substances are formed in several cultivated plants when subjected to fungal attack. It seems probable, however, that the changes observed can be due in part to further acylation of either polysaccharides or existing lignin in the cell walls by ferulic **(4)** and *p*-coumaric acids **(3).** These acids, besides rendering the polysaccharide components of the cell walls less susceptible to attack by microbial hydrolases, also presumably act as bactericidal and fungicidal compounds on release. It is tempting to speculate that they may also undergo transesterification reactions with the methyl glucuronate groups of the pectins and thus become even more effective barriers to pathogens since their methyl esters are up to 1000 times more effective antibiotics than the free acids (Friend, 1978).

The adverse effect of lignification of cell walls of pasture grasses on the nutrition of ruminant animals is well known. Recent evidence, however, indicates that here again the changes observed may well be due to increased acylation of cellulose and hemicelluloses by **3** and **4,** as outlined above (Hartley, 1972). Whatever the reason, there is no doubt that increases in lignin-like substances in the cell walls of grasses with age have a deleterious effect on their digestibility by herbivores, since the compounds reduce the availability of both carbohydrates and protein, in much the same way as do tannins. Furthermore, the two cinnamic acids themselves **(3** and **4)** when released by enzymic hydrolysis, have recently been shown to inhibit feeding by the insect *Locusta migratoria* on *Sorghum bicolor* (Woodhead and Cooper-Driver, 1979) and by the snail *Melampus bidentatus* and the amphipod *Orchestia grillus* on detritus of *Spartina alterniflora* (Valiela *et al.,* 1979).

Finally, consider the importance of lignins in the formation of fractions of soils containing humic acids. The latter are polymeric organic constituents that certainly contain elements of lignin, as shown by the fact that alkaline oxidation yields vanillin **(31)** and other benzaldehydes. Since, again, much carbohydrate and protein is complexed to humic acids, the reduction in their rate of breakdown is important in the overall nutrient cycle.

IV. CONCLUSIONS

It is hoped that this chapter has convinced the reader that tannins and lignins are the most important defensive components of plants—not only because of their widespread, almost universal distribution, but also because their evolutionary history is longer than that of any other class of secondary plant constituents.

In spite of this, the two groups have been sadly neglected by biochemical ecologists, who have generally concentrated on the chemically more glamorous alkaloids, glucosinolates, nonprotein amino acids, and the like. Yet, as has been pointed out, there are many more problems to be solved regarding the importance of these phenolic polymers. The methodology is there to be used, so as Joxer said in Sean O'Casey's "Juno and the Paycock," "Grasp it like a lad of mettle and soft as silk remains."

ACKNOWLEDGMENTS

I wish to thank Dr. E. C. Bate-Smith for his continual inspiration and Mrs. Audrey McNamara for her careful preparation of the manuscript.

REFERENCES

Bate-Smith, E. C. (1972a). *Phytochemistry* **11,** 1153–1156.
Bate-Smith, E. C. (1972b). *Phytochemistry* **11,** 1755–1757.
Bate-Smith, E. C. (1973). *Phytochemistry* **12,** 907–912.
Bate-Smith, E. C. (1974). *In* "Chemistry in Botanical Classification" (G. Bendz and J. Santesson, eds.), pp. 93–102. Academic Press, New York.
Bate-Smith, E. C. (1975). *Phytochemistry* **14,** 1107–1113.
Bate-Smith, E. C. (1977). *Phytochemistry* **16,** 1421–1426.
Bate-Smith, E. C. (1978). *Phytochemistry* **17,** 1945–1948.
Bate-Smith, E. C. and Ribereau-Gayon, P. (1974). *Qualitas Plantarum Mater. Veg.* **5,** 189.
Bland, D. E. (1960). *Biochem. J.* **75,** 195–201.
Cooper-Driver, G. A. and Balick, M. (1979). *Bot. Mus. Leaflets* (Harvard University.) **26,** 257.
Cooper-Driver, G. A., Finch, S., Swain, T., and Bernays, E. (1977). *Biochem. Syst. Ecol.* **5,** 211–218.
Côte, W. A. (1977). *Recent Adv. Phytochem.* **11,** 1–44.
Feeny, P. (1975). *In* "Coevolution of Animals and Plants" (L. E. Gilbert and P. H. Raven, eds.), pp. 3–19. Univ. of Texas Press, Austin.
Feeny, P. (1976). *Recent Adv. Phytochem.* **10,** 1–40.
Friend, J. (1976). *In* "Biochemical Aspects of Plant-Parasite Relationships" (J. Friend and D. R. Threlfall, eds.), pp. 291–304. Academic Press, New York.
Friend, J. (1978). *Recent Adv. Phytochem.* **12,** 557.

Glombitza, K. W. G., Coch, M., and Eckhardt, G. (1976). *Phytochemistry* **15,** 1082–1086.
Goldstein, J. L., and Swain, T. (1965). *Phytochemistry* **4,** 185–192.
Grant, W. D. (1976). *Science* **193,** 1137–1138.
Gross, G. G. (1977). *Recent Adv. Phytochem.* **11,** 141–184.
Gross, G. G. (1978). *Recent Adv. Phytochem.* **12,** 117–220.
Harborne, J. B. (1973). "Phytochemical Methods." Chapman & Hall, London.
Harborne, J. B. (1977). "Introduction to Ecological Biochemistry." Academic Press, New York.
Hartley, R. D. (1972). *J. Sci. Food Agric.* **23,** 1347–1354.
Hartley, R. D., and Jones, E. C. (1976). *Phytochemistry* **15,** 1157–1160.
Haslam, E. (1966). "The Chemistry of Vegetable Tannins." Academic Press, New York.
Haslam, E. (1974). "The Shikimic Acid Pathway." Halsted Press, New York.
Haslam, E. (1975). *In* "The Flavonoids" (J. B. Harborne, T. J. Mabry, and H. Mabry, eds.), pp. 505–559. Chapman & Hall, London.
Higuchi, T. M., Ito, Y., Shimada, M., and Kawamura, I. (1967). *Phytochemistry* **6,** 1551–1556.
Levin, D. (1971). *Am. Nat.* **105,** 157–181.
Levin, D. (1976). *Annu. Rev. Ecol. Syst.* **7,** 121–159.
Morrison, I. M. (1972). *J. Sci. Food Agric.* **23,** 455–463.
Morrison, I. M. (1974). *Phytochemistry* **13,** 1161–1165.
Neish, A. C., and Freudenberg, K. (1968). "Constitution and Biosynthesis of Lignin." Springer-Verlag, Berlin and New York.
Nimz, H. (1974). *Angew. Chem., Int. Ed. Engl.* **13,** 313–322.
Oates, J., Swain, T., and Zantovska, J. (1977). *Biochem. Syst. Ecol.* **5,** 317–321.
Rhoades, D. F. (1977a). *Biochem Syst. and Ecol.* **5,** 281–290.
Rhoades, D. F. (1977b). *In* "Creosote Bush" (T. J. Mabry, J. H. Hunziker, and D. R. Difeo, eds.), pp. 135–258. Dowden, Hutchinson and Ross, Stroudsburg, Pa.
Rhoades, D. F., and Cates, R. G. (1976). *Recent Adv. Phytochem.* **10,** 168–213.
Sakakibara, A. (1977). *Recent Adv. Phytochem.* **11,** 117–141.
Sarkanen, K. V., and Ludwig, C. H. (1971). "Lignins." Wiley (Interscience), New York.
Sieburth, J. M., and Conover, J. T. (1965). *Nature (London)* **208,** 52–53.
Stafford, H. A. (1974a). *Recent Adv. Phytochem.* **8,** 53–79.
Stafford, H. A. (1974b). *Annu. Rev. Plant Physiol.* **25,** 459–86.
Swain, T. (1974a). *In* "Chemistry in Botanical Classification" (G. Bendz and J. Santesson, eds.), pp. 81–92. Academic Press, New York.
Swain, T. (1974b). *Compr. Biochem.* **29,** Part A, 125–302.
Swain, T. (1976). *In* "Morphology and Biology of Reptiles" (A. d'A. Bellairs and C. B. Cox, eds.), pp. 107–122. Academic Press, New York.
Swain, T. (1977a). *Annu. Rev. Plant Physiol.* **28,** 479–501.
Swain, T. (1977b). *Proc. Int. Congr. Entomol., 15th, 1976* pp. 197–210.
Swain, T. (1978a). *In* "Biochemical Aspects of Plant and Animal Co-evolution" (J. B. Harborne, ed.), pp. 1–19. Academic Press, New York.
Swain, T. (1978b). *Recent Adv. Phytochem.* **12,** 617–640.
Swain, T. (1979). *Science* (to be published).
Synge, R. L. M. (1975). *Qual. Plant.—Plant Foods Hum. Nutr.* **24,** 337–350.
Towers, G. H. N., and Maass, W. S. G. (1965). *Phytochemistry* **4,** 57–66.
Valiela, I., Koumjian, L., Swain, T., Teal, J. M., and Hobbie, J. E. (1979). *Nature* (in press).
Woodhead, S., and Cooper-Driver, G. (1979). *Biochem. Syst. and Ecol.* **7** (in press)

Chapter **20**

Insect Hormones and Antihormones in Plants

KAREL SLÁMA

I. INTRODUCTION

Chemical regulation of growth and reproduction in insects is achieved by an enormously diversified neuroendocrine system, which does not show any comparable evolutionary counterpart in the plant kingdom. Yet, in spite of millions of years of independent evolution, there remain certain fundamental principles in the regulation of cellular growth that are common to both animals and plants. For example, similarities exist in the formation of some microstructural cell elements, and there are analogies in the process of mitotic division, similarities in the distribution of genetic information, common formation of the zygote, and common steps in early morphogenetic differentiation. It is not unlikely, therefore, that molecules involved in the regulation of certain basic biogenetic processes in plants could be specifically used as hormones to stimulate an analogous process in animals. Alternatively, it may be that some compounds that are present for various other reasons in plants may accidentally fit certain structural requirements and consequently act as animal hormones. Like some other secondary plant substances, the hormonally active compounds in plants could have been a factor of natural selection that modulated the coevolutionary relationships between plants and their insect herbivores.

ISBN 0-12-597180-X

A search for insect neurohormones or adenotropic hormones in plants would seem to be unwarranted. These hormones are mostly proteins or peptides with short biological half-lives. Even when accidentally present in the plant food, they should be rapidly broken down and inactivated by the digestive enzymes. In contrast to many other classes of secondary plant substances, which have been described in other chapters, identification of animal hormones in plants is sometimes very difficult. It depends mostly on well-elaborated methods for their detection. Our knowledge of endocrinological techniques in insects is most advanced for the sesquiterpenoid compounds of juvenile hormone (JH) and for polyhydroxylated steroidal compounds possessing molting hormone (MH) activity. As a result, a consideration of hormonal interactions between plants and insects is concerned mainly with these isoprenoid hormones and their pharmacobiological mimics.

Insects represent an economically important group of invertebrate herbivores. The control of larval growth, metamorphosis, and reproduction in this group is achieved by the combined action of neurohormones and two principal hormones acting at the peripheral target tissue. One is JH secreted by the corpora allata, and the other is MH produced by prothoracic and related glands. Secretion of these hormones from the endocrine glands is usually linked to nutritional factors derived from plant food. The main function of JH in insect ontogenesis is to inhibit selectively the morphogenetic process, which allows continued somatic growth. The hormone is absent during metamorphosis, but in the fully formed adult stage it is again used as a gonadotropic factor. The main function of MH is to stimulate development from one molting cycle to the next [for further information, see the reviews of Wigglesworth (1970) and Sláma *et al.* (1974)].

II. PLANT JUVENOIDS

During the past decade, endogenous JH from various insects has been isolated and identified (reviews by Menn and Beroza, 1972; Sláma *et al.*, 1974). The JH chemical structures are closely related to the esters of certain acyclic sesquiterpenoid acids, of which methyl 10,11-epoxy-3,7,11-trimethyl-2,6-dodecadienoate **(1)**, known as JH-III, and its homo **(2)** and dihomo analogs **(3)**, known as JH-II and JH-I, respectively, are the only JH compounds known to occur in insects to date.

The history of JH research is closely linked with secondary plant chemicals. Shortly after the development of a bioassay method for JH by Williams (1956), JH-active lipids were found in extracts of microorganisms and higher plants (Schneiderman *et al.*, 1960). The first compound found to cause JH activity was an acyclic sesquiterpenoid alcohol, farnesol **(4)**. It was isolated from yeasts and excrement of the beetle *Tenebrio molitor* (Schmialek, 1961) but was previously known as a common constituent of various flower oils.

A new search for JH analogs (juvenoids) among plants was stimulated by an accidental finding that certain American paper products contained an unknown factor that caused JH effects in a bug, *Pyrrhocoris apterus* (Sláma and Williams, 1965). The activity was traced to the lipid-soluble materials contained in the wood of certain pulp trees, mainly that of Canadian balsam fir, *Abies balsamea* (Sláma and Williams, 1966). Active extracts were also obtained from the wood of other, but not all, coniferous trees (see Sláma, 1969; Williams, 1970).

The compound responsible for the JH activity of balsam fir wood was identified by Bowers *et al.* (1966) as an alicyclic sesquiterpenoid ester of (+)-todomatuic acid and was named juvabione **(5)**. Černý *et al.* (1967) subsequently isolated two active compounds from a tree growing in a balsam fir grove in a Czechoslovakian arboretum. One active compound was assumed to be juvabione and the other dehydrojuvabione **(6)**. In contrast to farnesol and its synthetic derivatives, the juvabione-type compounds were selectively active only on insects of the family Pyrrhocoridae (Sláma, 1969; Williams, 1970).

In the original papers, Bowers *et al.* (1966) and Černý *et al.* (1967) did not establish the exact stereoconfigurations of juvabione and dehydrojuvabione, expecting that they would be identical with the *R, R* stereochemistry ascribed to the naturally occurring (+)-todomatuic acid by Nakazaki and Isoe (1963). Later, Pawson *et al.* (1970) synthesized all four diastereoisomers of juvabione. By measuring a sample of juvabione from the Czechoslovakian tree supplied by Černý *et al.* (1967), and not realizing the possibility that there could be two naturally occurring isomers of juvabione, they concluded that the stereochemistry of natural juvabione was *R,S* and not *R,R* as previously expected. Manville (1975) reconfirmed the *R,R* stereoconfiguration of juvabione from native Canadian *Abies balsamea* and deduced that the compound isolated by Černý *et al.* (1967) ought to be a geometric isomer of (+)-juvabione. Moreover, Manville *et al.* (1977) performed detailed analysis of the terpene constituents of the trees growing in the Czechoslovakian arboretum. They found that each of the three trees considerably differed from the native *A. balsamea* in terpene constituents and in some morphological patterns. One of the trees, which was the sample tree used by Černý *et al.* (1967), indeed contained the *R, S* isomers of juvabione and dehydrojuvabione, i.e., (+)-epijuvabione and (+)-dehydro-

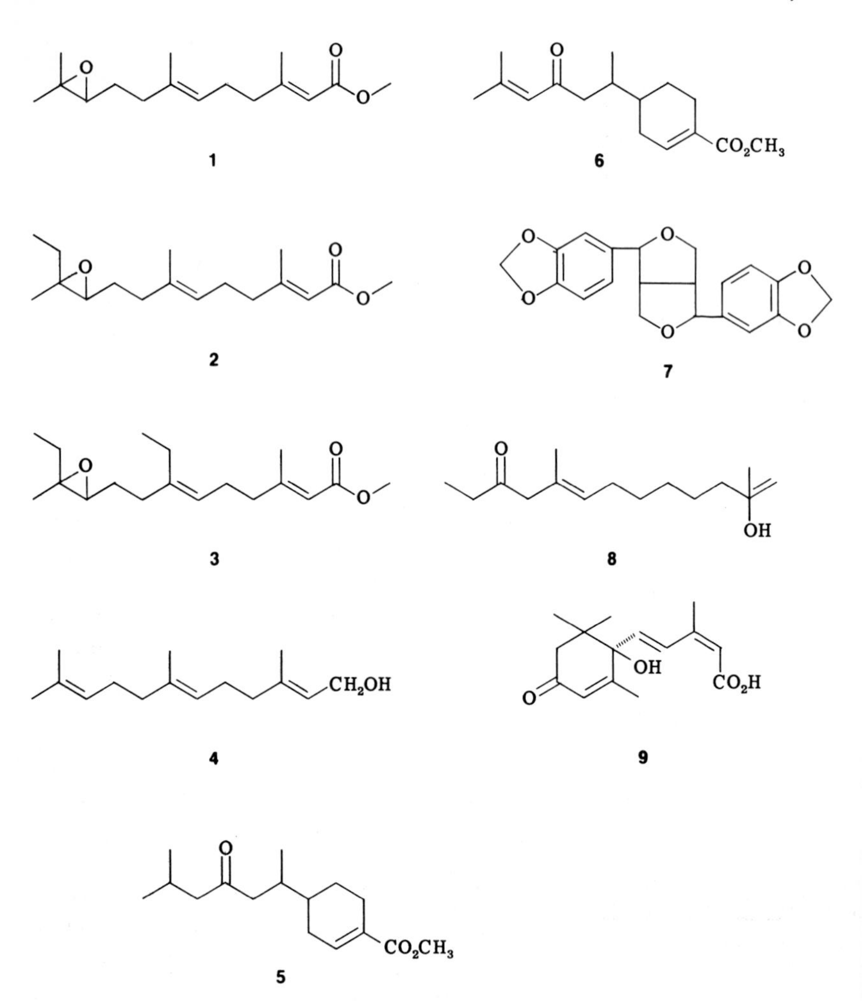

epijuvabione. They also found that the wood of balsam fir contains several other JH-active compounds of the juvabione type.

The woody portions of stems and branches are the main source of juvabione (Sláma, 1969). Very little or no JH activity is present in extracts of the foliage or bark. This explains why Staal (1967) was unable to find JH activity in the leaf extracts of a large number of conifers. The expected biosynthetic precursors of juvabione and dehydrojuvabione, i.e., (+)-todomatuic acid and dehydrotodomatuic acid, were also found in the wood of *Abies grandis* and *A. amabilis* (Puritch and Nijholt, 1974). Their presence was related to the infestation of the trees by the balsam woolly aphid (*Adelges piceae*). These acids were absent from the wood of noninfested trees and, in the infested trees, they occurred only in the wood adjacent to the aphid infestation. This suggests that the phytoalexin responses (see Harborne, 1977) of plants may also be induced in response to the feeding activity of herbivorous insects.

After the discovery of JH activity in balsam fir, there appeared several reports on the presence of hormonal mimics in plants (see Williams and Robbins, 1968). Some naturally occurring insecticide synergists, namely, sesamin **(7)** and sesamolin, which occur in sesame oil, showed low JH activity (Bowers, 1968). Some further reports of JH activity in plants included, for instance, isolation of JH materials from Douglas fir, red cedar (Wellington, 1969; Mansingh *et al.*, 1970), and various Indian species of plants (Saxena and Srivastava, 1973; Prabhu and John, 1975). Perhaps due to the limited specificity of certain assays for JH, some authors claim to have shown this activity for secondary plant substances that do not qualify as true analogs of JH (see Primo Yufera *et al.*, 1976; Ruscoe, 1972; Barton *et al.*, 1972).

Jacobson (1971) screened extracts from 200 higher plants for JH activity. Hormonal activity was assayed on larvae of a hemipteran insect, *Oncopeltus fasciatus,* and on pupae of the beetle *Tenebrio molitor,* using topical application. Only one extract, that of American coneflower (*Echinacea angustifolia*), showed reasonably high JH activity. More recently, Jacobson *et al.* (1975a) repeated this screening. They found that extracts of *Echinacea angustifolia* roots and *Chamaecyparis lawsoniana* seeds were highly active in *Tenebrio,* whereas extracts of *Clethra alnifolia* stems, leaves, and fruits, *Sassafras albidum* roots and root bark, *Eucalyptus camaldulensis* stems and bark, *Pinus rigida* twigs and leaves, and *Iris douglasiana* roots, stems, and fruits showed high activity on *Oncopeltus.* An extensive list of 326 species of plants that failed to provide JH-active ether extracts was also included. Finally, Jacobson *et al.* (1975b) isolated and identified the active compound from the *Echinacea* extracts and named it echinolone **(8)**.

The inhibitory action of JH on morphogenesis and differentiation of reproductive organs in insects seems to have its physiological counterpart in the effects of dormine or abscisic acid **(9)** in higher plants. This compound, which is structurally related to certain alicyclic sesquiterpenoid juvenoids, has a low but definite JH activity in pyrrhocorid bugs. Stowe and Hudson (1969) observed that farnesol and other juvenoids promoted the growth of isolated pea stem sections, as did other lipophilic compounds. They concluded that these effects in plants and insects might be mutually related. Other examples of well-expressed interactions between plants and insects that are mediated by isoprenoid compounds have been mentioned by Herout (1970).

Some authors have assumed that JH activity may contribute to the capacity of a plant to resist massive attacks by herbivorous insects (Williams, 1970; Sláma, 1969; Herout, 1970). However, examples of the JH-linked insect–plant interactions were found merely by chance. They depend on an accidental susceptibility of a few laboratory insects to a given plant juvenoid. It would be more important to ascertain whether the JH activity would also be involved in some special ecological situations in which particular phytophagous insects carefully avoid certain selected plants (see Sláma *et al.*, 1974). Our present information on the possible ecological importance of these JH-linked interactions is based exclusively on the results of topical assays. We have recently found in our laboratory that various classes of the isoprenoid juvenoids are far more effective when ingested by insects. This fact increases enormously the probability that plants may contain many other, still unknown juvenoid compounds, because feeding represents the most common means by which plants interact with herbivores.

The existence of over 2000 synthetic juvenoids [many of them related structurally to naturally occurring plant chemicals; see Sláma *et al.* (1974)] suggests that the receptors of JH within insect cells tolerate a wide range of structural analogs of the endogenous hormones. A somewhat similar situation exists with the receptors of estrogens accepting nonsteroidal phytoestrogens in mammalian cells (see Doecke, 1975). Further analysis of the presence of JH-active materials in plants by chemical ecologists offers real promise in the search for a practical method of controlling the reproduction of harmful insects by nontoxic natural products or their analogs.

III. PHYTOECDYSONES

The term *molting hormone* is used here to indicate the humoral factor that stimulates initiation of molting cycles in insects. In the absence of JH, it stimulates morphogenetic changes associated with metamorphosis. It is an animal hormone the study of which is most closely related to studies of secondary plant chemicals. In 1935, Fraenkel developed a sensitive bioassay for MH activity; this later became a general technique for identifying MH activity and is known as the *Calliphora, Musca*, or *Sarcophaga* bioassay. Using this assay, Butenandt and Karlson (1954) isolated the active substance from silkworm pupae and named it ecdysone. Its complete structure has been finally resolved as $2\beta,3\beta,14\alpha,$-$22R$,25-pentahydroxy-5β-cholest-7-en-6-one **(10)** (for review, see Karlson, 1966). Another slightly more polar compound, previously referred to as β-ecdysone, has been identified as 20-hydroxyecdysone and is now generally known as ecdysterone **(11)** (see Rees, 1971).

The presence of some polyhydroxylated isoprenoids in plants was noted by phytochemists before the structure of ecdysone became known, but it was Nakanishi *et al.* (1966) who first demonstrated that compounds isolated from *Podocarpus* trees were related by structure and biological activity to ecdysone. Among other reports that almost immediately followed, that of Jizba *et al.* (1967a,b), for instance, showed that the previously isolated polypodine A from rhizomes of the fern *Polypodium vulgare* was identical with ecdysterone. The dried *P. vulgare* rhizomes contained impressive amounts of the MH-active material, of the order of 1%. Thus, as pointed out by Williams (1970), 2.5 gm of these rhizomes contained the equivalent MH activity of 500,000 gm of silkworm pupae.

The simplicity of the *Calliphora* bioassay made possible an extensive screening of plants for their MH activity on insects. Takemoto *et al.* (1967) obtained active extracts from 22 of 39 species of ferns and from 27 of 74 species of higher plants. They were unable to detect MH activity in certain fungi and algae. Using a new topical assay (*Chilo* dipping test), Imai *et al.* (1969) and Matsuoka *et al.* (1969) conducted an extensive screening of plant extracts for analogs of MH (often called phytoecdysones, ecdysteroids, or ecdysoids). Some of the results have indicated a wide distribution of phytoecdysones, especially among Pteridophyta and Gymnospermae (Imai *et al.*, 1969). Among Angiospermae, most of the active extracts were found in the Amaranthaceae and Verbenaceae families. Plants that possess active extracts are mostly perrennial and woody plants rather than annual, herbaceous plants.

Hikino *et al.* (1973) evaluated the presence of phytoecdysones in 283 species of Japanese ferns, representing 76 genera and 20 families of plants. Positive responses were obtained with extracts from 170 species; some extracts contained up to seven different phytoecdysone compounds. Analogous screening of Formosan ferns revealed the presence of MH activity in 64 of 115 tested species (Yen *et al.*, 1974).

During one decade of research, several dozen phytoecdysones were isolated and identified in more than 80 plant families (Hikino and Takemoto, 1974). The details concerning their isolation and chemical structures can be found in numerous review articles (Hikino and Hikino, 1970; Herout, 1970; Nakanishi, 1971; Rees, 1971; Achrem *et al.*, 1973; Sláma *et al.*, 1974; Takemoto *et al.*, 1976). A few examples are given here to illustrate only the basic structural types of these compounds.

The C_{27} phytoecdysones containing a cholestane-type side chain are widely distributed in plants. Ecdysterone **(11)** is the most commonly occurring phytoecdysone. Other compounds of this series differ from ecdysterone in number, position, and stereochemistry of the hydroxyl groups in the side chain, e.g., ponasterone A **(12),** inokosterone, and pterosterone; in positions of the hydroxyl groups in the nucleus, e.g., polypodine B **(13),** ajugasterone C, and 2-deoxyecdysterone; or in different location of the double bonds, e.g., podecdysone B **(14),** cheilanthone A **(15),** stachysterone C, and calonysterone **(16).**

The C_{28} series of phytoecdysones, possessing an ergostane-type side chain, are relatively rare [see makisterone A **(17)**]. On the other hand, the C_{29} series of stigmastane-type phytoecdysones is again represented by numerous derivatives. Some examples are amarasterones A and B, lemmasterone **(18),** and makisterone D. This group includes many compounds containing a cyclic ether structure in the side chain, such as cyasterone **(19),** sengosterone, and capitasterone. Also, derivatives of phytoecdysones are known in which certain hydroxyl groups are replaced by an acetyl, alkoxyl, glucosyl, or cinnamoyl group. Ajugalactone **(29)** is also a compound related to phytoecdysones and is assumed to have MH-antagonist effects.

In addition to the above-named sterolic types, plants also contain C_{21} pregnane-type and C_{19} androstane-type compounds possessing an ec-

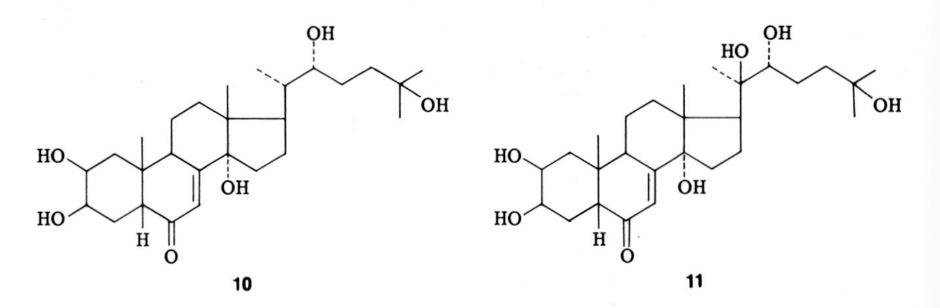

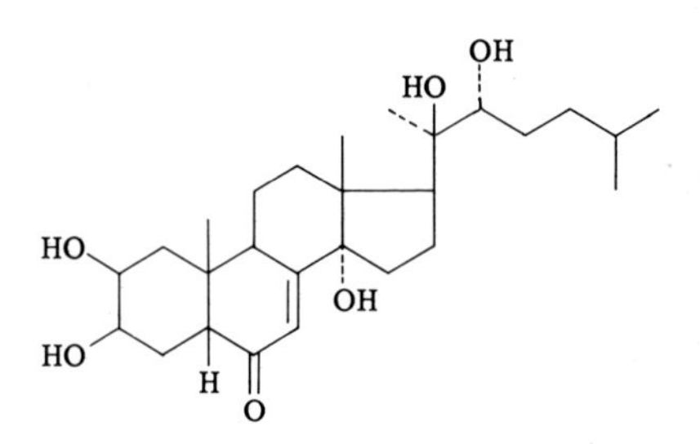

12

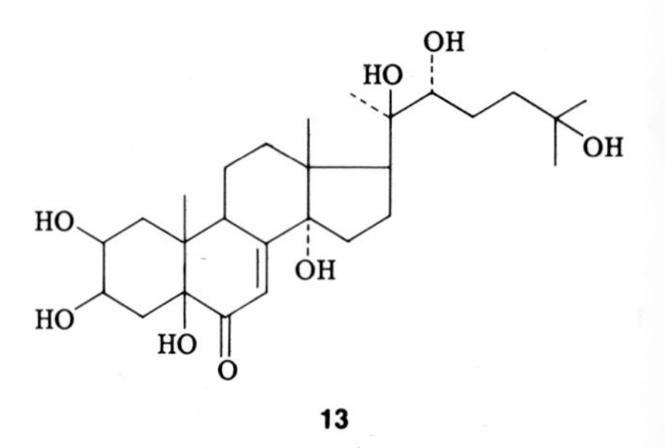

13

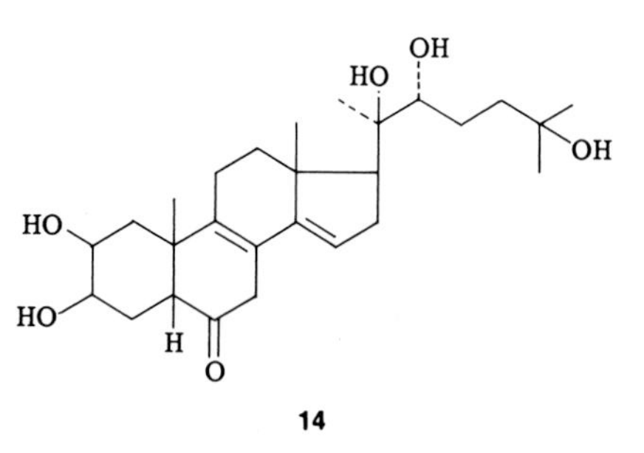

14

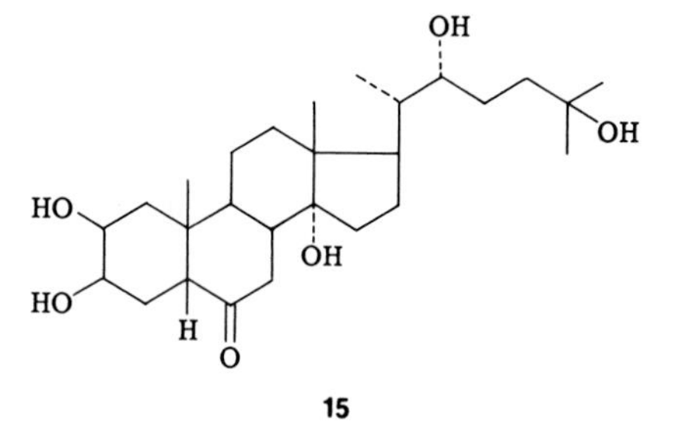

15

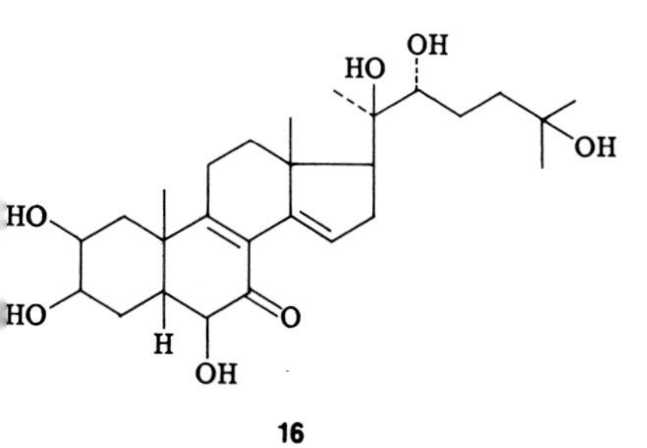

16

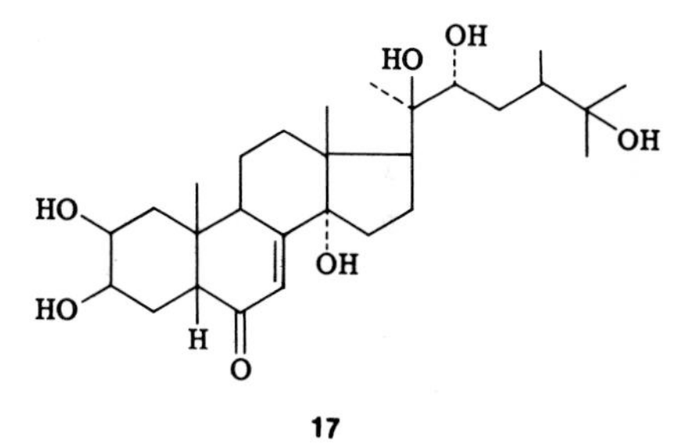

17

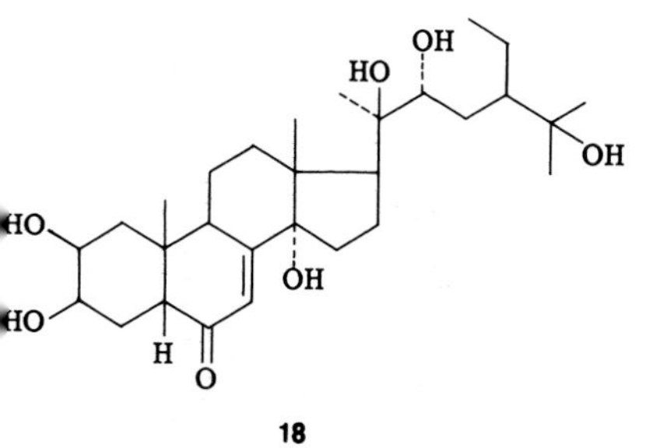

18

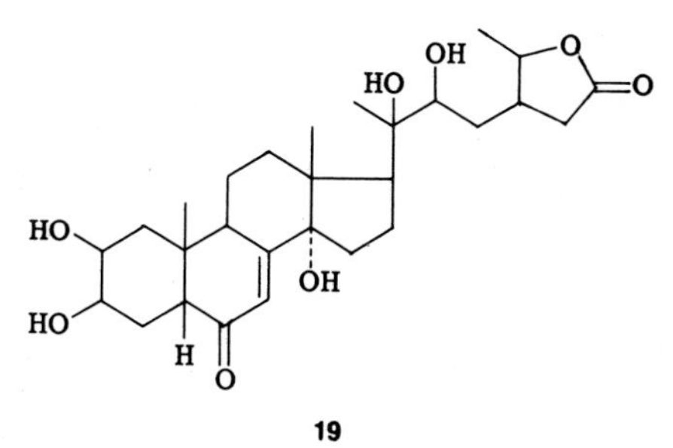

19

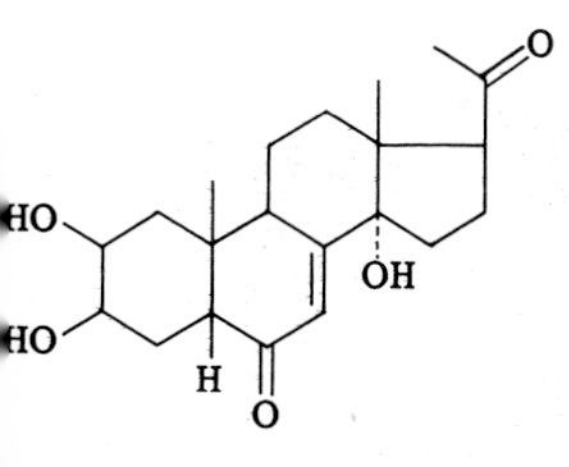

20

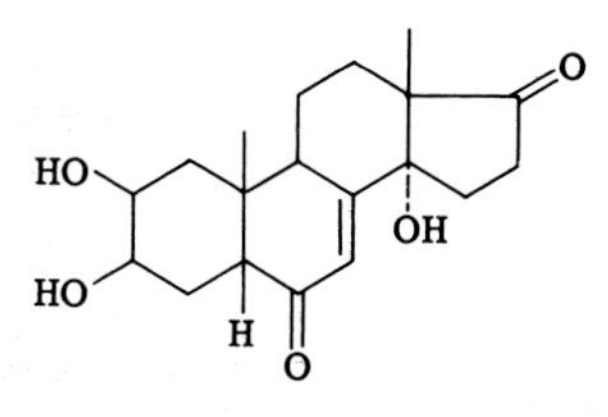

21

dysone nucleus. The first is represented by poststerone **(20)** and the second by rubrosterone **(21).** They are thought to be degradation products of other phytoecdysones (Rees, 1971). The occurrence of these metabolites in plants confirms that plant tissues can oxidize sterolic side chains as do animal tissues in the biogenesis of steroidal hormones. Although poststerone and rubrosterone are more closely related to vertebrate hormones by molecular size, their MH activity in insects is far less than that of the corresponding ecdysones containing the proper side chain.

It is generally expected that plants synthesize phytoecdysones from cholesterol. Incorporation of labeled cholesterol into ecdysterone or conversion of mevalonate to phytoecdysones was demonstrated in seedlings of various plants by a number of workers (see Heftmann *et al.*, 1968; de Souza *et al.*, 1970; Rees, 1971; Rees and Goodwin, 1974; Morgan and Poole, 1977). When compared to other classes of hormonal mimics, such as phytoestrogens (Doecke, 1975) or plant juvenoids, phytoecdysones show far more conservative structure–activity relationships. Chemical analysis of numerous plant extracts did not reveal a single nonsteroidal mimic exhibiting MH activity. Moreover, among so many steroids in plants, only compounds possessing the necessary structural features of ecdysone were able to cause MH-type responses in insects. According to Sláma *et al.* (1974), these essential structural features are the following: (a) cis fusion of A/B rings, (b) β-hydroxylic function at C-3, and (c) conjugated keto group at C-6 with an appropriately *R*-oriented hydroxylic function at C-22. Further restrictions in these structural requirements for biological activity would indicate the primary importance of the α,β-unsaturated keto group. This grouping is also essential for the biological activity of other steroid animal hormones, such as androgens, gestagens, and corticoids.

The content of phytoecdysones in plants containing hormonally active compounds usually varies between 0.01 and 0.05% of the dry weight. In some cases, concentrations as low as 0.0001% or as high as 2% have been detected (Achrem *et al.*, 1973), and the contents undergo considerable changes with respect to seasonal variations (Hikino *et al.*, 1973; Yen *et al.*, 1974). The plant *Serratula inermis* was reported to contain up to 2% ecdysterone on a dry weight basis in flowers and fruits, 0.25% in the leaves, 0.01% in the stem but practically none in the roots (see Achrem *et al.*, 1973). The *Calliphora* unit of MH activity of ecdysone or ecdysterone is approximately 0.01 μg per larval abdomen. The biological activity of most other phytoecdysones that possess the essential structural features does not differ significantly from this range.

The biological basis for so many plants having developed the ability to alaborate steroids with insect MH activity is still unknown (Morgan and Poole, 1977). There is a possibility that the presence of phytoecdysones may be a result of an evolutionary adaptation in plants that interferes with the growth processes of insect herbivores (Herout, 1970; Williams, 1970; Rees, 1971; Takemoto *et al.*, 1976). Indeed, Robbins *et al.* (1968) observed disturbances in insect development after ingestion of ecdysone analogs. However, other evidence suggests that this action cannot be the sole reason for the presence of phytoecdysones in plants. Many insects successfully develop and reproduce while feeding on plants that contain considerable amounts of phytoecdysones. Thus, at least certain species of insects became adapted to the dietary supply of MH activity by evolving efficient methods of excreting and metabolizing phytoecdysones. Chemotaxonomical conclusions reached by Hikino *et al.* (1973) and Yen *et al.* (1974) suggest that phytoecdysones cannot be viewed as general plant growth hormones. These compounds are among the most polar naturally occurring steroids ever known. It is thus possible that among other functions they may be involved in the transport of sterol within the plant system.

IV. ANTIHORMONAL FACTORS

Activation and inactivation of the neuroendocrine system of insects is dependent largely on nutrition. It thus appears that various secondary plant substances that function as repellents, feeding deterrents, or toxicants (see Munakata, 1975) can interfere with the hormonal control of growth and development in insects. Extracts of almost any plant can exhibit some type of inhibitory action on insect development, as shown by the screening of 325 plant extracts on *Aedes* mosquitoes (Patterson *et al.*, 1975).

The mode of action of insect hormones is still incompletely understood. Consequently, a reliable and fully specific bioassay for antihormonal agents does not exist. Sláma (1978) has enumerated the following types of antihormonal compounds: (a) antineurotropic factors, (b) antiadenotropic or antineurohormonal factors, (c) antihormonal factors in the strict sense, and (d) antihomeostatic factors. As is the case with antigonadotropins and antiestrogens in mammals (see Dorfman, 1962; Emmens and Martin, 1964a; Krause, 1970; Churý, 1971; Doecke, 1975), it is difficult to determine in insects whether the suspected antihormonal agent acts at the peripheral tissue level or at the level of adenotropic hormones.

In analogy with antiestrogenic action of the weakly active phytoestrogens (Emmens and Martin, 1964b; Krause, 1970; Doecke, 1975), attempts were made to induce antijuvenile effects in insects by 200 weakly active JH analogs (Sláma *et al.*, 1974). It was not possible to demonstrate such competitive inhibition of JH action. Due to physiological interactions within the neuroendocrine system, ecdysone and various phytoecdysones are able to cause antijuvenile-like prothetelic effects in immature insects (Sláma, 1978). Apart from this, however, the first antijuvenile-like effects from plant materials in insects were induced by kojic acid **(22).** This compound is produced as a sugar metabolite by various *Aspergillus* fungi. When fed in the diet to young larvae of the F_1 hybrid of *Bombyx* silkworms, it caused omission of one larval instar and caused premature metamorphosis (Murakoshi, 1972).

These prothetelic effects in silkworm larvae were later induced by synthetic derivatives related to kojic acid (Murakoshi and Ichimoto, 1972) and, more importantly, by a series of secondary plant substances. Compounds exhibiting marked "antijuvenile" action included diterpenes related to abietic or podocarpic acid, such as a compound referred to as A-11 **(23),** and various hydrofluorene compounds, such as B-3 **(24)** (see Murakoshi *et al.*, 1975). Other plant substances having this property were isolated from avocado, including dimethyl sciadinonate **(25)** and another new compound **(26)** (Murakoshi *et al.*, 1976). It seems, however, that the action of these substances is more or less limited to larvae of the F_1 hybrid of the silkworm (Sláma, 1978). Their action may result from a nutritional imbalance that deeply affects functions of the whole neuroendocrine system. It has also been observed (Elliger *et al.*, 1976) that some kaurenoid and trachylobanoic acids isolated from sunflower inhibit insect development by affecting the hormonal system.

A secondary plant substance that causes larger incidence of antijuvenile-like prothetelies in certain exopterygote insects is ageratochromene **(27),** isolated from *Ageratum* plants by Alertsen in 1955. Its prothetelic action on insects was first discovered by Bowers (1975). Topical treatment of young larvae of *Oncopeltus fasciatus* by *Ageratum* extracts or ageratochromene caused nutritional disproportions, which occasionally resulted in premature metamorphosis (Bowers *et al.*, 1976). The plant *Ageratum houstonianum* also contains another, less active 6-hydroxy-7-methoxy derivative of ageratochromene. Both of these compounds are now generally known by their trivial names, precocene II and precocene I, respectively. In addition to their effects on metamorphosis, they inhibit ovarian growth in adult females of some susceptible species and have a marked capacity to deter feeding and cause malnutrition (Bowers *et al.*, 1976). Prothetelic action of ageratochromene is not limited to hemipteran insects. For example, it causes premature metamorphosis in *Locusta migratoria* (Němec *et al.*, 1979).

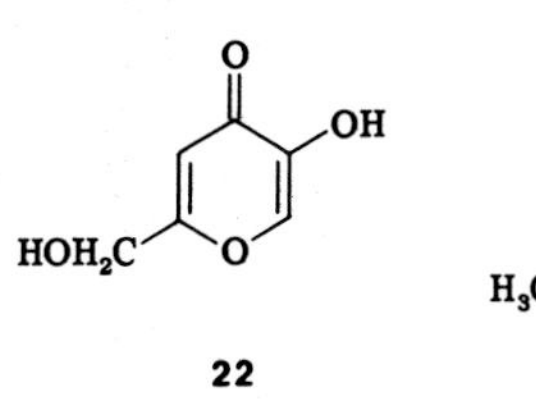

22

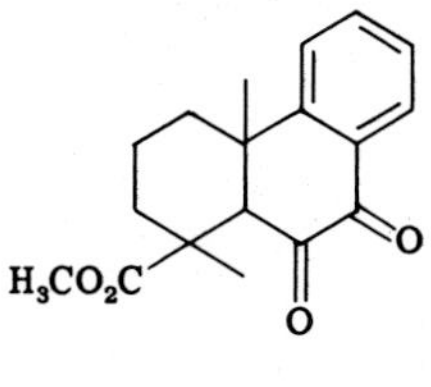

23

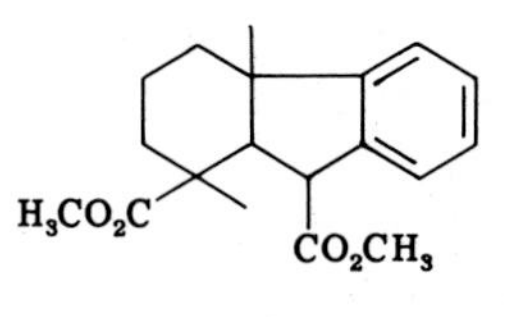

24

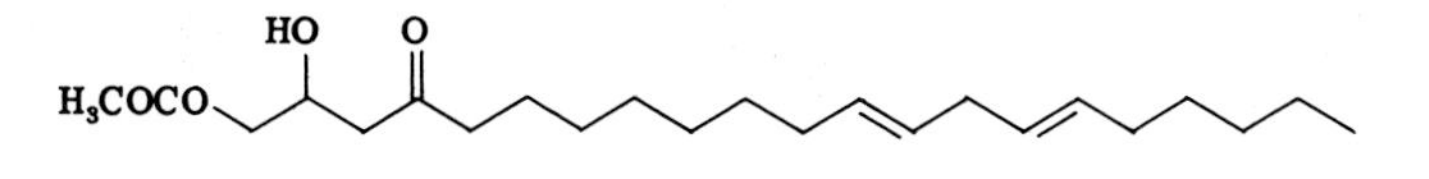

26

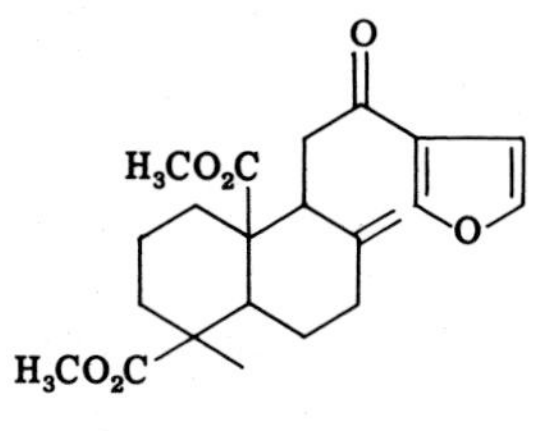

25

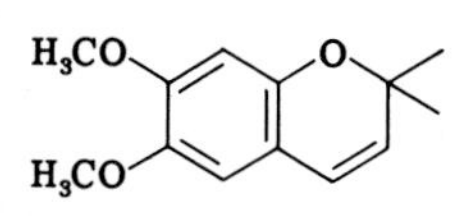

27

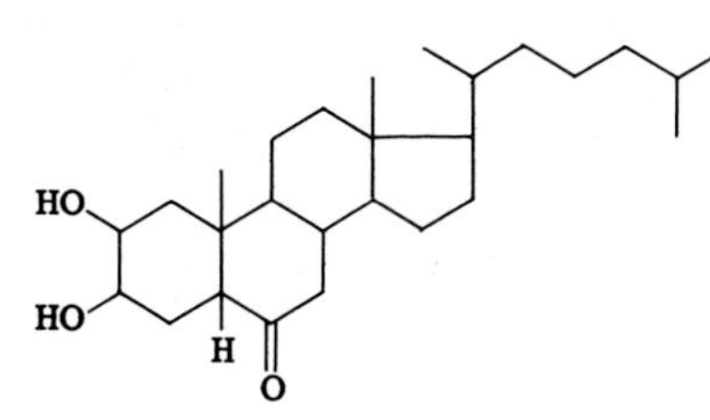

28

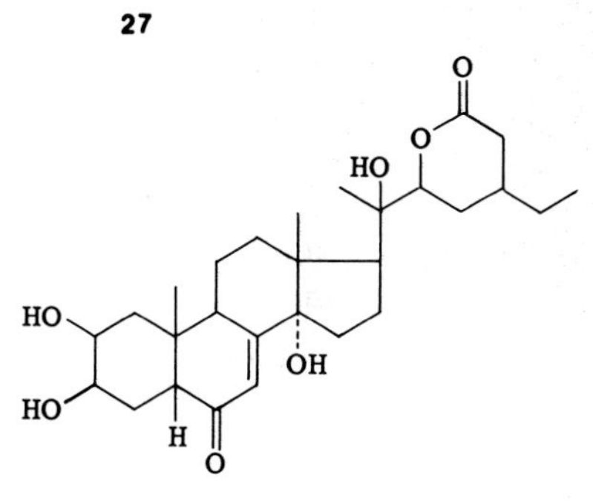

29

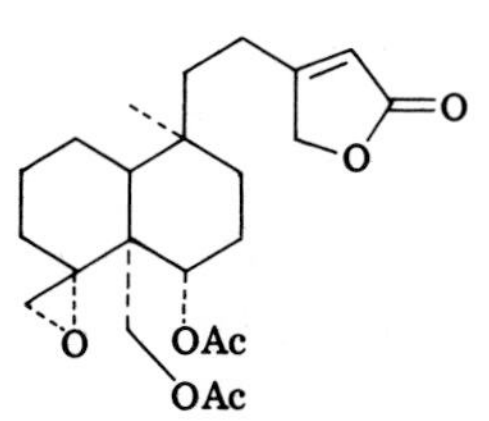

30

Ageratochromene (precocene) does not inhibit JH action at the peripheral tissue level. It is believed to interfere with the biosynthesis of JH in corpora allata, thereby exhibiting antiallatotropic properties (Bowers, 1977; Pratt and Bowers, 1977; Bowers and Martinez-Pardo, 1977). Due to a nutritionally linked activation of JH biosynthesis, however, this process can be nonspecifically inhibited by starvation, malnutrition, and various other secondary plant chemicals. Examples of the latter are quinine and ergometrine alkaloids (Chatterjee and Rajchandhuri, 1973).

Antijuvenile factors from plants are expected to have practical significance in the control of the development and reproduction of harmful insects [''fourth generation pesticides''; see Bowers *et al.* (1976)]. So far nothing is known about the distribution of other factors with antijuvenile or antiallatotropic activity in the plant kingdom. Assigning a possible role to antijuvenile action in the complex of coevolutionary plant interactions with insect herbivores is thus only speculative (Sláma, 1978).

Antihormonal factors with antiprothoractotropic and antiecdysone action should cause complete arrest of ontogenetic development, producing the symptoms of larval or pupal diapause. Similar inhibition of insect development can be achieved by a diversity of secondary plant chemicals ranging from repellents, feeding deterrents, antimetabolites, and antimitotic agents to plant toxicants such as pyrethrins and rotenone. Many of these compounds can be considered to be potential factors with antiecdysone or antiprothoracotropic properties. Because there is so far no way to determine specifically antihormonal activity among these developmental inhibitors (Sláma, 1978), all the effects discussed below must be viewed only as tentatively antihormonal.

The first compounds suspected of producing possible antiecdysone effects were derivatives of 2,3-dihydroxy-6-ketocholestane **(28)** (Hora *et al.*, 1966). Some of these compounds are structurally related to cheilanthones **(15)** occurring in plants. Robbins *et al.* (1968, 1970) demonstrated inhibitory effects on development by these and related ecdysone analogs in various insects. Another secondary plant substance, ajugalactone **(29)**, which occurs in *Ajuga decumbens* along with some phytoecdysones, showed ecdysone-antagonistic effects (Koreeda *et al.*, 1970). It inhibited the MH activity of a potent phytoecdysone, namely, ponasterone A. More recently, Kubo *et al.* (1976) isolated several substances from *Ajuga remota* that retard feeding and development. It was also suspected that these substances, ajugarins **(30)**, would exhibit antiecdysone effects. Under certain circumstances, as, for example, in larvae of

Anthonomus, developmental inhibitions can be produced by ingestion of phytoecdysones (Earle *et al.*, 1970; see also Robbins *et al.*, 1968). In the search for possible antiecdysone factors, Matolcsy *et al.* (1975) and Tóth *et al.* (1977) claimed to have obtained antiecdysone results with a fungicide, triarimol, which is an inhibitor of ergosterol synthesis.

The physiological reasons for the inhibition of insect development by various plant chemicals are little understood, except in the case of certain toxic compounds. It is likely that some of these effects are indeed associated with ecdysone insufficiency. This possibility has only recently been recognized, but adequate endocrinological methods for the determination of specific antiecdysone action are still lacking (Sláma, 1978). This shows that precautions must be taken so that antihormonal properties are not ascribed to some nonspecifically acting developmental inhibitors of plant origin.

V. CONCLUSIONS

This brief review provides considerable evidence that insect hormones and certain of their pharmacobiological mimics occur in plants. Various plants also contain the steroid hormones of vertebrate animals, mainly estrogens and their mimics (reviews by Heftmann, 1975; Doecke, 1975). Most of the hormonally active substances found in plants are related to isoprenoid animal hormones. The steroid hormones of vertebrates have been shown to affect the growth and differentiation of reproductive organs in plants (see Heftmann, 1975), but little is known about the possible growth regulatory functions of the other animal hormones in plants.

The presence of animal hormones and their mimics in plants occasionally leads to economically important consequences in the development of certain herbivores. For example, estrogens in forage plants may cause infertility in sheep (review by Bradbury and White, 1954); estrogen contained in spoiled grain induces pathophysiological hyperestrogenic syndromes in domestic animals, affecting both milk and meat production (Stob, 1973). The economic importance of insect hormones and antihormones from plants is underlined by their ability to inhibit the development and reproduction of insect herbivores. Some of these hormonal mimics of plant origin have potent biological activity, which can compete with the most effective toxicants. In addition, they commonly exhibit the desired selective activity toward one or a limited number of families of phytophagous insects. Their use offers an alternative method for controlling populations of harmful insects by means of nontoxic, selectively acting, and environmentally safer natural products.

In contrast to toxicants, the immediate effectiveness of the hormonal mimics on a single generation of a phytophagous insect species is rather small. It would be enormously increased by continuous exposure of the whole population to the hormone-containing plants. Thus, the hormonally active plant chemicals could very well contribute to modulating the coevolutionary interactions between plants and herbivores. On the basis of available evidence, it is not possible to state whether the presence of animal hormones in plants is a result of the above causal evolutionary interactions, or whether it is merely a matter of coincidence. Nevertheless, further studies on these hormonally linked relationships between plants and herbivores will be rewarding; they could well indicate methods for utilizing these chemicoecological interactions in integrated pest control programs.

REFERENCES

Achrem, A. A., Levina, I. S., and Titov, Yu. A. (1973). "Ecdysones-Steroidal Hormones of Insects." Nauka i Technika, Minsk (in Russian).

Alertsen, A. R. (1955). *Acta Chem. Scand.* **9,** 1725–1726.

Barton, G. M., MacDonald, B. F., and Sahota, T. S. (1972). *Bi-Mon. Res. Notes, Can. For. Serv.* **28,** 22–23.

Bowers, W. S. (1968). *Science* **161,** 895–897.

Bowers, W. S. (1975). *In* "The Juvenile Hormones" (L. I. Gilbert, ed.), pp. 394–408. Plenum, New York.

Bowers, W. S. (1977). *ACS Symp. Ser.* **37,** 271–275.

Bowers, W. S., and Martinez-Pardo, R. (1977). *Science* **197,** 1369–1371.

Bowers, W. S., Fales, H. M., Thompson, M. J., and Uebel, E. C. (1966). *Science* **154,** 1020–1021.

Bowers, W. S., Ohta, T., Cleere, J. S., and Marsella, P. A. (1976). *Science* **193,** 542–547.

Bradbury, R. B., and White, D. E. (1954). *Vitam. Horm. (N.Y.)* **12,** 207–233.

Butenandt, A., and Karlson, P. (1954). *Z. Naturforsch. Teil B* **9,** 389–391.

Černý, V., Dolejš, L., Lábler, L., Sorm, F., and Sláma, K. (1967). *Collect. Czech. Chem. Commun.* **32,** 3926–3933.

Chatterjee, N. B., and Rajchandhuri, D. N. (1973). *Labdev, Part B* **11,** 87–88.

Churý, J. (1971). *Dtsch. Tieraerztl. Wochenschr.* **78,** 332–337.

de Souza, N. J., Chisalberti, E. L., Rees, H. H., and Goodwin, T. W. (1970). *Phytochemistry* **9,** 1247–1252.

Doecke, F. (1975). *Veterinaermed. Endokrinol.* pp. 583–591.

Dorfman, R. I. (1962). *Methods Horm. Res.* **2,** 113–126.

Earle, N. W., Padovani, I., Thompson, M. J., and Robbins, W. E. (1970). *J. Econ. Entomol.* **63,** 1064–1069.

Elliger, C. A., Zinkel, D. F., Chan, B. G., and Waiss, A. C., Jr. (1976). *Experientia* **32,** 1364–1366.

Emmens, C. W., and Martin, L. (1964a). *Methods Horm. Res.* **3,** 1–80.

Emmens, C. W., and Martin, L. (1964b). *Methods Horm. Res.* **3,** 81–125.

Fraenkel, G. (1935). *Proc. R. Soc. London, B* **118,** 1–12.

Harborne, J. B. (1977). *Pure Appl. Chem.* **49,** 1403–1421.

Heftmann, E. (1975). *Phytochemistry* **14,** 891–901.

Heftmann, E., Sauer, H. H., and Bennett, R. D. (1968). *Naturwissenschaften* **55,** 37–38.

Herout, V. (1970). *Prog. Phytochem.* **2,** 143–202.

Hikino, H., and Hikino, Y. (1970). *Fortschr. Chem. Org. Naturst.* **28,** 256–312.

Hikino, H., and Takemoto, T. (1974). *In* "Invertebrate Endocrinology and Hormonal Heterophylly" (W. J. Burdette, ed.), pp. 185–203. Springer-Verlag, Berlin and New York.

Hikino, H., Okuyama, T., Jin, H., and Takemoto, T. (1973). *Chem. Pharm. Bull.* **21,** 2292–2302.

Hora, J., Lábler, L., Kasal, A., Černý, V., Šorm, F., and Sláma, K. (1966). *Steroids* **8,** 887–914.

Imai, S., Toyosato, T., Sakai, M., Sato, Y., Fujioka, S., Murata, E., and Goto, M. (1969). *Chem. Pharm. Bull.* **17,** 335–339.

Jacobson, M. (1971). *Mitt. Schweiz. Entomol. Ges.* **44,** 73–77.

Jacobson, M., Redfern, R. E., and Mills, G. D., Jr. (1975a). *Lloydia* **38,** 455–472.

Jacobson, M., Redfern, R. E., and Mills, G. D., Jr. (1975b). *Lloydia* **38,** 473–476.

Jizba, J., Herout, V., and Šorm, F. (1967a). *Tetrahedron Lett.* pp. 1689–1691.

Jizba, J., Herout, V., and Šorm, F. (1967b). *Tetrahedron Lett.* pp. 5139–5143.

Karlson, P. (1966). *Naturwissenschaften* **53,** 445–453.

Koreeda, M., Nakanishi, K., and Goto, M. (1970). *J. Am. Chem. Soc.* **92,** 7512–7513.

Krause, E. (1970). *Monatsh. Veterinaermed.* **25,** 148–157.

Kubo, I., Lee, Y.-W., Balogh-Nair, V., Nakanishi, K., and Chapya, A. (1976). *J. Chem. Soc., Chem. Commun.* **22,** 949–950.

Manville, J. F. (1975). *Can. J. Chem.* **53,** 1579–1585.

Manville, J. F., Greguss, L., Sláma, K., and von Rudloff, E. (1977). *Collect. Czech. Chem. Commun.* **42,** 3658–3666.

Mansingh, A., Sahota, T. S., and Shaw, D. A. (1970). *Can. Entomol.* **102,** 49–53.

Matolcsy, G., Varjas, L., and Bordás, B. (1975). *Acta Phytopathol. Acad. Sci. Hung.* **10,** 455–463.

Matsuoka, T., Imai, S., Sakai, M., and Kamada, M. (1969). *Annu. Rep. Takeda Res. Lab.* **28,** 221–235.

Menn, J. J., and Beroza, M. (1972). "Insect Juvenile Hormones." Academic Press, New York.

Morgan, E. D., and Poole, C. F. (1977) *Comp. Biochem. Physiol. B* **57,** 99–109.

Munakata, K. (1975). *Pure Appl. Chem.* **42,** 57–66.

Murakoshi, S. (1972). *Jpn. J. Appl. Entomol. Zool.* **16,** 111–113.

Murakoshi, S., and Ichimoto, I. (1972). *Jpn. J. Appl. Entomol. Zool.* **16,** 159–161.

Murakoshi, S., Nakata, T., Ohtsuka, Y., Akita, H., Tahara, H., and Tamura, S. (1975). *Jpn. J. Appl. Entomol. Zool.* **19,** 267–272.

Murakoshi, S., Isogai, A., Chang, C.-F., Kamikado, T., Sakurai, A., and Tamura, S. (1976). *Jpn. J. Appl. Entomol. Zool.* **20,** 87–91.

Nakanishi, K. (1971). *Pure Appl. Chem.* **25,** 167–195.

Nakanishi, K., Koreeda, M., Sasaki, S., Chang, M. L., and Hsu, H. Y. (1966). *Chem. Commun.* pp. 915–917.

Nakazaki, H., and Isoe, S. (1963). *Bull. Chem. Soc. Jpn.* **34,** 741–742.

Němec, V., Chen, T. T., and Wyatt, G. R. (1978). *Acta Entomol. Bohemoslov.* **74,** 285–286.

Patterson, B. D., Wahba-Khalil, S. K., Schermeister, L. J., and Quaraishi, M. S. (1975). *Lloydia* **38,** 391–403.
Pawson, B. A., Cheung, H. C., Gurbaxani, S., and Saucy, G. (1970). *J. Am. Chem. Soc.* **92,** 336–343.
Prabhu, V. K. K., and John, M. (1975). *Experientia* **31,** 913.
Pratt, G. E., and Bowers, W. S. (1977). *Nature (London)* **265,** 548–550.
Primo Yufera, E., Tadeo Lluch, L., Ribó Canutt, J., and Sendra Sena, J. (1976). *Rev. Agroquim. Tecnol. Aliment.* **16,** 69–78.
Puritch, G. S., and Nijholt, W. W. (1974). *Can. J. Bot.* **52,** 585–587.
Rees, H. H. (1971). *In* "Aspects of Terpenoid Chemistry and Biochemistry" (T. W. Goodwin, ed.), pp. 181–222. Academic Press, New York.
Rees, H. H., and Goodwin, T. W. (1974). *Biochem. Soc. Trans.* **2,** 1027–1033.
Robbins, W. E., Kaplanis, J. N., Thompson, M. J., Shortino, T. J., Cohen, C. F., and Joyner, S. C. (1968). *Science* **161,** 1158–1159.
Robbins, W. E., Kaplanis, J. N., Thompson, M. J., Shortino, T. J., and Joyner, S. C. (1970). *Steroids* **16,** 105–125.
Ruscoe, C. N. E. (1972). *Nature (London), New Biol.* **236,** 159–160.
Saxena, B. P., and Srivastava, J. B. (1973). *Indian J. Exp. Biol.* **11,** 56–58.
Schmialek, P. (1961). *Z. Naturforsch., Teil B* **16,** 461–464.
Schneiderman, H. A., Gilbert, L. I., and Weinstein, M. J. (1960). *Nature (London)* **188,** 1041–1042.
Sláma, K. (1969). *Entomol. Exp. Appl.* **12,** 721–728.
Sláma, K. (1978). *Acta Entomol. Bohemoslov.* **75,** 65–82.
Sláma, K., and Williams, C. M. (1965). *Proc. Natl. Acad. Sci. U.S.A.* **54,** 411–414.
Sláma, K., and Williams, C. M. (1966). *Biol. Bull. (Woods Hole, Mass.)* **130,** 235–246.
Sláma, K., Romaňuk, M., and Šorm, F. (1974). "Insect Hormones and Bioanalogues," Springer-Verlag, Wien and New York.
Staal, G. B. (1967). *Proc. K. Ned. Akad. Wet., Ser. C* **70,** 409–418.
Stob, M. (1973). *In* "Toxicants Occurring Naturally in Foods," 2nd ed., pp. 550–557. Natl. Acad. Sci., Washington, D.C.
Stowe, B. B., and Hudson, V. W. (1969). *Plant Physiol.* **44,** 1051–1057.
Takemoto, T., Ogawa, S., Nishimoto, N., Arihara, S., and Bue, K. (1967). *J. Pharm. Soc. Jpn.* **87,** 1414–1418.
Takemoto, T., Ogawa, S., and Nishimoto, N. (1976). *Kagaku (Kyoto)* **30,** 288–296.
Tóth, B., Tüske, M., Matolcsy, G., and Varjas, L. (1977). *Acta Phytopathol. Acad. Sci. Hung.* **12,** 375–381.
Wellington, W. G. (1969). *Can. Entomol.* **101,** 1163–1172.
Wigglesworth, V. B. (1970). *"Insect Hormones"*. Oliver & Boyd, R. & R. Clark Ltd., Edinburgh.
Williams, C. M. (1956). *Nature* **178,** 212–213.
Williams, C. M. (1970). *Chem. Ecol.,* 1970, 103–132.
Williams, C. M., and Robbins, W. E. (1968). *Bio Science* **18,** 791–799.
Yen, K. Y., Yang, L. L., Okuyama, T., Hikino, H., and Takemoto, T. (1974). *Chem. Pharm. Bull.* **22,** 805–808.

Subject Index

D

E

Q

R

T

U

V

W

X

Z